Serengeti II

Dynamics, Management, and Conservation of an Ecosystem

Serengeti II

Dynamics, Management, and Conservation of an Ecosystem

Edited by
A. R. E. Sinclair & Peter Arcese

THE UNIVERSITY OF CHICAGO PRESS
CHICAGO AND LONDON

A. R. E. Sinclair is professor of ecology at the University of British Columbia. His books include *The African Buffalo* and *Serengeti: Dynamics of an Ecosystem* (with M. Norton-Griffiths). **Peter Arcese** is assistant professor of wildlife ecology at the University of Wisconsin—Madison.

The University of Chicago Press, Chicago 60637
The University of Chicago Press, Ltd., London

Printed in the United States of America

04 03 02 01 00 99 98 97 96 95   1 2 3 4 5

ISBN 0-226-76031-6 (cloth)
ISBN 0-226-76032-4 (paper)

Library of Congress Cataloging-in-Publication Data

Serengeti II : dynamics, management, and conservation of an ecosystem / edited by A. R. E. Sinclair and Peter Arcese.
p. cm.
Based on papers presented at a workshop held in December 1991 at the Serengeti Research Institute, Tanzania.
Includes bibliographical references and index.
1. Animal ecology—Tanzania—Serengeti National Park Region—Congresses. 2. Wildlife conservation—Tanzania—Serengeti National Park Region—Congresses. 3. Ecosystem management—Tanzania—Serengeti National Park Region—Congresses. I. Sinclair, A. R. E. (Anthony Ronald Entrican) II. Arcese, Peter. III. Title: Serengeti 2.
QL337.T3S42 1995
574.5′264—dc20   94-45542
CIP

∞ The paper used in this publication meets the minimum requirements of the American National Standard for Information Sciences—Permanence of Paper for Printed Library Materials, ANSI Z39.48-1984.

CONTENTS

ACKNOWLEDGMENTS

We thank the boards of trustees of Tanzania National Parks and the Serengeti Wildlife Research Institute (SWRI), and the Tanzania Commission for Science and Technology for their permissions to conduct research over the years; also the directors of Tanzania National Parks, particularly Mr. David Babu, and of SWRI, F. Kirji and Mr. C. Mlay. We thank the directors of the Serengeti Wildlife Research Centre at Seronera, especially Mr. H. Nkya, and the chief park wardens of Serengeti National Park, Mr. B. Maragese and Mr. W. Summay. Many other wardens and rangers have helped over the years, too many to mention, and we thank them all.

Many people have helped in the production of this book, particularly as outside referees for chapters: M. Adamson, S. A. Albon, M. V. Ashley, M. Bekoff, J. Belsky, S. Boutin, J. Coleman, D. Covell, M. Dehn, L. Dill, R. D. Estes, R. W. Garrott, J. L. Gittleman, D. Hik, D. Houston, D. Huggard, A. Ives, L. Keller, N. Larter, N. R. Liley, J. Malcolm, K. McComb, F. Messier, M. O'Donaghue, R. Pech, K. Ralls, D. Read, R. Turkington, and D. Ward. Two anonymous reviewers read the entire manuscript and made many valuable suggestions. We appreciate the time they took to give critical reviews.

Tina Raudzus and Susan Brown helped with preparation of manuscripts. Anne Sinclair and Gwen Jongejan have helped us in many ways throughout, in fieldwork, organizing the workshop, and preparation of this book.

We would like to thank George Schaller, Mrs. Jorie Kent, and Tim Corfield, and the following organizations for their generous support of the editors over the years:

Canadian Natural Sciences and Engineering Research Council
Committee for Research and Exploration, National Geographic Society, Washington, D.C.
Friends of Conservation, Oak Brook, Illinois
Frankfurt Zoological Society
International Union for the Conservation of Nature and Natural Resources
Ker and Downey Safaris, Tanzania, Ltd., Arusha

National Science Foundation, U.S.A.
The Wildlife Conservation Society, New York
University of British Columbia
University of Wisconsin-Madison
World Wide Fund for Nature

I Overview

ONE

Serengeti Past and Present

A. R. E. Sinclair

Serengeti is a natural laboratory whose history dates back at least 4 million years to the beginnings of human evolution. The main feature of this system is the annual migration of large herds of ungulates, single herds of which are larger than any other on earth. The diversity of ungulates is also extraordinarily high, even for Africa, there being 28 species. The same may be said for the carnivores, the birds, and even the dung beetles. How does this system work? Why is it so diverse? The last 35 years of studies have been focused on answering these questions, with the primary purpose of conserving this unique system. An understanding of the system is necessary for conservation because without it we have no way of telling whether our efforts to manage and protect are correct, in much the same way that we cannot look after an orphaned animal if we have no idea what it feeds on.

The hominid footprints at Laetolil and the habitation on the ancient lake edge at Olduvai Gorge in the Serengeti plains (Leakey and Hay 1979) occurred in a savanna environment not much different from that of today. The climate was seasonal, and there is evidence of seasonal movement of ungulates, indicating a migration. The wildebeest of those days was not too different from the modern form, which appeared in this system 1.5 million years ago (Sinclair 1979c). Although there have been some changes in fauna—especially the disappearance of Pleistocene megafauna such as the white rhinoceros *(Ceratotherium simum)* and the giant bovine *Pelorovis*—the fossil record suggests that Serengeti is a natural ecosystem with a long history.

The Serengeti ecosystem, however, is not static. There have been many changes, both natural and human-induced, and our understanding has come from studying these changes. The first volume on the Serengeti-Mara area, *Serengeti: Dynamics of an Ecosystem* (abbreviated hereafter to *Serengeti*) (Sinclair and Norton-Griffiths 1979), covered research undertaken in the area through 1977. The years following have seen major changes in public perceptions of conservation, in the politics and populations of Africa, and in the dynamics of the Serengeti ecosystem. We know that it is no longer sufficient to save individual species; it is instead necessary to preserve intact assemblages of species and their habitats as functioning ecosystems. The latter is the essence of what is now known as the conservation of biodiversity. Understanding the ecological principles that promote biological diversity in natural ecosystems was a main goal of the work presented in the first Serengeti volume. In this second volume, we consolidate the recent results of this work and advise managers who plan for, and respond to, the social, political, and environmental perturbations that affect conservation policy. Their policy decisions will ultimately affect the longevity of the Serengeti-Mara ecosystem.

After a short description of the area as background, I summarize the perturbations that have taken place, their main results relating to the dynamics of the system, and the management actions that have been derived from these studies.

THE SERENGETI-MARA ECOSYSTEM

The Serengeti-Mara ecosystem (fig. 1.1) is an area of some 25,000 km^2 on the border of Tanzania and Kenya, East Africa (34° to 36° E, 1° to 3°30′ S) defined by the movements of the migratory wildebeest. The eastern boundary is formed by the Crater Highlands and the Rift Valley. An arm called the "western corridor" stretches west almost to Speke Gulf of Lake Victoria. The remaining western boundary is formed by dense cultivation. The northern boundary is formed by the Isuria escarpment and the Loita plains in Kenya. The southern and southwestern boundary runs along an area of rocks and dense woodland.

The ecosystem covers several different conservation administrations (see chaps. 25–28 for details). These are Ngorongoro Conservation Area (8,288 km^2), Serengeti National Park (14,763 km^2), Maswa Game Reserve (2,200 km^2) and Grumeti, Ikorongo, and Loliondo Game Controlled Areas in Tanzania, and the Masai Mara National Reserve (1,672 km^2) and adjoining Group Ranches in Kenya. Further details are given in Schaller (1972), Sinclair (1977), and Sinclair and Norton-Griffiths (1979).

History

An area of 2,286 km^2 was established in 1929 as a game reserve in what is now southern and eastern Serengeti. In 1940 Protected Area status was conferred, and a national park was established in 1951, covering southern Serengeti and Ngorongoro. After a commission of enquiry in 1956 (Pearsall 1957), the boundaries were realigned in 1959 to include the area between Banagi and the Kenya border (see fig. 1.1). At the same time

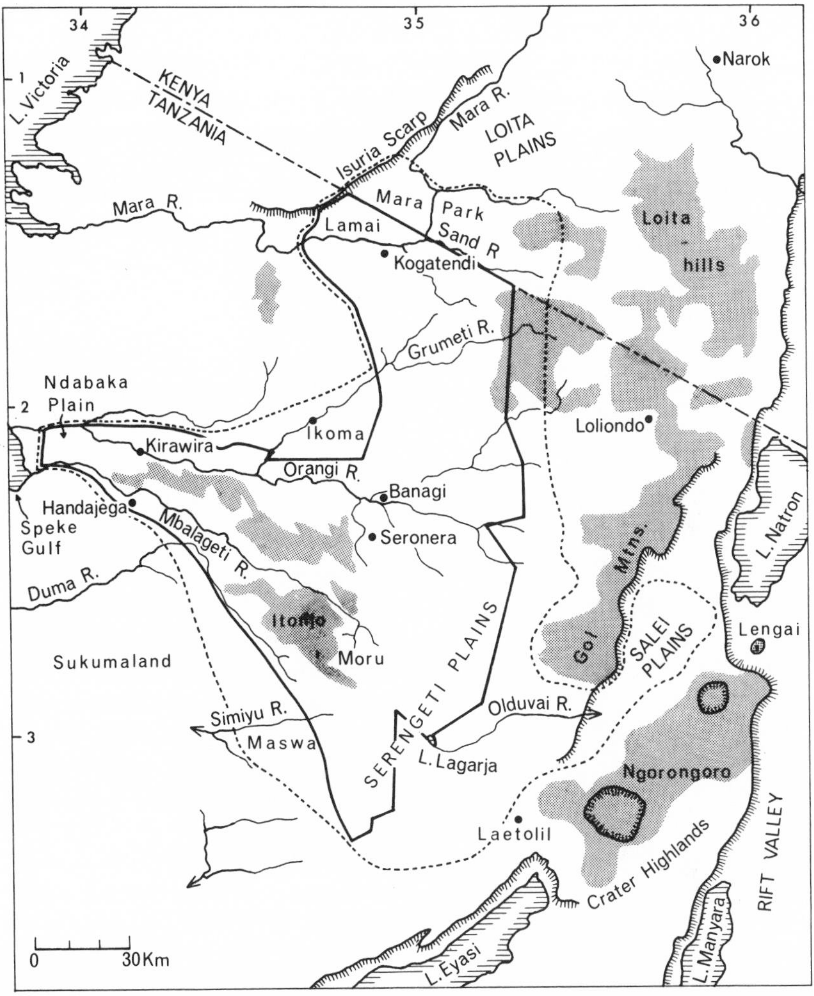

Figure 1.1 The Serengeti-Mara ecosystem is the area used by the wildebeest (even dashed line). The Serengeti National Park is shown by the heavy solid line. Hills are shaded.

Ngorongoro Conservation Area was excised from Serengeti National Park. In 1965 the Lamai Wedge between the Mara River and the Kenya border was added, and the Mara Reserve was formed in Kenya. A small area north of the Grumeti River in the corridor was added in 1967. The Serengeti National Park was one of the first areas to be proposed as a World Heritage Site by UNESCO in 1972, and it was formally gazetted as such and as a Biosphere Reserve, along with Ngorongoro Conservation Area, in 1981 (Campbell, Huish, and Kajuni 1991).

Climate

The temperature shows a relatively constant mean monthly maximum of 27°–28°C at Seronera. The minimum temperature varies from 16°C in the hot months of October–March to 13°C during May–August. Rain typically falls in a bimodal pattern, with the long rains during March–May and the short rains in November–December. However, the rains can fuse into one long period, particularly in the north, or the short rains can fail entirely, especially in the southeast. There is a rainfall gradient from the dry southeastern plains (500 mm per year) to the wet northwest in Kenya (1,200 mm per year) (Sinclair 1979c; Campbell, Huish, and Kajuni 1991).

Geology

The Serengeti is part of the high interior plateau of East Africa. It slopes from its highest part (1,850 m) on the eastern plains near the Gol Mountains toward Speke Gulf (920 m). West of a line going through Seronera, the underlying geology is very old Precambrian (2500 my) volcanic rock and banded ironstone. Late Precambrian sedimentary rocks unconformably overlie the shield and form the central and southern hills. A late Precambrian orogenic event produced outcrops of granitic gneisses and quartzite east of Seronera, forming the eastern hills and the kopjes. The Crater Highlands are volcanoes of Pleistocene age, and aerial debris from these was blown westward to form the Serengeti plains. Eruptions of one volcano, Lengai, continue to the present, the latest occurring in 1966.

Soils on the eastern plains are highly saline and alkaline, and are also shallow as a result of their recent volcanic origin. The soils become progressively deeper and less alkaline toward the northwestern plains and into the woodlands. The soil catena, which is the gradient of soil types from ridge top to drainage sump, is characterized by shallow, sandy, well-drained soil at the top changing to deep, silty, poorly drained soil at the bottom. Nutrient flow through the grazing food web determines the movement of migrants, the distribution of resident ungulates, and the biomass of herbivores (McNaughton and Banyikwa, chap. 3).

Vegetation

The southeastern plains are treeless except along Olduvai Gorge. The grasses are alkaline-tolerant and there are many small dicots (McNaughton and Banyikwa, chap. 3; Sinclair, chap. 5). With the deeper soils grass species change, the dominant ones being *Themeda triandra* and *Pennisetum mezianum*. These taller species continue into the woodlands.

The woodlands start at a sharp boundary running south of Seronera in one direction and east of Seronera in the other. The woodlands are dominated by *Acacia* species in all areas except for a small region south and west of Kogatende, where *Terminalia-Combretum* takes over (Herlocker 1976). Canopy cover of the trees varies between 2% and 30%. Along the main rivers, Mbalageti, Grumeti, and Mara and its tributaries, there is gallery forest with closed canopy.

Fauna

The Serengeti supports not only the largest herds of migrating ungulates but also one of the highest concentrations of large predators in the world. Estimates put wildebeest at about 1.3 million, zebras at 200,000 and Thomson's gazelles at 440,000. Hyenas are the most numerous of the large carnivores at about 7,500, with lions at about 2,800 (Caro and Durant, chap. 21). Other data are given by Campbell and Borner (chap. 6) and Hofer and East (chap. 16). A list of the medium and large mammals present in the system is given in appendix A.

Systematic studies of other groups of animals have been far fewer. There are some 517 species of birds (appendix B; Schmidl 1982). An incomplete collection of some 80 species of grasshoppers has been made (pers. obs.; A. Harvey, pers. comm.). Dung beetles have also been studied, and in one small area of the plains near Lake Lagarja over 100 species have been recorded (Foster and Bresele 1992).

PERTURBATIONS OF THE SERENGETI-MARA ECOSYSTEM

Research has documented several changes in the ecosystem. This information was instrumental in setting up conservation and management plans for the whole area, but it was also vital in furthering our understanding of the dynamics of this ecosystem. I document the major perturbations that have taken place since 1960.

Rinderpest

Rinderpest is a viral disease of cattle that occurs naturally in Asia. Its introduction to Africa in the 1880s and the ensuing epizootic in the 1890s is described in detail elsewhere (Ford 1971; Sinclair 1977, 1979a; Dobson, chap. 23). Rinderpest remained in the Serengeti region until the early 1960s (Talbot and Talbot 1963), when it disappeared from the wildlife populations as a result of a cattle vaccination campaign (Sinclair 1977; Plowright 1982). This vaccination program had the effect of protecting wildlife from infectious yearling cattle, so that the disease died out among the wildlife. Only ruminants are affected by rinderpest, the greatest morbidity being in species most closely related to cattle. Thus, buffalo were affected the most, followed by wildebeest. There are reports of giraffe and warthog being affected, but other ruminants appear to be less influenced by the disease.

The most important consequence of the rinderpest disappearance was a sixfold increase in the wildebeest population between 1963 and 1977 (fig. 1.2). Buffalo numbers also increased fivefold, but their effects on the ecosystem were less marked than those produced by the wildebeest eruption (Sinclair 1979b). The early consequences of these eruptions were documented in *Serengeti;* their later consequences are described in subsequent chapters of this volume.

After 1977 the wildebeest population leveled off, and it has remained at approximately 1.3 million animals since that time; the various censuses since 1977 are not significantly different from one another (Dublin et al. 1990; Campbell, Huish, and Kajuni 1991; Campbell and Borner, chap. 6). Zebra, which as nonruminants are not affected by rinderpest, have remained at constant numbers of about 200,000 over the 30 years from 1960 to 1989.

The Border Closure

Political events have also had a major impact on the ecosystem in the past 15 years. In April 1977 the international border between Tanzania and Kenya was closed, and it remained so until about 1986, when it was opened partially to tourism. However, the main tourist route between the Mara Reserve and Serengeti National Park (via the Sand River and Bologonja gates) remains closed.

The immediate effect of the border closure was a precipitous drop in the number of foreign visitors to Serengeti National Park from about 70,000 in 1976 to 10,000 in 1977 (fig. 1.3a). Visitor numbers remained at these very low levels throughout the 1980s, and only in the 1990s have they started to rebound. The Mara Reserve, in contrast, experienced an increase in tourist numbers during the same period of border closure.

One consequence of the border closure for Serengeti National Park was a drop in income from visitors. Since the operating budget of the park is linked to some extent to income, this budget did not show any overall increase through the 1980s (fig. 1.3b). In particular, the operating budget dropped continuously from 1982 to 1985 and remained low until 1987. In several of those years the park showed a net loss. As a result, the anti-poaching effort, in terms of patrol days, dropped by the mid-1980s to 60% of that prior to border closure.

Poaching

A marked increase in settlement along the western boundaries of both the Serengeti National Park and Mara Reserve (Dublin 1986) has been documented over the past 20 years; a detailed analysis is presented by Campbell and Hofer in chapter 25. In some areas the increase in popula-

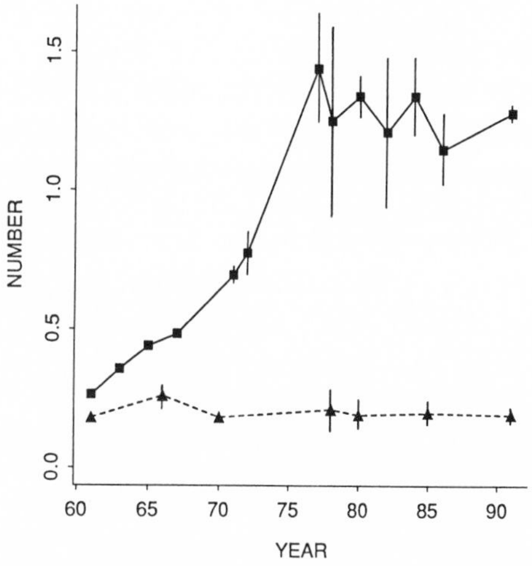

Figure 1.2 Population trends for Serengeti migrant wildebeest (squares) and zebra (triangles). (Data from Dublin et al. 1990; chap. 6). Numbers in thousands.

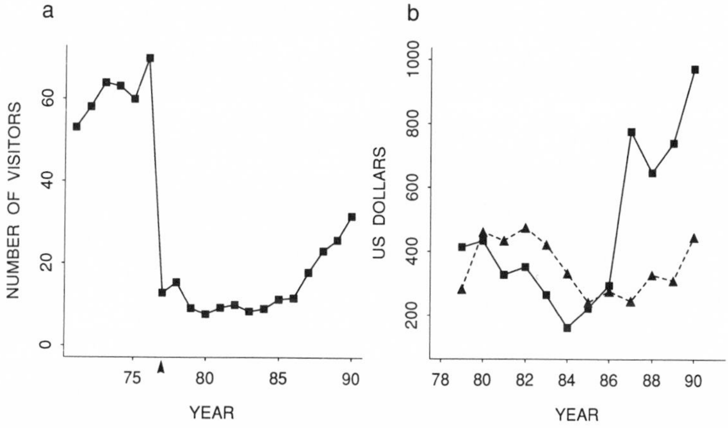

Figure 1.3 (*a*) Annual foreign visitor numbers and (*b*) income (solid line) and operating budget (dashed line) in U.S. dollar equivalents for Serengeti National Park. Arrow indicates border closure. (From Campbell, Huish, and Kajuni 1991). Numbers in thousands.

tion has approached 15% per year, mostly from immigration but partly from the 3% annual population growth rate due to demographic factors.

The combined effects of the human population increase and the marked drop in anti-poaching patrols resulted in an invasion of northern and western Serengeti by poachers. The first species to feel their effects was the rhinoceros, which lost 52% of its population in one year alone,

1977, the first year of border closure (fig. 1.4). By 1980 the population was effectively extinct, with only occasional sightings subsequently of animals that ventured in from Kenya. These did not survive long, for Serengeti was no longer a sanctuary for this species.

Elephant and buffalo were the next two species to show the effects of poaching. Data on elephant carcasses found in Serengeti (fig. 1.4) (Arcese, Hando, and Campbell, chap. 24) show a first, smaller, peak in 1977, coincident with the drop in rhinoceros numbers, and a larger, more prolonged peak during 1982–1986, when anti-poaching patrols were at their lowest. Some 50% of the population disappeared during 1984–1986. In 1986 the elephant population in Serengeti National Park was only 20% of that in 1976 (Dublin and Douglas-Hamilton 1987). Some 400–500 of the missing animals moved to the Mara Reserve, where poaching, although present, was less severe. The rest of the animals were killed. Elephant poaching in the Serengeti area slowed considerably in 1987, when legal ivory sales were disallowed in Burundi. It came to an abrupt stop for the region as a whole in 1989, when the world ban on the ivory trade was imposed.

Data on buffalo numbers are good until 1976, but there is a gap in census data through most of the period of border closure, until 1984 (fig. 1.5). By this time the northwest of Serengeti National Park was devoid of buffalo; this area had previously supported the highest densities in the park. In all some 50% of the park area suffered major reductions in buf-

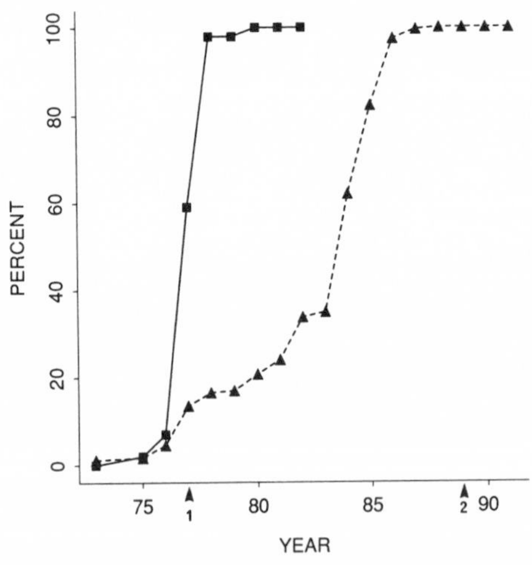

Figure 1.4 Cumulative % of total number killed by poachers and found by Serengeti National Park Staff, for rhinoceros (squares) and elephant (triangles) (data from pers. obs. and SNP records). Arrow 1 indicates border closure; arrow 2 indicates ivory trade ban.

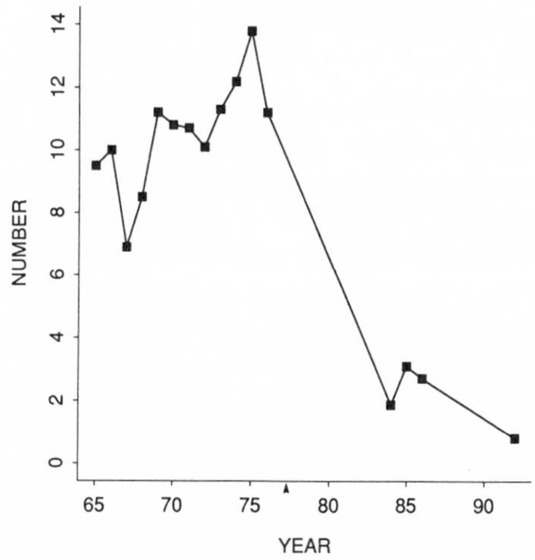

Figure 1.5 Censuses of African buffalo in northern Serengeti. No censuses were conducted during 1977–83. Arrow indicates border closure. (Data from Sinclair 1977; Dublin et al. 1990; Campbell & Borner 6). Numbers in thousands.

falo numbers, mainly in the north and west (Dublin et al. 1990; Campbell and Borner, chap. 6). Since this species was previously the dominant non-migratory herbivore, its removal constituted a marked change in the species composition of the area. Other anecdotal information (Sinclair, chap. 9) suggests that most of the removal took place in the late 1970s coincident with the removal of rhinoceros.

The consequences of these perturbations are of primary interest for conservation, and they have been the focus of much research. The rinderpest perturbation was the main theme of *Serengeti,* and its continuing effects are reflected in much of the present work. The aftereffects of the border closure and human population expansion, in particular through poaching, are examined in the sections on management. First, however, I review some of the results of the research with respect to population and ecosystem dynamics.

ECOSYSTEM DYNAMICS

Plant Dynamics

Heterogeneity, both spatial and temporal, was one of the main themes of *Serengeti.* Above all, heterogeneity allows the migration of herbivores around which the whole ecosystem is structured. One of the processes related to migration is the "grazing succession" (Vesey-Fitzgerald 1960) whereby some herbivores provide a feeding niche for others, an interaction termed "facilitation," a form of commensalism. R. H. V. Bell

(Gwynne and Bell 1968; Bell 1970, 1971) applied this idea to resident ungulates that move seasonally up and down the soil catena in the western corridor. He proposed that facilitation is also the underlying process for the synchronized migration of zebra, wildebeest, and Thomson's gazelle. Evidence for this process was provided by McNaughton (1979), who demonstrated both the ephemeral pattern of grass regrowth across the ecosystem (due to rainfall) that is tracked by the migration, and that grazing by wildebeest at certain seasons increases the food supply for Thomson's gazelle. Subsequent analysis of the microscale associations of zebra and wildebeest, however, shows that facilitation may not account for their relationship; rather, it is a combination of multispecies antipredator herding in the wet season and interspecific avoidance through competition in the dry season (Sinclair 1985).

In this volume, McNaughton and Banyikwa (chap. 3) elaborate on the theme of heterogeneity. Both spatial and temporal heterogeneity can be seen at all scales. The ecosystem is characterized by the major geographic, edaphic, and vegetation differences between the plains and woodlands. Within these there are differences on the scale of landscapes. For example, McNaughton discovered "hot spots" where the soils are rich in nutrients and where high densities of resident ungulates have persisted over the past 30 years. Examples of such areas are at Lobo and Lamai, and along the Talek River. Both within and outside of these "hot spots" there is yet a finer scale of association of ungulates determined in part by antipredator behavior (Sinclair 1985; Mduma and Sinclair 1994).

McNaughton's analysis of plant communities down the soil catena shows that they differ in both plant species and cation exchange capacity. There is a fine-scale heterogeneity underlying the herbivore movements observed by Bell (1970). In turn, grazing intensity and soil texture both determine community structure. This demonstrates the dominant role of grazers in determining plant community pattern. On a temporal scale there are patterns of plant growth at the scale of the whole ecosystem, the growing season, and the grazing season. In addition, in the dry season and at the beginning of the rains, stochastic rainfall events (storms) produce a finer-scale heterogeneity.

During the 1960s and 1970s elephants were considered the major determinants of woodland structure in African savannas (Dublin 1991). In *Serengeti,* however, Norton-Griffiths (1979) proposed a new hypothesis, based on a detailed study of burning patterns in the Serengeti ecosystem: that burning determines both structure and change in savanna tree populations. In chapter 4, Dublin describes her experimental test of this hypothesis in the Mara Reserve, where changes in tree populations have been greatest. She concludes not only that fire is the overriding cause of tree population collapse through the inhibition of seedling regeneration, but also that elephants act to maintain the grassland state even in the absence of fire by removing seedlings. There is evidence, therefore, that there are two stable states, woodland and grassland, and that elephants at high enough densities are capable of maintaining the grassland state, while fire acts as the perturbation to change the system from woodland to grassland.

A test of this idea is presented in chapter 5, where it is predicted that for woodland to return, both fire and elephant browsing of seedlings have to be reduced. These conditions have been met in Serengeti south of the international border in the past 15 years because high wildebeest grazing has reduced fire and high poaching mortality has reduced elephants to very low numbers (see fig. 1.4). The results show an exponential increase in small trees as predicted, in direct contrast to the Mara, where there is still no regeneration because there is a high density of elephants. The next prediction is that once sufficient numbers of trees have escaped predation, elephants will not be able to change this woodland back to a grassland state. This situation is being monitored as elephants move back into the Serengeti from the Mara in the 1990s (Campbell and Borner, chap. 6).

Herbivore Dynamics

Population Dynamics. The change in the wildebeest population and its effect on the rest of the ecosystem was the focus of *Serengeti* (Sinclair 1979a,b). The volume included some information on the other migrant species, the African buffalo, and other resident ungulates (Grimsdell 1979). The initial work on wildebeest and African buffalo concluded that intraspecific competition for food during the dry season regulated these populations (Sinclair 1977; Sinclair, Dublin, and Borner 1985). It now appears that these species represent special cases. Wildebeest are migrants and as such are capable of escaping population regulation by predators (Fryxell, Greever, and Sinclair 1988). African buffalo in this particular habitat are large enough to avoid capture by all predators except lions. Thus buffalo, along with the other very large herbivores—hippopotamus, rhinoceros, and elephant—are regulated by food rather than by predators. Subsequent work on the smaller, resident ungulates suggested that these populations are regulated by predators, in contrast to the results for the larger herbivores (Sinclair 1985). Little direct evidence was available up to the mid-1980s to test this hypothesis.

In the present book a large body of information has been collated for many ungulate species in Serengeti (Campbell and Borner, chap. 6; Sinclair, chap. 9; Mduma, chap. 10), Ngorongoro Crater (Runyoro et al., chap. 7) and Mara (Broten and Said, chap. 8). Campbell and Borner (chap. 6) examine ungulate populations and their distribution in the Tanzanian section of the Serengeti ecosystem. Populations of the three main migrant species are all constant, the zebra having been so for almost 30 years. In contrast, buffalo numbers have collapsed and are continuing to decrease through poaching. Elephant numbers also collapsed but are now increasing through immigration from the Mara. In the past we have had no information on the populations of very small ungulates. Mduma (chap. 10) documents a surprisingly high population of oribi, and it is possible that this species has increased since the 1960s. In terms of distribution, roan antelope are almost extinct, occurring now only in two small areas. Loss of habitat through burning may be a cause, but competition from wildebeest or human disturbance could be other reasons. Outside of the western Serengeti boundary, resident ungulate densities have declined by 95% in the 5 years from 1988 to 1992, and there are also areas inside the park in the west and north where there are few ungulates left because of poaching (Campbell and Borner, chap. 6).

For Ngorongoro Crater, Runyoro et al. (chap. 7) document ungulate numbers over 30 years and examine whether the removal of Masai livestock from the Crater in 1974 had any effects on the wildlife. In general, total herbivore biomass has remained constant, indicating a stable multispecies assemblage. However, in contrast to Serengeti, buffalo, which were absent in the 1960s, have increased in the Crater, possibly as a result of cattle removal and lower grazing competition. Another possibility is that buffalo that lived in the surrounding montane forest have been forced into the Crater by extensive deforestation, woodcutting, and poaching in the past 20 years. The ruminants, including wildebeest, have declined in numbers, but nonruminants such as zebra and ostrich have not. Runyoro et al. suggest that a decline in burning frequency after the Masai left has altered the grassland community to a taller and coarser structure more suitable for buffalo and zebra than for wildebeest. Among the concerns for conservation of the whole area is the recent doubling of the human population surrounding the Crater, the expansion of agriculture, and the increased use of water sources—all of which will affect the Crater wildlife.

For the Kenya Mara, Broten and Said (chap. 8) also emphasize the effects of accelerating human encroachment on the surrounding cattle ranches in the 1970s. Relict montane forests and natural grasslands have been reduced and wheat fields have replaced them. Remaining natural areas outside the Mara Reserve have been overgrazed by livestock, and wildlife populations have declined in these areas. Within the reserve, browsing species such as giraffe and eland have declined as a result of the long-term disappearance of trees and shrubs. More puzzling is the decline of warthog and topi, grazing species that have increased in the adjacent area in Serengeti (Sinclair, chap. 9).

Previous research has demonstrated that both buffalo and their main predators, lions, were removed by poachers in the period 1978–1985 (Dublin et al. 1990; Packer 1990; see fig. 1.5). These removals have been used as a perturbation to test hypotheses on predator limitation of resident ungulates (Sinclair, chap. 9; Sinclair 1985). Censuses in northern and western Serengeti, where buffalo and lions were removed, have been compared with those in Mara and eastern Serengeti, where buffalo and their predators have remained constant. In northern Serengeti, where buffalo removal was nearly complete, both topi (which are possible competitors with buffalo) and impala, and possibly oribi (which are not competitors), have increased relative to nonremoval areas. Although decreased competition cannot be excluded, it is possible that these species are experiencing a release from predation. This may happen through "apparent competition" (Holt 1977), since lions depend primarily on buffalo in this region (Scheel and Packer, chap. 14) and they may have dispersed when their primary prey disappeared.

The Migration. The annual cycle of movements by wildebeest, zebra, Thomson's gazelle, and eland is known as "the migration." The earliest anecdotal accounts of the migration go back to the 1910s (White 1915), but the first scientific accounts appeared in the 1950s (Grzimek and Grzimek 1960). Later studies filled in some of the gaps (Talbot and Talbot 1963; Watson 1967), but it was not until the early 1970s that M. Norton-Griffiths initiated a comprehensive systematic aerial survey to give the complete picture reported in *Serengeti* (Maddock 1979).

The general pattern of movements was understood by the 1970s. Migrants used the plains in the wet season (December–June). If there were dry periods during this time, the animals moved west into Maswa and the Mbalageti Valley until the rains returned. In late May or June the migrants moved northwest of the plains along the Simiyu, Mbalageti, Seronera, and Nyabogati Rivers to the western corridor. A part of the population moved directly north toward Lobo. As the dry season progressed (July–October) an increasing fraction of the migrants shifted north, appearing on the Mara watershed by early August. In early dry seasons, however, these events could occur as early as June; in late seasons they might not occur until September. When the wildebeest population was low (1950s and 1960s), the migrants barely crossed the border into the Mara, but by the 1970s, when numbers were high, a large proportion were using the Mara each dry season (Maddock 1979). Since 1973 there has been no systematic study of movements, and we have only fragmented accounts. The most significant change is that a fraction of the migrants now move directly north, east of Serengeti National Park through Loliondo. This is the area where there is potential conflict with wheat projects and fences proposed to protect them, because such fences could block the movements. The other important conservation factor is the progressive decline in area available to wildebeest during the dry season due to human settlement and agriculture. This decline is particularly severe in the Maswa and Grumeti areas in the west, Ikorongo in the northwest, and Loita in the Mara (Campbell and Hofer, chap. 25; Norton-Griffiths, chap. 27).

The study of the migration has led to understanding in two areas of human ecology. First, it suggests the selection pressure leading to the evolution of bipedalism and scavenging in hominids (Sinclair, Leakey, and Norton-Griffiths 1986): by becoming bipedal, carrying their offspring, and following the migration, the earliest hominids (pre-4 my B.P.) would have been able to make use of intact fresh carcasses (from animals dying of disease and starvation) that are two orders of magnitude more abundant than those in nonmigratory systems. Other carnivores cannot follow the migration throughout the year because they cannot take their young with them, so this was an empty niche waiting to be filled. Since protein is of major importance in the diet of large primates, there would be strong selection for adaptations to make use of this empty niche. Use of tools would not have been a necessary prerequisite, but selection would promote such an adaptation later because tools would speed up the dismembering of carcasses before larger carnivores could steal the food. Such a scenario predicts that for efficient long-distance walking on the savanna, there would have been a rapid and simultaneous evolution of the pelvis and foot. An alternative hypothesis (Lovejoy 1981) proposed that bipedalism arose from plant gathering within the forest first and that only later was there movement to the plains. This hypothesis predicts that upright stance (changes in the pelvis) could have preceded long-distance walking (changes in the foot); if such a sequence is found in future fossil discoveries, then this would be sufficient evidence to disprove the migration hypothesis.

The second applied aspect of migration is concerned with human seasonal migrations (transhumance) in the semiarid region of sub-Saharan Africa called the Sahel. This work indicates that the Sahel famines in North Africa are human-induced through prevention of migration: since migration allows higher populations than sedentary systems, the settlement of peoples around artificial wells automatically resulted in overpopulation, overgrazing, and habitat destruction (Sinclair and Fryxell 1985; Fryxell and Sinclair 1988).

The underlying causes of migration in the Serengeti ecosystem are still not fully established. By the late 1970s we knew that movements were tied to rainfall: in the dry season migrants move toward areas of higher rainfall because that is where there is the most food (Sinclair 1979b), but in the wet season they move to areas of lower rainfall, that is, to the plains (Maddock 1979). The reason for the latter movement was not clear; it seemed paradoxical, for food in the woodlands appeared as plentiful as on the plains. Hypotheses to explain why animals moved to the plains included their need to obtain more protein or energy (as Albon and Langvatn [1992] have proposed for other ungulates) or minerals such as calcium (Kreulen 1975); to avoid waterlogged ground in the woodland on which wildebeest, at least, develop foot diseases; to avoid tsetse flies; and to escape predation in the tall-grass areas. The mineral and nutrient hypotheses have been examined by Murray (chap. 11) and Fryxell (chap. 12). Murray measured the available nutrients in woodlands and plains and the dietary requirements of wildebeest. Although he does not reject the protein hypothesis, he finds that calcium is universally available and superabundant and cannot account for the movements. Instead, he proposes that phosphorus is the driving variable. This element is deficient on the dry-season range but is found on the plains at levels sufficient for lactation and the growth of young, which take place there. He also suggests that this is the reason why resident ungulates concentrate on the "hot spots" (McNaughton and Banyikwa, chap. 3).

Fryxell (1991; chap. 12) explores these same questions using models of grass maturation and herbivore functional responses. Such models indicate that because grass nutrient quality declines at high biomass (see fig. 4.5), herbivores should obtain less protein or energy in high-biomass areas such as the woodlands in the wet season than in intermediate-biomass areas such as the plains. Furthermore, grazers are more likely to keep grasses at immature, high-nutrient growth stages in low-productivity areas (plains) than in higher-productivity regions. This nutrient hypothesis is consistent with McNaughton's (1984, 1988) and Hobbs and Swift's (1988) proposition of intraspecific facilitation by which grazers stimulate the growth of their own food supply, and provides a mechanism for herbivore aggregation on the plains. It now needs testing.

Antipredator Adaptations. In his classic paper Jarman (1974) outlined the evolutionary adaptations that link feeding, social structure, and antipredator behavior in African antelopes, and he expanded on these ideas in *Serengeti* (Jarman and Jarman 1979). Since then many studies have concentrated on antipredator adaptations. Antipredator behavior in ungulates has a number of consequences for population size through differences in vulnerability among animals of various age, sex, and condition classes and reproductive status. FitzGibbon and Lazarus (chap. 13) illustrate these differences using the example of Thomson's gazelle. Females, for example, flee from predators before males do. Stotting, an antipredator display, varies with age, season, and condition. Males are caught by predators at a high frequency, so there is a female-biased sex ratio in the population. Some predators, such as wild dogs, select prey in poor condition, which suggests that prey have to trade off food intake against predation risk. Thus, both predation and intraspecific competition affect the gazelle population. FitzGibbon and Lazarus suggest ways in which this interaction can be incorporated into population models of these species.

Predator Dynamics

The study of lion populations on the plains began with Schaller (1972) in 1966, and a virtually unbroken record has been maintained since then. The first 11 years were summarized in *Serengeti* by Hanby and Bygott (1979) and Bertram (1979). They documented an increase in numbers of lion prides on the plains, not as a result of the wildebeest population increase, but because resident prey increased. This finding highlighted the main conclusion that predators with altricial young must maintain territories to protect them, and so cannot move with the migration. Thus, lion populations are limited by resident prey availability.

Subsequent information is summarized by Scheel and Packer (chap. 14) and Hanby, Bygott, and Packer (chap. 15). Due to a decline in rainfall, and hence prey biomass, lion numbers leveled off on the plains in the 1980s. New work on woodland lions suggests that these populations have a high biomass of resident prey, particularly buffalo, and are buffered from annual changes in food availability caused by wildebeest movements. These woodland populations are, therefore, net exporters of dispersers, and thus are "source" populations. In contrast, plains populations, subject to the vagaries of weather and wildebeest movements, cannot maintain themselves on the low resident prey numbers. They are "sink" populations.

These conclusions are particularly important for conservation. Woodlands supply the rest of the Serengeti ecosystem with lions, but it is in the woodlands (west and north) that they are most likely to be killed by poachers. Moreover, in the adjacent woodlands of Maswa and Loliondo they are being shot by hunters (Sinclair and Arcese, chap. 2), so that these areas draw out Serengeti animals, especially males, to fill the vacuum and be further exploited. Populations on the edge of the park cannot stand such heavy exploitation and are now showing extremely skewed sex ratios with few adult males present. This problem is discussed further by Caro and Durant in chapter 21.

The other main predator, the spotted hyena, was initially studied by Kruuk (1972). After a hiatus in the 1970s, research began again in the Mara in 1979 (Frank, Holekamp, and Smale, chap. 17) and recommenced in Serengeti in 1987 (Hofer and East 1993a,b,c; chap. 16). The hyena is the most abundant large predator, with a population of about 9,000 in the ecosystem as compared with 3,000 lions. Less is known about hyena population trends, but hyenas appear to have increased on the plains. Hofer and East have demonstrated that hyenas are more tied to the migrant populations than are lions because of their ability to commute for 40 km or more. They have a social system that allows commuters to pass through adjacent territories. Because of these adaptations hyena numbers may have increased due to higher wildebeest numbers. Frank et al. (chap. 17) concentrate on social behavior, female dominance, and reproductive success. In the Mara, breeding female numbers have remained stable, and predation by lions is the primary cause of mortality.

The finding that predators limit predators appears repeatedly and is one of the important general conclusions in this book (chaps. 17–20). Laurenson (chap. 18) demonstrates this clearly in cheetahs, which generally live at low density and are limited largely by predation by other carnivores. Thus, 92% of juveniles die before independence, and lions account for 73% of the mortality. Leopards and hyenas also kill cheetahs. In terms of population dynamics, the increase in lion numbers on the plains and their year-round residence there (lions used the plains only in the wet season in the 1960s) has resulted in higher mortality for cheetahs and almost certainly lower numbers.

Wild dog numbers have collapsed from over 100 in the 1960s to 2 (and a few transient groups) in the 1990s. The reasons for this decline are in debate (Burrows 1992, chap. 19; Burrows, Hofer, and East 1994; McDonald et al.1992; Creel 1992; Gascoyne et al. 1993) and remain to be resolved. Burrows (chap. 19) has shown that a major decline in wild dog numbers occurred in the early 1970s because several packs were shot outside Serengeti National Park. If this was the only cause of mortality, with no other underlying cause, then the population should have rebounded subsequently. The fact it did not do so, but remained low, is conclusive evidence that there was an underlying cause producing a decline and that the shooting merely accelerated it. Thus, there was a decline in numbers throughout the 1970s.

The most parsimonious approach to understanding the cause of the decline in wild dog numbers is to assume that the same ultimate cause has been present throughout the 30 years, so we must look at what has changed during that period. The three major changes of relevance to wild dogs that have occurred in the ecosystem are the increases in wildebeest, major predator (lion and hyena), and human populations. Wildebeest, which form a major food supply, are clearly not directly responsible. The increase in predator numbers has already been proposed as a likely cause in *Serengeti* (Sinclair 1979a), and has not yet been discounted. The observations of predation on other predators such as the cheetah lend credibility to this hypothesis. The most likely effect of the human population increase is increased transmission of diseases such as distemper and rabies from domestic dogs. Disease outbreaks in the wild dogs have been observed throughout the 30 years, so the disease hypothesis remains a strong candidate. The 1994 canine distemper outbreak in unhandled lions throughout Serengeti reinforces this (pers. obs., V. Morrell 1994 *Science* 264, 1664). Burrows has proposed a special hypothesis, namely, that the stress due to handling dogs for rabies vaccinations or to attach radio collars has resulted in increased mortality. Because handling has been a significant factor only since 1985, it cannot account for the general decline, and at most can only be a contributory factor. Whether handling in fact caused mortality remains to be confirmed.

Future work is needed urgently to sort out which of these hypotheses is most likely, not only because wild dogs are endangered throughout Africa, but also because some of these possible causes could be relevant to the whole complex of predator species. Because radiotelemetry is the only means by which the cause of decline will be established, I suggest that radio collars be used on some members of each pack in the areas surrounding Serengeti, and that monitoring of the packs be intensive. One of the lessons from past work is that monitoring was not sufficiently frequent to observe the fate of individuals or even whole packs; they just disappeared.

Most of the work on predators has focused on large carnivores. Since 1974, however, there has been a long-term study of small carnivores, particularly of mongooses, by J. Rood initially and subsequently by P. Waser and colleagues (chap. 20). Their population estimates of dwarf, banded, and slender mongooses are 94,000, 43,000, and 30,000 respectively. These numbers are an order of magnitude greater than those of the larger carnivores. Populations of the dwarf and banded may be increasing. Since these species feed on dung beetles, which depend on wildebeest, they may be increasing as an indirect result of the wildebeest increase (Waser et al., chap. 20). The slender mongoose feeds on small mammals and birds. Small mammals such as the grass rat *(Arvicanthis niloticus)* show population eruptions every few years, and both carnivores (e.g., wild cats) and raptors (black-shouldered kite, *Elanus caerulus;* long-crested hawk-eagle, *Spizaetus occipitalis*) respond with their own population increases. Whether the slender mongoose also responds remains unknown. However, Waser et al. (chap. 20) suggest that these mongoose populations may be limited by their own predators, such as raptors and other, larger mongoose species.

INTEGRATED RESEARCH AND MANAGEMENT

Since the opening of the international border with Kenya at Namanga and the resumption of full-scale research in Serengeti, there has been increased emphasis on applied ecology with a view to providing information on immediate problems of management and conservation.

Much of the research on predator species has been concerned with behavior. Caro and Durant (chap. 21) describe the link between behavior and conservation. Predator species usually live at low density, in small populations, and are sensitive to changes in prey populations. They are sensitive indicators of ecosystem disturbance. Two species in Serengeti, wild dog and cheetah, are probably below the minimum viable population levels. Caro and Durant recommend for all of these reasons that hunting of all predators around the edge of the park be stopped.

National parks and reserves are fast becoming islands in a matrix of humanity. Can we look at these protected areas as natural islands, or are they unnatural fragments isolated from an original larger area? The answer to this question lies in the degree to which there has been gene flow between populations. Georgiadis has examined this question by using the pattern of genetic subdivision among populations of wildebeest in East Africa (chap. 22). His results show that despite the ability of wildebeest to travel long distances, the current populations in different parks have been naturally isolated since long before the parks were demarcated. Gene flow was either very high or very low depending on whether the intervening habitat was suitable or not. Thus, isolation of these populations has not markedly altered gene flow because wildebeest are habitat specialists. Georgiadis suggests that the same result may not apply to habitat generalists.

The effects of disease on wildlife and domestic animals are becoming increasingly apparent (Burrows, chap. 19; Dobson, chap. 23). Dobson (chap. 23) points out that conservation depends on a high level of cattle inoculation against rinderpest. When veterinary services break down, as they did during the border closure, inoculations decline and disease breaks out. This was seen with an outbreak of rinderpest in 1982 in both Ngorongoro and Serengeti after an absence of 20 years. Tick-borne diseases of cattle have also increased in Ngorongoro. I have already referred to the problems in the wild dogs and the possible increase in transmission of distemper, rabies, or other canine diseases. Dobson concludes that we urgently need to understand how pathogens operate with two hosts in wildlife—domestic animal situations.

The majority of applied research has been on the effects of poaching. Arcese, Hando, and Campbell (chap. 24) describe the time dynamics of poaching while Campbell and Hofer (chap. 25) examine the spatial components. Snaring with wire nooses in thickets is the most common technique for killing animals, and this method disproportionately catches thicket-dwelling species such as buffalo, giraffe, and waterbuck. This was the main method for eliminating buffalo in northern Serengeti, and I suspect it is related to waterbuck and giraffe population declines. Other species that are caught frequently are impala, warthog, and topi; these need to be monitored carefully in the future.

Since the 1960s there has been a sixfold increase in the number of arrests of poachers. The number of arrests is determined by the number of patrols, as elsewhere in Africa (Leader-Williams, Albon, and Berry 1990). A combination of vehicle and foot patrols proved to be 3–5 times more effective in catching poachers than other methods, and there are several possible ways these could be improved in the future with funds from donors.

Although the actual number of illegal hunters is unknown, a rough extrapolation suggests that up to 30,000 illegal hunters could service about 1 million people within a radius of 45 km of the western boundary of Serengeti National Park (Campbell and Hofer, chap. 25). A very crude estimate of illegal harvest offtake is 200,000 animals per year, of which about 75,000 is of resident large herbivores both inside and outside of the park. This harvest is quite unsustainable and some areas within the park are already devoid of animals. The present and proposed Game Reserves (see fig. 25.1) will not prevent poaching, either in those reserves or in the park (Campbell and Hofer, chap. 25).

The widespread occurrence of illegal hunting within reserve boundaries and the difficulties that park wardens face in enforcement (Arcese, Hando, and Campbell, chap. 24; Campbell and Hofer, chap. 25) have caused conservationists to turn their attention to other methods, namely, Integrated Conservation and Development Projects. The value of these projects and their problems are discussed in chapter 2, so I mention only that two such projects are described for the Serengeti. One project, the Ngorongoro Conservation Area, is probably the first multiple land use scheme to have been set up in Africa, having started in 1959 (Perkin, chap. 26). Perkin describes the system and the lessons that have been learned in the past 30 years, and points out some of the major difficulties in maintaining control of the resources. The other project is the Serengeti Regional Conservation Strategy (Mbano et al., chap. 28), which has been in operation for only a few years. This project is concerned with the agricultural peoples west of Serengeti and is attempting to replace the illegal hunting inside the park with a legalized cull outside the park.

One of the problems with encouraging a legal cull is that it ties the human population, which is increasing at an extremely high rate, to a resource, the wildlife biomass, whose production cannot increase past a certain level. Thus, local peoples will either obtain a decreasing per capita food quota or will resort to illegal hunting again and exceed the maximum yield. Given this, it would perhaps be a safer strategy for conservation of the Serengeti ecosystem if the human population were encouraged to exploit another resource. This could be done by supplying domestic meat at a price that would undercut the price of illegal wildlife meat. Poaching, like all economic enterprises, requires that benefits exceed costs, with benefits coming from selling meat and costs from being caught by rangers. So far, anti-poaching efforts have concentrated on trying to raise the costs, with little success—presumably the chances of capture are low. Thus, a new strategy of lowering the benefits by undercutting the illegal market may be worth considering. Once the illegal trade dwindles, development agencies should concentrate on building an animal husbandry system, which currently does not exist. This is one of the recommendations put forward by Norton-Griffiths (chap. 27).

Researchers and park staff conducted a workshop at Seronera in December 1991 to integrate research knowledge with park management and conservation requirements (Hilborn et al., chap. 29). The overwhelming conclusion from these exercises is that the future level of poaching will determine what happens to the system. Since poaching offtake is approaching maximum sustained yield for the wildebeest (and is well beyond this for some of the resident species), changes in the level of poaching in the next few years will determine whether the wildebeest population will collapse or not. The wildebeest is the keystone species for this ecosystem, as outlined in *Serengeti,* and what happens to this population will determine the viability of the Serengeti ecosystem.

GENERAL CONCLUSIONS ON ECOSYSTEM STRUCTURE AND FUNCTION

The combined results reported in *Serengeti* and this volume lead to some general conclusions on how this ecosystem works.

1. Stochastic disturbances, such as burning and rainfall, and rare events, such as the conditions for tree regeneration, are major determinants of community structure, primarily on the vegetation but indirectly on the fauna.

2. Spatial heterogeneity in geology, soils (e.g., soil catena, "hot spots"), climate (e.g., rainfall distribution) and, therefore, vegetation (e.g., plains versus woodland) determines the migration.

3. Migratory behavior allows large populations of herbivores.

4. The large migratory wildebeest population determines nearly all other components of the system. Therefore, the wildebeest acts as a keystone species.

5. Pulse germination of seedlings produces even-aged stands of trees; this is a feature of tropical savanna ecosystems. Recruitment into the tree

population is determined by both stochastic events and herbivore feeding. Grassland structure and composition is determined by soils, edaphic features, and herbivore feeding.

6. Migratory herbivores and very large herbivores are regulated by their food supply. (However, it is not yet clear whether this applies to zebra.) In contrast, smaller resident herbivores may be regulated by predation.

7. Major predators (lion, hyena) are limited by their food supply. Smaller predators appear to be limited by their own predators rather than by food supply.

8. In terms of top-down or bottom-up effects on the system, we have in the component that involves the large mammals the complex result shown schematically in figure 1.6. The part of the system that involves migrants has largely bottom-up regulatory effects, whereas the part involving residents has mainly top-down effects. These differences are due not so much to differences in productivity and numbers of trophic levels, but rather to sensitivity in the scale of heterogeneity and disturbance (Menge and Olson 1990; Murray and Brown 1993). Migrants respond to large-scale heterogeneity in soils, landscapes, and habitats, and are regulated from below. Residents are insensitive to the large scale but respond to finer scales of habitat heterogeneity, and they appear to be regulated from above.

PRIORITIES FOR FUTURE RESEARCH

Despite the considerable amount of information that has been obtained, our understanding of the system is incomplete. Research includes both monitoring and baseline studies, and there are several major areas in which both kinds of work are needed. The basic research does not cover all areas of natural history. Many of the gaps in the first volume remain in this one, but this is inevitable so long as funding for projects is brought in by individual researchers who are obliged to meet other requirements. This process for funding research will inevitably miss the basic descriptive and natural history components of groups such as the insects, soil organisms, most of the plants, and even the birds. With the renewed interest in biodiversity research we hope that new research projects designed to document biodiversity will fill these gaps. I discuss these in a suggested order of priority for conservation and management.

1. The illegal killing of the migrant ungulates by poachers is potentially the most serious threat to the Serengeti system. Since the migration determines the structure and function of the system, overharvesting of the migrants with a collapse of their populations will result in the collapse of the whole system. There is an urgent need both to document this offtake

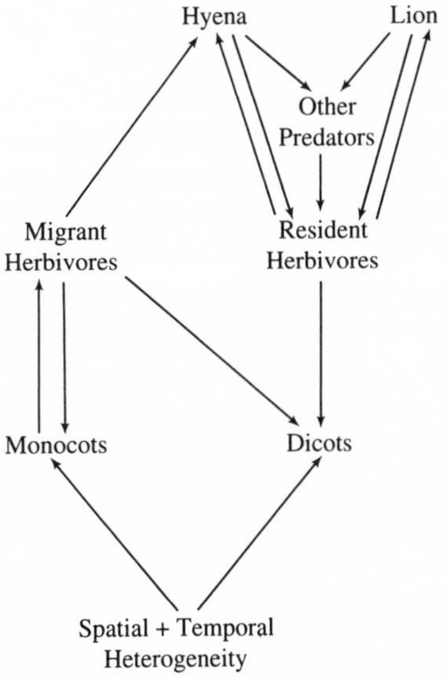

Figure 1.6 Schematic representation of the major trophic effects in the large mammal component of the ecosystem.

and to find ways to bring it under control. Furthermore, it is 20 years since the movements of the migrants were fully described, and there is some indication that both different routes and different areas are now being used, particularly outside Serengeti National Park, where animals are subject to poaching. Thus, the migration needs to be documented again.

2. There is a need to study the threatened habitats and species of the ecosystem. Unless we protect the habitats and their spatial arrangement, animal populations will decline. The riverine forests of the northern Serengeti and Mara are unique relict habitats and are fast disappearing. We have little knowledge of what lives in them, let alone why they are changing. An understanding of what influences regeneration is needed. We also need to understand the causes of patchiness and heterogeneity in the savanna vegetation.

3. The predator populations of the woodlands appear to be particularly important as source populations for the rest of the system. Little research has been done on these populations, and this should now be a priority for predator research.

4. Causes of mortality in the wild dog population are still unknown. To rescue this endangered species an intensive radiotelemetry study is required.

5. There has been a change in the woodland structure of Serengeti. We need to know how this affects the browsing ungulates. Little is known

about the population ecology of any browser. These include the migrant species, eland and Grant's gazelle, and residents such as giraffe, impala, and the several very small antelopes.

6. The *Terminalia* and *Combretum* habitats of the northwest and Maswa, with their different plant, bird, and mammal assemblages, are virtually unknown and warrant more attention.

7. The small carnivores are virtually unknown, and their basic population ecology needs description. Thus, although two of the three jackal species have been studied on the plains (Moehlman 1983), little is known about them in the western and northern woodlands. Even less is known about the small cats and the mongooses not covered in this volume.

8. The ostrich is under threat all over Africa. Serengeti provides an opportunity to study recruitment and mortality for baseline information.

9. Consistent with the appreciation that overall biodiversity is important in ecosystems, we need more basic information on the bird community to find out which species are vulnerable to natural and human-induced perturbations.

10. The invertebrate fauna is effectively untouched. Apart from being the major food supply of birds, it also has a dominant browsing impact on the vegetation (Sinclair 1975), and at least a basic description of phenology should be carried out.

These are some of the components of the ecosystem that urgently require study if we are to advise park managers. It is not an exhaustive list. Such studies should be placed in the context of increased human encroachment outside the system and increased tourism inside. Human impacts must be regarded as perturbations. They should be studied through adaptive management (Walters and Holling 1990) a method in which scientists and managers work together: managers alter their management actions in a planned and agreed-upon fashion, and scientists monitor the consequences of these actions.

The Serengeti is unique as an ecosystem. As one of the first proposed World Heritage Sites, and with its high profile, the Serengeti provides an example of applied research and conservation biology to the world. As the needs for conservation of biodiversity become increasingly clear, old-fashioned attitudes that parks and reserves should be left to look after themselves without research monitoring must be changed. The late Myles Turner, a warden in Serengeti during 1956–1972, wrote about research thus: "One thing is sure: it was a great confidence trick, and virtually nothing has ever come out of it to help the hard-pressed animals of East Africa" (Turner 1987, 163). He has thrown down the challenge. We have accepted that challenge, and this book is part of the answer. It is not complete; we shall build on it and continue to work toward syntheses of research that allow its beneficial application. In the process, we hope to show that research is fundamental to management and conservation.

REFERENCES

Albon, S. D., and Langvatn, R. 1992. Plant phenology and the benefits of migration in a temperate ungulate. *Oikos* 65:502–13.

Bell, R. H. V. 1970. The use of the herb layer by grazing ungulates in the Serengeti. In *Animal populations in relation to their food resources,* ed. A. Watson, 111–23. Oxford: Blackwell.

———. 1971. A grazing system in the Serengeti. *Sci. Am.* 224:86–93.

Bertram, B. C. R. 1979. Serengeti predators and their social systems. In *Serengeti: Dynamics of an ecosystem,* ed. A. R. E. Sinclair and M. Norton-Griffiths, 221–48. Chicago: University of Chicago Press.

Burrows, R. 1992. Rabies in wild dogs. *Nature* 359:277.

Burrows, R., Hofer, H., and East, M. L. 1994. Demography, extinction, and intervention in a small population: The case of the Serengeti wild dogs. *Proc. R. Soc. Lond. B* 256:281–92.

Campbell, K. L. I., Huish, S. A., and Kajuni, A. R., eds. 1991. *Serengeti National Park management plan.* Arusha: Tanzania National Parks.

Creel, S. 1992. Cause of wild dog deaths. *Nature* 360:633.

Dublin, H. T. 1986. Decline of the Mara woodlands: The role of fire and elephants. Ph.D. dissertation, University of British Columbia.

———. 1991. Dynamics of the Serengeti-Mara woodlands: An historical perspective. *For. Conserv. Hist.* 35:169–78.

Dublin, H. T., and Douglas-Hamilton, I. 1987. Status and trends of elephants in the Serengeti-Mara ecosystem. *Afr. J. Ecol.* 25:19–33.

Dublin, H. T., Sinclair, A. R. E., Boutin, S., Anderson, E., Jago, M., and Arcese, P. 1990. Does competition regulate ungulate populations? Further evidence from Serengeti, Tanzania. *Oecologia* 82:283–88.

Ford, J. 1971. *The role of trypanosomiases in African ecology.* Oxford: Clarendon Press.

Foster, R., and Bresele, L. 1992. *Dung beetles: Important invertebrates worth considering.* Conservation Monitoring News, no. 5, 6–7. Frankfurt: Frankfurt Zoological Society.

Fryxell, J. M., Greever, J., and Sinclair, A. R. E. 1988. Why are migratory ungulates so abundant? *Am. Nat.* 131:781–98.

Fryxell, J. M., and Sinclair, A. R. E. 1988. Causes and consequences of migration by large herbivores. *Trends Ecol. Evol.* 3:237–41.

Gascoyne, S. C., Laurenson, M. K., Lelo, S., and Borner, M. 1993. Rabies in African wild dogs *(Lycaon pictus)* in the Serengeti Region, Tanzania. *J. Wildl. Dis.* 29:396–402.

Grimsdell, J. J. R. 1979. Changes in populations of resident ungulates. In *Serengeti: Dynamics of an ecosystem,* ed. A. R. E. Sinclair and M. Norton-Griffiths, 353–59. Chicago: University of Chicago Press.

Grzimek, B., and Grzimek, M. 1960. *Serengeti shall not die.* London: Hamish Hamilton.

Gwynne, M. D., and Bell, R. H. V. 1968. Selection of vegetation components by grazing ungulates in Serengeti National Park. *Nature* 220:390–93.

Hanby, J. P., and Bygott, J. D. 1979. Population changes in lions and other predators. In *Serengeti: Dynamics of an ecosystem,* ed. A. R. E. Sinclair and M. Norton-Griffiths, 249–62. Chicago: University of Chicago Press.

Herlocker, D. J. 1976. *Woody vegetation of the Serengeti National Park.* College Station: Texas A&M University.

Hobbs, N. T., and Swift, D. W. 1988. Grazing in herds: When are nutritional benefits realized? *Am. Nat.* 760–64.

Hofer, H., and East, M. L. 1993a. The commuting system of Serengeti spotted hyaenas: How a predator copes with migratory prey. I. Social organization. *Anim. Behav.* 46:547–57.

———. 1993b. The commuting system of Serengeti spotted hyaenas: How a predator copes with migratory prey. II. Intrusion pressure and commuter's space use. *Anim. Behav.* 46:559–74.

———. 1993c. The commuting system of Serengeti spotted hyaenas: How a predator copes with migratory prey. III. Attendance and maternal care. *Anim. Behav.* 46:575–89.

Holt, R. D. 1977. Predation, apparent competition and the structure of prey communities. *Theor. Popul. Biol.* 12:197–229.

Jarman, P. J. 1974. The social organisation of antelope in relation to their ecology. *Behaviour* 48:215–67.

Jarman, P. J., and Jarman, M. V. 1979. The dynamics of ungulate social organization. In *Serengeti: Dynamics of an ecosystem,* ed. A. R. E. Sinclair and M. Norton-Griffiths, 185–220. Chicago: University of Chicago Press.

Kreulen, D. A. 1975. Wildebeest habitat selection on the Serengeti plains, Tanzania, in relation to calcium and lactation: A preliminary report. *E. Afr. Wildl. J.* 13:297–304.

Kruuk, H. 1972. *The spotted hyena: A study of predation and social behavior.* Chicago: University of Chicago Press.

Leader-Williams, N., Albon, S. D., and Berry, P. S. M. 1990. Illegal exploitation of black rhinoceros and elephant populations: Patterns of decline, law enforcement and patrol effort in Luangwa valley, Zambia. *J. Appl. Ecol.* 27:1055–87.

Leakey, M. D., and Hay, R. L. 1979. Pliocene footprints in the Laetolil beds at Laetoli, northern Tanzania. *Nature* 278:317–23.

Lovejoy, C. O. 1981. The origin of man. *Science* 211:341–50.

MacDonald, D. W., Artois, M., Aubert, M., Bishop, D. L., Ginsberg, J. R., King, A., Kock, N., and Perry, B. D. 1992. Cause of wild dog deaths. *Nature* 360:633–34.

Maddock, L. 1979. The "Migration" and grazing succession. In *Serengeti: Dynamics of an ecosystem,* ed. A. R. E. Sinclair and M. Norton-Griffiths, 104–29. Chicago: University of Chicago Press.

McNaughton, S. J. 1979. Grassland-herbivore dynamics. In *Serengeti: Dynamics of an ecosystem,* ed. A. R. E. Sinclair and M. Norton-Griffiths, 46–81. Chicago: University of Chicago Press.

———. 1984. Grazing lawns: Animals in herds, plant form, and coevolution. *Am. Nat.* 124:863–86.

———. 1988. Mineral nutrition and spatial concentrations of ungulates. *Nature* 334:343–45.

Mduma, S. A. R., and Sinclair, A. R. E. 1994. The function of habitat selection by oribi in Serengeti, Tanzania. *Afr. J. Ecol.* 32:16–29.

Menge, B. A., and Olson, A. M. 1990. Role of scale and environmental factors in regulation of community structure. *Trends Ecol. Evol.* 5:52–57.

Moehlman, P. D. 1983. Socioecology of silverbacked and golden jackals *(Canis mesomelas, Canis aureus).* In *Recent Advances in the Study of Mammalian Behavior,* ed. J. F. Eisenberg and D. G. Kleiman, 42–438. Special publication no. 7. Lawrence, Kans.: American Society of Mammalogists.

Murray, M. G., and Brown, D. 1993. Niche separation of grazing ungulates in the Serengeti: An experimental test. *J. Anim. Ecol.* 62:380–89.

Norton-Griffiths, M. 1979. The influence of grazing, browsing, and fire on the vegetation dynamics of the Serengeti. In *Serengeti: Dynamics of an ecosystem,* ed. A. R. E. Sinclair and M. Norton-Griffiths, 310–52. Chicago: University of Chicago Press.

Packer, C. 1990. Serengeti lion survey. Report to TANAPA, SWRI, MWEKA and the Game Department. Tanzania National Parks, Arusha, Tanzania.

Pearsall, W. H. 1957. Report on an ecological survey of the Serengeti National Park, Tanganyika. *Oryx* 4:71–136.

Plowright, W. 1982. The effects of rinderpest and rinderpest control on wildlife in Africa. *Symp. Zool. Soc. London* 50:1–28.

Schaller, G. B. 1972. *The Serengeti lion: A study of predator-prey relations.* Chicago: University of Chicago Press.

Schmidl, D. 1982. *The birds of the Serengeti National Park, Tanzania.* Checklist no. 5. London: British Ornithologists Union.

Sinclair, A. R. E. 1975. The resource limitation of trophic levels in tropical grassland ecosystems. *J. Anim. Ecol.* 44:497–520.

———. 1977. *The African buffalo: A study of resource limitation of populations.* Chicago: University of Chicago Press.

———. 1979a. Dynamics of the Serengeti ecosystem: Process and pattern. *In Serengeti: Dynamics of an ecosystem,* ed. A. R. E. Sinclair and M. Norton-Griffiths, 1–30. Chicago: University of Chicago Press.

———. 1979b. The eruption of the ruminants. In *Serengeti: Dynamics of an ecosystem,* ed. A. R. E. Sinclair and M. Norton-Griffiths, 82–103. Chicago: University of Chicago Press.

———. 1979c. The Serengeti environment. In *Serengeti: Dynamics of an ecosystem,* ed. A. R. E. Sinclair and M. Norton-Griffiths, 31–45. Chicago: University of Chicago Press.

———. 1985. Does interspecific competition or predation shape the African ungulate community? *J. Anim. Ecol.* 54:899–918.

Sinclair, A. R. E., Dublin, H., and Borner, M. 1985. Population regulation of Serengeti wildebeest: A test of the food hypothesis. *Oecologia* 65:266–68.

Sinclair, A. R. E., and Fryxell, J. M. 1985. The Sahel of Africa: Ecology of a disaster. *Can. J. Zool.* 63:987–94.

Sinclair, A. R. E., Leakey, M. D., and Norton-Griffiths, M. 1986. Migration and hominid bipedalism. *Nature* 324:307–8.

Sinclair, A. R. E., and Norton-Griffiths, M., eds. 1979. *Serengeti: Dynamics of an ecosystem.* Chicago: University of Chicago Press.

Talbot, L. M., and Talbot, M. H. 1963. *The wildebeest in western Masailand, East Africa.* Wildlife Monographs, no. 12. Washington, D.C.: The Wildlife Society.

Turner, M. 1987. *My Serengeti years.* Ed. by B. Jackman. London: Elm Tree Books/Hamish Hamilton.

Vesey-Fitzgerald, D. F. 1960. Grazing succession among East African game animals. *J. Mammal.* 41:161–72.

Walters, C. J., and Holling, C. S. 1990. Large-scale management experiments and learning by doing. *Ecology* 71:2060–68.

Watson, R. M. 1967. The population ecology of the wildebeest *(Connochaetes taurinus albojubatus)* in the Serengeti. Ph.D. thesis, Cambridge University, Cambridge.

White, S. E. 1915. *The rediscovered country.* New York: Doubleday.

TWO

Serengeti in the Context of Worldwide Conservation Efforts

A. R. E. Sinclair and Peter Arcese

The Serengeti-Mara ecosystem is one of the great natural wonders of the world. It was one of the first areas nominated as a World Heritage Site, and together with the Ngorongoro Conservation Area, it forms one of the world's largest Biosphere Reserves. Yet since the early 1900s, the Serengeti-Mara has lost over 50% of its area as a natural ecosystem. Most of this loss in the last 30 years has been within the legally protected boundaries. What has occurred in the Serengeti-Mara provides an indication of what is also happening in many other natural areas of worldwide significance. A once vast natural area bordered by undeveloped lands, the Serengeti-Mara ecosystem is fast becoming an insular assemblage of native species in a sea of humanity. As a result, the area is now severely threatened by the detrimental effects of human encroachment, the overexploitation and loss of its wildlife species, and the progressive loss of the natural system within its boundaries. We must ask in the face of these events, summarized in chapter 1, whether the current approaches to conservation will achieve the intended results for the protection of biodiversity, not just for Serengeti but for the world as a whole.

This chapter puts the events, research, and management in the Serengeti into the context of world conservation. To do so, we first outline three levels of questions that conservation ecologists currently ask in their research. We then discuss the roles of research in conservation, and in particular how these are applicable to management initiatives designed to integrate the natural areas of Serengeti with the surrounding agricultural and pastoral areas, and their human populations. Finally, we examine some of the problems and pitfalls confronting not only the Serengeti but also conservation communities of the world as a whole.

THREE LEVELS OF QUESTIONS IN CONSERVATION

Despite increasing awareness of the problems of conserving biological diversity, the conservation community—international bodies, governments and nongovernmental organizations, and the academic community—is still primarily responding to conservation problems in an ad hoc way, addressing each crisis individually as it occurs. A long-term conservation policy is needed, but this can only develop from an understanding of the types of problems involved. We suggest that there are three levels of problems that need to be addressed by researchers who wish to apply the results of their efforts to solving problems in conservation. We discuss these from the specific to the general and provide some examples of these levels of inquiry.

Third-Order Questions

The majority of research in conservation ecology is directed at questions at the level of individual populations, species, areas, or networks of populations or areas. Examples are presented in several chapters of this volume (e.g., FitzGibbon and Lazarus, chap. 13; Caro and Durant, chap. 21; Georgiadis, chap. 22). These questions focus on practical problems such as (1) the minimum population size that must be preserved for demographic or genetic reasons, (2) the minimum size and number of reserves necessary to maintain viable populations of individual species, (3) the shape of such reserves or reserve networks, and (4) the degree of dispersal shown by a species and hence whether or not corridors between reserves are necessary. These are only some of the issues at this level, and they absorb most of the academic attention and funding.

At this same level are the efforts of most governments and nongovernmental organizations. For the most part, these focus on the preservation of particular endangered species or habitats. These efforts aim primarily to describe the status of species or areas: for example, the population size of a rare species or the degree of human encroachment on a threatened habitat. The species that attract the most attention in such efforts are often those that have a high profile for political or emotive reasons, such as the giant panda *(Ailuropoda melanoleuca)*, the tiger *(Panthera tigris)*, or the African elephant. Usually, such species are among the higher "charismatic" vertebrates, and the higher their profile, the larger the organization that deals with them.

Second-Order Questions: Species Triage

Third-order questions beg a number of more general questions: for example, on what basis are particular species chosen for protection? Or, why should one area be preserved over another? It is estimated that of the 30 to 100 million species of organisms on this planet, at least 1 million will be extinct by the year 2000 (Norman 1981; Ehrlich and Wilson 1992; Raven and Wilson 1992). As a world community, we have neither the resources nor the political will to save each of these species (as was clear at the "Earth Summit" conference in 1992 at Rio). We are therefore faced with hard decisions concerning which species we should attempt to save and which must be abandoned (Soulé et al. 1986; McNeely et al. 1990; Vane-Wright, Humphries, and Williams 1991).

Given limited resources and time, we are obliged to adopt "species triage" ("triage" is a term originated during the Crimean War by doctors who were trying to save lives with insufficient resources). We use the concept here in the context of saving species or habitats: some species or areas can be left unattended because they can tolerate human encroachment; others are too far gone to warrant an extensive investment of time and resources because extinction in the near future is certain. A third group of species or areas are those that are possible to save with some help.

How do we decide into which species and habitats to put our efforts? The present ad hoc approach, although it sometimes offers political expediency, is highly unlikely to optimize the distribution of resources (money, land, time) in ways that maximize the number of species saved. Optimization underlies the fact that there is always an opportunity cost in conservation: resources made available for one species or habitat result in less being available for another. While this suggests that we must have a rational method for the optimal distribution of conservation resources, no method for the optimal partitioning of effort and resources to maximize conservation on a worldwide basis is currently available. Hence, we are left with the uncomfortable conclusion that the habitats and species currently receiving most of the resources and attention may not be the appropriate ones. However, a start has been made in developing a hierarchy for decisions concerning species (Noss 1990; Tisdell 1990; Vane-Wright, Humphries, and Williams 1991; Faith 1992) and reserves (Margules 1989; Margules, Nicholls, and Pressey 1988; Margules, Pressey, and Nicholls 1991), and these approaches must be used further to develop a world system. One point is clear from the above references: the approach of prioritizing areas with the highest diversity before proceeding to the less diverse ones is not necessarily an optimum procedure, and it could ultimately result in more extinctions than would occur under a procedure based on optimization.

Optimization procedures for deciding how to allocate resources become increasingly important as fewer resources are available for conservation. Thus, developing countries, such as many of those in Africa with annual rates of increase in human populations averaging near 3% (Sinclair and Wells 1989), are under more pressure to develop such procedures than their richer neighbors.

First-Order Questions: Habitat Renewal

An underlying premise of second- and third-order questions is that species and their associated communities and habitats will be conserved by reserving land. There is little information available to test this assumption, however, and it begs a higher-level question: is the preservation of natural habitats sufficient to promote conservation? Although reserving land is a necessary component of conservation, it is unlikely to be sufficient because all protected areas will suffer attrition over time through both deterministic and stochastic processes. Therefore, we must consider habitat renewal to counteract this erosion of land (Sinclair et al. 1995). In the Serengeti-Mara, conservation biologists and policymakers have a major challenge ahead in addressing this issue.

Loss of Ecological Integrity in Serengeti. The Serengeti is a prime example of the way many natural ecosystems are being eroded through the incremental loss of integrity of their natural communities, as judged by the local extinction of dominant or keystone species. The approximate extent of the Serengeti-Mara ecosystem from the beginning of the twentieth century through the 1950s is known from various early reports (White 1915; A. Moore, R. O. Sinclair, and M. I. M. Turner, pers. comm.) and during the 1960s from published data (Sinclair 1977; Dublin et al. 1990; Hilborn et al., chap. 29). This information allows us to trace the proportion of the original natural area in about 1910 that remained by 1992 (fig. 2.1), and it shows a progressive decline in the area of the "undisturbed" community (which included indigenous hunters using traditional weapons). This decline has occurred through the expansion of the human population. As a result, most of the area outside Serengeti National Park has been lost to cultivation, and a large proportion of the area inside the park has been significantly altered by poaching (Sinclair, chap. 1; Campbell and Borner, chap. 6). We see, first, that some 40% of the natural ecosystem has been lost; second, that there is no sign of this loss abating, and it may be accelerating; third, that this loss has taken place largely within the legal boundaries of the park and despite the expansion of the legal boundaries; and fourth, that the greatest loss has occurred during the period when the greatest attention has been given to the area by researchers and conservationists (1960s–1990s).

The Serengeti is not only progressively losing area as an integrated natural ecosystem, it is also losing species and habitats. Thus, rhinoceroses, once abundant, have been effectively exterminated from the ecosystem, and elephants have been reduced by 80%, both by poaching (Dublin and Douglas-Hamilton 1987). Roan antelope have disappeared from many areas, possibly due to the disappearance of their *Combretum*-dominated habitat on hills (Campbell and Borner, chap. 6; Sinclair, chap.

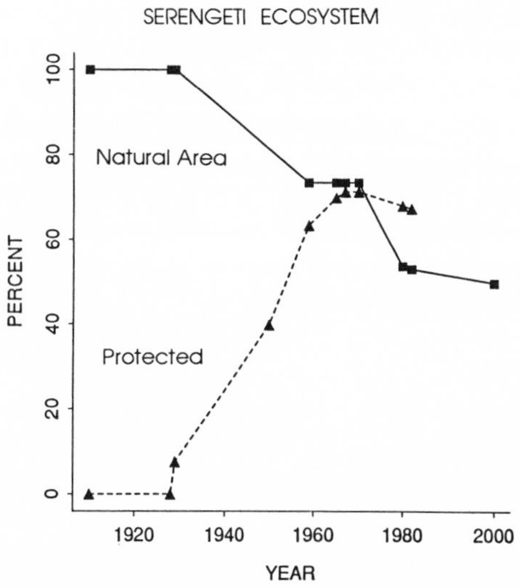

Figure 2.1 Natural, undisturbed area of the Serengeti-Mara ecosystem as a percentage of the original ecosystem (30,143 km^2) around the year 1910 (solid line), and percentage of the total area given legal protection as either park or reserve (dashed line). The estimate for year 2000 is the expected result of the expansion of wheat farms north of the Mara Reserve. (From Sinclair et al. 1995).

9). Wild dogs have also declined through the 1970s and 1980s from approximately 100 to two known animals (Burrows, chap. 19; Georgiadis, chap. 21). The cause of this decline remains unknown, but there are several hypotheses (MacDonald 1992; Creel 1992; Burrows 1992; Gascoyne et al. 1993; Sinclair, chap. 1; Burrows, chap. 19). A recent example likely to result in further attrition within the ecosystem is the 10-year hunting concession in Loliondo, adjacent to Serengeti National Park, let to the Sultan of Dubai in 1992. Unfortunately, this concession provides unlimited quotas and no specific protection for endangered species such as cheetah and wild dog. Unregulated hunting of even common species in areas adjacent to Serengeti may have dramatic negative effects. As explained in chapter 1, for example, overhunting of male lions alters local adult sex ratios, draws males out from the park, and thus disrupts populations within it. Whether this particular concession is allowed to proceed for its full term is perhaps a moot point, for if it can happen once, it can happen again if the fee is high enough (Newby 1990; Cloudsley-Thompson 1992).

The Serengeti example, then, highlights the principle that we cannot just protect land, habitat, and species and expect biodiversity to remain constant. We must assume that there will be an inevitable erosion of ecosystems through human impact irrespective of legal boundaries. We suggest that over the long-term a policy of active habitat restoration will be necessary to counteract this.

THE ROLES OF RESEARCH IN PROTECTED AREAS

To address the above questions in the conservation of biodiversity, research in Serengeti and other protected areas has had three major objectives differing in levels of generality. Since these objectives are not often stated explicitly, we outline them here.

Ecological Baselines

Of the three roles of research in conservation, the monitoring of natural areas as ecological baselines is the most important in the long term and the most general in application. As the human population changes the environment and habitats throughout the world, we must have a method of distinguishing the effects of humans from those of other (nonhuman) factors. The best way of doing this is to set up scientific controls for the effects of humans in areas that are maintained in as natural a state as is possible. These areas are the ecological baselines. Although they do not have to be national parks and reserves, in many countries these become, de facto, the baselines for lack of other protected areas. Such natural areas may be viewed as the insurance policy for our planet because they provide living examples of functioning natural communities and potentially, therefore, offer blueprints for restoration efforts in the future.

The rationale underlying the need for ecological baselines has been elaborated elsewhere (Sinclair 1981, 1983). Two important points underlie this argument. First, the degrading effects of humans on ecosystems often, perhaps usually, begin slowly and imperceptibly. At the same time, humans become accustomed to changes in their environment and so do not express concern. Second, there is evidence to suggest that ecosystems, including the Serengeti, do not change linearly in response to disturbance, but rather have breakpoints and move rapidly from one domain of stability to another (Holling 1973; May 1977; Peterman, Clark, and Holling 1979; Walker et al. 1981; Estes, Duggins, and Rathbun 1989; Sinclair 1989; Westoby, Walker, and Noy-Meir 1989). Such radical changes in the ecosystem can reduce biodiversity and may not be easily reversed. In the Serengeti, for example, the effects of fires set by humans have caused a change in vegetation from woodlands to grasslands and some areas (the Mara) appear to be locked into the grassland state (Dublin, Sinclair and McGlade 1990; Dublin, chap. 4; Sinclair, chap. 5).

At present, long-term monitoring efforts in the Serengeti region are being maintained under collaborative arrangements involving nongovernmental conservation organizations, the Serengeti Wildlife Research Institute, and Tanzania National Parks. However, the resources available to these groups severely limit the number of projects under way and the frequency with which recurring surveys can be conducted. History shows that funding for long-term monitoring in Serengeti has been subject to interruption, with near-catastrophic effects for even highly conspicuous species (Sinclair, chap. 1; see below).

Monitoring for Biodiversity

On a more specific level, research in the form of monitoring is required in any particular area for the conservation of its biodiversity. Ecosystems are dynamic, and changes in species abundances should be expected. However, most natural areas are arbitrarily curtailed by legal boundaries, and in these cases fluctuations in populations that once would have been tolerated by the system may now result in local extinctions. The best way to provide warning of this outcome is to collect information on the system through monitoring.

One of the clearest benefits of monitoring in the Serengeti was the detection of severe poaching on buffalo. In 1984 a census of buffalo in northern Serengeti showed a 90% drop in numbers in comparison with the long-term records from the period 1965–1976. Despite the magnitude of this drop, it had gone unnoticed previously. Subsequent monitoring data have shown other significant changes in Serengeti National Park (Sinclair, chaps. 5 and 9; Campbell and Borner, chap. 6), in Mara (Dublin, chap. 4; Broten and Said, chap. 8), and in Ngorongoro (Runyoro et al., chap. 7). Monitoring data have also shown where management efforts have succeeded in maintaining natural processes and viable populations within these areas (Sinclair, chaps. 5 and 9; Campbell and Borner, chap. 6; Runyoro et al., chap. 7; Broten and Said, chap. 8; Dobson, chap. 23).

Finally, monitoring data gathered in different areas managed under different mandates provide opportunities for researchers to treat these as experiments in order to understand better the outcomes of alternate management decisions (Walters 1986). Several opportunities for this type of research await future workers in the Serengeti region because of the diverse management goals and policies employed in the various protected areas comprised by the ecosystem (e.g., game reserves, conservation areas, controlled areas, and national parks; Perkin, chap. 26; Mbano et al., chap. 28).

Park Planning

The most specific type of research is that required for planning in reserves and national parks. This includes, for example, research aimed at providing information required for the siting of roads, hotels, and other tourist facilities, as well as research aimed at increasing the efficiency of management operations (e.g., Arcese, Hando, and Campbell, chap. 24). For example, information on sensitive areas where roads should not be placed must be made available. Roads sited in sensitive areas require costly maintenance, heavy equipment, and skilled engineers in order to prevent the proliferation of parallel tracks that are otherwise a common sight in areas with high rainfall or friable soils. In Serengeti, such a sensitive area is the short-grass plains, where tracks once made remain for decades: fourteen sets of parallel tracks are spread over about 400 m on the route from Naabi Hill to Lake Lagarja in 1993. Along the eastern boundary of the park, from the main road north to Barafu Kopjes, there are currently ten sets of parallel tracks, which have developed since the boundary was established in the late 1950s. These developed rapidly with the increased use of heavy vehicles associated with development in the Loliondo area, and they further illustrate the indirect effects of outside influences on the park. In Serengeti, research into ways of ameliorating such impacts has been limited, but it will be of increasing importance as development of the park accelerates.

INTEGRATED CONSERVATION AND DEVELOPMENT

Conflicts between human populations and the protected areas of the Serengeti-Mara ecosystem have led, as elsewhere in the world, to new initiatives known as Integrated Conservation and Development Projects (ICDPs). These have developed because traditional approaches to park management and enforcement have not kept up with competing pressures from outside the system. Thus, it has become clear that protected areas can no longer isolate themselves from their surrounding areas. The three main objectives of ICDPs are listed below, but little coordinated research is being undertaken to ascertain whether the policies espoused are having their desired effects overall. We suggest that management policies implemented as part of these new initiatives should be viewed as hypotheses to be tested (Walters 1986). This testing should be undertaken scientifically, via monitoring and temporal and comparative analyses, so that managers can adapt quickly if data suggest that the policies are failing to meet their objectives. In this section, we discuss some areas for consideration by researchers interested in these questions.

Integrated Conservation and Development Projects

The three main objectives of ICDPs are: (1) protected area management, (2) establishment of "buffer zones" where human activity is phased down to allow a blending with conservation objectives, and (3) local social and economic development (Wells, Brandon, and Hannah 1992).

Some of the results of such projects in Africa are described by Kiss

(1990). In the Serengeti ecosystem the best-established program is that of the Ngorongoro Conservation Area, which has been in existence since the late 1950s (Runyoro et al., chap. 7; Perkin, chap. 26). The Mara Reserve, in Kenya, is not a national park, but is administered by a local municipality together with the surrounding cattle ranches and wheat farms (Broten and Said, chap. 8; Norton-Griffiths, chap. 27).

The area surrounding Serengeti National Park is included in the Serengeti Regional Conservation Strategy (SRCS) (Mbano et al., chap. 28). This region contains some 1.7 million people to the west of the park. Poachers within a radius of 45 km commute to the park to obtain animals and supply an area containing 1 million people (Campbell and Hofer, chap. 25). Ironically, although a considerable number of people obtain direct and indirect economic benefit by such exploitation of the ecosystem, local people have a generally negative attitude toward Serengeti National Park (Mbano et al., chap. 28). They normally vote for political representatives who work against the park, a direct result of the historical antagonism between park authorities and administrators of the surrounding area. It is encouraging, however, that such attitudes are not general in Tanzania. Newmark et al. (1993; see also Newmark and Leonard 1990) found that a majority (slim in some areas) of residents living adjacent to several protected areas in Tanzania oppose their dissolution. These authors suggested that the future of protected areas depends in part on identifying and fostering those aspects that are perceived as valuable by local residents (see also Perkin, chap. 26; Mbano et al., chap. 28).

The three objectives of ICDPs listed above are also those of the SRCS. Protected area management includes the study of poaching offtake within and outside of the park (Arcese, Hando, and Campbell, chap. 24), and buffer zones have been proposed in several areas (Campbell and Hofer, chap. 25; Perkin, chap. 28). Local social and economic benefits include (1) natural resource management outside the area, particularly in the form of legalizing the poaching offtake through quotas to villages, (2) community social services such as schools and health clinics, (3) road construction, and (4) the possibility of employment through development projects, hotels, and other tourist-related activities.

These integrated development projects are only just under way in Serengeti and it is too early to assess their effects. Historical experience has shown that the traditional approach of isolating protected areas is not working, and the natural integrity of the ecosystem is being compromised, so the ICDP approach seems the better way to go. Nevertheless, there are some serious problems with the ICDP approach, and solutions to these have yet to be found. We discuss these problems not as a criticism but rather as a caution about what must be overcome.

Development without Conservation

The fundamental objective of ICDPs is to ensure that the development components result in the overall conservation of biodiversity. Wells, Brandon, and Hannah (1992) pose the question that has to be asked at the design stage: "What are the anticipated linkages between the planned realization of social and economic benefits by people living outside the park or reserve boundaries and the necessary behavioral response the project seeks to achieve to reduce pressure inside the boundaries?"

ICDPs are popular among development agencies and have been implemented around the world. Wells, Brandon, and Hannah (1992) reviewed twenty-three such projects in Africa, Asia, and Latin America. They concluded that virtually all the projects failed to meet the objective—namely, conservation—because "the critical linkage between development and conservation is either missing or obscure." The projects failed for several reasons, among which are: (1) although all projects provided local benefits, they lost sight of the need to translate this into conservation in the process; (2) they assumed naively that simple enhancement of living standards would reduce pressures on the nature reserves without law enforcement—results have disproved this assumption; (3) different organizations—government, conservation organizations, development organizations—working independently as they usually do could not effectively implement ICDPs; and (4) some factors leading to the erosion of biodiversity do not originate at the local level, but far from park boundaries. Such factors are the inability of central governments to manage public lands and powerful financial incentives that encourage overexploitation of wildlife, grazing, timber, and crops.

One of the more consistent findings is the need to provide adequate law enforcement inside the reserve at the same time positive incentives for conservation are provided to people outside (Milner-Gulland and Leader-Williams 1992; Wells, Brandon, and Hannah 1992; Newmark et al. 1993). This conclusion is in contrast to the suggestion that reserve boundaries and law enforcement should be contracted to the point where poaching incursion stops (Leader-Williams and Albon 1988). Since encroachment and attrition are inevitable, irrespective of the boundary, as mentioned earlier (see fig. 2.1), such a policy would lead to the complete elimination of the reserve. To counteract the inward pressure of poaching, law enforcement through cooperation of different agencies must be expanded, rather than contracted, to areas outside the reserve.

Uses of Wildlife

Direct consumptive exploitation of wildlife for meat, called "game cropping," is an old idea (Dassman 1964) still favored uncritically in some areas of southern Africa (Child 1990). The success of this approach to providing an economic incentive for conservation has usually been confined to special cases in which tourists or foreign hunters pay a higher purchase price than local people. There are many arguments against this approach, and in general such cropping projects have failed in eastern Africa (Parker 1984; Macnab 1991).

The Serengeti is the exception that proves this rule. For example, humans have been hunting in this area with bow and arrow, pitfall traps, and snares for decades, probably for centuries. The meat is dried and sold locally at prices affordable to the local community. Transport of products is minimal, the migrant wildebeest and zebra coming in some cases to the village gates. Initially, there was even a natural "closed season" when the migrants returned to the plains during the rains, an area too far away for the hunters to travel to and too exposed to allow effective hunting by traditional methods. This situation was probably sustainable several decades ago. It stands in stark contrast to recent attempts at harvesting using modern weapons, veterinary inspections, and cold-storage vehicle transport, all of which put the price of meat out of reach of local people (Parker 1984, 1987).

Thus, we can say that the original harvesting by native peoples was a successful example of "game cropping" for all the reasons that modern attempts are not. The original balance, however, is now clearly in danger of being lost with the expanding demands of an increasing human population and the consequent intensive harvest that takes place when the migrants are off the plains. Hence, although the legalization of harvesting wildlife outside the park and the setting of quotas for villages looks reasonable in the short term, there remains the question of whether the harvest can be controlled within sustainable limits in the long term.

Legalized harvest policies for animals that primarily dwell inside the protected portions of Serengeti must also overcome problems that are expected due to human demographic and economic factors. Historical experience shows that virtually all open-access, common property stocks, such as marine fish stocks and seal and whale populations, become overexploited due to economic incentives in the absence, and often even in the presence, of strict regulation and enforcement (Clark 1973a,b, 1981, 1990). The wildebeest and the other species of the Serengeti are a common property stock, not owned by the local human residents and subject to open access in the absence of effective anti-poaching operations. Since an increasing human population is likely to demand increasing quotas irrespective of biological constraints, the setting, monitoring, and enforcement of quotas will have to be done by an independent authority if stocks are to remain at sustainable levels. Following this scenario through, however, one can see that this situation is bound to encourage illegal hunting once demand exceeds supply, and at present illegal hunting is not effectively controlled. Calculations for the illegal harvest in the early 1990s suggest that it is already close to maximum sustained yield (Hilborn et al., chap. 29), so that a collapse of the wildebeest population in the near future is a real possibility. The question therefore remains: how does one sustain a harvest of a common property stock subject to economic incentives and in the face of insufficient enforcement?

An alternative solution, which may have some merit, is to promote domestic animal husbandry rather than wildlife harvesting outside of the reserve boundaries (Campbell and Hofer, chap. 25; Norton-Griffiths, chap. 27). Contrary to earlier literature, domestic ungulates are found in all habitats in Serengeti, form by far the highest ungulate densities, and considerably outweigh native wildlife on comparable areas (Watson, Graham, and Parker 1969; Coe, Cumming, and Phillipson 1976; Macnab 1991; Broten and Said, chap. 8). Since domestic stocks are privately owned they can be harvested economically, and as wealth accumulates there is decreasing incentive to harvest the wildlife (Norton-Griffiths, chap. 27). This approach provides a mechanism for unlinking increasing human demands from a stationary wildlife resource. At the same time, benefits from wildlife can be in a form in which overexploitation is less likely—that is, in nonconsumptive uses such as tourism (Western and Henry 1979)—with some portion of these benefits being passed to local communities through revenue-sharing agreements.

Stopgap versus Long-Term Solutions

A major problem that has arisen from development projects in Africa funded by the developed countries of the Northern Hemisphere, "the North," is that projects that help local people while they are operating leave them stranded and worse off when they stop. Most aid projects are not designed to last forever. During a project's operation the great infusion of funds encourages local people to change their activities and participate in the project, and in so doing to become dependent on it. When the project stops, it leaves people without their previous sources of support.

Kiss (1990, iv) states that self-sufficiency has not yet been achieved in any African wildlife project. As a result, ICDPs may in some cases be in danger of exacerbating the threat to wildlife and biodiversity. If ICDPs are encouraging increased dependence on consumptive uses of wildlife, then funding agencies must be very careful to ensure that the support systems do not collapse, because a collapse would cause a greater dependence on the wildlife resource and hence overexploitation. The extent to which ICDPs will work for the benefit of wildlife in the Serengeti region thus depends on the long-term commitment of the funding agencies involved. In 1993, however, funding for SRCS was reduced to one-seventh of its planned operating budget.

Economic Opportunity Costs of Protected Areas

One of the traditional approaches to the conservation of protected areas is to provide an economic rationale so as to encourage governments to continue support of these areas. Such economic arguments usually point to tourism, but sport hunting and commercial harvesting are often cited (Macnab 1991).

Economic reasons for conservation, however, only succeed so long as there are no other economic pressures that can outcompete them. This problem is outlined by Norton-Griffiths in chapter 27, where it is argued that the economic opportunity cost of the Serengeti-Mara region (that is, the financial benefit that would have accrued if the area had been used for some other purpose) is considerably greater than even the most optimistic projections of income from tourism. Furthermore, this discrepancy will become greater rather than less as human populations build, food prices go up, and demand on the land increases.

One of the conclusions one must draw from this argument, if it proves to stand up to closer inspection, is that it is unwise to base the rationale of conservation primarily on economic grounds (Clark 1973a,b, 1990). Conservation does not have to have an economic basis, or at least it does not have to depend entirely on an economic justification. Conservation can also be justified in terms of its existence, aesthetic, scientific, and other values to the world community (Orians et al. 1990), and we have discussed some of these here. To be successful, however, a strategy based on these latter values clearly requires that the world community set up a "Heritage Fund" to maintain such areas as Serengeti in perpetuity.

CONCLUSION

We have outlined the rationale for the existence of biological reserves, whatever their legal status, in order to address a range of questions. Such questions can be classified as tertiary (habitat fragmentation, minimum viable populations, etc.), secondary (optimization procedures for saving species or habitats), and primary (how to conserve habitats). We suggest that almost all effort by conservation agencies has been devoted to the tertiary questions, little to secondary ones, and almost none to the primary questions.

Within protected areas such as national parks, there are three broad areas of research questions that need to be addressed, ranging from general questions concerning world biodiversity to specific questions concerning park planning. To answer such questions we must understand the processes of the ecosystem, and this is one of the purposes of the present book.

We outline some of the dangers and pitfalls that have developed in some similar programs elsewhere, particularly those arising from Integrated Conservation and Development Projects. The extent to which these problems can be avoided in the Serengeti ecosystem will depend on (1) good research information, (2) awareness by the implementers of aid projects of the pitfalls, and (3) the long-term commitment of the donors.

REFERENCES

Burrows, R. 1992. Rabies in wild dogs. *Nature* 359:277.

Child, B. 1990. Assessment of wildlife utilization as a land use option in the semi-arid rangeland of southern Africa. In *Living with Wildlife,* ed. A. Kiss, 155–76. World Bank Technical Paper no. 130, Africa Technical Dept. Series. Washington, D.C.: The World Bank.

Clark, C. W. 1973a. The economics of over-exploitation. *Science* 181:630–34.

———. 1973b. Profit maximization and the extinction of animal species. *J. Polit. Econ.* 81:950–61.

———. 1981. Bioeconomics of the ocean. *BioScience* 31:231–37.

———. 1990. *Mathematical bioeconomics: The optimal management of renewable resources.* 2d ed. New York: John Wiley & Sons.

Cloudsley-Thompson, J. L. 1992. Wildlife massacres in Sudan. *Oryx* 26:202–4.

Coe, M. J., Cumming, D. M., and Phillipson, J. 1976. Biomass and production of large African herbivores in relation to rainfall and primary production. *Oecologia* 22:341–54.

Creel, S. 1992. Cause of wild dog deaths. *Nature* 360:633.

Dassman, R. F. 1964. *African game ranching.* Oxford: Pergamon Press.

Dublin, H. T. 1986. *Decline of the Mara woodlands: The role of fire and elephants.* Ph.D. dissertation, University of British Columbia.

Dublin, H. T., and Douglas-Hamilton, I. 1987. Status and trends of elephants in the Serengeti-Mara ecosystem. *Afr. J. Ecol.* 25:19–33.

Dublin, H. T., Sinclair, A. R. E., Boutin, S., Anderson, E., Jago, M., and Arcese, P. 1990. Does competition regulate ungulate populations? Further evidence from Serengeti, Tanzania. *Oecologia* 82:283–88.

Dublin, H. T. Sinclair, A. R. E., and McGlade, J. 1990. Elephants and fire as causes of multiple stable states in the Serengeti-Mara woodlands. *J. Anim. Ecol.* 59:1147–64.

Ehrlich, P. R., and Wilson, E. O. 1992. Biodiversity studies: Science and policy. *Science* 253:758–62.

Estes, J. A., Duggins, D. O., and Rathbun, G. B. 1989. The ecology of extinctions in kelp forest communities. *Conserv. Biol.* 3:252–64.

Faith, D. P. 1992. Conservation evaluation and phylogenetic diversity. *Biol. Conserv.* 61:1–10.

Gascoyne, S. C., Laurenson, M. K., Lelo, S., and Borner, M. 1993. Rabies in African wild dogs *(Lycaon pictus)* in the Serengeti Region, Tanzania. *J. Wildl. Dis.* 29:396–402.

Holling, C. S. 1973. Resilience and stability of ecological systems. *Annu. Rev. Ecol. Syst.* 4:1–23.

Kiss, A. ed. 1990. *Living with Wildlife.* World Bank Technical Paper no. 130, Africa Technical Dept. Series. Washington, D.C.: The World Bank.

Leader-Williams, N., and Albon, S. D. 1988. Allocation of resources for conservation. *Nature* 336:533–35.

MacDonald, D. W. 1992. Cause of wild dog deaths. *Nature* 360:633–34.

Macnab, J. 1991. Does game cropping serve conservation? A reexamination of the African data. *Can. J. Zool.* 69:2283–90.

Margules, C. R. 1989. Introduction to some Australian developments in conservation evaluation. *Biol. Conserv.* 50:1–11.

Margules, C. R., Nicholls, A. O., and Pressey, R. L. 1988. Selecting networks of reserves to maximize biological diversity. *Biol. Conserv.* 43:63–76.

Margules, C. R., Pressey, R. L., and Nicholls, A. D. 1991. Selecting nature reserves. In *Nature conservation: Cost-effective biological surveys and data analysis,* ed. C. R. Margules and M. P. Austin, 90–97. Melbourne: CSIRO.

May, R. M. 1977. Thresholds and breakpoints in ecosystems with a multiplicity of stable states. *Nature* 269:471–77.

McNeely, J. A., Miller, K. R., Reid, W. V., Mittermeir, R. A., and Werner, T. B. 1990. *Conserving the world's biological diversity.* Washington, D.C.: The World Bank.

Milner-Gulland, E. J., and Leader-Williams, N. 1992. A model of incentives for the illegal exploitation of black rhinos and elephants: Poaching pays in Luangwa Valley, Zambia. *J. Appl. Ecol.* 29:388–401.

Newby, J. E. 1990. The slaughter of Sahelian wildlife by Arab royalty. *Oryx* 24:6–8.

Newmark, W. D., and Leonard, N. L. 1991. Attitudes of local people toward Kilimanjaro National Park and Forest Reserve. In *The conservation of Mount Kilimanjaro,* ed. W. D. Newmark, 87–96. Gland, Switzerland: IUCN.

Newmark, W. D., Leonard, N. L., Sariko, H. I., and Gamassa, D.-G. M. 1993. Conservation attitudes of local people living adjacent to five protected areas in Tanzania. *Biol. Conserv.* 63:177–83.

Norman, C. 1981. The threat to one million species. *Science* 214:1105–7.

Noss, R. F. 1990. Indicators for monitoring biodiversity: A hierarchical approach. *Conserv. Biol.* 4:355–64.

Orians, G. H., Brown, G. H., Kunin, W. E., and Sweirbinski, J. E., eds. 1990. *The preservation and valuation of biological resources.* Seattle: University of Washington Press.

Parker, I. S. C. 1984. Perspectives on wildlife cropping or culling. In *Conservation and wildlife management in Africa,* ed. R. H. V. Bell and E. McShane-Caluzi, 233–53. Washington, D.C.: U.S. Peace Corps, Office of Training and Program Support, Forestry and Natural Resources Sector.

———. 1987. Game cropping in the Serengeti region. *Parks* 12:12–13.

Pearsall, W. H. 1957. Report on an ecological survey of the Serengeti National Park, Tanganyika. *Oryx* 4:71–136.

Peterman, R. M., Clark, W. C., and Holling, C. S. 1979. The dynamics of resilience: Shifting stability domains in fish and insect systems. In *Population dynamics,* ed. R. M. Anderson, B. D. Turner, and L. R. Taylor, 321–41. Symp. Brit. Ecol. Soc. 20. Oxford: Blackwell Scientific Publications.

Raven, P. H., and Wilson, E. O. 1992. A fifty-year plan for biodiversity surveys. *Science* 258:1099–1100.

Sinclair, A. R. E. 1977. *The African buffalo: A study of resource limitation of populations.* Chicago: University of Chicago Press.

———. 1981. Environmental carrying capacity and the evidence for overabundance. In *Problems in management of locally abundant wild mammals,* ed. P. A. Jewell and S. Holt, 247–57. New York: Academic Press.

———. 1983. Management of conservation areas as ecological baseline controls. In *Management of large mammals in African conservation areas,* ed. R. N. Owen-Smith. Pretoria, South Africa: Haum Educational Publ.

———. 1989. Population regulation in animals. In *Ecological concepts,* ed. J. M. Cherrett, 197–241. Symp. Brit. Ecol. Soc. 26.

Sinclair, A. R. E., Hik, D., Schmitz, O. J., Scudder, G. G. E., Turpin, D., and Larter, N. C. 1995. Biodiversity and the need for habitat renewal. *Ecol. Appl.* In press.

Sinclair, A. R. E., and Wells, M. P. 1989. Population growth and the poverty cycle in Africa: Colliding ecological and economic processes? In *Food and Natural Resources,* ed. D. Pimentel and C. Hall, 439–84. New York: Academic Press.

Soulé, M., Gilpin, M., Conway, W., and Foose, T. 1986. The millennium ark: How long a voyage, how many staterooms, how many passengers? *Zoo Biol.* 5:101–13

Tisdell, C. 1990. Economics and the debate about preservation of species, crop varieties and genetic diversity. *Ecol. Econ.* 2:77–90.

Vane-Wright, R. I., Humphries, C. J., and Williams, P. H. 1991. What to protect: Systematics and the agony of choice. *Biol. Conserv.* 55:235–54.

Walker, B. H., Ludwig, D., Holling, C. S., and Peterman, R. M. 1981. Stability of semi-arid savanna grazing systems. *J. Ecol.* 69:473–98.

Walters, C. 1986. *Adaptive management of renewable resources.* New York: Macmillan.

Watson, R. M., Graham, A. D., and Parker, I. S. C. 1969. A census of the large mammals of Loliondo Controlled Area, northern Tanzania. *E. Afr. Wildl. J.* 7:43–59.

Wells, M., Brandon, K., and Hannah, L. 1992. *People and parks: Linking protected area management with local communities.* Washington, D.C.: The World Bank.

Western, D., and Henry, W. 1979. Economics and conservation in Third World National Parks. *BioScience* 29:414–18.

Westoby, M., Walker, B., and Noy-Meir, I. 1989. Opportunistic management for rangelands not at equilibrium. *J. Range Mgmt.* 42:265–73.

White, S. E. 1915. *The rediscovered country.* New York: Doubleday.

II Plants and Herbivory

THREE

Plant Communities and Herbivory

S. J. McNaughton and F. F. Banyikwa

The Serengeti ecosystem, operationally defined by the movements of migratory wildebeest, zebra, and eland herds, is a vast territory spilling beyond the reaches of core protected areas (Sinclair, chap. 1). Throughout the year, the populations of large migratory herbivores engage in system-wide movements, traveling and pausing incessantly as the seasons change. Spatial variation, from the level of entire landscapes to the level of local grassland swards, both influences and is influenced by the movements of the animals (McNaughton 1989).

The theme of this chapter is the coupling of spatial patterns and processes in grazing ecosystems. It focuses on spatial and temporal variation of ecosystem properties and the significance of process-pattern interactions for conservation of the Serengeti ecosystem. Its conclusions are based on 20 years of research on ecosystem properties throughout the Serengeti, from the short-grass plains of the southeast to the far reaches of the western corridor to the expansive landscapes of the northwest.

Serengeti trophic levels are united in a complex, integrated functional web of energy and nutrient flow (McNaughton, Ruess, and Seagle 1988). Much of the fundamental character of the Serengeti is linked to functional interdependence that is spatially and temporally variable (McNaughton 1983, 1985). Therefore, conservation policy in the region must recognize this linkage and its changeable qualities; attempts to impose uniformity are efforts to overpower the Serengeti's essential characteristics.

A WILDEBEEST'S VIEW OF THE ECOSYSTEM

Consider for a moment a migratory wildebeest's view of its world, the Serengeti. It spends much of its time with head bent down, walking along, taking bite after bite after bite, continually hearing the cacophony of the surrounding herd, and, we presume, reassured and protected from the Serengeti's predators by that dissonance. It bites and swallows, bites and swallows, until its rumen signals "full." Then it lies down, regurgitates and remasticates the previously swallowed forage, swallows it again, and thus accelerates the extraction of nutrients as the forage is processed by the complex of microbes and chemicals in its digestive tract. From time to time come a few days when it and its companions cannot fill their rumens fast enough to meet their needs, so the herd moves to another location. As the weeks proceed, the herd moves again and again—less during the wet season when foraging on the Serengeti short-grass plains, incessantly during the dry season when forage quality and quantity are both meager (Sinclair 1977).

As it walks along feeding, taking several bites per minute, our wildebeest, moving through visually homogeneous grasslands, in fact encounters a constantly varying environment (McNaughton 1989). It encounters different individual grass plants at different stages of growth, different genotypes of the same grass species, different grass species, a mixture of grasses, forbs, shrubs, and, off the Serengeti plains, tree seedlings, saplings, and adults. As it moves from area to area, it encounters grasslands that are genetically short, medium, and tall. All may be ungrazed, lightly grazed, moderately grazed, or heavily grazed. Some landscape regions it moves through are flat to gently undulating, others are highly dissected by ravines, gorges, and steep hills.

At all of the levels that it experiences, our wildebeest and its companions confront a highly variable and complex environment (McNaughton 1983, 1985, 1989, 1991). This complexity at many levels presents herbivores with numerous options (Senft et al. 1987; McNaughton 1989): Do I take a bite of this or not? Do I stay here or do I move? If I move, in what direction do I go? How far should I move in this direction before I change direction? Should I stop here, or are there more profitable areas ahead? The first consideration occurs on a time scale of seconds, others on scales of weeks or months. The responses to some of these alternatives are probably, to a certain extent, genetically programmed, but many also undoubtedly may be learned to a considerable extent. How the herbivores respond to these alternatives is influenced by, and influences, the entire character of the Serengeti ecosystem (McNaughton 1989, 1991).

COMMUNITY PATTERNS IN SPACE

An Extensive Approach: Grassland Community Organization

Inclusive studies of grasslands sampled at 105 locations throughout Serengeti National Park and Masai Mara National Reserve identified a core of grassland community types (table 3.1), each characterized by a similar species composition, that recur in different locations (McNaughton 1983). The most widespread of these are various mid-grasslands dominated by red oat grass *(Themeda triandra)* throughout the regions above a mean annual rainfall of about 700 mm. Short grasslands dominated by species of dropseed *(Sporobolus)* are abundant on the short-grass plains

Table 3.1 Grassland community classification based on species.

Code	Dominant species	Relative abundance
1	*Sprobolus kentrophyllus*	.275
	Kyllinga nervosa	.256
2	*Sporobolus ioclados*	.296
	Digitaria macroblephara	.235
3	*Digitaria macroblephara*	.308
	Pennisetum mezianum	.251
4	*Andropogon greenwayi*	.577
5	*Chrysochloa orientalis*	.339
	Sporobolus ioclados	.201
6	*Chloris pycnothrix*	.567
7	*Sporobolus fimbriatus*	.265
8	*Heteropogon contortus*	.285
9	*Themeda triandra*	.461
	Eustachys paspaloides	.153
10	*Themeda triandra*	.337
	Panicum coloratum	.253
11	*Themeda triandra*	.407
	Eragrostis tenuifolia	.204
12	*Themeda triandra*	.552
	Sporobolus pyramidalis	.146
13	*Themeda triandra*	.521
	Pennisetum mezianum	.062
14	*Themeda triandra*	.519
	Bothriochloa insculpta	.068
15	*Hyparrhenia filipendula*	.262
16	*Cymbopogon excavatus*	.475
17	*Echinochloa haploclada*	.822

Note: Community numerical code corresponds with numbers in fig. 3.1. Relative abundance is the average proportion that the species contributes to peak biomass.

and top catenas in the western corridor, and tall grasslands of russet grass *(Loudetia)* and thatching grass *(Hyperthelia)* occur in the far northwest and the corridor.

Ordination of the communities based on species compositions revealed two intergrading clusters of short (ungrazed canopy height < 50 cm) to tall (> 150 cm) grasses (fig. 3.1a) and outlying tall *Echinochloa* grasslands, which occur in frequently to regularly submerged marshlands, most extensively in the Grumeti and Mbalageti drainages near Lake Victoria (McNaughton 1983). The last is an extremely low diversity community, with over 80% of the standing crop typically contributed by the dominant species and only a few other species present at low abundance. Therefore, the Serengeti grasslands are organized as a continuum of vegetation with covarying species composition and stature.

These communities are not entities determined by the physical environment independently of the herbivores. Instead, the herbivores exert a powerful controlling influence on grassland composition (McNaughton 1983). The first axis of the ordination was most closely related to grazing intensity (fig. 3.1b). Mean annual grazing intensity is greatest in the *Sporobolus-Digitaria* grasslands that occur on the Serengeti plains and

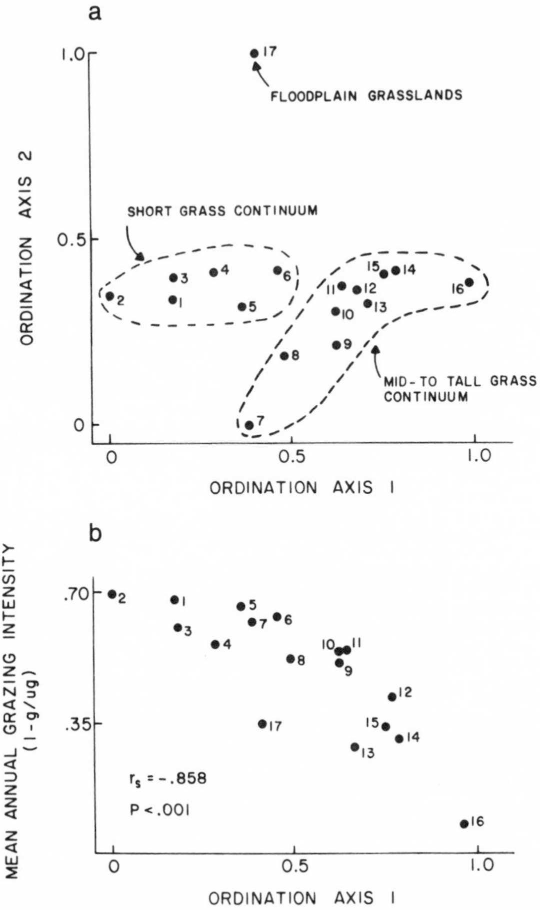

Figure 3.1 Ordination of grassland communities characterized in table 3.1 and their relation to two principal environmental variables. (*a*) Grassland ordination with communities coded as in table 3.1. (*b*) Relationship between position of community on ordination axis 1 and mean annual grazing intensity the community experiences. Grazing intensity is measured as $(1-g/ug)$, where g is biomass of unfenced plots and ug is biomass of fenced plots, averaged over the year. (*c*) Relationship between position of community on ordination axis 2 and soil clay content in the upper 10 cm. (From McNaughton 1983.)

on top catenas in the western corridor. At the opposite extreme are the grasslands dominated by citronella grass *(Cymbopogon)*, a very low palatability grass because of the fragrant and, presumably, distasteful chemicals that it contains (Vesey-Fitzgerald 1973).

The second axis of the ordination was most closely related to soil texture, here indexed by clay content (fig. 3.1c). The soils with the lowest clay content support short grasslands dominated by *Chrysochloa* and *Sporobolus*, and the soils with the highest clay content are the floodplain grasslands dominated by *Echinochloa*. Clay content is probably an ap-

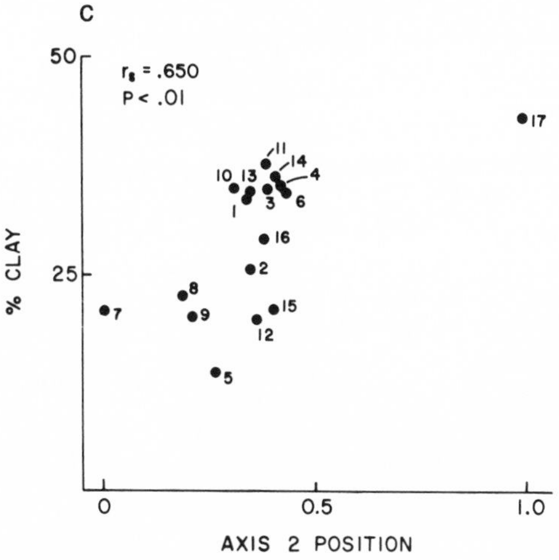

proximate measure of nutritional capacity because more finely divided soil particles generally retain more cations.

An Intensive Approach: Fuzzy Set Ordination

To provide more detail at a lower level of spatial organization, the short grasslands of the eastern Serengeti plains were partitioned by fuzzy set ordination. Fuzzy set theory (Zadeh 1965; Zimmerman 1984) is an extension of classic set theory that has been introduced into ecology only recently (Marsili-Libelli 1986; Roberts 1986; Feoli and Zuccarello 1986, 1988). It combines classification and ordination to provide direct ordinations with environmental fuzzy sets. Data for short-grass community analyses were from two matrices of 32 stands by 21 plant species and 7 soil variables (Banyikwa, Feoli, and Zuccarello 1990). Classification of stands was by sum of squares clustering using the chord distance, while species classification was by average linkage clustering using the correlation coefficient. The clusters of stands defined were then used to obtain four fuzzy sets:

a. community fuzzy sets of stands
b. community fuzzy sets of species
c. environmental fuzzy sets of stands
d. environmental fuzzy sets of species

Community fuzzy sets were derived from community data while environmental fuzzy sets were obtained by combining the community sets with environmental data (Feoli and Zuccarello 1988). Classification produced five clusters of stands (table 3.2), which represent short grassland communities. They are arranged sequentially along the topographic gradient

Table 3.2 Mean percentage cover of species and average soil properties for stand clusters and species clusters from short grasslands of the eastern Serengeti plains.

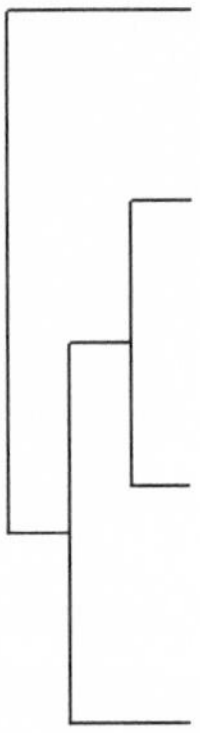

	Species/Soil factors	Code
A	Sporobolus ioclados	SIOD
	Kyllinga nervosa	KYNE
	Eragrostis papposa	ERPA
B	Eustachys paspaloides	EUPA
	Microchloa kunthii	MIKU
	Sporobolus fimbriatus	SFIM
	Digitaria macroblephara	DIMA
	Digitaria abyssinica	DIAB
C	Sporobolus kentrophyllus	SKEN
	Harpachne schimperi	HARS
	Andropogon greenwayi	AGRE
	Chloris pycnothrix	CHPY
D	Undifferentiated dicots	UNDI
	Cynodon dactylon	CYDA
	Sporobolus spicatus	SPSP
	Chloris gayana	CHGA
	Pennisetum clandestinum	PECL
	Cenchrus ciliaris	CECI
	Psilolema jaegeri	PSJA
	Sporobolus sanguineus	SPSA
	pH	pH
	Sodium	Na
	Potassium	K
	Calcium	Ca
	Magnesium	Mg
	Cation exchange capacity	CEC
	Moisture factor	MF

Source: Banyikwa, Feoli, and Zuccarello 1990.
Note: This is a detailed analysis of the same region covered by community 1 in table 3.1. Stand group relationships are indicated in the top dendrogram by sum of squares and chord distance coupling and in the basal dendrogram by cosine and average linkage. The left dendrogram is species group classification (capital letters) based on average linkage clustering and correlation coefficient. Topographic positions of stands are coded as I = hilltops; II = upper, III = middle, IV = lower hillslopes; V = drainage areas. *Hypoestes forskalei* was combined with other dicots in the table but dominated the forbs in stand cluster V.

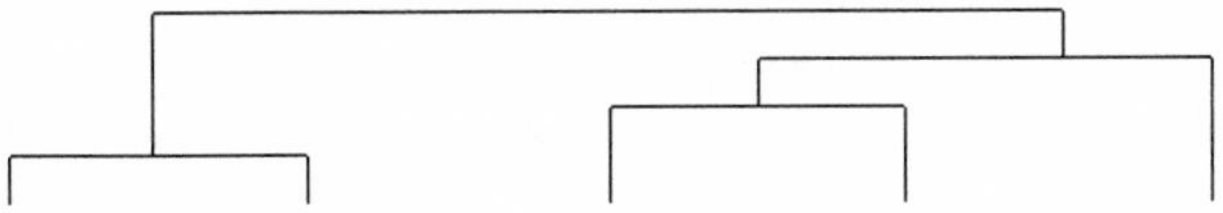

I	II	III	IV	V	Mean
16.8	10.2	10.1	3.0	2.0	9.4
14.6	18.7	4.0	4.1	2.0	8.7
3.5	3.4	2.8	2.1	1.0	2.6
9.4	3.7	14.9	5.2	3.4	7.3
0.6	0.5	1.5	0.9	0.2	0.7
2.9	3.6	4.8	4.6	4.2	4.0
1.6	1.4	2.4	2.3	1.0	1.7
2.7	5.4	4.7	3.3	1.6	3.5
2.4	6.7	7.6	19.0	1.1	7.4
0.7	0.5	1.9	2.1	1.2	1.2
0.0	0.0	0.8	1.5	0.0	0.5
0.0	0.0	0.0	0.5	0.1	0.1
3.3	3.8	1.8	1.7	14.9	5.0
1.6	1.6	1.2	0.9	2.8	1.8
0.1	0.5	0.9	0.9	4.9	1.5
0.0	0.0	0.0	0.0	7.7	1.5
0.0	0.0	0.0	0.0	1.8	0.4
0.0	0.0	0.0	0.0	0.1	0.0
0.0	0.0	0.0	0.0	1.3	0.0
0.0	0.0	0.0	0.5	0.1	0.1
8.3	8.6	8.1	7.9	8.6	8.3
1.6	1.4	1.5	1.5	1.5	1.0
16.7	15.5	13.3	16.6	13.9	15.0
71.4	68.2	61.7	60.7	61.2	64.6
8.5	10.2	5.0	7.9	8.7	8.1
70.3	47.6	68.2	51.1	34.0	54.3
4.9	6.2	5.8	6.2	6.1	5.8

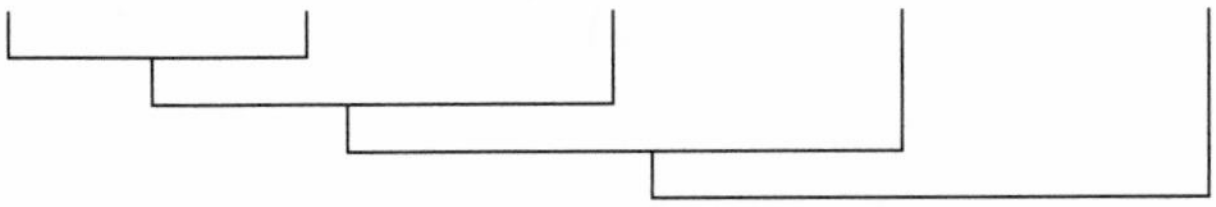

from hilltops (clusters I and II) through hillslopes (III and IV) to drainage lines (V).

Community fuzzy set ordination (fig. 3.2a) indicates that topography accounted for by fuzzy set I has an overriding effect on the arrangement of stands and that it is the most generalized environmental factor. The ordination of species according to community fuzzy sets I and V (fig. 3.2b), and superimposition of figure 3.2b on figure 3.2a, indicate that the arrangement of species groups follows the topographic gradient. Group A species occur on hilltops, group B and C species on hillslopes, and group D species in drainage areas.

Soil factor fuzzy set ordination for calcium and cation exchange capacity, which were the most discriminating factors in the fuzzy set system space (fig. 3.3), indicates that the arrangement of stands along the topographic gradient is not fully retained. Clusters I and II overlap completely while V shows high heterogeneity. The cluster I and II overlaps suggest that the species composition differentiation is not reflected in these two major environmental factors.

This intensive approach to the first three communities in the extensive community classification suggests a somewhat different organization within those clusters:

1. A *Sporobolus ioclados—Kyllinga nervosa* community that occurs on the hilltops and upper hillslopes and is characterized by the two most abundant group A species. This is the most extensive community within the Serengeti short grasslands (Herlocker and Dirschl 1972; Banyikwa 1976; McNaughton 1983). The soils are saline and alkaline because they have been least affected by leaching (de Wit 1976).

2. A previously unrecognized short-grass *Eustachys paspaloides* community characterized by the group B species combination, in contrast to the mid-grass *Themeda-Eustachys* community in higher-rainfall regions (see table 3.1). It occurs on relatively steep slopes and around inselbergs where erosion is relatively high.

3. A *Sporobolus kentrophyllus* community on lower hillslopes where the topography is relatively flat. It is characterized by group C species, especially *S. kentrophyllus, Andropogon greenwayi,* and *Chloris pycnothrix* (the only abundant annual grass). It is associated with heavy grazing and animal disturbance.

4. A forb-dominated *Hypoestes forskalei* community characterized by group D species occurring in drainage channels. It is a mixture of forbs (*H. forskalei, Oropetium capense)* and grasses (*M. kunthii, S. spicatus, S. sanguineus*) associated with salt accumulation due to surface water run-on during the wet season and subsequent drying during the dry season.

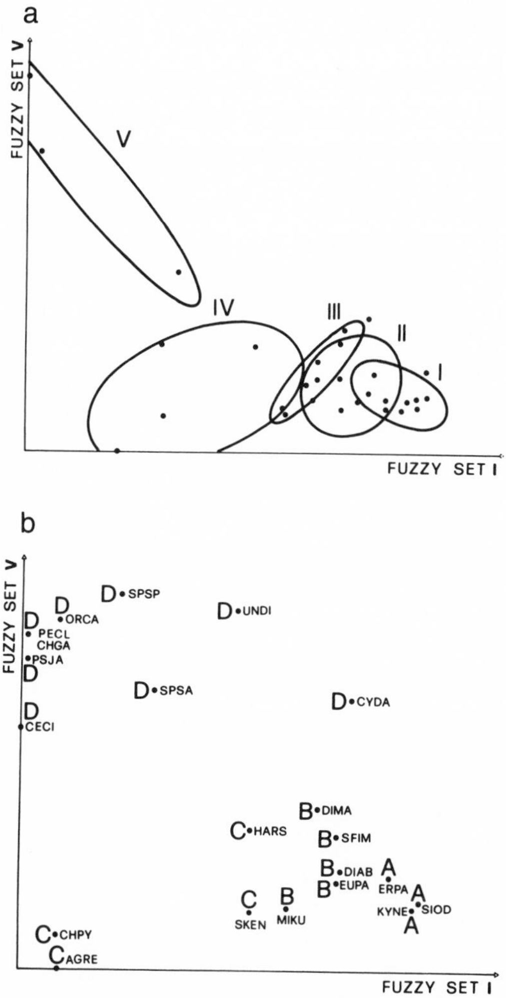

Figure 3.2 Fuzzy set ordination of Serengeti short grasslands on the eastern Serengeti plains. (*a*) Ordination of stand clusters identified in table 3.2. (*b*) Ordination of species with species clusters identified with capital letters and species identifications abbreviated as in table 3.2. ORCA is *Oropetium capense*. (From Banyikwa, Feoli, and Zuccarello 1990).

Thus, this fine-scale examination of pattern within the short grasslands of the Serengeti plains reveals levels of resolution below those of the system-wide survey. Evidence indicates that large mammals respond to scale at all levels, from the bites they select as they forage to the landscapes they move to at certain times of year (Senft et al. 1987; McNaughton 1989, 1991).

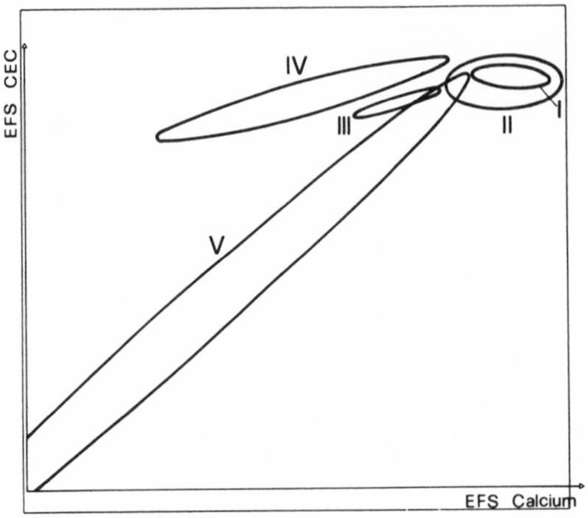

Figure 3.3 Environmental fuzzy set ordination of eastern Serengeti plains short grasslands based on the two principal soil properties related to stand groupings. The abscissa is the calcium fuzzy set axis and the ordinate is the cation exchange capacity fuzzy axis. Elipses of equal concentration at the .05 probability level are drawn around centroids of the five stand clusters identified in table 3.2. (From Banyikwa, Feoli, and Zuccarello 1990).

SPATIAL PATTERNS OF ECOSYSTEM FUNCTION

Extensive Patterns

The gradient of rainfall across the Serengeti is well characterized (Norton-Griffiths, Herlocker, and Pennycuick 1975), and it is common to refer to a rainy season from November to May, and a dry season from June to October. However, that generalization is only very approximately reliable.

Associated with the annual rainfall gradient from 400 mm in the southeast to over a meter in the northwest are gradients of growing season and grazing season (McNaughton 1985). Grassland growing season, as measured by the occurrence of live forage biomass above 20 g/m^2, varies from about 75 days in the driest areas of the Serengeti plains to almost continuous in the tall grasslands on deep, sandy soils in the northwest (fig. 3.4a). A particularly interesting property of this relationship is its sigmoid form. At both the upper and lower rainfall levels, there are regions of relatively minor change in growing season with change in precipitation, but between about 650 and 800 mm of annual rainfall, there is a more abrupt increase in growing season with increased rain. This transitional region is associated with the occurrence of resident animals in the Serengeti; they are, by and large, restricted to regions with an annual rainfall above 700 mm and, therefore, a grassland growing season of about 250 days.

At lower rainfall levels, the growing season and the grazing season are identical (fig. 3.4b). In fact, this is true of short grasslands even up to

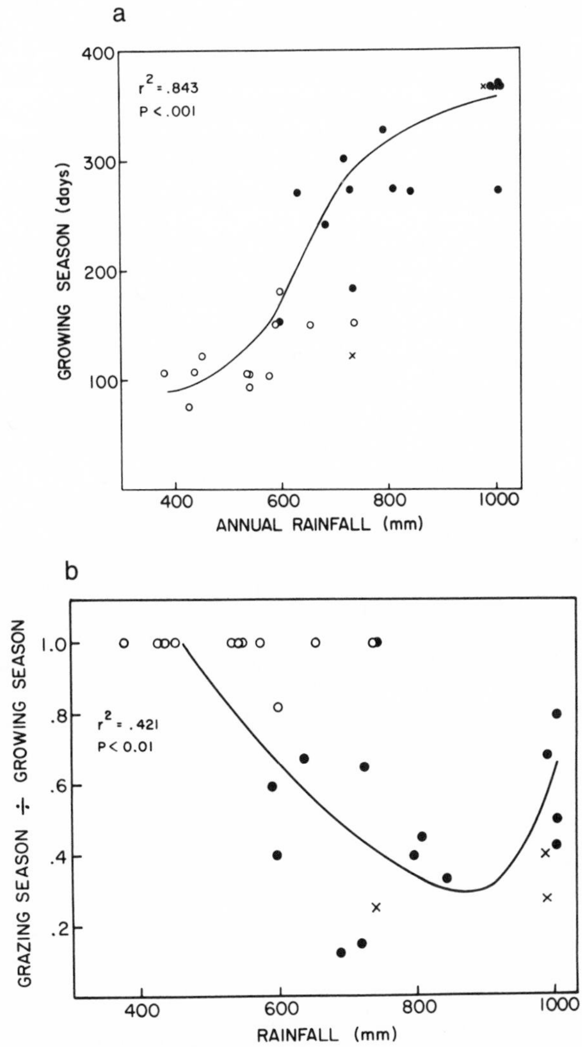

Figure 3.4 Relationship between annual rainfall and (*a*) grassland growing season and (*b*) grazing season as a proportion of the growing season. Open circles are short grasslands, solid circles are medium-height grasslands, and x indicates tall grasslands. (From McNaughton 1985).

the upper limit of their occurrence near 800 mm annually. Above about 600 mm, however, mid-grasslands begin to predominate, and the proportion of the growing season during which they experience grazing declines to a minimum of about 10% near 700 mm of annual rain. Above this rainfall level, there is a tendency for the period of grazing to increase again.

A variety of different factors contribute to this pattern (McNaughton 1985). However, they can be summarized in the following generalizations: (1) At low mean annual rainfalls, which occur on the Serengeti plains, continuous grazing throughout the growing season is due to concentration of migrant herds during the rainy season. (2) At intermediate rainfalls, resident grazers create short grasslands due to continuous grazing on top catenas throughout the growing season, while less continuously grazed medium and tall grasslands occur on mid- and lower catenas. (3) Higher grazing at the highest rainfall levels is due to both the presence of resident animals and the tendency of migratory herds to concentrate in these grasslands at the peak of the dry season when they are the only ones that have the capacity to continue growing (e.g., fig. 3.4a). The wet season spatial concentration of herbivores produces grazing lawns of low stature but high nutritional quality (McNaughton 1984).

Grazing Systems: Sustained Yield and Rotational Passage

There are two major types of herbivores in the Serengeti, as there are in most other ecosystems that support abundant populations of large herbivorous mammals. Migratory herds dominate animal biomass (Fryxell, Greever, and Sinclair 1988), and resident herds occupy distinct home ranges throughout the year and contribute the most to biodiversity. Residence or migratory behaviors obviously are not a species-specific trait because all of the major Serengeti migrants are also characterized by resident herds in some localities within the ecosystem.

Two fundamentally different types of grazing behaviors, distinguished by their occurrence in the wet and dry seasons, characterize both migratory and resident herds. Sustained-yield grazing is characteristic of the wet season, rotational-passage grazing of the dry season.

Sustained-Yield Grazing. Grazing herds engage in a sustained-yield grazing system during the wet season, in which they occupy a limited area for an extended period of time, moving back and forth through the occupied area as the rains continue (McNaughton 1985). Members of the migratory herds proceed rapidly to the Serengeti plains at the onset of the rains, oscillate spatially over areas receiving continuing rainfall, then leave the plains abruptly when the grassland dries out (Inglis 1976). Resident herds commonly concentrate on top catenas during the wet season and may reach very high densities due to the co-occurrence of several species of herbivores (Bell 1970). These grasslands are typically of short stature, even when fenced (McNaughton 1984), and are characterized by grazing seasons that equal the growing season (see fig. 3.4b).

Rotational-Passage Grazing. During the dry season, in contrast, herbivores engage in a rotational-passage grazing system, moving rapidly through tall and medium grasslands, but returning to them when there is rain behind the herd or when residual soil water at the time of grazing is sufficient to generate new grass growth (McNaughton 1985). The transition from a sustained-yield to a rotational-passage grazing system can be very abrupt at seasonal transitions (Inglis 1976). Although they move over a much less expansive area, resident herds also move more during the dry season than during the wet season, tending to follow sporadic dry-season showers (McNaughton 1985).

Local Functional Heterogeneity

There is pronounced functional heterogeneity at intermediate spatial scales associated with stochastic rainfall events (McNaughton 1985). At three experimental plot locations in the short-grass plains separated by distances of 3–10 km, green biomass in fenced plots protected from grazers fluctuated asynchronously during the first 60 days of the rainy season (fig. 3.5). Productivity at the two closest sites was only weakly correlated ($r = .486$; $P < .1$), and there was no correlation in the two other comparisons. Therefore, herbivores during this period of buildup to the peak of the rainy season are confronted with a forage resource that can be extremely productive, approaching 30 g/m^2/day, but which can also collapse, as it dries out at a similar rate. The herbivores encounter a constantly shifting mosaic of productivity driven by spatially localized, stochastic rains during this period.

Grazers track these rainfall-driven pulses of primary productivity with considerable accuracy (McNaughton 1985). During periods of increasing primary productivity on localized patches, grazer densities increase. Then, as grassland productivity declines when rain does not resume for a period and soil water is depleted, grazer densities decline.

In situ mineralization rates (Raison, Connell, and Khanna 1987) measured at a northern Serengeti plains mid-grassland site were highly variable in sixteen fenced plots separated by a maximum distance of 10 m (fig. 3.6). Although there was a highly significant mineralization rate of 0.4 gN/m^2/week (averaged across all plots by the best fit regression), there also was substantial heterogeneity, which increased with time. At the beginning of the experiment, mineral N (both NO_3 and NH_4) ranged from near zero to 2.8 gN/m^2. By the end of the experiment, values ranged from 1.8 to 11.3 gN/m^2 for the sixteen plots. The data indicate, at the very least, substantial local variation in plant-available N. This variation can produce considerable local patchiness in grass growth, with high growth rates in patches where mineralization rates are high and low rates in nearby patches.

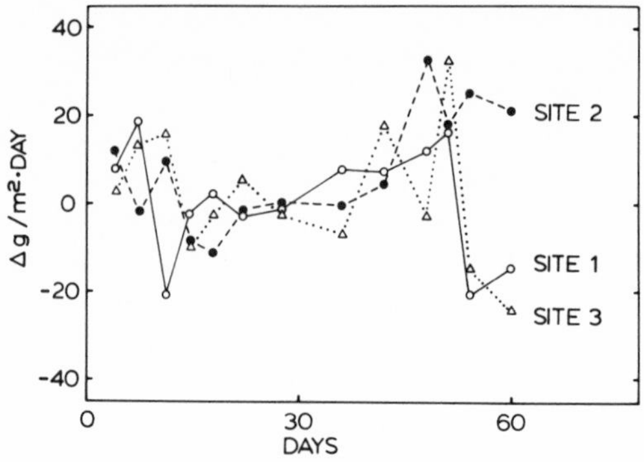

Figure 3.5 Rates of change in green biomass in fenced plots on the Serengeti plains separated by a maximum linear distance of 10 km during the first 60 days of the rainy season.

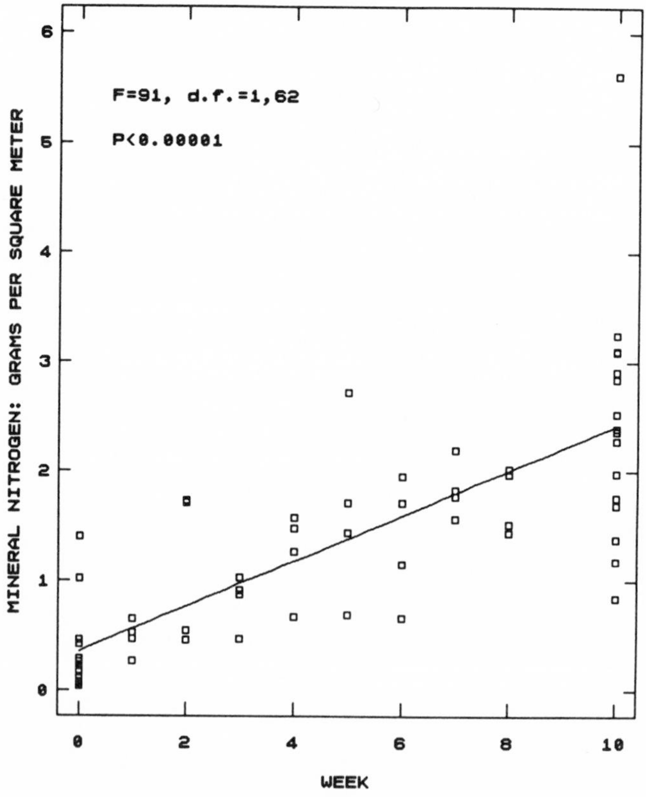

Figure 3.6 Nitrogen mineralization (as total mineral NO_3 and NH_4) inside mineralization tubes at a site on the north central Serengeti plains. Maximum distance between plots was 10 m.

The Micropattern: Causes and Consequences. Recent experimental studies document a pronounced, and very rapid, effect of different perennial grass species on N mineralization rates in initially identical soils (Wedin and Tilman 1990). After only 3 years, annual net mineralization had diverged up to tenfold in soils supporting monocultures of different species. These data document a strong feedback from grassland species composition to soil processes and suggest that local vegetation spatial patchiness, a characteristic of Serengeti grasslands closely associated with grassland biodiversity (McNaughton 1983; Banyikwa, Feoli, and Zuccarello 1990), can produce equally heterogeneous patterns of soil processes.

Species distribution patterns on very localized spatial scales can often be related to underlying soil patterns (Snaydon 1962). Such patterns are not solely independent of species occurrence, but are coupled to species nutritional patterns: Ca-accumulating species may be preferentially associated with, and perform better in, high-Ca patches within the underlying soil. Furthermore, the tendency of such species to accumulate Ca in their tissues will tend to progressively enrich the spots that they occupy. There also can be rapid intrapopulation genetic differentiation of plants in mosaic environments (Snaydon 1970): that is, species can differentiate into a variety of local genotypes, some favored in certain habitat patches, others favored in different patches.

Wildebeest foraging through a visually uniform grass sward, then, can be confronted with a bewildering variety of nutritional conditions (McNaughton 1989, 1991). Some grasses will be high in Ca, others low; some high in Na, others low; and so on. This heterogeneity will depend upon forage species, genotype, growth stage, and microhabitat. Some factors will be independent of the grazers themselves; others may be affected in both the short and the long term by the grazers (McNaughton 1983, 1985).

In the short term, for instance, urination can have a rapid, promotive effect on plant nitrogen content (Jarmillo and Detling 1988). This results in a marked preference of grazers for grass regrowth on patches that are urine-affected (Day and Detling 1990).

In the longer term, over periods of decades, grazers can affect plant community composition and soil properties in ways promoting higher nutritional status of available forages (Georgiadis and McNaughton 1990). Forages in areas of high herbivore usage are characterized by higher fiber nutritional value, even during periods when there has been no grazing; this pattern carries across soil types. So, too, forage mineral properties are influenced by the intensity of herbivore use. Therefore, herbivores can create areas of nutritional sufficiency due to high usage even in the absence of intrinsic soil differences.

Grazers have a strong promotive effect on biodiversity of the Serengeti grasslands (McNaughton 1979, 1983). Fencing almost immediately results in a diversity decline as a few taller-growing species that invest heavily in stems overtop other species. Within a few years, diversity plummets, and even those species present both inside and outside fences are genetically differentiated: tall-growing genotypes occur in the fenced vegetation, short genotypes outside fences (McNaughton 1984).

NUTRITIONAL HETEROGENEITY AND ANIMAL DISTRIBUTION AND ABUNDANCE

Resident Herds: Hot Spots

Resident herds of many species of herbivorous mammals are widely distributed throughout the Serengeti ecosystem at locations with rainfalls above about the 700 mm isohyet. Those herds represent the principal reservoir of large mammal diversity in the ecosystem. Topi, buffalo, kongoni, impala, waterbuck, reedbuck, roan antelope, oribi, resident gazelles, and others constitute this storehouse of biodiversity.

Resident animals, however, are not uniformly distributed in space. Instead, they are concentrated in localized "hot spots." Hot spots of resident animal distribution are temporally stable through periods exceeding two decades, are characterized by the occurrence of several intermixed animal species, and are associated with grazing lawns similar to the Serengeti short-grass plains, although occurring in much higher rainfall regions (see fig. 3.4).

Two principal explanations offered for local distribution concentrations are (1) that different species of herbivores facilitate one another's foraging and (2) that association protects animals from predation (Vesey-Fitzgerald 1960; Bell 1970; Sinclair 1985). The stable occurrence of hot spots argues against both hypotheses. First, if facilitation of foraging were a principal factor leading to association, herds should move in concert throughout the growing season as rainfall shifts in space and, therefore, the base of primary productivity shifts. Second, if predation were the principal factor leading to species association, predation risk could be further reduced by rapid, unpredictable movements by the herds, reducing the predictability of their occurring in any one location.

Sampling of grass leaf blades at the same stage of growth on hot spots and adjacent, more lightly utilized areas revealed substantial nutritional differences (McNaughton 1988). Particularly important in discriminating between hot spot and control swards (fig. 3.7) were Na and P. Sodium is generally found at a very low concentration in plant tissues but is required by animals; the concentration was much higher in hot spot vegetation than in control grasslands. Similarly, P concentration was higher in hot

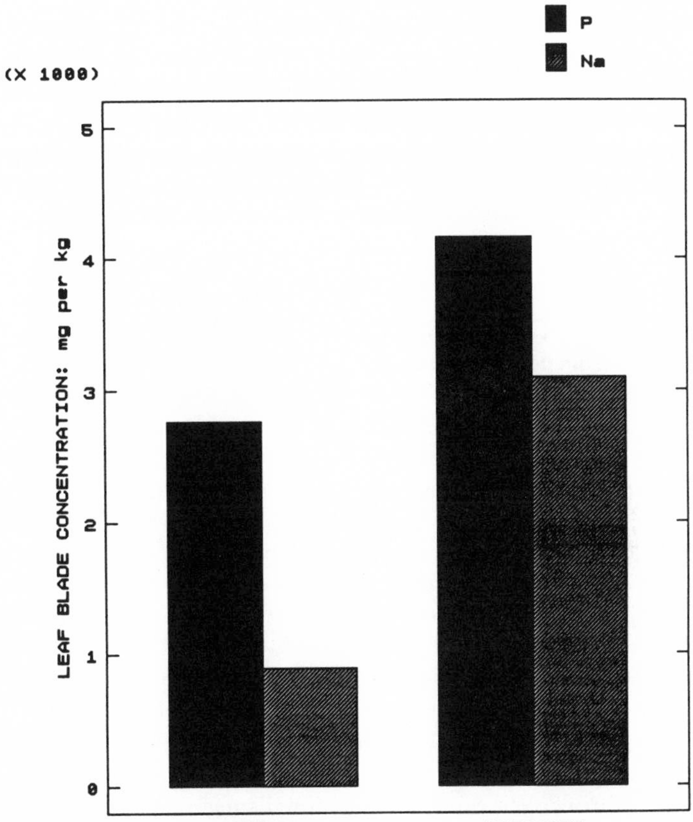

Figure 3.7 Grass leaf blade concentrations of P and Na on hot spots (animal concentration areas) and nearby, control locations. (Data from McNaughton 1988.)

spot swards. Both nutrients are particularly important during late pregnancy and lactation. Sodium concentration was over three times and P about 50% greater in forages on animal concentration areas.

These data suggest that nutritional sufficiency, particularly during late pregnancy and lactation for females and early growth for calves, is an important factor governing the distribution of resident herds in the Serengeti. This does not indicate that facilitation of foraging or protection from predation do not also contribute to such animal concentrations (see also Fryxell, chap. 12). In particular, prey often concentrate in buffer zones between territorial predators (Mech 1977; Rogers et al. 1980), so there could be a tendency for hot spots to develop in such zones.

Perhaps surprisingly, underlying soil properties, as indexed by total nutrient concentrations rather than mineralization rates, did not differ between hot spots and control areas (McNaughton 1988). This finding suggests that underlying geologic factors or general differences in physical environments do not govern mineralogical differences in forages associated with animal concentration areas.

It is possible that animal concentration on areas predisposed to higher mineralization rates by subtle physical differences could lead to the development of hot spots. Our current research is investigating this possibility.

Migratory Herds: Temporal Synchronization of Requirements and Spatially Heterogeneous Resources

To determine whether the seasonal movements of migratory herds, like the distributions of resident herds, might be related to nutritional heterogeneity, forage samples were collected from 115 locations throughout Serengeti National Park (McNaughton 1990). As in the hot spot study, the youngest fully expanded leaf blades of dominant species and paired 10 cm depth soil samples were collected during the rainy season, when the migrants are concentrated in the southeast.

Two discriminant functions separated grasslands used in different seasons with high resolution. The first axis separated wet season grasslands from all others, and the second axis separated dry season grasslands from those used principally in transitional periods between wet and dry seasons. Particularly important were minerals required by pregnant and lactating cows and by growing young: Ca, Cu, N, Mg, Na, P, and Zn. All were higher in wet season and transitional ranges than in dry season ranges.

These data indicate that the migratory herds move along a nutritional gradient that is counter to the rainfall gradient. Although forage and water are everywhere abundant during the wet season, the huge migratory herds concentrate in the most arid portion of the ecosystem during the wet season, when calving is most frequent. In contrast to the localized distribution patterns of residents, this migratory pattern is also related to underlying soil properties. Of the nutrients required at high levels by cows and calves, and which are higher in wet season and transitional range forages, all but Mg were correlated with soil levels.

The concurrent calving periods of the major migrants suggest that there has been natural selection for a migratory pattern synchronizing reproduction with that period of the year when there is a high probability of sufficient rain, and resultant grass growth, on the nutritionally superior southeastern Serengeti plains (Kruelen 1975): that is, that the animals most successful in this ecosystem, as measured by their contribution to total herbivore biomass (Fryxell, Greever, and Sinclair 1988), have evolved birth periods coincident with high rainfall on the Serengeti plains, a high probability of that rainfall occurring, and the spatial distribution of nutritionally suitable forages (see also Murray, chap. 11).

CRITICAL HABITATS: A CONSERVATION STANDARD

This research indicates that nutritionally critical habitats are identifiable for both resident and migratory herds in the Serengeti ecosystem. Those habitats are characterized by nutrient-accumulating grasses, usually short in stature, that are concentrated in certain localities. Those habitats are critical to reproduction, meeting the requirements of that biological process when no other areas can do so. However, those other areas utilized at other times of the year do supply the animals with energy needed for maintenance.

It seems likely that seasonal patterns of landscape utilization in the Serengeti, and in other ecosystems that support large populations of herbivorous mammals, are distinguishable by two factors. First, there are spatially localized areas where the crucial nutritional needs of reproduction are fulfilled. The processes leading to this fulfillment must be related both to soil processes governing nutrient availability and to plant uptake properties, which may be regulated by feedback from the animals to the plants (Day and Detling 1990; Georgiadis and McNaughton 1990). Second, there are spatially extensive areas where animals meet their energy requirements but which are nutritionally marginal for many mineral nutrients. Animal condition may gradually decline during the period when animals exploit these nutritionally marginal, but nevertheless essential, habitats. Third, transitional areas also are likely to be important in providing an improvement in the dietary balance available to pregnant females during mid-pregnancy before rains begin in the southeastern plains.

Spatial variability is also an essential property of the Serengeti grazing ecosystem. From the contribution of local pattern differentiation to community diversity (McNaughton 1983), through the topographically related diversification of grasslands on the Serengeti plains, to the spatially extensive distribution of grassland types throughout the region, heterogeneity is crucial to biodiversity. The nutritional and process heterogeneity that accompany pattern diversity reveal an important functional contribution of spatial and temporal variation. Preservation of this variation is important to preserving the ecosystem's essential character.

RESERVE DESIGN, CRITICAL HABITATS, AND LARGE HERBIVORES

The data available suggest that analytical tools are now available to accomplish a concerted habitat mapping of reserve areas designed to support large herbivorous mammals. A working hypothesis is that current reserves lack critical habitats if animal populations move far beyond reserve boundaries, particularly during the wet season when animals may freely choose habitats unconstrained by water or forage availability.

Because the boundaries of the Serengeti and surrounding multiple-use areas were more or less continuously adjusted over several decades, they may more clearly encompass the full range of habitats required by herds than those of other reserves with more arbitrarily drawn perimeters. However, should animal reintroduction become an important conservation policy in the future, as it has been for American bison, the approach outlined here, focusing on the temporal and spatial variability of potential reserves, could focus attention on delimiting borders that are ecologically sound (McNaughton 1988, 1990).

A potentially important application of this research to the Serengeti ecosystem would be the expansion of nutritional mapping to the Maswa Game Reserve. That region is extensively utilized by the migratory herds whenever rain temporarily fails on the Serengeti plains, and there have been years when the herds have calved there (McNaughton 1985). Therefore, we predict that the Maswa may contain critical habitats to which more rigorous protection should be extended than currently applies. The practice of removing portions of the Maswa region from the protected area, which occurred during the 1970s and 1980s, also could have a significant long-term deleterious effects upon the animals of the Serengeti if this area is a critical source of nutrients when the Serengeti plains are dry.

ACKNOWLEDGMENTS

The Serengeti Ecosystem Processes Project has been supported by the U.S. National Science Foundation Ecosystem Studies Program and by Syracuse University since 1974. The second author expresses appreciation for support from the Third World Academy of Science and the University of Dar es Salaam. Margaret McNaughton was instrumental to the execution of much of the original research reviewed here.

REFERENCES

Banyikwa, F. F. 1976. A quantitative study of the ecology of the Serengeti short grasslands. Ph.D. thesis, University of Dar es Salaam.

Banyikwa, F. F., Feoli, E., and Zuccarello, V. 1990. Fuzzy set ordination and classification of Serengeti short grasslands, Tanzania. *J. Veg. Sci.* 1:97–106.

Bell, R. H. V. 1970. The use of the herb layer by grazing ungulates in the Serengeti. In *Animal populations in relation to their food sources*, ed. A. Watson, 111–24. Oxford: Blackwell.

Day, T. A., and Detling, J. K. 1990. Grassland patch dynamics and herbivore grazing preference following urine deposition. *Ecology* 71:180–88.

de Wit, H. 1976. Soils and grassland types of the Serengeti plains (Tanzania). Ph.D. thesis, University of Wageningen.

Feoli, E., and Zuccarello, V. 1986. Ordination based on classification: Yet another solution? *Abstr. Bot.* 10:203–19.

———. 1988. Syntaxonomy: A source of useful fuzzy sets for environmental analysis? *Coenoses* 3:141–47.

Fryxell, J. M., Greever, J., and Sinclair, A. R. E. 1988. Why are migratory ungulates so abundant? *Am. Nat.* 131:781–98.

Georgiadis, N. G., and McNaughton, S. J. 1990. Elemental and fibre contents of savanna grasses: Variation with grazing, soil type, season, and species. *J. Appl. Ecol.* 27:623–34.

Herlocker, D. J., and Dirschl, H. J. 1972. *Vegetation of Ngorongoro Conservation Area, Tanzania.* Canadian Wildlife Service Report ser. 19. Ottawa: Canadian Wildlife Service.

Inglis, J. M. 1976. Wet season movements of individual wildebeests of the Serengeti migratory herd. *E. Afr. Wildl. J.* 14:17–34.

Jarmillo, V. J., and Detling, J. K. 1988. Grazing history, defoliation, and competition: Effects on shortgrass production and nitrogen accumulation. *Ecology* 69:1599–1608.

Kruelen, D. 1975. Wildebeest habitat selection on the Serengeti plains, Tanzania, in relation to calcium and lactation: A preliminary report. *E. Afr. Wildl. J.* 13:297–304.

Marsili-Libelli, S. 1986. Crop growth and stress assessment through fuzzy clustering. In *Proc. 2nd Eur. Simulation Congr.*, Antwerp, 31–37.

McNaughton, S. J. 1979. Grassland-herbivore dynamics. In *Serengeti: Dynamics of an ecosystem,* ed. A. R. E. Sinclair and M. Norton-Griffiths, 46–81. Chicago: University of Chicago Press.

———. 1983. Serengeti grassland ecology: The role of composite environmental factors and contingency in community organization. *Ecol. Monogr.* 53:291–320.

———. 1984. Grazing lawns: Animals in herds, plant form, and coevolution. *Am. Nat.* 124:863–86.

———. 1985. Ecology of a grazing ecosystem: The Serengeti. *Ecol. Monogr.* 55:259–94.

———. 1988. Mineral nutrition and spatial concentrations of African ungulates. *Nature* 334:343–45.

———. 1989. Interactions of plants of the field layer with large herbivores. In *The biology of large African mammals in their environment,* ed. P. A. Jewell and G. M. O. Maloiy, 15–29. Zool. Soc. London Symp. no. 61. Oxford: Clarendon Press.

———. 1990. Mineral nutrition and seasonal movements of African migratory ungulates. *Nature* 345:613–15.

———. 1991. Evolutionary ecology of large tropical herbivores. In *Plant-animal interactions: Evolutionary ecology in tropical and temperate regions,* ed.

P. W. Price, T. M. Lewinsohn, G. W. Fernandes, and W. W. Benson, 509–22. New York: Wiley.

McNaughton, S. J., Ruess, R. W., and Seagle, S. W. 1988. Large mammals and process dynamics in African ecosystems. *BioScience* 38:794–800.

Mech, L. D. 1977. Wolf pack buffer zones as prey reservoirs. *Science* 198:320–21.

Norton-Griffiths, M., Herlocker, D., and Pennycuick, L. 1975. The pattern of rainfall in the Serengeti ecosystem, Tanzania. *E. Afr. Wildl. J.* 13:347–74.

Raison, R. J., Connell, M. J., and Khanna, P. K. 1987. Methodology for studying fluxes of soil mineral-N in situ. *Soil Biol. Biochem.* 19:521–30.

Roberts, D. W. 1986. Ordination on the basis of fuzzy set theory. *Vegetatio* 66:123–43.

Rogers, L. L., Mech, L. D., Dawson, D. K., Peek, J. M., and Korb, M. 1980. Deer distribution in relation to wolf pack territory edges. *J. Wildl. Mgmt.* 44:253–56.

Senft, R. L., Coughenour, M. B., Bailey, D. W., Rittenhouse, L. R., Sala, O. E., and Swift, D. W. 1987. Large herbivore foraging and ecological hierarchies. *BioScience* 37:789–99.

Sinclair, A. R. E. 1977. *The African buffalo: A study of resource limitation of populations.* Chicago: University of Chicago Press.

———. 1985. Does interspecific competition or predation shape the African ungulate community? *J. Anim. Ecol.* 54:899–918.

Snaydon, R. W. 1962. Micro-distribution of *Trifolium repens* L. and its relation to soil factors. *J. Ecol.* 50:133–43.

———. 1970. Rapid population differentiation in a mosaic environment. I. The response of *Anthoxanthum odoratum* populations to soil. *Evolution* 14:257–69.

Vesey-Fitzgerald, D. 1960. Grazing succession among East African game animals. *J. Mammal.* 41:161–72.

———. 1973. *East African grasslands.* Nairobi: E. Afr. Publ. House.

Wedin, D. A., and Tilman, D. 1990. Species effects on nitrogen cycling: A test with perennial grasses. *Oecologia* 85:433–41.

Zadeh, L. A. 1965. Fuzzy sets. *Inf. Control* 8:338–58.

Zimmerman, H. J. 1984. *Fuzzy set theory and its applications.* Dordrecht: Kluwer Nijhoff.

FOUR

Vegetation Dynamics in the Serengeti-Mara Ecosystem: The Role of Elephants, Fire, and Other Factors

Holly T. Dublin

Incomplete and comparatively poor documentation exists on the history of vegetation changes in Africa's major plant communities over the past century. Reconstructing the factors and processes involved in these vegetation shifts, given the paucity of data, requires a blend of qualitative and quantitative approaches. Recent attempts to analyze these changes in detail have added a great deal to our understanding of the nature and complexity of vegetation dynamics. Many vegetation communities in Africa have long been viewed as examples of "climax" communities resulting from conditions of long-term stability, but "climax" is not an appropriate description of the dynamics of savanna woodland ecosystems.

The Serengeti-Mara ecosystem, like many of Africa's savanna woodlands, has experienced major vegetation changes in its recent history, alternating between open grassland and dense woodland and back over the past century. There have been reports of similar changes in other ecosystems within the region over the same period. These patterns of vegetation change suggest that the Serengeti-Mara ecosystem is dynamic and may be subject to long-term vegetation cycles or transitions between stable states following ecological perturbations (Caughley 1976; Carson and Abbiw 1990; Dublin, Sinclair, and McGlade 1990).

At the turn of the century, the Serengeti-Mara area was described by explorers, traders, and hunters as an open grassland with lightly wooded patches, much as it is in many areas today. By the time the colonial administrators arrived in the 1930s and early 1940s, the area had become densely wooded. The land set aside as the Serengeti National Park and the Masai Mara National Reserve was characterized by dense woody vegetation and stayed that way for over 20 years. In the 1950s, however, the woodlands and thickets began a rapid decline and reverted to grasslands.

Norton-Griffiths (1979) predicted that if the wildebeest population remained high, fire frequency and severity would decrease and lead to a recovery of woody vegetation throughout the ecosystem over time. This prediction has been borne out in the Serengeti woodlands south of the border over the past 5–10 years (Sinclair, chap. 5), but the opposite has taken place in Masai Mara. One obvious difference between the two areas is the presence of elephants at high densities in the Mara and their virtual absence in the Serengeti over the same period. Poaching for ivory has played a significant role in determining elephant distributions in the ecosystem. These distributions have, in turn, influenced the pattern of woody loss and recovery throughout the northern Serengeti and Mara.

This chapter provides an overview of vegetation changes over the past century in the Serengeti-Mara ecosystem. The dynamics of the Serengeti-Mara woodlands are complex and continually changing, perhaps more rapidly than previously assumed. Our research has provided insights into the many factors that are involved in the dynamics of this system.

METHODS

Reconstruction of 100 years of vegetation dynamics involved qualitative and quantitative assessment of historical and experimental information coupled with simple modeling. The early history, from the late 1880s to the early 1900s, was compiled from numerous written sources, including the records of early slave traders, explorers, hunters, and naturalists. Information on the 1930s and 1940s came from both written reports and personal interviews with colonial administrators, journalists, and hunters who had worked extensively in the Serengeti-Mara region (Dublin 1991). Changes in woody vegetation cover over the 30 years from 1950 through the early 1980s were described using a series of five complete aerial photographic surveys flown in 1950, 1961, 1967, 1974, and 1982. Woodland changes were estimated using a "dot-grid" analysis of these photographs (Norton-Griffiths 1979; Lamprey 1985; Dublin 1991).

Experimental work conducted in the Mara between 1982 and 1986 provided specific data on the factors involved in vegetation changes, particularly the annual rates of seedling and tree damage and mortality caused by fire, elephants, wildebeest, and resident antelopes (Dublin 1986). These data were then combined in a simple model to simulate the effects of varying elephant densities and burning rates on woodland dynamics in the Serengeti-Mara ecosystem (Dublin, Sinclair, and McGlade 1990). Predictions from this model were compared with estimates of woodland change derived from aerial photography and fixed photopoints in the Serengeti and Mara to analyze the biological forces driving woodland declines in the 1960s and their failure to recover in the 1980s.

RESULTS

From Grasslands to Woodlands: The 1890s

Following the great rinderpest epidemic of 1890, human and animal populations are thought to have been reduced to negligible numbers in the Serengeti-Mara region (Sinclair and Norton-Griffiths 1979). Fires were presumably infrequent due to low human populations, and elephant numbers suffered from heavy ivory poaching in the previous decades (Spinage 1973; Dublin 1986). Explorers and hunters of the early 1900s encountered a Serengeti-Mara characterized by broad, open expanses of grassland studded by occasional *Acacia* trees, much as we see the Mara Reserve today (Woosnam 1913; White 1915; Buxton 1927; Eastman 1927).

Over the next 30 to 50 years the prevailing conditions saw the establishment of dense woodlands and thickets. This dense, woody vegetation provided suitable habitat for heavy infestations of the tsetse fly (Lewis 1935; Swynnerton 1936; Beaumont 1944; Buxton 1955; Ford and Clifford 1968; Langridge, Smith, and Wateridge 1970; Ford 1971). This factor alone prevented any significant human settlement within the Serengeti and Mara at that time and was one reason why the area was set aside for protection in the 1930s. Ironically, the colonial administrators who worked so diligently to protect these "pristine" woodlands did not recognize that this area had been open grassland less than 50 years earlier.

From Woodlands to Grasslands: The 1960s

It was the general failure to recognize the dynamic nature of the extant vegetation that led to an extreme reaction by scientists and managers when the woodlands began their decline in the late 1950s and early 1960s (Lamprey et al. 1967; Watson and Bell 1969; Glover and Trump 1970). During this same period human populations, recovering from the secondary effects of the rinderpest epidemic, increased dramatically (Morgan and Shaffer 1966; Kurji 1976). Two primary pathways stemming from the increase in human populations may have initiated woodland declines. The first was fire. Ungulate populations, still sparse as a result of the rinderpest, were unable to reduce significantly the standing crop of dry grass resulting from unusually high rainfall in the late 1950s and early 1960s (fig. 4.1). As the dry season progressed, fires became widespread (M. Turner, pers. comm.). Some of these fires were intentionally set by Masai herdsmen to improve grazing pastures and to clear tsetse-infested bush, while some burns facilitated hunting by neighboring tribes, some were set by park authorities under fire management schemes, some were inadvertently lit by wandering honey hunters, and still others, by European hunters. Under conditions of normal rainfall, woodland regeneration would have been impeded by this sudden increase in fire occurrence. Under the

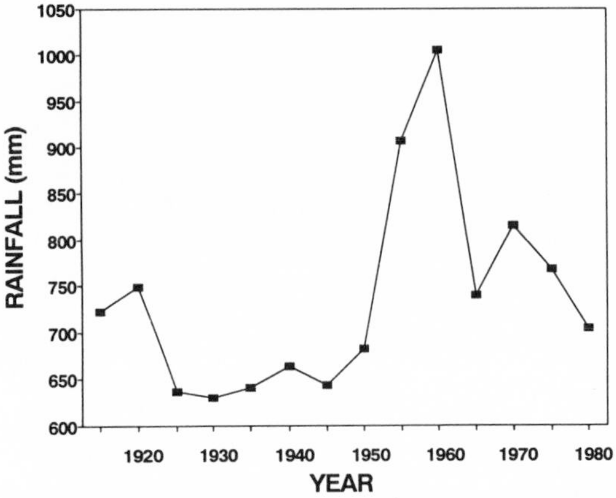

Figure 4.1 Five-year rainfall averages for Narok, Kenya, for the period 1914–1980.

high-rainfall conditions of the early 1960s and the subsequent high fuel production, these fires were devastating. They served to clear the area of bush and attract ungulates to the lush grazing "lawns" created by the fires.

The increase in humans also opened a second pathway to woodland loss in protected areas: the compression of elephant populations into smaller areas. The role of elephants in dramatic woodland declines during the 1960s had gained much notoriety and attention among ecologists throughout Africa (Buechner and Dawkins 1961; Brooks and Buss 1962; Glover 1963; Darling 1964; Buss and Savidge 1966; Field 1971; Anderson and Walker 1974; Caughley 1976). Like many other areas in Africa, the Serengeti-Mara woodlands felt the effects of increased elephant densities as the animals moved in from surrounding areas such as Loliondo, the Isuria (Oloololo or Siria) escarpment, the Chepalunga Forest, and the Lambwe Valley (M. Turner, K. Smith, and R. Elliot, pers. comm.) These areas were experiencing rapid human settlement, increasing livestock numbers, and cultivation, and therefore elephants were being forced out (Narok District Commissioner's Report 1955). Subsequent change in the woodlands was inevitable.

An analysis of the aerial photographs shows a steady loss of cover in woodlands (fig. 4.2). Absolute and relative rates of woodland cover loss in the Mara Reserve were highest from 1961 to 1967, though declines continued into the 1980s (Dublin 1986). During the 1960s, the mean cover density of *Acacia* woodlands and *Croton* thickets dropped significantly throughout the Mara Reserve (Dublin 1991). These losses corresponded significantly to the period of unusually high rainfall (Dublin 1991), the subsequent fires, which occurred twice and sometimes three

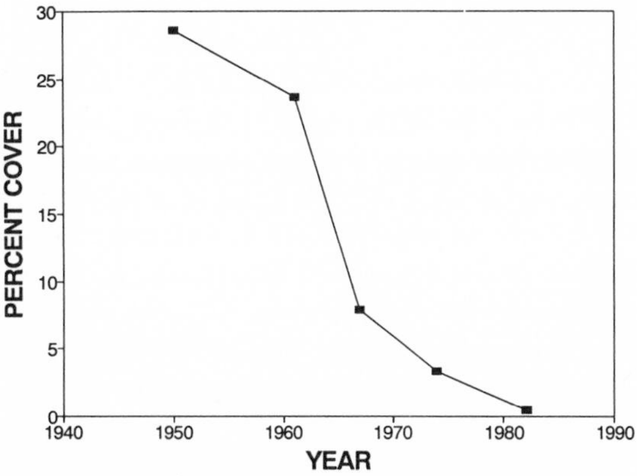

Figure 4.2 Mean percentage cover in *Acacia* woodlands in the Mara Reserve. Percentage cover densities are derived from dot-grid analyses of five sets of aerial photographs taken between 1950 and 1982.

times per year (Langridge, Smith, and Wateridge 1970), and increasing elephant densities throughout the Serengeti-Mara ecosystem (Dublin and Douglas-Hamilton 1987).

A Stable State: Grasslands of the 1980s

Between the 1960s and the 1980s dramatic changes occurred in the Serengeti-Mara ecosystem (Sinclair and Norton-Griffiths 1979). By the mid-1970s wildebeest populations, freed from the limitations of rinderpest, had increased by a factor of five, and they have remained at approximately 1.5 million ever since (Dublin et al. 1990). Each year these wildebeest travel to the northern Serengeti and Mara in search of dry season forage and remove the majority of the available standing crop (Onyeanusi 1989), which would otherwise burn. Despite Norton-Griffiths's (1979) predictions for woodland recovery under these conditions, woodland declines continued in the Serengeti and Mara well into the 1980s.

Elephant Numbers and Distribution. Survey work on elephant numbers and distribution, observational work on elephant habitat choice and feeding patterns, and experimental work on the effects of elephants, fire, wildebeest, and browsing antelope on woody vegetation elucidated the reasons behind the persistence of grasslands over the recovery of woodlands.

Significant changes have taken place in both the numbers and distribution of elephants in the Serengeti-Mara ecosystem over the past 20–25 years (Dublin and Douglas-Hamilton 1987). These changes can largely be attributed to an increase in poaching within the Serengeti National Park during the late 1970s and early 1980s (Douglas-Hamilton 1983; Dublin et al. 1990). The recent but extreme difference in elephant distribution and densities between the Serengeti and Mara played a significant role in the divergent pattern of woodland changes seen today.

The Serengeti elephant population declined by 81% from 2,460 in 1970 to 467 in 1986. Of these, over 1,500 were apparently killed by poachers, while the remaining 400 to 500 sought safe refuge in the Mara. A combination of high tourist exposure and increased anti-poaching effort provided a secure home in the Mara, and elephant numbers have increased significantly there since 1970. Today, over 1,300 elephants move back and forth between the Mara and the adjacent pastoral lands to the north (fig. 4.3). This population increase is presumably due to a combination of immigration and natural population growth.

Elephant Feeding Patterns and Habitat Choice. The effects of these elephants, which are now year-round residents of the Mara, on vegetation are pronounced. Elephants in the Mara eat woody species of all sorts, and their use of shrubs and trees increases significantly when dry conditions prevail (Dublin 1986). Facing an already reduced availability of browse forage and shade trees, elephants have begun concentrating their time within the *Croton* thickets. With the significant loss of other woodland habitats in the Mara over the past three decades, these thickets now provide one of the last wooded refuges available to elephants. Here they are able to find shade, and also to forage on woody species and herbs that thrive in the moist, shady conditions within these thickets.

This constant use of *Croton* thickets for food and shade has caused severe damage to their internal structure and opened large pathways through the vegetation (Norton-Griffiths 1979; Dublin 1991). In subsequent rainy seasons, these light gaps grow thick swards of grass (Norton-Griffiths 1979). Most grazers avoid the risk of hidden predators in thickets, so this grass is frequently left to dry and form litter in the herb layer. When fires occur in the reserve, these grass pathways through the thickets burn very hot and destroy the trees and bushes along their boundaries. As the years progress the thickets become fragmented—like pies being cut and removed in ever smaller pieces.

Acacia woodlands have been subjected to similar pressures, primarily from elephant bulls, but occasionally from cow-calf groups as well. Most *Acacia* woodlands in the Mara are composed of trees that are too small to be used for shade. In unusually dry seasons, however, bulls spend a lot of time in these woodlands and feed heavily on the trees. In the drought of 1984, a herd of six bull elephants visited a 2 km^2 mature *Acacia gerrardii* stand near Emarti. Within 24 hours they had left 34% of the trees dead or fatally damaged and another 22% with multiple broken branches. Many other stands experienced similar damage during this and other very dry periods.

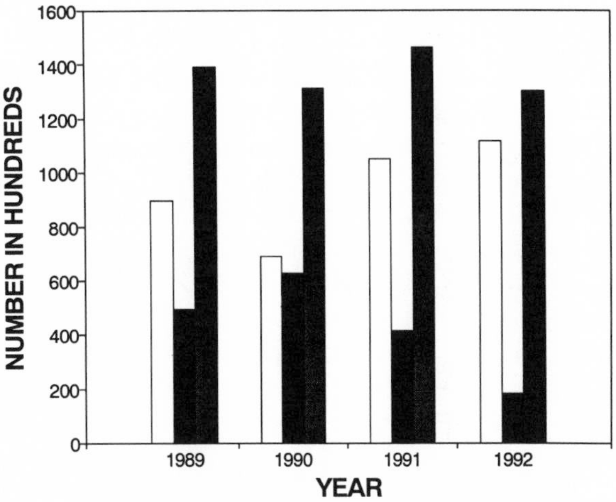

Figure 4.3 Elephant numbers within the Mara Reserve (solid bars), its adjacent pastoral areas (shaded bars), and the Lamai Wedge of northern Serengeti (open bars).

In addition to damaging large trees, elephants spend significant amounts of time feeding on seedlings and saplings under 1 m in height. Norton-Griffiths (1979) and Belsky (1984) inferred from the earlier observations of Croze (1974b) that elephants "largely ignored" trees under 1 m, and therefore had little effect on seedling survivorship in the Serengeti. In contrast, the results from this study demonstrate that elephant browsing in the 1980s was a primary factor in seedling mortality and the inhibition of seedling growth in the Mara (Dublin 1986).

There are several possible explanations for this difference in findings. First, Croze (1974a,b) concluded that Serengeti elephants in the late 1960s and early 1970s browsed trees in approximate proportion to their height availabilities, but avoided trees under 1 m. Trees above 1 m were abundant then, constituting up to 60% of the total population in *Acacia* woodlands (Lamprey et al. 1967; Glover 1968; Croze 1974b; Norton-Griffiths 1979). In the Mara there has been a major change in the height distribution of trees following a progressive loss of the taller height classes (Dublin 1984, 1986). Unlike the height distribution of trees in the 1960s, the current height structure of the Mara woodlands is heavily biased toward the 0–1 m height class (fig. 4.4). Dublin (1986) showed that over 80% of the browse diets of elephants in the Mara came from trees shorter than 1 m, which is in proportion to the current frequency distribution of tree heights in *Acacia* woodland stands. Therefore, the observed difference in the feeding behavior of elephants in the Mara may result from this obvious change in the population structure of trees.

Second, elephants browsing on seedlings may appear to be feeding on

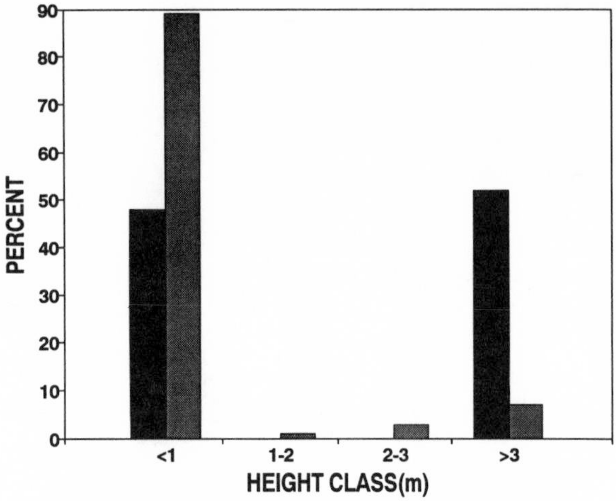

Figure 4.4 Frequency distribution by height class for *Acacia* in the Seronera woodlands (dark shading) (data from Croze 1974b) and in the Mara woodlands (light shading) (data from Dublin 1984).

nearby grasses. Seedlings occur in their highest densities in open grassland areas of the Mara. From a distance, it is possible to mistake feeding on seedlings for grazing on grasses. Forage selection by elephants was determined in the Mara through direct, close-up observations (Dublin 1986), whereas earlier studies often determined the forage preferences of elephants through their direct effects on vegetation and their distribution among habitats. These earlier, indirect methods can be misleading. When elephants forage on seedlings there is no obvious sign left after the seedling has been completely removed. However, the patterns of elephant foraging recorded in the Mara are consistent with findings from other elephant populations, which reported that elephants utilized regenerating seedlings where they were available (Brooks 1957; Buss 1961; Field and Ross 1976; Guy 1976; Jachmann and Bell 1985; Okula and Sise 1986; Weyerhaeuser 1985). It is, therefore, possible that earlier studies of elephant feeding in the Serengeti simply failed to notice the extent to which elephants utilized seedlings.

This heavy utilization of seedlings, of almost any size, has important consequences for woodland dynamics. Caughley (1976) suggested that a "stable limit cycle" existed because seedlings escape predation by elephants and so, over time, provide the stock for woodland regeneration. Seedlings were assumed to be in a "height refuge." In the Mara, however, where so many seedlings are removed by elephants, this height refuge does not exist, and the potential for woodland recovery is effectively diminished. Furthermore, the general tendency for elephants to overutilize woodlands in the dry season may be a direct consequence of competition with the migratory wildebeest that are present in the Mara throughout that time. These migrants remove over 90% of the standing crop during their stay (Sinclair 1975; Onyeanusi 1989). Immediately following cessation of the rains and prior to the arrival of the wildebeest, both sexes of elephants show an obvious preference for grasses (Dublin 1986). At this time, the newly germinated grasses are very abundant and at their peak of digestible protein levels (Sinclair 1975; Onyeanusi 1989) (fig. 4.5). However, upon the arrival of the wildebeest, a distinct change in diet choice occurs. Elephants immediately decrease their feeding on grasses and in the latter stages of the wildebeest occupancy stop feeding on grasses altogether, despite the production of new grass following intermittent local thunderstorms. Wildebeest graze the grasses down to lawn height (McNaughton 1984). At this height grasses may become difficult for elephants to eat.

This notable change from grasses to woody species probably reflects a decline in both grass quantity and grass quality. Woody species remain relatively higher in crude protein than herbaceous species as the dry season progresses. Pellew (1981) reported that some preferred browse species, such as *Acacia gerrardii, A. senegal,* and *Commiphora trothae,* maintained crude protein levels above 13% even in older leaves. Dougall (1963) and Dougall, Drysdale, and Glover (1964) found even higher values for the mature leaves of *Acacia brevispica* (17%), *Boscia angustifolia* (33%), and *Solanum incanum* (30%), which are also used by Mara elephants in the dry season. The grasses eaten by elephants, such as *Themeda triandra* and *Cynodon dactylon,* drop to as low as 5% and 8% crude

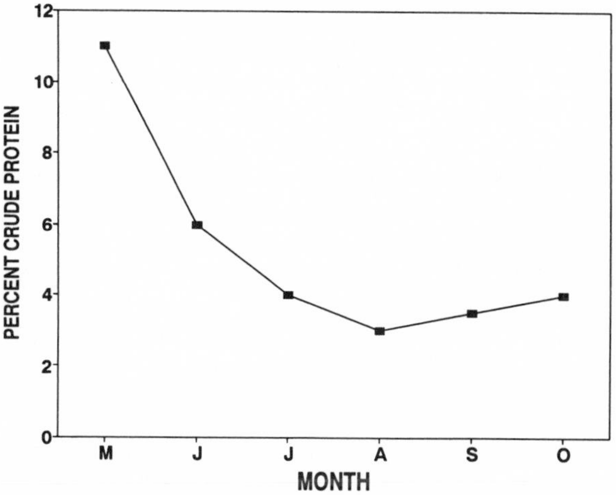

Figure 4.5 Crude protein percentage of long grasses over the course of the dry season. (Redrawn from Sinclair 1975.)

protein during the dry season (Dougall 1963; Dougall and Glover 1964; Field 1971; Sinclair 1975). By the height of the dry season, elephants spend the majority of their feeding time on the available browse species (fig. 4.6).

Throughout Africa elephants are considered the primary agents of habitat change (Laws 1970; Thomson 1975). Although Field (1971) suggested that giraffes may compete with elephants under certain circumstances, no other species is known to alter habitats as significantly as elephants, or to compete with them for grazing resources. The evidence provided above, however, suggests that the sheer numbers of migratory wildebeest coming into the Mara each year do alter the habitat during their annual stay. Through trampling and inadvertent browsing they reduce the availability of small trees used by elephants late in the dry season. Although elephant browsing activity increases significantly during any period of low rainfall, the added effect of the wildebeest migration may exacerbate this pattern and result in elephants placing even greater pressure on available browse species in the dry season.

Croze (1974a) studied the feeding behavior of bull elephants in an area of the central Serengeti where wildebeest were not present in the dry season. He found that bulls spent approximately 70% of their time in the *Acacia* woodlands and 30% of their time in vegetation along permanent water sources. In the two habitats combined, browsing took up 34% of the elephants' feeding time. In contrast, elephants in the Mara spent 70% of their foraging time on browse during the dry season.

This extreme dependence on browse in the Mara is also exacerbated by the poaching pressure that kept elephants in the Mara on a year-round basis rather than allowing them to follow their seasonal migratory routes as before.

Effects of Elephants, Fire, Wildebeest, and Other Browsers on Woodlands

The number of mature trees in the Masai Mara has been reduced over the past 30 years. The remaining mature trees are now being lost at a rate of over 8% per year. The future of these disappearing woodlands is dependent on the potential for replacement by regenerating seedlings.

In the Mara, the growth potential of seedlings and coppicing rootstocks is being severely inhibited. Dublin (1986) established that 4% of all seedlings were killed annually by elephants, 4% by fire, 1% by wildebeest, and another 1% through other natural causes. Seedlings experienced the greatest impacts from elephants and other browsers in the dry season. Wildebeest, elephants, and other browsers removed up to 60% of all surviving stems in unburned plots and an even greater proportion in burned plots. While wildebeest effects were similar in both burned and unburned plots, elephants and other browsers showed a distinct prefer-

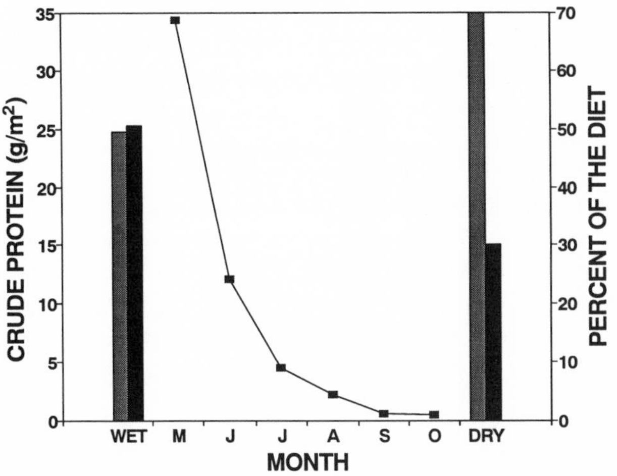

Figure 4.6 Available grams crude protein per m[2] of long grasses during the dry season (left axis, line graph) and the proportion of elephant diet composed of grass (solid bar) and browse (shaded bar) in the wet and dry seasons (right axis).

ence for seedlings in burned plots. Browsing effects on seedlings varied accordingly and were significantly higher in burned plots than in unburned plots. The majority of seedlings removed at ground level by wildebeest or fire resprouted within 6 months. Those taken by elephants, however, experienced much greater delays in recovery (fig. 4.7). This suggests that elephants did more severe damage to the plants than did fire or wildebeest.

Seedlings exposed to browsing showed significant decreases in height but were significantly taller than those seedlings that had been both burned and browsed (fig. 4.8). Those seedlings that were browsed but not burned showed a steady decline in stem numbers. Burning coupled with browsing stimulated an increase in stem numbers. In exclosure plots, where seedlings were neither burned nor browsed, they grew a maximum of 10–15 cm per year while the average number of stems remained more or less the same.

Multiple-burn experiments demonstrated that seedling survivorship was inversely related to the level of fire intensity. Seedling survivorship remained high after repeated cool burns (fuel loads $\leq$ 150 g/m^2) but dropped significantly after repeated hot burns (fuel loads $>$ 300 g/m^2), are often characteristic of Serengeti-Mara grasslands (fig. 4.9).

A MODEL OF TREE POPULATION DYNAMICS

The data described above were combined in a simple model (Dublin, Sinclair, and McGlade 1990) to describe the separate and synergistic effects

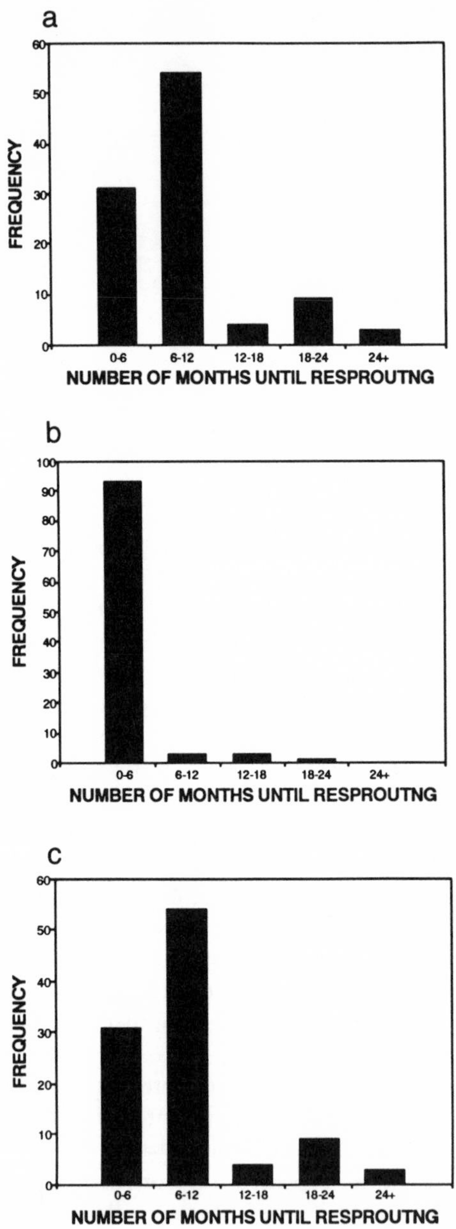

Figure 4.7 Frequency distribution of return times for resprouting "regenerates" originally removed by (*a*) fire, (*b*) wildebeest, and (*c*) elephants.

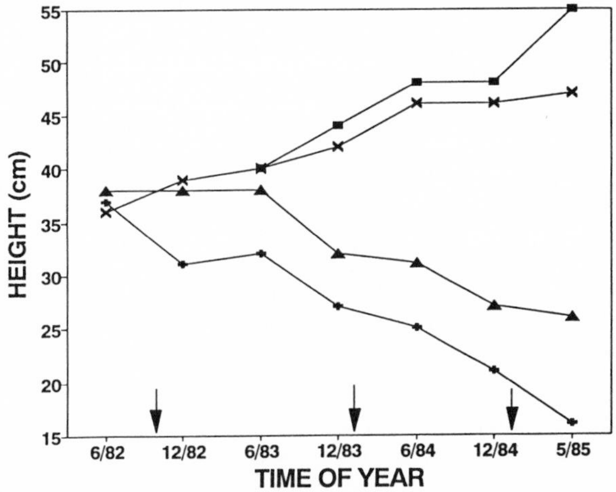

Figure 4.8 Change in the average heights of marked stems over the study period. Arrows indicate experimental burns. Squares, clipped exclosure; Xs, unclipped exclosure; triangles, browsed only; plus signs, browsed and burned.

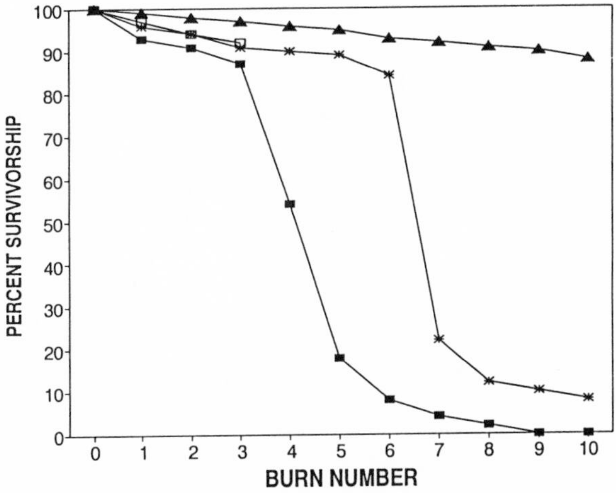

Figure 4.9 Seedling survivorship for multiple-burn experiments under three burning treatments: Triangles, 150 g/m²; asterisks, 300 g/m²; solid squares, 600 g/m². Open squares, 200–400 g/m². This last treatment experienced only three sequential large-scale experimental burns.

of fire, elephants, and wildebeest and to answer two questions: What caused the decline of woodlands in the 1960s, and what prevents their recovery in the 1980s? The model was based on a series of constant and variable parameters (table 4.1) based on data collected in the field. These parameters were set to levels measured in both the 1960s and the 1980s to "re-create" the conditions of those periods. Recruitment rates generated by the model were compared with known rates of decline derived from serial aerial photography of the area.

By varying the individual factors of fire, elephants, wildebeest, and browsing antelopes while holding others constant, it was possible to analyze the effects of each factor individually. The differential and synergistic effects of fire and elephants most affected the dynamics of the woodlands.

Findings from the period of greatest woodland losses, the 1960s, supported the hypothesis that fire alone was able to drive the decline (fig. 4.10). Elephants, acting with fire, further exacerbated these declines but were not in and of themselves responsible for the losses. The early 1960s was a time of unusually high rainfall, and the subsequent increase in hot and frequent fires acted as the primary force in woodland loss.

In the 1980s the dynamics of the northern Serengeti and Mara woodlands diverged. Mara woodlands continued their decline, resulting in a

Table 4.1 Constants and variables built into the model of tree dynamics with elephant and fire effects.

Constants

For all plants

1. Seedling densities: 850/ha
2. Adult tree (>3 m) densities: 32/ha
3. New seedlings: 17/ha/year
4. Growth of seedlings: 15 cm/year
5. Fire escapement height: 3 m
6. Natural mortality rate of adult trees and seedlings: 1%/year

For burned plants

1. Number of plants under 1 m killed by fire: 5%
2. Number of plants in the 0–1 m class that revert to the 0–1 m class: 95%
3. Number of plants in the 1–2 m class that revert to the 0–1 m class: 90%
4. Number of plants in the 2–3 m class that revert to the 0–1 m class: 5%
5. Number of plants in the 2–3 m class that revert to the 1–2 m class: 29%

Variables

1. Burning rate
2. Elephant mortality rate on seedlings
3. Elephant reversal rate on seedlings
4. Wildebeest inhibition rate on seedlings
5. Wildebeest reversal rate on seedlings
6. Browser inhibition rate on seedlings
7. Mortality rate of mature, adult trees

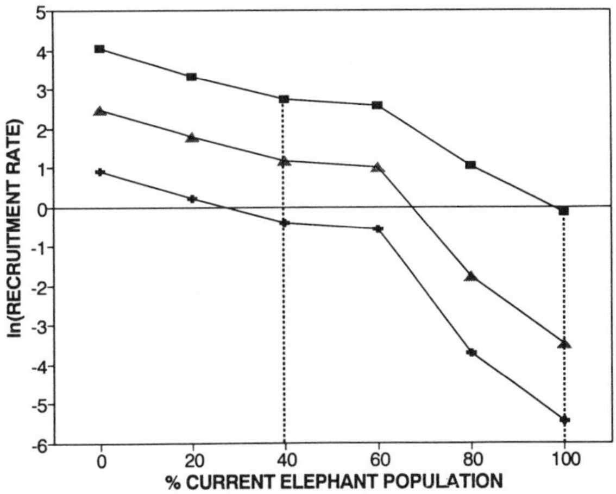

Figure 4.10 The natural log of tree recruitment rates (R) for varying elephant populations under the model: squares, without wildebeest or browsers; triangles, with browsers but no wildebeest; plus signs, with both wildebeest and browsers. Those values less than zero indicate woodland loss, while those values greater than zero indicate woodland increases. The dotted line at 40% represents elephant population sizes of the 1960s; that at 100% represents population size in the 1980s. Burning rates are held at zero. (Adapted from Dublin, Sinclair, and McGlade 1990.)

perpetuation of grasslands throughout the area. Conversely, woodlands of the northern Serengeti showed a tremendous regeneration of woody species, particularly *Acacia clavigera* and *Acacia gerrardii* (Sinclair, chap. 5). The model suggests that this differential recruitment between the Serengeti and the Mara resulted from the differential use of these areas by elephants during the 1980s. During this time, the Mara woodlands received increased browsing pressure from elephants, while the northern Serengeti experienced just the opposite. For a number of years in the mid- to late 1980s, elephants were virtually absent from the northern Serengeti.

Permanent photopoints established in the Mara and the northern Serengeti provided evidence that woodland recruitment very closely resembled that predicted by the model. In the absence of elephants, woodlands should recover even with the pressure of current wildebeest impacts and the effects of browsing antelopes (fig. 4.11). This result has been verified in the field. While the Mara Reserve continued to lose its remaining *Acacia* woodlands, the Serengeti woodlands began a rapid recovery (Sinclair, chap. 5). Despite Norton-Griffiths's (1979) prediction that the woodlands would recover under conditions of high wildebeest numbers and low fire, this has not been the case in the Mara. Supporting over 1,000 elephants on a year-round basis since the mid-1980s, the Mara Reserve appears to be locked into a grassland phase.

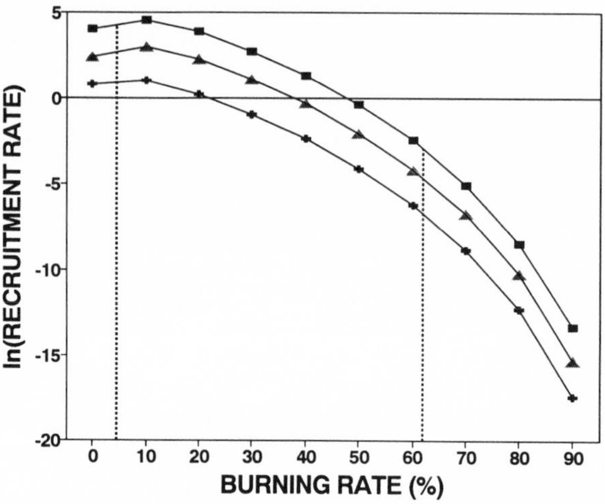

Figure 4.11 The natural log of tree recruitment rates (*R*) for varying burning rates under the model: squares, without wildebeest or browsers; triangles, with browsers but no wildebeest; plus signs, with both wildebeest and browsers. Those values less than zero indicate woodland loss, while those values greater than zero indicate woodland increases. The dotted line at 62% represents the burning rates of the 1960s, that at 5% represents the burning rates in the 1980s. Elephant effects are held at zero. (Adapted from Dublin, Sinclair, and McGlade 1990a.)

LOOKING AHEAD: THE ROLE OF FIRE AND ELEPHANTS IN HABITAT DYNAMICS

Historical, experimental, and simulated findings suggest that vegetation communities in the Serengeti-Mara ecosystem are capable of shifting from one state to another following ecological perturbations. These states are identifiable when the system does not return to the original state following removal of the original factor that caused the perturbation. Rather, one factor may cause the perturbation while another factor maintains the vegetation in its new state. In this way, the vegetation may remain in a specific state for varying periods of time depending on the predominating conditions. Likewise, it may rapidly switch back to the old state or to another new state when new factors come into play.

Under current conditions, mature trees will not be replaced and *Acacia* woodlands in the Mara will not expand or even persist. A significant change in one or more of the prevailing inhibitory factors must take place for woodlands within the reserve to regenerate.

A decrease in fire frequency or severity is unlikely, as fire forms an integral part of the Serengeti-Mara ecosystem. Although burning rates are not as severe as they have been in the past, fires still occur in the Mara on an annual basis. Due to the current high levels of offtake by the migratory wildebeest, the most severe fires do not occur at the height of the dry season (September—October) but are more frequently restricted to the short dry season (January–February) when the migratory herbivores are not present. These fires inhibit the growth and survivorship of seedlings. If disease and/or poaching ever reduced wildebeest numbers to their previous levels of 250,000 or less, more fuel would be available in the dry season and fire would once again play an important role in the inhibition of woodland recovery.

In addition to the effect of fire, the large number of resident elephants currently in and around the reserve does not bode well for the future of the woodlands. The only possible reprieve from a scenario of ever-declining woodlands would be the emigration of significant numbers of elephants back to the Serengeti. This will very likely occur when the elephants feel safe enough to cross back into their previous haunts in the northern Serengeti. Since early 1990, there has been evidence that this process is beginning to take place.

If elephants do reestablish themselves in the northern Serengeti, the monitoring of woodlands in both the Serengeti and the Mara will be extremely important over the next decade. If elephant numbers in the Mara are significantly reduced as a result of this emigration, there may be some chance for seedlings in the Mara to escape and for woodland recovery to proceed. The opposite may be true in the northern Serengeti.

The future of the Serengeti-Mara woodlands involves many individual and interacting factors, which may themselves be governed by prevailing conditions of the times. As Norton-Griffiths (1979) stated, "stability has no place in systems such as these." This reality reflects serious doubts on the applicability of the once widely accepted "climax" theory of vegetation dynamics and has important implications for the management of such ecosystems.

ACKNOWLEDGMENTS

I thank the wardens and staff of the Masai Mara National Reserve and the Wildlife Department of Kenya for their help. This work was supported by the Canadian Natural Sciences and Engineering Research Council, the New York Zoological Society, and the East African Wildlife Society. I was supported by a Fulbright Fellowship and a graduate fellowship from the University of British Columbia.

REFERENCES

Anderson, G. D., and Walker, B. H. 1974. Vegetation composition and elephant damage in the Sengwa Wildlife Research Area, Rhodesia. *J. S. Afr. Wildl. Mgmt. Assoc.* 4:1–14.

Beaumont, E. 1944. Control of east coast fever in the Masai District: A general survey of the Lemek Valley, Narok District. Kenya Archives Document, DC/NRK.1/12/2.

Belsky, A. J. 1984. Role of small browsing mammals in preventing woodland regeneration in the Serengeti National Park, Tanzania. *Afr. J. Ecol.* 22:271–79.

Brooks, A. C. 1957. *Notes on some ecological studies in the Queen Elizabeth National Park with particular reference to grasses.* Game and Fisheries Dept., Entebbe, Uganda. (Mimeographed.)

Brooks, A. C., and Buss, I. O. 1962. Past and present status of elephant in Uganda. *J. Wildl. Mgmt.* 26:38–50.

Buechner, H. K., and Dawkins, H. C. 1961. Vegetation change induced by elephants and fire in Murchison Falls National Park, Uganda. *Ecology* 42:752–66.

Buss, I. O. 1961. Some observations on food habits and behavior of the African elephant. *J. Wildl. Mgmt.* 25:131–48.

Buss, I. O., and Savidge, J. M. 1966. Change in population number and reproductive date of elephants in Uganda. *J. Wildl. Mgmt.* 30:791–809.

Buxton, A. M. 1927. *Kenya Days.* London: Edward Arnold.

Buxton, P. A. 1955. *The natural history of tsetse flies.* London: H. K. Lewis.

Carson, W. P., and Abbiw, D. K. 1990. The vegetation of a fire protection site on the northern Accra Plains, Ghana. *Afr. J. Ecol.* 28:143–46.

Caughley, G. 1976. The elephant problem: an alternative hypothesis. *E. Afr. Wildl. J.* 14:265–83.

Croze, H. 1974a. The Seronera bull problem. I. The bulls. *E. Afr. Wildl. J.* 12:1–27.

———. 1974b. The Seronera bull problem. II. The trees. *E. Afr. Wildl. J.* 12:29–47.

Darling, F. F. 1964. Conservation and ecological theory. *J. Ecol.* 52 (suppl.):36–45.

Dougall, H. W. 1963. Average chemical composition of Kenya grasses, legumes, and browse. *E. Afr. Wildl. J.* 1:120–21.

Dougall, H. W., Drysdale, V. M., and Glover, P. E. 1964. The chemical composition of Kenya browse and pasture herbage. *E. Afr. Wildl. J.* 2:86–121.

Dougall, H. W., and Glover, P. E. 1964. On the chemical composition of *Themeda triandra* and *Cynodon dactylon. E. Afr. Wildl. J.* 2:67–70.

Douglas-Hamilton, I. 1983. Elephants hit by arms race: Recent factors affecting elephant populations. *African Elephant and Rhino Specialist Group Newsletter* 2:11–13.

Dublin, H. T. 1984. The Serengeti-Mara ecosystem. *Swara* 7:8–13.

———. 1986. Decline of the Mara woodlands: The role of fire and elephants. Ph.D. dissertation, University of British Columbia.

———. 1991. Dynamics of the Serengeti-Mara woodlands: An historical perspective. *For. Conserv. Hist.* 35:169–78.

Dublin, H. T., and Douglas-Hamilton, I. 1987. Status and trends of elephants in the Serengeti-Mara ecosystem. *Afr. J. Ecol.* 25:19–33.

Dublin, H. T., Sinclair, A. R. E., and McGlade, J. 1990. Elephants and fire as causes of multiple stable states in the Serengeti-Mara woodlands. *J. Anim. Ecol.* 59:1147–64.

Dublin, H. T., Sinclair, A. R. E., Boutin, S., Anderson, E., Jago, M., and Arcese, P. 1990. Does competition regulate ungulate populations? Further evidence from Serengeti, Tanzania. *Oecologia* 82:283–88.

Eastman, G. E. 1927. *Chronicles of an African trip.* Privately printed.

Field, C. R. 1971. Elephant ecology in Queen Elizabeth National Park, Uganda. *E. Afr. Wildl. J.* 9:99–123.

Field, C. R., and Ross, I. C. 1976. The savanna ecology of Kidepo Valley Park. II. Feeding ecology of elephant and giraffe. *E. Afr. Wildl. J.* 14:1–15.

Ford, J. 1971. *The role of the trypanosomiases in African ecology.* Oxford: Clarendon Press.

Ford, J., and Clifford, H. R. 1968. Changes in the distribution of cattle and of bovine trypanosomiasis associated with the spread of tsetse flies (*Glossina*) in south-west Uganda. *J. Appl. Ecol.* 5:301–37.

Glover, J. 1963. The elephant problem at Tsavo. *E. Afr. Wildl. J.* 1:30–39.

Glover, P. E. 1968. The role of fire and other influences on the savannah habitat, with suggestions for further research. *E. Afr. Wildl. J.* 6:131–37.

Glover, P. E., and Trump, E. C. 1970. *An ecological survey of the Narok District of Kenya Masailand.* Part II: *The vegetation.* Nairobi: Kenya National Parks.

Guy, P. R. 1976. The feeding behaviour of elephant (*Loxodonta africana*) in the Sengwa Area, Rhodesia. *S. Afr. J. Wildl. Res.* 6:55–63.

Jachmann, H., and Bell, R. H. V. 1985. Utilization by elephants of *Brachystegia* woodlands of the Kasungu National Park, Malawi. *Afr. J. Ecol.* 23:245–48.

Kurji, F. 1976. *Human ecology.* Serengeti Research Institute Annual Report, 1974–1975, 12–31. Arusha: Tanzania National Parks.

Lamprey, H. F., Glover, P. E., Turner, M. I. M., and Bell, R. H. V. 1967. The invasion of the Serengeti National Park by elephants. *E. Afr. Wildl. J.* 5:151–66.

Lamprey, R. H. 1985. Masai impact on Kenya savanna vegetation: A remote sensing approach. Ph.D. dissertation, University of Birmingham.

Langridge, W. P., Smith, J. A., and Wateridge, L. E. D. 1970. *Some observations on the ecology of* Glossina swywnnertoni *(Austen) in the Mara region of Kenya.* Proc. 12th Meeting of the International Scientific Council for Trypanosomiasis Research, Bangui.

Laws, R. M. 1970. Elephants as agents of habitat and landscape change in East Africa. *Oikos* 21:1–15.

Lewis, E. A. 1935. Tsetse flies in the Maasai Reserve, Kenya colony. *Bull. Entomol. Res.* 25:439–55.

McNaughton, S. J. 1984. Grazing lawns: Animals in herds, plant form, and coevolution. *Am. Nat.* 124:863–86.

Morgan, W. T. W., and Shaffer, N. M. 1966. *Population of Kenya: Density and distribution.* Nairobi: Oxford University Press.

Narok District Commissioner's Report. 1955. Kenya Archives Documents, DC/NRK.1/1/1–7.

Norton-Griffiths, M. 1979. The influence of grazing, browsing, and fire on the vegetation dynamics of the Serengeti. In *Serengeti: Dynamics of an ecosystem* ed. A. R. E. Sinclair and M. Norton-Griffiths, 310–52. Chicago: University of Chicago Press.

Okula, J. P., and Sise, W. R. 1986. Effects of elephant browsing on *Acacia seyal* in Waza National Park, Cameroon. *Afr. J. Ecol.* 24:1–6.

Onyeanusi, A. E. 1989. Large herbivore grass offtake in Masai Mara National Reserve: Implications for the Serengeti-Mara migrants. *J. Arid Environ.* 16:203–9.

Pellew, R. A. P. 1981. The giraffe (*Giraffa camelopardalis tippelskirchi* Matschie) and its *Acacia* food resource in the Serengeti National Park. Ph.D. thesis, Cambridge University, Cambridge.

Sinclair, A. R. E. 1975. The resource limitation of trophic levels in tropical grassland ecosystems. *J. Anim. Ecol.* 44:497–520.

Sinclair, A. R. E., and Norton-Griffiths, M., eds. 1979. *Serengeti: Dynamics of an ecosystem.* Chicago: University of Chicago Press.

Spinage, C. A. 1973. A review of ivory exploitation and elephant population trends in Africa. *E. Afr. Wildl. J.* 11:281–89.

Swynnerton, C. F. M. 1936. The tsetse flies of East Africa. *R. Ent. Soc. Lond.* 84:1–579.

Thomson, P. J. 1975. The role of elephants, fire, and other agents in the decline of a *Brachystegia boehmii* woodland. *J. S. Afr. Wildl. Mgmt.* 5:11–18.

Watson, R. M., and Bell, R. H. V. 1969. The distribution, abundance, and status of elephant in the Serengeti region of northern Tanzania. *J. Appl. Ecol.* 6:115–32.

Weyerhaeuser, F. J. 1985. Survey of elephant damage to baobabs in Tanzania's Lake Manyara National Park. *Afr. J. Ecol.* 23:235–43.

White, S. E. 1915. *The Rediscovered Country.* London: Hodden & Stotten.

Woosnam, R. B. 1913. Report on a search for *Glossina* on the Amala (Engabei) River, southern Masai Reserve, East African protectorate. *Bull. Entomol. Res.* 4:271–78.

FIVE

Equilibria in Plant-Herbivore Interactions

A. R. E. Sinclair

The vegetation pattern in tropical African savanna is a mosaic of habitat patches that appear repetitively in physical units called landscapes (Gerrescheim 1972; Belsky 1993). The pattern and structure of the vegetation, it has been argued by Belsky (1995), are determined largely by environmental and edaphic factors and hardly at all by large mammals. Thus rainfall, geology, soil moisture, soil characteristics on the slope (the soil catena), and soil alkalinity, salinity, and sodicity are the dominant influences on pattern in these African systems. Termites appear to be the major animal influence through the formation of mounds and redistribution of nutrients (Belsky 1988, 1993). Large mammals, however, have a strong influence on species composition, decomposition rates, nutrient cycling, and microsite structure, as in the effects of herbivory on the height of the herb layer (Botkin, Mellilo, and Wu 1981; McNaughton 1983; Ruess and McNaughton 1987; McNaughton, Ruess, and Seagle 1988). Thus, the presence of a calcareous hardpan and high soil nutrients (de Wit 1978) prevent trees from establishing themselves on the Serengeti plains even after prolonged protection from large herbivores, but herb species composition and height does change with protection (McNaughton 1983).

Two exceptions to this generality have been proposed. First, elephants seem to determine both the structure and the dynamics of the tree community in Africa (Laws 1970; Dublin, Sinclair, and McGlade 1990). Second, it has been proposed that the pattern and dynamics of the herb layer on the Serengeti plains is determined by large mammal grazers (McNaughton 1983).

In this chapter I present tests of these two hypotheses by taking advantage of the changes in populations of the major ungulate species of Serengeti—wildebeest, buffalo, and elephant—in the past 25 years.

Elephants in Africa

Initial ideas on elephant-tree interactions in Africa developed from research in Uganda (Laws 1969, 1970; Laws, Parker, and Johnstone 1975), where it appeared that elephants in national parks could reduce and sometimes eliminate tree populations. These observations suggested that there was no stable state of coexistence between herbivores and trees, possibly because in the high rainfall of Uganda there was an alternative "prey" available to elephants, namely, the very tall (3–4 m) grasses that occurred on the banks of the Nile. In the late 1960s Laws (1969) proposed that the same outcome (i.e., a single state of elephants, grass, and no trees) would be seen in the semiarid region of Tsavo, in southeastern Kenya. However, this region differed markedly from Uganda in that there were no tall grasses, and the herb layer in general was sparse, as is common in semiarid scrubland. A more plausible alternative prediction for this habitat, therefore, was that as tree populations declined, so also would that of elephants due to the absence of an alternative food source. Either a cyclical situation would develop (Caughley 1976) or some sort of stable coexistence of elephants and trees would pertain. Part of the confusion in the debate on elephant-tree interactions stems from differences in the preferred foods of elephants in different areas. Elephants are primarily grazers in areas where the grass layer is abundant (as in Uganda); they switch to browsing only when grass is unavailable, as in the dry season (Laws 1970; Croze 1974a; Dublin 1986). Where grass is scarce, however, elephants browse most of the time, as they do in Tsavo.

As it turned out, the alternative prediction in Tsavo was confirmed when the elephant population collapsed in 1970–1971 (Corfield 1973) and there followed a rebound in tree numbers. This regeneration of trees took place before the subsequent decline of elephant numbers due to poaching in the mid-1970s, indicating that the initial response of the trees was due to the starvation of elephants and not to poaching mortality.

The situation in Serengeti in the mid-1960s developed along similar lines. Elephant numbers within the park boundary were about 2,500 in 1965 and remained more or less constant until 1976, when poaching caused a decline in numbers (Sinclair, chap. 1; Dublin and Douglas-Hamilton 1987). An early idea, the "nonequilibrium hypothesis," suggested that woodlands in Serengeti would disappear due to elephant over-browsing, as in Uganda, because elephants were not found in the region historically and the habitat was unsuitable for them (Lamprey et al. 1967; Watson and Bell 1969). In contrast, H. A. Fosbrooke (1968, pers. comm.) found historical records from the 1860s and 1870s (Wakefield 1870, 1882; Farler 1882) pointing to the presence of elephants in the Serengeti at that time. These subsequently disappeared due to the ivory trade (Spinage 1973). In addition, Croze (1974a,b; Croze, Hillman, and Lang 1981) suggested that elephants were not the ultimate cause of the decline in tree numbers in Serengeti and that there could be stability in the elephant-tree interaction (the "equilibrium hypothesis"). Norton-

Griffiths (1979) then proposed the "fire hypothesis," namely, that fire caused the mortality of large trees as well as preventing the regeneration of small trees, and that mortality of trees due to elephants was incidental. In other savanna communities in Africa, fire either inhibits growth or kills trees (Trapnell 1959; Hopkins 1965; Thomas and Pratt 1967; Pratt and Knight 1971; Afolayan 1978; Gillon 1983; Stark 1986; Frost and Robertson 1987). A further effect of fire was that mature trees would occur in the population as a similar-aged cohort (having escaped fire at some point in the past) and that widespread mortality would occur due to senescence, as has been proposed for Amboseli, Kenya (Young and Lindsay 1988). The fire hypothesis predicts (1) that if fire frequency were reduced, tree regeneration would take place and woodland would return to Serengeti. Pellew (1981, 1983) supported these predictions, but added that giraffe browsing slows down the rate of escapement of small trees into large size classes.

Dublin, Sinclair, and McGlade (1990) and Dublin (chap. 4) have elaborated on the equilibrium hypothesis by proposing a "multiple stable state" (MSS) hypothesis: that there are not just one but two equilibria, one with many trees and few elephants, the other with few trees and many elephants. The prediction (2) from this hypothesis differs from prediction 1 in that both fire and elephant numbers must be reduced for trees to regenerate. The MSS hypothesis also predicts (3) that if there is little fire but many elephants, then trees will not regenerate. This MSS hypothesis differs from Caughley's (1976) "stable limit cycle" hypothesis in that the former predicts that elephants by themselves are unable to cause a collapse in the tree population, and that the tree population cannot escape spontaneously from predator regulation by elephants (Sinclair 1989), whereas the latter predicts that both events can occur without external perturbations such as fire.

The predictions of the "nonequilibrium hypothesis" failed both in Tsavo (Corfield 1973) and Serengeti-Mara (Dublin, chap. 4). The "stable limit cycle" hypothesis appears to be supported in Tsavo, but not in Serengeti (Dublin, Sinclair, and McGlade 1990). With respect to the "equilibrium hypothesis," prediction 3 is examined in chapter 4 by H. Dublin. In this chapter I will explore predictions 1 and 2.

Grazers and the Herb Layer

During the 1960s and 1970s the wildebeest population exhibited a five-fold increase from 250,000 to about 1.3 million. In 1977 the population stabilized, and it has remained approximately constant for the past 15 years. Wildebeest are regulated by food in the dry season (Sinclair, Dublin, and Borner 1985). At this time they eat either dry, dormant grass or the short regrowth that appears after thunderstorms in northern Serengeti. In either case, growth of the grass is influenced mostly by rainfall and very little by grazing, so that growth is not the result of a reciprocal herbivore-plant interaction. Therefore, one does not predict changes in the plant community as a result of the increased grazing pressure brought about by the increase in the wildebeest population. This prediction is supported by observations of the grass community in the northern woodlands by Belsky (1992).

On the plains in southern Serengeti, however, wildebeest graze the swards in the wet season, when growth is occurring, and the potential for a reciprocal interaction between herbivore population and plant community exists. The plains are true grasslands, trees being excluded because of shallow alkaline soils (de Wit 1978; Belsky 1990). However, the structure of the grassland is strongly influenced by grazing, particularly on the eastern plains (McNaughton 1983; Belsky 1986a,b, 1992); perennially short-grass areas (10 cm) grow to heights of 50 cm when protected from grazing. It has been suggested by park wardens (Belsky 1985) that the increased grazing pressure might be resulting in "overgrazing" (i.e., a decrease in perennial grasses and a concomitant increase in dicots). Both McNaughton (1983), on the basis of changes in grass species, and Sinclair (1979), on the basis of increases in the abundance of browsers (as a putative response to increased dicots) have suggested that this hypothesis was plausible. However, Belsky (1985) did not detect species changes in permanent vegetation plots monitored over the period 1972–1982. This period covered the increase phase of the wildebeest and so the full effect of grazing may not yet have been felt. There is also a third possibility that grazers may promote and maintain grasslands through a positive feedback process: grazers may raise soil nutrients locally through urination and defecation and so enhance the growth of grasses at the expense of dicots (McNaughton 1988; McNaughton, Ruess, and Seagle 1988; Georgiadis and McNaughton 1990).

I examine here whether the species composition of the southeastern plains herbaceous community has changed as a result of the increased grazing pressure. The definition and detection of "overgrazing" is confused in the literature (Macnab 1985). I define "overgrazing" in this context as a state in which there is no equilibrium between herbivore and food plant populations, so that under herbivory the food plant species disappears and the herbivore ceases to use the area. A necessary (but not sufficient) prediction (4) of this "nonequilibrium" hypothesis is that under conditions of high herbivore populations, as have occurred with wildebeest on the Serengeti plains during 1977–1993, there should be a progressive change in plant species composition toward unpalatable or inedible dicots.

THE ELEPHANT-TREE INTERACTION

Fire and Wildebeest

The test of predictions 1 and 2 requires a reduction in fire frequency. Norton-Griffiths (1979) has shown that fire frequency declined in the Serengeti through the 1960s and 1970s concomitant with the increase in wildebeest numbers. He suggests that wildebeest caused the change in fire frequency by grazing the grass that would have provided fuel for the fires. However, Stronach (1989) has shown that climatic changes may also have influenced the fire regime. Stronach (1989) conducted a detailed study of fire and its effects on the herb and tree communities in the Serengeti during 1985–1987. The two most important factors that determined the frequency of fires were the quantity of dry grass available for burning at the beginning of the dry season, and the moisture content of this fuel (Stronach and McNaughton 1989). The quantity of fuel is largely determined by the rainfall in the previous wet season, with more rain producing more grass (Sinclair 1977; McNaughton 1985). Moisture content is determined by rainfall in the dry season, with more rain producing more moisture and lower fire frequency. Stronach (1989) proposed that the wet season/dry season rainfall ratio is an index of fire frequency: the higher the ratio, the higher the incidence of fire. This index (fig. 5.1) shows a decline from the early 1960s to the late 1970s coincident with the drop in area burnt reported by Norton-Griffiths (1979). The index was also low in the early 1980s, consistent with the low fire incidence in the Mara (Dublin, Sinclair, and McGlade 1990). There was a 2-year peak in the index in 1979–1980, and I observed unusually widespread and hot burns in both of those years, although no exact figures are available.

It is likely that the effects of the pattern of rainfall and wildebeest grazing are interrelated. When dry seasons start early (June) and have low rainfall the wildebeest do not remain long in the Serengeti woodlands but move rapidly north to the Mara. They leave much of the grass ungrazed and thus available for hot burns in central Serengeti. I observed this in 1969, 1970, and 1980, when wildebeest reached the Mara in June and burning farther south was extensive. Conversely, when there is much rain in the dry season, not only is the grass damp (and so does not burn), but the wildebeest linger in the central Serengeti, graze down the standing crop of grass, and so reduce the fuel. This occurred in 1971, 1972, and 1973; in 1972 they inhabited the Mara for only 6 weeks in September and October. Therefore, we can restate prediction 2, such that regeneration of trees should occur when wildebeest numbers are high, fire incidence is low, and elephant numbers are decreasing. This situation occurred in Serengeti after 1976 because wildebeest numbers were high and constant, fire incidence was low, and elephant numbers were reduced by 80%

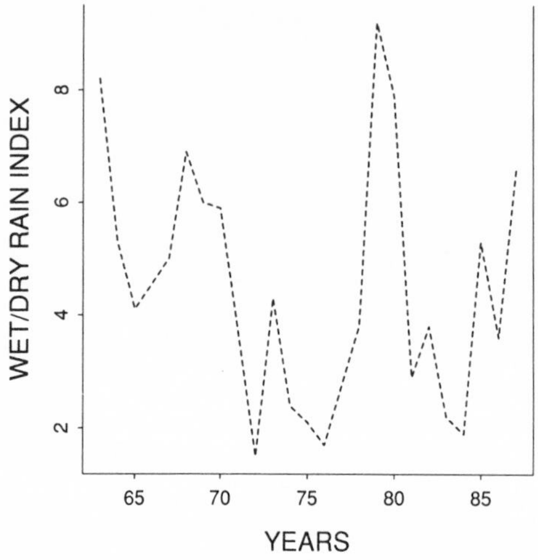

Figure 5.1 The ratio of wet season (Nov.–May) to dry season (June–Oct.) rainfall for the central Serengeti woodlands, 1960–1989. (From Stronach 1989.)

through poaching between 1976 and 1986 (Dublin and Douglas-Hamilton 1987). Therefore, regeneration should be seen during the 1970s and 1980s when the above conditions prevailed. In contrast, regeneration should not be observed during the period of high fire frequency, namely, the 1950s and 1960s.

Methods

Photopoints. Beginning in 1980, twenty-two photopoints were established on the tops of hills and kopjes in the woodlands. Oblique photographs were taken at set compass bearings of the woodland vegetation below. These were repeated at intervals ranging from 3 to 11 years up to 1991. In addition, I repeated some earlier photographs taken during the late 1960s and 1970s.

A much earlier set of photographs was taken by Osa and Martin Johnson. These two American adventurers and filmmakers made three expeditions to Serengeti in 1926, 1928, and 1933 (their first expedition of 1922 did not provide photographs of Serengeti as far as I could ascertain). The last expedition used an aircraft, and some aerial photographs are available. Some of the photographs from these expeditions reside in the Martin and Osa Johnson Safari Museum, Chanute, Kansas. In September 1981 I visited this museum, and with the assistance of the curators, obtained copies of photographs whose locations I thought were identifiable. In 1982 I rephotographed the ten identifiable Johnson pictures taken in the woodlands. There were several others that I could not identify, and despite much searching in subsequent years, I added only two more. The rest remain a mystery. I also used one aerial photograph of a valley in northern Serengeti taken by Burtt (1942) in about 1935 when conducting a botanical survey of Tanganyika. This valley, now known as Burtt's Valley, was rephotographed in 1968, 1978, and 1988.

Photographs were examined in pairs, and each one compared with the next in the time sequence. Pairs of oblique photographs were analyzed by drawing a line around identifiable points in the foreground and middle distance so that the same area was identified on each photograph. Two size classes of trees were identified: large mature trees (adults) and small (juvenile) trees. Most of the trees were *Acacia clavigera* E. Mey (= *A. robusta* Burch.) ssp. *usambarensis* (Taub) Brenan, with *Terminalia mollis* Laws. in some northern photographs. The other common tree was *A. tortilis* (Forsk) Hayne ssp. *spirocarpa* (Hochst ex A. Rich) Brenan, and trees that appeared occasionally were *A. xanthophloea* Benth, *A. mellifera* (Vahl) Benth spp. *mellifera, A. nilotica* (L) Del., *Balanites aegyptiaca* (L) Del., *Combretum molle* R. Br. ex G. Dom, and *Commiphora trothae* Engl. By visiting sites on the ground, I established that small trees below 50–100 cm in height were not usually visible because long grass obscured visibility. Adult and juvenile trees were marked as different-colored dots on transparent acetate overlay. The dots were then counted and compared over time.

Because absolute area was unknown for these oblique photographs, I measured relative change of tree numbers using instantaneous rates of increase *(r)*. Thus,

$$r = \ln (N_t/N_o)/t$$

where N_o and N_t are the numbers of either adult or juvenile trees at the beginning and end of the time period, and t is the number of years elapsed between the pair of photographs.

Figures 5.2a, b, and c illustrate hypothetical survivorship curves for a cohort of adult trees alive in the early 1900s and dying at different age-specific rates up to the present day. Thus, figure 5.2a suggests higher mortality in the early part of the century and less later. Conversely, figure 5.2c represents the situation in which mortality has been severe in recent decades, but slight during earlier periods. Figure 5.2b shows a period of severe mortality in the middle, and little at the beginning and end periods. Figures 5.2d, e, and f are the respective instantaneous rates of change for the hypothetical survivorship curves shown in figures 5.2a, b, and c when change is measured from a set point A in the present (1991) backward to different times in the past. The instantaneous rates are all negative in this figure because the decline in population is being examined. However, similar diagrams for regeneration give the converse positive instantaneous

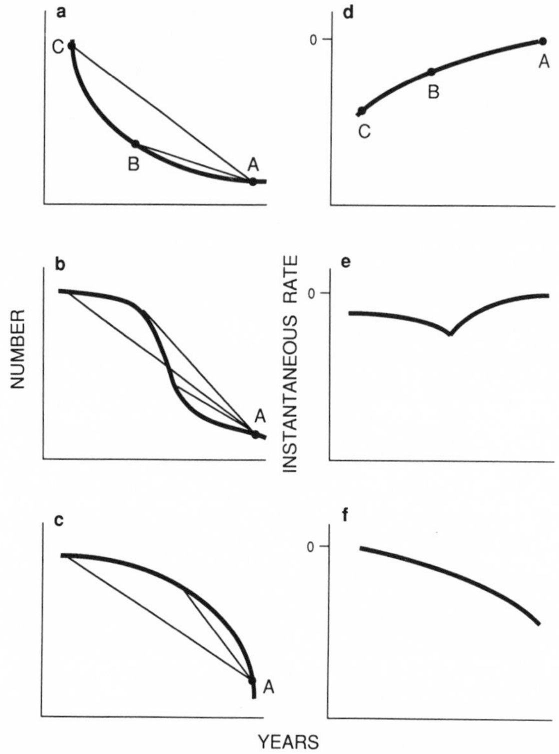

Figure 5.2 Hypothetical survivorship curves for adult trees in Serengeti through the twentieth century (a,b,c). The corresponding instantaneous rates of decline (d,e,f) are calculated from a set point A backward over progressively longer time periods B, C, etc., and plotted against the start of the time period (the year for B, C). The slope of the lines BA, CA in (*a*) is an index of the instantaneous rates shown as B, C in (*d*).

rates. The curves for the instantaneous rates (fig. 5.2) provide the predictions with which to compare the results from the photopoints.

Tree Ages. I concentrated on *A. clavigera* because it is the most widespread tree species in the woodlands and is regenerating extensively. It is also browsed by elephants but uneaten by giraffe (Pellew 1981) and other browsers (pers. obs.). Thus the effects of browsing by smaller mammals noted by Belsky (1984) can be ignored for this *Acacia* species. Basal sections from a selected sample of 25 *A. clavigera* trees were collected from live regenerating stands in the west, center, and northern woodlands: Kogatende, Quartz Hill, a point 10 km south of Lobo Kopjes, Seronera, and Kirawira. At Seronera a section was collected from a tree of known age (22 years) at one of the houses in the Serengeti Wildlife Research Centre (SWRC). This was one of several *A. clavigera* seedlings that I protected from fire in January 1971. From this tree I was able to identify and confirm the annual growth rings. These rings are broad (about 3 mm thick) and can be seen with the naked eye after some practice. The broad rings are composed of many finer rings that can be seen under the microscope, but the nature of those rings remains unknown.

The selected sample of sections was chosen to cover a range of diameters, the largest being 27 cm. Trees much larger than this could not be examined for rings because they rot in the center, although they continue to grow on the outside. Basal circumference and diameter were measured for these samples and for other known-age trees of *A. clavigera, A. tortilis, A. senegal* (L) Willd., and *Dichrostachys cinerea* (L) Wight and Arn that had been protected at Seronera since 1971.

Tree Densities. There are two size ranges for *A. clavigera,* small trees ($\leq$ 60 cm circumference), and large trees ($>$ 80 cm) which are sparsely scattered. The density of small trees was measured in dense stands using seventeen 30 m by 2 m belt transects spaced systematically through the stand. The density of large trees was estimated using T-square sampling (Krebs 1989). Circumference was measured for all trees, about 15 cm from the base in small trees, 100 cm up in large trees.

Results

Photopoints. Figure 5.3 shows the instantaneous rates of change of adult (large) trees over varying lengths of time backward from 1991. There has been an overall decrease in large trees since the 1920s, and the rates of decline have been much greater in the second half of the century than in the first half. Although the scatter of points suggests a continuing acceleration in the rate of decline (conforming to figs. 5.2c and f), the weighted smoothed mean line indicates that the highest rate of decline was during the 1960s, suggestive of the curves in figures 5.2b and e.

Figure 5.4 shows the instantaneous rates of change for young trees. In general, there is an opposite trend to that for large trees. Thus, there were few young trees in the earlier part of this century (1930–1960). Instantaneous rates of increase jumped in the late 1970s and have remained high since then, conforming to the positive version of figure 5.2f.

Tree Ages. Table 5.l gives the mean circumference for the known-age trees of four species for future reference, although only *A. clavigera* is examined here. The relationship between growth rings of *A. clavigera* (as an index of age), and basal diameter is shown in figure 5.5. Also shown are the diameters of known-age trees. The regression explains 95% of the variance. The linear regression of circumference on rings explained as much of the variance, but that for basal area on rings was not as good

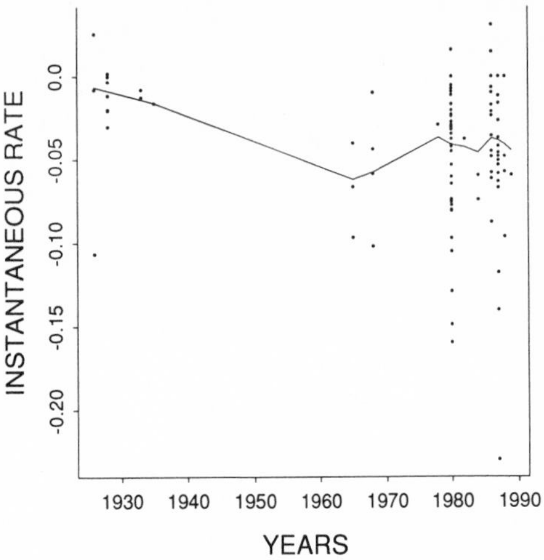

Figure 5.3 The instantaneous rate of decrease for adult trees in Serengeti calculated from pairs of oblique photographs. Each point represents one pair of photographs plotted against the year of the earlier picture. The line is calculated from a locally weighted and smoothing function. (From Becker, Chambers, and Wilks 1988.)

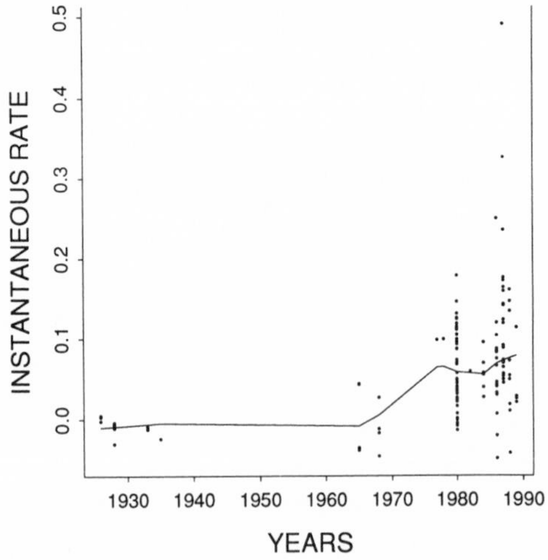

Figure 5.4 The instantaneous rate of increase for juvenile trees in Serengeti calculated from pairs of oblique photographs as for figure 5.3.

Table 5.1 Circumference of various tree specis of known age 22 years collected from SWRC in July 1992.

Species	*N*	Mean circumference (cm)	S.E.
A. clavigera	4	63.8	3.6
A. tortilis	8	68.8	3.6
A. senegal	1	20	—
D. cinerea	1	44	—

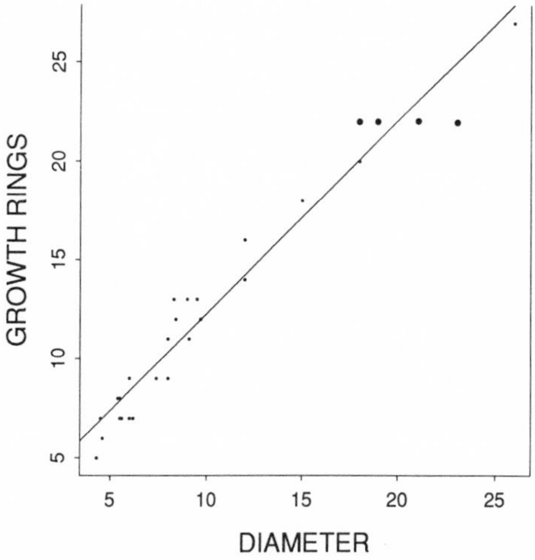

Figure 5.5 The relationship of number of growth rings as an index of age in years and the basal diameter (cm) of *A. clavigera* trees. Large dots indicate known-age trees.

(87% of the variance explained) since the relationship was curvilinear.

The linear regression equation for figure 5.5 where age in years is predicted by basal diameter *(d)* in cm is

$$\text{age} = 2.49 + 0.977\, d$$

This equation was used to estimate the ages of all measured trees. For fieldwork this is convenient because age in years is approximately the diameter in cm plus 2. Although this sample came from a wide range of sites, I recognize that size-age relationships depend on site characteristics, climate, and tree density, so this relationship can be used only as a rough approximation for age distributions until much larger samples are obtained.

Tree Densities. The density of small *A. clavigera* trees (< 60 cm circumference) was measured on the Kogatende Ridge in northern Serengeti.

These regenerating stands appear similar to other stands in the rest of the Serengeti woodlands. The density in July 1992 was 3,490 (± 146 SE) trees/ha. This is similar to the densities of 3,000–5,000/ha measured by Stronach (1989) in the east near Lobo.

The density of large trees (> 80 cm circumference) measured from twenty-nine T-square points in the Kogatende area was three orders of magnitude less than that for small trees, being 1.29/ha with a standard error range of 1.04–1.70.

Size and Age Frequencies. The circumferences of 412 small trees and 110 large trees recorded in northern Serengeti were divided into 10 cm size classes. The mean density for small and large trees was split into these size classes according to the proportions found (fig. 5.6). As expected, there were two distinct size classes, a small one 10–30 cm in circumference and a large one spread more evenly over 100–200 cm in circumference. The largest circumference was 334 cm. No trees were recorded in the size range 57–82 cm.

Figure 5.7 shows the frequency distribution of this sample for dates of regeneration. (Regeneration refers to the start of growth above ground, not necessarily germination, since previously burnt or browsed seedlings may have germinated earlier). These dates are reliable back to 1950 but are less reliable earlier because they are extrapolations of the diameter-age regression. However, these dates give an indication of the time scale

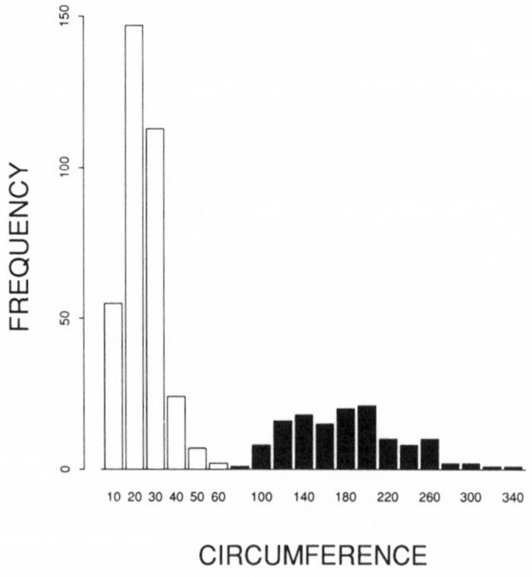

Figure 5.6 The frequency distribution of circumferences of two cohorts of *A. clavigera*. The smaller cohort (open bars) is in units of stems/1,000 m^2 with 10 cm divisions; the larger cohort (solid bars) is in units of stems/km^2 with 20 cm divisions. There is no overlap in the distributions.

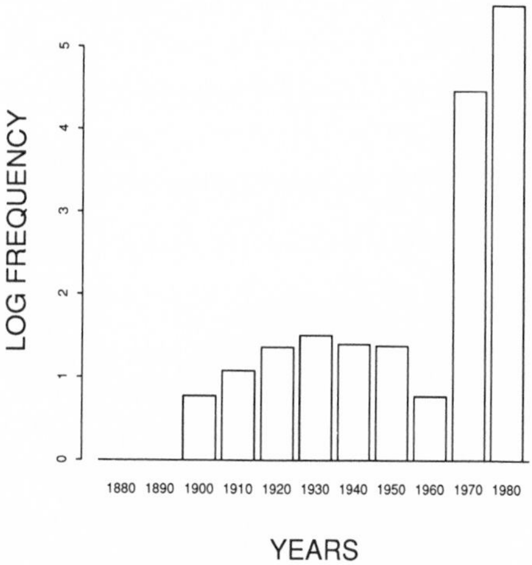

Figure 5.7 The log frequency distribution of dates of regeneration of *A. clavigera* (stems/km^2) grouped into decades starting with the year indicated.

for the two periods of tree regeneration. The more recent period of regeneration began in the late 1970s and is still continuing (fig. 5.8), which is consistent with figure 5.4 and with Stronach's (1989) observations. This period of regeneration began when the wildebeest reached their maximum numbers.

Most of the larger trees have died in the past 20 years (see fig. 5.3), some from elephant damage but many from other undetermined causes, possibly related to senescence. Ruess and Halter (1990) provide similar data for sites near Seronera. The age distribution of the remaining trees does suggest that a major period of regeneration took place in the period 1900–1920. In contrast, the period 1950–1975, when fires were most prevalent, showed almost no regeneration.

The oldest *A. clavigera* tree measured (on the track 6 km east of the Tabora-Wogakuria turnoff) would have started growth in 1886. An *A. tortilis* tree in a Johnson photograph of 1928, near the research center at Seronera, is still alive. This tree was rephotographed in 1992 from the same point and the circumference measured. From the ratio of the diameters on the photographs I estimated the basal diameter in 1928 to be 43 cm. From table 5.1 it appears that *A. tortilis* grows at about the same rate as *A. clavigera*. If this is so, then the same equation for age predicts the Johnson tree at SWRC to have started growth in 1884.

WILDEBEEST-HERB INTERACTIONS

If the increase in wildebeest has resulted in an unstable plant-herbivore interaction on the plains, then as predicted above by the "nonequilib-

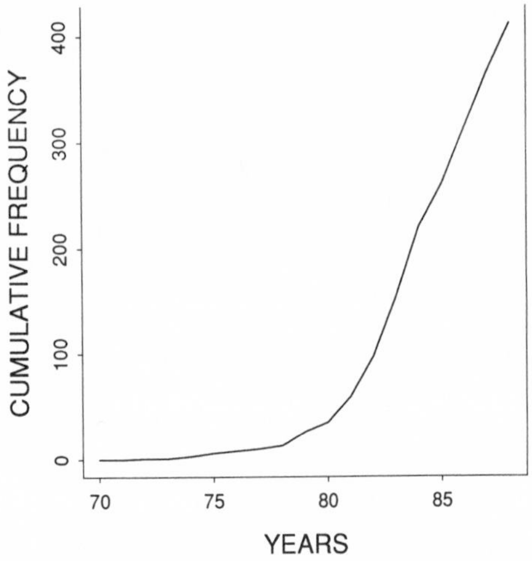

Figure 5.8 The cumulative frequency of dates of regeneration by year since 1970 for the recent cohort of *A. clavigera* (sample $n = 412$).

rium" hypothesis, one should see a continuing change in the plant community of the herb layer, with unpalatable and ungrazed species increasing during the period when wildebeest have been at highest density, that is, since 1977. On the other hand, if the interaction is stable, no trend should be seen in species composition.

Methods

Ten 30 m transects were set up along a baseline that follows the eastern park boundary, running from the main road northward until it crosses the Ngare Nanyuki River near Barafu Kopjes. This baseline was about 35 km long and the transects were roughly 3–5 km apart. Each transect started from one of the oildrums that act as permanent markers for the park boundary. A 30 m tape was attached to a T-shaped hole in the drum and laid on a compass bearing of magnetic east to the northwest corner of a cement block buried in the ground. The plant species that lay under the tape at each 20 cm mark was recorded. Grass species were lumped as one category (because usually they were not flowering), but the sedge *Kyllinga nervosa* Steud. was distinguished. Dominant grasses in these swards are *Digitaria scalarum* (Schweinf.) Choiv., *Andropogon greenwayi* Napper, *Sporobolus ioclados* (Trin.) Nees, *S. kentrophyllus*, and *S. fimbriatus* (Trin.) Nees (Belsky 1983). Other small flowering herbs were identified where possible; these are listed in table 5.2. Identification of dicot species was made with the help of those at SWRC, and a reference collection was sent to the East African Herbarium for identification. The transects were monitored following good rain as the herbs were

Table 5.2 Dicot species recorded on the southeastern plains transects 1980–1993, in order of abundance.

Solanum incanum L.
Oldenlandia wiedemannii K. Schum.
Indigofera microcharoides Taub.
Pluchea monocephala E. A. Bruce
Melhania ovata (Cav.) Spreng.
Becium obovatum (E. Mey.) N. E. Br.
Euphorbia inaequilatera Sond.
**Hirpicium diffusum* (Oliv.) Roess
Leucas neuflizeana Courb.
Ipomea longituba Hall. f.
Heliotropium longiflorum (A. DC.) Jaub and Spach.
Athroisma psylloides (Oliv.) Mattf.
Medicago laciniata (L.) Mill.
Hypoestes forskalei (Vahl.) Roem and Schult
Justicia uncinulata Oliv.
Crotolaria sp. nr. keniensis Bak. f.
Felicia abyssinica A. Rich.
Gutenbergia fischeri R. E. Fries
Phylanthus aspericaulis Pax
Cycnium tubulosum (L.f.) Engl. ssp. *montanum* (N.E. Br.) O. J. Hansen
Craterostigma spp.

*Annual.

flowering, but there were always some specimens that were unidentified due to lack of flowering, particularly in 1986.

Results

The mean proportions for the main groups of plants in each transect calculated over all years are given in table 5.3. Grass made up about 56% of the sward, *Kyllinga* 13%, and dicots other than *Solanum* 18% on average. *Solanum* showed a geographic trend, being commonest in the southernmost transect (7.8%) and declining progressively northward. *Solanum* was effectively absent in the northern half of the transects. Transect 10 lay on a slope leading into the Ngare Nanyuki River where there is erosion and much bare ground. The microhabitat of this transect is different from the others.

Transects were recorded in 1980, 1984, 1986, 1989, 1991, and 1993. The records of the commonest species in the combined set of transects are given in table 5.4. There was no overall change in the species composition over the 13 years using values for all transects combined. In particular, one of the important indicators of overgrazing, *Solanum incanum,* has shown no detectable change over time ($F_{9,45} = 2.40$, $P > .05$). Bare ground varied with both location ($F_{9,45} = 10.1$, $P < .001$) and time ($F_{9,45} = 6.25$, $P < .001$), but there was no consistent trend. In contrast, the sedge *Kyllinga,* which is a preferred food for grazers, has increased from 6.5% to 18.0%. The regression of total *Kyllinga* records on year is

Table 5.3 Percentage of each transect on the eastern plains in each category as a mean over all years 1980–1993.

Transect	Grass	*Kyllinga*	*Solanum*	Other dicots	Bare ground
1	46.9	19.2	7.8	13.7	12.4
2	52.9	13.8	5.7	18.7	8.9
3	62.6	11.0	2.9	14.7	8.8
4	49.3	19.7	2.5	18.4	10.1
5	55.2	11.2	1.0	17.4	15.2
6	56.9	13.1	0.1	25.7	4.2
7	53.0	10.5	0.2	28.7	7.6
8	61.1	18.9	0	10.4	9.6
9	64.8	8.4	0	17.1	9.7
10	59.3	4.1	0	11.3	25.3
Mean	56.2	13.0	2.0	17.6	11.2
SE	1.8	1.6	0.7	1.8	1.8

Table 5.4 Frequency of occurrence of herbaceous species for each year in total of 1,500 points on the ten line transects, eastern Serengeti plains.

Species	1980	1984	1986	1989	1991	1993
Grass	870	946	813	814	824	788
Kyllinga nervosa	98	187	179	230	262	272
Solanum incanum	53	30	4	34	34	26
Oldenlandia wiedemannii	36	8	27	38	51	47
Indigofera microcharoides	54	6	7	21	31	38
Pluchea monocephala	25	31	7	21	10	3
Melhania ovata	17	23	21	15	39	40
Becium obovatum	15	14	6	28	12	18
**Hirpicium diffusum*	6	4	4	2	29	—
Hypoestes forskalei	8	8	4	—	11	8
Other dicots	236	57	199	131	63	71
Bare ground	72	186	229	166	134	189

*Annual.

significant ($p < .001$), and analyses of variance showed both location ($F_{9,45} = 3.82$, $P < .001$) and year effects ($F_{9,45} = 5.00$, $P < .001$). Grass showed significant variation with location ($F_{9,45} = 3.46$, $P < .01$) and year ($F_{9,45} = 2.46$, $P < .05$). *Kyllinga* appears to have increased at the expense of grass, so that when both *Kyllinga* and grass records are combined there is no trend with year.

DISCUSSION

Browsers

The equilibrium hypothesis of Norton-Griffiths (1979), that fire is the cause of the disappearance of *Acacia* trees due to the inhibition of regeneration, could be tested by the removal of fire (prediction 1). The multiple stable state hypothesis of Dublin, Sinclair, and McGlade (1990) added the condition that elephants must also be in low numbers (prediction 2). These conditions prevailed in Serengeti south of the Kenya border during the 1970s and 1980s because *(a)* wildebeest numbers were increasing and the animals were eating more of the dry season standing grass so that there was less fuel to burn, *(b)* there were poor climatic conditions for burning (see fig. 5.1), and *(c)* poachers were removing up to 80% of the elephants (Dublin et al. 1990).

The results from the photopoints on regeneration rates in Serengeti since the 1920s (fig. 5.4) and from age distributions (figs. 5.7, 5.8) indicate that there was a marked increase in tree regeneration in the 1970s and 1980s. In contrast, there was little if any regeneration in the adjacent Mara area during the same period (Dublin, chap. 4). These observations are not consistent with prediction 1 (equilibrium hypothesis), that regeneration occurs merely by removing fire, but are consistent with prediction 2 (MSS hypothesis), that the effects of both fire and elephants have to decrease for regeneration to occur. Dublin shows in chapter 4 that, in the presence of high elephant density in the Mara, regeneration does not take place, consistent with prediction 3 (MSS hypothesis). This result also was observed in Uganda (Harrington and Ross 1974; Lock 1985; Sabiiti and Wein 1988). The results from both this chapter and chapter 4 are, therefore, consistent with the MSS hypothesis.

The MSS hypothesis also proposed that fire was the cause of the reduction in adult trees. Therefore, adult tree densities should have declined at a faster rate before the wildebeest increased significantly in the 1970s and 1980s because wildebeest removed the grass fuel for fires and so caused a reduction in fire prevalence. The results from the photopoints (fig. 5.3) are consistent with this prediction. In support of this conclusion, the change in tree cover measured by Dublin (chap. 4) from sets of vertical aerial photographs shows that the greatest decline in adult trees took place in the 1960s. Other studies in other parts of Africa have concluded that, as in Serengeti, the absence of certain size cohorts is due to a period of severe burns (Thomson 1975; Spinage and Guiness 1971).

Dublin, Sinclair, and McGlade (1990) and Dublin (1991) proposed that the higher densities of trees observed in photographs of the 1930s and 1940s, both in Serengeti and across savanna Africa, were not (contrary to general belief) the shape of pristine savanna Africa before the arrival of Europeans. Instead, they proposed that these trees were the consequence of a special and probably highly unusual set of conditions near the turn of the century produced by the absence of elephants and fire brought about by the ivory trade, rinderpest, and trypanosomiasis (Ford 1971; Spinage 1973; Sinclair 1979). If this is correct, then tree regeneration should have begun after 1890 and built up in the first decades of this century. Although my techniques of age determination for older trees are inevitably imprecise (because trees 40 or more years old rot in the center), the results (fig. 5.7) do suggest that regeneration started late in the nineteenth century and built up in subsequent decades in a pattern consistent with the above scenario.

Grazers

The nonequilibrium hypothesis that "overgrazing" could be occurring on the plains as a result of the high numbers of wildebeest predicted that species composition should change in favor of unpalatable herb species after wildebeest numbers had leveled out (prediction 4). Monitoring of herb species composition on the southeastern plains along the border of the national park indicated that no detectable changes were occurring. In addition, Belsky (1985) did not detect changes in the composition of one of K. Gerresheim's plots on the southeastern plains over the period 1972–1982, and Watson (1967) recorded grasses constituting 60% of the sward during 1962–1966, similar to the present values. Therefore, it seems there is no obvious trend toward unpalatable dicots in the past 30 years. Nevertheless, anecdotal accounts by M. D. Leakey and A. Root (pers. comm.) record the grass layer as being tall (over 50 cm) in the 1950s on the eastern plains where the present transects are situated. Thus, a change in the height of the grassland has undoubtedly taken place, but this occurred before or during the early stages of the wildebeest population increase. The reason for this change remains obscure. The suggestion of an increase in the palatable sedge *Kyllinga* conforms to McNaughton's nutrient facilitation hypothesis, but more data are needed.

CONCLUSION

Two general ideas on plant-herbivore interactions have been suggested: first, that these are unstable unless the herbivores are controlled by predators or by humans; and second, that such interactions can be stable (Caughley 1976; Macnab 1985). Examination of the elephant-tree and wildebeest-herb interactions shows that in neither case has the herbivore resulted in a destabilized plant community changing monotonically toward a different species composition. In the case of trees, the results confirm that fire was the major disturbance resulting in a decline in tree populations. When fire was reduced as a result of climatic changes and wildebeest grazing, tree seedling regeneration expanded rapidly, and *A. clavigera* woodland of the Serengeti south of the Kenya border is now supporting dense thickets of young trees. This change has occurred in areas of few elephants. In contrast, in the Mara Reserve north of the border, where elephants are numerous, there is little or no regeneration of young *Acacia* trees, as discussed by Dublin in chapter 4. These results are consistent with the multiple stable state hypothesis (Dublin, Sinclair, and McGlade 1990), and inconsistent with either a single stable state or a nonequilibrium hypothesis.

The higher densities of trees seen across Africa in the first half of this century may have been due to an extraordinary outbreak of trees precipitated by the rinderpest epizootic late in the nineteenth century. This also suggests that these areas of Africa may have been dominated by grassland rather than by savanna before the arrival of Europeans.

Monitoring of the herb layer on the eastern plains shows no systematic change in plant species composition. The implication from these results is that (1) while wildebeest are regulated by food shortage in the dry season (Sinclair, Dublin, and Borner 1985), they have abundant food on the plains in the wet season, and (2) the degree of feeding on the herb layer is not so great as to destabilize the plant-herbivore interaction. There is some evidence that the palatable plants have increased.

ACKNOWLEDGMENTS

I thank Neil Stronach and Holly Dublin for information they supplied. Dianne Goodwin helped set up the monitoring program in 1980, but her death in January 1992 prevented her from seeing the result. The 1984 plains data were collected by Trudy and Steve Chatwin, while Stan Boutin and Holly Dublin helped in 1986, Anne Sinclair in 1991. Simon Mduma and Anne Sinclair helped collect the tree data. Stan Boutin, Barbara Thomas, and Holly Dublin helped with photopoint photographs. Barbara Henshall and Sondra Alden found albums and arranged for prints at the Martin and Osa Johnson Safari Museum. I thank the museum for their permission to obtain copies. Janet Winbourne, Helen Davis, Susan Brown, and Tina Raudzus helped with data analysis. The East African Herbarium, Nairobi, and G. Amworo kindly identified plant species. Joy Belsky and Doug Houston provided helpful comments on an earlier draft.

REFERENCES

Afolayan, T. A. 1978. Savanna burning in Kainji Lake National Park, Nigeria. *E. Afr. Wildl. J.* 16:245–55.

Becker, R. A., Chambers, J. M., and Wilks, A. R. 1988. *The new S language*. Pacific Grove, Calif.: Wadsworth & Brooks/Cole.

Belsky, A. J. 1983. Small-scale pattern in grassland communities in the Serengeti National Park, Tanzania. *Vegetatio* 55:141–51.

———. 1984. The role of small browsing mammals in preventing woodland regeneration in the Serengeti National Park, Tanzania. *Afr. J. Ecol.* 22:271–79.

———. 1985. Long-term vegetation monitoring in the Serengeti National Park, Tanzania. *J. Appl. Ecology* 22:449–60.

———. 1986a. Population and community processes in a mosaic grassland in the Serengeti, Tanzania. *J. Ecol.* 74:841–56.

———. 1986b. Revegetation of artificial disturbances in grasslands in the Serengeti National Park, Tanzania. II. Five years of successional change. *J. Ecol.* 75:937–51.

———. 1988. Regional influences on small-scale vegetational heterogeneity within grasslands in the Serengeti National Park, Tanzania. *Vegetatio* 74:3–10.

———. 1990. Tree/grass ratios in East African savannas: A comparison of existing models. *J. Biogeogr.* 17:483–89.

———. 1992. Effects of grazing, competition, disturbance, and fire on species composition and diversity in grassland communities. *J. Vegetation Sci.* 3:187–200.

———. 1995. Spatial and temporal landscape patterns in arid and semi-arid African savannas. In *Mosaic landscapes and ecological processes,* ed. L. Hansson, L. Fahrig, and G. Merriam, 31–56. New York: Chapman and Hall.

Botkin, D. B., Mellilo, J. M., and Wu, L. S. Y. 1981. How ecosystem processes are linked to large mammal population dynamics. In *Dynamics of large mammal populations,* ed. C. W. Fowler and T. D. Smith, 373–87. New York: John Wiley & Sons.

Burtt, B. D. 1942. Some East African vegetation communities. Ed. C. H. N. Jackson. *J. Ecol.* 30:65–146.

Caughley, G. 1976. The elephant problem: an alternative hypothesis. *E. Afr. Wildl. J.* 14:265–83.

Corfield, T. 1973. Elephant mortality in Tsavo National Park, Kenya. *E. Afr. Wildl. J.* 11:339–68.

Croze, H. 1974a. The Seronera bull problem. I. The bulls. *E. Afr. Wildl. J.* 12:1–27.

———. 1974b. The Seronera bull problem. II. The trees. *E. Afr. Wildl. J.* 12:29–47.

Croze, H., Hillman, A. K. K., and Lang, E. M. 1981. Elephants and their habitats: How do they tolerate each other? In *Dynamics of large animal populations,* ed. C. W. Fowler and T. D. Smith, 297–316. New York: John Wiley & Sons.

de Wit, H. A. 1978. Soils and grassland types of Serengeti plains (Tanzania). Ph.D. dissertation, Agricultural University, Wageningen.

Dublin, H. T. 1986. Decline of the Mara woodlands: The role of fire and woodlands. Ph.D. thesis, University of British Columbia.

———. 1991. Dynamics of the Serengeti-Mara woodland. *For. Conserv. Hist.* 35:169–78.

Dublin, H. T., and Douglas-Hamilton, I. 1987. Status and trends of elephants in the Serengeti-Mara ecosystem. *Afr. J. Ecol.* 25:19–33.

Dublin, H. T., Sinclair, A. R. E., and McGlade, J. 1990. Elephants and fire as causes of multiple stable states in the Serengeti-Mara woodlands. *J. Anim. Ecol.* 59:1147–64.

Dublin, H. T., Sinclair, A. R. E., Boutin, S., Anderson, E., Jago, M., and Arcese, P. 1990. Does competition regulate ungulate populations? Further evidence from Serengeti, Tanzania. *Oecologia* 82:283–88.

Farler, J. P. 1882. Native routes in East Africa from Pangani to the Masai country and the Victoria Nyanza. *Proc. R. Geogr. Soc.,* new series 4:730–42.

Ford, J. 1971. *The role of trypanosomiases in African ecology.* Oxford: Clarendon Press.

Fosbrooke, H. A. 1968. Elephants in the Serengeti National Park: An early record. *E. Afr. Wildl. J.* 6:150–52.

Frost, P. G. H., and Robertson, F. 1987. The ecological effects of fire in savanna. In *Determinants of tropical savannas,* ed. B. H. Walker, 93–140. Oxford: IRL Press.

Georgiadis, N. J., and McNaughton, S. J. 1990. Elemental and fibre contents of savanna grasses: Variation with grazing, soil type, season and species. *J. Appl. Ecol.* 27:623–34.

Gerrescheim, K. 1972. Die Landschaftsgliederung als ökologischer Datenspeicher—angewandte Landschaftsökologie im Serengeti National Park, Tanzania. *S. Natur. Landschaft* 47:35–45.

Gillon, D. 1983. The fire problem in tropical savannas. In *Tropical savannas,* ed. F. Bourliere, 617–41. Amsterdam: Elsevier.

Harrington, G. N., and Ross, I. C. 1974. The savanna ecology of Kidepo Valley National Park. I. The effects of burning and browsing on the vegetation. *E. Afr. Wildl. J.* 12:93–106.

Hopkins, B. 1965. Observations on savanna burning in the Olokemeji Forest Reserve, Nigeria. *J. Appl. Ecol.* 2:367–81.

Krebs, C. J. 1989. *Ecological Methodology.* New York: Harper and Row.

Lamprey, H. F., Glover, P. E., Turner, M., and Bell, R. H. V. 1967. Invasion of the Serengeti National Park by elephants. *E. Afr. Wildl. J.* 5:151–66.

Laws, R. M. 1969. The Tsavo research project. *J. Reprod. Fert.,* suppl. 6:495–531.

———. 1970. Elephants as agents of habitat and landscape change in East Africa. *Oikos* 21:1–15.

Laws, R. M., Parker, I. S. C., and Johnstone, R. C. B. 1975. *Elephants and their habitats.* London: Clarendon Press.

Lock, J. M. 1985. Recent changes in the vegetation of Queen Elizabeth National Park, Uganda. *Afr. J. Ecol.* 23:63–65.

Macnab, J. 1985. Carrying capacity and related slippery shibboleths. *Wildl. Soc. Bull.* 13:403–10.

McNaughton, S. J. 1983. Serengeti grassland ecology: The role of composite environmental factors and contingency in community organization. *Ecol. Monogr.* 53:291–320.

———. 1985. Ecology of a grazing system: The Serengeti. *Ecol. Mongr.* 55:259–94.

———. 1988. Elemental nutrition and spatial concentrations of African ungulates. *Nature* 334:343–45.

McNaughton, S. J., Ruess, R. W., and Seagle, S. W. 1988. Large mammals and process dynamics in African ecosystems. *BioScience* 38:794–800.

Norton-Griffiths, M. 1979. The influence of grazing, browsing, and fire on the vegetation dynamics of the Serengeti. In *Serengeti: Dynamics of an ecosystem,* ed. A. R. E. Sinclair and M. Norton-Griffiths, 310–52. Chicago: University of Chicago Press.

Pellew, R. A. P. 1981. The giraffe (*Giraffa camelopardalis tippelskirchi* Matschie) and its *Acacia* food resource in the Serengeti National Park. Ph.D. dissertation, University of London.

———. 1983. The impacts of elephant, giraffe, and fire upon the *Acacia tortilis* woodlands of the Serengeti. *Afr. J. Ecol.* 21:41–74.

Pratt, D. J., and Knight, J. 1971. Bush-control studies in the drier areas of Kenya. V. Effects of controlled burning and grazing management on *Tarchonanthus/Acacia* thicket. *J. Appl. Ecol.* 8:217–37.

Ruess, R. W., and Halter, F. L. 1990. The impact of large herbivores on the Seronera woodlands, Serengeti National Park, Tanzania. *Afr. J. Ecol.* 28:259–75.

Ruess, R. W., and McNaughton, S. J. 1987. Grazing and the dynamics of nutrient and energy regulated microbial processes in the Serengeti grasslands. *Oikos* 49:101–10.

Sabiiti, E. N., and Wein, R. W. 1988. Fire behavior and the invasion of *Acacia sieberiana* into savanna grassland openings. *Afr. J. Ecol.* 26:301–14.

Sinclair, A. R. E. 1977. *The African buffalo: A study of resource limitation of populations.* Chicago: University of Chicago Press.

———. 1979. Dynamics of the Serengeti ecosystem: Process and pattern. In *Serengeti: Dynamics of an ecosystem,* ed. A. R. E. Sinclair and M. Norton-Griffiths, 1–30. Chicago: University of Chicago Press.

———. 1989. Population regulation of animals. In *Ecological concepts,* ed. J. M. Cherrett, 197–241. Oxford: Blackwell Scientific Publications.

Sinclair, A. R. E., Dublin, H. T., and Borner, M. 1985. Population regulation of Serengeti wildebeest: A test of the food hypothesis. *Oecologia* 65:266–68.

Spinage, C. A. 1973. A review of ivory exploitation and elephant population trends in Africa. *E. Afr. Wildl. J.* 11:281–89.

Spinage, C. A., and Guiness, F. E. 1971. Tree survival in the absence of elephants in the Akagera National Park, Rwanda. *J. Appl. Ecol.* 8:723–28.

Stark, M. A. 1986. The relationship between fire and basal scarring in *Afzelia africana* in Benoue National Park, Cameroun. *Afr. J. Ecol.* 24:263–71.

Stronach, N. R. H. 1989. Grass fires in Serengeti National Park, Tanzania: Characteristics, behavior and some effects on young trees. Ph.D. dissertation, Cambridge University, Cambridge.

Stronach, N. R. H., and McNaughton, S. J. 1989. Grassland fire dynamics in the Serengeti ecosystem, and a potential method of retrospectively estimating fire energy. *J. Appl. Ecol.* 26:1025–33.

Thomas, D. B., and Pratt, D. J. 1967. Bush control studies in the drier areas of Kenya. IV. Effects of controlled burning on secondary thicket in upland *Acacia* woodland. *J. Appl. Ecol.* 4:325–35.

Thomson, P. J. 1975. The role of elephants, fire, and other agents in the decline of a *Brachystegia boehmii* woodland. *J. S. Afr. Wildl. Mgmt. Assoc.* 5:11–18.

Trapnell, C. G. 1959. Ecological results of woodland burning experiments in Northern Rhodesia. *J. Ecol.* 47:129–68.

Wakefield, T. 1870. Routes of native caravans from the coast to the interior of East Africa. *J. R. Geogr. Soc.* 11:303–38.

———. 1882. New routes through Masai country. *Proc. R. Geogr. Soc.* 4:742–47.

Watson, R. M. 1967. The population ecology of the wildebeest *(Connochaetes taurinus albojubatus)* in the Serengeti. Ph.D. dissertation, Cambridge University, Cambridge.

Watson, R. M., and Bell, R. H. V. 1969. The distribution, abundance, and status of elephant in the Serengeti region of northern Tanzania. *J. Appl. Ecol.* 6:115–32.

Young, T. P., and Lindsay, W. K. 1988. Disappearance of *Acacia* woodlands: The effect of size structure. *Afr. J. Ecol.* 26:69–72.

III Herbivores and Predation

SIX

Population Trends and Distribution of Serengeti Herbivores: Implications for Management

Ken Campbell and Markus Borner

Over the last 15 years significant changes have occurred within the Serengeti ecosystem. The dramatic increase in the wildebeest population that started in the 1960s (Sinclair 1979) effectively ceased during the late 1970s (Sinclair and Norton-Griffiths 1982), and wildebeest numbers have since stabilized (Sinclair, Dublin, and Borner 1985; Borner et al. 1987; Dublin et al. 1990). In some parts of the ecosystem nonmigratory wildlife have been subjected to severe illegal hunting pressures. Total off-take of wildlife through illegal game meat hunting may be in the region of 200,000 animals annually (chap. 25). Rhinoceroses have been eliminated from much of the ecosystem by poachers (Borner 1981) and the elephant distribution has also been affected (Dublin and Douglas-Hamilton 1987). Buffalo populations in northern Serengeti have declined due to illegal meat hunting (Sinclair 1987b; Dublin et al. 1990; Sinclair, chap. 1). Outside the protected areas, settlements and human populations have increased, and agriculture is contiguous with extensive sections of the national park and game reserve boundaries. This chapter summarizes present population estimates of large herbivores and demonstrates some population trends and patterns of species distribution that have particular relevance to the management of the protected areas.

METHODS

Study Area

The study area (fig. 6.1) comprised some 27,000 km² of *Acacia* woodland and grassland and included those protected areas within the Tanzanian part of the Serengeti ecosystem: Serengeti National Park, Maswa Game Reserve (GR), proposed Ikorongo and Grumeti GRs, and Ngorongoro Conservation Area (excluding the Crater Highlands). The western boundary was defined by the limits of cultivation and settlement (the sur-

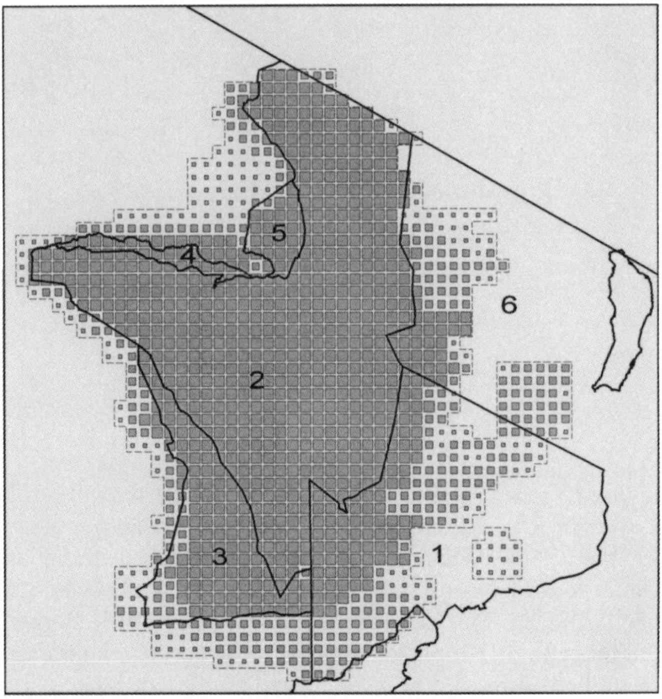

Figure 6.1 Boundaries of protected areas and the area surveyed by three aerial censuses in 1988, 1989, and 1991 (gridded line). Graduated symbols indicate the number of surveys covering each 5 by 5 km UTM grid square. Strata indicated are: 1, Ngorongoro Conservation Area; 2, Serengeti National Park; 3, Maswa GR; 4, proposed Grumeti GR; 5, proposed Ikorongo GR; 6, Loliondo. Areas 5 and 6 are currently Game Controlled Areas.

vey aircraft typically flew some distance beyond the first settlements). In the north the survey area was limited by the international border; in the northeast, by the Loliondo highlands; in the southeast, by the Rift Valley escarpment and Crater Highlands and by Lake Eyasi to the south. In the west, therefore, survey limits broadly corresponded to the change to an agricultural land use, while in the east they were determined by the physical barriers of mountainous terrain.

Historical Data

Surveys of resident ungulates were carried out in July 1971 (Sinclair 1972) and August 1976 (Grimsdell 1979). The 1971 census covered three strata totaling 9,576 km^2, while the 1976 census covered two of these strata (7,349 km^2). Both surveys used stratified random sampling (STRS: Sinclair 1972; Clarke 1986). Prior to these dates, surveys of resident ungulates covered smaller areas and were largely confined to the western Serengeti. These earlier data are quoted by Grimsdell (1979) together with an analysis of population trends prior to 1976. Estimates of wildebeest and zebra population size between 1957 and 1984 are available

(Swynnerton 1958; Norton-Griffiths 1973; Sinclair 1973, Sinclair and Norton-Griffiths 1982; Sinclair, Dublin, and Borner 1985). Buffalo and elephant have been censused by total counts since the mid-1960s (Sinclair 1973; Dublin, Sinclair, and McGlade 1990). Outside the protected areas, Watson, Graham, and Parker (1969) estimated wildlife densities in the Loliondo area to the northeast of the national park.

Recent Surveys

More recently, aerial sample surveys gathered data from the whole of the protected area, with the exception of the Crater Highlands. An aerial survey of 18,300 km^2 of the protected area, carried out during 1985 (Borner, Serengeti Ecological Monitoring Programme records), was the first Systematic Reconnaissance Flight (SRF: Norton-Griffiths 1978) to cover the whole of Serengeti National Park and Maswa GR. SRF surveys during the wet seasons of 1988, 1989, and 1991 covered increased areas up to 27,500 km^2 (Campbell 1988, 1989, and see below). During May 1986 2,313 km^2 of northern Serengeti were flown (A. R. E. Sinclair, pers. comm.). A total count of buffalo and elephant was carried out during May 1986 (Sinclair 1987b) and another during May 1992 (see below), when the survey area was extended to cover Maswa GR and parts of Ngorongoro Conservation Area. Wildebeest censuses were carried out in 1986 (Sinclair 1987a, Dublin et al. 1990), in 1989 (Campbell 1989), and in 1991 (see below). The 1988, 1989, and 1991 ecosystem-wide SRF surveys and 1992 total count of buffalo and elephant form the basis for an analysis of current wildlife distribution patterns and trends.

Sample Survey Methods. Since 1985, all sample censuses with the exception of the 1986 count in northern Serengeti have used SRF methods. The choice of the SRF, rather than the STRS methods used earlier, was based on a requirement for information on both distribution and densities, as well as the inherent lack of flexibility in STRS for incorporating data from external sources (Clarke 1986). East-west transects were flown at a spacing of 5 km. Subunits were defined as 30 seconds of flying time. Left and right rear seat observers (RSO) dictated into hand-held tape recorders all wildlife (except for wildebeest on the Serengeti plains) observed within sample strips defined by pairs of fiberglass rods attached to the wing struts of the Cessna 182 or 185 aircraft. Target strip width was 150–170 m for each RSO. Strip widths under census conditions were calculated by regression of observed strip widths against radar altimeter readings obtained by flying over a series of white cement blocks spaced 20 m apart.

An extension of SRF, Aerial Point Sampling (APS: Norton-Griffiths 1988) was used for the wildebeest censuses in 1989 and 1991. Vertical sample 35 mm color transparencies were exposed on a systematic basis throughout the survey area at the start of each 30-second subunit using 18 or 24 mm lenses. During the April 1989 wildebeest survey the sampling intensity was increased in areas with high concentrations of wildebeest (Campbell 1989). Sampling times were recorded to the nearest second on each frame by the Nikon F3 camera's 250-exposure databack and used to estimate distance between photos and area sampled by each frame.

Analysis. Transects and subunits were geo-referenced and assigned to strata (administrative boundaries, previous survey areas, and 5 by 5 km UTM grid squares). Strip widths were calculated from recorded radar altimeter readings and linear regressions of each RSO strip width against height above ground. Observed strip areas were derived for each subunit from subunit lengths and strip widths. Subunit areas were derived from subunit length and transect spacing. Wildlife densities were calculated for each subunit using combined RSO data. Stratum and grid square areas were calculated by combining subunit areas. Population estimates and their confidence limits were calculated for each stratum by combining data from subunits within each transect and applying a formula for unequal-sized sample units (Jolly 1969; Norton-Griffiths 1978). Strata used were: 5 by 5 km UTM grid squares, administrative areas, 1971 census boundaries, modeled zones of human and wildlife interaction, and, for topi, regions of high, medium, and low topi density.

APS data were analyzed by vertical projection of the full 35 mm frame at 10 times magnification onto a desktop grid, resulting in an interpretation scale of about 1:600. All wildebeest in sample photos were counted. Population estimates were obtained by extrapolation of each sample density. A frame area, $a_{N,}$ for each photo was calculated from image scale and dimensions, modified by a factor based on the area obscured by the camera's databack information. The area sampled by each frame, $A_{N,}$ was calculated as transect length × (sample interval/total transect time). Total surveyed area, $\Sigma A_{n,}$ was the sum of individual sample areas. Population estimates for each sample area were calculated from sample densities × sampled area. Total population estimates were the sum of the estimates for each sample, Σ [(count/a_N) × A_N). The variance of sample estimates was calculated by summing data within each transect and treating transects as primary sample units (Norton-Griffiths 1973; Sinclair and Norton-Griffiths 1982). Additional data from small wildebeest populations in the western corridor and Loliondo were obtained by SRF techniques and added to APS data to obtain a final population estimate.

A separate flight was made during May 1989 to test for possible bias relating to a change from black-and-white print to color transparency film. Six passes with both types of film were carried out over the same group of wildebeest: three with an 18 mm lens at a nominal flying height of 400 feet above ground level (a.g.l.) (range 350–450) and three with a 50 mm lens at a nominal height of 1,200 feet a.g.l. (range 1,100–1,300). A total of 184 black-and-white and 179 color photos were exposed, all including wildebeest. Color transparencies were projected at 14 times magnification. Black-and-white negatives were examined as prints at similar final magnification using a binocular microscope. No difference was observed in density estimates between film types (mean density on black-and-white 45.44 animals/ha, on color 41.56/ha, $P = .55$) or lens and flying height combinations (18 mm 45.22/ha, 50 mm 41.85/ha, $P = .63$).

Total counts were analyzed using standard techniques (Sinclair 1973, Norton-Griffiths 1978). In 1992, color transparencies replaced black-and-white prints. Global Positioning Systems (GPS) in each of the five aircraft participating in the 1992 count (Campbell 1992) enabled coordinates for each observation of buffalo and elephant to be recorded. Prior to the census, observers participated in training flights and were shown a series of color slides of cattle and buffalo taken from the air and asked to estimate herd size before being told the real number. Observers not already experienced in SRF surveys participated in more than one training flight. Larger groups of buffalo ($N \geq 50$) either not photographed or for which the film was damaged were corrected for observer counting bias by the use of linear regressions of observer estimates and photo counts derived independently for each of the survey crews. Observer estimates of less than 50 showed no detectable counting bias ($R^2 = .3$) and no corrections were made.

Distribution maps were prepared by summarizing information on wildlife densities within 5 by 5 km UTM grid squares, and densities were represented by graduated symbols for display purposes. Densities of all nonmigratory wildlife combined were averaged over the three surveys of 1988 to 1991 and used to calculate a density contour map using Surfer (Golden Software 1990) by kriging using three nearest data points in a quadrant search pattern.

RESULTS

Population Numbers

Resident Herbivores. In the ecosystem as a whole, few significant changes in herbivore populations occurred between 1988 and 1991. Changes on a smaller spatial scale are considered below, under "Population Trends." Data for administrative strata are presented using estimates merged from the 1988, 1989 and 1991 surveys (table 6.1). Separate annual estimates are given for species in which significant changes occurred ($d > 1.96$, Norton-Griffiths 1978) Estimates for the ecosystem as a whole were calculated by summing merged estimates of the different administra-

Table 6.1 Population estimates of resident herbivores in: A. Serengeti National Park and the Tanzanian part of the Serengeti ecosystem, B. Maswa GR, and Ikorongo and Grumeti GRs combined with areas outside, and C. Ngorongoro Conservation Area, excluding the Crater Highlands and the western parts of Loliondo.

A.	Serengeti National Park			Serengeti Ecosystem		
	Estimate	SE	95% c.l.	Estimate	SE	95% c.l.
Giraffe	6,673	446.6	13.1	8,832	485.8	10.8
Grant's gazelle	14,886	479.8	6.3	26,019	693.9	5.2
Impala	59,289	5,686.9	18.8	74,560	5,739.1	15.1
Kongoni	10,156	547.0	10.6	13,771	620.9	8.8
Warthog	4,666	371.9	15.6	5,872	397.1	13.3
Waterbuck	1,185	120.2	19.9	1,635	206.2	24.7
Topi (1988)	42,645	2,978.3	13.7	—		
Topi (1989)	74,338	8,052.9	21.2	—		
Topi (1991)	92,119	23,875.4	50.8	95,184	23,877.0	49.2
B.	**Maswa Game Reserve**			**Ikorongo and Grumeti GR**		
	Estimate	SE	95% c.l.	Estimate	SE	95% c.l.
Grant's gazelle	16	13.4	161.1	1,073	396.1	72.3
Impala	3,204	42.2	2.6	6,097	734.6	23.6
Kongoni	275	114.5	81.6	57	30.0	103.1
Roan	42	25.5	119.7	0	—	—
Topi	899	211.4	46.1	1,806	152.1	16.5
Warthog	341	61.9	35.6	508	85.3	32.9
Waterbuck	200	59.2	58.0	54	32.7	118.7
Giraffe (1988)	262	110.8	82.9	—		
Giraffe (1989)	396	123.2	61.0	—		
Giraffe (1991)	987	167.9	33.3	—		
Giraffe merged (88–91)	—			100	48.5	95.3
C.	**Ngorongoro Conservation Area (excluding the Crater Highlands)**			**Loliondo**		
	Estimate	SE	95% c.l.	Estimate	SE	95% c.l.
Giraffe	567	114.4	39.6	1,042	125.3	23.6
Grant's gazelle	7,097	99.3	2.7	2,946	290.5	19.3
Impala	624	220.3	69.2	5,345	75.4	2.8
Kongoni	205	97.3	93.2	3,078	250.8	16.0
Topi	0	—	—	360	101.0	55.0
Warthog	139	68.5	96.7	220	59.7	53.2
Waterbuck	0	—	—	196	153.3	152.9

Note: 1988, 1989, and 1991 survey estimates are merged where no significant changes occurred during this period and are presented for each survey separately where changes occurred between survey estimates ($d > 1.96$). Estimates are uncorrected for survey bias (see text).

tive strata. Population estimates of buffalo and elephant from the 1992 total count (Campbell 1992) are given in tables 6.2 and 6.3. Buffalo numbers within the ecosystem were approximately 46,000 (1992 census), but within comparable census areas have declined from 43,456 in 1986 to 37,823 in 1992. Elephant numbers have increased since 1986, largely in an area around Lobo in the northeast of the park. Figure 6.2 shows

Table 6.2 Buffalo numbers in the Serengeti ecosystem.

	Census block	Block name	Area (km²)	1970 population	1986 population	1992 population
Northern Serengeti	T0	Lamai	387	341	60	510
	T1	Mara-Tabora	493	1,861	195	2
	T2	Nyamalumbwa	900	8,887	2,521	812
	T3	Lobo-Tabora	898	6,912	2,917	1,603
	T4	Tagora-Lobo	1,147	6,146	5,511	3,215
	T4w*	Ikorongo north	285	—	—	0
	Subtotal		4,110	24,147	11,204	6,142
Central Serengeti	T5	Banagi-Ikoma	1,177	8,502	3,224	3,436
	T6	Grumechen	1,335	5,802	6,174	7,090
	T7	Musabi	1,235	5,773	7,933	7,005
	Subtotal		3,747	20,077	17,331	17,531
Western Serengeti	T8	W. Corridor	1,036	4,111	1,895	2,683
	T8n*	Grumeti GR (part)	370	—	—	6
	T9n&s	Dutwa and Ndoha	1,454	7,272	5,218	(4,352)
	T9n**	Dutwa	617	—	—	1,957
	T9s**	Ndoha	837	—	—	2,395
	Subtotal		2,860	11,383	7,113	7,041
Plains, Southwest, and Maswa	T10	Moru-Mamarehe	1,380	5,768	6,054	2,662
	T11 (1986)	Simiyu-Makao	1,114	1,769	1,754	(4,353)
	T11 (1992)	Simiyu-Ndutu	1,951	—	—	2,743
	T12	Eastern plains	1,205	—	0	354
	T13*	Maswa North	610	—	—	944
	T14*	Maswa Central	1,136	—	—	887
	T15*	Maswa South	1,206	—	—	2,431
	Subtotal (1986)		3,699	7,537	7,808	7,369
	Subtotal (1992)		7,487	—	—	10,021
Loliondo	T16*	Loliondo North	583	—	—	1,758
	T17	Loliondo South	897	***	—	407
	Subtotal		1,480			2,165
Ngorongoro Conservation Area	T18*	Endulen	438	—	—	308
	T19*	Makao-Kakesio	1,016	—	—	345
	T20*	Ngorongoro Crater	275	—	—	2,388
	Subtotal		1,729			3,041
Totals administrative areas	Serengeti National Park		13,063	63,144	43,456	36,396
	Maswa Game Reserve		2,879	—	—	4,173
	Makao Open Area (part)		334	—	—	112
	Ngorongoro (part)		2,204	—	—	2,929
	Grumeti and Ikorongo GRs		1,020	—	—	166
	Loliondo		1,695	—	—	2,165
Ecosystem totals	TOTAL (in 1986 Blocks only)		13,761	63,144	43,456	37,823
	TOTAL (in all 1992 Blocks)		21,413	—	—	45,941

Note: May 1992 total count compared with 1970 and 1986 data (Sinclair 1987b; Dublin et al. 1990). New census blocks added in 1992 are indicated by * and subdivided blocks by **. Block T17, Loliondo South (***), was last flown in 1968 when 432 buffalo were counted in an equivalent area (raw data held at SWRC). Block T11 was modified for

continued

Table 6.2 (continued)

the 1992 census by the addition of T13, T14, and T15 to cover the whole of Maswa GR. Figures in parentheses refer to 1992 buffalo numbers within 1986 census block boundaries where these differ from 1992 boundaries. Areas of census blocks and administrative areas were calculated from digitized maps using Idrisi (Eastman 1990). Census block boundaries are shown in figure 6.2.

Table 6.3 Elephant numbers within the Serengeti ecosystem.

Census block	Block name	1986 elephant	1992 elephant
T0	Lamai	0	13
T2	Nyamalumbwa	56	38
T3	Lobo-Tabora	18	19
T4	Tagora-Lobo	91	491
T5	Banagi-Ikoma	6	3
T6	Grumechen	144	140
T7	Musabi	28	0
T10	Moru-Mamarehe	69	79
T11 (1992)	Simiyu-Ndutu	55	189
T15*	Maswa South	—	27
T16*	Loliondo North	—	8
T18*	Endulen	—	169
T19*	Makao-Kakesio	—	90
T20*	Ngorongoro Crater	—	29
TOTAL		467	1,295

Note: 1992 total count data compared with 1986 census data (Sinclair 1987b). Blocks with no elephant in either census are omitted. New census blocks added in 1992 are indicated by *.

boundaries of census blocks used during these counts. Some species such as reedbuck, bushbuck, and the smaller antelopes are affected by extreme visibility bias during SRF surveys, making survey results unreliable for anything but order of magnitude estimates. Waterbuck are similarly affected by visibility bias, but since comparable earlier data are available, population estimates are given. Due to their restricted distribution and preference for areas with relatively high woody canopy along the major rivers, population estimates for waterbuck usually have high standard errors. High standard errors were also noted for estimates of topi. While SRF surveys are suitable for sampling smaller topi groups, total count techniques are better suited to estimating large herds. Total count data are unavailable, but SRF estimates of topi indicate an increase (tables 6.4 and 6.5), emphasizing the necessity to combine total counts of large topi herds with future SRF censuses to obtain more accurate population estimates.

Highly Mobile or Migratory Herbivores. Many species show considerable seasonal movement, while others, such as the wildebeest, are considered truly migratory (Maddock 1979) and must therefore be considered within an ecosystem-wide context. Census estimates derived from the two

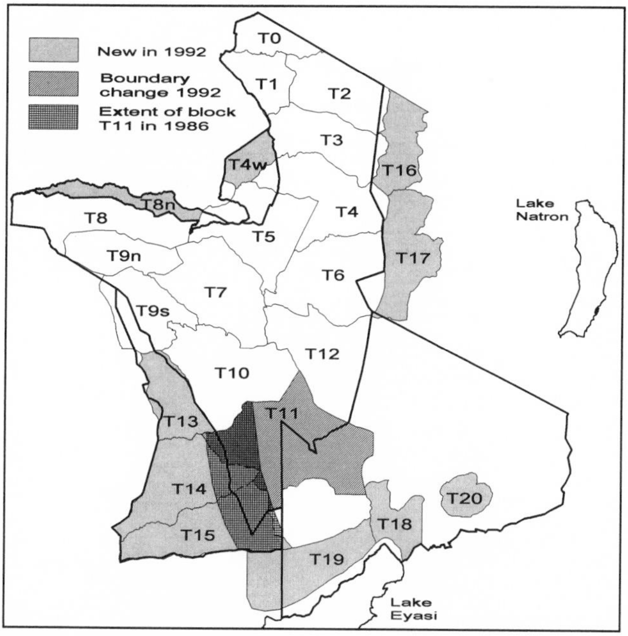

Figure 6.2 Total count census blocks with block IDs, 1986 and 1992. Unshaded blocks were flown in both censuses without boundary modifications. New census blocks added for the 1992 count are shaded. The extent of block T11 as flown at the time of the 1986 census is indicated by cross-hatching, while a fine diagonal shading shows the extent of block T11 in 1992. Boundaries were digitized from 1:250,000 and 1:50,000 scale flight maps, and that of T11 in 1986 from a copy of the 1986 flight map.

SRF surveys covering the greatest area (May 1989 and April 1991) indicate population numbers of approximately 4,300 ostrich, 12,000 eland, 191,000 zebra, and 355,500 Thomson's gazelle (figures rounded, see table 6.6 for confidence limits). An estimate of 1,278,603 wildebeest (± 95% c.l. of 2.6%) was derived by merging the April 1989 and April 1991 survey estimates. This figure is strongly influenced by the low sample variance calculated for the 1991 survey.

Spatial Distribution

Wildlife densities from the 1988, 1989, and 1991 surveys were summarized by 5 by 5 km UTM grid squares and average densities of each nonmigratory wildlife species were calculated. Figures 6.3–6.6 show distributions of impala, giraffe, topi, and warthog. Topi show an increase in density along the southeast-northwest rainfall gradient, the area occupied by the large herds, and an absence from the southern parts of Maswa. Roan antelope are now restricted to parts of Maswa and a small area outside the reserve, having disappeared from their former range within

Table 6.4 Wildlife population estimates and standard errors, 1971 to 1991, within three strata in northern, central, and western Serengeti.

Stratum 1 — Western corridor

Year	1971		1976		1988		1989		1991	
Area (km^2)	4,045		3,862		3,616		3,642		3,837	
	Est.	S.E.	Est.	S.E.	Est.	S.E.	Est.	S.E.	Est.	S.E.
Impala	6,759	870	30,440	4,550	24,279	3,927	33,347	7,997	21,106	5,324
Topi	17,072	961	51,633	7,281	37,849	5,979	35,864	8,146	81,930	23,188
Kongoni	461	163	934	209	497	135	1,089	235	538	265
Giraffe	2,855	209	2,240	474	3,015	470	1,955	178	1,690	185
Waterbuck	594	196	1,554	417	668	139	241	144	1,489	306

Stratum 2 — Central Woodlands

Year	1971		1976		1988		1989		1991	
Area (km^2)	3,687		3,487		3,295		3,203		3,246	
	Est.	S.E.	Est.	S.E.	Est.	S.E.	Est.	S.E.	Est.	S.E.
Impala	39,255	3,155	33,738	4,503	16,559	4,059	33,888	7,595	16,500	3,710
Topi	6,946	585	9,651	1,530	4,951	983	11,332	1,598	6,916	1,054
Kongoni	6,865	599	6,591	1,042	5,070	1,024	7,256	522	4,084	757
Giraffe	2,780	307	4,970	436	2,407	482	3,736	593	1,392	259
Waterbuck	365	185	352	176	324	120	339	196	207	99.7

Stratum 3 — North of Grumeti (excluding Lamai)

Year	1971		1985		1986		1988		1989		1991	
Area (km^2)	1,844		1,844		1,921		2,022		1,955		1,944	
	Est.	S.E.	Est.	S.E.	Est.	S.E.	Est.	S.E.	Est.	S.E.	Est.	S.E.
Impala	10,577	2,786	8,251	1,689	11,383	4,807	10,335	2,056	11,754	2,592	5,767	1,195
Topi	3,378	622	5,706	885	12,123	4,393	9,718	2,064	23,211	2,564	5,720	1,439
Kongoni	1,738	315	2,820	897	1,505	967	1,340	333	6,156	1,650	1,665	117.8
Giraffe	1,672	298	496	163	1,254	294	670	189	856	250	251	94.9
Waterbuck	336	60	132	85	211	94.2	212	110	19	18	17	14.4
Warthog		—	1,355	255	758	198.6	740	150	430	95	280	112
Grants		—	430	183		—	91	85	188	76	86	90

Sources: July 1971, Sinclair 1972; August 1976, Grimsdell 1979; 1985, Borner, SEMP records; 1986, Sinclair, pers. comm.; March 1988 and May 1989, Campbell 1988, 1989; April 1991, this chapter.
Note: Stratum boundaries are indicated in figure 6.1. Stratum areas are calculated from individual aerial survey parameters.

Table 6.5 Changes in topi density in low-, medium-, and high-density strata.

	Density strata		
	Low	Med	High
Densities (/km²)			
1988	1.55	7.88	57.34
1989	2.15	14.51	74.61
1991	1.40	9.02	115.04
% of total population			
1988	15.19	22.73	62.07
1989	16.31	30.27	53.64
1991	8.57	16.71	74.71

Table 6.6 Population estimates and standard errors of the estimates (SE) for mobile and migratory wildlife within the Serengeti ecosystem.

	1989		1991		Merged Estimates 1989 and 1991		
	Estimate	S.E.	Estimate	S.E.	Estimate	S.E.	95% c.l.
Eland	13,709	984	9,486	1,225	12,053	767	12.5
Ostrich	6,058	799	3,362	592	4,317	475	21.6
Thomson's gazelle	440,844	42,454	325,769	25,053	355,493	21,576	11.9
Wildebeest	1,407,287[a]	109,904	1,274,728	17,130	1,278,603	17,049	2.6
Wildebeest	1,686,079[b]	175,658					
Zebra	256,562	18,470	148,942	14,802	191,028	11,550	11.9

Note: Population estimates are merged from 1989 and 1991 SRF surveys. Observer data for Thomson's gazelle are corrected using a sightability factor, $k = 1.41$, derived by comparing observer and vertical photographic data (Campbell 1987) to facilitate comparison with earlier data. 1989 wildebeest census results are from Campbell (1989).
[a]26 February, [b]29–30 April

the national park. From 1969 to 1972 the Serengeti Ecological Monitoring Programme recorded wildlife presence in 10 by 10 km grid squares covering most of the ecosystem. During 35 flights roan were recorded from 17 grid squares in Serengeti National Park, 25 in Maswa, and 3 in Kenya. Recent reports of a small population of roan (ca. five individuals) outside and to the northwest of the park are not reflected by current aerial survey data. Their presence, but not their status, is confirmed by a sighting of five roan on the park boundary 10 km north of Tabora ranger post (A. R. E. Sinclair, pers. comm.) and also by village leaders at workshops held by the Serengeti Regional Conservation Strategy. Two roan (out of an estimated total of 11,871 of all species) were reported killed by hunters from eleven villages (SRCS 1991).

Distributions and densities of buffalo and elephant from 1992 are summarized on 5 by 5 km UTM grid squares (figs. 6.7 and 6.8). Distributions and densities of all nonmigratory wildlife species recorded by the 1988, 1989, and 1991 SRF censuses are summarized as zones of very low, low, medium, and high density (fig. 6.9). Densities of resident species are

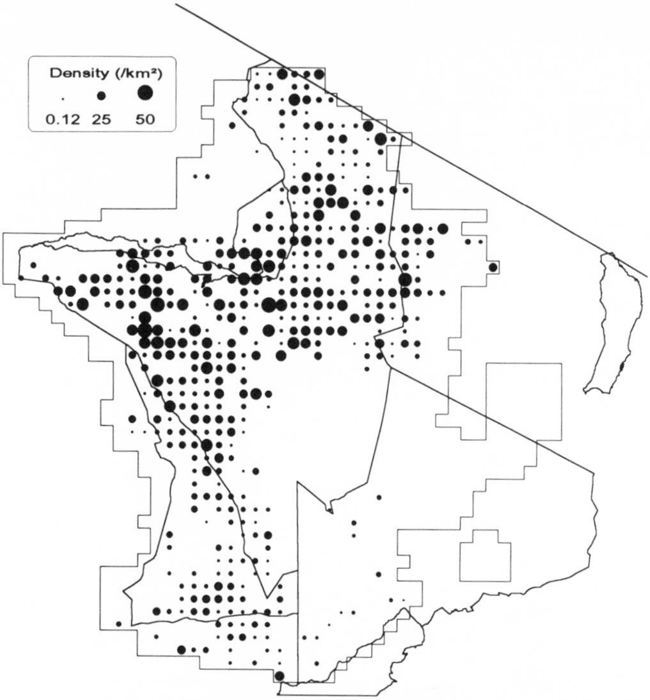

Figure 6.3 Spatial distribution and densities of impala over three aerial surveys in March 1988, April 1989, and April 1991. Densities are averaged within each 5 by 5 km UTM grid square. A gridded line indicates the total area covered during the three surveys.

lower on the drier southeastern plains and generally higher in the wetter, wooded areas of the west and north. Important exceptions to this pattern include areas of very low wildlife density ($< 1/km^2$) inside the northwestern border of the park and zero to very low densities inside the western boundary of Maswa. Both of these areas should, in the absence of human disturbance, contain higher densities of wildlife. Available historical data on one species, buffalo, from the northwest of the Serengeti show that greater numbers were formerly present in that area.

Population Trends

Long-term trends in population numbers are indicated in table 6.4 by survey estimates from 1971 survey blocks in northern, central, and western Serengeti. These roughly corresponded to total count census blocks T1, T2, and that portion of T3 north of the Grumeti River (northern survey area); blocks T4, T5, T6, and the remainder of T3 (central); and blocks T8, T8n, T9n, T9s and T7 (western survey area) respectively (see fig. 6.2). In these areas, for species enumerated during the 1971 census,

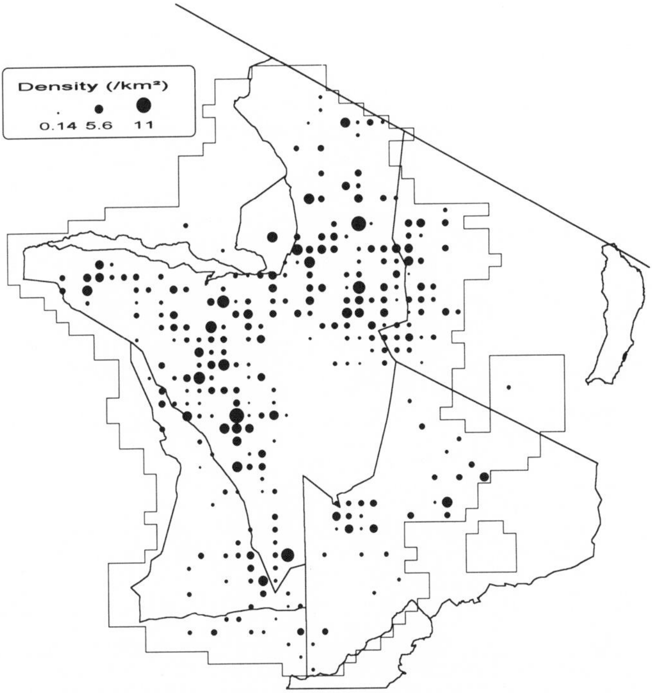

Figure 6.4 Spatial distribution and densities of giraffe over three aerial surveys in March 1988, April 1989, and April 1991. Densities are averaged within each 5 by 5 km UTM grid square. A gridded line indicates the total area covered during the three surveys.

there are few significant long-term changes in wildlife density. Estimates of both impala and topi population numbers in stratum 1 increased dramatically between 1971 and 1976. Both changes are significant (d = 5.1 for impala, d = 4.7 for topi). Since then there have been no significant changes in impala estimates. A recent increase in topi estimates between 1989 and 1991 is not significant at the 95% level due to the high standard errors of the 1991 estimate (d = 1.87). The 1971 and 1976 estimates indicate an annual rate of increase for impala of about 35% and for topi a rate of about 25% per annum. These high rates of increase may not be feasible, and the observed changes may be partly due to survey biases.

More recent survey data cover the greater part of the Serengeti ecosystem, and the SRF methods used facilitate examination of population trends within a variety of different strata. Since 1988, giraffe numbers have increased in Maswa, while estimates of the topi population in the national park indicate an upward trend (see table 6.1). Human influences (in the form of illegal meat hunting) are likely to have resulted in the zero and very low wildlife densities currently recorded in parts of the ecosystem. To test for effects due to illegal meat hunting, estimates of wildlife

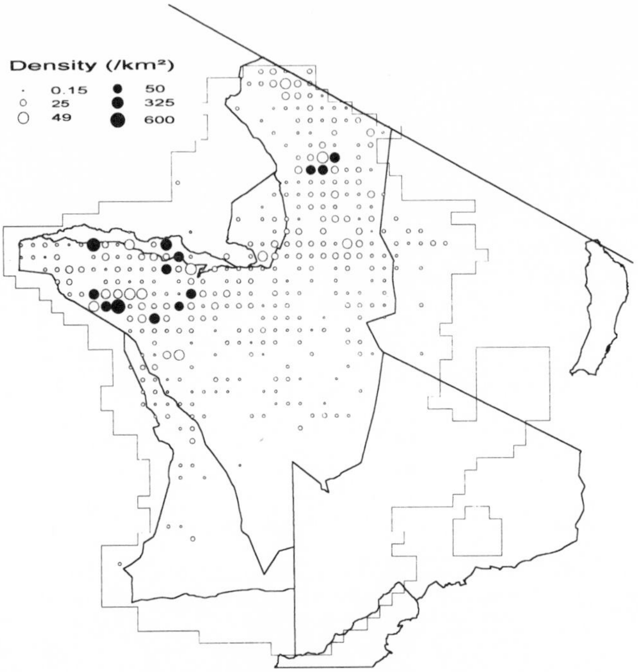

Figure 6.5 Spatial distribution and densities of topi over three aerial surveys in March 1988, April 1989, and April 1991. Densities are averaged within each 5 by 5 km UTM grid square. Solid circles indicate densities recorded as a result of encounters with large herds of topi, while shaded circles show the densities of topi in smaller groups. A gridded line indicates the total area covered during the three surveys.

density were calculated within zones corresponding to an increasing gradient of human/wildlife interaction (see table 6.7). These zones were modeled using a combination of environmental and wildlife data, human settlement patterns, and information from anti-poaching patrols (Campbell and Hofer, chap. 25). For the purposes of this analysis resident wildlife densities were calculated for three of these risk zones—(1) overexploited, (2) endangered, and (3) escalation of conflict (fig. 25.12, chap. 25)—the ones most likely to be affected by illegal meat hunting. The results of this analysis showed a decline in density within zone 1 but increased densities in zones 2 and 3.

DISCUSSION

Survey Bias

The calculation of observer and survey bias is complicated by a wide variety of factors, including size of the animal; group size and complexity; background color, contrast, and illumination; observer experience; visual acuity and color vision deficiency; observer fatigue; strip width; flying

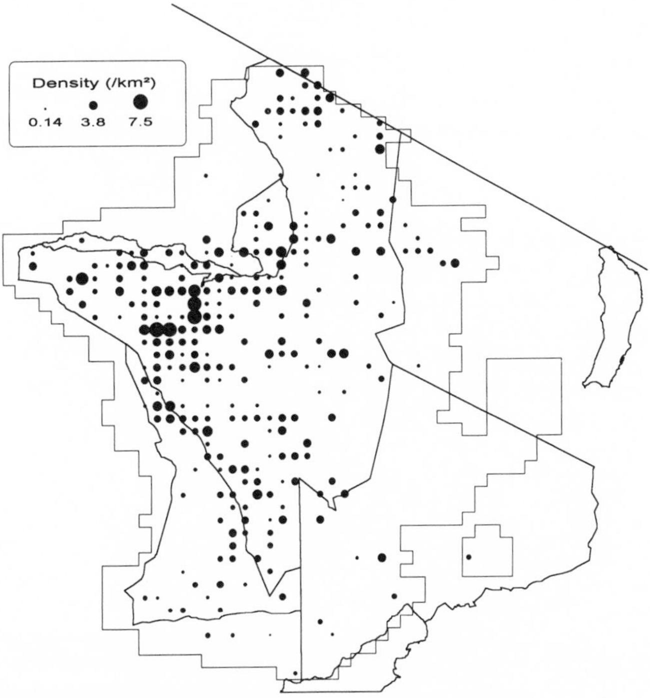

Figure 6.6 Spatial distribution and densities of warthog over three aerial surveys in March 1988, April 1989, and April 1991. Densities are averaged within each 5 by 5 km UTM grid square. A gridded line indicates the total area covered during the three surveys.

height; relative ground speed; time of day, temperature, and cloud cover; woody canopy cover; the overall density of observations and numbers of different species within the strip at a single site; and the numbers of different tasks required of the observers (for discussion of some of these see Graham and Bell 1969; Watson, Freeman, and Jolly 1969; Norton-Griffiths 1978; Pollock and Kendall 1987; Drummer and McDonald 1987; Otto and Pollock 1990). Many of these parameters are at any given time likely to vary considerably within a large survey area such as the Serengeti ecosystem. Attempts at correcting observer bias have used photographs taken by observers during the census. Using this method, Sinclair (1972) estimated that, for the 1971 census, impala numbers counted by observers were some 30% too low. This survey was flown during July at a dry time of year (June and July 1971 rainfall at Seronera meteorological station = 0). Under such conditions brown species, such as impala, are more difficult to observe against a dry background than during a wetter and greener time of year. Photographic records of the larger groups enable some sources of bias to be estimated, but at the same time also introduce additional biases. Photointerpretation itself may suffer from

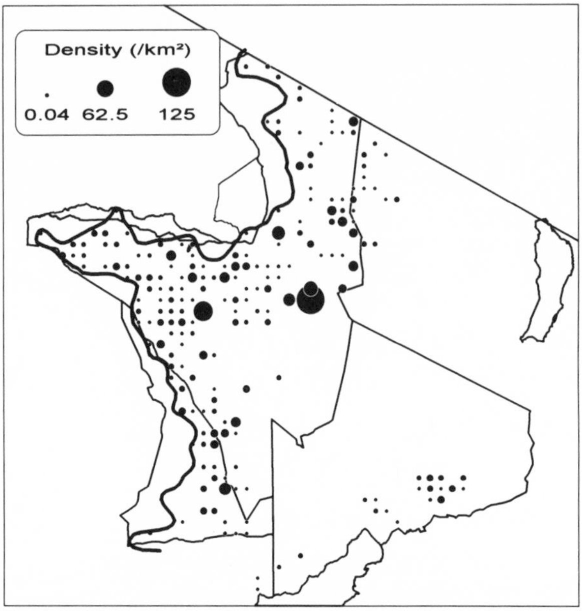

Figure 6.7 Spatial distribution and density of buffalo, May 1992. Total count data are summarized within each 5 by 5 km UTM grid square. A solid line, drawn by eye, indicates the current western limits of buffalo distribution and clearly shows areas along the western boundary of the protected areas with an absence of buffalo.

visibility bias. For example, photographs present the viewer with a single perspective, whereas observers are able to use a range of different viewing angles as the aircraft passes any given location and can usually see behind obstructing objects. Other factors such as contrast and shadow may affect photographic media to a greater extent than they do the human eye. Additionally, observers tend to miss, and fail to record, individuals or groups of animals within the sample strip while occupied by photographing others. The latter source of bias is more likely to occur among the relatively high wildlife densities found in the Serengeti, especially in "hot spots" (McNaughton 1988), where multispecies assemblages are common. Observers may also fail to notice individuals or small groups of one species among larger groups of a different species, the chances of this being reduced with more experienced observers who are not required to both count and photograph. Successful use of hand-held cameras during SRF surveys demands considerable expertise and experience on the part of the observers, and since few trained observers were also fully confident with cameras, oblique photography was not used during the 1988, 1989, and 1991 SRF surveys. While some of the above sources of bias can be measured (e.g., flying height and ground speed) and others can be guarded against (all observers were tested for color vision deficiency), the magni-

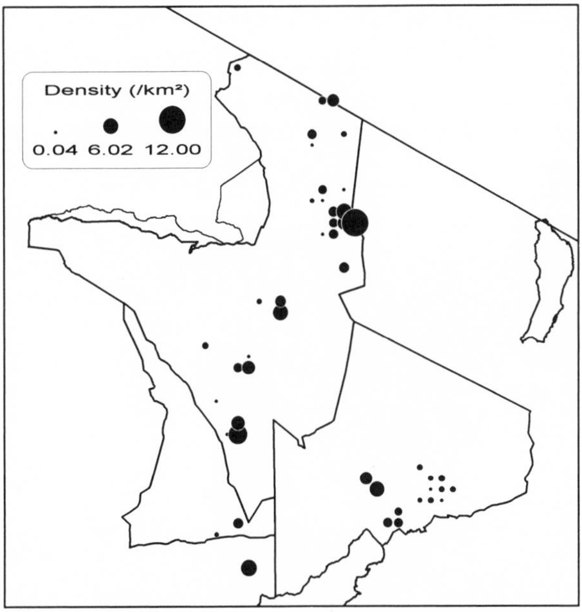

Figure 6.8 Spatial distribution and density of elephant, May 1992. Total count data are summarized within each 5 by 5 km UTM grid square.

tude and spatial variation of others remain unknown. Since during these surveys there was no practical means of accurately quantifying all of these factors, data in this chapter are presented as raw data uncorrected for survey bias, except where stated.

Spatial Patterns

Spatial distribution patterns of migratory wildlife are highly seasonal (Maddock 1979) and likely to change in emphasis from year to year due to local variations in rainfall and growing conditions. Since migratory wildlife are not present in any one area on a year-round basis, they are not constrained by the same factors that govern resident wildlife densities; that is, they periodically escape from different set of constraints. Non-migratory wildlife densities, being governed by parameters effective at a local scale, represent a more robust set of indicators of conservation status than do migratory wildlife, which only provide an indicator of the status of the entire ecosystem. This discussion therefore focuses on recent spatial patterns of resident wildlife species.

Roan Antelope. Although they were never considered common within the ecosystem, the former distribution of roan antelope, as recorded by the 1969 to 1972 series of monthly reconnaissance flights, included virtually the whole of Maswa GR and significant parts of the southwest and

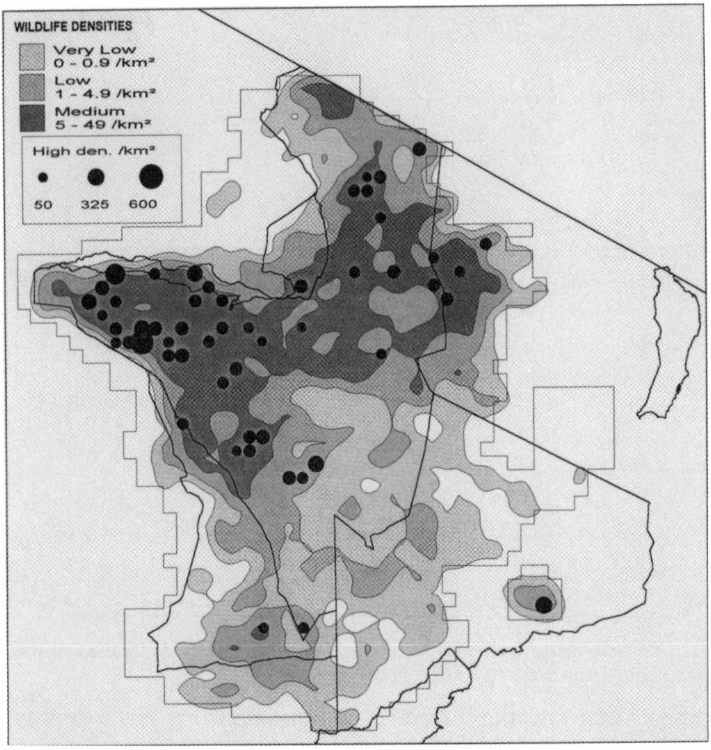

Figure 6.9 Nonmigratory wildlife densities, estimated from three aerial surveys in March 1988, April 1989, and April 1991. A gridded line shows the extent of the survey area. Unshaded areas inside this line indicate zero recorded wildlife density. Species included are: baboon, buffalo, giraffe, Grant's gazelle, impala, kongoni, roan, topi, warthog, and waterbuck.

northwest of the park, as well as the Banagi and Nyaraswiga hills (M. Norton-Griffiths, pers comm.; A. R. E. Sinclair, pers. comm.; SEMP records). Roan were also recorded in parts of the Masai Mara in Kenya. The current distribution of roan, recorded by three aerial surveys, is limited to isolated areas in and to the west of Maswa GR. They may also still occur in very small numbers in the northwest. Since these refuge areas are close to human settlement, the disappearance of roan from their former range must be partly linked to factors other than illegal meat hunting. Possible factors include: (1) changes in habitat—the *Combretum* vegetation formerly occurring on many hills in the Serengeti has now largely disappeared; (2) the prevalence of fire, including that caused by early burning campaigns conducted by park management (these act to prevent later, and theoretically more destructive, higher-temperature burns [Stronach 1988], but hillslopes are often burnt by uncontrolled fires as a result of early burning), which may be linked to the decline in *Combretum* and in roan; (3) a scarcity of food caused by fire early in the dry season, which may affect some herbivore species to a greater extent than others; and (4) because the current distribution of roan is close to some of the greatest human disturbance in the ecosystem, it is possible that in the Serengeti, roan are unable to compete successfully against other herbivores in the absence of a certain level of human disturbance.

Boundary Effects and Conservation Status. Current spatial patterns of nonmigratory wildlife are the result of interactions of each species with a wide variety of physical and biological factors inside the protected areas as well as with human disturbance originating from outside the protected areas. Several parameters affecting resident wildlife species are examined by Campbell and Hofer (chap. 25). Factors such as rainfall, woody canopy, and the nature of the terrain were shown to be important in determining broad-scale species distribution patterns inside the protected areas. Such environmental gradients do not, as a rule, change abruptly as one crosses the boundaries of the protected areas, but interactions between people and natural resources form an important factor determining wildlife densities in areas close to protected area boundaries. Such interactions lead not only to a reduction in the overall size of the protected area, through, for example, the repositioning of the western boundary of Maswa GR (Borner and Maragesi 1985; W. J. Ngowo, pers. comm.; M. Maige, pers. comm.), but also to reduced wildlife densities through continued and long-term illegal overexploitation of the resource by meat hunters.

The extent of cross-boundary human influence on wildlife densities varies in different parts of the ecosystem, as is evident from the distribution of nonmigratory wildlife (see fig. 6.9). Three types of boundaries can be seen:

1. The distribution of resident wildlife closely follows the protected area boundary and steep density gradients follow the boundary.
2. The distribution of wildlife extends a significant distance beyond the protected area boundary.
3. Very low density ($< 1/km^2$) or empty areas occur inside the boundary.

The first type of boundary effect occurs along the northern and southern edges of the park's western corridor. The second occurs in the Loliondo area to the east of the park, along the border between Maswa GR and Makao Open Area and in a smaller region near Ikoma gate. The third boundary effect represents zones where overexploitation of the wildlife resource has led to the current situation in which large tracts of land inside the protected areas contain either no wildlife or very low densities (recorded by SRF). Two overexploited areas were identified, one in the northwest of the Park and the other along the western boundary of Maswa. The width of both zones is broadly similar and roughly corresponds to a single day's walk by a hunter from settlements close to the boundary (J. Hando, pers. comm.).

What factors are responsible for these different boundary conditions? Levels of management infrastructure are similar in northern Serengeti and the western corridor (Tanzania National Parks 1991), with an average prior to 1992 of one ranger per 115 km^2 in the north (total count census blocks T0–T4) and one per 123 km^2 in the west (T8, T9n and s; see fig. 6.2). Differences in the current conservation status of these two areas are therefore not simply related to staffing levels. In the north, access and patrolling are made difficult by the nature of the terrain, which includes numerous drainage lines, while the vegetation consists of a mosaic of open woodland and grassland patches. The west has a generally flatter terrain, facilitating greater access by patrols to much of the area, and a woodland/grassland mosaic in which open grassland plays a more dominant role. These factors combine to reduce access by law enforcement patrols in the north but facilitate it in the west and, as a result, may increase the chance of arrest in the western corridor relative to the northwest. The complete absence of an effective buffer zone in the northwest also plays an important role. Within Ikorongo patrolling has been sparse since there is no law enforcement presence. To the north of this, cultivation stretches to the border of the park, facilitating access to the park by meat hunters. In contrast, much of the western corridor is bordered by a de facto buffer zone that, although narrow, increases the distance that a hunter is required to travel. The problem of zero and low wildlife densities in western Maswa is similar to that of the northern Serengeti, except that the lack of infrastructure, nature of the terrain, drainage lines, and high woody canopy cover all tend to favor an increased probability of unhindered access by illegal meat hunters. *Wildlife distribution and densities in these areas indicate that an unsustainable consumptive utilization has already occurred over a period of time sufficient to cause a severe degradation of large mammal biodiversity.*

In the Lobo/Loliondo and South Maswa boundary zones, much of the wildlife extends well beyond the protected area. A major contributory factor in maintaining this high biodiversity must surely be the attitudes and cultural traditions of the local people toward wildlife. Cultural differences exist between settlements adjacent to the protected areas in the north, west, and southwest, and between these and the pastoral settlements to the east of the park, but adequate socioeconomic data from the relevant areas are lacking. Finally, there is undoubtedly a wide range of additional parameters related to environmental, social, cultural, economic, natural resource, and management-related factors that play a role in determining the extent and nature of wildlife/human interactions along the boundaries of protected areas. Clearly we have little understanding of many of these parameters. There is an urgent requirement for a rigorous spatial analysis of these and other factors contributing to the observed differences in wildlife density gradients along protected area boundaries.

Population Trends

Trends in wildlife densities depend on the proximity of different boundary categories rather than on simpler functions such as distance to the nearest settlements. Those parts of protected areas that are close to human settlements and lack an intermediate buffer zone (equivalent to boundary condition 3 above) show a decrease in wildlife densities. These trends are found in long-term and more recent data.

Buffalo. While population estimates of buffalo (see table 6.2) have essentially remained steady since 1970 in central and southwestern Serengeti National Park (average annual rate of change 1970 to 1992, −0.5% per annum) and have declined marginally in the west (−2.1% per annum), this species has suffered a dramatic decline in the north of the park (−6% per annum over the same period). Prior to the 1986 census there were small declines in population numbers in western and central Serengeti but no change in buffalo numbers in the southwest (fig. 6.10). Disease is one of a number of possible causes of the observed declines in buffalo numbers. Dublin et. al. (1990) concluded that although rinderpest was present within the population, the disease was not responsible for any observed population decline. Data on other diseases are lacking, but a continued lack of population change throughout much of the rest of the park since 1986 indicates that other factors are primarily responsible. The southwestern census blocks and block T6 (see fig. 6.2) lie at a greater distance from settlements outside the park than the remaining blocks. Buffalo numbers in these areas show either an increase or no change. In the north the majority of the remaining buffalo are currently found well away from the park's western boundary (see fig. 6.7), indicating that changes in numbers have occurred largely in western areas nearest to human settlement. These data indicate that illegal meat hunting has been the primary cause of the observed decline in buffalo numbers.

Other Resident Herbivores. Recent SRF data on densities of nonmigratory herbivores within zones of increasing risk are shown in table 6.7. In the high poaching risk, or overexploited, zone, average nonmigratory wildlife densities declined from 5.84/km^2 to 0.09/km^2 between 1988 and 1991. With over 1 million people within the 45 km wildlife meat demand

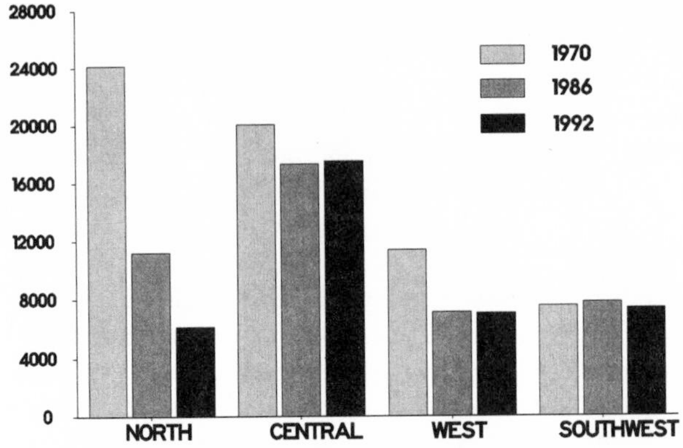

Figure 6.10 Long-term changes in buffalo population numbers, 1970 to 1992, within four regions of Serengeti National Park. Strata are combined from individual total count census blocks and correspond to the subdivisions in table 6.2

zone to the west of the protected area boundaries, it is not surprising that wildlife numbers have been affected in this manner (Campbell and Hofer, chap. 25). For Maswa (boundary condition 3) the lack of data from before the mid-1980s precludes long-term comparison. However, the current distribution of nonmigratory wildlife in this area suggests that densities of larger herbivores have sharply declined, especially along and inside the western boundary.

In strong contrast, wildlife densities in the Loliondo area to the northeast of the park (boundary condition 2) have not changed appreciably, and in some instances (e.g., giraffe) may have increased since the late 1960s when the first estimates of wildlife densities were made (Watson, Graham, and Parker 1969). Similarly, buffalo census data from 1968 show very little difference from the 1992 result (432 and 407 buffalo) in block T17.

In areas identified as having a lesser risk of poaching by meat hunters (zones 2 and 3), wildlife densities sampled since 1988 indicate recent population increases of several nonmigratory species, notably impala and topi, while buffalo numbers have remained stable. The increase in zone 2 is largely the result of increased numbers of impala, but in zone 3 both impala and topi contribute to an overall increase in wildlife density (table 6.7).

To examine long-term changes, census results from July 1971 (Sinclair 1972) were compared with recent data from the same stratum in western Serengeti, (see table 6.4), the greater part of which corresponds to boundary condition 1. In this area the only significant changes have been the early increase in impala numbers between 1971 and 1976 and the decrease in buffalo between 1970 and 1986 (see table 6.2). The trend

Table 6.7 Mean population densities of resident wildlife, 1988, 1989, and 1991, within zones of increasing risk from meat hunting.

	Wildlife densities (/km^2)			
Zone 1	1988	1989	1991	Average
Grant's gazelle	0.34	0.02	0	0.12
Giraffe	0.33	0.83	0.01	0.39
Impala	4.35	2.74	0.05	2.38
Kongoni	0.23	0.13	0	0.12
Topi	0.39	0.71	0.01	0.37
Warthog	0.15	0.09	0.01	0.09
Waterbuck	0.05	0.01	0.01	0.02
TOTAL	5.84	4.53	0.09	3.49
Zone 2				
Grant's gazelle	0.23	1.00	1.12	0.78
Giraffe	0.77	0.38	0.82	0.66
Impala	3.48	5.19	16.25	8.31
Kongoni	0.85	0.16	0.95	0.65
Topi	3.07	3.64	3.80	3.50
Warthog	0.58	0.56	0.41	0.52
Waterbuck	0.07	0.00	0.45	0.17
TOTAL	9.05	10.93	23.80	14.59
Zone 3				
Grant's gazelle	1.72	1.45	3.27	2.15
Giraffe	0.47	0.79	0.83	0.69
Impala	5.87	10.15	10.88	8.97
Kongoni	0.54	1.44	0.92	0.97
Topi	5.60	14.62	21.51	13.91
Warthog	0.23	0.58	0.78	0.53
Waterbuck	0.07	0.04	0.34	0.15
TOTAL	14.50	29.07	38.53	27.37

Note: Zones are: 1, *Overexploited* (2,205 km^2); *Endangered* (1,502 km^2); 3, *Escalating conflict* (2,875 km^2) (Campbell and Hofer, chap. 25).

toward increased numbers of topi and a possible recent decline in giraffe numbers are masked by the high standard errors of the survey estimates. If giraffe have declined in numbers in the western corridor, one possible cause is indicated by their scarcity close to the boundaries of the protected areas (see fig. 6.4) where illegal meat hunting is more likely. In areas where illegal meat hunting is uncommon (e.g., Loliondo), giraffe are found up to the boundaries of the protected area and there is little evidence of decline in density outside of the protected area.

Population estimates of topi are subject to high standard errors because the population consists of a small number of very large herds (up to about 2,000 individuals) and a large number of small groups. Whether or not an individual large herd is included in an observer's sample strip during an SRF census makes an appreciable difference to the final survey estimate and its variance. Estimates of topi within the park, where the bulk of the population occurs, increased from 42,645 (SE ± 2,978) in 1988 to 92,119 (SE ± 23,875) in 1991. What part of the population accounted for the increase? To answer this question, the survey area was stratified into low-, medium-, and high-density topi areas on the basis of the average numbers recorded during this period. The increase in topi population estimates was shown to result from increased numbers in the high-density areas (see table 6.5). Such an increase can be accounted for either by (1) an increasing proportion of the large herds being recorded within the sampling strips while the actual population remained stable, or (2) a real increase in the topi population centered around the large herds. While the former remains a possibility, the second explanation is considered more likely.

Management Resources and Infrastructure

The Demand for Wildlife Meat. Resources available for law enforcement work in the Serengeti have increased in recent years (Arcese, Hando, and Campbell, chap. 24). This increase has undoubtedly had an effect in reducing the levels of illegal meat hunting that would otherwise have occurred, given the increase in human settlement close to the boundary. Increased patrolling effort is more likely to have occurred preferentially in more accessible areas, corresponding to risk zones 2 and 3, but not, in general, to zone 1. A likely outcome of this scenario would be decreased mortality due to poaching in zones 2 and 3, resulting in increased wildlife density there. This has been observed by the SRF surveys. In order to compensate, and since there is no evidence to suggest that demand for game meat products has decreased, hunters may have been forced to make greater efforts in lower-density or otherwise less suitable areas. This would result in accelerated rates of decline, especially in areas that already had low wildlife densities. By providing greater accessibility for law enforcement patrols in these overexploited zones, the profitability of poaching could be reduced to such a level that illegal meat hunting in these zones would no longer be regarded as a suitable source of food and money. Concentrated enforcement effort in the low-density wildlife areas, while in the long run likely to increase wildlife numbers in those zones, may in turn increase illegal meat hunting efforts in the higher-density wildlife areas (e.g., zone 2, endangered).

Wildlife Utilization. Nonmigratory wildlife account for approximately 33% of the total wildlife recorded killed by game meat hunters (Arcese, Hando, and Campbell, chap. 24) or approximately 75,000 animals annually (Campbell and Hofer, chap. 25). Game reserves adjacent to the western border of the national park contain an estimated 21,000 nonmigratory animals (see tables 6.1 and 6.2). The entire resident wildlife population of these areas is therefore equivalent to only a small part of the estimated annual demand for nonmigratory wildlife within the ecosystem. If a legalized 10% offtake within game reserves and other buffer zones could be considered sustainable (and this is highly questionable in those areas already with very low densities), this amount would satisfy less than 5% of the current demand for wildlife meat. Moreover, the demand is likely to grow at a rate comparable to the rate of increase of the human population. In zones adjacent to the protected area boundaries this rate approaches an average of 4% per annum. Within 15 km of the protected area boundaries, where demand for wildlife meat is greatest and where wildlife utilization schemes are most likely to be targeted, there was a total human population of 363,800 (Bureau of Statistics 1988). Within this zone the estimated demand for game meat is 86% of the total demand (Campbell and Hofer, chap. 25). Assuming current wildlife population levels and a maximum of 10% per annum offtake as being sustainable, the game reserves can satisfy less than 5% of this demand.

It is clear that in the absence of significant measures designed both to increase effective law enforcement presence in targeted areas and to reduce dependence on wildlife meat as a source of both protein and funds, illegal offtake from the national park will continue. Ways must be found to reduce the demand for wildlife meat and increase demand for domestic sources of meat, as through livestock development programs (cattle, poultry, etc.). The development of alternative and sustainable forms of natural resource exploitation (e.g., beekeeping), provision of employment opportunities, and provision of adequate transport facilities for domestic sources of meat and fish must also play a role in the future conservation of the Serengeti ecosystem.

CONCLUSION

The spatial distribution and densities of nonmigratory wildlife species within the Serengeti ecosystem were examined in relation to different administrative areas and to varying levels of human-wildlife interaction. Within the ecosystem as a whole there is little evidence of significant changes in resident wildlife densities over the last 20 years. Exceptions to this include rhino (poached for its horn and now more or less absent), roan antelope (now restricted to a small fraction of its former range), and buffalo. Buffalo have suffered a serious decline in the northwest of the park; on the other hand, buffalo numbers in other areas show either little change or a slight increase. A recent trend toward increasing density was recorded for topi.

In areas within which law enforcement activities have increased since

1988, densities of several resident wildlife species have also increased. Conversely, in areas that are both close to the protected area boundaries and less accessible to patrols (especially vehicle patrols), the already low wildlife densities have declined still further. Along the borders of the western corridor of the national park wildlife densities are generally high, and a steep gradient from high to low wildlife densities more or less coincident with the boundary is evident. It is suggested that greater access to much of this area has enabled the existing, relatively low, levels of staffing and infrastructure to maintain an effective offensive against the demand from illegal meat hunters. Similar staff levels in the northwest have been unable to halt an observed decline in numbers. It is suggested that much of this decline has occurred in areas that are, for a variety of different reasons, still relatively inaccessible to law enforcement patrols.

Very low densities of wildlife, together with a continuing decline in numbers along parts of the western edge of the protected areas, were shown to be partly correlated with the presence of an agricultural or agro-pastoral form of land use outside the protected area and with the nature of the intervening buffer zone. Conversely, a pastoral land use to the east of the national park has resulted in significant wildlife populations and no marked changes in density on either side of the protected area boundaries.

An expansion of human populations adjacent to the western boundaries of the protected areas is likely to be reflected in a similar increase in the demand for wildlife meat. Various strategies have been proposed to combat this problem, including the introduction of a legalized and sustainable system of offtake quotas. This study, together with that of Campbell and Hofer (chap. 25), shows that the demand for wildlife meat far outstrips the current capacity of the game reserves to supply it. The implications for regional conservation are important. Development programs aimed at improving the quality, quantity, and availability of domestic sources of meat and other protein (e.g., fish) must begin to play an important role within the overall conservation effort in the Serengeti ecosystem.

ACKNOWLEDGMENTS

The authors would like to thank the many people who assisted with aerial survey work. Front and rear seat observers in aerial surveys between 1988 and 1992 were: H. T. Dublin, B. N. Mbano, J. Hando, S. Kihaule, J. ole Kuwai, S. Lelo, B. Lyimo, J. K. Magombe, G. K. Makumbule, C. Mdoe, G. B. Mngon'go, C. Mufungo, B. Mwasaga, H. Mwinyijuma, H. M. Nkya, M. Obed, V. Runyoro, E. Severre, F. Silkiluwasha, and W. Summay, all of whom have spent many uncomfortable hours in light aircraft during the course of this work. Invaluable assistance was provided by the Serengeti National Park ecologist, A. R. Kajuni, while others also assisted in many ways.

Much of the data was entered on computer by C. Mufungo. S. A. Huish spent many days checking and validating all of the many thousands of data points collected during these surveys and commented on the first draft of this chapter. Special mention must be made of the pilots, G. Bigurube, J. Hando, R. Lamprey, B. N. M. Mbano, S. Tham, and C. Trout, who undoubtedly had the most difficult and demanding task of all. We also thank the Director General, Tanzania National Parks, the Director of the Wildlife Division, the Director General of the Serengeti Wildlife Research Institute, the Principal Park Warden of Serengeti National Park, and the management and wardens of Serengeti National Park. The work was financed by the Frankfurt Zoological Society, whose support and long-term funding of conservation monitoring enabled these data to be gathered.

REFERENCES

Borner, M. 1981. Black Rhino disaster in Tanzania. *Oryx* 16:59–66.

Borner, M., Fitzgibbon, C. D., Borner, M., Caro, T. M., Lindsay, W. K., Collins, D. A., and Holt, M. E. 1987. The decline of the Serengeti Thomson's gazelle population. *Oecologia* 73:32–40.

Borner, M., and Maragesi, B. 1985. Land-use pressure around the Serengeti National Park. In *Towards a regional conservation strategy for the Serengeti.* Nairobi: IUCN.

Bureau of Statistics. 1988. *Population census: Preliminary report.* Dar es Salaam, Tanzania: Ministry of Finance, Economic Affairs and Planning.

Campbell, K. L. I. 1987. *Quarterly Report, June 1987.* Arusha, Tanzania: Serengeti Ecological Monitoring Programme.

———. 1988. *Programme report, March 1988.* Arusha, Tanzania: Serengeti Ecological Monitoring Programme.

———. 1989. *Programme report, September 1989.* Arusha, Tanzania: Serengeti Ecological Monitoring Programme.

———. 1992. *Census of buffalo and elephant in the Serengeti ecosystem, May 1992.* Arusha, Tanzania: Tanzania Wildlife Conservation Monitoring.

Campbell, K. L. I., S. A. Huish, and A. R. Kajuni. 1991. *Serengeti National Park management plan, 1991–1995.* Arusha, Tanzania: Tanzania National Parks.

Clarke, R. 1986. *The handbook of ecological monitoring.* Oxford: Oxford University Press.

Drummer, T. D., and McDonald, L. L. 1987. Size bias in line transect sampling. *Biometrics* 43:13–21.

Dublin, H. T., and Douglas-Hamilton, I. 1987. Status and trends of elephants in the Serengeti-Mara ecosystem. *Afr. J. Ecol.* 25:19–33.

Dublin, H. T., Sinclair, A. R. E., Boutin, S., Anderson, E., Jago, M., and Arcese,

P. 1990. Does competition regulate ungulate populations? Further evidence from the Serengeti, Tanzania. *Oecologia* 82:283–88.

Dublin, H. T., Sinclair, A. R. E., and McGlade, J. 1990. Elephants and fire as causes of multiple stable states in the Serengeti-Mara woodlands. *J. Anim. Ecol.* 59:1147–64.

Eastman. R. 1990. *Idrisi, a grid-based geographic analysis system, version 3.2, November 1990.* Clark University, Graduate School of Geography.

Golden Software. 1990. *Surfer, version 4.* Golden, Colo.: Golden Software Inc.

Graham, A., and Bell, R., 1969. Factors influencing the countability of animals. *E. Afr. Agric. For. J.* 34 (special issue):38–43.

Grimsdell, J. J. R. 1979. Changes in populations of resident ungulates. In *Serengeti: Dynamics of an ecosystem,* ed. A. R. E. Sinclair and M. Norton-Griffiths, 353–59. Chicago: University of Chicago Press.

Jolly, G. M. 1969. Sampling methods for aerial censuses of wildlife populations. *E. Afr. Agric. For. J.* 34:46–49.

Maddock, L. 1979. The "migration" and grazing succession. In *Serengeti: Dynamics of an ecosystem,* ed. A. R. E. Sinclair and M. Norton-Griffiths, 104–29. Chicago: University of Chicago Press.

McNaughton, S. J. 1988. Mineral nutrition and spatial concentration of African ungulates. *Nature* 334:343–45.

Norton-Griffiths, M. 1973. Counting the Serengeti migratory wildebeest using two-stage sampling. *E. Afr. Wildl. J.* 11:135–49.

———. 1978. *Counting animals.* 2d ed. Handbook no. 1. Techniques in African Wildlife Ecology. Nairobi: African Wildlife Foundation.

———. 1988. Aerial point sampling for land use surveys. *J. Biogeog.* 15:149–56.

Otto, M. C., and Pollock, K. H. 1990. Size bias in line transect sampling: A field test. *Biometrics* 46:239–45.

Pollock, K. H., and Kendall, W. L. 1987. Visibility bias in aerial surveys: A review of estimation procedures. *J. Wildl. Mgmt.* 51:502–10.

Sinclair, A. R. E. 1972. Long-term monitoring of mammal populations in the Serengeti: Census of non-migratory ungulates, 1971. *E. Afr. Wildl. J.* 10:287–97.

———. 1973. Population increases of buffalo and wildebeest in the Serengeti. *E. Afr. Wildl. J.* 11:93–107.

———. 1979. The eruption of the ruminants. In *Serengeti: Dynamics of an ecosystem,* ed. A. R. E. Sinclair and M. Norton-Griffiths, 82–103. Chicago: University of Chicago Press.

———. 1987a. *Long-term monitoring in the Serengeti-Mara: Trends in wildebeest and gazelle populations.* Typed manuscript. Arusha, Tanzania: SEMP, February 1987.

———. 1987b. *Long-term monitoring of mammal populations in the Serengeti-Mara, 1986.* Typed manuscript. Arusha, Tanzania: SEMP, February 1987.

Sinclair, A. R. E., Dublin, H., and Borner, M. 1985. Population regulation of the Serengeti wildebeest: A test of the food hypothesis. *Oecologia* 65:266–68.

Sinclair, A. R. E., and Norton-Griffiths, M. 1982. Does competition or facilitation regulate migrant ungulate populations. *Oecologia* 53:364–69.

SRCS. 1991. *Serengeti Regional Conservation Strategy, a plan for conservation and development in the Serengeti region: Phase II final report and Phase III action plan.* United Republic of Tanzania, Ministry of Tourism, Natural Resources and Environment.

Stronach, N. 1988. *The management of fire in Serengeti National Park: Objectives and prescriptions.* Arusha, Tanzania: Tanzania National Parks.

Swynnerton, G. H. 1958. Fauna of the Serengeti National Park. *Mammalia* 22:435–50.

Watson, R. M., Freeman, G. H., and Jolly, G. M. 1969. Some indoor experiments to simulate problems in aerial censusing. *E. Afr. Agric. For. J.* 34 (special issue):56–59.

Watson, R. M., Graham, A. D., and Parker, I. S. C. 1969. A census of large mammals of Loliondo Game Controlled Area, northern Tanzania. *E. Afr. Wildl. J.* 7:43–59.

SEVEN

Long-Term Trends in the Herbivore Populations of the Ngorongoro Crater, Tanzania

Victor A. Runyoro, Heribert Hofer,
Emmanuel B. Chausi, and Patricia D. Moehlman

The Ngorongoro Crater is renowned for the abundance and diversity of its wildlife. Although a geographically distinct ecological unit, it is not an isolated system but is linked to the adjacent Serengeti plains and Crater Highlands by the seasonal migration of several species of herbivores (Estes and Small 1981) and by emigration and immigration of predators (Kruuk 1972; Pusey and Packer 1987). The Ngorongoro Conservation Area Authority (NCAA), the College of African Wildlife Management, and research scientists have monitored the wildlife populations of Ngorongoro Crater since 1963 (Turner and Watson 1964; Ngorongoro Conservation Unit 1967; Rose 1975; Estes and Small 1981). Since its inception in 1987, the Ngorongoro Ecological Monitoring Programme (NEMP) has been responsible for conducting wet and dry season censuses. The complete data set covers 30 years (1963–1992), permitting us to examine long-term population trends and assess the stability of a multispecies wild herbivore community. Analyses of wet and dry season fluctuations in numbers reveal those species that are principally resident in the Crater and those that seasonally utilize the Highlands or move down to the short-grass plains of the Serengeti (Estes and Small 1981; Boshe 1988). We also assess the long-term consequences for the wild herbivore community of the removal of resident pastoralists and their livestock from the Crater in 1974 (NCAA 1985; Fosbrooke 1986). If the presence of livestock had depressed populations of wild herbivores prior to 1974, wild herbivore population sizes should have increased after the removal of pastoralists. Our analysis demonstrates that the removal of pastoralist livestock and concomitant changes in rangeland management had complex consequences.

STUDY AREA

The Ngorongoro Crater (3°10' S, 35°35' E) is a caldera covering an area of approximately 300 km^2, of which 250 km^2 forms the floor; the remainder is on the steeply rising slopes (Herlocker and Dirschl 1972). The Crater walls are steep except in the northeast and generally rise 500 m above the Crater floor (1,737 m above sea level). The vegetation and the geology of the crater are described by Herlocker and Dirschl (1972) and Anderson and Herlocker (1973). The Crater has its largest catchment basin in the Crater Highlands (Homewood and Rodgers 1991) and receives water from Lalratati and Edeani streams and Lerai spring from Oldeani Mountain. Lerai spring has been measured as producing 1 million gallons of water per day in the dry season (Estes and Small 1981). Olmoti crater provides runoff for two rivers in the north, the Laawanay and Lemunga, which feed the Mandusi swamp and Lake Magadi. Seneto spring feeds Seneto swamp and Lake Magadi. Lljoro Nyuki River flows into the Crater from Losirua in the northeast and into Gorigor swamp. Dry season water availability is therefore dependent on the Highlands' catchment system.

The wildlife of the Ngorongoro Crater has had a protected status since 1921, and the area has been administrated by the Ngorongoro Conservation Unit since 1959 and by the Ngorongoro Conservation Area Authority since 1975 as part of a larger (8,292 km^2) multiple land use area.

METHODS

The Ngorongoro Crater has been divided into six blocks (fig. 7.1) that have been used since the 1960s for censusing total numbers of large animals. The blocks cover the entire Crater floor, except for the inaccessible areas of Lake Magadi in the southwest, Lerai Forest in the south, and Maandusi and Gorigor swamps in the northwest and southeast. Ground censuses are conducted by one censusing team per block, consisting of one driver, one recorder, and one observer, in a vehicle driving along line transects 1 km apart. Since 1987 the six teams have been provided with a 1:50,000 map marked with transects, a compass, binoculars, and a mechanical counter. Each block takes 6 to 8 hours to complete, and all blocks are censused simultaneously to minimize counting errors due to movement of animals (NEMP 1987, 1988a,b, 1989, 1990, 1991). In 1986, 200 m wide strip samples were incorporated into the total ground census for counting Thomson's gazelle, Grant's gazelle, and warthog, and analyzed with Jolly's method number 2 (Norton-Griffiths 1978). Strip counts were discontinued for Grant's gazelle in 1987 due to unacceptably large confidence limits. The drivers of census vehicles had serious prob-

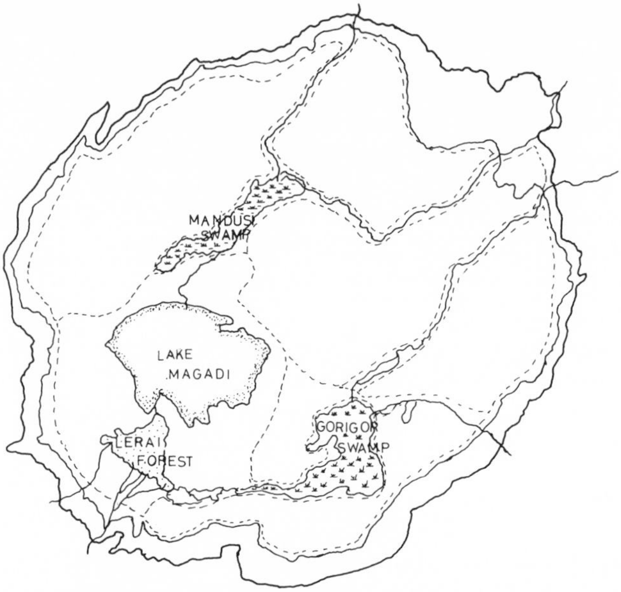

Figure 7.1 Ngorongoro Crater, showing census blocks (dashed lines).

lems with maintaining a straight compass course during the April wet season counts, and therefore wet season strip counts were discontinued after 1989. In 1992 it was decided to continue driving the transects but to count all species as total counts. Aerial counts were conducted as total counts with the exception of one systematic reconnaissance flight (SRF, a strip sample count technique) in 1980 (Ecosystems Ltd. 1980). The methodology of earlier aerial total counts is described in detail in Norton-Griffiths (1978) and SEMP (1986).

Unpublished records of NCAA and NEMP (1987–1992) provided the bulk of the data on animal numbers. Information on early censuses was taken from Estes and Small (1981), cross-checked against original records in the NCAA archives, and corrected where necessary. An unpublished report by Ecosystems Ltd. (1980) provided additional aerial census information. Data for the first census in 1963 were taken from the annual report of the Ngorongoro Conservation Unit (1967). A major gap in population estimates occurs between 1978 and 1986, with the exception of the count by Ecosystems Ltd. (1980). Body masses of mammalian herbivores were taken from Sinclair and Norton-Griffiths (1979), mass for ostrich from Brown, Urban, and Newman (1980). Total wild herbivore biomass was calculated as the sum of wildebeest, zebra, buffalo, Grant's gazelle, Thomson's gazelle, kongoni, eland, waterbuck, elephant, black rhinoceros, and ostrich. Warthog, hippopotamus, and bohor reedbuck were not included in total wild herbivore biomass because of small samples or lack of information. Warthogs have been counted only from 1980 onward. Ground counts were inappropriate for hippos, and there were only seven aerial counts that provided information on their population (see appendix 7.1). Elliott and McTaggart Cowan (1978) state that there were several hundred reedbuck in the swamps, but there is no information available to confirm this. Excluding warthogs and hippos from biomass is of little consequence, as the sum of their mean biomass constitutes only 1.3% of total herbivore biomass. We classified the herbivores into four functional categories representing the key species and the variation in feeding ecology encountered: (1) wildebeest, (2) buffalo, (3) other grazers (Thomson's gazelle, kongoni, eland, zebra, ostrich), and (4) mixed grazers/browsers (Grant's gazelle, waterbuck, black rhinoceros, elephant). Classification of species followed the categories developed for the nearby Serengeti (Jarman and Sinclair 1979), Tarangire (McNaughton and Georgiadis 1986), and Lake Manyara (Prins and Douglas-Hamilton 1990).

Masai pastoralists occupied permanent bomas on the Crater floor until the beginning of the dry season 1974. Because data on their livestock holdings (cattle, goats, donkeys) were available only for 1963–1967, we had to estimate livestock biomass for 1963–1974. Between 1963 and 1966 total livestock biomass constituted on average 12.72 ± 2% ($n = 4$) of the total wildlife biomass in terms of stock units (Ngorongoro Conservation Unit 1967), almost all cattle. Combined total biomass for wildlife and livestock was therefore estimated for 1963–1974 as 112.72% of total wild herbivore biomass.

Long-term rainfall data were not available for the Crater floor. We therefore used the long-term rainfall data for the Crater rim measured at Ngorongoro headquarters. Statistical analyses were performed using SYSTAT 5.0 (Wilkinson 1990), following the procedures recommended by Conover (1980) and Sokal and Rohlf (1981). Statistics are given as means ± standard errors, and probabilities are for two-tailed tests unless indicated otherwise. If more than one census occurred during a season, data were averaged to provide a mean estimate. Data from the 1963 census were excluded from statistical analysis as details on counting methods were not available. The final sample size thus comprised data for 39 seasons. For Thomson's gazelle and Grant's gazelle, only total counts (ground or aerial) were included because differences between sample and total counts were unacceptably high. We first checked on possible temporal interdependencies by treating each sequence of population numbers as a time series and calculating correlograms for each species. None of the species exhibited any higher-order temporal patterns. First-lag autocorrelation coefficients were significant for kongoni, buffalo, eland, ostrich, Thomson's gazelle, and waterbuck, and not significant for elephant,

Grant's gazelle, black rhinoceros, wildebeest, and zebra. Thus for half the species no interdependencies were found; for the others only consecutive numbers (mostly seasonal changes) showed any resemblance to each other. We then checked for independence of data points by testing the residuals of general linear models (see below) for autocorrelation with the Durbin-Watson statistic. With the exception of kongoni, residuals were not autocorrelated. As the autocorrelation in kongoni is moderate and the uncorrected *P* values of the significant effects in kongoni are $< .0001$, adjustment of *P* values would change little in the significance of effects. We therefore report unadjusted *P* values below. We assessed the significance of changes in population size by using general linear models that incorporated as effects (1) season (dry or wet), (2) the presence/absence of permanent Masai bomas (and their livestock), (3) an interaction term expressing the change in strength of seasonal differences before and after the removal of the Masai in 1974, (4) an interaction term expressing the significance of changes in linear trends before and after the removal of the Masai, and (5) rainfall. In the models for wildebeest and total herbivore biomass we also included an interaction between season and year, as preliminary analyses showed that there was a strong decline during the dry but not the wet season. In all general linear models we confirmed the normal distribution of residuals with the Lilliefors test as implemented (with corrected *P* values: Wilkinson 1990) in SYSTAT.

We calculated a stability index for the wet and dry season herbivore communities following McNaughton (1978). The index is $s = i/(nc)^{-1/2}$, where n is the number of species (11), i is the mean value of significant (at $P = .05$) Spearman rank correlations (r_s) between the population numbers of all possible pairs of species, and c is the proportion of all r_s values that were significant.

RESULTS

Between 1963 and 1992 a total of 50 counts of the herbivores in the Ngorongoro Crater were undertaken, 28 during the wet and 22 during the dry season. Population estimates are available for most years except 1980–1986. Regular wet and dry season counts started in 1968 (appendix 7.1). There were 24 censuses prior to the removal of Masai pastoralists (period I) and 26 censuses since (period II).

Long-Term Population Trends

Wildebeest and buffalo exhibited the most dramatic changes in population numbers during the 30-year period of wildlife censuses (fig. 7.2). Wildebeest increased until approximately 1974 but then declined after

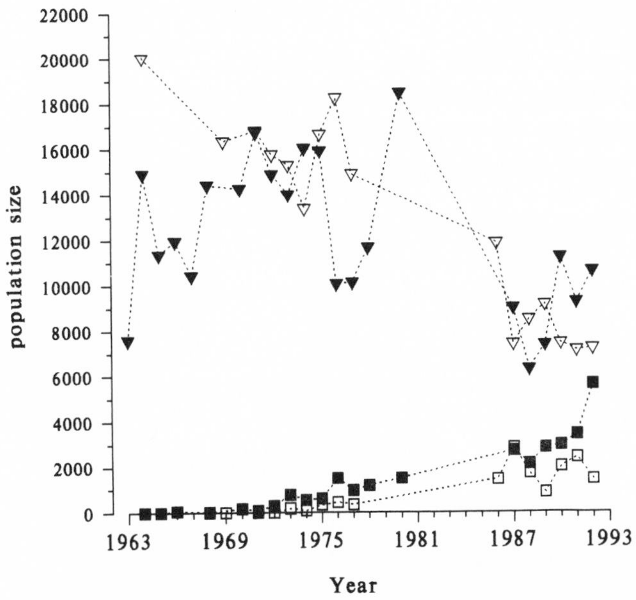

Figure 7.2 Numbers of wildebeest (triangles) and buffalo (squares) in wet (solid symbols) and dry seasons (open symbols).

the Masai were removed from the Crater (table 7.1). The wildebeest decline was pronounced during the dry season but nonsignificant during the wet season. Wildebeest were the only species for which rainfall had a significant, negative effect ($F_{1,31} = 7.22$, $P = .01$). Rainfall did not affect numbers of any other species, nor the dynamics of any functional biomass category. The migratory wildebeest population in the Serengeti experienced a threefold increase between 1961 and 1973 (Sinclair 1979). In order to provide a comparable estimate of Ngorongoro wildebeest numbers for the same period, we regressed the Ngorongoro wildebeest numbers on year during period I and compared the regression estimate for 1961 with that for 1973. The estimated 1973 figure of 15,227 was 38% higher than the 1961 figure of 11,068, a comparatively modest increase. The average number of wildebeest during period II was significantly lower than during period I (table 7.1). Buffalo were virtually absent from the Crater prior to 1960 and appeared in slowly increasing numbers, reaching a substantial level by 1973. The increase in buffalo numbers significantly accelerated after the removal of the Masai (table 7.1), breaching the 5,000 mark during the wet season 1992 (appendix 7.1).

All other species except ostrich, waterbuck, and elephant showed significant population changes in one or both periods. During period I, spe-

Table 7.1 Factors influencing population changes of herbivores in the Ngorongoro Crater.

		Difference in numbers before and after 1974?		Change in slope of linear trend from before to after 1974?		Season			
Species	r^2 of model	Direction of change	*P*	Trends[a]	*P*	Season with high population	*P*	Interaction?[b]	*P*
Wildebeest[c]	.71	Decline	***	B: increase A: decline	***	Dry	*	No	NS
Buffalo	.76	Increase	***	B: slight increase A: sharp increase	***	Wet	*	Yes[d]	*
"Other grazers"									
Kongoni	.51	Increase	***	B: increase A: decline	***	—	NS	No	NS
T. gazelle	.68	Decline	***	B: increase A: decline	***	—	NS	No	NS
Eland	.72	Decline	***	B: constant A: decline	***	Wet	**	No	NS
Zebra	.32	Decline	.054	B: constant A: decline	.053	Dry	.057	Yes[e]	.051
Ostrich	.21	—	NS	No	NS	Wet	*	No	NS
"Mixed grazers/browsers"									
G. gazelle	.57	Decline	***	B: increase A: decline	***	—	NS	No	NS
Rhino	.29	Decline	NS	B. constant A: decline	***	Wet	***	No	NS
Waterbuck	.50	—	NS	No	NS	Wet	**	No	NS
Elephant	.37	—	NS	No	NS	Wet	**	No	NS

Note: Significance assessed by general linear models (*P* values: *< .05; **< .01; ***< .001).
[a]B/A, before/after 1974
[b]Did strength of seasonal effect change from period I to period II?
[c]Also a significant ($P = .01$) interaction between time and season: numbers decline strongly through time during dry season but weakly (and not significantly) during wet season.
[d]Seasonal differences moderate before 1974, substantial after 1974.
[e]Before 1974 dry season higher than wet season, after 1974 both equal.

cies either increased in number (kongoni, Thomson's gazelle, Grant's gazelle) or showed no trend (eland, zebra, black rhinoceros). Apart from buffalo, kongoni was the only other species with a higher average during period II. All other grazers had a significantly lower average during period II, except ostrich. Among the mixed grazers/browsers, waterbuck and elephant showed no change in mean population size, while Grant's gazelle and black rhino had a lower average during period II (table 7.1). During period II, kongoni, Thomson's gazelle, eland, zebra, Grant's gazelle, and black rhinoceros declined significantly.

The seven aerial total counts of hippopotamus during 1964–1988 showed a highly significant increase in population size ($Y = -3{,}613.6 + 1.852X$, $F_{1,5} = 29.9$, $P < .003$, $r^2 = .86$) with a mean of 38 ± 7. There was no discernible trend in warthog population size between 1980 and 1992 (mean 115 ± 24).

Seasonal Population Trends

Our results on seasonal changes mostly confirmed Estes and Small's (1981) earlier analysis for 1964–1978. Buffalo, eland, ostrich, waterbuck, rhino, and elephant had larger populations in the Crater during the wet season. Wildebeest and zebra were the only species with a larger population during the dry season (table 7.1). In contrast to Estes and Small (1981), we did not find significant seasonal changes in Thomson's gazelle. When we ran a general linear model on the Thomson's gazelle data of Estes and Small (1981), seasonal change was also not a significant factor. Although more wildebeest and zebra were present during the dry season, it was during this season that the long-term declines in their populations were most pronounced. After 1990, the seasonal pattern in wildebeest and zebra numbers appears to be reversed, with higher numbers during the wet season. Only in buffalo and zebra did the magnitude of seasonal changes vary significantly between periods I and II (table 7.1). In buffalo, there were few seasonal differences during period I but substantial differences during period II, while in zebra the seasonal differences during period I vanish in period II.

Long-Term Biomass Trends

Despite marked changes in the population sizes of individual species, both total wild herbivore biomass (fig. 7.3) and total herbivore biomass (including livestock during period I) showed considerable variation but no significant trend over time, and the two estimates varied little from each other. Total wild herbivore biomass was not significantly affected by the removal of pastoralists (table 7.2). Wet season biomass was significantly higher than dry season biomass. Furthermore, dry season biomass declined significantly over time while wet season biomass showed a slight trend to increase (table 7.2). Mean total wild herbivore biomass was 3,146,210 ± 95,228 kg (N = 27). Mean total herbivore biomass was 3,294,450 ± 103,615 kg (N = 27), equivalent to 10,982 kg/km^2. This figure is lower than that for nearby Lake Manyara National Park but higher than the figure for the Serengeti in the mid-1970s (table 7.3). There was a positive association between biomass (kg/km^2) and rainfall in fourteen East African wildlife and pastoral areas ($r_s = 0.81$, $P < .01$, table 7.3), with Lake Manyara National Park a significant outlier with a higher than expected biomass, given its precipitation.

The stability index, s, was 0.39 for the wet season and 0.75 for the dry season. Both values are well below 1, indicating that the herbivore community was stable in both seasons.

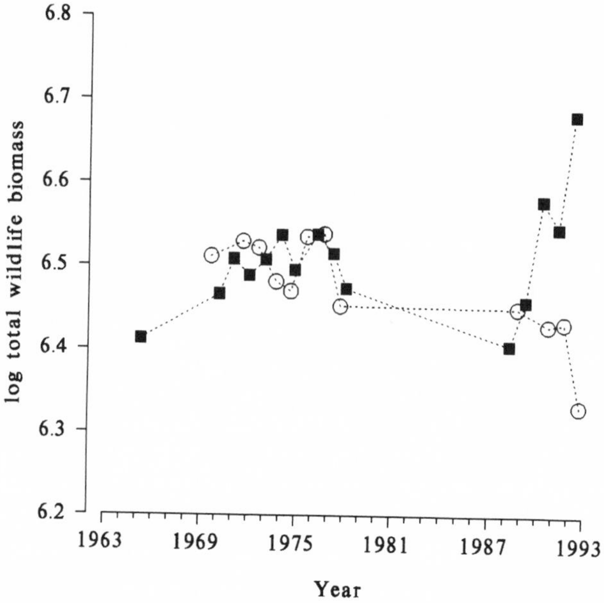

Figure 7.3 Log_{10} of total herbivore biomass (kg) during wet (solid squares) and dry (open circles) seasons.

Table 7.2 Factors influencing biomass changes of herbivores in the Ngorongoro Crater.

		Difference in numbers before and after 1974?		Change in slope of linear trend from before to after 1974?		Season			
Category	r^2 of model	Direction of change	*P*	Trends[a]	*P*	Season with high biomass	*P*	Interaction?[b]	*P*
Wildebeest[c]	0.71	Decline	***	B: increase A: decline	***	Dry	*	No	NS
Buffalo	0.76	Increase	***	B: slight increase A: sharp increase	***	Wet	*	Yes[d]	*
Other grazers[e]	0.47	Decline	**	B: increase A: decline	**	—	NS	No	NS
Mixed grazers[f]	0.51	—	NS	No	NS	Wet	***	No	NS
Total wild herbivores[g]	0.44	—	NS	No	NS	Wet	**	No	NS

Note: Significance assessed by general linear models (*P* values: *< .05; **< .01; ***< .001).
[a]B/A, before/after 1974
[b]Did strength of seasonal effect change from period I to period II?
[c]Also a significant (P = .01) interaction between time and season: biomass declined strongly through time during dry season but weakly (and not significantly) during wet season.
[d]Seasonal differences moderate before 1974, substantial after 1974.
[e]Kongoni, Thomson's gazelle, eland, zebra, ostrich.
[f]Grant's gazelle, black rhinoceros, waterbuck, elephant.
[g]Also a significant (P = .01) interaction between time and season: biomass declined strongly through time during dry season but increased weakly (and not significantly) during the wet season.

Table 7.3 Biomass of large herbivores and mean annual rainfall in wildlife and pastoral areas in East Africa.

Locality	Country	Biomass (kg/km²)	Rainfall (mm/year)	Reference[a]
Wildlife areas				
Rwenzori National Park	Uganda	19,928	1,010	1
Lake Manyara National Park	Tanzania	16,933	581	2
Bunyoro North	Uganda	13,261	1,150	1
Ngorongoro Crater	Tanzania	10,982	630	3
Serengeti National Park	Tanzania	8,268	803	4
Amboseli Game Reserve	Kenya	4,848	350	1
Nairobi National Park	Kenya	4,824	844	1
Samburu-Isiolo	Kenya	2,018	375	1
Pastoral areas				
Kaputei District	Kenya	7,884	710	5
Samburu District	Kenya	6,514	500	5
Garissa District	Kenya	3,818	398	5
Turkana District	Kenya	2,406	330	5
Mandera District	Kenya	1,901	228	5
Wajir District	Kenya	1,151	218	5

[a]1, Coe, Cumming, and Phillipson 1976; 2, calculated from Prins and Douglas-Hamilton 1990; 3, this study; 4, Maddock 1979; 5, Watson 1972, in Jewell 1980.

Trends in Biomass of Functional Categories of Herbivores

Before the removal of pastoralists, all categories of grazers (wildebeest, buffalo, and "others") showed a significant increase (table 7.2). During period II, however, wildebeest and other grazers declined while buffalo increased rapidly. In contrast to pure grazers, mixed grazer/browser biomass remained constant. This group showed pronounced seasonal changes with a significantly higher biomass in the Crater during the wet season (table 7.2).

During the past 30 years, wildebeest moved from being the dominant species (50%–65% of total herbivore biomass) to second position (35%–45%; fig. 7.4). Other grazers of similar or smaller body size followed a similar downward trend. By contrast, buffalo biomass was initially low (0%–20%), but rapidly gained in importance during period II and is now the largest biomass component during the wet season (55%). Mixed grazers/browsers contributed a small, constant proportion to the overall biomass throughout the period (fig. 7.4).

DISCUSSION

During the past 30 years there have been significant changes in the population sizes of many herbivores in the Ngorongoro Crater. Wildebeest declined significantly and are no longer the dominant species in terms of biomass. Most other ruminant grazers also declined significantly. The exception is buffalo, which, virtually absent from the Crater 30 years ago,

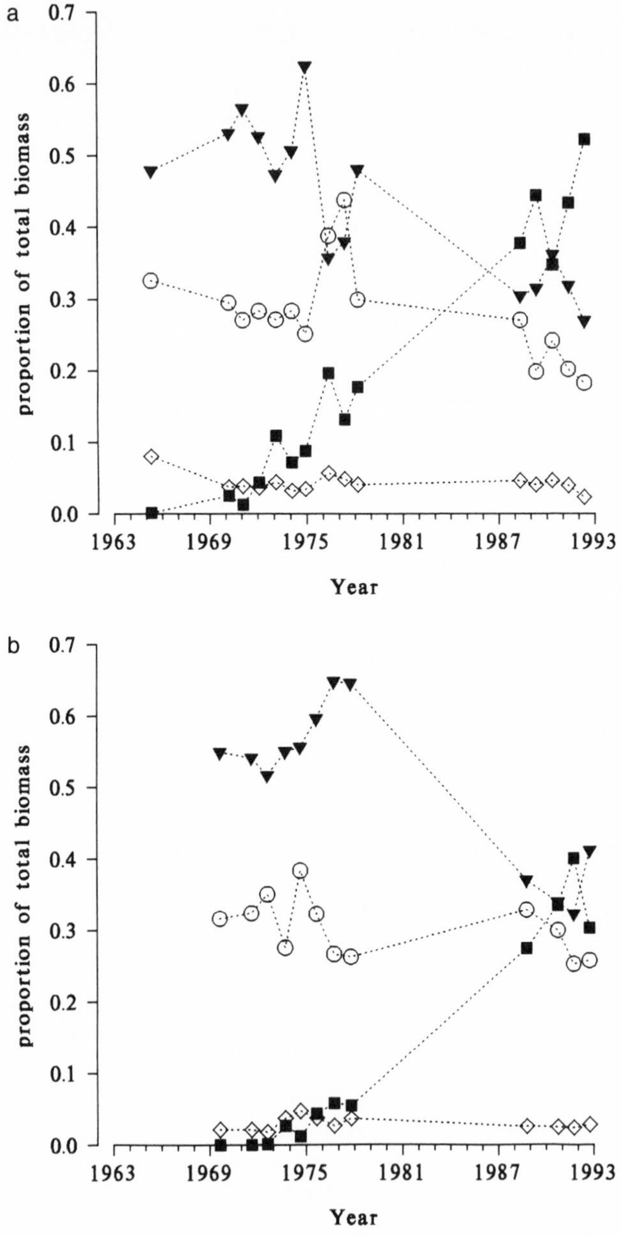

Figure 7.4 Proportion of total herbivore biomass of several functional categories of herbivores during (*a*) wet and (*b*) dry seasons: wildebeest, solid triangles; buffalo, solid squares; other grazers, open circles; mixed grazers/browsers, open diamonds.

is now the dominant species in terms of biomass. At the same time, nonruminant grazers (zebra and ostrich) changed comparatively little. Mixed grazers/browsers also changed little in biomass, although Grant's gazelle and black rhino experienced significant declines in population size. Despite these changes in population sizes of single species, total herbivore biomass of the Crater showed no significant change. Neither interactions with adjacent ecosystems (exception: buffalo), nor disease, poaching (exception: black rhinoceros), or predation are sufficient to explain the changes in population numbers since the removal of the Masai from the Crater, as we will discuss below.

Interactions with Adjacent Ecosystems

Buffalo were rarely seen in the Crater until the early 1970s, and were thought to be intermittent visitors from the Olmoti highlands (Rose 1975; NCAA 1985). Their arrival in the Crater may have been due to increasing pressure from agricultural communities to the east of the conservation area in Oldeani and Karatu (NCAA 1985). When buffalo visited farmland adjacent to the Northern Highlands Forest Reserve in the early 1970s, the NCAA took crop protection measures, and many buffalo were killed (NCAA 1985). Other buffalo presumably moved out of the forest reserve into the Crater, where they were protected, and stayed.

The importance and destinations of seasonal movements are not known for any of the herbivore species present in the Crater, although the significance of the seasonal changes in several species indicates that such movements take place. It is conceivable that the Crater could be populated each season by a random selection of individuals from species in the massive Serengeti "pool." However, the relatively high degree of habituation by Crater herbivores to tourist vehicles, in marked contrast to the behavior of the Serengeti migratory herds, suggests that most individuals of at least some species are residents.

Kabigumila (1988) cautioned that if elephant movement corridors were blocked and elephants were unable to move out of the Crater, increased numbers could have an adverse affect on Lerai Forest in the Crater. Although the present Crater population of black rhinoceros is sedentary (Goddard 1967; Kiwia 1983, 1986, 1989; NEMP 1991), the population was once contiguous with animals found in Olduvai Gorge and to the west in Serengeti National Park (Frame 1980).

Factors Affecting Herbivore Populations

Vegetation Structure and Fire. As large-bodied herbivores, buffalo require a lot of food, but its quality can be lower and can contain a higher proportion of fiber than what is required by the smaller wildebeest and Thomson's gazelle (Sinclair and Norton-Griffiths 1979; Demment and van Soest 1985). Buffalo prefer long grass, select for a high ratio of leaf to stem (Sinclair 1977), and are as tolerant as zebra (a nonruminant) of tall, fibrous grasses (Bell 1970). Before 1974, pastoralists managed the available forage in the Crater with a combination of livestock grazing and fire. This type of rangeland management favors shorter grasses with a larger proportion of palatable species (Gichohi 1990; Mwalyosi 1992). The removal of Masai pastoralists and their livestock must have affected vegetation structure and species composition, as forage has become increasingly rank and of lower quality, particularly in medium-height grassland (NEMP 1989). Wildebeest prefer shorter, less fibrous grasses than do buffalo, and as the structure and quality of grasses changed, wildebeest would have been at a disadvantage. Data on the changes in vegetation species composition and structure in the Crater from 1963 to 1987 are not available. The Ngorongoro Ecological Monitoring Programme started collecting such data in 1987, which will permit a more comprehensive exploration of the factors underlying herbivore population changes in the future.

If changes in forage quality were important, then we would expect (1) smaller-bodied ruminant grazers to decline during period II and (2) nonruminant grazers to be less affected. Our results provide some, but not unequivocal, support for this hypothesis. Thomson's gazelle and wildebeest declined strongly during period II, while zebra showed a moderate decline approaching statistical significance. Because these three species together form a substantial component of total herbivore biomass, the significant dry season decline in total herbivore biomass would follow from the decline in their numbers. As nonruminants, zebras have the option of utilizing larger quantities of lower-quality forage to meet their nutritional requirements (Demment and van Soest 1985). On a seasonal basis, the smallest-bodied ruminant (Thomson's gazelle, 15 kg) declined significantly in both seasons, while in wildebeest (123 kg, ruminant) and zebra (200 kg, nonruminant) the decline was restricted to the dry season, when limitation in high-quality forage is most pronounced. In contrast to the prediction, kongoni were the only grazers with higher numbers during period II; the reason for this is unknown.

Elephant numbers remained constant in the Crater over the last 30 years. Only male elephants, mostly large older bulls, were present (Kabigumila 1988). During the dry season they fed mainly on tree browse and sedges, while during the wet season they ate forbs and grasses (Kabigumila 1988). Female elephants did not use the Crater, presumably because food availability was insufficient to meet their higher nutritional needs for reproduction (Kabigumila 1988). The Crater probably acted mainly as a spillover area for males from the surrounding Highlands. Although Lerai Forest was the preferred habitat for elephants, their destruction of

Acacia xanthophloea, the dominant tree species, did not appear to exceed recruitment rates of trees (Kabigumila 1988).

Black rhinoceros mostly grazed in open grasslands during the wet season and browsed in swamps and in Lerai Forest during the dry season (Kiwia 1983, 1986). The significant decline in both elephant and black rhinoceros numbers in the Crater from wet to dry season may in part be due to their greater utilization of swamp and forest habitat during the dry season. Black rhinoceros and elephants move in and out of the Crater, as documented by Kiwia (1983) and Kabigumila (1988), who followed individually recognized animals.

Disease. An outbreak of rinderpest in 1958 affected buffalo, wildebeest, and eland in the Crater Highlands, and in 1961 a serious outbreak affected yearling buffalo adjacent to the Crater (Machange 1988). The Ngorongoro Conservation Area Authority started the inoculation of cattle against rinderpest in the 1950s and eradicated the disease in cattle by the 1960s (Machange 1988). Inoculations continued as a yearly "rinderpest campaign." Since then the only outbreak was in 1982, affecting buffalo, eland, and giraffe, but not cattle (NEMP 1989). Unfortunately there were no animal counts during 1980–1986, but despite the losses from rinderpest during 1982, the buffalo population doubled. Rinderpest had also been responsible for heavy mortality in wildebeest and buffalo in the Serengeti ecosystem, but died out there in wildebeest in 1963 and in buffalo in 1964 (Sinclair 1979). This release may have contributed to the appearance of buffalo in the Crater and the adjacent highlands.

The parallel eradication of rinderpest in the Serengeti ecosystem and the Crater Highlands apparently released wildebeest populations. From 1963 to 1974 the Serengeti migratory wildebeest population increased by a factor of three (Sinclair 1979). During the same period, the sedentary wildebeest population centered in Ngorongoro Crater increased by only 38%. If both populations had been kept low by rinderpest prior to 1961, then this difference in rate of population change supports Fryxell, Greever, and Sinclair's (1988) hypothesis for fundamental differences between migratory and sedentary herbivores in the way their populations are regulated.

Predation. Population sizes and feeding ecology of the two main predators, spotted hyenas and lions, were studied by Kruuk (1972), Elliott and McTaggart Cowan (1978), Pusey and Packer (1987) and Hanby, Bygott, and Packer (chap. 15). Kruuk (1972) estimated the total spotted hyena population between 1964 and 1968 to be 430 animals. Recent information on spotted hyena population size is unavailable. Elliott and McTaggart Cowan (1978) estimated a resident population of 65 lions for 1970–

1972. Since 1975 the lion population has been stable, and numbers have fluctuated around 85 (Pusey and Packer 1987; Hanby, Bygott, and Packer, chap. 15).

In 1970–1972 the estimated annual percentage of wildebeest killed or scavenged by lions and hyenas was 14.6% (Elliot and McTaggart Cowan 1978), approximately equal to the recruitment rate (Kruuk 1972). Whether this ratio has changed during period II, when wildebeest have declined, is presently unknown. In the early 1970s Elliott and McTaggart Cowan (1978) estimated that the resident lion population would kill 15 buffalo per year. Therefore the effect of lions on the buffalo population appears to be negligible. The estimated annual kill of Ngorongoro Crater lions in 1970–1972 was 233 Thomson's gazelle and 238 zebra, and the annual percentage of those populations killed was 7.8% and 10.8% respectively (Elliot and McTaggart Cowan 1978). Zebra reproductive biology (Klingel 1969) is similar to that of feral equids in North America. Population modeling of those species indicates that equids have a potentially low rate of population growth; thus predation rates on the order observed by Elliot and McTaggart Cowan may be an important source of zebra mortality (Wolfe 1980).

Poaching. The Ngorongoro Crater has the highest density of naturally occurring black rhinoceros in Tanzania (Cumming, du Toit, and Stuart 1990). Poaching on black rhinoceros in the Crater began in 1972 and was a major source of mortality until the mid-1980s (Kiwia 1989). Kiwia's study indicated that rhinoceros sex and age class ratios in the 1980s were similar to those recorded in the 1960s. Intercalf intervals and reproduction have been good in the last 10 years, with no poaching since 1988 (NEMP 1991).

Long-Term Biomass Trends

Although there have been dramatic changes in the population sizes of individual species, total herbivore biomass has not changed significantly over the past 30 years. Our stability analysis using McNaughton's measure of community stability showed that the Crater wild herbivore community is an essentially stable multispecies assemblage. The computed values of the stability index were well below the critical threshold of 1.0 and lower than the value of 0.94 recorded in a long-term study of large mammalian herbivores in Lake Manyara National Park (Prins and Douglas-Hamilton 1990). Both localities are small in size and subject to semiarid rainfall regimes. In both studies annual precipitation fluctuated widely but showed no consistent trends during the study periods (NCU 1967; Chausi 1985; Prins and Douglas-Hamilton 1990; NEMP 1991). Both areas receive substantial water flow from a wide catchment area of adjacent forest highlands that experience higher rainfall. For instance, the

Crater floor received only 47% of the annual precipitation on the rim in 1989–1991. It is therefore not surprising that neither our study nor Prins and Douglas-Hamilton (1990) found any correlation of total herbivore biomass with annual rainfall. Both ecosystems contrast with the Serengeti, where Sinclair (1979) demonstrated significant changes in forage availability and herbivore numbers with dry season rainfall.

Despite this general picture of stability, the developments since 1990 indicate that major changes in the dynamics of the system are possible, and they emphasize the importance of long-term monitoring. Wet season biomass and the difference between wet and dry season biomass have increased dramatically since 1990 because (1) buffalo continued to increase and (2) wildebeest and zebra switched from a peak in biomass in the dry season to a peak in the wet season. We are not sure about the cause of this change. One hypothesis is that the increase in wet season numbers of wildebeest and zebra is due to a temporary influx of Serengeti migratory animals. This hypothesis would predict that the relative constancy of their numbers between successive dry seasons indicates the true size of the resident populations of the two species.

CONCLUSIONS

The Ngorongoro Crater, while to some extent a self-contained ecological unit, is clearly an integral part of the Crater Highlands catchment area (Estes and Small 1981). However, the importance of movements in and out of the Crater is unknown for most species. We need to know what factors influence these movements and which corridors and adjacent areas are vital to the Crater herbivores. In the last 30 years, the pastoralist population has doubled within the NCA (NEMP 1989), and as more people inhabit the Highlands surrounding the Crater, habitat will inevitably be lost.

The removal of pastoralists from the Crater may have been the key factor in the population changes observed in many herbivores, as the absence of their pasture management, including controlled burning, probably led to changes in forage quality. Experimental burning could be used to investigate how important this traditional tool of pasture management is for creating and maintaining grassland communities suitable for smaller-sized ruminants (Gichohi 1990) and wildebeest (Estes and Small 1981). Even so, we currently have no indication that the productivity of the Crater grasslands has either decreased or increased since the removal of pastoralists from the Crater, as we found no change in overall herbivore biomass.

There are two kinds of processes currently under way that could threaten the integrity of the Crater ecosystem. First, diversion of water from the Crater for human utilization—for example, for tourist lodges—may diminish the water supply to the Crater. Second, with the high utilization of the Crater floor by tourist vehicles, vegetation damage needs to be carefully assessed and restrictions on off-road driving enforced.

ACKNOWLEDGMENTS

We gratefully acknowledge the support of Juma Kayera, the former Conservator of the NCAA, and the hard work and substantial contributions of Sebastian Chuwa, Allan Kijazi, Mohamed Fadhili, Laban Maruo, and the staff of the Ngorongoro Ecological Monitoring Programme (NEMP). Samson Mkumbo, Paul J. Mshanga, and Michael Shoo provided assistance for the Crater counts. NEMP is supported by New York Zoological Society—Wildlife Conservation Society. Assistance for this study was provided by the Ngorongoro Conservation Area Authority (V. A. R., E. B. C.), NYZS/WCS (P. D. M.) and the Fritz-Thyssen-Stiftung and the Max-Planck-Gesellschaft (H. H.). We are particularly grateful for fruitful and informative discussions with Bjørn Figenschou and constructive comments by two anonymous referees.

Appendix 7.1 Population sizes of herbivores in the Ngorongoro Crater 1963–1992 from general Crater counts.

Year	Month and season	Source[a]	Wildebeest	Zebra	Thomson's gazelle	Grant's gazelle	Gazelles combined
1963	?	NCA	7,600 t	2,500 t	1,500 t	800 t	
1964	2 wet	TWA	14,922 a	5,038 a			2,310 a
1964	9 dry	WAT	20,038 a	6,078 a			2,373 a
1965	4 wet	WAT	11,352 a	4,145 a	1,229 a	1,273 a	
1966	3 wet	TBE	11,944 t	3,935 t			2,377 t
1966	11 wet	TLA	10,438 a	4,040 a			2,100 a
1966	11 wet	NCA					
1968	3 wet	DML	14,417 t	3,058 t			
1968	9 dry	MWE			4,269 t	1,376 t	
1969	6 dry	MWE	14,452 t	3,948 t	3,812 t	1,927 t	
1969	10 dry	MWE	18,238 t	5,734 t	3,769 t	1,478 t	
1970	1 wet	MWE	14,091 t	3,104 t	3,862 t	1,620 t	
1970	4 wet	MWE	14,422 t	4,267 t	2,576 t	1,798 t	
1970	12 wet	NCA	17,597 t	2,596 t	2,860 t	1,376 t	
1971	1 wet	NCA	15,853 t	5,405 t	3,800 t	1,024 t	
1971	8 dry	NCA	16,797 t	5,523 t	5,166 t	1,492 t	

Note: a, aerial total count; f, systematic reconnaissance flight (strip sample count, Norton-Griffiths 1978); s, strip sample ground count; t, total ground count.

[a]Count organizers: DML: Des Meules and Lemieux (NCAA); ECO: Ecosystems Ltd. Nairobi; MWE: Mweka College of African Wildlife Management; NCA: Ngorongoro Conservation Area Authority; NEM: Ngorongoro Ecological Monitoring Programme; SEM: Serengeti Ecological Monitoring Programme; TBE: Turner and Bell (NCAA); TLA: Turner and Lamprey (NCAA); TWA: Turner and Watson (1964); WAT: Watson. Sources of data: Estes and Small (1981) for MWE and WAT; Turner and Watson (1964) for TWA: Ngorongoro Conservation Area Authority files and Ngorongoro Conservation Unit (1967) for DML, NCA, NEM, TBE, TLA; Serengeti Ecological Monitoring Programme (1986) for SEM; Ecosystems Ltd. (1980) for ECO.

REFERENCES

Anderson, G. D., and Herlocker, D. J. 1973. Soil factors affecting the distribution of the vegetation types and their utilization by wild animals in Ngorongoro Crater, Tanzania. *J. Ecol.* 61:627–51.

Bell, R. H. V. 1970. The use of the herb layer by grazing ungulates in the Serengeti. In *Animal populations in relation to their food sources,* ed. A. Watson, 111–23. Oxford: Blackwell Scientific Publications.

Boshe, J. I. 1988. *Wildlife ecology.* Technical Report no. 3, Ngorongoro Conservation and Development Project. Nairobi: IUCN.

Brown, L. H., Urban, E. K., and Newman, K. 1980. *The birds of Africa.* vol 1. London: Academic Press.

Chausi, E. B. 1985. *Range management and ecology in Ngorongoro Conservation Area, Tanzania: An integrated range resources management prospective.* M.Sc. thesis, University of Idaho.

Coe, M. J., Cumming, D. H., and Phillipson, J. 1976. Biomass and production of large African herbivores in relation to rainfall and primary production. *Oecologia* 22:341–54.

Conover, W. J. 1980. *Practical nonparametric statistics.* New York: John Wiley.

Eland	Elephant	Black rhino	Buffalo	Kongoni	Waterbuck	Hippo	Warthog	Ostrich
300 t		20 t		20 t	100 t	30		
342 a	28 a	27 a	11 a	49 a	35 a	23 a		37 a
77 a	0 a	6 a	0 a	35 a	12 a	25 a		25 a
267 a	87 a	24 a	8 a	54 a	85 a	25 a		34 a
488 t	10 t	46 t	67 t	52 t	47 t			44 t
320 a						34 a		
				67 t	85 t			
355 t	11 t	3 t	25 t	19 t	26 t			30 t
213 t	10 t	7 t	1 t	139 t	62 t			52 t
324 t	0 t	0 t	0 t	148 t	91 t			34 t
329 t	0 t	0 t	0 t	140 t	37 t			40 t
490 t	0 t	0 t	135 t	167 t	130 t			35 t
478 t	18 t	52 t	197 t	136 t	114 t			34 t
240 t	20 t	25 t	18 t	114 t	63 t			32 t
418 t	49 t	32 t	163 t	167 t	64 t			55 t
98 t	7 t	8 t	0 t	154 t	20 t			13 t

Appendix 7.1 (continued)

Year	Month and season	Source[a]	Wildebeest	Zebra	Thomson's gazelle	Grant's gazelle	Gazelles combined
1972	1 wet	MWE	14,876 t	4,003 t	2,736 t	1,317 t	
1972	8 dry	NCA	15,754 t	5,782 t	3,216 t	1,105 t	
1972	11 wet	NCA	14,458 t	3,482 t	3,023 t	1,702 t	
1973	1 wet	MWE	13,422 t	3,286 t	2,778 t	1,833 t	
1973	4 wet	MWE	14,089 t	4,446 t	4,567 t	1,679 t	
1973	8 dry	NCA	16,484 t	3,764 t	2,291 t	1,825 t	
1973	10 dry	MWE	14,089 t	3,881 t	2,749 t	1,365 t	
1974	1 wet	NCA	16,025 t	4,573 t	3,559 t	1,383 t	
1974	8 dry	NCA	13,366 t	4,768 t	4,134 t	1,797 t	
1974	11 wet	NCA	15,917 t	3,099 t	3,999 t	1,333 t	
1975	8 dry	NCA	16,642 t	4,951 t	4,584 t	2,037 t	
1976	4 wet	NCA	10,059 t	5,306 t	3,407 t	2,346 t	
1976	9 dry	NCA	18,240 t	4,139 t	3,419 t	1,450 t	
1977	4 wet	NCA	10,133 t	6,172 t	2,804 t	1,278 t	
1977	9 dry	NCA	14,451 t	3,312 t	2,827 t	1,507 t	
1977	10 dry	ECO	15,335 a	3,038 a			
1978	2 wet	ECO	13,736 a				
1978	2 wet	NCA	9,587 t	3,623 t	3,125 t	1,376 t	
1980	2 wet	ECO	18,450 f	4,258 f			5,883 f
1986	7 dry	SEM	11,847 t	4,297 t	3,392 s	2,136 s	
1987	4 wet	NEM	9,011 t	3,127 t	4,342 s	3,588 s	
1987	9 dry	NEM	7,415 t	4,332 t	4,677 s	1,135 t	
1988	4 wet	NEM	6,305 t	2,810 t	5,877 s	991 t	
1988	9 dry	NEM	8,633 t	4,214 t	7,830 s	1,254 t	
1988	9 dry	NEM	8,689 a	3,905 a			
1988	10 dry	MWE	8,179 t	4,000 t	4,452 s	868 t	
1989	4 wet	NEM	7,373 t	2,642 t	890 t	985 t	
1989	9 dry	NEM	9,152 t	3,313 t	2,175 s	756 t	
1990	4 wet	NEM	11,221 t	4,330 t	1,826 t	1,043 t	
1990	9 dry	NEM	7,439 t	3,788 t	1,493 s	601 t	
1991	4 wet	NEM	9,254 t	3,314 t	1,216 t	927 t	
1991	9 dry	NEM	7,111 t	2,993 t	4,160 s	853 t	
1992	4 wet	NEM	10,608 t	4,185 t	1,237 t	1,386 t	
1992	9 dry	NEM	7,192 t	2,540 t	1,285 t	834 t	

Eland	Elephant	Black rhino	Buffalo	Kongoni	Waterbuck	Hippo	Warthog	Ostrich
340 t	32 t	7 t	303 t	188 t	103 t			36 t
288 t	8 t	7 t	12 t	74 t	23 t			23 t
452 t	16 t	34 t	526 t	226 t	83 t			42 t
499 t	56 t	32 t	1,279 t	150 t	49 t			37 t
2 t	37 t	18 t	539 t	161 t	38 t			44 t
201 t	39 t	18 t	184 t	161 t	54 t			28 t
477 t	13 t	11 t	176 t	170 t	25 t			28 t
336 t	20 t	33 t	550 t	139 t	58 t			34 t
266 t	20 t	37 t	80 t	188 t	19 t			30 t
203 t	14 t	34 t	610 t	283 t	27 t			34 t
55 t	18 t	17 t	329 t	207 t	9 t			44 t
574 t	42 t	34 t	1,508 t	249 t	20 t			45 t
61 t	8 t	23 t	447 t	156 t	7 t			30 t
428 t	48 t	29 t	960 t	111 t	18 t			43 t
176 t	5 t	16 t	109 t	176 t	6 t			23 t
109 a	31 a	10 a	582 a	115 a	10 a	34 a		3 a
	17 a		1,169 a			47 a		31 a
284 t	29 t	25 t	1,172 t	166 t	48 t			32 t
325 f		67 f	1,498 f	215 f	33 f		56 f	56 f
59 t		2 t	1,455 t	72 t	19 t			26 t
64 t		9 t	2,714 t	70 t	25 t		147 s	23 t
7 t		8 t	2,855 t	112 t	31 t		239 s	40 t
56 t	32 t	17 t	2,139 t	120 t	61 t		336 s	53 t
21 t	12 t	5 t	2,339 t	103 t	7 t		121 s	30 t
21 a	15 a	6 a	2,114 a	126 a	7 a	76 a		31 a
49 t		4 t	722 t	107 t	25 t			52 t
27 t	36 t	10 t	2,851 t	126 t	53 t		35 t	46 t
15 t		9 t	859 t	94 t	32 t		146 s	23 t
43 t	74 t	10 t	2,946 t	98 t	8 t		49 t	37 t
35 t	18 t	8 t	2,005 t	108 t	30 t		123 s	24 t
43 t	56 t	5 t	3,405 t	104 t	23 t		68 t	40 t
20 t	14 t	5 t	2,417 t	106 t	15 t		54 s	35 t
27 t	27 t	12 t	5,608 t	160 t	8 t		50 t	33 t
9 t	13 t	3 t	1,445 t	154 t	11 t		75 t	33 t

Cumming, D. H. M., du Toit, R. F., and Stuart, S. N. 1990. *African elephants and rhinos: Status survey and conservation action plan.* Gland, Switzerland: IUCN.

Demment, M. W., and van Soest, P. J. 1985. A nutritional explanation for body-size patterns of ruminant and non-ruminant herbivores. *Am. Nat.* 125:641–72.

Ecosystems Ltd. 1980. *The status and utilisation of wildlife in Arusha Region, Tanzania: Final report.* Nairobi: Ecosystems Ltd. Unpublished report.

Elliott, J. P., and McTaggart Cowan, I. 1978. Territoriality, density, and prey of the lion in Ngorongoro Crater, Tanzania. *Can. J. Zool.* 56:1726–34.

Estes, R. D., and Small, R. 1981. The large herbivore populations of Ngorongoro Crater. *Afr. J. Ecol.* 19:175–85.

Fosbrooke, H. 1986. Fire! Master or servant? *Swara* 9(2):12–16.

Frame, G. W. 1980. Black rhinoceros (*Diceros bicornis* [L.]) subpopulation on the Serengeti plains, Tanzania. *Afr. J. Ecol.* 18:155–66.

Fryxell, J. M., Greever, J., and Sinclair, A. R. E. 1988. Why are migratory ungulates so abundant? *Am. Nat.* 131:781–98.

Gichohi, H. W. 1990. *The effects of fire and grazing on grasslands of Nairobi National Park.* M.Sc. thesis, University of Nairobi.

Goddard, J. 1967. Home range, behaviour and recruitment rates of two black rhinoceros (*Diceros bicornis* [L.]) populations. *E. Afr. Wildl. J.* 5:133–50.

Herlocker, D. J., and Dirschl, H. J. 1972. *Vegetation of the Ngorongoro Conservation Area, Tanzania.* Canadian Wildlife Service Report ser. 19. Ottawa: Canadian Wildlife Service.

Homewood, K. M., and Rodgers, W. A. 1991. *Maasailand ecology.* Cambridge: Cambridge University Press.

Jarman, P. J., and Sinclair, A. R. E. 1979. Feeding strategy and the pattern of resource partitioning in ungulates. In *Serengeti: Dynamics of an ecosystem,* ed. A. R. E. Sinclair and M. Norton-Griffiths, 130–63. Chicago: University of Chicago Press.

Jewell, P. A. 1980. Ecology and management of game animals and domestic livestock in African savannas. In *Human ecology in savanna environments,* ed. D. Harris, 353–82. London: Academic Press.

Kabigumila, J. D. L. 1988. *The ecology and behaviour of elephants in Ngorongoro Crater.* M.Sc. thesis, University of Dar es Salaam.

Kiwia, H. D. 1983. *The behaviour and ecology of the black rhinoceros (*Diceros bicornis* L.) in Ngorongoro Crater.* M.Sc. thesis, University of Dar es Salaam.

———. 1986. Diurnal activity pattern of the black rhinoceros (*Diceros bicornis* [L.]) in Ngorongoro Crater, Tanzania. *Afr. J. Ecol.* 24:89–96.

———. 1989. Black rhinoceros (*Diceros bicornis* [L.]): Population size and structure in Ngorongoro Crater, Tanzania. *Afr. J. Ecol.* 27:1–6.

Klingel, H. 1969. Reproduction in the plains zebra *Equus burchelli boehmi:* Behavior and ecological factors. *J. Reprod. Fert.,* suppl. 6:339–45.

Kruuk, H. 1972. *The spotted hyena: A study of predation and social behavior.* Chicago: University of Chicago Press.

Machange, J. 1988. *Livestock/wildlife interactions.* Ngorongoro Conservation and Development Project Technical Report 4. Nairobi: IUCN.

Maddock, L. 1979. The "migration" and grazing succession. In *Serengeti: Dynamics of an ecosystem*, ed. A. R. E. Sinclair and M. Norton-Griffiths, 104–29. Chicago: University of Chicago Press.

McNaughton, S. J. 1978. Stability and diversity of ecological communities. *Nature* 274:251–53.

McNaughton, S. J., and Georgiadis, N. J. 1986. Ecology of African grazing and browsing mammals. *Annu. Rev. Ecol. Syst.* 17:39–65.

Mwalyosi, R. B. B. 1992. Influence of livestock grazing on range condition in southwest Masailand, Northern Tanzania. *J. Appl. Ecol.* 29:581–88.

Ngorongoro Conservation Unit. 1967. *Annual report 1967*. Dar es Salaam, Tanzania: Ministry of Agriculture and Co-operative, Tanzania.

Ngorongoro Conservation Area Authority. 1985. *Annual reports 1968–1985*. Dar es Salaam, Tanzania: Ministry of Agriculture and Co-operative, Tanzania.

Ngorongoro Ecological Monitoring Programme. 1987. *Semi-annual report*. Ngorongoro: Ngorongoro Conservation Area Authority. Unpublished report.

———. 1988a. *Semi-annual report*. Ngorongoro: Ngorongoro Conservation Area Authority. Unpublished report.

———. 1988b. *Annual report*. Ngorongoro: Ngorongoro Conservation Area Authority. Unpublished report.

———. 1989. *Annual report*. Ngorongoro: Ngorongoro Conservation Area Authority. Unpublished report.

———. 1990. *Annual report*. Ngorongoro: Ngorongoro Conservation Area Authority. Unpublished report.

———. 1991. *Annual report*. Ngorongoro: Ngorongoro Conservation Area Authority. Unpublished report.

Norton-Griffiths, M. 1978. *Counting animals*. 2d ed. Handbook no. 1. Techniques in African Wildlife Ecology. Nairobi: African Wildlife Foundation.

Prins, H. H. T., and Douglas-Hamilton, I. 1990. Stability in a multi-species assemblage of large herbivores in East Africa. *Oecologia* 83:392–400.

Pusey, A. E., and Packer, C. 1987. Philopatry and dispersal in lions. *Behaviour* 101:275–310.

Rose, G. A. 1975. Buffalo increase and seasonal use of Ngorongoro Crater. *E. Afr. Wildl. J.* 13:385–87.

Serengeti Ecological Monitoring Programme. 1986. *Ngorongoro Crater wildlife census 26–27 July 1986*. Arusha, Tanzania: Serengeti Ecological Monitoring Programme. Unpublished report.

Sinclair, A. R. E. 1977. *The African buffalo: A study of resource limitation of populations*. Chicago: University of Chicago Press.

———. 1979. The eruption of the ruminants. In *Serengeti: Dynamics of an ecosystem*, ed. A. R. E. Sinclair and M. Norton-Griffiths, 82–103. Chicago: University of Chicago Press.

Sinclair, A. R. E., and Norton-Griffiths, M. 1979. *Serengeti: Dynamics of an ecosystem*. Chicago: University of Chicago Press.

Sokal, R. R., and Rohlf, F. J. 1981. *Biometry*. 2d ed. San Francisco: Freeman.

Turner, M., and Watson, R. M. 1964. A census of game in Ngorongoro Crater. *E. Afr. Wildl. J.* 2:165–68.

Wilkinson, L. 1990. *SYSTAT: The system for statistics.* Evanston, Ill.: SYSTAT Inc.

Wolfe, M. L. 1980. Feral horse demography: A preliminary report. *J. Range Mgmt.* 33:354–60.

EIGHT

Population Trends of Ungulates in and around Kenya's Masai Mara Reserve

Michael D. Broten and Mohammed Said

The Mara area of Kenya's Narok District forms part of the northern portion of the Serengeti ecosystem. It encompasses an area of about 5,560 km^2, of which about 1,680 km^2 forms the Masai Mara National Reserve (MMR); the rest consists of private group ranches that are owned by different Masai clans. *Mara* is "a Masai word referring to the patchy distribution of forests and bush within the study area especially along the major drainage systems" (Douglas-Hamilton 1990).

Although the Mara constitutes only a small portion of the Serengeti migrant wildlife's dispersal range, this area is critical beyond its size. Its high rainfall, permanent water, and high grassland productivity make it an important dry season refuge for most of the migrants (wildebeest and zebra) for about 4 months (July–October) each year. The area is also rich in resident wildlife species. Thus, through wildlife-based tourism, the Mara is presently one of Kenya's highest foreign exchange—earning areas (Douglas-Hamilton 1990). For example, in 1987 alone, the Mara Reserve received 18% of all visits to national parks and reserves in Kenya. The revenue it generated for the year represented 8% of gross tourism revenues for the whole country.

The major issues of concern in Narok District are conflicting land use practices, increasing human population growth, environmental degradation, and land ownership. Livestock and wildlife grazing was historically the most important land use in the district. Forest also forms an important land-cover type, particularly in the Mau uplands to the north. However, agriculture is fast gaining in importance over pastoral activities. Originally, the district was inhabited mainly by Masai, whose livelihood has always been livestock keeping. Wildlife species have also had access to the entire area and have traditionally coexisted in relative harmony with Masai livestock grazing activities.

Agricultural practices increased with the influx of people from other parts of the country into the district starting in the early 1960s. Intensification of this influx in the 1970s sparked the current conflicts over land use. Now, the natural forests and woodlands are being cleared to provide room for agricultural expansion and settlements. All these activities are leading to drastic constriction of livestock and wildlife ranges and consequently to increased overgrazing problems and general degradation of the environment. Although livestock keeping is still an important activity in the district, more emphasis is now being accorded to agricultural production, at the expense of forest and wildlife conservation as well as pastoral grazing (Republic of Kenya 1980, 1984, 1989).

The Department of Resource Surveys and Remote Sensing (DRSRS; formerly Kenya Rangeland Ecological Monitoring Unit—KREMU) has been conducting yearly ecological surveys in Narok District (particularly in the Mara area) since 1977. The main objectives of these surveys are to monitor, on a long-term basis, the ecological changes taking place in the district and to provide resource data to decision makers and development planners. Most of the data collected in the district are on animal numbers and their seasonal distributions, and on land use/land cover parameters.

In this chapter, we use these long-term monitoring data to analyze ungulate population trends, their seasonal habitat utilization, and their distributions.

STUDY AREA

The study area is composed of the MMR and the surrounding ranches (Siana, Koyaki, Dagurugurueti, Olkinyei, Lemek, and Ol Chorro Orogwa). The study area lies completely within Narok District, located in southwestern Kenya (fig. 8.1). Narok District lies at approximately 1°–2° south latitude and 34°–36° degrees east longitude and covers a land area of about 17,821 km^2. Physiographically, it consists of a series of hills in the extreme north, the Loita and Siana hills in the southeast, and the Mara and Loita plains in the southern and north-central parts. To the north, the study area is bounded by the Loita plains, which rise to an altitude of 1,800 meters and slope southward. To the east are the Siana plains and hills, which rise to 2,650 meters. To the west is the Isiria escarpment (Esoit Oloololo), which rises to 1,850 meters. To the south lies the Serengeti National Park of Tanzania.

The Mara, Sand, and Talek Rivers and their numerous tributaries form the main drainage systems in the study area. These surface water features are of great importance with regard to wildlife habitat in the area.

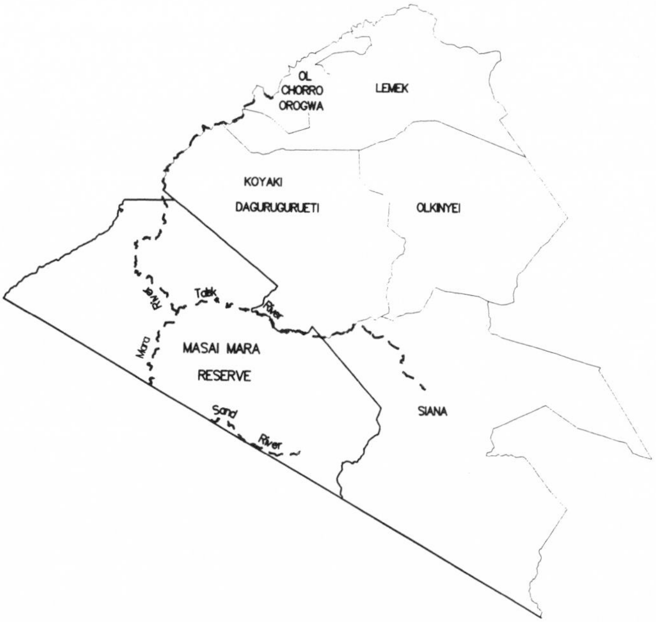

Figure 8.1 The Masai Mara National Reserve, surrounding group ranches, and major rivers (dashed lines).

Climate

The rainfall pattern in Narok District is loosely associated with the movement of the low-pressure intertropical convergence zone (ITCZ), which alternates annually between mid-Sudan and northern Zimbabwe (Brown and Cocheme 1973). The ITCZ shifts according to the zenith sun angle but has a lag of approximately 1 month before reaching its northern boundary in late July and its southern boundary in late January. This cycle brings about the two seasons in the equatorial regions. The annual distribution of rainfall is bimodal in pattern, characterized by two rainy seasons as well as two dry seasons. The long rains are generally from March to May and the short rains occur in November and December. The main dry period is from mid-June to mid-October, with a lesser dry spell in January and February.

In addition to the very general influence of the ITCZ, local variations in topography, such as high mountains and escarpments, plus orographic and diurnal effects play a major role in rainfall patterns within the study area (Griffiths 1972).

Both the climate and soils are closely related to the landforms of the area. The lowland Loita and Siana plains receive a mean annual rainfall of about 750 mm. The southeastern part of the study area receives between 375 and 500 mm, while the Mara plains receive between 500 and 700 mm annually. Aore (1988) noted that (1) there is a general increase of rainfall from east to west and northward; (2) there is a general decrease of rainfall southward, but with an increase again in the Loita hills; and (3) rainfall occurs in heavy downpours with limited infiltration into the soil.

For the purposes of this study, rainfall variability was based on four meteorological stations: Narok, Naikarra, Lemek, and Aitong (table 8.1). Mean annual rainfall values were calculated for each of the four stations for the following periods: annual total (November–October), wet season total (November–May), dry season total (June–October), and the total for each of the 12 months. The year-to-year variations of the annual total rainfall (November–October) and the wet season total (November–May) are fairly large, while those of the dry season are considerably smaller (Amuyunzu 1984). Due to the seasonality of the rainfall, it is more realistic to classify the rainfall year into months that are moist, moderately dry, and dry (table 8.2). The grouping is done by comparing the relationship between the monthly rainfall and evaporation rates (Brown and Cocheme 1973).

Soils

The soils of Narok have been described by Sombroek, Braun, and Van der Pouw (1982). The soils are mainly of volcanic origin, the fertility ranging from high to very low. They include the red friable clays of the Loita hills, which were developed from intermediate igneous rocks, with volcanic ash mixture. The soils of the southern plains, and in particular, parts of the Mara Reserve, were developed on biotite gneiss, also with volcanic ash mixture. These soils are brown to dark gray clay, seasonally waterlogged but better drained than the clay of moderate to low-fertility soils. The central Loita plains have deep friable silt-loam to clay-loam, derived from biotite gneiss, with volcanic ash mixture.

Land Use

Regulations exclude cultivation and livestock from the Mara Reserve. Those livestock that appear to be located within the reserve boundary are largely the result of map generalization, trespasses, and recent boundary changes. Cultivation was not recorded anywhere in the reserve. However, livestock and wildlife utilize the group ranches freely with no serious legal or physical barriers. This has been the case for centuries.

Cultivation of wheat and small-scale farming is increasing north of the Loita plains and Aitong areas. This activity has been a new development in the last 10 to 15 years and has brought wildlife and humans into

Table 8.1 Annual and monthly rainfall at four stations.

	Station (Altitude)							
	Narok (1,890 m)		Lemek (1,980 m)		Aitong (1,760 m)		Naikarra (2,030 m)	
Time period	Mean (mm)	Var. (%)	Mean (mm)	Var. (%)	Mean (mm)	Var. (%)	Mean (mm)	Var. (%)
Annual Total (Nov.–Oct.)	737	30	664	22	1,029	37	716	22
Wet season total (Nov.–May)	620	28	473	29	789	25	544	25
Dry season total (June–Oct.)	117	20	191	33	240	23	172	42
November	64	89	43	64	119	109	63	108
December	72	80	49	100	113	72	70	102
January	70	94	76	100	89	67	71	87
February	80	73	71	60	109	71	77	54
March	98	76	78	72	110	84	74	96
April	144	59	99	68	170	72	100	76
May	92	72	60	73	79	137	65	59
June	29	98	50	50	58	78	52	63
July	17	99	48	124	41	98	28	108
August	21	123	24	111	56	66	30	77
September	23	100	42	91	53	83	46	109
October	27	95	26	63	32	91	16	80
Mean monthly rainfall[a]	61	64	56	40	86	47	58	43

Note: Variability (Var.) expresses the year-to-year variation by taking the standard deviation as a percentage of the mean.

[a]Mean monthly rainfall is equal to the annual total divided by 12.

Table 8.2 Distribution of wet and dry months

		Number of months		
Station	Probability	Moist	Moderately Dry	Dry
Narok	75%	2	4	6
	50%	7	—	5
Lemek	75%	1	10	1
	50%	7	5	—
Niakarra	75%	2	8	2
	50%	9	3	—
Aitong	75%	3	8	1
	50%	10	2	—

Note: Moist months are those in which rainfall is greater than 50% of the monthly evapotranspiration (ET). Moderately dry months are those in which monthly rainfall is less than 50% but greater than 25% of ET. Dry months are those in which monthly rainfall is less than 10% of ET. (Adapted from Brown and Cochene 1973).

direct confrontation. Wildlife is being displaced at the same time that fence barriers are being erected along wildlife migration routes.

METHODS

Aerial Survey

Systematic reconnaissance flights (SRF) for estimation of animal populations are fully described by Norton-Griffiths (1978). This method as practiced by DRSRS is briefly described below:

1. High-winged aircraft equipped with global navigation systems, intercoms, and radar altimeters were used for aerial censuses. The crew for each flight consisted of a pilot, a front seat observer (FSO), and two rear seat observers (RSO). The RSO were responsible for animal counts, while the FSO gathered information on general land use and environmental parameters.
2. The aircraft followed systematic flight lines or transects that were oriented either east-west or north-south, depending on the terrain. Survey of Kenya topographic sheets of scale 1:250,000 were used for flight planning. All transects followed the Universal Transverse Mercator (UTM) coordinate grid lines shown on those topographic base maps. A transect spacing of 5 km was used.
3. Each transect was divided into equal sample intervals of 5 km length. Consequently, with transect spacing of 5 km, a final sample unit of 5 km by 5 km was defined.
4. During the surveys, standard flying height of either 91 or 122 m (300 or 400 ft.) and flying speed of between 160 and 190 km/hour were maintained.
5. A calibrated survey strip was defined by two parallel rods mounted on the wings of the aircraft and by corresponding window markings. Only animals observed within this strip were recorded during the survey.
6. During the survey all visual observations by RSO of animals within the strip width were recorded orally using tape recorders. For herds of greater than ten animals, a photo was also taken.
7. After every survey, the tape-recorded observations were summarized for each sample unit and transcribed to data sheets by the RSO. The photos were processed and photo counts made. The photo count was then entered on the RSO data sheet.

The photo-corrected data sets were then key-entered into a computer-based geographic information system (GIS) for final processing, mapping, and calculation of population estimates. Population estimates were calculated according to the procedure of Jolly II for transects of unequal length

(Jolly 1969). The final statistical processing can be summarized as follows:

1. Data for sample units found within the study area were selected.
2. Characteristics of each flight/transect (height deviation, etc.) and animal counts for units along these transects were combined to calculate population estimates for each observed species. Correction factors for the aircraft altitude variation were made to actual strip width according to:

Actual strip width = Expected strip width × (actual flying height/ expected flying height)

3. Population estimates based on Jolly's (1969) method of unequal transect lengths were calculated and a summary report was generated. The sample data set was then saved for other analyses.

For the purposes of this study, no correction was made for partial units along the study area boundary; each unit of 5 by 5 km (an area of 25 km^2) was used. Additionally, sample variances and corresponding confidence limits were not analyzed on time for this study.

Vegetation/Habitat Classification

The vegetation map used for this study is illustrated in figure 8.2. It is based on the work of Epp and Agatsiva (1982), who mapped an area of 7,500 km^2 of the Mara, Siana, and Loita ecosystems. The method used in delineating the habitat types was based on the interpretation of satellite images from 1973 to 1976, aerial photography, and ground checks. The classification system used in mapping was a modification of Pratt, Greenway, and Gwynne (1966). The physiognomic characteristics of height and canopy cover were the two main criteria used in classifying the habitat types. The classification has forty-six habitat types with twelve main categories covering all the rangelands in Kenya. For the purposes of this study, the vegetation maps of Epp and Agatsiva (1982) and Msafiri (1984) were used.

There are seventeen habitat types in the study area. Generalization of these detailed classes into fewer classes became necessary for two reasons: (1) some vegetation polygons formed very small or narrow areas on the map, and matching them with wildlife sampling units (25 km^2) was not possible; and (2) vegetation classification differed between authors. A listing of the generalized vegetation categories and areas covered by each is presented in table 8.3.

The most extensive habitat category in the study area is shrubland, which covers an area of approximately 2,527 km^2 (about 45% of the study area). Grassland is the second most extensive type, covering an area of 1,467 km^2 (26%). These grasslands are found mainly between Talek

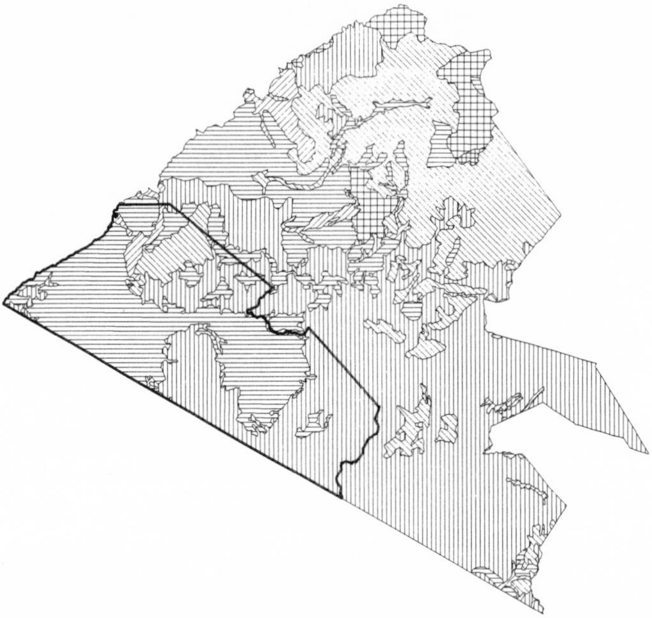

Figure 8.2 The major vegetation categories in the area shown in figure 8.1. Horizontal hatching, grassland; vertical hatching, shrubland; diagonal hatching (left up), dwarf shrubland; diagonal hatching (right up), woodland; small squared hatching, bushland; large squared hatching, agriculture. Dark line delimits the Mara Reserve.

Table 8.3 Generalized vegetation categories for the study area.

General vegetation type	Map symbol	Area (km²)	% of total	Detailed types comprising general category
Grasslands	GL	1,467	26.4	Grasslands, wooded grasslands, shrubby grasslands, grassland with isolated trees
Shrublands	SL	2,527	45.4	Shrublands, grassy shrublands, shrubby riverine, thicket shrublands
Dwarf shrublands	DS	1,165	21.0	Dwarf grassy shrublands, dwarf shrubby grasslands
Woodlands	WL	156	2.8	Wooded riverine, woodlands
Bushlands	BL	8	0.1	Wooded shrublands, shrubby woodlands
Agricultural land	AG	238	4.3	Agricultural or settled areas, wheat farms
TOTAL		5,561	100.0	

River and Bardamant hills. Dwarf shrublands are mainly found in the Loita plains and cover an area of approximately 1,165 km^2 (about 21%). Riverine gallery forest borders the Mara, Talek, and Ewaso Ngiro Rivers. Wooded grassland (merged with the grasslands category) is found only in the Mara triangle, in the western part of the Mara Reserve. Agriculture is found mainly in the northernmost part of the study area. This category is made up of wheat farms, abandoned wheat farms, and other settled areas. Agriculture covers about 4% of the total study area. However, it is known that agricultural activities in this area have expanded since 1988.

RESULTS AND DISCUSSION

Population Trends 1977–1990

A simple analysis of wildlife and livestock trends was completed by graphing trends for each species. The livestock species that are considered of importance to the Mara are cattle, sheep and goats ("shoats"), and donkeys. With respect to wildlife, fifteen species of herbivores are considered to be the most important to the general ecology of the Mara: wildebeest, elephant, giraffe, rhinoceros, buffalo, Grant's gazelle, Thomson's gazelle, impala, kongoni, topi, zebra, eland, ostrich, warthog, and waterbuck. Population estimates for each species are listed in table 8.4. Estimates are given for the Mara Reserve (Mara), the group ranch areas (Ranch), and the entire study area. Some of these data are also presented in graphical form in the following discussion.

Wildlife Trends. Based on DRSRS aerial survey data, the overall number of resident wild herbivores in the study area is currently about 150,000 head (fig. 8.3), with 100,000 in the ranch areas and 50,000 in the Mara Reserve. The resident wildebeest and zebra populations are included in this estimate. Wildlife trends within the reserve appear to be fairly stable, while resident wildlife species for the surrounding ranch areas may be on a general decline, although further statistical analyses are required.

As part of this analysis, the sizes of zebra and wildebeest populations were estimated and graphed (fig. 8.4). The population estimates over the years were sorted by month and graphed on a log scale to help differentiate between the resident and migrant (those migrating across the Tanzania border) populations. The migrant animals show up clearly as a large seasonal swell in the population estimates. It is clear that wildebeest immigrants swell the overall populations in the Mara Reserve more than in the surrounding ranches. However, the migration appears to have no effect on the numbers of zebra found in the ranch areas. It can also be seen that the resident populations of both wildebeest and zebra reside mostly on the ranches.

Based on the estimates, the migration appears to bring in an average of between 200,000 and 600,000 wildebeest, while the resident wildebeest population in the ranch areas appears relatively stable at about 10,000 to 20,000. However, the long-term wildebeest estimates indicate a slight downward trend for resident wildebeest in the reserve. Resident populations in the reserve, which appear to have been in the 1,000 to 10,000 range in the 1970s, appear to be an order of magnitude smaller (in the range of 100 to 1,000) in the 1990s. For zebra, the fluctuation is

Table 8.4 Population estimates for ungulates in the study area.

		Month-Year of Survey						
Species	Area	Nov. 77	Oct. 80	Jan. 81	Feb. 83	Aug. 83	Apr. 85	Aug. 85
Buffalo	Mara	14,581	11,450	11,234	15,723	14,342	13,547	6,733
	Ranch	9,243	2,192	—	—	3,675	783	1,740
	Total	23,824	13,642	—	—	18,017	14,330	8,473
Cattle	Mara	13,923	9,688	26,347	30,698	7,238	14,675	11,512
	Ranch	99,733	144,314	—	—	75,027	149,661	84,911
	Total	113,656	154,002	—	—	82,265	164,336	96,423
Donkey	Mara	219	1,057	97	281	401	231	44
	Ranch	9,351	2,546	—	—	3,616	4,106	2,700
	Total	9,570	3,603	—	—	4,017	4,337	2,744
Eland	Mara	2,154	1,198	2,432	265	382	14	794
	Ranch	2,410	1,130	—	—	197	1,744	1,002
	Total	4,564	2,328	—	—	579	1,758	1,796
Elephant	Mara	1,496	1,163	1,197	1,466	1,413	1,446	118
	Ranch	0	118	—	—	0	288	264
	Total	1,496	1,281	—	—	1,413	1,734	382
Giraffe	Mara	1,436	1,022	772	1,373	1,031	578	970
	Ranch	3,309	2,681	—	—	2,063	2,746	1,559
	Total	4,745	3,703	—	—	3,094	3,324	2,529
Grant's gazelle	Mara	1,436	1,392	1,023	920	573	708	2,014
	Ranch	4,963	6,120	—	—	9,393	4,984	10,036
	Total	6,399	7,512	—	—	9,966	5,692	12,050
Impala	Mara	14,083	7,011	5,771	15,583	16,863	11,595	10,674
	Ranch	37,548	24,699	—	—	55,239	41,467	28,174
	Total	51,631	31,710	—	—	72,102	53,062	38,848
Kongoni	Mara	2,533	1,180	1,872	1,763	1,738	1,272	2,514
	Ranch	1,942	3,406	—	—	2,771	3,707	2,116
	Total	4,475	4,586	—	—	4,509	4,979	4,630
Ostrich	Mara	239	106	135	265	57	231	191
	Ranch	0	118	—	—	138	55	390
	Total	239	224	—	—	195	286	581
Rhino	Mara	180	0	0	0	0	0	0
	Ranch	36	0	—	—	0	0	0
	Total	216	0	—	—	0	0	0

Month-Year of Survey								
Dec. 85	May 86	Aug. 86	Nov. 86	Apr. 87	May 89	Aug. 89	Aug. 90	Apr. 91
11,178	9,179	2,879	3,263	7,473	8,929	10,720	9,972	9,228
2,134	1,114	1,315	1,536	4,241	848	6,624	3,314	482
13,312	10,293	4,194	4,799	11,714	9,777	17,344	13,286	9,710
1,268	2,497	12,401	14,063	4,782	27,683	27,479	10,204	15,043
70,650	133,068	106,004	78,669	96,340	150,678	162,444	136,515	130,712
71,918	135,565	118,405	92,732	101,122	178,361	189,923	146,719	145,755
177	0	107	276	169	412	371	0	601
3,208	2,953	2,559	2,346	3,695	4,792	1,974	1,201	4,213
3,385	2,953	2,666	2,622	3,864	5,204	2,345	1,201	4,814
944	0	137	230	123	0	680	609	15
1,145	599	57	642	1,036	1,131	343	663	633
2,089	599	194	872	1,159	1,131	1,023	1,272	648
752	1,270	168	475	1,292	1,074	479	681	132
481	891	325	559	112	269	601	470	220
1,233	2,161	493	1,034	1,404	1,343	1,080	1,151	352
988	823	366	306	784	309	309	275	615
1,781	2,382	1,697	1,131	2,561	1,456	1,445	1,284	1,212
2,769	3,205	2,063	1,437	3,345	1,765	1,754	1,559	1,827
1,165	462	1,615	2,007	1,215	1,427	664	1,928	952
8,027	5,585	5,811	7,624	8,958	7,972	7,740	5,316	4,833
9,192	6,047	7,426	9,631	10,173	9,399	8,404	7,244	5,785
13,980	15,500	2,681	8,778	22,850	16,872	9,221	11,349	9,111
33,325	39,709	28,120	23,598	35,846	32,454	25,968	28,336	27,523
47,305	55,209	30,801	32,376	58,696	49,326	35,189	39,685	36,634
1,489	5,051	564	1,563	2,337	1,795	556	1,232	1,684
1,328	4,373	947	1,257	1,960	3,124	787	773	1,473
2,817	9,424	1,511	2,820	4,297	4,919	1,343	2,005	3,157
192	404	213	138	477	44	139	130	337
28	404	99	182	238	184	114	290	344
220	808	312	320	715	228	253	420	681
0	0	0	0	0	0	0	0	0
0		0	14	0	0	0	0	0
0	0	0	14	0	0	0	0	0

Table 8.4 (continued)

Species	Area	Month-Year of Survey						
		Nov. 77	Oct. 80	Jan. 81	Feb. 83	Aug. 83	Apr. 85	Aug. 85
Sheep and goats	Mara	4,847	4,580	6,235	5,288	5,252	9,080	912
	Ranch	195,582	104,459	—	—	73,494	107,237	106,724
	Total	200,429	109,039	—	—	78,746	116,317	107,636
Thomson's Gazelle	Mara	8,497	9,635	20,362	12,884	16,806	17,104	14,143
	Ranch	46,288	26,047	—	—	62,097	53,330	34,577
	Total	54,785	35,682	—	—	78,903	70,434	48,720
Topi	Mara	9,275	9,265	15,770	12,479	10,599	12,492	10,556
	Ranch	9,387	9,087	—	—	12,714	11,863	12,584
	Total	18,662	18,352	—	—	23,313	24,355	23,140
Waterbuck	Mara	1,197	1,092	483	328	1,127	564	74
	Ranch	504	34	—	—	531	481	0
	Total	1,701	1,126	—	—	1,658	1,045	74
Warthog	Mara	4,787	1,057	3,455	2,480	2,330	2,212	1,073
	Ranch	3,848	421	—	—	1,788	2,742	2,144
	Total	8,635	1,478	—	—	4,118	4,954	3,217
Wildebeest[a]	Mara	2,314	1,000	1,000	343	1,000	810	1,000
	Ranch	20,000	20,000	—	—	20,000	18,784	20,000
	Total	22,314	21,000	—	—	21,000	19,594	21,000
Zebra[a]	Mara	12,227	5,000	5,000	1,045	5,000	2,487	5,000
	Ranch	25,176	22,372	—	—	40,000	39,449	22,369
	Total	37,403	27,372	—	—	45,000	41,936	27,369
Overall wildlife	Mara	76,435	51,571	70,506	66,917	73,261	65,060	55,854
	Ranch	164,654	118,425	—	—	210,606	182,423	136,955
	Total	241,089	169,996	—	—	283,867	247,483	192,809
Overall livestock	Mara	18,989	15,325	32,679	36,267	12,891	23,986	12,468
	Ranch	304,666	251,319	—	—	152,137	261,004	194,335
	Total	323,655	266,644	—	—	165,028	284,990	206,803

[a]Wildebeest and zebra numbers adjusted to show approximate resident populations only.

Month-Year of Survey								
Dec. 85	May 86	Aug. 86	Nov. 86	Apr. 87	May 89	Aug. 89	Aug. 90	Apr. 91
27,326	130	9,415	3,615	5,259	9,826	2,332	8,160	8,100
84,132	89,836	109,115	92,101	107,915	57,077	87,889	80,962	75,015
111,458	89,966	118,530	95,716	113,174	66,903	90,221	89,122	83,115
24,967	21,807	8,181	14,951	18,890	20,593	12,959	12,596	13,652
42,624	45,433	23,073	30,272	49,758	41,288	29,745	21,542	27,897
67,591	67,240	31,254	45,223	68,648	61,881	42,704	34,138	41,549
15,307	14,706	10,573	7,767	14,731	7,958	7,893	3,348	4,043
9,709	15,195	6,489	3,644	10,973	6,248	6,767	6,394	6,045
25,016	29,901	17,062	11,411	25,704	14,206	14,660	9,742	10,088
870	621	46	689	338	177	62	203	513
693	864	28	140	98	85	501	373	165
1,563	1,485	74	829	436	262	563	576	678
3,185	2,800	1,021	1,241	646	618	417	217	864
2,332	2,368	877	461	1,008	735	630	110	330
5,517	5,168	1,898	1,702	1,654	1,353	1,047	327	1,194
2,964	1,010	1,000	1,000	92	382	1,000	1,000	190
20,000	12,452	20,000	20,000	13,815	17,245	20,000	20,000	14,348
22,964	13,462	21,000	21,000	13,907	17,627	21,000	21,000	14,538
5,000	1,948	5,000	5,000	2,491	4,192	5,000	5,000	3,310
26,499	25,265	16,612	11,645	28,273	31,309	21,146	20,216	20,282
31,499	27,213	21,612	16,645	30,764	35,501	26,146	25,216	23,592
82,981	75,581	34,444	47,408	73,739	64,370	50,099	48,540	44,646
150,106	156,634	105,450	102,705	158,877	144,348	122,411	109,081	105,787
233,087	232,215	139,894	150,113	232,616	208,718	172,510	157,621	150,433
28,771	2,627	21,923	17,954	10,210	37,921	30,182	18,364	23,744
157,990	225,857	217,678	173,116	207,950	212,547	252,307	218,678	209,940
186,761	228,484	239,601	191,070	218,160	250,468	282,489	237,042	233,684

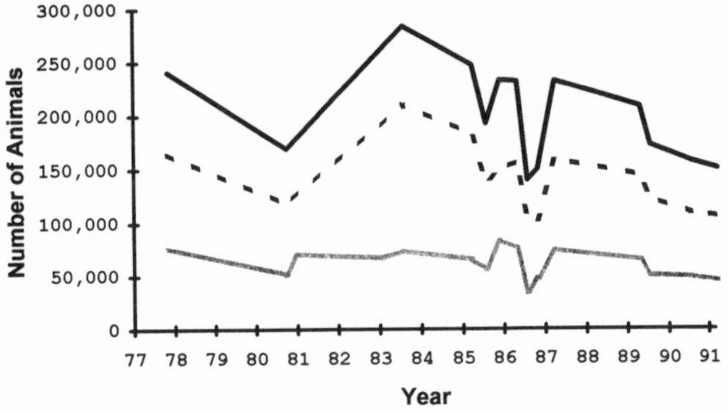

Figure 8.3 Overall wildlife trends based on population estimates for the fifteen species covered in this study. Gray line, population within Mara Reserve; dashed line, surrounding group ranches; solid line, combination of reserve and ranches.

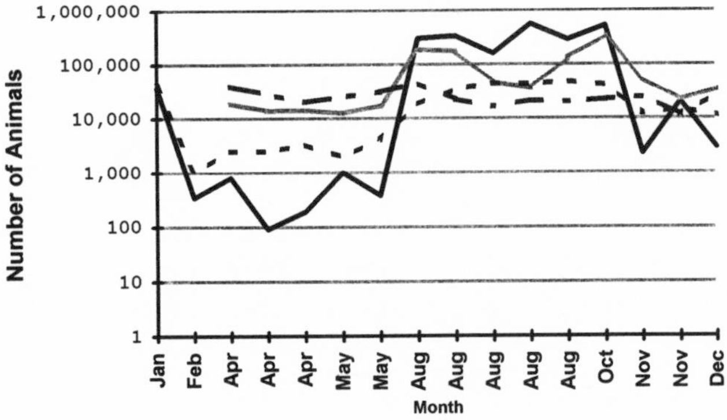

Figure 8.4 Seasonal shifts in wildebeest and zebra populations. Estimates from various surveys are plotted by month with no regard to year. Solid line, wildebeest-Mara; dashed line, zebra-Mara; gray line, wildebeest-ranch, dotted and dashed line, zebra-ranch.

less pronounced. About 10,000 to 50,000 appear to migrate into the reserve, and the resident population estimate stays between 1,000 and 5,000 animals. Zebra populations appear to be stable.

An examination of population trends for the other wildlife species delineated five general species-based groupings, each with a characteristic trend pattern. These groupings and the species falling within each are:

1. *Migrant species.* This group consists of wildebeest and zebra that move between the Serengeti of Tanzania and the Mara region of Kenya. They comprise two populations, the migrants and the resident populations. Their trends are described above.

2. *Stable populations.* Species in this group do not exhibit detectable population changes either in the Mara Reserve or on the ranches. This group includes impala and Grant's gazelle.

3. *Apparently stable populations.* Species in this group appear stable, but show erratic changes in numbers over the years. These fluctuations may be caused by counting biases rather than reflecting real changes. The species included in this category are elephant (fig. 8.5a), kongoni, ostrich (fig. 8.5b), and Thomson's gazelle. With respect to ostrich, there is strong indication of an increase in population.

4. *Apparently declining populations.* This group seems to indicate a downward trend in population but shows erratic changes in numbers over the years. These changes may be caused by counting biases rather than reflecting real changes. The species included in this category are buffalo and waterbuck (fig. 8.5c).

5. *Declining populations.* Species in this group have shown an apparent slow decline in numbers since 1977. This category includes the warthog (fig. 8.6a), topi, rhinoceros, giraffe (fig. 8.6b), and eland (fig. 8.6c). Except for the rhinoceros and warthog, the changes are fairly gradual.

Livestock Trends. Since livestock grazing is generally not allowed within the Mara Reserve, there was no need to compare livestock populations in the reserve and in the ranch areas, although there is known trespassing by livestock along the borders of the reserve. Overall livestock population estimates are quite stable at about 250,000 head. The decline seen in 1985–1986 may be explained by the 1984 drought. Cattle populations appear to be stable with a slight increasing trend. The population estimate for cattle in 1991 was about 150,000. Sheep and goats show a population estimate of around 80,000 in 1991 with an apparent decreasing trend. Donkeys appear to be in the 2,000 to 6,000 range in 1991 and show an apparent decrease in numbers, which may be explained by a decreasing reliance on donkeys as beasts of burden.

Seasonal Stocking Rates

Major changes in seasonal stocking rates are observed in the study area during any one year, mainly due to the migration of wildebeest and zebra, which starts in late May and ends in late October. Maximum numbers are normally attained by July or August. This seasonal change is illustrated clearly in figure 8.4. Based on a 1986 survey (Stelfox et al. 1986), the northward thrust of these migrants has the greatest impact in the Mara Reserve and just north of it. In 1986, however, the numbers of wildebeest that migrated into Kenya were small, reaching just about 160,000, as opposed to the more typical levels of 300,000 or more.

This seasonal fluctuation is further illustrated in table 8.5. The numbers show that the overall biomass, expressed as tropical livestock units, or TLUs (1 TLU = 250 kg live weight), increases from the wet season (November–May) to the dry in-migration season (June–October). This

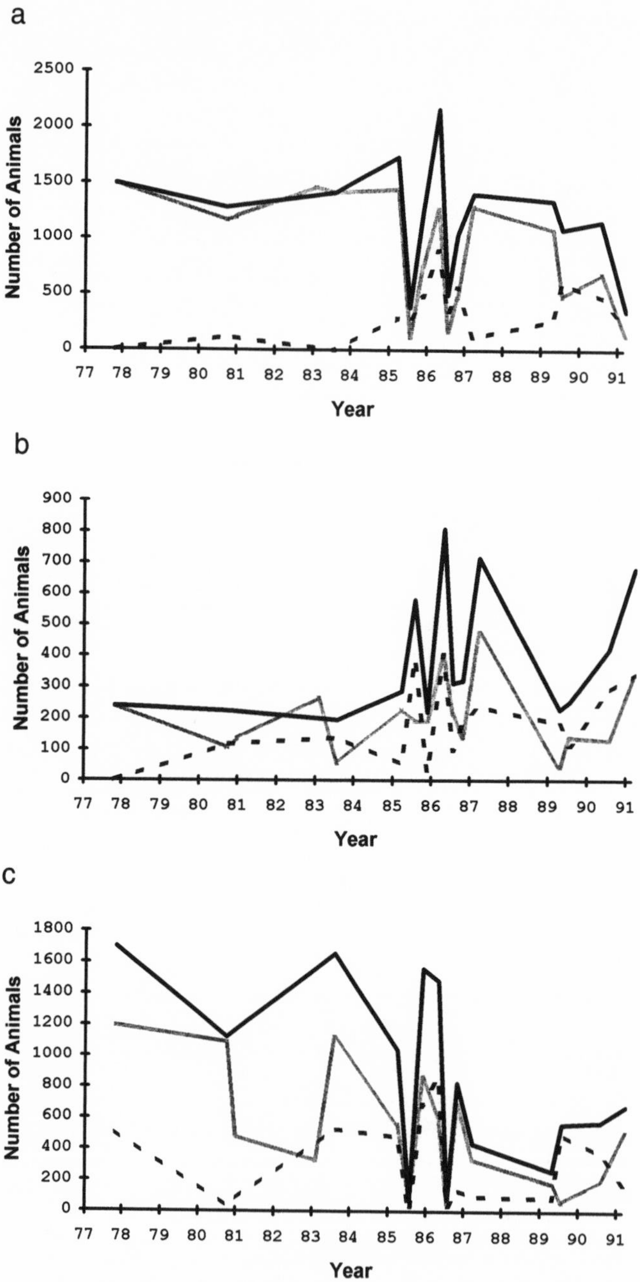

Figure 8.5 Population trends for (*a*) elephant, (*b*) ostrich, and (*c*) waterbuck. Gray lines, Mara Reserve; dashed lines, surrounding group ranches; solid lines, combination of reserve and ranches.

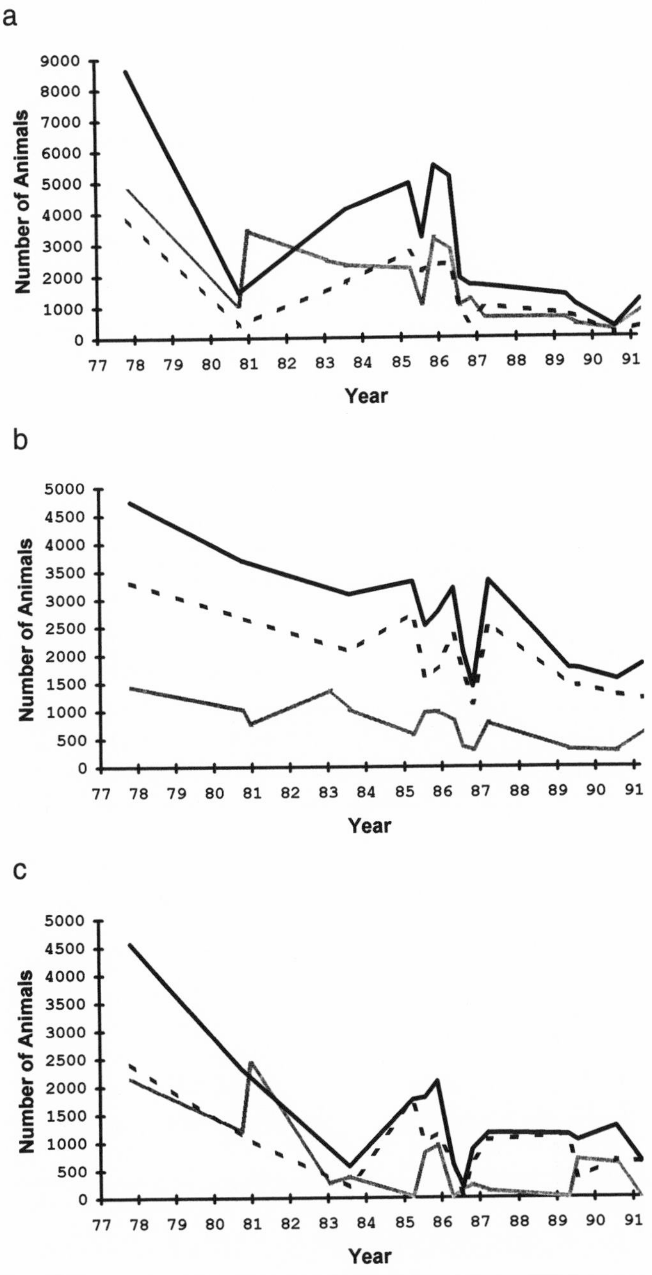

Figure 8.6 Population trends for (*a*) warthog, (*b*) giraffe, and (c) eland. Gray lines, Mara Reserve; dashed lines, surrounding group ranches; solid lines, combination of reserve and ranches.

Table 8.5 Average seasonal herbivore densities in the Mara study area (TLU/km²).

	Mara Reserve[a]		Ranches[b]	
	Out-migration Nov.–May	In-migration June–Oct.	Out-migration Nov.–May	In-migration June–Oct.
Livestock	7.1	8.0	32.3	29.6
Wildlife	32.7	142.2	15.2	23.6
Total	39.8	150.2	47.5	53.2
Overall change (%)	277		12	

Note: 1 tropical livestock unit (TLU) is equivalent to 250 kg live weight. Species-specific TLUs based on Stelfox et al. 1986.
[a]Averaged for eight surveys
[b]Averaged for four surveys

seasonal increase is pronounced in the Mara Reserve, where the change is about 277%, while the increase on the ranches is only about 12%. This means that usage of ranch areas is relatively well spread throughout the year, except for the ranch area immediately north of the reserve, which is heavily utilized by wildebeest and zebra between August and September.

The use of the two areas by resident populations is illustrated by TLU densities between November and May. In the Mara Reserve the total TLU density is about 40 TLU/km², of which about 33 comes from wildlife species and the remaining 7 from livestock along the reserve border. These livestock densities within the reserve are attributed both to averaging of data for units along the reserve border and to trespassing livestock (a historical problem that was largely corrected in 1984 by the strict exclusion of livestock grazing within the reserve). On the ranches, the total density is about 47 TLU/km², with livestock contributing about 32 and wildlife 15.

Habitat Use

Distribution and movement of animal populations was analyzed with respect to different habitat types (see fig. 8.2). The twenty-one detailed vegetation cover types were lumped into six broad habitat types (see table 8.3). Animal density was then overlaid on the general vegetation map using GIS software to calculate animal densities by vegetation category. These results were then used to indicate habitat use by species and by season (table 8.6).

Of the six general vegetation community types, only four were extensive enough to qualify for habitat utilization analysis. The woodland and bushland categories were both excluded because they are not extensive and because only a few aerial survey units could be completely assigned

Table 8.6 Animal densities by habitat type and season.

	Shrubland			Grassland			Dwarf shrubland			Agricultural		
	May	Aug.	Nov.	May	Aug.	Nov.	May	Aug.	Nov.	May	Aug.	Nov.
Buffalo	1	0.6	0.8	1.9	0.8	0.9	0.9	0.2	0.1	0	0.1	0.6
Cattle	27	27	24	28	23	20	25	21	24	37	27	23
Elephant	0.3	0.1	0.1	0.3	0.1	0.2	0.3	0.5	0.1	0.1	0.1	0
G. gazelle	0.8	1.1	1.6	1.1	1.2	1.8	1.4	2	1.7	0.6	1.6	0.7
Giraffe	0.5	0.4	0.3	0.6	0.3	0.3	0.4	0.5	0.3	0.1	0.1	0.1
Impala	9.1	5.3	4.8	9.9	6.4	5.6	6.9	4.6	4.3	2.1	1	1.3
Sheep/goats	18	22	18	17	19	16	18	28	24	18	29	11
T. gazelle	8	3.9	6.2	12	5.6	8.6	7.6	4.3	4.8	4.6	6.9	4.6
Topi	3.8	2.3	1.5	5.4	3.3	2.3	2.5	1	0.7	0.2	0.8	0.5
Wildebeest	1.4	32	6.4	1.9	42	8.6	4.7	10	3.9	6.1	34	0.4
Zebra	3.1	8.7	3.5	3.9	11	4.7	6.9	2.8	2.9	6.3	3	1
TOTAL	73	103	67	82	113	69	75	75	67	75	104	43
Wildlife %	38	53	37	45	63	48	42	35	28	27	46	21
Livestock %	62	47	63	55	37	52	58	65	72	73	54	79

Note: Densities are expressed as number of animals/km^2. Densities were calculated based on population estimates from surveys completed in 1986. Percentage values for wildlife and livestock are given as a proportional percentage by adding the densities of the species in question and dividing by the total density for the given habitat type and month.

to one of these categories. The four vegetation classes used in the analysis were grasslands, shrublands, dwarf shrublands, and agricultural areas.

Species densities were calculated for each vegetation type for three surveys done in May, August, and November 1986. These three surveys were used to represent seasonal changes, and the results are presented in table 8.6 for each species. Several important points emerge:

1. In general, cattle are spread out in all habitat types in high densities during all the seasons. In comparison with other species, they form the highest densities in all vegetation types. The sheep and goats show a pattern similar to that of cattle and, in terms of numbers in most habitat types, are the second most dominant species after cattle. Since the highest livestock densities are recorded in the agricultural areas, it is evident that cultivation and livestock can mutually coexist.

2. Conversely, wildlife populations are shrinking in the agricultural areas, where the proportional density of wildlife is already low compared with that in other vegetation communities. However, wildebeest still cross into agricultural areas during the August migration, and this movement inevitably brings them into direct conflict with humans.

3. In the period of June through October the migrant wildebeest and Burchell's zebra distort the overall distribution and densities of wildlife in the study area. However, when just the May and November data are considered, wildlife constitutes about 46%, 37%, 35%, and 24% of the animals in the grasslands, shrublands, dwarf shrublands, and agricultural areas respectively.

4. Before the in-migration in May, the resident wildebeest population constitutes about 9%, 5%, 2%, and 2% of the animals in agricultural lands, dwarf shrublands, grasslands, and shrublands respectively. Immediately after the out-migration in November, they constitute about 10%, 12%, 6%, and 1% in shrublands, grasslands, dwarf shrublands, and agricultural areas respectively. In May, it appears that the resident wildebeest show increased use of dwarf shrubland and agricultural areas.

5. Zebra occur in higher densities in dwarf shrublands and agricultural areas than in grasslands and shrublands in May. The reverse is true in August. In November they are evenly spread throughout all habitats except for the agricultural areas, where they exhibit very low densities.

6. In August, Thomson's gazelle occur in relatively lower densities (compared with May and November) in the grasslands, dwarf shrublands, and shrublands. The reverse is true for the agricultural areas, where they appear to increase in density in August.

7. Impala appear to occur in very low densities in the agricultural areas compared with the other habitats. In all habitats, the densities appear lowest in the months of August and November. There is an apparent strong buildup of the population in the month of May that is most pronounced in the shrublands and grasslands.

8. Buffalo are found in shrublands and grasslands most of the time, but rarely in dwarf shrublands and agricultural areas, except in May, when they concentrate in the dwarf shrublands, and in November, when they concentrate in the agricultural areas.

To examine habitat use by species further, maps were prepared that show animal densities plotted on top of the vegetation/habitat map. Simplified examples are given in figures 8.7–8.9 for wildebeest, cattle, and Thomson's gazelle, respectively. These maps show seasonal shifts of resident wildebeest between habitats (fig. 8.7) and the higher densities of Thomson's gazelle noted in the May 1986 survey (fig. 8.9).

CONCLUSION

In general, wildlife trends within the Mara Reserve between 1977 and 1990 appear to be fairly stable, while resident wildlife species in the surrounding ranch areas appear to be in a general decline. It is during this same period that the area has experienced rapid land use changes, especially in the north, where large wheat fields were opened up and are slowly spreading southward. This farming activity has resulted in constriction of the area available for wildlife. Alteration of the vegetation habitat and the fact that wildlife species are in direct conflict with farmers have probably been the major factors leading to their decline within the agricultural areas. The general decline of wildlife numbers in other areas

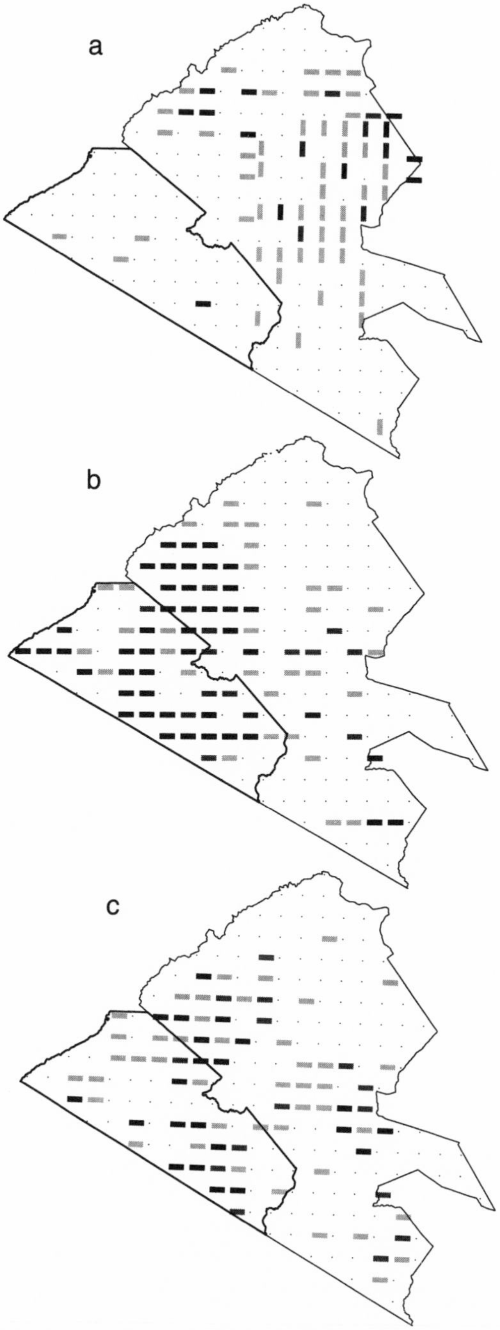

Figure 8.7 Seasonal shifts in habitat utilization for wildebeest: (*a*) May 1986; (*b*) August 1986; (*c*) November 1986. Animal densities (animals/km^2) are plotted with different shading intensity: darkest represents densities above 50; 2d darkest, densities between 20 and 50; 3d darkest, densities between 10 and 20; lightest, below 20. A dot represents a surveyed cell with zero animals. The orientation of shaded bars shows the direction of transects.

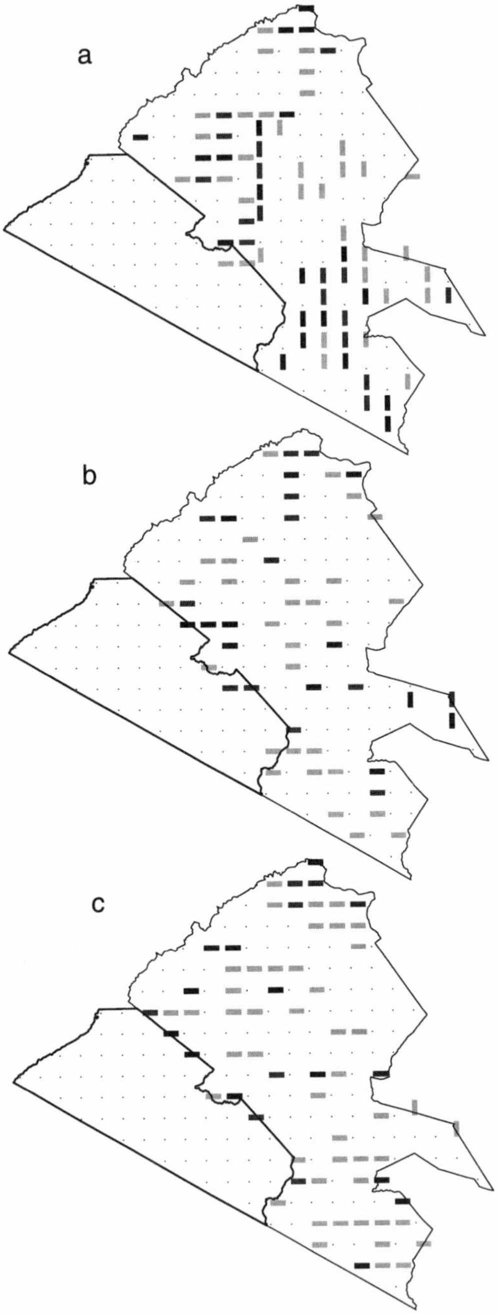

Figure 8.8 Seasonal shifts in habitat utilization for cattle: (*a*) May 1986; (*b*) August 1986; (*c*) November 1986. Animal densities (animals/km^2) are plotted with different shading intensity: darkest represents densities above 200; 2d darkest, densities between 100 and 200; 3d darkest, densities between 50 and 100; lightest, below 50. See fig. 8.7 for other symbols.

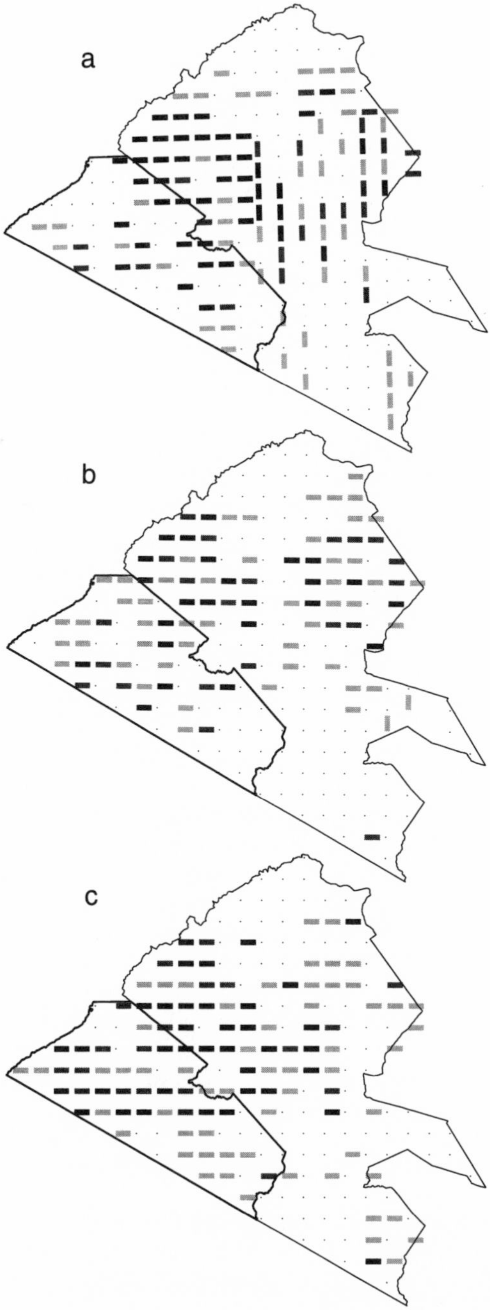

Figure 8.9 Seasonal shifts in habitat utilization for Thomson's gazelle: (*a*) May 1986; (*b*) August 1986; (*c*) November 1986. Animal densities (animals/km^2) are plotted with different shading intensity: darkest represents densities above 50; 2d darkest, densities between 20 and 50; 3d darkest, densities between 10 and 20; lightest, below 20. See fig. 8.7 for other symbols.

outside of the reserve may be partly explained by competition with livestock grazing. Further research will be required to document and explain the reasons for such declines.

The analysis of habitat use presented here illustrates the importance of seasonal changes in animal populations and provides some baseline observations that require further research.

There is still a very rich reservoir of wildlife species and numbers in the private ranching areas of the Mara region. The ranches also form the northernmost migratory route for Serengeti-Mara migrants. We need to identify management solutions for the conservation of the area in order to respond to today's environmental and economic pressures and preserve these species for posterity.

We have presented the results of an ongoing long-term monitoring program and illustrated some preliminary analyses of the collected data. The most important conclusion is that there is no substitute for reliable long-term monitoring data. Variations and trends in populations can be established only by using consistent and accurate data that are gathered in the same way over a long period of time.

REFERENCES

Amuyunzu, C. L. 1984. Land resource inventory as a basis of land evaluation and rural development: The role of remote sensing techniques. M.Sc. thesis, ITC, Netherlands.

Aore, W. W. 1988. An assessment of effects of wheat cultivation on land qualities in Angata Plains, Kenya. M.Sc. thesis, ITC, Netherlands.

Brown, L. H., and Cocheme, J. 1973. *A study of the agroclimatology of the highlands of eastern Africa.* Technical Note no. 125 (WMO 339). Geneva, Switzerland: World Meteorological Organization.

Douglas-Hamilton, I. 1990. *Identification study for the conservation and sustainable use of the Kenya portion of the Mara-Serengeti ecosystem.* Nairobi: European Development Fund of the European Economic Community.

Epp, H., and Agatsiva, J. 1982. *Habitat types of Mara-Narok area, western Kenya.* KREMU Technical Report no. 20. Ministry of Environment and Natural Resources, Kenya Rangeland and Ecological Monitoring Unit.

Griffiths, J. F. 1972. *Climates of Africa.* World Survey of Climatology, vol. 10.

Jolly, G. M. 1969. Sampling methods for aerial censuses of wildlife populations. *E. Afr. Agric. For. J.* 34:46–49.

Msafiri, F. 1984. A vegetation map of Lolgorien Area of Narok District. KREMU Technical Report no. 119. Ministry of Environment and Natural Resources, Kenya Rangeland and Ecological Monitoring Unit.

Norton-Griffiths, M. 1978. *Counting animals.* 2d ed. Handbook no. 1. Techniques in African Wildlife Ecology. Nairobi: African Wildlife Foundation.

Pratt, D. J., Greenway, P. J., and Gwynne, M. D. 1966. A classification of East

African rangeland with an appendix on terminology. *J. Appl. Ecol.* 3:369–82.

Republic of Kenya. 1980, 1984, 1989. Narok District Development Plans. Nairobi: Ministry of Planning and National Development.

Sombroek, W. G., Braun, H. M. H., and Van der Pouw, B. J. 1982. Exploratory soil map and agroclimatic zone map of Kenya, 1980, scale: 1:1,000,000. Republic of Kenya, Ministry of Agriculture, National Agricultural Laboratories, Kenya Soil Survey.

Stelfox, J. G., Peden, D. G., Epp, H., Hudson, R. J., Mbugwa, S. A., Agatsiva, J. L., and Amayunzu, C. L. 1986. Herbivore dynamics in southern Narok, Kenya. *J. Wildl. Mgmt.* 50:339–47.

NINE

Population Limitation of Resident Herbivores

A. R. E. Sinclair

In chapters 6 and 8, Campbell and Borner and Broten and Said present information on the population trends of various ungulate species in the whole Serengeti-Mara region, and discuss management and conservation concerns for threatened species. In this chapter I look at a part of this area in detail to consider possible causes for the natural limitation of some of the larger resident ungulate populations. To do this I use census information covering 27 years (1967–1993) collected by myself from a study area in northern Serengeti and examine the consequences of a major perturbation due to poaching partway through this period. These data are supplemented by information on age and sex ratios, collection of which began in 1980 during the perturbation period, and by census data from other areas of Serengeti, which have been collated from various sources. Both methods and observers have changed for these other censuses, and reliability is variable; I consider this in the text. The most complete data come from topi and impala, with that for kongoni, giraffe, and waterbuck being less complete. Data on warthog, ostrich, and Thomson's gazelle (in the north only) are fragmentary, but I present them for the record.

HYPOTHESES OF POPULATION LIMITATION

Previous work on the causes of natural regulation of ungulate populations in Serengeti has pointed to lack of food during the dry season in both the migratory wildebeest population (Sinclair 1979; Sinclair, Dublin, and Borner 1985) and the nonmigratory buffalo population (Sinclair 1977). However, migratory populations may escape predator regulation because of their movements, which implies that resident populations could be more vulnerable to the effects of predators (Fryxell, Greever, and Sinclair 1988; Fryxell and Sinclair 1988). Similar observations have been made in Namibia (Mills and Shenk 1992) and the Northwest Territories of Canada (Heard and Williams 1992). These observations lead to the hypothesis that resident ungulate populations, especially those of smaller species, may be regulated by predators. Exceptions to this hypothesis are produced by the very large herbivores, such as buffalo, hippopotamus, rhinoceros, and elephant, which may have outgrown their predators. The hypothesis of predator regulation of smaller resident ungulates is supported by circumstantial evidence that antelope species tend to coexist in closer proximity, and with greater ecological overlap, than would be expected either by chance or from interspecific competition (Sinclair 1985; Mduma and Sinclair 1994).

The alternative hypothesis that resident populations are resource-limited is supported by the several studies of niche partitioning in African ungulates (Lamprey 1963; Gwynne and Bell 1968; Field 1972; Jarman 1971; Ferrar and Walker 1974; Murray and Brown 1993). This circumstantial evidence implies that resources are in short supply, so that interspecific competition results and coexistence develops by the use of different resources. In addition, increases of ungulate populations following removal of the dominant herbivore species provide compelling evidence for resource competition. For example, the removal of hippopotamus through culling in Queen Elizabeth National Park, Uganda, resulted in increases in buffalo and waterbuck (Eltringham 1974), and the die-off of wildebeest through drought in Nairobi National Park, Kenya, produced an increase in kongoni (Foster and Kearney 1967; Foster and McLaughlin 1968). However, experimental removals of waterbuck in Umfolozi Reserve, South Africa, pointed more to limitation by parasites than to limitation by competition (Melton 1987).

EXPERIMENTAL PERTURBATION

Researchers are testing the predation and interspecific competition hypotheses by observing the consequences of an artificial (human-caused) perturbation in the wildlife populations of northern Serengeti. Sometime between 1976 and 1982, approximately 85% of buffalo (ca. 9,000 animals) in northern Serengeti were removed by poachers (Sinclair 1977; Dublin et al. 1990) (Sinclair, chap. 1, fig. 1.5). Figure 9.1 shows the distribution of buffalo herds in northern Serengeti in 1970 and 1984, before and after this removal, in comparison with the Mara, where removals have not occurred. In a 182 km^2 study area (fig. 9.1) to the west of the Wogakuria-Kogatende road, on the south bank of the Mara River, there were 1,700 animals during 1967–1970. In 1978 there were only about 500, in 1980 there were 150, and by 1984 there were a mere 50, only 3% of the original number. For the following 5 years (1984–1989) heavy poaching continued in this study area, and the buffalo population was limited to a single herd of only 100–200 animals that made occasional forays into the area from the eastern boundary. Since 1989 increased anti-

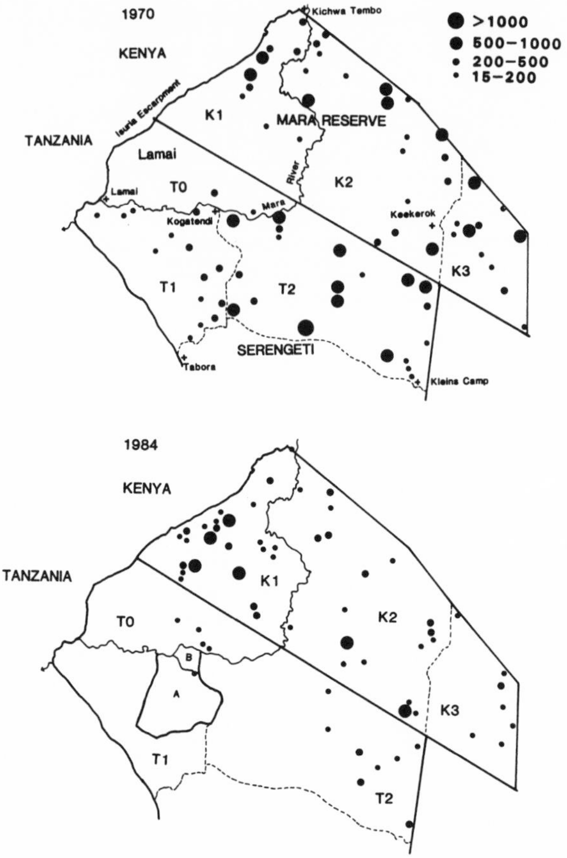

Figure 9.1 The distribution of buffalo herds in 1970, before removal began in 1976, and in 1984. The Mara areas K1–K3 constitute the control where no removals occurred; the Serengeti areas T0–T2 experienced the removals. A is the northern study area, and B is the Kogatende block.

poaching activity by Tanzanian authorities has led to less disturbance for buffalo, and a herd of 300 is seen more frequently on the eastern boundary. However, few other buffalo occur in the study area. Thus, the previously dominant herbivore in this area (Sinclair 1977) has been effectively absent for the 14 years 1980–1994. While the removal of buffalo has been the best-documented consequence of poaching, there is some circumstantial evidence that lions, which are the main predators in the removal area (Schaller 1972), also disappeared: roar counts were considerably lower in the study area in 1990 compared with other parts of Serengeti (Packer 1990), and I have noticed a decline in sightings since the 1970s. There are two ways lion numbers could have declined. One is through poaching. The other is a natural decline consequent on the reduction of buffalo, the main food of lions in this area. The other potential predator, the hyena, is very uncommon in the north (Kruuk 1972; pers. obs.).

A second removal area in the western part of Serengeti (fig. 9.2) experienced a 50% reduction of buffalo (Dublin et al. 1990) and reductions of predators. Counts of lion roars were also low here (Packer 1990), and I observed lions less frequently in 1992 than in the 1960s. Piles of hyena skulls, clearly killed by humans, were observed in 1978, 1980, and 1986 (pers. obs.). Hofer and East (chap. 16) report poaching of hyenas in the west. Therefore, I have used this area as a replicate for the northern Serengeti in terms of census information.

This removal experiment can be compared with control treatments in two other areas (fig. 9.2), one immediately adjoining the northern removal area in the Mara Reserve, Kenya, and the other on the eastern side of Serengeti National Park. In both areas poaching on buffalo has been minimal (Ottichilo, Kufwafwa, and Stelfox 1987; Dublin et al. 1990) and lions have not suffered measurable poaching mortality. In terms of habitat, the Mara control is similar to the northern removal area, and the eastern control is similar to the western removal area for all species except kongoni. The west is an unsuitable removal treatment for kongoni because this species does not occur over most of it. While young woodland has been increasing there (Sinclair, chap. 5) this has occurred in all areas. In terms of other effects, elephant and rhinoceros have been removed from all areas. Their diet (Dublin 1986; pers. obs.) overlaps only that of giraffe in the dry season.

The effects of the removal experiment are most likely to be observed in topi and impala because of their larger numbers and the more complete

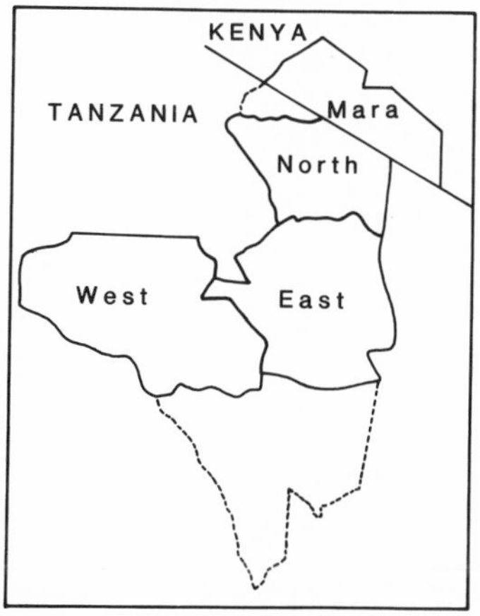

Figure 9.2 The Serengeti-Mara, showing the aerial survey blocks. North and West are removal blocks, Mara and East are control blocks.

and reliable data. Topi, kongoni, warthog, and waterbuck are pure grazers (Bell 1970; Field 1972; Duncan 1975; Spinage 1982; Murray 1991; Campbell and Borner, chap. 6); impala and Thomson's gazelle are wet season grazers, but dry season browsers (Bell 1970; Jarman and Sinclair 1979; Underwood 1983; Sinclair 1985); and giraffe are pure browsers (Hall-Martin 1974; Leuthold and Leuthold 1975). Ostrich are both grazers and browsers (Sinclair 1978; Brown, Urban, and Newman 1982). The removals of a major herbivore competitor (buffalo) and a major predator (lion) allow predictions from the two hypotheses on how the populations of other resident ungulates should respond. Since buffalo are pure grazers (grass eaters) in the Serengeti region, (Sinclair 1977; Sinclair and Gwynne 1972) the competition hypothesis predicts (1) that other populations of grazing species should increase in the absence of buffalo, while those of browsing species should not be affected; and (2) that the competitive effect of buffalo should be most apparent in the survival of juveniles of other grazers during their first dry season because that is when resources are limiting (Sinclair 1977; Sinclair, Dublin, and Borner 1985). Thus yearling ratios (i.e., after the first dry season) should be higher for grazers in the removal area. In contrast, the predation hypothesis predicts (3) that both grazer and browser populations should increase as a result of predator removal; and (4) that since predators affect early calf survival, both calf (those alive before the first dry season) and yearling ratios should be higher in the removal area.

METHODS

Census

Detailed descriptions of census methods and extensive analyses of sample errors for aerial transect counts and bias errors for total counts may be found in Sinclair (1972, 1973, 1977), Norton-Griffiths (1973, 1978), and Sinclair and Norton-Griffiths (1982), so I will not repeat those details here. I report what type of census method was used except where a variation of method requires elaboration. Censuses were conducted at three different scales. First, broad-scale aerial transect surveys covered the Mara and most of Serengeti, including the two removal and two control areas described above. Second, aerial total counts covered a 182 km² "northern study area." This area is situated on the south bank of the Mara River in northern Serengeti (area A, fig. 9.1) and lies between Nyamburi Ridge on the west and Kogatende Ridge on the east. The third level of census was conducted in a subsection of the northern study area, called the "Kogatende block." This is an area of some 10 km² on the northern part of Kogatende Ridge and includes Rhino plain (area B, fig. 9.1). Censuses in this block were total counts also, sometimes by air, other times from the ground.

Aerial transect censuses for resident ungulates are described in Sinclair (1972) and Grimsdell (1979) for the counts of 1971 and 1976 respectively. These censuses were designed for resident ungulates and used transects oriented in a specific direction but randomly spaced apart. During monthly aerial surveys in 1969–1972 and aerial transect censuses since 1985, east-west—oriented transects spaced systematically 10 km apart were employed using a standardized survey grid. Details of this grid are given in Maddock (1979). These surveys covered the whole Serengeti-Mara region at monthly intervals from August 1969 to August 1972 (5 months were omitted). Until 1971 the surveys did not use transect markers on the wing struts and so the transect width was unmeasured. However, calibration of this technique in February 1971 with the subsequent approach of a set transect width defined by strings attached to the strut showed that at 91 m (300 ft.) flying height, an effective counting width of 150 m either side of the aircraft, or approximately 3% of the area, was covered (Norton-Griffiths 1978). In some of these early survey flights, particular species were counted properly (rather than classified in group size categories), and it is these counts that I have used. The 1985 census was conducted under difficult conditions with less experienced observers, so the results may be less reliable.

Aerial total counts of the northern study area for all resident ungulates were conducted monthly from August 1967 to September 1969 and again from July 1971 to November 1972 (Sinclair 1977). The census was repeated in March 1989. The area was also surveyed by air for buffalo only during May 1965–1976, 1978, 1984, and 1986, and from the ground in June 1980. In November 1988 a ground count of topi was conducted. Only wet season counts (November–June) were used to avoid confusion caused by the presence of migratory wildebeest and zebra. The several wet season surveys allowed an estimate of variation in the counts to be calculated.

The Kogatende block was censused during the northern study area aerial total counts described above. In addition, ground counts were conducted in June 1980, as well as in December 1990–1993 for some species and in June 1986 for topi. As with the aerial counts, animals were plotted on maps. All were counted with binoculars. Visibility was good, so herds were easily identified and double counting or missed herds were unlikely. The repeated monthly censuses in the 1960s and 1970s give some indication of variability in bias error. Since the aerial survey was conducted in the same way each time, the results may be regarded as indices of population trends.

The three different scales of census can address different scales of movement. For example, subpopulations, or herds, of both topi and giraffe have been observed to move 10–20 km between seasons. This means that the northern study area and Kogatende block have high variability in counts between seasons, which is not reflected in the larger-scale aerial transect censuses. To avoid problems of seasonality, only wet season counts have been used. Conversely, changes in the highly resident impala and Thomson's gazelle populations can be monitored more closely in the intensively covered Kogatende block, while such changes may go undetected in the less intensive aerial transect surveys.

Age and Sex Ratios

Since 1980, age and sex ratios were obtained for resident species in the north, east and Mara areas of Serengeti. Ages of juveniles were classified into categories based on body size, horn length, horn shape, coat color, or combinations of all these. Ages were assigned to these categories on the basis of birth seasons and growth rates. Once animals reached the adult stages, ages were not assigned.

In most species the sexes of juveniles up to 1 year old were not distinguished. In both impala and waterbuck, however, females do not have horns at all, while 6-month-old males already have horns up to the length of their ears. Thus in these species, age categories older than 6 months were estimated separately for the two sexes. In most species the sexes of adults could be distinguished by the presence of horns, shape and size of horns, or other secondary sexual characters. In warthogs, however, small size precluded consistent identification of the sexes (by presence or absence of testes), so both sexes were lumped as adults.

Data were obtained by counting animals from a vehicle and recording the information either on data sheets or on tape. Earlier data were obtained intermittently, but more recent samples were obtained monthly. The open habitat and good visibility meant that, with the possible exception of waterbuck, biases in recording age or sex categories were unlikely (Samuel et al. 1992). Standard errors of the ratios were computed using standard formulae (Zar 1984).

RESULTS

Censuses

Topi. Estimates of topi numbers are collated in table 9.1. The estimate by R. H. V. Bell in 1966 for the west was from a total count. All other counts were from transects, and standard errors are included where published. In removal areas topi numbers have increased substantially. In the north, density increased by a factor of 4, from about 2/km^2 in the early 1970s to about 8/km^2 in the mid-1980s. In the west, overall densities were

Table 9.1 Aerial transect censuses for topi.

Date of survey (Source)	North (Removal)				Mara (Control)			
	Area	Total	SE	D	Area	Total	SE	D
Pre-removal								
1970–71 (2)	1,844	2,448	400	1.30	2,600	31,983	2,706	12.30
1971 (3)	1,844	3,378	672	1.83	—	—	—	—
Post-removal								
1977 (4)	—	—	—	—	1,588	17,900	—	11.27
1978–79 (4)	—	—	—	—	2,350	30,832	7,495	13.12
1985 (5)	1,844	6,987	1,208	3.80	—	—	—	—
1986 (6)	1,844	11,637	4,209	6.31	1,588	20,343	7,159	12.81
1987 (7)	—	—	—	—	2,350	28,642	8,794	12.19
1988–89 (8)	1,989	18,167	972	9.13	—		—	

	West (Removal)				East (Control)			
	Area	Total	SE	D	Area	Total	SE	D
Pre-removal								
1966 (1)	2,000	11,909	—	5.95	—	—	—	—
1970–71 (2)	—	—	—	—	3,687	7,691	1,345	2.09
1971 (3)	4,045	17,072	988	4.49	3,687	6,946	604	1.88
1972 (1)	4,045	26,384	—	6.52	—	—	—	—
Post-removal								
1976 (1)	4,045	51,633	7,281	13.40	3,687	9,651	1,530	2.76
1985 (5)	3,616	25,226	3,829	6.98	4,121	13,489	1,566	3.27
1988–89 (8)	3,629	37,154	4,820	10.24	3,249	6,703	837	2.06

Sources: (1) Grimsdell 1979; (2) Serengeti Ecological Monitoring Programme, unpublished data; (3) Sinclair 1972; (4) Stelfox et al. 1986; (5) Campbell 1988; (6) this study; (7) Department of Resource Surveys and Remote Sensing, Nairobi, 1989; (8) Campbell 1989.
Note: Area is in km^2. D = density as number per km^2.

much higher than in the north because of the difference in habitats. Density increased in the west by a factor of about 2, from about $5/km^2$ during 1966–1972 to about $10/km^2$ during 1976–1989. The increase was probably not as sudden as it appears in the data because the estimates may have been too high in 1976 and too low in 1985, so that a more gradual increase occurred through the 1970s with a leveling out in the 1980s. In contrast to the two removal areas, the two controls showed no consistent trend in the last two decades. Thus, the Mara, which resembles the north in habitat, has remained at high density ($11–13/km^2$) since 1970. The east, which is dry savanna, less suitable for topi, has remained at lower densities ($2–3/km^2$).

There were nine total counts in the northern study area between July 1971 and July 1972 (table 9.2). The mean count of topi there was 141 ($\pm$ 16 SE); in the Kogatende block it was 76 ($\pm$ 13) (table 9.3). In contrast, estimates of 1,200 in November 1988 and 467 in March 1989 in the northern study area were significantly higher ($P < .01$) than the nine counts in the early 1970s. Similarly, five counts of the Kogatende block from May 1986 to December 1992 (249 $\pm$ 33) were significantly above

Table 9.2 Total counts in the northern study area.

	Pre-removal					Post-removal	
Year	1967	1968	1969	1971	1972	1988	1989
Topi							
Mean	73	75	71	124	146	1,200	467
SE	—	15	12	34	19	—	—
Max	73	127	114	158	215	1,200	467
Impala							
Mean	92	150	136	162	196	—	527
SE	29	22	32	—	34	—	—
Max	120	249	198	162	410	—	527
Kongoni							
Mean	—	51	66	—	28	—	42
SE	—	6	8	—	4	—	—
Max	—	74	93	—	49	—	42
Giraffe							
Mean	37	46	24	41	32	—	—
SE	12	11	9	32	10	—	—
Max	62	126	57	73	78	—	—
Waterbuck							
Mean	7.7	16	10	7	30	—	21
SE	0.9	9	5	2	8	—	—
Max	9	50	26	9	60	—	21
Warthog							
Mean	—	14	—	—	—	—	15
SE	—	10	—	—	—	—	—
Max	—	30	—	—	—	—	15
Ostrich							
Mean	—	6.5	12	4	5.6	—	7
SE	—	0.5	—	1	1.8	—	—
Max	—	7	12	5	9	—	7

those in the 1970s. In general, these counts show that topi numbers have increased by a factor of 2–8 in the removal areas, while they have remained constant in the control areas (fig. 9.3).

Impala. The first transect census of impala in 1971 (table 9.4) in the west was an underestimate because orientation of the transects was parallel to the riverine thickets where the animals occurred, and many groups were missed. The repeat census in 1972 by P. Duncan, with better orientation of transects, is probably more reliable. The early transect surveys in Mara during 1970–1971 were corrected for underestimates of herd sizes using a photographic calibration (Sinclair 1973).

In the two removal areas, no changes in impala densities could be detected from the aerial surveys (but see the ground census data for the north below). In the control areas, the Mara population has also remained constant. The eastern population appeared to show a steady decline until 1988, but then rebounded to original densities in 1989. However, none of the censuses since 1976 are significantly different from one another, so the population appears constant.

Table 9.3 Total counts in the Kogatende Block.

	Pre-removal					Post-removal							
	1967	1968	1969	1971	1972	1980	1986	1988	1989	1990	1991	1992	1993
Topi													
Mean	57	33	22	86	73	107	300	—	340	153	247	207	*
SE	—	10	3	38	15	—	—	—	—	—	—	—	*
Max	57	74	29	124	126	107	300	—	340	153	247	207	*
Impala													
Mean	92	77	81	63	110	—	—	—	445	291	404	270	555
SE	29	21	28	39	23	—	—	—	—	—	—	—	—
Max	120	182	128	101	209	—	—	—	445	291	404	270	555
Kongoni													
Mean	—	8.6	20.8	—	4.3	—	—	—	16	5	8	11	0
SE	—	3.4	5.1	—	1.7	—	—	—	—	—	—	—	—
Max	—	24	41	—	12	—	—	—	16	5	8	11	0
Giraffe													
Mean	4.5	9.8	10.0	16	9.4	13	—	10	7	10	9	4	4
SE	1.5	4.9	2.5	14	2.3	—	—	—	—	—	—	—	—
Max	6	34	17	30	18	13	—	10	7	10	9	4	4
Waterbuck													
Mean	26	5	—	5	7	11	—	—	15	2	3	2	9
SE	18	1.2	—	—	2.7	—	—	—	—	—	—	—	—
Max	43	7	—	5	21	11	—	—	15	2	3	2	9
Warthog													
Mean	—	4.0	—	—	—	4	—	—	—	22	19	16	22
SE	—	0.7	—	—	—	—	—	—	—	—	—	—	—
Max	—	5	—	—	—	4	—	—	—	22	19	16	22
Ostrich													
Mean	—	3.3	6	—	4.3	—	—	—	—	2	0	1	2
SE	—	0.3	—	—	1.1	—	—	—	—	—	—	—	—
Max	—	4	6	—	6	—	—	—	—	2	0	1	2
Thomson's gazelle													
Mean	—	—	30	—	20	—	—	—	—	170	180	332	455
SE	—	—	—	—	—	—	—	—	—	—	—	—	—
Max	—	—	30	—	20	—	—	—	—	170	180	332	455

*Drought conditions in December 1993 caused dispersal out of the census area.

In the northern study area, total counts (table 9.2) showed 147 (± 17 SE) impala for 1967–1972, compared with 527 in March 1989, which was significantly ($p < .01$) above the earlier distribution. In the Kogatende block (fig. 9.3, table 9.3), there were 85 (± 18) pre-removal, compared with 393 (± 52) in five later counts ($p < .001$). Thus, although impala increases were not detected in the wider aerial survey of the northern removal area, impala numbers have increased by a factor of 3–5 in its western portion (the northern study area) where poaching has been most severe, as indicated by the buffalo removals mentioned earlier.

Kongoni. In the Mara during the 1970s, there appears to have been a significant decline in kongoni density by a factor of 0.5, but since then the population has remained constant (table 9.5). The northern removal area had constant kongoni densities from 1971 to 1988, there being no

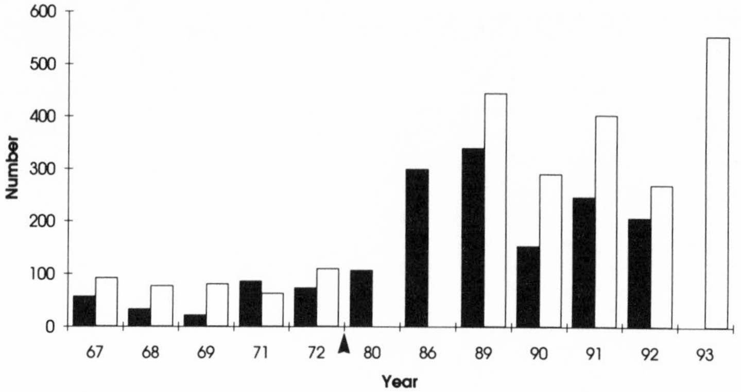

Figure 9.3 Topi (solid bars) and impala (open bars) numbers in the Kogatende block of northern Serengeti. Arrow separates the pre- and post-removal periods.

Table 9.4 Aerial transect censuses for impala.

	North (Removal)				Mara (Control)			
	Area	Total	SE	D	Area	Total	SE	D
Pre-removal								
1970–71	—	—	—	—	2,600	30,751	4,216	11.83
1971	1,844	10,577	3,008	5.74	—	—	—	—
Post-removal								
1978–79	—	—	—	—	2,350	29,960	8,250	12.75
1985	1,844	9,513	1,737	5.16	—	—	—	—
1986	1,844	10,927	4,606	5.92	1,588	24,278	5,609	15.29
1987	—	—	—		6,400	81,792	14,211	12.78
1988	2,022	10,335	2,056	5.11	—	—	—	—
1989	1,955	11,754	2,592	6.00	—	—	—	—
	West (Removal)				**East (Control)**			
	Area	Total	SE	D	Area	Total	SE	D
Pre-removal								
1966	2,000	5,181	—	2.59	—	—	—	—
1971	4,045	6,759	894	1.67	3,687	39,255	3,256	10.65
1972	4,045	20,422	—	5.05	—	—	—	—
Post-removal								
1976	3,862	30,440	4,550	7.88	3,687	33,738	4,503	9.15
1985	3,616	15,248	1,458	4.22	4,121	30,507	5,385	7.40
1988	3,616	24,279	3,927	6.71	3,295	16,559	4,059	5.03
1989	3,642	33,347	7,997	9.16	3,203	33,888	7,595	10.58

Note: Area is in km^2. D = density as number per km^2. Sources as for table 9.1.

significant difference in the estimates from different aerial surveys. The 1989 aerial survey estimate, however, is higher by a factor of 6 than the previous year's, and this difference is not feasible from reproduction. Whether the 1989 estimate is due to immigration from the Loliondo area east of the park or is merely a result of sampling error must be decided by future censuses. Total counts of the northern study area (table 9.2) in

Table 9.5 Aerial transect censuses for kongoni.

	North (Removal)				Mara (Conrol)			
	Area	Total	SE	D	Area	Total	SE	D
Pre-removal								
1970–71	—	—	—	—	2,600	7,838	1,439	3.01
1971	1,844	1,738	340	0.94	—	—	—	—
Post-removal								
1978–79	—	—	—	—	2,350	3,293	835	1.40
1985	1,844	2,871	800	1.56	—	—	—	—
1986	1,844	1,445	926	0.78	1,588	2,580	870	1.62
1987	—	—	—	—	4,200	5,218	—	1.24
1988	2,022	1,340	333	0.66	—	—	—	—
1989	1,955	8,790	1,945	4.50	—	—	—	—
	West (Removal)				East (Control)			
	Area	Total	SE	D	Area	Total	SE	D
Pre-removal								
1966	2,000	197	—	0.10	—	—	—	—
1971	4,045	461	168	0.11	3,687	6,865	619	1.86
Post-removal								
1976	3,862	934	209	0.24	3,687	6,591	1,042	1.79
1985	3,616	1,102	96	0.30	4,121	8,768	1,229	2.13
1988	3,616	497	135	0.13	3,295	5,070	1,024	1.56
1989	3,642	1,089	235	0.30	3,203	7,256	522	2.27

Note: Area is in km^2. D = density as number per km^2. *Sources* as for table 9.1.

the 1970s show a mean of 28 (± 4) and a maximum of 49. The estimate of 42 in March 1989 is not significantly different from the earlier counts. Results from the Kogatende block are similar (table 9.3). There is a possibility that kongoni populations in the eastern part of the northern removal area have been increasing, for I have recorded large groups of bachelor males on its eastern border in the 1980s (e.g., 52 in 1982, 89 in 1988) but only small groups in the 1960s and early 1970s.

In eastern Serengeti, which supports higher kongoni densities than elsewhere, the population has remained constant. The west, with its wide-open floodplains, is unsuitable habitat for this species; densities of kongoni were very low, so this area does not represent a good removal treatment. Densities may have increased after 1976. The density post-removal of 0.24/km^2 is double that of the two earlier estimates but it is not yet significantly different.

Giraffe. In the northern removal area, the mean density from the four censuses post-removal (0.41 ± 0.09) is significantly lower ($p < .05$) than the pre-removal estimate (table 9.6). In the west there appears to have been little change post-removal. None of the censuses differ significantly from one another. Although total counts in the Kogatende block (table 9.3) show no significant decline, these numbers are very small and little

Table 9.6 Aerial transect censuses for giraffe.

	North (Removal)				Mara (Control)			
	Area	Total	SE	D	Area	Total	SE	D
Pre-removal								
1970–71	—	—	—	—	2,600	2,002	344	0.77
1971	1,844	1,672	322	0.91	—	—	—	—
Post-removal								
1978–79	—	—	—	—	2,350	1,995	501	0.85
1985	1,844	444	149	0.24	—	—	—	—
1986	1,844	1,204	281	0.65	1,588	1,454	416	0.92
1987	—	—	—	—	2,350	1,716	235	0.73
1988	2,022	670	189	0.33	—	—	—	—
1989	1,955	856	250	0.44	—	—	—	—
	West (Removal)				East (Control)			
	Area	Total	SE	D	Area	Total	SE	D
Pre-removal								
1971	4,045	2,855	215	0.71	3,687	2,780	317	0.75
Post-removal								
1976	3,862	2,240	474	0.58	3,687	4,970	436	1.35
1985	3,616	3,621	617	1.00	4,121	2,040	428	0.50
1988	3,616	3,015	470	0.83	3,295	2,407	482	0.74
1989	3,642	1,955	178	0.54	3,203	3,736	593	1.17

Note: Area is in km^2. D = density as number per km^2. Sources as for table 9.1.

weight can be placed on them. For this species, the more extensive aerial surveys are probably more reliable.

In the Mara control area, densities appear to have remained constant in these surveys, but Broten and Said (chap. 8) report a decline in numbers. In the east, Grimsdell (1979) reported a significant increase from 1971 to 1976, but subsequently densities have been similar to that for 1971. The mean density post-removal does not differ significantly from that in 1971.

Waterbuck. Waterbuck numbers are low in all areas because these animals are confined to narrow strips of riverine grassland. This habitat is nowhere extensive in the Serengeti region, but is most prevalent in the Mara, and here waterbuck densities are highest (table 9.7).

In the northern removal area there has been a significant decline in population post-removal if the 1989 census result is included. However, more waterbuck were recorded in the total count (table 9.2) of the northern study area (182 km^2) than in the whole northern removal area census zone (1,844 km^2) in 1989, indicating that the latter result is probably underestimated through sampling error. Without the 1989 result the decline is not significant. Total counts in the northern study area and in the Kogatende block (tables 9.2, 9.3) do not show a change in number. In the west there is a significant decline ($p < .05$) between 1976 and 1989.

Table 9.7 Aerial transect censuses for waterbuck.

	North (Removal)				Mara (Control)			
	Area	Total	SE	D	Area	Total	SE	D
Pre-removal								
1970–71	1,844	252	—	0.14	2,600	1,332	—	0.51
1971	1,844	336	65	0.18	—	—	—	—
Post-removal								
1986	1,844	202	90	0.11	1,588	798	374	0.50
1988	2,022	212	110	0.10	—	—	—	—
1989	1,955	19	18	0.01	—	—	—	—
	West (Removal)				**East (Control)**			
	Area	Total	SE	D	Area	Total	SE	D
Pre-removal								
1971	4,045	594	201	0.14	3,687	365	191	0.10
Post-removal								
1976	3,862	1,554	417	0.40	3,687	352	176	0.10
1988	3,616	668	139	0.18	3,295	324	120	0.10
1989	3,642	241	144	0.07	3,203	339	196	0.11

Note: Area is in km². D = density as number per km². Sources as for table 9.1.

Both control areas show no change in waterbuck densities. However, the early 1970s estimate in the Mara is based on two surveys (19 March 1971 and 18 November 1971) with no calculated sampling error, so this result may not be reliable.

Warthog. The only early estimate of warthog comes from one of the first aerial surveys (24 September 1969). Because there is only one estimate and because observers were inexperienced, the result is probably too low. Nevertheless, the data (table 9.8) indicate that there has been a marked increase in both the northern removal and the Mara control areas. This is consistent with anecdotal records of major increases in warthog numbers throughout the Serengeti system during the 1970s.

Roan Antelope. Nineteen systematic aerial surveys between 1969 and 1971 were combined for the Mara control and northern removal areas. These combined data produce a mean estimate of 45 (± 17 SE) roan in the Mara and 23 (± 14 SE) animals in the northern removal area. Although estimates from aerial surveys have not been repeated, it is thought that few, and perhaps no, roan remain in the Mara and possibly 10 occur in the northern removal area (H. Dublin, pers. comm.; pers. obs.).

Ostrich. Ostrich occur in very low numbers in the northern removal area, but judging from total counts in the northern study area and Kogatende block (tables 9.2, 9.3), these numbers have not changed noticeably over the 20 years.

Table 9.8 Aerial transect censuses for warthog.

	North (Removal)				Mara (Control)			
	Area	Total	SE	D	Area	Total	SE	D
Pre-removal								
1970–71	1,844	175	—	0.10	2,600	146	—	0.04
Post-removal								
1978–79	—	—	—	—	2,350	3,300	—	1.40
1986	1,844	722	190	0.39	1,588	1,953	372	1.23
	West (Removal)				East (Control)			
	Area	Total	SE	D	Area	Total	SE	D
Post-removal								
1985	9,305	6,362	912	0.68	4,121	2,763	444	0.67
1988	8,537	6,847	1,162	0.80	3,535	1,162	102	0.33

Note: Area is in km^2. D = density as number per km^2. Sources as for table 9.1.

Thomson's Gazelle. Little information is available on the resident populations of this species because most work is concerned with the migratory population (Dublin et al. 1990), which does not reach the northern removal area (Maddock 1979). In the early 1970s, this gazelle was uncommon in the Kogatende block (table 9.3), with approximately 20–30 present during the wet season. In contrast, during 1991–1993 numbers increased from 170 to 455. In general, the resident Thomson's gazelle population in the northern Serengeti appears to have increased by a factor of 10.

Recruitment

The age ratios have been analyzed in two ways. First, all records for the Serengeti-Mara for the periods October–February, March–July, and August–September were summed to produce mean calves per female (1 year or older) for each period. For species with defined birth peaks in October–December (topi, warthog), these ratios give newborn, 6-month calf, and 12-month calf proportions. Similarly, the ratio of yearlings (18 months) per female in March–July indicates losses in the first dry season between the ages of 6 and 18 months.

The second approach looks at the monthly ratios of juveniles (< 6 months) for the northern removal area and the two control areas of the Mara and eastern Serengeti.

Topi. Table 9.9 gives the age ratios for topi since 1980 for the Serengeti-Mara as a whole. The calf ratio is for calves per adult female plus yearling females during the March–July period. The yearling ratio is the mean of 12–15-month juveniles per female in the October–February period. From observations of pregnant females, it appears that all adult females give

Table 9.9 Calves and yearlings per female (March–July) for the whole Serengeti-Mara.

	Topi		Impala	Kongoni		Giraffe		Waterbuck	Warthog[a]
Year	Calves	Yearlings	Calves	Calves	Yearlings	Calves	Yearlings	Calves	Calves
1980	.330	—	—	.406	—	—	—	—	—
1982	.364	—	—	.421	.107	—	—	.391	—
1983	.179	—	—	.310	.106	—	—	—	—
1984	.279	—	—	—	—	—	—	—	—
1986	.382	—	.200	.380	.135	.177	.502	.190	.447
1987	.287	—	.153	.262	.173	.194	.336	.253	—
1988	.317	.265	.174	.396	—	.229	.366	.172	.681
1989	.499	.104	.177	.390	.081	.175	.267	.224	.970
1990	.417	.234	.180	.300	.130	.179	.194	.340	.996
1991	—	.149	—	—	—	.334	.263	.276	.653

[a]For warthog the ratio is young per adult in the October–February period.

birth. With this assumption of pregnancy rate, the mortality of calves in the first 6 months is about 63% (table 9.10), while that in the second 6 months (which covers the dry season) is 50%.

The monthly calf ratios for the three areas, north, east, and Mara, are shown in figure 9.4. The regressions of this ratio against time, starting at birth in October, indicate that the northern removal area has a significantly higher calf ratio than the two control areas (analysis of covariance, $P < .05$). From the regressions, calf mortality through June (before the dry season) is 65.1%, 70.1%, and 72.4% for the north, east, and Mara respectively. The dry season mortalities (July–September) are 20.6%, 16.4%, and 24.6% respectively. Thus, the Mara control area has the highest mortality rate; it is also the area with the highest density of predators. The northern removal area has the lowest early calf mortality, and also the lowest density of predators (Packer 1990).

Impala. Impala births occur in most months. The annual ratios of calves less than 6 months old per female are given in table 9.9. Yearlings could not be identified with accuracy. Again, assuming that all females produce a calf each year, the yearly calf mortality through 6 months is 79% (table 9.10).

The monthly ratios of 6-month calves per female are relatively constant through the year, except for a small peak in December–January. The three areas of north, east, and Mara show similar ratios of 0.175 (± 0.017 SE), 0.223 (+ 0.030), and 0.176 (+ 0.014) respectively, and regression lines do not differ from one another. Early calf mortalities for these areas are estimated as 82.5%, 77.7%, and 82.4%.

Kongoni. As with impala, kongoni births occur in most months. There is a peak in August–September in the middle of the dry season and another in November–December in the short rains. The ratio of calves per

Table 9.10 Estimates of calf and yearling recruitment per female and calf and yearling mortalities for the Serengeti-Mara area (SE in parentheses).

Species	Calves/female	Yearling/female	% calf mortality	% yearling mortality
Topi	0.371 (.033)	0.188 (.037)	62.9	49.3
Impala	0.215 (.009)	—	78.5	—
Kongoni	0.393 (.045)	0.139 (.012)	60.7	64.6
Giraffe[a]	0.225 (.030)	0.155 (.028)	59.0	31.1
Waterbuck	0.359 (.041)	—	64.1	—
Warthog[b]	1.131	0.467	71.7	58.7

Notes: Calf and yearling ratios are averaged for the whole year except for topi, for which calves are for the period Mar.–July, and yearlings for Oct.–Feb. The calf ratio was adjusted for the presence of yearlings. Thus, if y was the ratio of yearlings to females, and c' was the ratio of calves to females plus yearlings, then the ratio of calves to females (c) is $c = c'\ (1 + y)$. Newborn mortality (M_n) assumes that all adult females give birth, so $M_n = (1 - c)$. Yearling mortality (M_y) = $(c - y/c)$.

[a]For giraffe, assume 0.55 females give birth per year.

[b]For warthog, assume litter size = 4 and adult sex ratio is 2 female/1 male, similar to other ungulates.

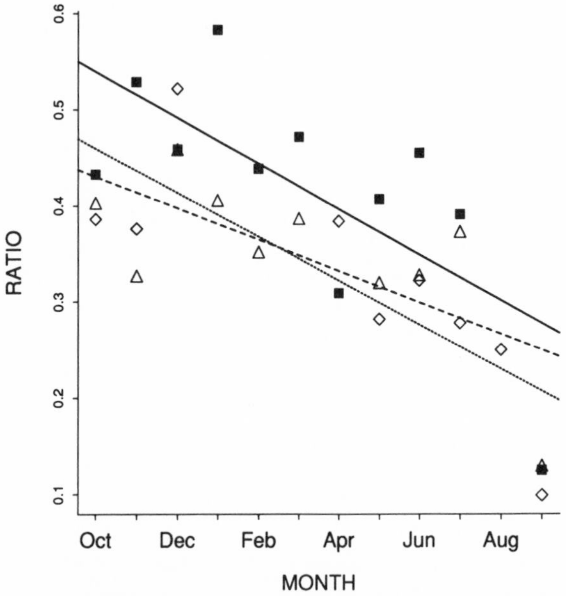

Figure 9.4 Monthly ratios of topi calves per female in the northern removal area (solid squares, solid line) and in the eastern control (open triangles, dashed line) and Mara control (open diamonds, dotted line) areas.

female over the years 1980–1990 show no trend (table 9.9). Assuming all females give birth, early calf mortality (up to 6 months) and yearling mortality (6–18 months) are roughly 61% and 64% respectively (table 9.10).

The mean monthly ratios for calves and yearlings for the north, east, and Mara are shown in figure 9.5. There is no clear trend over the year, but the north may have a higher calf ratio (0.463 [0.033 SE]/female) than

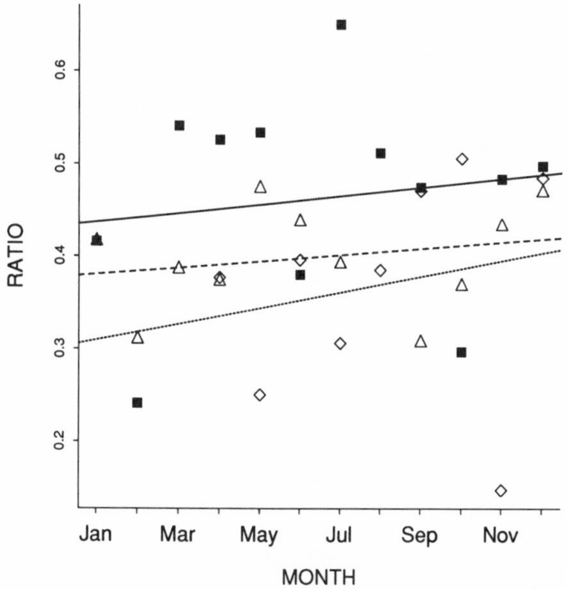

Figure 9.5 Monthly ratios of kongoni calves per female in the northern removal (solid squares, solid line), eastern control (open triangles, dashed line), and Mara control (open diamonds, dotted line) areas.

the east (0.399 [0.017]/female) or Mara (0.369 [.039]/female) (analysis of covariance $P = .06$). If all females give birth, then early calf mortality is 53.7% in the north, which is significantly lower than in the east (60.1%) or Mara (63.1%). Conversely, yearling mortality. which covers the dry season, shows the reverse relationship, being 72.1%, 68.4%, and 57.5% in the three areas respectively. Thus, the north, with the lowest predator density, has the lowest early calf mortality and the highest dry season mortality.

Giraffe. The annual ratios for the 6-month and 18–30-month age categories (table 9.9, fig. 9.6) show a distinct downward trend in the yearlings, suggesting that survival during the dry season (6–12 months old) may be declining. Since this trend is not reflected in the early calf survival, it suggests that intraspecific competition for resources during the dry season is the major limiting factor for this species.

Young are born throughout the year and there is no clear peak of births. At 6 months of age there are 0.225 calves per female, and at 18–30 months there are 0.310 per female. I have assumed that half of this ratio (0.155) represents the yearling ratio, so that mortality from 6 to 18 months is 31% (table 9.10). The gestation length of the giraffe is 15 months (Hall-Martin, Skinner, and Van Dyk 1975) and inter-calf interval

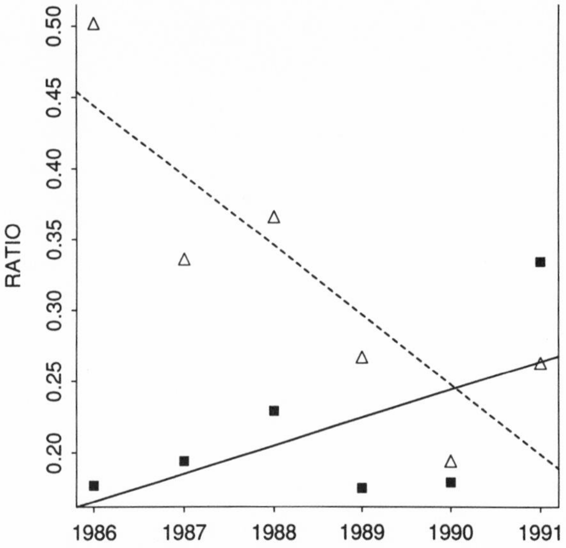

Figure 9.6 Annual ratios for giraffe of 6-month calves (solid squares, solid line) and 18-month yearlings (open triangles, dashed line) per female over the whole Serengeti-Mara.

is 22 months (Foster 1987), so that 55% of females give birth each year. This estimate is supported by the mean monthly proportion of newborns (0.045/female), which represents 54% of females giving birth per year. With a 55% pregnancy rate, calf mortality in the first 6 months is 59%.

Waterbuck. The annual ratio of calves per female (table 9.9) shows no long-term trend. The mean ratio of 0.359 per female (table 9.10) gives an early calf mortality of 64% if all females reproduce in the year, similar to that of topi and kongoni. Although there may be a peak of births during the long rains in May–June, young appear throughout the year.

Warthog. Warthog, unlike other ungulates in Serengeti, give birth to more than one offspring, the average litter size being about four. Births are highly seasonal and fall in the period August–November. Because the sexes of adults are not easily distinguishable in the field, young are recorded as a ratio per adult. The ratio of 1.145 young per adult in September declines by 72% to 0.311 per adult in the following August, when young are nearly a year old. In table 9.10 I have assumed an adult sex ratio of two females to one male, similar to that of other ungulates, so that mortality from birth to 6 months is 72%, and for 6–12 months, over the dry season, it is 59%. There is a suggestion of increasing survivorship for very young ages (October–February period) from 1986 to 1990 (table 9.9).

DISCUSSION

These results can be compared with the predictions of the interspecific competition and predation hypotheses. The competition hypothesis predicts that as a result of the removal of buffalo, (1) grazing species but not browsing species populations should increase and (2) dry season mortality of calves should decrease in the removal areas. In contrast, the predation hypothesis predicts that as a result of the disappearance of predators, (3) both grazers and browsers should increase and (4) newborn as well as calf mortality should decrease in the removal areas.

Topi. Numbers have increased in both removal areas but not in the control areas. In addition, survival of newborn animals increased in the one removal area where it was measured as compared with the control areas. While the census results can be explained by both the interspecific competition and predation hypotheses, the survival of newborn animals is better explained by the predation hypothesis.

Impala. Numbers increased in the northern removal area but not in the west in comparison with controls. Since this species is a browser in the dry season, competition with buffalo would not occur, and the interspecific competition hypothesis should not explain the result. The predation hypothesis is, therefore, a likely possibility (prediction 3 above). This result could have come about either through a direct reduction in predators or through the process of "apparent competition" (Holt 1977; Holt and Lawton 1994). The latter occurs when predator numbers decline as their main prey, in this case buffalo, decline, reducing incidental predation on other prey species.

One other possible explanation for the increase in numbers is a change in habitat due to the increase in young trees (Sinclair, chap. 5) producing thickets. These thickets are a preferred habitat for this species. However, this habitat change has occurred throughout Serengeti, including both removal areas and the eastern control area, but not the Mara control area. Therefore, habitat changes are not consistent with population changes of impala and are an unlikely explanation.

Kongoni. Numbers have not changed significantly in either removal or control areas. Newborn survival has increased in the northern removal area, as predicted by the predation hypothesis, but numbers have not increased because of higher yearling mortality in the dry season. In general, there has been no strong reaction to the perturbations in Serengeti, and this suggests that other processes, such as intraspecific competition in the dry season, may be limiting the population. This species is better adapted to the tall dry grasslands of eastern Serengeti (Murray and Brown 1993) where perturbations have not occurred.

Giraffe. The aerial surveys indicate a decline in numbers in the north. Although newborn survival has been either constant or increasing, yearling survival has been decreasing in the past 10 years. This finding suggests that mortality during the first dry season is the limiting factor, and this could be caused either by intraspecific competition or by poaching.

Waterbuck. Waterbuck numbers in the west have declined, and there is a suggestion this may also have occurred in the northern removal area, relative to the controls. These results suggest that poaching has affected this species as well. Waterbuck are confined to riverine vegetation during the dry season in Serengeti, and they inhabit thickets and forest. This is where poachers set snares for wildebeest and zebra, so it is highly likely that waterbuck will also be caught. This incidental catch by snaring, if it remains a constant number, would become an increasing proportion of a declining population, and so could eradicate the population in those areas.

Warthog. Warthog have increased in the northern removal area but have declined in the Mara (Broten and Said, chap. 8). They are unlikely to be competitors of buffalo because the species' diets do not overlap. In contrast, they are one of the preferred prey of lions, which suggests that population changes are due to the disappearance of predators.

An alternative hypothesis suggests that warthog numbers have increased as a result of the disappearance of rinderpest, as proposed for the population increases of buffalo and wildebeest in the 1960s and 1970s (Sinclair 1979). Warthog are susceptible to the disease, and the recent evidence of an increase in this species on the Serengeti plains suggests that they may also have responded to the absence of rinderpest. Why they should have increased more slowly than the other species remains unknown. If lions switched to warthog as they became more common (i.e., showed a type III functional response) then predation could have slowed the rate of increase of warthog, and it may eventually regulate the population.

Roan Antelope. Roan antelope are now extremely rare in the northern part of the Serengeti-Mara, although they are more abundant (but still not common) in the far south and in the adjacent Maswa region. Small groups of 5–10 used to occur in the 1960s on the central hills of Nyaraswiga and Banagi. I saw one lone yearling at Banagi in 1980 and none since. Poaching could have caused the roan's extinction, as may be occurring with waterbuck. There is another possible cause, however, through loss of habitat. Roan inhabit *Combretum* woodlands on hills. With the intensive fires since the 1950s (Dublin, chap. 4; Sinclair, chap. 5) these woodlands have suffered a severe reduction in central Serengeti and above the Isuria escarpment in the Mara. There is still much of this woodland in the south, and it is regenerating in the north (Sinclair, chap. 5); these are the two areas where roan are still found. A study of these woodlands and the species that occupy them is needed.

Thomson's Gazelle. The Thomson's gazelle has increased in the northern study area but not in the Mara (Broten and Said, chap. 8). Since it is a browser in the dry season it is unlikely to be a competitor of buffalo, and so may be responding to a drop in predation rate. Alternatively, other observations suggest that in the wet season in the north, when the grass is very tall (1–2 m) and unsuitable for gazelle, this species associates very closely with topi, as it does in the Mara (Sinclair 1985). Gazelle use the short-grass swards created by the large topi herds that have appeared in the 1980s. This raises the hypothesis that Thomson's gazelle were previously limited in northern Serengeti by suitable short-grass habitat, and that their subsequent increase may have been caused by facilitation from topi through production of new habitat. This hypothesis needs testing experimentally.

CONCLUSION

I have treated the removals of buffalo and large predators in northern and western Serengeti as a perturbation experiment to test the predictions of the predation and interspecific competition hypotheses for the coexistence of ungulate species in African savannas. In terms of population responses, topi have increased in the removal areas while they have remained constant in the controls. Impala, Thomson's gazelle, and warthog have all increased in the northern removal area but not in the Mara control area. Kongoni have not responded to removals. Waterbuck have declined in the west, and giraffe have declined in the north. Warthog have increased in both the north and Mara.

For the decade 1980–1990, both topi and kongoni early juvenile recruitment appear to be higher in the northern removal area than in the controls. In kongoni, higher yearling mortality counteracts the juvenile recruitment, which could explain the lack of population response. Warthog recruitment has also increased. Giraffe yearling recruitment has declined over the Serengeti as a whole.

The results for topi, impala, Thomson's gazelle, and warthog are consistent with the hypothesis that these species are predator-limited. Their increase in numbers may be either a direct result of lower predation if predators were killed by poachers, or a consequence of "apparent competition" if predators declined from lack of their main food species, the buffalo. At present the data do not distinguish between these two interpretations.

The results for kongoni suggest that intraspecific competition may be operating. Both giraffe and waterbuck appear to be declining through poaching. Thomson's gazelle may also have increased in the north through facilitation by topi.

ACKNOWLEDGMENTS

I would like to thank all those, too many to mention, who have helped with the censuses in the past quarter century. My colleague, Mike Norton-Griffiths, has helped through most of this time. In the past decade, I have been greatly helped in the collection of age and sex classification data by P. Arcese, S. Boutin, H. Dublin, G. Jongejan, S. Mduma, and Anne Sinclair. I thank D. Schluter, D. Hik, and two anonymous referees for many constructive comments on a preliminary draft.

REFERENCES

Bell, R. H. V. 1970. The use of the herb layer by grazing ungulates in the Serengeti. In *Animal populations in relation to their food resources,* ed. A. Watson, 111–23. Oxford: Blackwell.

Brown, L. H., Urban, E. K., and Newman, K. 1982. *The birds of Africa.* vol. 1. New York: Academic Press.

Campbell, K. L. I. 1988. *Programme report, September 1988.* Seronera: Serengeti Ecological Monitoring Programme.

———. 1989. *Programme report, September 1989.* Seronera: Serengeti Ecological Monitoring Programme.

Department of Resource Surveys and Remote Sensing. 1989. Wildlife/Livestock Survey. Nairobi, Kenya.

Dublin, H. T. 1986. Decline of the Mara woodlands: The role of fire and elephants. Ph.D. dissertation, University of British Columbia.

Dublin, H. T., and Douglas-Hamilton, I. 1987. Status and trends of elephants in the Serengeti-Mara ecosystem. *Afr. J. Ecol.* 25:19–23.

Dublin, H. T., Sinclair, A. R. E., Boutin, S., Anderson, E., Jago, M., and Arcese, P. 1990. Does competition regulate ungulate populations? Further evidence from Serengeti, Tanzania. *Oecologia* 82:283–88.

Duncan, P. 1975. Topi and their food supply. Ph.D. dissertation, University of Nairobi.

Eltringham, S. K. 1974. Changes in the large mammal community of Mweya peninsula, Rwenzori National Park, Uganda, following removal of hippopotamus. *J. Appl. Ecol.* 11:855–66.

Ferrar, A. A., and Walker, B. H. 1974. An analysis of herbivore/habitat relationships in Kyle National Park, Rhodesia. *J. S. Afr. Wildl. Mgmt. Assoc.* 4:137–47.

Field, C. R. 1972. The food habitats of wild ungulates in Uganda by analyses of stomach contents. *E. Afr. Wildl. J.* 10:17–42.

Foster, J. B. 1987. Walking tall. *Equinox* 6 (34):24–35.

Foster, J. B., and Kearney, D. 1967. Nairobi National Park game census, 1966. *E. Afr. Wildl. J.* 5:112–20.

Foster, J. B., and McLaughlin, R. 1968. Nairobi National Park game census, 1967. *E. Afr. Wildl. J.* 6:152–54.

Fryxell, J. M., Greever, J., and Sinclair, A. R. E. 1988. Why are migratory ungulates so abundant? *Am. Nat.* 131:781–98.

Fryxell, J. M., and Sinclair, A. R. E. 1988. Causes and consequences of migration by large herbivores. *Trends Ecol. Evol.* 9:237–41.

Grimsdell, J. J. R. 1979. Changes in populations of resident ungulates. In *Serengeti: Dynamics of an ecosystem,* ed. A. R. E. Sinclair and M. Norton-Griffiths, 353–59. Chicago: University of Chicago Press.

Gwynne, M. D., and Bell, R. H. V. 1968. Selection of vegetation components by grazing ungulates in the Serengeti National Park. *Nature* 220:390–93.

Hall-Martin, A. J. 1974. Food selection by Transvaal lowveld giraffe as determined by analysis of stomach contents. *J. S. Afr. Wildl. Mgmt. Assoc.* 4:191–202.

Hall-Martin, A. J., Skinner, J. D., and Van Dyk, J. M. 1975. Reproduction in the giraffe in relation to some environmental factors. *E. Afr. Wildl. J.* 13:237–48.

Heard, D. C., and Williams, T. M. 1992. Distribution of wolf dens on migratory caribou ranges in the Northwest Territories, Canada. *Can. J. Zool.* 70:1504–10.

Holt, R. D. 1977. Predation, apparent competition and the structure of prey communities. *Theor. Popul. Biol.* 12:197–229.

Holt, R. D., and Lawton, J. H. 1994. The ecological consequences of shared natural enemies. *Ann. Rev. Ecol. Syst.* 24:495–520.

Jarman, P. J. 1971. Diets of large mammals in the woodlands around Lake Kariba, Rhodesia. *Oecologia* 8:157–78.

Jarman, P. J., and Sinclair, A. R. E. 1979. Dynamics of ungulate social organization. In *Serengeti: Dynamics of an ecosystem,* ed. A. R. E. Sinclair and M. Norton-Griffiths, 185–220. Chicago: University of Chicago Press.

Kruuk, H. 1972. *The spotted hyena: A study of predation and social behavior.* Chicago: University of Chicago Press.

Lamprey, H. F. 1963. Ecological separation of the large mammal species in the Tarangire Game Reserve, Tanganyika. *E. Afr. Wildl. J.* 1:63–92.

Leuthold, W., and Leuthold, B. M. 1975. Temporal patterns of reproduction in ungulates of Tsavo East National Park, Kenya. *E. Afr. Wildl. J.* 13:159–69.

Maddock, L. 1979. The "migration" and grazing succession. In *Serengeti: Dynamics of an ecosystem,* ed. A. R. E. Sinclair and M. Norton-Griffiths, pp. 104–29. Chicago: University of Chicago Press.

Mduma, S. A. R., and Sinclair, A. R. E. 1994. The function of habitat selection by oribi in Serengeti, Tanzania. *Afr. J. Ecol.* 32:16–29.

Melton, D. A. 1987. Waterbuck *(Kobus ellipsiprymnus)* population dynamics: The testing of an hypothesis. *Afr. J. Ecol.* 25:133–45.
Mills, M. G. L., and Shenk, T. M. 1992. Predator-prey relationships: The impact of lion predation on wildebeest and zebra populations. *J. Anim. Ecol.* 61:693–702.
Murray, M. G. 1991. Maximizing energy retention in grazing ruminants. *J. Anim. Ecol.* 60:1029–45.
Murray, M. G., and Brown, D. 1993. Niche separation of grazing ungulates in the Serengeti: An experimental test. *J. Anim. Ecol.* 62:380–89.
Norton-Griffiths, M. 1973. Counting the Serengeti migratory wildebeest using two-stage sampling. *E. Afr. Wildl. J.* 11:135–49.
Norton-Griffiths, M. 1978. *Counting animals.* 2d ed. Handbook no. 1. Techniques in African Wildlife Ecology. Nairobi: African Wildlife Foundation.
Ottichilo, W. K., Kufwafwa, J. W., and Stelfox, J. G. 1987. Elephant population trends in Kenya: 1977–1981. *Afr. J. Ecol.* 25:9–18.
Packer, C. 1990. *Serengeti lion survey.* Report to TANAPA, SWRI, MWEKA, and the Game Department. Arusha, Tanzania: Tanzania National Parks.
Samuel, M. D., Steinhorst, R. K., Garton, E. O., and Unsworth, J. W. 1992. Estimation of wildlife population ratios incorporating survey design and visibility bias. *J. Wildl. Mgmt.* 56:718–25.
Schaller, G. B. 1972. *The Serengeti lion: A study of predator-prey relations.* Chicago: University of Chicago Press.
Sinclair, A. R. E. 1972. Long-term monitoring of mammal populations in the Serengeti: Census of non-migratory ungulates, 1971. *E. Afr. Wildl. J.* 10:287–97.
———. 1973. Population increases of buffalo and wildebeest in the Serengeti. *E. Afr. Wildl. J.* 11:93–107.
———. 1977. *The African buffalo: A study of resource limitation of populations.* Chicago: University of Chicago Press.
———. 1978. Factors affecting the food supply and breeding season of resident birds and movements of Palaearctic migrants in a tropical African savannah. *Ibis* 120:480–97.
———. 1979. The eruption of the ruminants. In *Serengeti: Dynamics of an ecosystem,* ed. A. R. E. Sinclair and M. Norton-Griffiths, 82–103. Chicago: University of Chicago Press.
———. 1985. Does interspecific competition or predation shape the African ungulate community? *J. Anim. Ecol.* 54:899–918.
Sinclair, A. R. E., Dublin, H., and Borner, M. 1985. Population regulation of the Serengeti wildebeest: A test of the food hypothesis. *Oecologia* 65:266–68.
Sinclair, A. R. E., and Gwynne, M. D. 1972. Food selection and competition in the East African buffalo (*Syncerus caffer* Sparrman). *E. Afr. Wildl. J.* 10:77–89.
Sinclair, A. R. E., and Norton-Griffiths, M. 1982. Does competition or facilitation regulate migrant ungulate populations in the Serengeti? A test of hypotheses. *Oecologia* 53:364–69.
Spinage, C. A. 1982. *A territorial antelope: The Uganda waterbuck.* London: Academic Press.

Stelfox, J. G., Peden, D. G., Epp, H., Hudson, R. J., Mbugwa, S. A., Agatsiva, J. L., and Amuyunzu, C. L. 1986. Herbivore dynamics in southern Narok, Kenya. *J. Wildl. Mgmt.* 50:339–47.

Underwood, R. 1983. The feeding behaviour of grazing African ungulates. *Behaviour* 84:195–243.

Zar, J. H. 1984. *Biostatistical analysis.* 2d ed. Englewood Cliffs, N.J.: Prentice-Hall.

TEN

Distribution and Abundance of Oribi, a Small Antelope

Simon A. R. Mduma

Although the abundance of most of the large ungulates in Serengeti National Park has been documented (Campbell and Borner, chap. 6; Dublin et al. 1990), little is known about the small antelopes, such as steinbok, dikdik, klipspringer, oribi, and common duiker, that occur in Serengeti. These species are difficult to census because they are small and secretive. We need knowledge of their abundance and distribution for conservation purposes, because they appear frequently in the kill by poachers around the borders of the Serengeti. This chapter describes the distribution and abundance of oribi in the northern extension of Serengeti National Park and compares this with the results of studies elsewhere. The methods used may be adapted to some of the other small species in the park.

Oribi are distributed from Senegal to Ethiopia and south to Cape Province of South Africa (Dorst and Dandelot 1970; Smithers 1983; Estes 1990). Little is known about oribi ecology (East 1988), possibly because of the species' small size (ca. 16 kg), sparse distribution, and tendency to remain concealed in tall vegetation, but considerable information is now available for the Serengeti population (Jongejan, Arcese, and Sinclair 1992; Mduma and Sinclair 1994; Arcese, Jongejan, and Sinclair 1995).

METHODS

Observations were made between October 1988 and November 1989, and were confined to an area of 398 km^2 in the northwestern corner of Serengeti National Park (see fig. 1.1). The altitude of the area is about 1,400 m above sea level, and annual precipitation typically reaches 1,100 mm (Norton-Griffiths, Herlocker, and Pennycuick 1975). Seasonal rainfall shows a major peak in April–May and a lesser one in November–December (Jongejan, Arcese, and Sinclair 1992).

The study area was surveyed to map the extent of oribi distribution.

Three counting techniques were used to estimate oribi population size, and these allowed me to compare the counting biases between methods.

Determining the Census Zone

The census zone was determined by the availability of potential oribi habitat. Accessible areas in the northwestern extension of Serengeti National Park, south of the Mara River, were searched by vehicle and any oribi seen were recorded. Oribi were abundant in *Combretum-Terminalia* woodland habitat, but rare in *Acacia* woodlands, thus areas with the latter habitat type were excluded from the census zone.

The census zone covered about 700 km^2, approximating the area of the broad-leaved *Combretum-Terminalia* woodlands. Few oribi were sighted in the *Acacia clavigera* relict woodlands that form a transition zone between *Combretum-Terminalia* woodlands and the other *Acacia* woodlands. These *A. clavigera* areas were considered less suitable habitats for oribi, but were, nevertheless, included in the census zone.

Boundaries of the census zone were delineated according to Herlocker's (1974) Serengeti National Park Vegetation Map. The census zone was bounded by the Mara River on the north and the park boundary on the west. The southern boundary followed the edge of the *Combretum-Terminalia* woodlands as far east as the Bologonja River, which formed the remaining eastern boundary.

The sample zone was confined to areas accessible by vehicle and with a grass height low enough to allow visibility ($<$ 40 cm). Grass height changed seasonally, so the sample zone also changed. The sample zone, which included all transect tracks (length by 200 m width) and census blocks used for population estimates, covered 398 km^2. Areas were measured from 1:250,000 and 1:50,000 scale maps using a planimeter (Norton-Griffiths 1978).

Counting Techniques

I used three techniques for estimating animal population size: (1) total counts of known individuals; (2) total counts of animals in census blocks; and (3) sample counts using ground transects. Except in a few areas that were inaccessible to vehicles, all counts were made from a vehicle (Suzuki Samurai). The locations of counted oribi were recorded on 1:50,000 scale maps kept separately for each census.

Total Counts of Known Individuals. In one intensive study area (Kogatende Ridge, fig. 1.1), individuals were identified in their respective territories and/or home ranges. Identification was made by a combination of several natural physical features (e.g., facial marks, ear shape, body size, and horn patterns for males) and artificial marks (colored ear tags and radio collars; Jongejan, Arcese, and Sinclair 1992). The area was repeatedly searched throughout the study period (October 1988–November 1989).

Features of all sighted individuals were recorded on identification cards kept separately for each individual. Resightings were recorded, and individuals that were not seen were considered to have emigrated or died. The total number of known individuals alive at any one time was used to estimate the density (total number of animals/area) of oribi in the intensive study area. This technique is equivalent to that known as "Minimum Number Alive."

Total Counts in Blocks. Total counts by searching and counting all sighted groups of oribi within a short period of time (at most 5 hours) were conducted in two other areas: the Wogakuria Ridge and East Ridge census blocks. These blocks lay respectively west and east of the Kogatende intensive study area. They comprise flat-topped ridges bounded by drainage lines (seasonal rivers) and dense *Acacia clavigera* thickets. All parts of the catena, from ridge top to valley bottom, were searched.

Blocks were searched by driving across the ridge in a systematic zigzag fashion. More time was spent in areas with dense shrubs or rock outcrops. All flushed animals were counted and their group composition recorded. The searching exercise was designed to avoid double counting, and brief identification notes of counted individuals were taken for this purpose. The drive paths and locations of all counted groups were indicated on a map.

Sample Counts by Ground Transects. Three censuses were conducted: (1) October–November 1988, (2) August 1989, and (3) October–November 1989. A preliminary survey determined the distribution and relative abundance of oribi in the northern Serengeti, and this determined the census area. Fourteen transects were established. All except one originated from the main road that runs from Kogatende to Tabora B guard post. Most transects followed ridge tops, but some cut across drainage lines and different vegetation types.

All large mammal species sighted on both sides of the transects were counted. Sighting distances from the transect were estimated and recorded in meters, and grass height where an animal was sighted was recorded. Total length of transects passing through different habitats (vegetation type and grass height) were recorded from the vehicle odometer. Oribi sex and age classes were also identified using binoculars. Ground counts were conducted in the morning (starting time about 09:00 depending on weather) when most individuals were actively feeding.

Data Analysis

Estimates of total population (Y) and its variance (var [Y]) for ground transects were obtained using Jolly's (1969) method for unequal-sized samples (Norton-Griffiths 1978):

$$Y = Z \text{ x } (\textstyle\sum y / \sum x)$$
$$= Z \times R$$
$$\text{var}(Y) = N[(N - n)/n] \times [Sy^2 - (2R \times Szy) + (R^2 \times Sz^2)]$$

where

y = individual transect total of animals
z = individual transect areas
Z = total census area
N = total possible number of transects
n = actual number of transects driven
Sy^2 = variance of transect totals of animals
Sz^2 = variance of transect areas
Szy = covariance between animals counted and transect area

RESULTS

Total Counts by Known Individuals

A total of 65 individuals were known to be alive in the Kogatende Ridge study area (8.1 km^2) in November and December 1988 and November 1989, and the density of oribi was estimated as 8.02 km^2 (table 10.1).

Although there was turnover of the individual animals in the population, total numbers remained more or less the same throughout the study period. In December 1989 there were 62 resident oribi, giving a density of 7.7/km^2 (P. Arcese and G. Jongejan, pers. comm.).

Total Counts in Blocks

Total counts in the Wogakuria and East Ridge blocks are given in table 10.1. A total of 4 hours 40 minutes and 4 hours were spent searching the two blocks respectively. Sixty-one individuals were recorded in the Wogakuria Ridge block and 58 in the East Ridge block, giving densities of 3.0/km^2 and 4.1/km^2 respectively.

Sample Counts by Ground Transects

The factors that affected the visibility of oribi were grass height and sighting distance. The frequency distribution of oribi counted in different grass heights is shown in table 10.2. Only 4.7% of all oribi counted were seen in areas with grass height greater than 40 cm. Owing to their small body size (60 cm at the shoulder) and their antipredator "freezing" behavior (Jarman 1974; Estes 1990) oribi are difficult to see in long grass

Table 10.1 Total counts of oribi from known individuals (*a*) and total block counts (*b, c*) in three census blocks, northwestern Serengeti (November 1989).

Census block	Area (km²)	Total (*n*)	Density (km²)
(a) Kogatende Ridge	8.1	65	8.02
(b) East Ridge	14.1	58	4.11
(c) Wogakuria Ridge	20.5	61	2.98

Table 10.2 The number of oribi counted on ground transects in different grass height classes, combined over all three census periods.

	Oribi (*n*) Census				
Height (cm)	1	2	3	Total	%
<20	90	45	200	335	51.9
21–40	79	51	150	280	43.4
41–60	14	3	10	27	4.2
61–80	1	0	2	3	0.5
>80	0	0	0	0	0
Total	184	99	362	645	100.0

(> 40 cm), and hence they are considerably undercounted in these conditions. To reduce this bias, only transect segments with grass height of less than 40 cm were used in the calculations, approximately 80% of the total transect length.

Figure 10.1 shows the frequency distribution of oribi counted at different sighting distances. I used this information to decide on the effective transect width by using Kelker's method as described in Norton-Griffiths's (1978) census handbook.

Of the 645 oribi counted, 32.2% were within 50 m of the observer (table 10.3). A further 50.9% were seen at 50 to 100 m. It is probable that those animals originally within 50 m of the vehicle were running away before they were sighted beyond the 50 m boundary. If there were no bias errors due to distance, then the number of oribi counted between 100 m and 150 m should be similar to the average for 0–50 and 50–100 m, namely 268, (fig. 10.1). The actual number sighted (99) (table 10.3) was considerably below the expected number, implying that many oribi (63%) that were beyond 100 m were missed, either because of their small size or because they remained hidden in cover. To reduce this bias, I discarded those counts made beyond 100 m on either side of the transect. Transect width was, therefore, 200 m.

Density Estimates. Densities per km² on each transect were 10.8 (4.4 SE) in November 1988, 7.8 (1.3) in August 1989, and 9.9 (1.5) in November 1989. The merged density of oribi over all censuses across the study

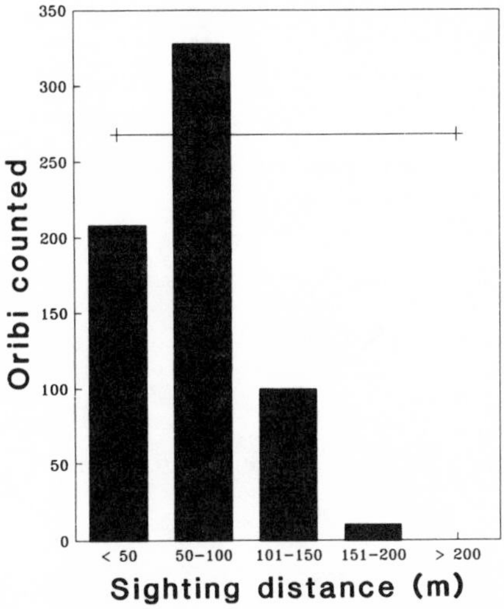

Figure 10.1 Frequency distribution of oribi counted at different sighting distances. The horizontal line shows the average number of oribi seen at the 0–50 and 50–100 m distances from the observer. This average is the expected number that should have been seen at the 100–150 m distance if there was no visibility bias.

Table 10.3 The number of oribi counted on ground transects at different sighting distances, combined over all three census periods.

	Oribi (*n*) Census				
Distance (m)	1	2	3	Total	%
<50	50	28	130	208	32.2
50–100	117	48	163	328	50.9
101–150	9	23	67	99	15.3
151–200	8	0	2	10	1.6
>200	0	0	0	0	0
Total	184	99	362	645	100.0

area was 9.59 oribi/km^2 (± 2.83; 95% c.l.). Individual transect densities varied considerably (e.g., in November 1988 from 2.5 to 31.5 animals/km^2). Anecdotal observations and repeated counts suggest that these local differences in density are real.

Independent Census Estimates of Population Size. Results from each ground transect census with their 95% confidence limits are shown in table 10.4 for the sample zone (398 km^2) and the census zone (700 km^2).

All three censuses show similar population estimates, with that of August–November 1989 having the smallest sampling error due to the

Table 10.4 Oribi population estimates (Y) from ground transects, using Jolly's (1969) method for unequal-sized sampling units, in the sample zone (398 km²) and census zone (700 km²).

	Nov. 1988	Aug. 1989	Nov. 1989
Oribi counted	159	76	277
Transect area (km²)	15.84	9.82	29.22
Sample zone			
(Y)	3,995	3,080	3,773
95% c.l.	3,677	1,322	1,030
Census zone			
(Y)	7,026	5,418	6,636
95% c.l.	6,525	2,334	1,852

larger number of transects. For the census zone the most precise estimate was 6,636 ± 1,852.2 (95% c.l.).

Merging Independent Estimates of Population Size. Merging independent estimates has the advantage of reducing the sampling error (Eltringham and Din 1977; Norton-Griffiths 1978), but is allowed only if no detectable population change takes place between censuses. The total counts of known individuals at Kogatende Ridge indicated that the population was not changing. The merged estimate, by the method of Norton-Griffiths (1978), gave a population total of 5,986 ± 1,086 (95% c.l.).

DISCUSSION

Sources of Error

Sample Design. Because the positions of transects were determined by accessibility to vehicle and by visibility, rather than according to a random design, the estimated density may have been biased. Since transects were along ridge tops favored by oribi, there may have been a positive bias. This possible bias can be examined by comparing the mean transect density with the independent estimate from known animals on Kogatende Ridge, which is bias-free in this respect. Density from transects (9.59/km² + 2.83, 95% c.l.) was not significantly different from that found through total counts of known individuals (8.02/km²). This suggests that the bias from nonrandom placement of transects was relatively small.

Counting Bias. A counting bias occurs when observers either over- or underestimate the number of individuals in large groups of animals. Since groups had a maximum of six individuals, this bias was negligible.

A more important bias could arise from missing animals that are resting in shade or concealed places. Undercounting was reduced in the second and third transect censuses by counting when animals were active. In addition, bias was reduced by (1) using a narrow strip width that encompassed oribi flushing distance (< 100 m), (2) driving slowly (20 to 25 km/hour), and (3) having at least two observers. Nevertheless, some animals along the transects, and in census blocks, would have been missed in areas with dense bushes, long grass, termite mounds, rock outcrops, or small drainage lines that offer good cover.

Comparison of Census Techniques

Results from total counts of known individuals were the most reliable, being free from both sampling error and biases inherent in the other techniques. This technique could be used as an independent estimate with which to compare other methods. The stability of territorial groups, their affinity to particular areas, the permanence of territories, and the method used to identify animals meant that few individuals were unrecorded (Arcese, Jongejan, and Sinclair 1995)

The density of oribi in the intensive study area was approximately stable at 8.02 individuals/km^2 throughout the study period. This estimate is straddled by the 95% confidence range from the merged transect estimate (9.6 + 2.8). Similarly, the density from these results suggests that the technique of ground transect counts is relatively bias-free when conducted in short grass less than 40 cm in height.

Transect 1 went through the study area where the total count by known individuals was conducted. The density from transect 1 was 9.6 + 1.6 (95% c.l.).

The one-time total count of blocks appeared to provide the poorest density estimate. When compared with the mean density derived from ground transects, total counts underestimated density by 57.3% and 68.7% on the East Ridge and Wogakuria Ridge respectively.

Two aerial censuses of the large mammals in northern Serengeti were conducted in November 1988 and March 1989 (Sinclair, chap. 9). Both censuses covered 1,000 km^2 and included the oribi study area. Only two oribi were counted in the first census, none in the second. Hence oribi cannot be counted reliably in aerial surveys.

Comparison with Other Studies

This study estimated a population size of 6,635 ± 1,842 (95% c.l.) oribi in the northern Serengeti. This estimate, however, was based on an area of 700 km^2, of which 15% consisted of unsuitable forest habitat for oribi. By removing this 15%, an adjusted figure of 5,640 ± 1,566 oribi is obtained. The northern Serengeti clearly supports a higher density of oribi (9.6 ± 2.8 km^2) than other areas in Africa (table 10.5). Monfort and Monfort (1974) recorded an overall density of 12.84 within one transect in Akagera National Park, but the mean density in occupied habitats there was 2.7 ± 1.8 (c.l.) animals/km^2.

Table 10.5 Oribi densities recorded at different locations in Africa.

Location	Density (km^2)	Source
Akagera N.P., Rwanda		
Savanna highlands	12.8	Monfort and Monfort (1974)
Mean density	2.7	Monfort and Monfort (1974)
Hill glades	0.02	Monfort and Monfort (1974)
South Africa		
Highmoor State Forest	1.53	Oliver, Short, and Hanks (1978)
Piet Rief	4.5	Viljoen (1982)
Amsterdam	2.68	Viljoen (1982)
Mount Sheba	1.5	Shackleton and Walker (1985)
Northern Serengeti		
Highest density	31.48	This study
Mean density	9.6	This study
Lowest density	2.50	This study

Factors Influencing Abundance

Estimates from ground counts showed a west-east decline of oribi density in the census zone. This trend coincided with the rainfall gradient (Norton-Griffiths, Herlocker, and Pennycuick 1975), with higher rainfall in the west declining toward the east. High annual precipitation spread over most of the year supports the *Combretum-Terminalia* woodlands favored by oribi.

Several studies in South Africa concluded that nutritional quality of food might limit the oribi population, especially during the coldest and driest time of the year (Rowe-Rowe 1983; Mentis 1978; Oliver, Short, and Hanks 1978; Shackleton and Walker 1985). These studies suggested that fire stimulated new grass with higher nutritive value, and thus in turn supported a higher density of oribi. If fire is an important ecological component for maintaining a high density of oribi, then the annual burns set by poachers, wardens, and others in Serengeti National Park may be one reason for the high densities of oribi. Studies of habitat preference indicated that oribi preferred recently burned areas (Rowe-Rowe 1982; Mduma 1991), but these results may be confounded by lower visibility in tall grass (Mduma and Sinclair 1994).

Oribi were restricted to *Combretum-Terminalia* woodlands, which constituted less than 5% of the total Serengeti woodlands. This habitat may be shrinking due to the increased frequency of hot fires (Dublin, chap. 4; Sinclair, chap. 5) that inhibit tree recruitment. Therefore, to maintain the present oribi population (and populations of other antelopes), I recommend that Tanzania National Parks consider early burning as a high priority. Early burning could provide new pastures with high-quality food while reducing the accumulation of dry fuel that will eventually result in hot fires and cause damage to the woodlands from uncontrolled late burns.

CONCLUSION

Three census techniques were used to estimate the oribi population in northern Serengeti: (1) total counts of known individuals, (2) total counts in blocks, and (3) sample counts by ground transects. Counts by known individuals gave a relatively unbiased density of 8.02 individuals/km^2, which is comparable to 9.59/km^2 (+ 2.83, 95% c.l.) from ground transects. These transects in short grass are, therefore, a good method for censusing small antelopes in open savanna habitats. Total counts of blocks significantly underestimated the density by 63%, suggesting that many animals were missed. One-time total block counts were not a suitable method.

The oribi population in the census zone was estimated to be 5,986 ± 1,086 (95% c.l.) after the merging of the three independent ground transect censuses. The estimated density is the highest so far recorded in Africa.

ACKNOWLEDGMENTS

I would like to thank the Director of Tanzania National Parks and the Director of Serengeti Wildlife Research Institute for permission to carry out research in Serengeti National Park. This study was funded by Wildlife Conservation International (WCI), a division of the New York Zoological Society, with additional support from German Academic Exchange Program (DAAD) and Natural Sciences and Engineering Research Council of Canada (NSERC) grants to A. R. E. Sinclair. I thank A. R. E. Sinclair, R. B. M. Senzota, P. D. Moehlman, P. Arcese, and G. Jongejan for their help with this work.

REFERENCES

Arcese, P., Jongejan, G., and Sinclair, A. R. E. 1995. Behavioural flexibility in a small African antelope: Group size and composition in the oribi (*Ourebia ourebi,* Bovidae). *Ethology.* In press.

Dorst, J., and Dandelot, P. 1970. *A field guide to the larger mammals of Africa.* London: Collins.

Dublin, H. T., Sinclair, A. R. E., Boutin, S., Anderson, E., Jago, M., and Arcese, P. 1990. Does competition regulate ungulate populations? Further evidence from Serengeti, Tanzania. *Oecologia* 82:283–88.

East, R., ed. 1988. *Antelopes global survey and regional action plans.* Part 1, *East and North-East Africa.* IUCN/SSC Antelope Specialist Group/Allen Press.

Eltringham, S. K. and Din, N. A. 1977. Estimates of the population size of some ungulate species in the Ruwenzori National Park, Uganda. *E. Afr. Wildl. J.* 15:305–16.

Estes, R. D. 1990. *Behaviour guide to African mammals.* Los Angeles: University of California Press.

Herlocker, D. J. 1974. *Woody Vegetation of the Serengeti National Park.* Kleberg Studies in Natural Resources. College Station: Texas A&M University.

Jarman, P. J. 1974. The social organization of antelope in relation to their ecology. *Behaviour* 48:215–66.

Jolly, G. M. 1969. Sampling methods for aerial census of wildlife populations. *E. Afr. Agric. For. J.* 34:46–49.

Jongejan, G., Arcese, P., and Sinclair, A. R. E. 1992. Growth, size, and the timing of births in an individually identified population of oribi. *Afr. J. Ecol.* 29:340–52.

Mduma, S. A. R. 1991. Population ecology of oribi in Serengeti National Park, Tanzania. M.Sc. thesis, University of Dar es Salaam.

Mduma, S. A. R., and Sinclair, A. R. E. 1994. The function of habitat selection by oribi in Serengeti, Tanzania. *Afr. J. Ecol.* 32:16–29.

Mentis, M. T. 1978. Population limitation in grey rhebuck and oribi in the Natal Drakensberg. *Lammergayer* 26:19–28.

Monfort, A., and Monfort, N. 1974. Notes sur l'ecologie et le comportement des oribis (*Ourebia ourebi* Zimmermann 1783). *Terre et Vie* 28:169–208.

Norton-Griffiths, M. 1978. *Counting animals.* Handbook No. 1. Techniques in African Wildlife Ecology. Nairobi: African Wildlife Foundation.

Norton-Griffiths, M., Herlocker, D. and Pennycuick, L. 1975. The patterns of rainfall and climate in the Serengeti ecosystem. *E. Afr. Wildl. J.* 13:347–74.

Oliver, M. D. N., Short, N. R. M., and Hanks J. 1978. Population ecology of oribi, grey rhebuck, and mountain reedbuck in Highmoor State Forest Land, Natal. *S. Afr. J. Wildl. Res.* 8:95–105.

Rowe-Rowe, D. T. 1982. Influence of fire on antelope distribution and abundance in the Natal Drakensberg. *S. Afr. J. Wildl. Res.* 12:124–29.

———. 1983. Habitat preference of five Drakensberg antelopes. *S. Afr. J. Wildl. Res.* 13:1–8.

Shackleton, C., and Walker, B. H. 1985. Habitat and dietary species selection by oribi antelope at Mount Sheba Nature Reserve. *S. Afr. J. Wildl. Res.* 15:49–53.

Smithers, R. H. N. 1983. *The mammals of the Southern Africa sub-region.* Pretoria: University of Pretoria.

Viljoen, P. C. 1982. Die gedragsekologie van die Oorbietjie (*Ourebia ourebi* Zimmermann 1783) in Transvaal. M.Sc. Thesis, University of Pretoria.

ELEVEN

Specific Nutrient Requirements and Migration of Wildebeest

Martyn G. Murray

With the arrival of heavy rainstorms, African ungulates seek out new pastures, leaving their dry season refuges by way of well-worn paths and trails. Usually they concentrate near permanent water during dry periods and disperse, often moving upward into neighboring "dry country," during wet periods. So the annual cycle of movement typically consists of contraction and concentration in the dry season followed by expansion and dispersion in the wet season (Dasmann and Mossman 1962; Lamprey 1964; Jarman 1972; Stanley Price 1974; Western 1975; Afolayan and Ajayi 1980; Sinclair 1983). In certain exceptional cases, the pattern of movement differs in that concentrations of game take place in both the wet and dry seasons (Bell 1969; Murray 1982). In many cases, it seems that some additional factor, related to breeding, concentrates these populations on wet season calving grounds (Sinclair 1983). Where the two concentration areas are sufficiently far apart, the movements are called seasonal migrations.

Despite the major influence of migrations on ecosystem dynamics (Sinclair and Norton-Griffiths 1979), the acute vulnerability of migrations to human developments in land use (Borner 1985; Williamson and Mbano 1988; Howell, Lock, and Cobb 1989), and the scientific interest in describing and understanding the phenomenon itself, the reasons for migratory movements of African ungulate populations remain inadequately known. In their study of migratory kob (*Kobus kob leucotis* Lichtenstein and Peters), Fryxell and Sinclair (1988) point out that movements onto the wet season concentration area (which they share with the migratory tiang, *Damaliscus lunatus tiang* Heuglin) cannot be explained by the need to avoid annual flooding, or by selection of habitats with greater forage abundance. They conclude that the grasslands of the wet season range must be attractive "for more subtle reasons." These more subtle reasons form the subject matter of this chapter, in which I investigate seasonal habitat selection by white-bearded wildebeest in the Serengeti-

Mara ecosystem of Tanzania and Kenya. In particular, I examine the possibility that wildebeest migration is driven by seasonal demands for specific nutrients.

Each year some 1 million wildebeest migrate across the Serengeti-Mara ecosystem, traveling 10 km per day averaged over a whole year (D. Kreulen, pers. comm., cited in Pennycuick 1979). The crude cost of this movement, relative to neighboring resident populations of wildebeest, is a 3% increment in mortality per year (Sinclair 1983), possibly combined with reduced fertility in young females (Watson 1969). The overall migratory pattern is thought to be related to food supply, which is itself dependent on an uneven distribution of rainfall (Grzimek and Grzimek 1960; Talbot and Talbot 1963; Anderson and Talbot 1965; Pennycuick 1975; Maddock 1979). It has been suggested that the principal northwesterly movement at the start of the dry season is in response to the need to find surface water (Sinclair and Fryxell 1985), but according to Watson (1967), in some years the movement may begin before the wet season has ended and despite continued growth of grass and abundant supplies of drinking water. The unique movement south and east that returns the wildebeest to their wet season calving ground is the most predictable feature of the migration, and its explanation is crucial to an understanding of the migration as a whole.

The wet season range of wildebeest, situated on the short and medium-length grasslands of the Serengeti plains, is thought to provide the best grazing in the ecosystem (Bell 1971; Braun 1973; Sinclair and Fryxell 1985). Usually taller grasses are more fibrous, have lower protein concentrations, and are less digestible (Van Soest 1982), so selection by wildebeest of short grassland could be due to the requirements of lactating females for metabolizable energy and protein (Watson 1967; McNaughton 1985). Grasses prone to heavy grazing have evolved a variety of specialized growth traits (McNaughton 1984). An alternative hypothesis for why wildebeest select their wet season habitat is that grasses on the dry season range are unable to sustain production of new growth if heavily grazed, as they lack appropriate genetic adaptations (R. Ruess, pers. comm.). Long-distance movements have also been attributed to the wildebeest's dislike of wet and sticky ground (Talbot and Talbot 1963; Anderson and Talbot 1965), escape from predation pressure by large predators confined to territories (Fryxell, Greever, and Sinclair 1988), reduction of competition for food with resident grazers, and avoidance of areas with tsetse fly (Maddock 1979).

Wildebeest movements could also be influenced by changes in requirements for specific nutrients. Kreulen (1975) noted that Serengeti wildebeest on their wet season range preferred a short-grass over a long-grass habitat, and that calcium concentrations were higher on the short grassland. By extrapolation from livestock data, he showed that the elevated requirement for calcium in lactating females could be met only by forage gathered from the short grassland, concluding that this could explain habitat selection. Recent surveys of element concentrations in Serengeti grasslands reveal that most minerals have substantially higher concentrations in short grasslands on the wet season range of migratory wildebeest than in other Serengeti grasslands (McNaughton and Banyikwa, chap. 3; McNaughton 1989). Movements of lactating wildebeest were related to forage levels of Cu, Mg, N, Na, and P, and to the Ca:P ratio (McNaughton 1990). These findings indicate that cyclic requirements for one or more elements by female wildebeest could account for localized movements and seasonal migration in the Serengeti-Mara ecosystem.

In this chapter, habitat preferences of wildebeest are investigated in relation to (1) the availability of specific nutrients in different habitats, (2) seasonal variation in dietary requirements of female wildebeest, and (3) the evidence for mineral deficiency in lactating females from assays of serum and urine electrolytes. These data are used to test among six hypothetical constraints that could account for seasonal movements. It is hypothesized that wildebeest select wet season habitats to (1) increase their daily intake of green leaf; (2) increase the concentration of metabolizable energy in their diet; (3) increase the concentration of protein in their diet; (4) meet minimum requirements for dietary sodium; (5) meet minimum requirements for dietary calcium; (6) meet minimum requirements for dietary phosphorus. The first two hypotheses are also considered by Fryxell (chap. 12).

STUDY REGION

Bounded to the north by the Isuria escarpment and Loita plains, on the east by the Loliondo highlands and the western wall of the Rift Valley, to the south by the Crater Highlands and Eyasi escarpment, and to the west by Lake Victoria and expanding human cultivation, the Serengeti-Mara ecosystem extends over some 25,000–35,000 km^2 (fig. 11.1). In April 1989, the population of migratory wildebeest was estimated to be 1.6 million, with a smaller population of 25,000 resident wildebeest on grassland plains at Kirawira in the west of Serengeti National Park (Campbell 1989). The ecosystem contains a diverse assemblage of other grazing herbivores, including substantial numbers of zebra, buffalo, kongoni, topi, and Thomson's gazelle. Major predators of these species are hyena, wild dog, lion, leopard, and cheetah.

The migratory wildebeest typically spend wet season months (November to May) on treeless plains to the southeast (fig.11.1), which in-

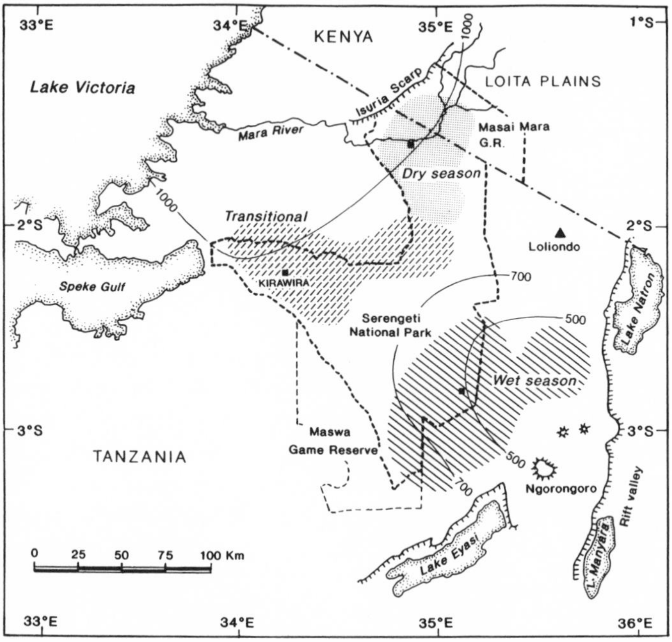

Figure 11.1 The seasonal ranges occupied by migratory wildebeest in the Serengeti-Mara ecosystem. Mean annual rainfall is shown by isohyets (in mm). The locations of sample sites are indicated by solid squares.

clude extensive areas of short grassland on saline and alkaline soils, dominated by *Sporobolus ioclados, S. kentrophyllus, S. fimbriatus, Digitaria abyssinica, D. macroblephara,* and *Kyllinga nervosa* (Cyperaceae) (McNaughton 1983). Rainfall is largely confined to the wet season, annually averaging 400 mm in the southeast to 800 mm in the northwest of the plains (Norton-Griffiths, Herlocker, and Pennycuick 1975).

In the early dry season, migratory wildebeest occupy a transitional range in the western part of the ecosystem that overlaps with areas used by the resident population at Kirawira. The range is wooded but is broken by extensive plains on which *Themeda triandra, Panicum coloratum, Chrysochloa orientalis,* and *Eriochloa fatmensis* are common grass species. Taller grasses (such as *Panicum maximum* and *Echinochloa hapoclada*) occur in swamps and along major rivers (Duncan 1975). Rainfall is more evenly distributed through the year, averaging 900–1,000 mm per annum. In wet months, the resident wildebeest at Kirawira prefer short grasslands on the well-drained catena top, dominated by *Digitaria macroblephara, Sporobolus ioclados, Cynodon dactylon,* and *Chrysochloa orientalis.*

In the late dry season, the migratory wildebeest move northeast to their dry season range in the northwest of Serengeti National Park, spilling over into the Masai Mara National Reserve of Kenya. The area contains forest and thicket patches in open, relatively tall grassland with scattered *Acacia* trees. Common grass species are *Themeda triandra, Setaria sphacelata, Sporobolus fimbriatus, Pennisetum mezianum,* and *Digitaria macroblephara* (Sinclair 1977). The dry season range has a comparatively high annual rainfall of 1,000–1,200 mm, often with appreciable rainfall in the dry season (June to October).

METHODS

Structural and Chemical Composition of Pastures

Aboveground plant biomass and the proportional representation of green leaf in the sward were studied from December 1988 to July 1989 at two sites, one located on the short-grass plains in the wet season range and the other on open *Themeda* grassland in the dry season range of the migratory wildebeest (see fig. 11.1). Each study site had three permanent exclosures (5 by 5 m), one each on catena top, mid-catena, and just above catena bottom. The sites were visited twice per calendar month (as close as possible to the 1st and 16th day of each month).

Three permanent plots of 60 by 40 cm were located within each exclosure and clipped to 2 cm twice per month to measure net primary aboveground production. One other plot of the same dimensions was moved at each visit onto an untreated area within the exclosure and also clipped to 2 cm, to measure the standing crop. Clipped samples were sorted by hand to separate green leaf from the remainder, and the sorted fractions were oven-dried at 65°C and weighed.

Samples of green leaf were analyzed for crude protein (CP), neutral cellulase digestibility (NCD), sodium, calcium, and phosphorus by the Agricultural Development and Advisory Service (ADAS) of the Ministry of Agriculture, Fisheries, and Food, U.K., using their methods (MAFF 1986); NCD was determined using method e of Dowman and Collins (1982). Degradable nitrogen (DN) was estimated from a regression equation for digestible crude protein (DCP) of tropical green forages (Minson 1982):

$$\text{DCP (g/kg)} = 0.96\text{CP(g/kg)} - 38,$$

($R^2 = 0.98$), and DN = DCP/6.25. Metabolizable energy content (ME) was estimated from a regression equation for spring-grown herbage in the U.K. (Givens, Everington, and Adamson 1990), after confirming similar relationships between NCD and CP in the U. K. and Serengeti forage samples (Murray 1991):

$$\text{ME (MJ/kg)} = 0.0111\text{NCD(g/kg)} + 3.24.$$

(SE = 0.65 MJ/kg). In addition to the samples collected from within exclosures, a single collection of forage species was gathered from unprotected swards in February 1989, when the growth stage of the grass was immature. These samples were collected along 2 km transects, from the upper to the lower catena, at the two study sites for migratory wildebeest and also at two sites within wet and dry season habitats of resident wildebeest at Kirawira (see fig. 11.1). Samples of green leaf were gathered from the two dominant grass species at 50 m intervals along each transect; these were oven-dried and analyzed for crude protein, sodium, calcium, and phosphorus, as described above.

Mineral Requirements

Given the uncertainties involved in calculating requirements for wildebeest based on those recommended for cattle, only two extreme cases are considered: (1) the female wildebeest in the early to middle stage of pregnancy, maintaining body weight and ingesting a low-quality forage, which is assumed to be typical of conditions in the dry season in an extended period without rainfall; and (2) the female wildebeest maintaining body weight at peak lactation while ingesting a high-quality forage, which is assumed to be typical of conditions in the wet season in a period with abundant rainfall.

During extended periods without rain, wildebeest graze pastures of mature or senescent growth with little green leaf; a common forage species is *Themeda triandra*. In recent feeding trials conducted in the Serengeti, two yearling male wildebeest, with body masses of 86 and 108 kg, were provided a forage of mature *Themeda* with metabolizable energy concentration of 7.20 MJ/kg (Murray 1993). The animals maintained constant body weight with an average daily intake of metabolizable energy of 0.512 MJ/kg $W^{0.75}$. Extrapolating from these results, the intake of metabolizable energy for wildebeest of 143 kg live weight (the average for adult females in the migratory population; Watson 1967) would be 21.17 MJ/day. If this "average" animal walked 3 km/day, an additional intake of 1.15 MJ would be required, or 22.32 MJ/day in all (Kreulen 1975). Dry matter intake for maintenance of body weight would then be 3.10 (22.32/7.20) kg/day.

Peak lactation in the Serengeti wildebeest occurs during the wet season when the animals graze on short pastures of young green leaf and stem. Under these conditions, the intake of dry matter by an adult female wildebeest of 143 kg live weight was estimated by Kreulen (1975) to be 4.54 kg/day.

There is a wide range in the recommended allowances of calcium and phosphorus for sheep and cattle (ARC 1965, 1980; INRA 1978; NRC 1985), which has stimulated debate and new research into the actual requirements for these minerals (MAFF 1984; Brodison et al. 1989). In revising previous estimates, AFRC (1991) introduced two new principles: that the obligatory component of endogenous fecal loss of calcium, E(Ca), and of phosphorus, E(P), is related to the level of food intake, and in the case of E(P) also to the proportion of roughage in the diet. Their new equations for cattle, on a diet containing at least 50% roughage, are applied here to wildebeest:

$$E(Ca)\ g/day = -0.74 + 0.0079W + 0.66DMI, \quad (11.1)$$

and

$$E(P)\ g/day = 1.6\ (-0.06 + 0.693DMI), \quad (11.2)$$

where W is live weight (kg) and DMI is dry matter intake (kg). By substitution into equations (11.1) and (11.2), estimates for endogenous fecal losses of calcium and phosphorus in wildebeest are E(Ca) = 2.44 g/day and E(P) = 3.34 g/day.

The minimum dietary intake of both calcium and phosphorus at maintenance is obtained by dividing E by the appropriate absorption coefficient (ARC 1980). Minimum concentrations of minerals in forage are determined by dividing the daily requirements by the estimated intake of dry matter per day (table 11.1).

The new factorial models recommended for estimating the dietary requirements of lactating cattle are

$$Ca\ (g/day) = (-0.74 + 0.0079W + 0.66DMI + m \times c)/0.68, \quad (11.3)$$

and

$$P\ (g/day) = 1.6(-0.06 + 0.693DMI + m \times p)/0.58, \quad (11.4)$$

where m is the milk yield (kg/day), and c and p are the Ca and P concentrations of milk (g/kg) respectively.

From equations (11.1) and (11.2), the endogenous fecal losses of Ca and P in lactating wildebeest are 3.39 g/day and 4.94 g/day respectively. The average composition of milk collected from fifteen lactating wildebeest on their wet season range in the Serengeti was 1.9 g Ca/kg and 1.4 g P/kg. Peak milk yield was estimated by Kreulen (1975) from secondary sources to be 3.77 kg/day, giving losses of 7.16 g Ca/day and 5.28 g P/day in milk.

Dietary requirements of sodium were estimated in a similar way, and once again the estimations for cattle were used for wildebeest. On a low dietary intake of sodium, beef cattle at maintenance are estimated to lose sodium at a daily rate of 6.8 mg/kg live weight (ARC 1980). The absorption coefficient of sodium is estimated to be 0.91, so the dietary requirement at maintenance is

$$0.0068W/0.91\ g/day. \quad (11.5)$$

The dietary requirement of sodium during lactation, with the animal maintaining body weight, is estimated to be

$$0.0068W + m \times s/0.91 \text{ g/day,} \tag{11.6}$$

where s is the concentration of sodium in milk (g/kg). For wildebeest sampled in the Serengeti, s = 0.31 g Na/kg, giving losses of 1.17g Na/day in milk.

Requirements for minerals at peak lactation (equations [11.3], [11.4], and [11.6]) are listed in table 11.1, together with the minimum concentrations in forage that would be necessary to meet dietary requirements assuming the DMI estimated above.

Serum and Urine Electrolytes and Proteins

In April 1989, seventeen adult female wildebeest were immobilized for sample collections on the wet season range of migratory wildebeest; in July, nine animals were immobilized on the transitional range and five more on the dry season range. Samples of serum, urine, and milk were taken from each animal and stored temporarily in the Serengeti at −10°C. In addition, the age of the mother's calf was estimated from its horn length, and the visual condition of the mother was estimated from the profile of pelvic bones, vertebrae, and ribs. Serum and urine electrolytes and proteins were analyzed by Rossdale and Partners, Newmarket, U.K. Sodium and potassium were assayed with a Corning Flame Photometer, model 435 (supplied by Corning Medical, Halstead, Essex, U.K.); all other assays were conducted with a Hitachi 705 Automatic Biochemistry Analyser (supplied by Boehringer Mannheim U.K., Lewes, East Sussex, U.K.). Milk samples were analyzed by the Agricultural Development and Advisory Service, U.K., who assayed the major minerals by atomic absorption spectrophotometry (MAFF 1986).

A general linear model procedure was used to determine what proportion of the variation in serum electrolytes and creatinine clearances was due to differences in location of sampled animals, and what proportion was due to differences in the reproductive stage and physical condition of the animals.

Table 11.1 Calcium, phosphorus, and sodium intake requirements, and minimum dietary concentrations in forage for an adult female wildebeest maintaining a constant live weight of 143 kg.

	Intake (g/day)			Minimum concentration (% DM)		
Reproductive status	Ca	P	Na	Ca	P	Na
Early–mid-pregnancy	3.59	5.76	1.07	0.12	0.19	0.035
Peak of lactation	15.51	17.61	2.35	0.34	0.39	0.052

RESULTS

Structure and Nutritional Composition of Pastures

Primary production peaked sharply on the wet season range of migratory wildebeest during February 1989 following heavy rain in January. It declined through the remainder of the growth season (fig. 11.2a). Production over the same period on the dry season range was less variable (fig. 11.2b). Over the entire growth period, overall production and green leaf production did not differ significantly between the wet and dry season ranges (table 11.2). Standing crop within the fenced exclosures increased steadily over the growth season in both areas, with a substantial rise on the dry season range in April and May (fig. 11.3). Over the entire growth

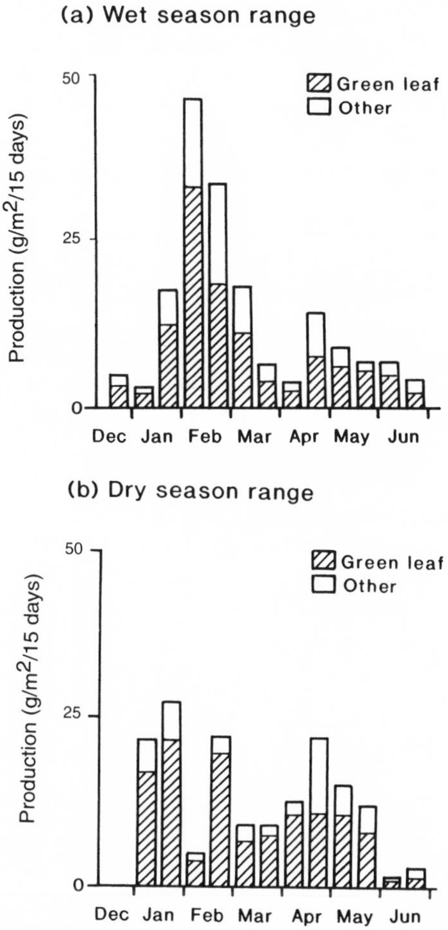

Figure 11.2 Total production and leaf production at bimonthly intervals on (*a*) the wet season range (calving ground) and (*b*) the dry season range of migratory wildebeest.

Table 11.2. Sward characteristics on the wet and dry season ranges of migratory wildebeest in the Serengeti.

	Standing crop			Production per 15 days		
Range	Green leaf (g/m^2)	Total (g/m^2)	Green leaf (%)	Green leaf (g/m^2)	Total (g/m^2)	Green leaf (%)
Wet season	31.2	100.4	39.7	8.7	13.6	63.3
	(5.2)	(20.2)	(4.2)	(2.4)	(3.5)	(2.4)
Dry season	75.0	207.0	45.9	10.9	14.7	71.8
	(12.2)	(42.7)	(4.7)	(2.0)	(2.5)	(3.7)

Note: Means and standard errors (in parentheses) taken from 14 bimonthly measures over the growing season.

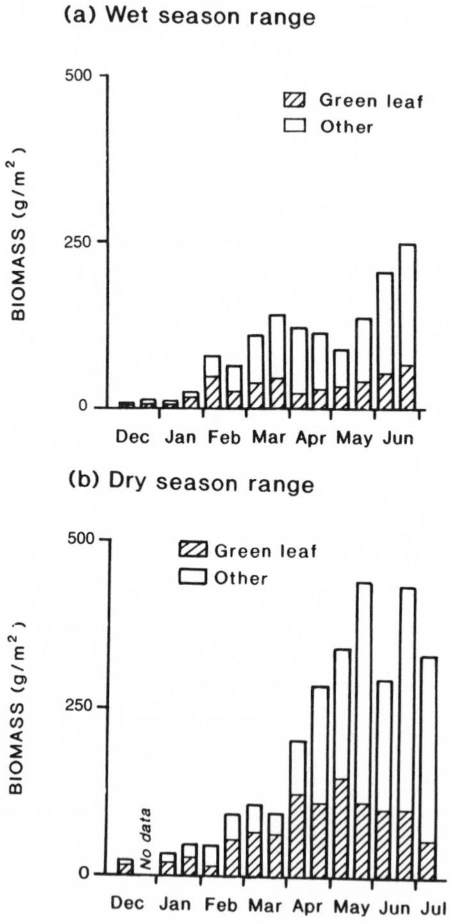

Figure 11.3 Standing crop and green leaf crop within exclosures situated in (*a*) the wet season range and (*b*) the dry season range of migratory wildebeest.

period, the total standing crop and the standing crop of green leaf were significantly higher in the dry season range ($P < .01$ and $P < .05$, respectively). The proportion of green leaf in wet and dry season ranges did not differ significantly, either in clipped or in control plots.

Digestibility of green leaf remained high over the growth season, and there was no consistent difference between samples gathered from the wet and from the dry season ranges of migratory wildebeest (fig. 11.4a; table 11.3). This result indicates that metabolizable energy was equally avail-

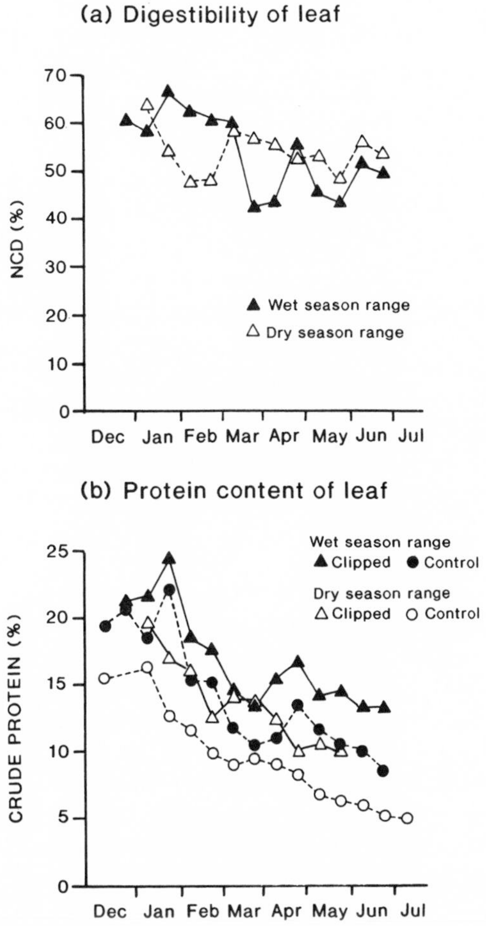

Figure 11.4 Bimonthly variation in (*a*) digestibility of clipped and (*b*) protein content of both clipped and control green leaf samples on seasonal ranges of migratory wildebeest.

Table 11.3 Nutritional composition of green leaf collected from within exclosures on the wet and dry season ranges of migratory wildebeest in the Serengeti.

Range	Treatment	NCD	CP	Sodium	Calcium	Phosphorus	Ca:P	n
Wet season	Clipped	53.93	16.77	0.13	0.56	0.54	1.15	12
		(8.26)	(3.65)	(0.06)	(0.08)	(0.13)	(0.57)	
Dry season	Clipped	53.91	13.58	0.13	0.40	0.26	1.53	10
		(4.66)	(3.20)	(0.06)	(0.05)	(0.05)	(0.18)	
		NS	*	NS	***	***		
Wet season	Control	52.31	14.19	0.22	0.50	0.52	1.01	14
		(5.98)	(4.42)	(0.08)	(0.13)	(0.09)	(0.43)	
Dry season	Control	50.60	9.30	0.09	0.37	0.19	2.06	14
		(7.03)	(3.61)	(0.06)	(0.06)	(0.06)	(0.39)	
		NS	**	***	**	***		

Note: All measures are expressed as a percentage of dry weight. Means and standard deviations (in parentheses) are given for the growth period (December through June). Clipped, plot clipped every 15 days; control, plot not previously clipped. NCD, neutral cellulase digestibility; CP, crude protein; Ca:P, calcium to phosphorus ratio; *n*, minimum sample size. Significance levels refer to Mann-Whitney tests of differences in nutritional composition between samples collected from the wet and dry season ranges.
$^*P < .05$; $^{**}P < .01$; $^{***}P < .001$; NS, not significant.

able in both areas. The crude protein content of green leaf declined throughout the growth period (fig. 11.4b), averaging 3% higher in clipped plots on the wet season range (table 11.3). In samples collected from unprotected sites, the concentration of crude protein was not significantly different between wet and dry season ranges (table 11.4), nor was there a significant difference between the wet and dry season ranges of resident wildebeest. The availability of protein in grass leaf for fermentation (g degradable nitrogen/MJ of metabolizable energy) declined during the growth period, but reached levels likely to constrain dry matter intake (< 1.0 gN/MJ; see "Discussion" below) only toward the end of the growth period, in the standing crop of the dry season range (fig. 11.5).

Mineral Concentrations in Grass Leaf

In samples collected from inside exclosures, major minerals were usually more concentrated on the wet than on the dry season range of migratory wildebeest (table 11.3). For example, green leaf from clipped plots on the wet season range contained twice as much phosphorus, 1.4 times as much calcium, and equivalent quantities of sodium. Sodium concentrations were above the required level for lactating wildebeest in all but three samples, which were collected from the standing crop on the dry season range toward the end of the growing season. Calcium concentrations fell just below the required level for lactating wildebeest in one clipped and six control samples taken from the dry season range (table 11.1; fig. 11.6a, b). Phosphorus concentrations were below the required level for lactating wildebeest in all 24 samples gathered from the dry season range, but in

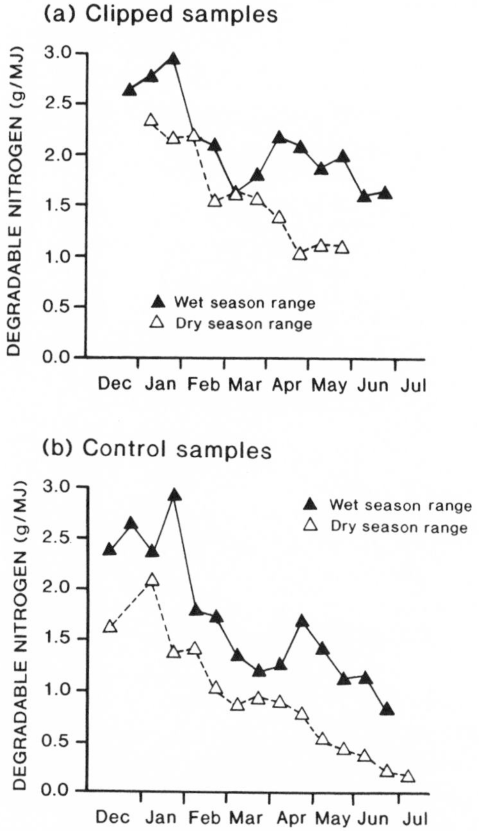

Figure 11.5 The availability of protein for rumen fermentation in green leaf collected inside exclosures on the wet and dry season ranges of migratory wildebeest: (*a*) clipped plots; (*b*) control plots.

only 4 of 26 samples from the wet season range (table 11.1; fig. 11.6c, d). There was also a seasonal trend, with calcium concentration declining over the growth period on both ranges, but phosphorus concentration declining on the dry season range and increasing on the wet season range.

In samples collected from unprotected sites, calcium and phosphorus were again more concentrated on the wet season range of the migratory population (table 11.4). Sodium was lower on the wet season range, but not significantly so. Differences in mineral concentrations on the wet and dry season ranges of resident wildebeest were less marked and failed to achieve statistical significance. However, sodium concentrations were several times greater than on the wet and dry season ranges of the migratory

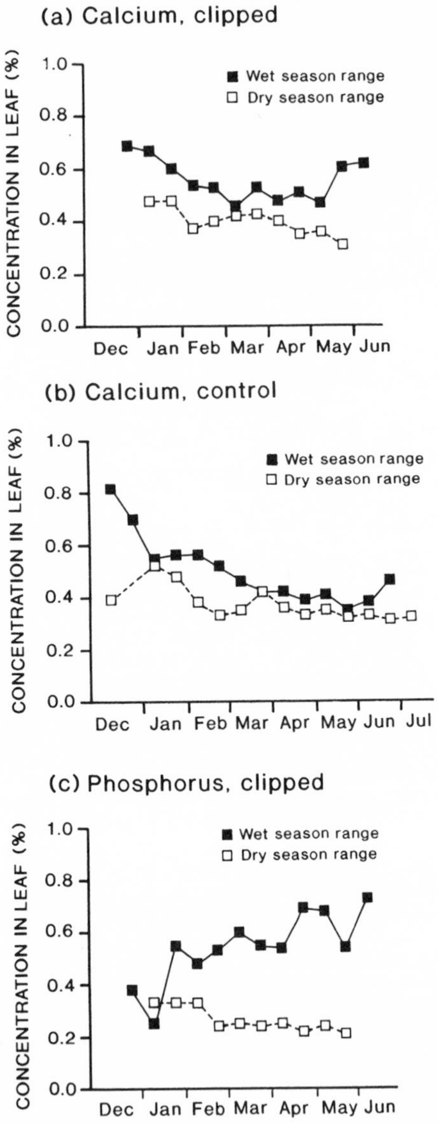

Figure 11.6 Bimonthly variation in concentration of calcium and phosphorus in green leaf collected within exclosures on the wet and dry season ranges of migratory wildebeest. (*a*) calcium, clipped sample; (*b*) calcium, control sample; (*c*) phosphorus, clipped sample; (*d*) phosphorus, control sample.

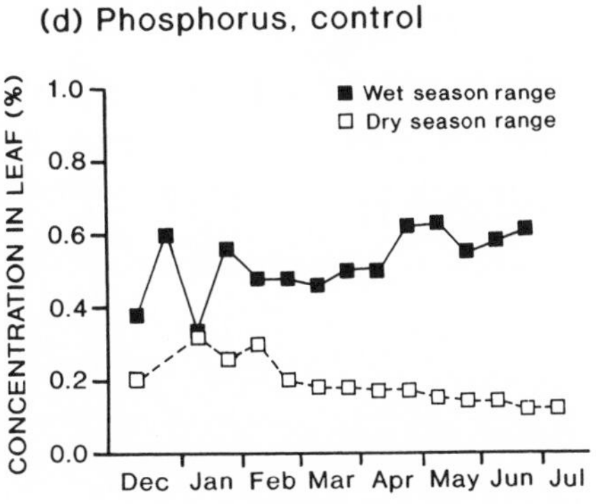

Table 11.4 Nutritional composition of green leaf collected from unprotected sites on the wet season and dry season ranges of migratory and resident wildebeest.

Population	Range	CP	Sodium	Calcium	Phosphorus	Ca:P	n
Migratory	Wet season	18.13	0.10	0.66	0.52	1.28	30
		(4.55)	(0.13)	(0.19)	(0.12)	(0.41)	
Migratory	Dry season	16.09	0.19	0.47	0.31	1.58	33
		(3.60)	(0.34)	(0.11)	(0.06)	(0.43)	
		NS	NS	**	***		
Resident	Wet season	15.97	0.55	0.41	0.45	0.99	23
		(3.42)	(0.38)	(0.09)	(0.14)	(0.38)	
Resident	Dry season	13.68	0.73	0.42	0.36	1.22	30
		(3.44)	(0.57)	(0.06)	(0.08)	(0.36)	
		NS	NS	NS	NS		

Note: All measures are expressed as a percentage of dry weight. Means and standard deviations (in parentheses) refer to single collections from each site. Significance levels refer to Kolgomorov-Smirnov two-sample tests of nutrient concentrations within populations and between ranges.
$^{**}P < .01$; $^{***}P < .001$; NS, not significant.

population (table 11.4). Average Ca:P ratios for all grassland habitats varied from 0.99:1 to 2.06:1 (tables 11.3 and 11.4).

Serum and Urine Electrolytes and Proteins

Total protein in sera of the Serengeti wildebeest was high by comparison with wildebeest kept at Whipsnade Zoo, U.K., and by comparison with normal values of cattle (table 11.5). The high level of total protein was associated with a high level of serum globulin, possibly reflecting high levels of immunoglobulins. Animals on the dry season and transitional ranges may have been dehydrated, since both the concentration of creatinine in their serum samples and the specific gravity of their urine were elevated.

Serum phosphate levels were generally low, in keeping with the trend noted for savanna-dwelling artiodactyls (S. Gascoyne, pers. comm.) but

Table 11.5 Protein concentrations in sera of adult female wildebeest in the Serengeti.

Range	Total protein (g/l)	Creatinine (mol/l)	Albumin (g/l)	Globulin (g/l)	n
Reference values (1)	78	—	35	43	
(2)	68.0–71.0	124–154	38.0–40.0	30.0–31.0	
Wet season	79.0 (4.5)	108.6 (12.9)	30.8 (2.4)	48.2 (3.2)	17
Transitional	79.1 (5.3)	165.7 (22.6)	30.0 (3.8)	49.1 (5.2)	9
Dry season	82.0 (3.7)	175.8 (9.8)	30.6 (3.0)	51.4 (3.8)	5
	NS	***	NS	NS	

Note: Mean values and standard deviations (in parentheses) are given. The locality of animals sampled is given by reference to the seasonal movements of the migratory population. Reference values are (1) means for cattle (Topps and Thompson 1984) and (2) those obtained from two healthy adult wildebeest kept at Whipsnade Zoo, U.K. (S. Gascoyne, pers. comm.). Significance of variation in concentrations between different ranges is indicated (see text).
***$P < .001$; NS, not significant.

Table 11.6 Electrolyte concentrations in sera of adult female wildebeest in the Serengeti.

Range	Sodium (mmol/l)	Potassium (mmol/l)	Calcium (mmol/l)	Inorganic phosphate (mmol/l)	Chloride (mmol/l)	n
Reference values (1)	141	5.6	2.48	1.94	—	
(2)	141–143	4.0–4.3	2.23–2.29	1.00–1.13	96.0–98.0	
(3)	—	—	2.0	1.45	—	
Wet season	133.0	4.75	1.91	1.02	98.7	17
	(3.3)	(0.41)	(0.14)	(0.26)	(2.2)	
Transitional	131.4	4.58	1.95	1.05	101.2	9
	(2.7)	(0.46)	(0.13)	(0.18)	(3.4)	
Dry season	132.2	4.72	1.96	0.60	101.8	5
	(1.5)	(0.26)	(0.06)	(0.18)	(1.5)	
	NS	NS	NS	**	NS	

Note: Mean values with standard deviations (in parentheses) are given. The locality of animals sampled is given by reference to the seasonal movements of the migratory population. Reference values are (1) means for cattle (Topps and Thompson 1984), (2) those obtained from two healthy adult wildebeest kept at Whipsnade Zoo, U.K. (S. Gascoyne, pers. comm.), and (3) critical minimum levels for cattle (McDowell 1985). Significance of variation in concentration between different ranges is indicated (see text).
**$P < .01$; NS, not significant.

levels were very low in animals on the dry season range (table 11.6). No phosphate was detected in urine. Serum calcium was low relative to the reference animals at Whipsnade Zoo, but not significantly lower than the critical minimum value for cattle (table 11.6). Lower albumin levels may have reduced the level of bound calcium. Sodium in sera was also low relative to the Whipsnade animals, but the absence of variation in serum sodium among samples collected from the different ranges of wildebeest, in association with a 30-fold variation in sodium clearance (table 11.7), signifies effective homeostatic control. Potassium clearances were particularly high on the wet season range of migratory wildebeest, suggesting a greater effort by these animals to conserve sodium. Conversely, the lower clearance of potassium on the dry season range suggests less need for

Table 11.7 Creatinine clearances (% Cr) of adult female wildebeest in the Serengeti.

Range	Sodium	Potassium	Chloride	n
Wet season	0.037 (0.023)	105.3 (36.3)	1.26 (0.54)	14
Transitional	1.135 (0.972)	52.0 (31.1)	1.26 (0.42)	9
Dry season	0.039 (0.038)	50.9 (9.4)	1.32 (0.42)	5
	**	***	NS	

Note: Mean values with standard deviations (in parentheses) are given. The locality of animals sampled is given by reference to the seasonal movements of the migratory population. Significance of variation in clearances between different ranges is indicated (see text).
$^{**}P < .01$; $^{***}P < .001$; NS, not significant.

sodium retention, which indicates a higher sodium content in the diet relative to requirements. Chloride in serum was normal.

The concentration of phosphate in serum and the clearances of sodium and potassium were strongly influenced by the location of wildebeest at the time of sampling (tables 11.5–11.7), but were not significantly associated with the calf's age or the mother's body condition.

DISCUSSION

In a pioneering study, Weir (1972) demonstrated that localized sources of sodium strongly affected the distribution of elephant in western Zimbabwe. Likewise, in the Serengeti, resident herbivores are found to concentrate on pastures with a high mineral content (McNaughton 1988, 1989), and it has been suggested that migratory herbivores select habitats on the basis of mineral availabilities (Kreulen 1975; McNaughton 1990). In reviewing the evidence presented here, I am aware of the uncertainty in extrapolating from a narrow database (8 months of data collection from four sites for vegetation; two time periods and three locations for animal sampling). Bearing this caveat in mind, several conclusions concerning the causation of wet season habitat selection in wildebeest emerge from a comparison of the results with the six hypotheses set out above.

Availability of Energy and Protein

The standing crop of grass and of green leaf was higher in the dry season range than in the wet season range of migratory wildebeest, but the clipped plots produced similar quantities of grass and of green leaf in the two areas, suggesting that grassland productivity contained a component related to grass biomass. In a similar comparison between the long and short grasslands of the wet season range (clipped every 2 weeks), Braun (1973) found higher production in the long grasslands and, as with this study, observed a similar proportion of leaf to stem in the two grassland types. Both sets of results are inconsistent with hypothesis 1. There is some evidence that the long grasslands cannot withstand continuing grazing pressure over several years. Plots that were clipped repeatedly for four consecutive growing seasons showed a steady decline in production (Sinclair 1977). Nevertheless, the results presented here, as well as those of Braun (1973), reveal that tall grasses of the dry season range are capable of withstanding heavy grazing pressure for at least one season, and that wildebeest do not migrate at the start of the wet season because of differences in the availability of green leaf. Fryxell (chap. 12) suggests that rotational grazing by migratory species may result in a mosaic of pastures of different biomass, with the result being a higher production overall than in ungrazed areas. While this scenario was not specifically investigated, it seems unlikely given the high intensity of grazing throughout the shortgrass plains.

Metabolizable energy concentration within green leaf was similar in different localities, providing no basis of support for hypothesis 2. Protein was more concentrated in leaf from the wet season range of migratory wildebeest than from the dry season range, but the difference was not great. In fact, CP on the dry season range of migratory wildebeest was as high as that on the wet season range of resident wildebeest. A similar result was obtained by Braun (1973) from experiments on the long grasslands of the wet season range. He found that crude protein in clipped leaf "remains constant at a remarkably high level," a finding that prompted Kreulen (1975) to reject crude protein availability as an explanation for wildebeest habitat selection.

Even small differences in crude protein content could be important if they affected the rate of rumen fermentation. The availability of protein for fermentation was lower on the dry season range of migratory wildebeest, dropping below 1.34 g N/MJ metabolizable energy, the level recommended as sufficient for the rumen microflora to make full use of fermentable carbohydrate (ARC 1984). Values as low as 0.2 g N/MJ ME in the standing crop suggest the potential for a severe constraint on food intake by livestock standards, but in the clipped plots the value remained above 1.0 g N/MJ ME, indicating the possibility of a mild constraint on intake by livestock standards (E. L. Miller, pers. comm.).

Thus, these data do not provide sufficient grounds for rejection of hypothesis 3. They suggest a possible advantage for wildebeest moving onto the wet season range on the basis of an increased protein intake.

Evidence for Mineral Deficiencies

Sodium-conserving mechanisms in the ruminant are so effective that incidents of deficiency on natural forages, even those low in sodium, are uncommon. Given a low intake of dietary sodium, clinical deficiency is most likely in rapidly growing animals, in dairy cattle, which have large losses of sodium in milk, and in any animals that have large losses of sodium in sweat (Underwood 1981; McDowell 1985). Unlike McNaughton (1990), who found low levels of sodium in the dry season range of the migratory population, this survey found few forage samples that were deficient in sodium relative to the minimum requirements of lactating wildebeest. In fact, the most active retention of sodium was observed in lactating females on the wet season range of the migratory population, but these animals maintained normal levels of serum sodium. The evidence from this study suggests that wildebeest can cope with a low sodium intake by minimizing losses and by selecting sodium-rich habitats within the dry season range. The data do not implicate sodium deficiency in long-distance movements of wildebeest, contrary to hypothesis 4.

Calcium deficiency is rare in ruminant livestock except in high-yielding dairy cows and in other livestock when feeding on quick-growing grasses in humid areas, which can contain very low concentrations of Ca (< 0.2%; Underwood 1981). In the Serengeti, average concentrations of calcium were highest in the wet season range of the migratory wildebeest and lower elsewhere (this study; McNaughton 1990); but even the lower values were usually above the estimate for the minimum requirements of wildebeest at peak lactation. Serum calcium was low relative to the Whipsnade Zoo animals, but this difference may simply reflect lower levels of bound calcium associated with the lower albumin levels. The absence of phosphate in urine samples (indicating that calcium was not being mobilized from bone reserves) and the presence of calcium carbonate crystals in urine also point to a sufficient dietary intake of calcium in all areas sampled. Thus, these data are not consistent with hypothesis 5.

On the strength of more precise livestock models and new information on the calcium concentration of milk from Serengeti wildebeest, Kreulen's (1975) estimate of the minimum requirements for dietary calcium of lactating wildebeest was downwardly revised. Most of the grasses sampled by Kreulen had higher concentrations than this revised minimum standard, which brings into question his explanation of habitat selection within the wet season range on the basis of calcium availability. Wildebeest may risk calcium deficiency in the wet season because of low Ca:P ratios in some habitats. Kreulen (1975) found that the habitat with the lowest Ca in grass leaf (open plains with long grassland) also had a low Ca:P ratio, noting that high concentrations of P may interfere with absorption of Ca in the digestive tract. Recent research has revealed a wider tolerance to the Ca:P ratio than previously suspected (AFRC 1991); nevertheless, ratios as low as those recorded by Kreulen may still be significant (Underwood 1981). The Ca:P ratios recorded in the dry season range of migratory wildebeest were above 1 (this study; McNaughton 1990), and so the problem of Ca absorption does not appear to explain the occurrence of long-distance migratory movements (hypothesis 5).

Natural pastures with a deficiency in phosphorus for cattle or sheep occur extensively throughout the world, including areas in eastern and southern Africa (Underwood 1981; McDowell 1985). Phosphorus deficiency causes severe clinical and pathological change in grazing livestock, with impairment of fertility, appetite, milk yield, and growth, as well as abnormalities of bones and teeth and an increased mortality rate (Underwood 1981; Read, Engels, and Smith 1986b). Short-term deficiency need not cause harmful effects, as ruminants withstand marginal and even moderately severe dietary deficiencies by drawing upon skeletal reserves (Read, Engels, and Smith 1986a; Brodison et al. 1989). Nonetheless, phosphorus is given as a supplement to grazing livestock more often than any other nutrient, excepting salt.

In the Serengeti, phosphorus concentration on the wet season range of migratory wildebeest was well above the minimum requirements for lactating wildebeest (hypothesis 6), but on the dry season range it remained below lactation requirements throughout the growing season, declining as the season progressed. In the last of the clipped samples, it approached the minimum concentration required for maintenance. In the standing crop, it dropped below this level early on in the wet season, declining to 0.12% by June. A low concentration of phosphorus in the dry season range of wildebeest was also recorded by McNaughton (1990). These data suggest that pregnant wildebeest on the dry season range could fail to maintain phosphorus balance while foraging on mature swards. Were the same animals to remain on the dry season range while lactating, they would fail to meet their phosphorus requirements even on growing swards in the wet season. The animals would therefore have difficulty in replenishing skeletal reserves of phosphorus. It is interesting that the level of serum phosphate in the five animals sampled on the dry season range averaged 0.6 mmol/l, the same level that induced a strong appetite for naturally occurring sources of available phosphate (bone and bird feces) in experiments with phosphorus-deficient cattle (Denton 1984). Thus both sets of data are consistent with hypothesis 6.

Causation of Migratory Movements

Geographic variation in the availability of dietary phosphorus within the Serengeti grasslands, combined with a reproductive cycle in requirements for phosphorus in female wildebeest, can provide an underlying explanation for long-distance migratory movements. Four of the five alternative hypotheses are rejected, but the third hypothesis—that female wildebeest select wet season habitats with higher concentrations of protein in green leaf—is not ruled out. Phosphorus and protein concentrations were only weakly correlated in this study ($r = .23$, $N = 116$), so more intensive sampling of protein and minerals among different pasture blocks, combined with analyses of fecal phosphorus levels (Belonje 1978; Belonje and van den Berg 1983), might discriminate between the two extant hypotheses.

The return movement of migratory wildebeest to their dry season range occurs with the advent of dry weather. Grass growth on the wet season range stops after a few days without rain, and there remains almost no standing crop as a food reservoir (McNaughton 1985). Freestanding water is also largely absent from this area. The wildebeest are thus forced to return to their dry season range, which maintains green leaf for a longer period and retains a substantial reservoir of grass swards with high biomass due to light grazing pressure in the wet season. Freestanding water is also available there in pools along major river systems. Lactating females may take advantage of swards in the transitional range (see fig. 11.1), which have higher concentrations of phosphorus and other minerals than those in the dry season range (McNaughton 1990).

Resident Ungulates

A requirement of any hypothesis seeking to explain wildebeest migrations is that it should also account for the nonmigratory habits of resident ungulates. If it is hypothesized that wildebeest migrate in order to avoid predators, why do not resident ungulates also migrate? Given the argument forwarded here that wildebeest migration is a behavioral strategy to avoid phosphorus deficiency, how is it that resident ungulates can remain year-round in the long grasslands, which are deficient in minerals? The answer to the latter question may be found in the mineral composition of certain localized pastures favored by resident grazers during the wet season. McNaughton (1988; McNaughton and Banyikwa, chap. 3) has shown that the magnesium, sodium, and phosphorus content of green leaf is substantially higher in these "hot spots" than in surrounding areas. Although any one of the preferred locations could support only a fraction of the migratory wildebeest population, why do so few wildebeest remain on hot spots in the long grasslands during the wet season?

Resident species could be better adapted to diets that are marginal in phosphorus content. One such possible adaptation is in reproductive strategy. Wildebeest breeding is highly synchronized, and the majority of adult females calve each year during the wet season. By contrast, some resident grazers (such as kongoni and waterbuck) breed throughout the year, which would provide postpartum females with the opportunity to replenish reserves of phosphorus by extending the anestrous period. Another way to supplement reserves of dietary phosphorus is by selective consumption of browse species (Pellew 1984). However, neither strategy is easily extended to topi, a resident grazer that rarely feeds on dicotyledons and in which the majority of females breed annually in a well-defined calving season. Possibly, the more selective grazing strategy of topi

(Murray and Brown 1993) provides the opportunity for sustained-yield grazing in mineral-rich hot spots (McNaughton and Banyikwa, chap. 3). Wildebeest graze swards down to a lower biomass than do topi, and this could reduce the productivity of pastures, forcing wildebeest to forage over wider areas.

Implications for Management

In the last century, game migrations in Africa were probably widespread (Houston 1979), but today, only three large-scale migrations remain relatively intact: those of tiang and white-eared kob in southern Sudan (Howell, Lock, and Cobb 1989; Fryxell and Sinclair 1988) and that of wildebeest in the Serengeti-Mara ecosystem. Until recently, migrations of wildebeest also occurred in Botswana (Williamson, Williamson, and Ngwamotsoko 1988), Namibia (Berry 1980), and South Africa (Whyte and Joubert 1988), but each of these populations has declined after the erection of game-control fences that severed traditional routes of migration. Smaller numbers of ungulates still move over long distances in parts of southern, central, and eastern Africa, but these populations are also under an increasing threat (e.g., Borner 1985; Prins 1987; Howell, Lock, and Cobb 1989). The findings reported here and elsewhere (Kreulen 1975; McNaughton 1989, 1990; McNaughton and Banyikwa, chap. 3), that African ungulates move seasonally between pastures to find specific nutrients, provide a new opportunity in the development of land units for conservation purposes. Identification of widespread mineral deficiencies in grasslands at the ecosystem level, and of mineral-rich pastures within the ecosystem, would be an important early step in formulating plans for protection or expansion of migratory herds. As recently burned areas can also be a direct source of minerals for alcelaphines (Messana 1993), there is also a potential for management of fire to provide mineral supplements.

ACKNOWLEDGMENTS

Field research and writing were supported by the Nuffield Foundation, U.K. I thank P. Arcese, L. J. Brandt, M. Jago, N. Kapinga, and S. Mduma for their substantial help with sample collection and preparation, and Rossdale and Partners for analyzing serum and urine samples at their own expense. I would especially like to thank Mrs. P. B. Allen for her practical help on many occasions and P. A. Jewell for providing much-needed backup in the U.K. Many of the concepts in this chapter were developed in discussion with colleagues, either in Serengeti or in Cambridge, and helpful comments on the manuscript were provided by P. Duncan, D. A. Kreulen, S. J. McNaughton, L. Phillips, A. R. E. Sinclair, N. Suttle, and an anonymous reviewer. Mrs. D. Hughes drew the Serengeti map.

REFERENCES

Afolayan, T. A., and Ajayi, S. S. 1980. The influence of seasonality on the distribution of large mammals in the Yankari Game Reserve, Nigeria. *Afr. J. Ecol.* 18:87–96.

Anderson, G. D., and Talbot, L. M. 1965. Soil factors affecting the distribution of the grassland types and their utilization by wild animals on the Serengeti plains, Tanganyika. *J. Ecol.* 53:33–56.

AFRC (Agricultural and Food Research Council)—Technical Committee on Responses to Nutrients. 1991. A reappraisal of the calcium and phosphorus requirements of sheep and cattle. Report 6. *Nutr. Abstr. Rev.,* ser B 61:573–612.

ARC (Agricultural Research Council). 1965. *The Nutrient Requirements of Farm Livestock. No. 2, Ruminants.* HMSO, London.

———. 1980. *The Nutrient Requirements of Ruminant Livestock.* Commonwealth Agricultural Bureaux, England.

———. 1984. *The Nutrient Requirements of Ruminant Livestock. Supplement no. 1.* Commonwealth Agricultural Bureaux, England.

Bell, R. H. V. 1969. The use of the herbaceous layer by grazing ungulates in the Serengeti National Park, Tanzania. Ph.D. thesis, University of Manchester.

———. 1971. A grazing ecosystem in the Serengeti. *Sci. Am.* 224:86–93.

Belonje, P. C. 1978. An investigation into possible methods of assessing the intake of calcium and phosphorus by grazing sheep. *Onderstepoort J. Vet. Res.* 45:7–22.

Belonje, P. C., and van den Berg, A. 1983. Estimating phosphorus intake by grazing sheep. *S. Afr. J. Anim. Sci.* 13:195–98.

Berry, H. H. 1980. Behavioural and eco-physiological studies on blue wildebeest *(Connochaetes taurinus)* at the Etosha National Park. Ph.D. thesis, University of Cape Town.

Borner, M. 1985. The isolation of Tarangire National Park. *Oryx* 19:91–96.

Braun, H. M. H. 1973. Primary production in the Serengeti: Purpose, methods, and some results of research. *Ann. Univ. Abidjan, ser. E (Ecologie)* 6:171–88.

Brodison, J. A., Goodall, E. A., Armstrong, J. D., Givens, D. I., Gordon, F. J., McCaughey, W. J., and Todd, J. R. 1989. Influence of dietary phosphorus on the performance of lactating dairy cattle. *J. Agric. Sci.* (Camb.) 112:303–11.

Campbell, K. L. I. 1989. *Programme report, September 1989.* Seronera: Serengeti Ecological Monitoring Programme.

Dasmann, R. F., and Mossman, A. S. 1962. Population studies of impala in Southern Rhodesia. *J. Mammal.* 43:375–95.

Denton, D. 1984. *The hunger for salt.* Berlin: Springer-Verlag.

Dowman, M. G., and Collins, F. C. 1982. The use of enzymes to predict the digestibility of animal feeds. *J. Sci. Food Agr.* 33:689–96.

Duncan, P. 1975. Topi and their food supply. Ph.D. thesis, University of Nairobi.

Fryxell, J. M., Greever, J., and Sinclair, A. R. E. 1988. Why are migratory ungulates so abundant? *Am. Nat.* 131:781–98.

Fryxell, J. M., and Sinclair, A. R. E. 1988. Seasonal migration by white-eared kob in relation to resources. *Afr. J. Ecol.* 26:17–31.

Givens, D. I., Everington, J. M., and Adamson, A. H. 1990. The nutritive value of spring-grown herbage produced on farms throughout England and Wales over 4 years. III. The prediction of energy values from various laboratory measurements. *Anim. Feed Sci. Technol.* 27:185–96.

Grzimek, B., and Grzimek, M. 1960. *Serengeti shall not die.* London: Hamish Hamilton.

Houston, D. C. 1979. The adaptations of scavengers. In *Serengeti: Dynamics of an ecosystem,* ed. A. R. E. Sinclair and M. Norton-Griffiths, 263–86. Chicago: University of Chicago Press.

Howell, P., Lock, J. M., and Cobb, S., eds. 1989. *The Jonglei canal: local impact and opportunity.* Cambridge: Cambridge University Press.

INRA (Institut National de la Recherche Agronomique). 1978. *Alimentations des ruminants.* INRA Publications, Versailles.

Jarman, P. J. 1972. Seasonal distribution of large mammal populations in the unflooded Middle Zambezi Valley. *J. Appl. Ecol.* 9:283–99.

Kreulen, D. K. 1975. Wildebeest habitat selection on the Serengeti plains, Tanzania, in relation to calcium and lactation: A preliminary report. *E. Afr. Wildl. J.* 13:297–304.

Lamprey, H. F. 1964. Estimation of the large mammal densities, biomass, and energy exchange in the Tarangire Game Reserve and the Masai Steppe in Tanganyika. *E. Afr. Wildl. J.* 2:1–46.

Maddock, L. 1979. The "migration" and grazing succession. In *Serengeti: Dynamics of an ecosystem,* ed. A. R. E. Sinclair and M. Norton-Griffiths, 104–29. Chicago: University of Chicago Press.

MAFF (Ministry of Agriculture, Fisheries, and Food, U.K.) 1984. Mineral, trace element and vitamin allowances for ruminant livestock. In *Recent advances in animal nutrition–1984,* ed. W. Haresign and D. J. A. Cole, 113–37. London: Butterworths.

———. 1986. *The analysis of agricultural materials.* 3d ed. Reference Book 427. London: HMSO.

McDowell, L. R. 1985. *Nutrition of grazing ruminants in warm climates.* New York: Academic Press.

McNaughton, S. J. 1983. Serengeti grassland ecology: The role of composite environmental factors and contingency in community organization. *Ecol. Monogr.* 53:291–320.

———. 1984. Grazing lawns: Animals in herds, plant form, and coevolution. *Am. Nat.* 124:863–86.

———. 1985. Ecology of a grazing ecosystem: The Serengeti. *Ecol. Monogr.* 55:259–94.

———. 1988. Mineral nutrition and spatial concentrations of African ungulates. *Nature* 334:343–45.

———. 1989. Interactions of plants of the field layer with large herbivores. *Symp. Zool. Soc. Lond.* 61:15–29.

———. 1990. Mineral nutrition and seasonal movements of African migratory ungulates. *Nature* 345:613–15.

Messana, G. H. 1993. The reproductive ecology of Swayne's hartebeest *(Alcelaphus buselaphus swaynei).* Ph.D. thesis, University of Cambridge.

Minson, D. J. 1982. Effect of chemical composition on feed digestibility and metabolizable energy. *Nutr. Abstr. Rev., ser. B,* 52:591–615.

Murray, M. G. 1982. Home range, dispersal and the clan system of impala. *Afr. J. Ecol.* 20:253–69.

———. 1991. Maximising energy retention in grazing ruminants. *J. Afr. Ecol.* 60:1029–45.

———. 1993. Comparative nutrition of wildebeest, hartebeest, and topi in the Serengeti. *Afr. J. Ecol.* 31:172–77.

Murray, M. G., and Brown, D. 1993. Niche separation of grazing ungulates in the Serengeti: An experimental test. *J. Anim. Ecol.* 62:380–89.

Norton-Griffiths, M., Herlocker, D., and Pennycuick, L. 1975. The patterns of rainfall in the Serengeti ecosystem, Tanzania. *E. Afr. Wildl. J.* 13:347–74.

NRC (National Research Council) 1985. *Nutrient requirements for sheep.* 6th rev. ed. Washington, D.C.: National Academic Press.

Pellew, R. A. 1984. The feeding ecology of a selective browser, the giraffe *(Giraffa camelopardalis tippelskirchi). J. Zool.* (Lond.) 202:57–81.

Pennycuick, L. 1975. Movements of the migrating wildebeest population in the Serengeti area between 1960 and 1973. *E. Afr. Wildl. J.* 13:65–87.

———. 1979. Energy costs of locomotion and the concept of "foraging radius." In *Serengeti: Dynamics of an ecosystem,* ed. A. R. E. Sinclair and M. Norton-Griffiths, 164–84. Chicago: University of Chicago Press.

Prins, H. H. T. 1987. Nature conservation as an integral part of optimal land use in East Africa: The case of the Masai ecosystem of northern Tanzania. *Biol. Conserv.* 40:141–61.

Read, M. V. P., Engels, E. A. N., and Smith, W. A. 1986a. Phosphorus and the grazing ruminant. 1. The effect of supplementary P on sheep at Armoedsvlakte. *S. Afr. J. Anim. Sci.* 16:1–6.

———. 1986b. Phosphorus and the grazing ruminant. 2. The effects of supplementary P on cattle at Glen and Armoedsvlakte. *S. Afr. J. Anim. Sci.* 16:7–11.

Sinclair, A. R. E. 1977. *The African buffalo: A study of resource limitation of populations.* Chicago: University of Chicago Press.

———. 1983. The function of distance movement in vertebrates. In *The ecology of animal movement,* ed. I. R. Swingland and P. J. Greenwood, 240–59. Oxford: Clarendon Press.

Sinclair, A. R. E., and Fryxell, J. M. 1985. The Sahel of Africa: Ecology of a disaster. *Can. J. Zool.* 63:987–94.

Sinclair, A. R. E., and Norton-Griffiths, M. 1979. *Serengeti: Dynamics of an ecosystem.* Chicago: University of Chicago Press.

Stanley Price, M. R. 1974. The feeding ecology and energetics of Coke's hartebeest. D.Phil. thesis, University of Oxford.

Talbot, L. M., and Talbot, M. H. 1963. *The wildebeest in western Masailand, East Africa.* Wildlife Monographs, no. 12. Washington, D. C. The Wildlife Society.

Topps, J. H., and Thompson, J. K. 1984. *Blood characteristics and the nutrition of ruminants.* London: HMSO.

Underwood, E. J. 1981. *The mineral nutrition of livestock.* 2d ed. Commonwealth Agricultural Bureaux, England.

Van Soest, P. J. 1982. *Nutritional ecology of the ruminant.* Corvallis, Ore.: O and B Books.

Watson, R. M. 1967. The population ecology of the wildebeest (*Connochaetes taurinus albojubatus* Thomas) in the Serengeti. Ph.D. thesis, Cambridge University.

———. 1969. Reproduction of wildebeest, *Connochaetes taurinus albojubatus* Thomas, in the Serengeti Region, and its significance to conservation. *J. Reprod. Fert.,* suppl. 6:287–310.

Weir, J. S. 1972. Spatial distribution of elephants in an African National Park in relation to environmental sodium. *Oikos* 23:1–13.

Western, D. 1975. Water availability and its influence on the structure and dynamics of a savanna large mammal community. *E. Afr. Wildl. J.* 13:265–86.

Whyte, I. J., and Joubert, S. C. J. 1988. Blue wildebeest population trends in the Kruger National Park and the effects of fencing. *S. Afr. J. Wildl. Res.* 18:78–87.

Williamson, D. T., and Mbano, B. 1988. Wildebeest mortality during 1983 at Lake Xau, Botswana. *Afr. J. Ecol.* 26:341–44.

Williamson, D., Williamson, J., and K. T. Ngwamotsoko. 1988. Wildebeest migration in the Kalahari. *Afr. J. Ecol.* 26:269–80.

TWELVE

Aggregation and Migration by Grazing Ungulates in Relation to Resources and Predators

John M. Fryxell

The Serengeti ungulate community, like those in other pristine savanna ecosystems, is dominated numerically by several species (wildebeest, zebra, and Thomson's gazelle) that exhibit extensive seasonal migrations (Fryxell, Greever, and Sinclair 1988). Moreover, these ungulate species are commonly highly aggregated within their seasonal ranges in Serengeti. Despite the fact that these features have long been recognized, the adaptive nature of migration and aggregation has been little considered.

In this chapter I use simple theoretical models to explore the ecological implications of migration and aggregation by grazing ungulates. In particular, I explore conditions under which energy gain might be enhanced or predation risk might be reduced by seasonal migration and spatial aggregation by Serengeti ungulates.

NATURAL HISTORY

In general terms, the Serengeti ecosystem can be divided into two components: (1) open southern grasslands with low annual rainfall (ca. 600 mm/year) that support a relatively low biomass of short-growing grasses and (2) wooded northern grasslands with higher rainfall (ca. 1,000 mm/year) that support tall, highly lignified grasses (Braun 1973; McNaughton 1979, 1985). Rainfall is a key factor influencing primary productivity in both grassland types (Braun 1973; Sinclair 1975; McNaughton 1979, 1985). Most rainfall is concentrated in the period from November to May, with occasional dry periods in January and February (Norton-Griffiths, Herlocker, and Pennycuick 1975). Rainfall shows a pronounced north-south gradient, with localized variation superimposed on this general pattern. As a result, there is considerable spatial and seasonal variation in grass productivity (McNaughton 1985; McNaughton and Banyikwa, chap. 3).

All of the migratory species (wildebeest, zebra, and Thomson's gazelle) show similar seasonal shifts in habitats, using the short grasslands in the south during the wet season and using the tall grasslands in the north during the dry season (Pennycuick 1975; Maddock 1979). Their long-range movements roughly correspond to seasonal transitions. Moreover, within each seasonal range, there are continual movements on a smaller spatial scale, well documented in wildebeest (Inglis 1976) and probably occurring as well in zebra and Thomson's gazelle (Maddock 1979). The general pattern, therefore, is of predictable seasonal shifts in range use, superimposed on extensive within-season migration or nomadism.

Within seasons, all three species are highly aggregated in space. For example, population densities of wildebeest during the wet season commonly exceed 1,000 individuals/km^2 in specific locations in the southern short grasslands. Although dry season distribution patterns are less marked, there is still a pronounced degree of spatial aggregation (Maddock 1979). Seasonal changes in the degree of aggregation by Serengeti ungulates are similar to those recorded in other African savannas, such as the Boma ecosystem in the southern Sudan (Fryxell and Sinclair 1988).

With these general descriptive patterns in mind, I now outline a series of general models that may help clarify some of the demographic implications of seasonal patterns of migration and aggregation. I particularly focus on the various possible effects of movement and grouping strategies on energy intake and the risk of predation.

MODELS

Forage Quality

One of the most straightforward benefits of seasonal migration and aggregation may be the opportunity to take advantage of seasonal and spatial variation in food availability. In order to consider these effects explicitly, it may be instructive to consider the contrasting ways in which forage biomass and nutritional quality of forage constrain rates of energy gain. The model I develop is a simplified version of more physiologically precise, but more detailed, models I have published elsewhere (Fryxell 1991).

Instantaneous consumption rates of grazing herbivores commonly decelerate monotonically with increasing forage biomass (Wickstrom et al. 1984; Renecker and Hudson 1985; Short 1985) and/or bite size attainable from individual plants (Spalinger, Hanley, and Robbins 1988; Gross et al. 1993). Hence, the rate of energy intake (I) can be estimated by a modified disc equation (Holling 1959):

$$I = Q \times [aV/(1 + ahV)] \quad (12.1)$$

where I = digestible energy intake over a short time interval, V = vegetation biomass per unit area, a = area searched per unit time, h = handling time (i.e., feeding plus digestion), and Q = digestible energy content per unit mass of forage ingested. At low levels of forage abundance, the rate of intake is constrained by the rate at which food is encountered, whereas handling time constrains intake at high forage abundance.

Both Q and h can be affected by spatial or temporal variation in forage quality. Of particular interest, forage quality commonly varies as a function of maturation stage and, therefore, of biomass of vegetation (Mowat et al. 1965; Raymond and Terry 1966; Wilmshurst, Fryxell, and Hudson 1995). For example, Braun's (1973) data suggest that crude protein content (which is commonly correlated with digestible energy; Van Soest 1982) declines steadily with maturation of grasses in several parts of the Serengeti. Taking the simplest case, assume that forage quality declines linearly with grass biomass,

$$Q = c - [(c - d) \times V/K] \qquad (12.2)$$

where c = maximum forage quality in immature swards, d = minimum forage quality in mature swards, and K = vegetative carrying capacity. Hence, digestible energy intake should decline at high grass biomass, despite high instantaneous rates of intake.

Moreover, the handling time of forage is also often related to forage quality (review in Hobbs 1990; Fryxell 1991). For simplicity, assume a direct correspondence between the two parameters, such that $h = b - fQ$, where b = maximum handling time and f = rate of decline in handling time with increasing forage quality. Hence a more realistic energy intake model incorporating maturational changes in forage quality might be

$$I = \frac{a\,[c - (c - d)\,V/K]\,V}{1 + abV - af[c - (c - d)\,V/K]\,V} = \frac{\alpha V - \beta V^2}{1 + \theta V + \phi V^2} \qquad (12.3)$$

where α, β, θ, and ϕ are aggregate parameters derived from parameters a–f.

Hence, inclusion of maturational changes in forage quality leads to higher-order effects (the squared terms of V in the numerator and denominator) that imply that energy gain could be maximized at intermediate levels of vegetation biomass (fig. 12.1). It is interesting to note that maturational effects on either Q or h can lead to declining energy intake at high grass biomass. Ruminants should be more sensitive to maturational effects on h because passage rates are often constraining (Belovsky 1978, 1986), whereas nonruminants are thought to compensate for poor forage quality by speeding up passage rates through the gut (Janis 1976). In the latter case, however, it is unlikely that faster passage rates can be achieved without compromising digestibility (but see Duncan et al. 1990). Hence, both ruminants and nonruminants might be expected to have complex

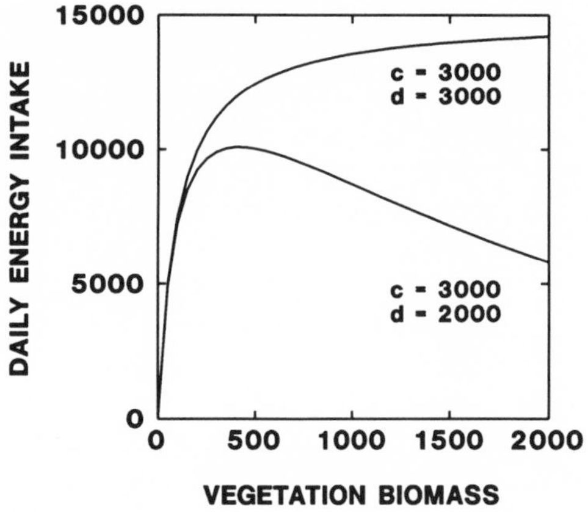

Figure 12.1 Daily rates of energy intake per herbivore (Kcal per day) in relation to vegetation abundance (kg/ha). Energy intake was modeled according to equation (12.3), where c = maximum forage quality in immature swards and d = minimum forage quality in mature swards. When there is no maturational change in forage quality (top line), energy intake is maximized at vegetative carrying capacity. Pronounced changes in forage quality with plant maturation (bottom line) imply that energy intake will be maximized at intermediate vegetation abundance.

energy intake functions that are affected by maturational changes in diet quality, albeit for different reasons.

Maturational changes in vegetation quality and their effect on herbivore foraging preferences have been recently tested using *Cervus elaphus* (commonly known as wapiti or red deer) at sites in Norway (Langvatn and Hanley 1993) and Canada (Wilmshurst, Fryxell, and Hudson 1995). At both sites, grasses experienced considerable changes in digestibility, fiber content, and protein content as they matured; hence there was a trade-off between plant abundance and plant nutritional quality. In the Canadian experiments, the effect of maturational changes in energy content on the maximum rate of forage processing (i.e., the reciprocal of handling time) and the functional response to changes in grass abundance were measured and the resulting parameters were used to model energy gain according to a modified version of equation (12.3). These measurements suggested that study animals should experience maximal rates of energy gain when feeding in grass swards with a biomass of 1,100 kg/ha (fig. 12.2).

In both studies, manipulative experiments were used to test whether the grazers preferred grass patches of intermediate biomass. The protocols involved creation of a mosaic of large grassland patches of varying biomass, either by staggered grass planting (Langvatn and Hanley 1993) or staggered mowing schedules (Wilmshurst, Fryxell, and Hudson 1995. In each case, grazers spent a disproportionate amount of time in the

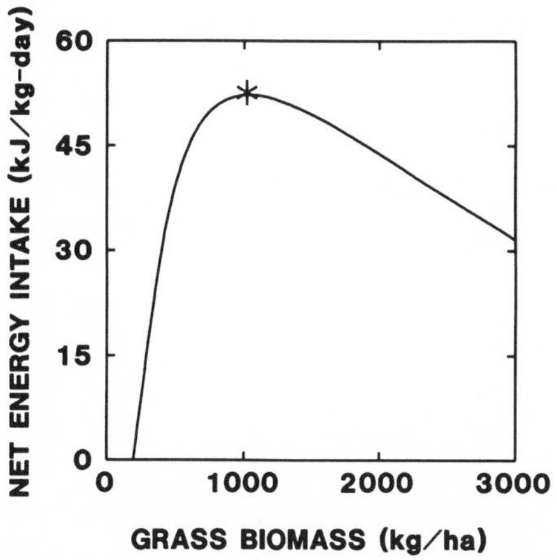

Figure 12.2 Net energy gain (kJ/kg-day) predicted for wapiti in relation to grass biomass (kg/ha). The optimal value is indicated by an asterisk. (From Wilmshurst, Fryxell, and Hudson 1995.)

patches with intermediate grass biomass. In the Canadian study, the preferred patch type (with a biomass of 1,212 kg/ha) was close to the value predicted by the model (fig. 12.3). These experimental results strongly suggest that herbivores are capable of responding behaviorally to the postulated trade-off between plant abundance and plant nutritional quality.

By embedding such responses in a plant-herbivore model, we can now consider how energy intake might vary in relation to herbivore density and nutritional characteristics of the vegetation. Assume that vegetation growth is logistic, such that per capita growth rates of vegetation (G) decline linearly with vegetation density,

$$G = r(V + s)\,[1 - (V + s)/(K + s)] \qquad (12.4)$$

where r = maximum per capita rate of growth of vegetation, s = ungrazable root or crown reserves, and K = carrying capacity of vegetation (Fryxell, Greever, and Sinclair 1988). Then $V_{t+1} = V_t + G - IH_t/Q$, where H_t = herbivore density at time t.

Three general results emerge from such simulations (Fryxell 1991). First, energy gain should be highest at intermediate herbivore densities (fig. 12.4) because of trade-offs between forage quality and the rate at which forage can be obtained. At high forage biomass, as would occur for a low density of herbivores, the rate of energy gain is depressed because forage is maintained at a relatively mature growth stage with low nutritional quality and a long retention time. At low forage biomass, as would occur for a high density of herbivores, the rate of energy gain is constrained by availability. Hence, energy gain should be highest at intermediate resource densities, as on "grazing lawns" where herbivores maintain the vegetation at an immature growth stage that optimizes the trade-

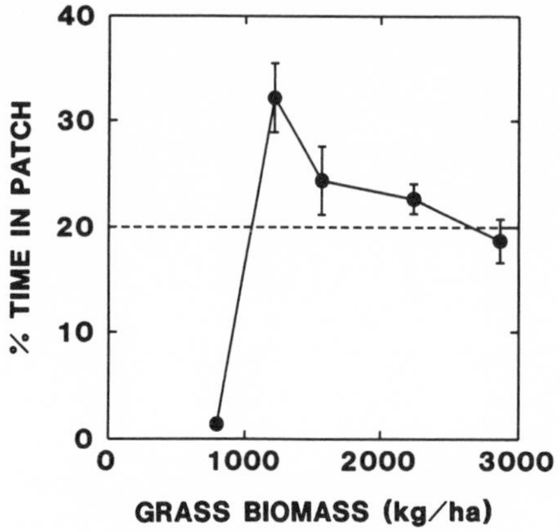

Figure 12.3 Percentage foraging time by wapiti in relation to grass biomass (kg/ha) within experimental patches. Vertical bars are 1 SE. The horizontal hatched line represents the expected value if animals foraged indiscriminately among patches. (Data from Wilmshurst, Fryxell, and Hudson 1995.)

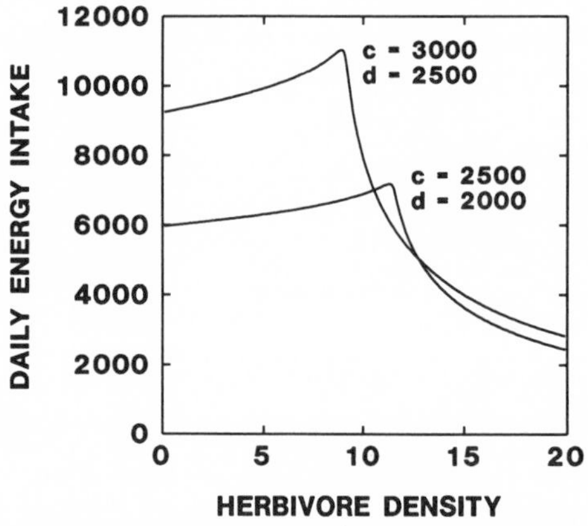

Figure 12.4 Daily energy intake (Kcal per day) in relation to herbivore density (per ha), for animals in high-quality (top line) and low-quality (bottom line) habitats. Energy intake was modeled as in figure 12.1.

off between availability and digestive constraints (McNaughton 1984; Georgiadis and McNaughton 1990; Fryxell 1991).

Second, herbivores at a given population density should obtain appreciably higher rates of energy gain feeding on vegetation types with high mean quality than ones with low mean quality, even with no difference in per capita growth rates or carrying capacity (fig. 12.4). Hence, pronounced spatial variation in forage quality, due to localized variation in rainfall, geomorphology, or nutrient cycling, should contribute to pronounced herbivore aggregation, provided that herbivores distribute themselves in an ideal free manner (Fretwell and Lucas 1970).

Third, given the choice between low-productivity and high-productivity grasslands (assuming forage quality is similar across habitats), herbivores could potentially realize higher rates of long-term energy gain in less productive habitats than in highly productive habitats, at least at low to intermediate herbivore densities (fig. 12.5), because the less productive grasslands would be maintained more readily by grazers at an immature growth stage that optimizes the trade-off between availability and quality constraints. However, forage quality is also often correlated with rainfall in African grasslands (Breman and DeWit 1983). Hence, productivity and forage quality should often negatively covary.

These relationships could help to explain the seasonal distribution patterns of the predominant Serengeti ungulates. As described earlier, there is a decreasing rainfall gradient in Serengeti from north to south (Norton-Griffiths, Herlocker, and Pennycuick 1975), leading to a similar gradient in grass productivity (McNaughton 1979, 1985). Although this pattern has not been studied in detail in Serengeti, there is some empirical evidence suggesting that more northerly habitats support poorer-quality grasses at a given growth stage than do the short grasslands of the southern Serengeti (Braun 1973; Murray, chap. 11). On this basis, migratory ungulates should be better off foraging on the less productive, but higher-quality, Serengeti plains during the wet season than elsewhere in the ecosystem. This general prediction is certainly supported by the seasonal distribution patterns of all three of the predominant migratory species (Maddock 1979). During the dry season, very little grass growth takes place in the south, and there is little available surface water there for grazing ungulates. Hence, migration back to the lower-quality northern grasslands should be expected during the dry season, which is indeed the case.

In summary, the observed seasonal migration pattern of Serengeti ungulates is consistent with the predictions of the forage quality model. If one assumes that mortality rates and/or reproduction are associated with long-term rates of energy gain, then seasonal migration should result in higher population growth rates at a given level of herbivore abundance than year-round residence in the north. More specific predictions are currently impossible because most of the relevant parameters on vegetation quality, constraints on intake, and herbivore demographic response to variation in energy gain are unavailable.

The forage quality hypothesis could also help to explain the pronounced degree of aggregation by the Serengeti migratory ungulates observed within seasons (Inglis 1976; Maddock 1979). If there is appreciable spatial variation in forage quality parameters and/or grassland productivity within the dry season or wet season ranges, then one would predict pronounced clumping of subpopulations of the migratory species. Local variation in the nutritional quality of soils and vegetation in Ser-

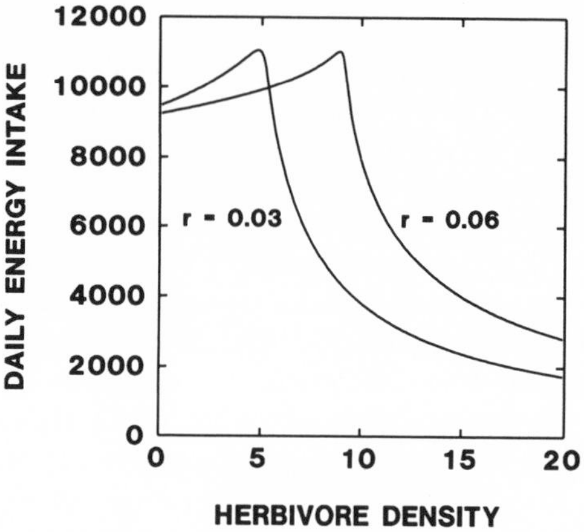

Figure 12.5 Daily energy intake (Kcal per day) in relation to herbivore density (per ha), for animals in slow-growing (left curve) and fast-growing (right curve) habitats. Energy intake was modeled as in figure 12.1.

engeti is pronounced (McNaughton 1990; McNaughton and Banyikwa, chap. 3). Moreover, local densities of both migratory and resident ungulates are correlated with several of these nutritional parameters (McNaughton 1990). A certain degree of aggregation would be predicted even in the absence of spatial variation in forage quality, simply because aggregated ungulates should theoretically be more capable of maintaining nutritious "grazing lawns" of vegetation than a dispersed population would be (McNaughton 1984). Hence, forage quality or vegetation structure could explain both the large-scale seasonal changes in home range (i.e., migration) and regional variation in distribution (i.e., aggregation) of the Serengeti ungulate community.

Rotational Grazing

A second possible advantage of migration could be that herbivores maximize the growth potential of the vegetation through rotational grazing. Let us assume that migratory herbivores occur in a large group that forages together, moving as a unit in a circuit over a large home range. Food resources in any given location are rapidly depleted by the group, which then moves on to exploit a new food patch. A mosaic of food patches should emerge over time, with local resource abundance depending on the time since the migratory group last visited. If we further assume that per capita growth of vegetation declines with vegetation abundance, then rotational grazing could conceivably lead to increased vegetation productivity. This would occur because local patches of vegetation would periodically attain the intermediate levels of abundance at which incremental growth is highest (Noy-Meir 1976, 1978; Fryxell, Greever, and Sinclair 1988).

For example, consider a tropical grassland whose per capita growth can be modeled by equation (12.4), with $r = 0.06$, $s = 2{,}000$, and $K = 5{,}000$. Assume further that no appreciable change in plant quality occurs with maturation (Q is constant), and that the herbivore functional response is of the form $I = aQV/(1 + ahV)$. The effects of rotational grazing can be simulated by having the herbivore population utilize discrete vegetation patches for a period t. Hence, the total rotation period (T) depends on the time per patch (t) and the total number of patches (p) visited during a foraging circuit, assuming that the size of each patch is the reciprocal of the number of patches used.

Food intake varies according to both time in each patch and number of patches, with various combinations of each set of parameters optimizing rates of food intake (fig. 12.6). For short patch times, the optimal number of patches visited is large, whereas the optimal patch number is smaller for long patch times. Hence, there is an optimum total rotation period (approximately 50 days in the example shown in fig. 12.6) that can be obtained either by visiting a small number of large patches for extended periods or by visiting a large number of small patches for a short time. In any case, rotational grazing could clearly improve the rate of food intake by individual herbivores in a cohesive herd, provided that overall herbivore density is high enough.

Maturational changes in forage quality should tend to diminish the advantages of rotational grazing because allowing vegetation to reach high biomass would actually reduce the rate of energy gain. Hence, rotational grazing advantages should be most pronounced when ungulates feed on high-quality grasses, such as those growing in arid regions. On this basis, then, one might predict that rotational grazing benefits would be most pronounced for Serengeti herbivores when they are on the wet season range, whereas these benefits would be slight or nonexistent in the northern wooded grasslands.

All other things being equal, rotational grazing should, however, be evolutionarily unstable. A solitary herbivore that does not migrate with the herd should always obtain equivalent or higher longer-term rates of energy gain than individuals in the herd (fig. 12.7). Although rotational grazing still increases the energy intake of group members, solitary foragers always obtain higher rates of intake because they infrequently suffer intense intraspecific competition for food (fig. 12.7). If other benefits of social cohesion outweigh this benefit of cheating, however, rotational grazing could be selectively advantageous.

Predation

Lions and hyenas are the most abundant natural predators of Serengeti ungulates. Reproductive individuals of both species have relatively fixed

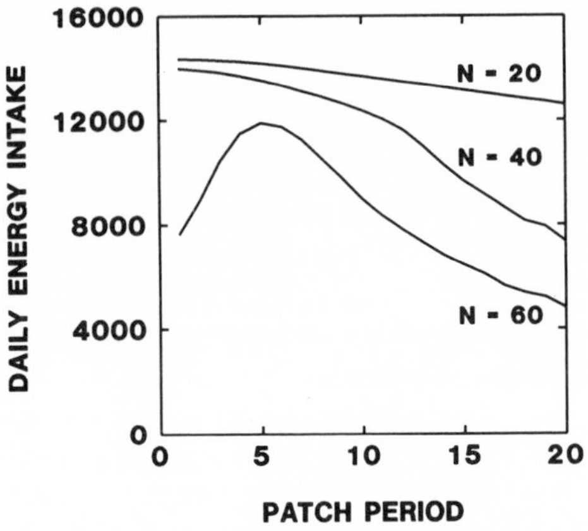

Figure 12.6 Daily energy intake (Kcal per day) in relation to patch period (days) by herd members in a cohesive foraging rotation. In all simulations there were 10 patches available. Energy gain is depicted for various herbivore densities ranging between 20 and 60 animals/ha.

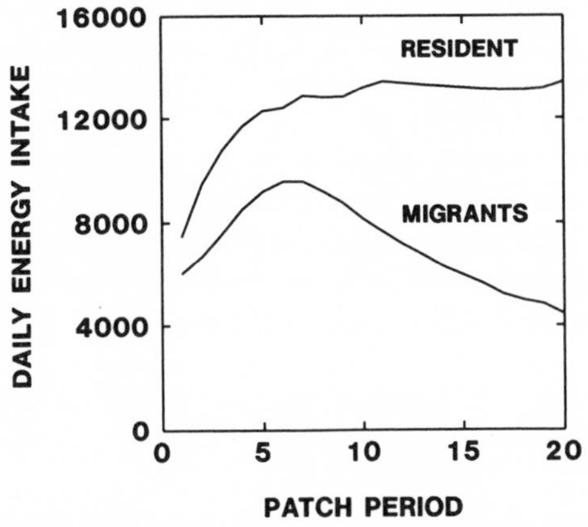

Figure 12.7 Daily energy intake (Kcal per day) by migrant herd members in a grazing rotation and by a solitary resident forager, in relation to patch period (days).

home ranges that are concentrated in the northern plains and wooded grasslands (Schaller 1972; Kruuk 1972). Because the offspring of both species require extended periods of parental care, many lions and hyenas are thought to be unable to follow the migratory herds continually (Schaller 1972; Hanby and Bygott 1979). This is a crude generalization, however, as both lions and hyenas can commute to obtain prey from areas 50 km distant (Hofer and East, chap. 16). In addition, nonreproductive predators, having no fixed home range, could readily follow the migratory herds. Nonetheless, migratory ungulates may be able to reduce the risk of predation by simply moving away from the northern grasslands for extended periods.

Simulation modeling of this process in Serengeti suggests that seasonal escape from predators could be sufficient to prevent predators from regulating their migratory prey (fig. 12.8). As a result, resident and migratory species could have different modes of population regulation, with residents commonly regulated by predators and migrants regulated by food availability (Fryxell, Greever, and Sinclair 1988).

Aggregation should also tend to reduce the risk of predation through simple dilution (Hamilton 1971; Bertram 1978). Hence, doubling the local density of herbivores while predator density remains unchanged would lead to a halving of the mortality risk. Of course, this assumes that predators do not seek out areas with high prey density, which we know does occur (Schaller 1972). Nonetheless, aggregation should be effective in reducing predation risk if social factors prevent predators from associating proportionately with their mobile prey (i.e., if a fivefold ratio between herbivore densities in two locales is not matched by a fivefold ratio in predator densities). Currently, we do not know the short-term numerical response by Serengeti predators to spatial and temporal variation in prey abundance, apart from cheetahs (Caro 1994).

If prey are sufficiently aggregated that they form a cohesive group, they might also benefit from increased probability of detection of predators (Estes 1974; Jarman 1974; Leuthold 1977). There is certainly evidence from other systems that individual members of foraging groups spend less time scanning for predators than do solitary individuals (Lipetz and Bekoff 1982; Underwood 1982; Berger and Cunningham 1988) and that hunting success for lions declines with group size of prey (Van Orsdol 1984).

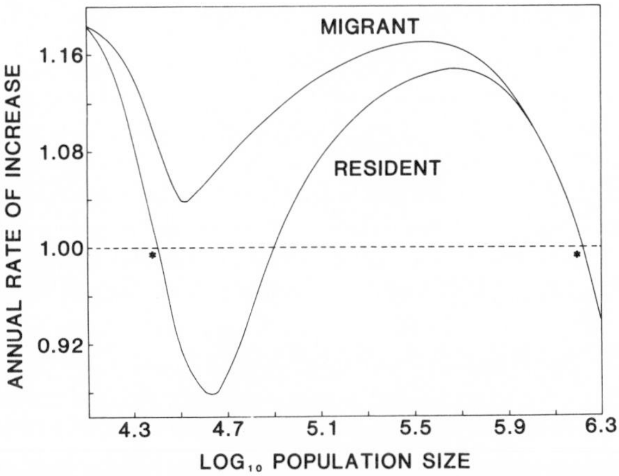

Figure 12.8 Predicted rates of population increase in relation to population size for a migratory versus a resident population of wildebeest in Serengeti, based on Fryxell, Greever, and Sinclair's (1988) model. Asterisks mark stable equilibria due to predation (left) and food availability (right).

Hence, seasonal escape from predators, dilution of predation risk, and improved surveillance could lead to diminished mortality due to predation for migratory and/or aggregated prey. The magnitude of these effects will depend on the degree of predator home range fidelity, the local numerical response by predators, and the effectiveness of group vigilance, none of which are known in detail for Serengeti predators.

Combined Effects

Many of the postulated advantages of ungulate migration and aggregation should be additive. Seasonal movements by Serengeti grazers could result in both appreciably higher energy intake and lower risk of predation. Hence, such additive effects should tend to reinforce the adaptive advantage of seasonal movements. For example, the coincident advantages of predator dilution could be sufficient to make rotational grazing an evolutionarily stable strategy. In turn, the energetic benefits of rotational grazing could further strengthen the antipredatory advantages of group formation. As a result, it may prove difficult to discriminate between the adaptive significance of energetic and of antipredatory benefits in a given population.

Perhaps more importantly, the adaptive advantages of seasonal migration and aggregation should contribute to pronounced spatial heterogeneity in both herbivores and their natural predators. An emerging theme in theoretical ecology is that such spatial heterogeneity can have a powerful stabilizing influence on trophic interactions (May 1978; Chesson and Murdoch 1986; Pacala, Hassell, and May 1990; Fryxell and Lundberg 1993). Hence, an understanding of the factors leading to herbivore spatial heterogeneity may be instrumental in understanding why herbivore/plant and herbivore/carnivore interactions have been relatively stable in Serengeti over the past 15 years.

DISCUSSION

In this chapter, I have tried to show that there are numerous ways in which migration and aggregation might enhance the survivorship of Serengeti ungulates. The combined effects of these factors for various behavioral lifestyles are summarized in table 12.1. While it is unlikely that any single species can be precisely accommodated within the categories given in table 12.1, it seems clear that a tendency toward aggregatedness by a migratory species offers multiple advantages compared with alternative lifestyles. Hence, it may not be particularly surprising that those species that show particularly pronounced variation in spatial and temporal distribution predominate in the Serengeti ecosystem.

The ultimate objective of science is to be able to predict empirical

Table 12.1 Hypothetical advantages of different behavioral lifestyles for Serengeti herbivores.

	Migration	Aggregation	Solitary
Spatial resource heterogeneity	+	+	0 or +
Forage quality enhancement	0	+	0
Forage productivity enhancement	+	0	0
Seasonal escape from predators	+	0	−
Predator dilution	+	+	−

Note: +, positive demographic effect; 0, no substantial effect; −, negative demographic effect.

outcomes. These simple models may provide a useful first step in understanding the implications of movement and grouping patterns in ungulates. But after more than 30 years of research, we are still far from being able to provide precise models of basic demographic processes directly applicable to the Serengeti system. In the remainder of this discussion I would like to highlight particular research topics that would allow further development of such models.

The basic conceptual element in understanding any trophic interaction is the functional response by consumers to variation in food resource density. Despite the abundance of field research in Serengeti, there are no basic descriptions of functional responses for any major herbivore or carnivore species. It seems unlikely that much progress will be made in predicting ecosystem processes in Serengeti until these critical data become available.

The effect of variation in forage quality on ungulate functional responses would seem particularly important. Moreover, although we understand in general terms how forage quality tends to vary in relation to maturational stage and local conditions, these parameters need to be estimated empirically to apply the functional response models to the Serengeti system.

Similarly, an understanding of predator functional responses should be a critical research priority. There is strong evidence that lions show density-dependent changes in prey preferences (Foster and Kearney 1967; Rudnai 1974), which could result in either type II (Hilborn and Sinclair 1979) or type III (Fryxell, Greever, and Sinclair 1988) functional responses. These different functional responses can produce considerably different long-term dynamics.

Data on numerical responses by both herbivores and carnivores to variation in resource density would also be enormously useful. The herbivore numerical response to variation in food abundance and food quality is needed to test the predictions of the forage quality model. The carnivore numerical response is needed to assess the effectiveness of migration and aggregation in diluting predation risk or apportioning predation unevenly across prey subpopulations.

Finally, there is a need for more detailed information on the movement patterns of ungulates within seasonal ranges. The positive effects of movement by groups on forage productivity and forage quality depend on repeated use of specific habitats. Hence, nomadic or random movements within seasonal ranges would reduce or eliminate some of the potential effects outlined in these models, whereas cohesive aggregations and small-scale migration within seasonal ranges would enhance these effects.

In conclusion, empirical studies in Serengeti have stimulated several alternative hypotheses that could explain the predominance of wildebeest, zebra, and Thomson's gazelle there. Further progress in testing these ideas will largely depend on organizing future research that focuses on processes affecting the outcome of key trophic interactions.

ACKNOWLEDGMENTS

I wish to thank the National Sciences and Engineering Research Council of Canada for financial support. I also thank P. Arcese, A. R. E. Sinclair, and two anonymous referees for their comments on an earlier draft of this chapter.

REFERENCES

Belovsky, G. 1978. Diet optimization in a generalist herbivore: The moose. *Theor. Popul. Biol.* 14:103–34.

———. 1986. Generalist herbivore foraging and its role in competitive interactions. *Am. Zool.* 26:51–69.

Berger, J., and Cunningham, C. 1988. Size-related effects on search times in North American grassland female ungulates. *Ecology* 69:177–83.

Bertram, B. C. R. 1978. Living in groups: Predators and prey. In *Behavioural ecology,* ed. J. R. Krebs and N. B. Davies, 64–96. Oxford: Blackwell Scientific Publications.

Braun, H. M. H. 1973. Primary production in the Serengeti: Purpose, methods, and some results of research. *Ann. Univ. Abidjan, ser. E (Ecologie)* 6:171–88.

Breman, H., and de Wit, C. T. 1983. Rangeland productivity and exploitation in the Sahel. *Science* 221:1341–47.

Caro, T. 1994. *Cheetahs of the Serengeti plains: Group living in an asocial species.* Chicago: University of Chicago Press.

Chesson, P. L., and Murdoch, W. W. 1986. Aggregation of risk: Relationships among host-parasitoid models. *Am. Nat.* 127:696–715.

Duncan, P., Foose, T. J., Gordon, I. J., Gakahu, C. G, and Lloyd, M. 1990. Comparative nutrient extraction from forages by grazing bovids and equids: A test of the nutritional model of equid/bovid competition and coexistence. *Oecologia* 84:411–18.

Estes, R. D. 1974. Social organization of the African Bovidae. In *The behaviour of ungulates and its relation to management,* ed. V. Geist and F. Walther, 166–205. Morges, Switzerland: IUCN.

Foster, J. B., and Kearney, D. 1967. Nairobi National Park game census, 1966. *East Afr. Wildl. J.* 5:112–20.

Fretwell, S. D., and Lucas, H. L. 1970. On territorial behavior and other factors influencing habitat distribution in birds. *Acta Biotheor.* 19:16–36.

Fryxell, J. M. 1991. Forage quality and aggregation by large herbivores. *Am. Nat.* 138:478–98.

Fryxell, J. M., Greever, J., and Sinclair, A. R. E. 1988. Why are migratory ungulates so abundant? *Am. Nat.* 131:781–98.

Fryxell, J. M., and Lundberg. P. 1993. Optimal patch use and metapopulation dynamics. *Evol. Ecol.* 7:379–93.

Fryxell, J. M., and Sinclair, A. R. E. 1988. Seasonal migration by white-eared kob in relation to resources. *Afr. J. Ecol.* 26:17–31.

Georgiadis, N. G., and McNaughton, S. J. 1990. Elemental and fibre contents of savanna grasses: Variation with grazing, soil type, season, and species. *J. Appl. Ecol.* 27:623–34.

Gross, J. E., Shipley, L. A., Hobbs, N. T., Spalinger, D. E., and Wunder, B. 1993. Foraging by herbivores in food-concentrated patches: Tests of a mechanistic model of functional response. *Ecology* 74:778–91.

Hamilton, W. D. 1971. Geometry for the selfish herd. *J. Theor. Biol.* 31:295–311.

Hanby, J. P., and Bygott, J. D. 1979. Population changes in lions and other predators. In *Serengeti: Dynamics of an ecosystem,* ed. A. R. E. Sinclair and M. Norton-Griffiths, 249–62. Chicago: University of Chicago Press.

Hilborn, R., and A. R. E. Sinclair. 1979. A simulation of the wildebeest population, other ungulates, and their predators. In *Serengeti: Dynamics of an ecosystem,* eds A. R. E. Sinclair and M. Norton-Griffiths, 287–309. Chicago: University of Chicago Press.

Hobbs, N. T. 1990. Diet selection by generalist herbivores: A test of the linear programming model. In *Behavioral mechanisms of food selection,* ed. R. N. Hughes, 395–414. Heidelberg: Springer-Verlag.

Holling, C. S. 1959. The components of predation as revealed by a study of small-mammal predation of the European pine sawfly. *Can. Entomol.* 91:293–320.

Inglis, J. M. 1976. Wet season movements of individual wildebeests of the Serengeti migratory herd. *E. Afr. Wildl. J.* 14:17–34.

Janis, C. 1976. The evolutionary strategy of the equidae and the origins of rumen and cecal digestion. *Evolution* 30:757–74.

Jarman, P. J. 1974. The social organisation of antelope in relation to their ecology. *Behaviour* 48:215–67.

Kruuk, H. 1972. *The spotted hyena: A study of predation and social behavior.* Chicago: University of Chicago Press.

Langvatn, R., and Hanley, T. A. 1993. Feeding-patch choice by red deer in relation to foraging efficiency. *Oecologia* 95:164–70.

Leuthold, W. 1977. *African ungulates.* Berlin: Springer-Verlag.

Lipetz, V. E., and Bekoff, M. 1982. Group size and vigilance in pronghorns. *Z. Tierpsychol.* 58:203–16.

Maddock, L. 1979. The "migration" and grazing succession. In *Serengeti: Dynamics of an ecosystem,* ed. A. R. E. Sinclair and M. Norton-Griffiths, 104–29. Chicago: University of Chicago Press.

May, R. M. 1978. Host-parasitoid systems in patchy environments: A phenomenological model. *J. Anim. Ecol.* 47:833–43.

McNaughton, S. J. 1979. Grassland-herbivore dynamics. In *Serengeti: Dynamics of an ecosystem,* ed. A. R. E. Sinclair and M. Norton-Griffiths, 46–81. Chicago: University of Chicago Press.

———. 1984. Grazing lawns: Animals in herds, plant form, and coevolution. *Am. Nat.* 124:863–86.

———. 1985. Ecology of a grazing ecosystem: The Serengeti. *Ecol. Monogr.* 55:259–94.

———. 1990. Mineral nutrition and seasonal movements of African migratory ungulates. *Nature* 345:613–15.

Mowat, D. N., Fulkerson, R. S., Tossell, W. E., and Winch, J. E. 1965. The *in vitro* digestibility and protein content of leaf and stem portions of forages. *Can. J. Plant Sci.* 45:321–31.

Norton-Griffiths, M., Herlocker, D., and Pennycuick, L. 1975. The patterns of rainfall in the Serengeti ecosystem, Tanzania. *East Afr. Wildl. J.* 13:347–74.

Noy-Meir, I. 1976. Rotational grazing in a continuously growing pasture: A simple model. *Agric. Syst.* 1:833–43.

Noy-Meir, I. 1978. Stability in simple grazing models: Effects of explicit functions. *J. Theor. Biol.* 71:347–80.

Pacala, S. W., Hassell, M. P., and May, R. M. 1990. Host-parasitoid associations in patchy environments. *Nature* 344:150–53.

Pennycuick, L. 1975. Movements of the migratory wildebeest population in the Serengeti area between 1960 and 1973. *East Afr. Wildl. J.* 13:65–87.

Raymond, W. F., and Terry, R. A. 1966. Studies of herbage digestibility by an *in vitro* method. *Outlook on Agriculture* 5:60–68.

Renecker, L. A., and Hudson, R. J. 1985. Estimation of dry matter intake of free-ranging moose. *J. Wildl. Mgmt.* 49:785–92.

Rudnai, J. 1974. The pattern of lion predation in Nairobi Park. *E. Afr. Wildl. J.* 12:213–25.

Schaller, G. B. 1972. *The Serengeti lion: A study of predator-prey relations.* Chicago: University of Chicago Press.

Short, J. 1985. The functional response of kangaroos, sheep, and rabbits in an arid grazing system. *J. Appl. Ecol.* 22:435–47.

Sinclair, A. R. E. 1975. The resource limitation of trophic levels in tropical grassland ecosystems. *J. Anim. Ecol.* 44:497–520.

Spalinger, D. E., Hanley, T. A., and Robbins, C. T. 1988. Analysis of the functional response in foraging in the Sitka black-tailed deer. *Ecology* 69:1166–75.

Underwood, R. 1982. Seasonal changes in African ungulate groups. *J. Zool.* (Lond.) 196:191–205.

Van Orsdol, K. G. 1984. Foraging behaviour and hunting success of lions in Queen Elizabeth National Park, Uganda. *Afr. J. Ecol.* 22:79–99.

Van Soest, P. J. 1982. *Nutritional ecology of the ruminant.* Corvallis, Ore.: O and B Books.

Wickstrom, M. L., Robbins, C. T., Hanley, T. A., Spalinger, D. E., and Parish, S. M. 1984. Food intake and foraging energetics of elk and mule deer. *J. Wildl. Mgmt.* 48:1285–1301.

Wilmshurst, J. F., Fryxell, J. M., and Hudson, R. J. 1995. Forage quality and patch choice by wapiti *(Cervus elaphus). Behav. Ecol.* In press.

THIRTEEN

Antipredator Behavior of Serengeti Ungulates: Individual Differences and Population Consequences

Clare D. FitzGibbon and John Lazarus

Although work in the Serengeti and elsewhere has revealed a rich and increasingly subtle story of the adaptive modulation of antipredator strategies—and, more recently, of predator responses to those strategies—the population consequences of antipredator behavior are poorly understood. Our aim here is to explore the implications of individual differences in antipredator behavior for population dynamics and to show, in consequence, the importance of behavioral studies for a full understanding of the Serengeti ecosystem.

Behavioral studies are important for an understanding of population dynamics for the obvious reason that fecundity and mortality are strongly influenced by behavioral responses to conspecifics, competitor species, resources, parasites, and predators (e.g., Sibly and Smith 1985; Ives and Dobson 1987; Abrams 1992). Individual differences in behavior influence population dynamics in the following way:

1. Individuals in populations fall into different classes of age, sex, reproductive status, and body condition, and class membership influences fecundity and mortality. The existence of such classes and the relative frequency of each class therefore influences the dynamics of the population.
2. Between-class differences in fecundity and mortality are partly a consequence of the different behavioral strategies adopted by each class.
3. Within-class differences in fecundity and mortality arise as a result of adaptive behavioral responses to various ecological and social factors.
4. Consequently, population dynamics is influenced by individual differences in behavior, and can be informed by ideas in behavioral ecology that seek to explain this individual variation in adaptive terms.

This line of argument is developed for the case of antipredator behavior in the following section. We then illustrate the relationship between individual differences in behavior and population dynamics, largely with data from the best-studied species, Thomson's gazelle. There are two stages to this analysis: first, we review the influences on antipredator behavior that are expected to have population consequences, and second, having demonstrated the importance of predation as a mortality factor, we examine the evidence that differential predation rates can be explained in terms of these influences. Finally, we consider the population consequences of the costs of antipredator behavior, and the role of antipredator behavior in the dynamics of predator-prey systems. In appendix 13.1 we calculate the relative strength of selection on an antipredator defense as a consequence of interaction with different predator species.

FROM INDIVIDUAL BEHAVIOR TO POPULATION DYNAMICS

The consequences of antipredator behavior for population dynamics are complex (fig. 13.1). Four aspects of population structure—age, sex, reproductive status, and body condition—influence antipredator behavior (1: see numbered arrows in fig. 13.1 for this and later references). For each aspect, the various classes of individuals (e.g., fawn, adolescent, adult; male and female; territorial and nonterritorial males, females with young of varying ages, etc.) vary in their risk of predation as a result of their behavioral and morphological characteristics. Such characteristics influence both the probability that an individual is selected as a target (2 and 3 respectively) and its chance of surviving the attack (4 and 5 respectively). These two effects work synergistically since predators tend to select just those individuals whose antipredator behavior renders them more vulnerable when attacked (FitzGibbon 1989).

Antipredator behavior—and consequently predation risk—is also influenced by two groups of extrinsic factors, ecological (6) and social (7), that alter the costs and benefits of the behavior. Some of these factors influence predation risk directly (8 and 9)—for example, cover—while others act through their effects on antipredator behavior (6 and 7, 4)—for example, food, due to its influence on the time budget.

Antipredator behavior also has implications for the competing demands of feeding (10) and reproduction (11), the requirements of which are generally in conflict with those of surviving attack (Lima and Dill 1990). Selection will tend to maximize reproductive success as a joint function of these three demands, and inevitably involves trade-offs between them.

Antipredator behavior has immediate effects on the feeding time bud-

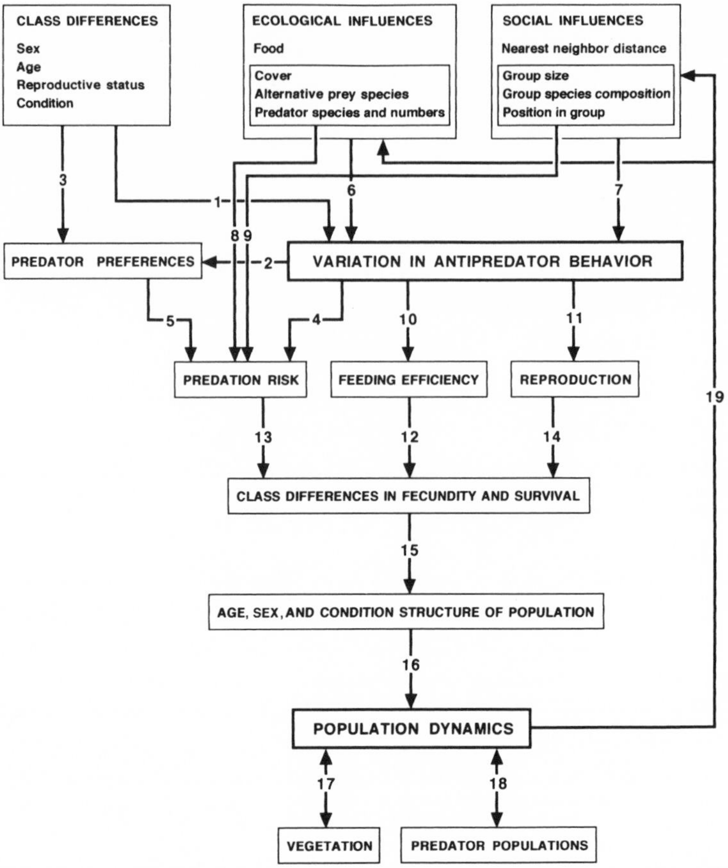

Figure 13.1 The relationship between antipredator behavior and population dynamics.

get, while escape responses may involve a longer-term shift to poorer feeding sites (10). The consequences of feeding deficits for fecundity and survival (12) are difficult to trace but act through loss of condition, which increases the chances of starvation and predation risk (Caro and FitzGibbon 1992), and may reduce mating success (e.g., Prins 1989).

Class differences in fecundity and survival are therefore influenced by antipredator behavior, both directly via predation risk (4, 13), and indirectly through its effects on feeding (10, 12) and reproduction (11, 14). Variance in fecundity and survival directly determines population structure (15) which, in turn, influences population dynamics (16).

Prey population dynamics interacts with lower (17) and higher (18) trophic levels and feeds back on many of the social and ecological factors influencing antipredator behavior (19).

CLASS DIFFERENCES IN ANTIPREDATOR DEFENSES AND THE INFLUENCE OF SOCIAL AND ECOLOGICAL FACTORS

The antipredator behavior of ungulates in general has been reviewed by Caro and FitzGibbon (1992), and that of African ungulates in particular by Kruuk (1972) and Leuthold (1977). We now review the influence of sex, age, reproductive status, and physical condition on various antipredator defenses, focusing first on Thomson's gazelle (summarized in table 13.1), and then adding information for other Serengeti species. For each defense we finally discuss the influence of ecological and social factors.

Grouping

The most abundant Serengeti ungulates—wildebeest, zebra, and Thomson's gazelle—all live in herds (Maddock 1979; Sinclair 1985). Gregariousness reduces the probability that a prey will be detected and then captured by a searching predator (e.g., Hamilton 1971; Treisman 1975a,b).

Class Differences. In Thomson's gazelle, herd size varies from 2 to over 1,000 individuals. In common with many other species (Leuthold 1977), male gazelles are more likely to remain on their own and less likely than females to be in large groups (FitzGibbon 1990c). This difference results primarily from the need for males to defend territories, even in the absence of other gazelles. Fawns and their mothers are more likely to be found in smaller groups than larger ones, while adolescents show the opposite trend (fig. 13.2). This might be because mothers lose contact with the preferred larger groups while they maintain contact with fawns in hiding.

Little is known about the effect of condition on grouping patterns, but in the African buffalo individuals in poor condition often leave the herd in order to increase their feeding time, because they are too weak to keep up, or because they are forced out by more dominant individuals (Sinclair 1977; Prins 1989; Prins and Iason 1989).

Ecological and Social Influences. In Thomson's gazelle, groups are larger on the short-grass plains than in tall-grass areas (mean group size 110.0 where vegetation height $\leq$ 30 cm, 62.1 where vegetation height $>$ 30 cm), probably because of a change in food distribution (Jarman 1974) and the difficulty of maintaining contact when visibility is restricted (LaGory 1986). Similarly, kongoni, impala, Grant's gazelle, and zebra form larger groups in more open country (Leuthold 1970; Leuthold and Leuthold 1975). Since lack of cover inhibits stalking and gives both predator and prey a better view, the antipredator implications of this relationship

Table 13.1 Summary of class, ecological, and social influences on antipredator behaviour in Thomson's gazelles.

	Group size	Vigilance	Monitor	Flight	Stotting
Class differences					
Sex	F>M	F>M	?	F>M[a]	F>M[a]
Age	adol>hg>f	adol>ad>f>hg	adol>hg>ad>f[b] f=hg=adol=ad[c]	f>hg=adol[b,c]	f>hg>ad[d]
Parental status	m/adol>m/hg>m/f	m/f>m/hg=m/adol	m/f=m/hg=m/adol>ad[b] m/f>m/hg=m/adol>ad[c]	m/f>m/hg=m/adol[b,c]	m/f>ad[d]
Condition	?	?	?	?	good>poor[a]
Social factors					
Group size		small>large	?	?	small>large[a]
Position in group		edge=center	?	?	predator near>predator far[a]
Presence of heterospecifics	P>A	P<A[e] P>A[f]	?	?	?
Ecological factors					
Grass height	high<low	high>low	?	?	high=low[b,d]
Season	wet=dry[g]	?	?	?	wet>dry[a]

Note: F, female; M, male; ad, adult; adol, adolescent; hg, half-grown; f, fawn; m, mother; P, present; A, absent.
[a]In response to wild dogs.
[b]In response to cheetahs.
[c]In response to hyenas.
[d]In response to an approaching vehicle.
[e]In groups of fewer than six Thomson's gazelles.
[f]In groups of more than twenty Thomson's gazelles.
[g]Source: C. FitzGibbon, unpublished data.

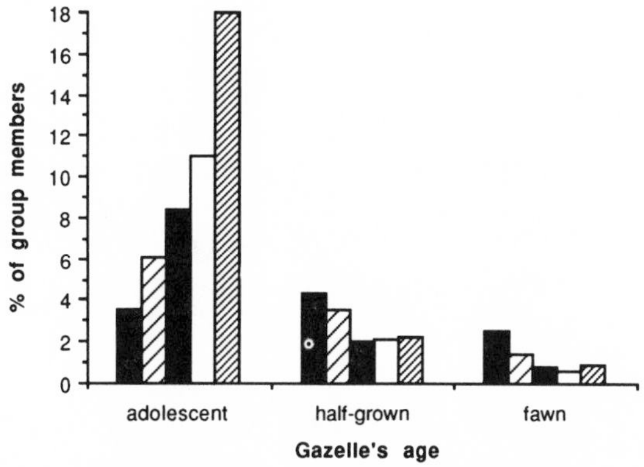

Figure 13.2 The age class composition of different-sized groups. Each histogram shows the percentage of a particular group size that is composed of a particular age class. Group sizes are, from left to right: 2–10; 11–50; 51–100; 101–500; >500. Since fawns out of hiding are always accompanied by their mothers, the distribution of fawns among different group sizes is equivalent to that of their mothers.

are not straightforward. Herds of Thomson's gazelle containing Grant's gazelles tend to be larger than monospecific groups, providing greater protection through the dilution effect (FitzGibbon 1990b).

Vigilance

Vigilance is crucial in providing early warning of predator approach; more vigilant gazelles detect approaching cheetahs at greater distances (FitzGibbon 1989, 1990b), and consequently have more time to take avoiding action.

Class Differences. Individuals adjust their vigilance levels, trading off the benefits of predator detection against those of competing activities. In Thomson's gazelle, males are less vigilant than females (FitzGibbon 1990c), probably because they need more food to support their greater body size and reproductive costs. Their vigilance may also be directed at competitors and potential mates. In females, vigilance is also influenced by the vulnerability of attendant offspring, with mothers of fawns spending more time vigilant than those of older infants (fig. 13.3), particularly when fawns are out of hiding (FitzGibbon 1994). This difference corresponds with the increasing ability of older infants to outrun predators once chased (table 13.2). In contrast, the fawns and half-growns are themselves less vigilant than adolescents and adults (fig. 13.4). The reduced utility of vigilance in young animals may be an explanation; despite a high risk of predation, they often do not recognize predators even if they detect them (C. F., pers. obs.). In addition, feeding for growth is a high priority for young animals, so they rely to some extent for early warning

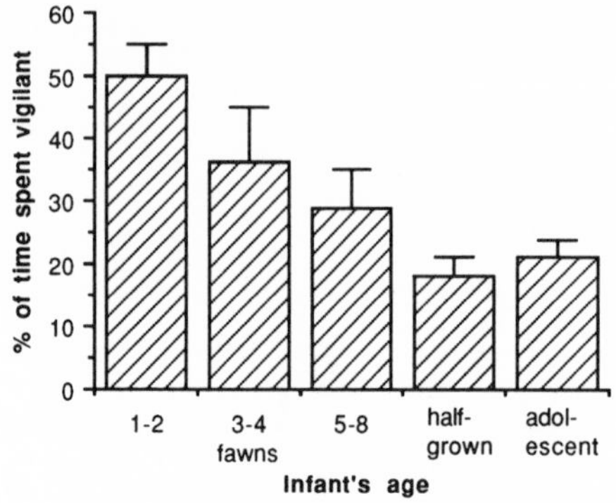

Figure 13.3 The mean percentage of time spent vigilant by gazelle mothers with attendant offspring of different ages. Fawns are divided into three age classes based on size relative to their mother: 1–2 weeks, 3–4 weeks, and 5–8 weeks. Bars indicate SE. ANOVA, $F = 8.41$, df = 4, 63, $P < .0001$. (From FitzGibbon 1988.)

Table 13.2 Percentage hunting success of cheetahs, wild dogs, spotted hyenas, and golden jackals against Thomson's gazelles of different ages.

	Adult (*N*)	Adolescent (*N*)	Half-grown (*N*)	Fawn (*N*)
Cheetah	20.6 (141)	37.5 (16)	48.0 (25)	70.5 (61)
Wild dog	42.7 (131)	50.0 (8)	85.7 (7)	100.0 (14)
Hyena	20.0 (10)	—	—	36.0 (33)
Jackal	—	—	—	50.0 (14)

Sources: Data on hyenas are from Kruuk 1972; on wild dogs, from Fanshawe and FitzGibbon 1993; on other predators, from FitzGibbon 1988.

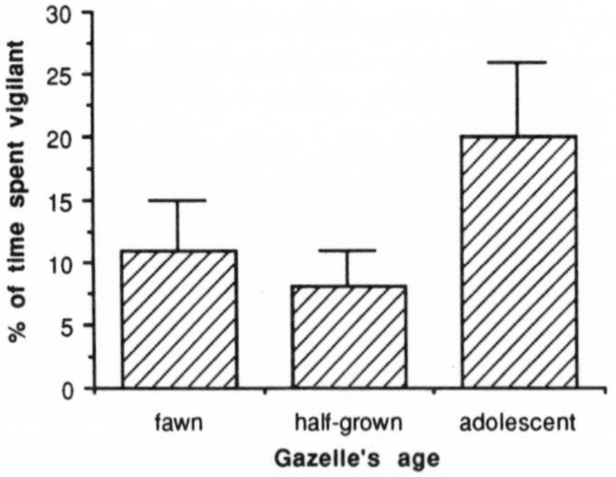

Figure 13.4 The mean percentage of time spent vigilant by young gazelles in three age classes. Bars indicate SE. ANOVA, $F = 3.37$, df = 2, 65, $P = .01$.

on the greater vigilance of their mothers and other adults (compare figs. 13.3 and 13.4).

In contrast to Thomson's gazelle, adult male African buffalo in mixed-sex herds have the same vigilance level as adult females, except for peripheral animals at night, when males are *more* vigilant (Prins and Iason 1989). The klipspringer lives in monogamous pairs in which the male is again the more vigilant sex, and is the first of the pair to detect approaching predators (Dunbar and Dunbar 1980).

Ecological and Social Influences. In Thomson's gazelles vigilance is more effective for escaping stalking predators than coursing predators, flight distance having a dramatic influence on the success rate of cheetahs (stalkers) (FitzGibbon 1988) but not of wild dogs (coursers) (Fanshawe and FitzGibbon 1993). This is presumably because stalking predators rely on surprise for their success, while coursers rely on stamina over long distances. The availability of alternative prey also influences risk, and consequently vigilance level. For example, warthogs, a favored prey of lions, are less vigilant in the dry season when zebras move into the area (Scheel 1992).

Individuals are more vigilant in cover-rich habitats (reedbuck, impala, topi, and wildebeest: Underwood 1982; Gosling and Petrie 1990; Thomson's gazelle: FitzGibbon 1988), reflecting the greater hunting success of stalking predators in these areas (Van Orsdol 1984). However, the assumption here—that individuals increase vigilance to compensate for a greater predation risk—does not necessarily hold, since the utility of vigilance must also be considered (Lima 1987; Lazarus and Symonds 1992); and indeed wildebeest are sometimes *less* vigilant in areas with more cover (Scheel 1992). Another important ecological factor is food, its poor quality in the dry season possibly explaining the lower vigilance levels at this time (Underwood 1982). Gazelles in larger groups are less vigilant than those in smaller groups (FitzGibbon 1988).

The presence of Grant's gazelles reduces vigilance in Thomson's gazelle groups of fewer than six, but increases it in groups of more than twenty (FitzGibbon 1990b). Although individual gazelles on the herd periphery are at greater risk than those in the center (FitzGibbon 1990c), they are no more vigilant (FitzGibbon 1988). Peripheral buffalo, however, are more vigilant than central individuals (Prins and Iason 1989).

Monitoring

Having detected a predator, prey face the problem of monitoring its movements and intentions, to prevent surprise attack, until the prey decides that the predator is no longer a threat.

Class and Ecological Differences. Stalking predators, such as cheetahs, are monitored for longer than coursers, such as spotted hyenas (fig. 13.5). This influence of predator species interacts with offspring age since vulnerability to the different predators is age-dependent (see table 13.2). Thus, while the period for which gazelle mothers monitor predators decreases with offspring age, the decline is particularly dramatic in response to golden jackals (fig. 13.5). These predators pose a threat primarily to fawns and rarely attack older gazelles. Adolescent gazelles monitor cheetahs for longer than younger animals (fig. 13.6), even though their risk of predation is lower (see table 13.2), probably because it provides an opportunity to learn about predators.

Flight

If the predator attacks, all but the largest ungulate species flee to avoid capture. Flight distance, reaction time, speed, and the time for which high speed can be sustained (i.e., stamina) are the important determinants of successful escape.

Class Differences. In small herds of blesbok *(Damaliscus dorcas),* mountain reedbuck, and gray rhebok *(Pelea capreolus)* (Rowe-Rowe 1974), and in monogamous klipspringer pairs, the female flees first, even though the male is the more vigilant sex in this last species (Dunbar and Dunbar 1980). Male Thomson's gazelles have shorter flight distances than females in response to wild dogs, but the sexes do not differ in their ability to outrun them (Fanshawe and FitzGibbon 1993). Older gazelles out-

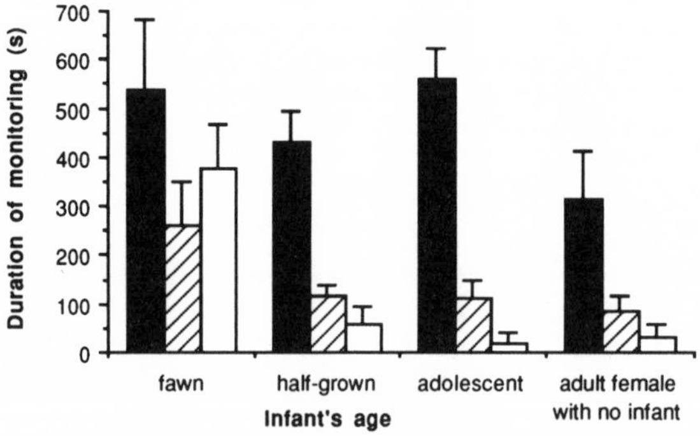

Figure 13.5 The mean time gazelle mothers with attendant offspring of varying ages spent monitoring three predator species, compared with an adult female with no offspring. Bars indicate SE. Predator species: cheetah, solid histograms; spotted hyena, hatched histograms; golden jackal, open histograms. There was no effect of infant's age on the time that mothers stared at cheetahs (ANOVA, $F = 0.66$, df = 2, 55, $P > .05$), but there was on the duration of staring at hyenas ($F = 4.28$, df = 2, 31, $P < .05$) and jackals ($F = 11.24$, df = 2, 30, $P < .0001$). (From FitzGibbon 1988.)

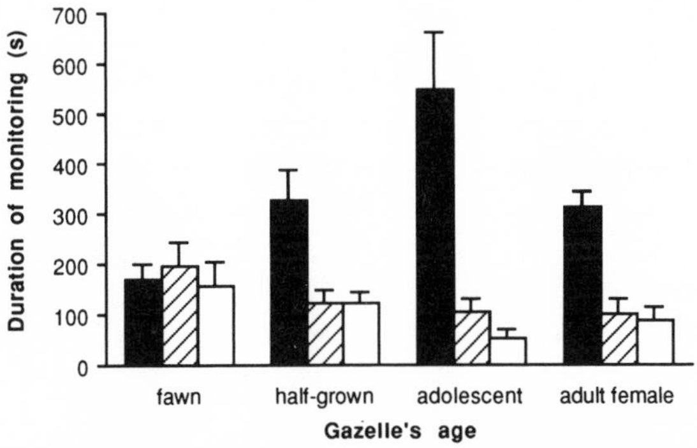

Figure 13.6 The mean time gazelles in different age classes spent monitoring three predator species. Bars indicate SE. Predator species: cheetah, solid histograms; spotted hyena, hatched histograms; golden jackal, open histograms. Within the three youngest age classes, older gazelles watched cheetahs for longer (ANOVA, $F = 3.53$, df = 2, 55, $P < .05$) and jackals for shorter periods ($F = 3.62$, df = 2, 30, $P < .05$) but there was no effect of gazelle age on the time for which hyenas were monitored ($F = 2.53$, df = 2, 31, $P > .05$). (From FitzGibbon 1988.)

run cheetahs, wild dogs, and spotted hyenas more successfully than younger animals do (see table 13.2), explaining why mothers of younger infants react to predators at greater distances than those of older infants (fig. 13.7).

Ecological and Social Influences. Group size has no effect on the flight distance of Thomson's gazelles responding to wild dogs (Fanshawe and FitzGibbon 1993).

Stotting

Stotting is defined as leaping off the ground with all four legs held stiff and straight, and occurs during flight (Walther 1969, Caro 1986a).

Class Differences. In response to coursing predators, stotting probably acts as an honest signal of a gazelle's escape ability, with individual differences in stotting height and rate in Thomson's gazelle reflecting differences in physical condition (FitzGibbon and Fanshawe 1988). Males stot less rapidly than females, and individuals stot less frequently in the dry season than in the wet (FitzGibbon and Fanshawe 1988), both effects probably resulting from differences in physical condition. In Thomson's gazelle, Grant's gazelle, topi, and kongoni, the probability of stotting in response to an approaching person is unaffected by condition, although in individuals that do stot, the number of stots and/or the rate of stotting increase with condition in topi and Grant's gazelle (T. M. Caro 1994b).

Although adult gazelles rarely stot in response to stalking predators,

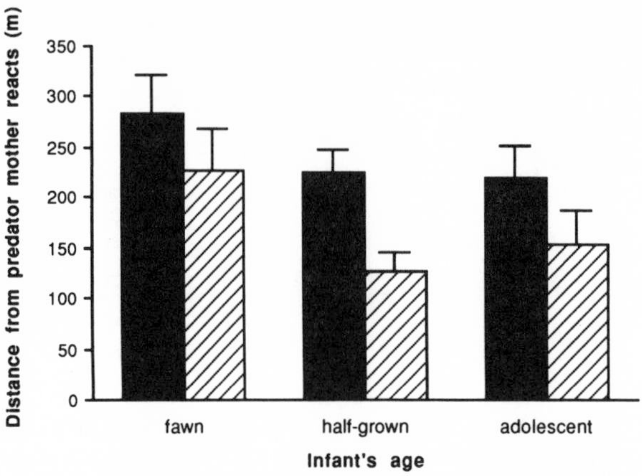

Figure 13.7 The mean distance at which mothers of different-aged young reacted to approaching predators (normally by staring and then fleeing). Bars indicate SE. Predator species: Cheetahs, solid histograms; spotted hyenas, hatched histograms. The distance is influenced by both infant's age (ANOVA, $F = 3.10$, df = 2, 73, $P < .05$) and predator species ($F = 7.66$, df = 1, 73, $P < .05$).

like cheetahs and lions, the behavior is more common in younger animals, particularly fawns frightened from hiding (Caro 1986b), and probably alerts mothers to assist them.

Ecological and Social Influences. In Thomson's gazelle, stots are generally performed when fleeing from coursing predators, such as wild dogs and spotted hyenas, which tend to select their prey during the chase (Caro 1986b; FitzGibbon and Fanshawe 1988). Stotting is also influenced by the risk of being selected as prey; in response to wild dogs, gazelles are less likely to stot when in larger groups and when further from the predator (FitzGibbon and Fanshawe 1988). Stotting in Thomson's gazelle does not vary with vegetation height (Caro 1986b).

EFFECTS OF PREDATION ON THE SIZE OF THE THOMSON'S GAZELLE POPULATION

How important is predation as a cause of mortality in Thomson's gazelles? Borner et al. (1987) and Caro (1994a) have reviewed the impact of predation on the gazelle population in terms of the estimated numbers of adult gazelles killed by each of the five main predator species (table 13.3), compared with the annual recruitment into the gazelle population. Using the best available data on predator hunting rates and prey choice from a variety of sources, Caro recalculated that predators kill between 51% and 82% of the estimated 73,000–86,000 adult gazelles recruited

Table 13.3 Estimated numbers of predators and of adult Thomson's gazelles (TG) killed or scavenged per annum in the Serengeti ecosystem.

	Predator numbers		Adult TG killed/ predator/year		Adult TG killed/year		
	Plains	Woodlands	Plains	Woodlands	Plains	Woodlands	Ecosystem
Lion	200	2,600	2.2	3.3	440	8,580	9,020
Cheetah	110	100	35.4	12.4	3,894	1,240	5,134
Leopard	0	800–1,000	0	4.9–5.9	0	3,920–5,900	3,920–5,900
Wild dog		60		6.7		402	402
Spotted hyena	5,200	3,500–4,000	5.6	1.3	29,120	4,550–5,200	33,670–34,320

Source: From Caro 1994a.
Note: Estimates assume no change in killing rates since the 1960s, except for cheetahs and lions, whose rates were calculated in the 1980s.

into the population each year, using the 1989 population estimate (Campbell et al. 1990).

If the Thomson's gazelle population is stable, as recent evidence suggests (Dublin et al. 1990), so that mortality equals recruitment, then either the above predation rates are too low or at least 13,000 gazelles are dying of other causes each year. Predation rates vary according to the age/sex class of the predator (e.g., Caro 1994a), and from area to area, depending on prey availability (e.g., table 36 in Schaller 1972). More data are required on intraspecific variation in hunting behavior to obtain more reliable estimates of predator offtake. However, estimates would have to be increased substantially to make up the shortfall. Alternatively, many gazelles may be dying of starvation in the late dry season in the woodlands (where little research work is carried out) and being consumed by predators not listed in table 13.3, such as vultures. It is also possible that the Thomson's gazelle population is increasing, although recent counts suggest otherwise (Campbell et al. 1990; Dublin et al. 1990).

DIFFERENTIAL PREDATION AND THE ROLE OF ANTIPREDATOR BEHAVIOR IN POPULATION DYNAMICS

To what extent can differences in predation rates on different sex, age, and condition classes be explained by the differences in the antipredator behavior of these classes described above?

Sex Ratio of Kills

In both Serengeti Thomson's gazelles (table 13.4) and a number of other African ungulates (references in Prins and Iason 1989) predation risk is higher for males than for females. The role of antipredator behavior in this effect is well illustrated by cheetah and wild dog predation of Thomson's gazelles (FitzGibbon 1990c).

Table 13.4 Relative predation risk for adult male and female Thomson's gazelles from five Serengeti predators.

Predator	Relative risk male:female	*N*	Source
Cheetah	1.8:1	25	FitzGibbon 1990c
	1.2:1	102	Schaller 1972
	4.8:1	27	Laurenson, unpub.
Wild dog	2.1:1	44	Fanshawe and FitzGibbon 1993
	3.0:1	25	Fuller and Kat 1990
	1.8:1	44	Schaller 1972
	3.0:1	46	Kruuk 1972
Spotted hyena	6.4:1	54	Kruuk 1972
Lion	2.4:1	125	Schaller 1972
Leopard	5.0:1	75	Schaller 1972

Note: Calculated from the observed sex ratio of kills and the biased sex ratio of the gazelle population (1 male:2 females, combining adults and subadults; Laurenson 1992).

Cheetahs prefer to hunt single animals, and more males than females are found on their own. In addition, males concentrate on the edges of groups and are less vigilant than females, two factors that also predispose them to attack by cheetahs. Compared with females, males flee from wild dogs at shorter distances, perhaps because they are reluctant to leave their territories (Estes and Goddard 1967; Fanshawe and FitzGibbon 1993). In addition, the poorer condition and lower stotting rates of males may make them more vulnerable (Fanshawe and FitzGibbon 1993).

Males have therefore accepted a greater risk of predation in order to increase their chances of obtaining a breeding territory and access to females. However, the relative vulnerability of different classes of males is not clear. While the vulnerability of lone animals and the lower flight distance of males suggest that territorial males are at more risk than bachelors, the selection by predators of animals in poor condition and of individuals from the edges of groups suggests the opposite, since subordinate males may be forced to the edges of groups (Hirth and McCullough 1977).

In the African buffalo the greater predation risk of adult males (3.3 times that of females: calculated from Prins and Iason 1989) probably arises from the time spent outside the mixed-sex herd in small bachelor groups. Males are either forced out of the mixed-sex herd by more dominant animals or move out to improve their body condition. Loss of condition probably results from lost feeding time, which in turn is associated with increased vigilance, probably directed at predators, competitors, and mates (Sinclair 1977; Prins 1989; Prins and Iason 1989).

Male-biased predation results in a heavily female-biased gazelle population, substantially affecting the reproductive rate of the population. Since Thomson's gazelles are polygynous (Walther 1978), a 1:1 population sex ratio is unnecessary to maintain female reproductive rates. At the current sex ratio (1 male:2 females; Laurenson 1992), and assuming that each female produces one fawn per year, the birth rate is 667 fawns/1,000 animals. If the sex ratio were 1:1, the birth rate would be only 500 fawns/year. Thus the sex ratio bias results in a 33% increase in reproductive rate, a dramatic consequence of a combination of differences between the sexes in morphology, behavior, and ecology.

Finally, in some circumstances the sex ratio of kills may be biased toward males as a result of female choice. In puku *(Kobus vardoni),* topi (Balmford, Rosser, and Albon 1992), and Uganda kob *(Kobus kob)* (Deutsch and Weeks 1992), females prefer male territories or leks with a lower predation risk and are therefore expected to suffer a lower average predation rate than males in this context.

Age Class Composition of Kills

A number of studies have examined predator selection of different gazelle age classes (fig. 13.8). All predators select a higher proportion of immatures than predicted from the composition of the population (Kruuk 1972; Schaller 1972; FitzGibbon and Fanshawe 1989), reducing recruitment rates and, in theory, favoring rapid early growth. Antipredator behavior is important in reducing predation on these classes. For example, hiding by gazelle fawns considerably reduces their risk of cheetah predation, increasing survivorship by 2.6% per year (appendix 13.2), and the increased vigilance of mothers reduces the success of cheetah attacks on fawns out of hiding (FitzGibbon 1990a). Thus antipredator behavior by mothers and immatures reduces juvenile mortality, thereby increasing recruitment rates.

While predators do not obviously select a high proportion of gazelles in the oldest age class (fig. 13.8), the lack of information about the age class composition of the population prevents us from determining whether predators prefer older animals.

While coursing predators have been predicted to take a greater proportion of old, young, and sick animals than stalking predators (Schaller 1972; Kruuk 1972, 1986; FitzGibbon and Fanshawe 1989), Greene (1986) argues that ambush (sit and wait) predators bias their prey selection toward larger and older prey, whereas cruising predators (which move to locate prey) take population components more in relation to their abundance. There is little evidence for any difference in prey selection between predators using different hunting techniques in the Serengeti (Schaller 1972; FitzGibbon and Fanshawe 1989). For any one predator species, prey selection can vary dramatically according to season, habitat, and the age and sex of the individual predator. For example, lactating female cheetahs select more adult gazelles than nonlactating females, which hunt a greater proportion of fawns (Laurenson 1992).

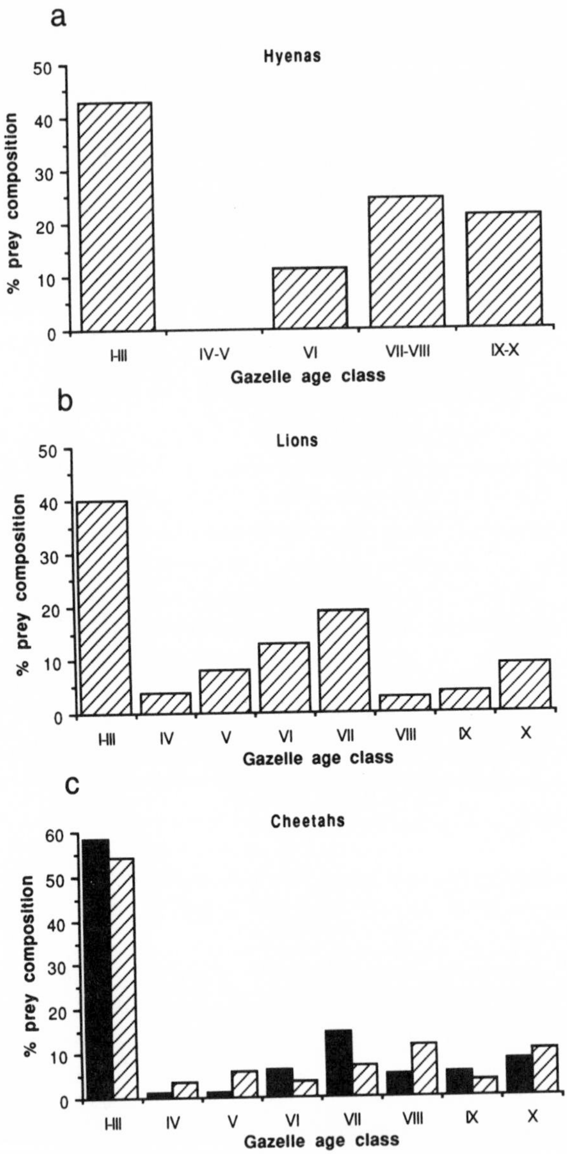

Figure 13.8 The age class composition of Thomson's gazelles killed by (*a*) spotted hyenas (Kruuk 1972), (*b*) lions (Schaller 1972), (*c*) cheetahs, and (*d*) wild dogs (*c* and *d:* solid histograms, Schaller 1972; hatched histograms, FitzGibbon and Fanshawe 1989; open histograms, Fuller and Kat 1990). Fuller and Kat recorded an additional seven adult gazelles (18% of total) killed by wild dogs but not aged, while Schaller recorded fifteen (18% of total). Except for gazelles killed by hyenas (Kruuk 1972), age classes are based on tooth eruption and wear (Walther 1973).

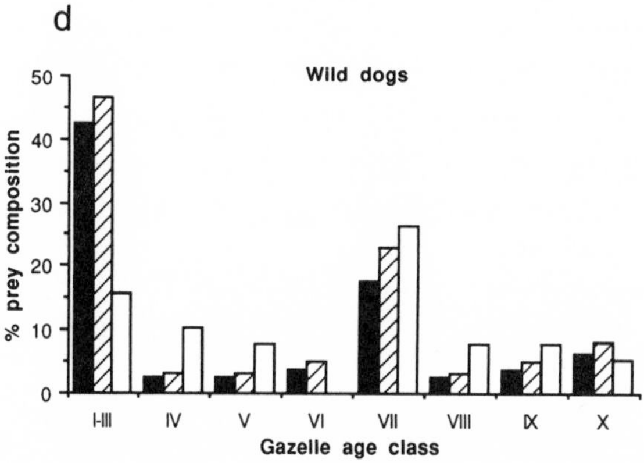

Condition

Although predators probably select prey in poor condition (Kruuk 1972; Schaller 1972), there is little conclusive evidence for this (Caro and FitzGibbon 1992). On the basis of bone marrow fat indices, wild dogs select more poor-condition gazelles than do cheetahs (FitzGibbon and Fanshawe 1989), although a comparison with the condition of individuals available in the population has not been possible.

Wild dogs' selection of Thomson's gazelles with low stot rates (FitzGibbon and Fanshawe 1988) could result in coursing predators selecting less healthy prey. Similarly, since animals in poor condition may reduce vigilance in favor of feeding, cheetahs' preference for animals with low vigilance rates (FitzGibbon 1989) could have the same effect. However, animals in poor condition may, alternatively, increase vigilance to compensate for their greater risk of capture once chased.

Behavioral correlates of condition probably provide predators with more sensitive cues for assessing body condition than do morphological indicators alone. In any case, the result is fewer old and sick animals in the population, increasing the reproductive rate per individual. Since males tend to be in poorer condition than females (Bradley 1977; FitzGibbon and Fanshawe 1989), condition-based selection will contribute to the female bias in the population, again producing a greater per capita recruitment rate.

POPULATION CONSEQUENCES OF THE COSTS OF ANTIPREDATOR BEHAVIOR

Adult gazelles spend about 10% of their time vigilant (FitzGibbon 1990c), and a further 1% responding to cheetahs. Since cheetahs constitute only 2% of large predators in Serengeti (see table 13.3), the percentage of time spent responding to all predators will be far higher. Time spent in antipredator behavior constrains the time available for feeding and other essential activities. In addition, predators may prevent gazelles from utilizing prime food resources where the risk of predation is high (e.g., in thick cover near water).

Unlike the direct costs of predation, the costs of antipredator behavior are paid by all prey individuals and do not translate into increases in the growth and reproductive rates of predators. Consequently, as a result of antipredator behavior, fewer predators can be supported by a given prey population (Ives and Dobson 1987).

ANTIPREDATOR BEHAVIOR AND PREDATOR-PREY SYSTEMS

We have shown how variation in antipredator behavior due to age, sex, reproductive status, and condition influences the structure and dynamics of prey populations. These effects result from variation in predation risk and the benefits and costs of competing activities. Some of our conclusions have necessarily been qualitative in nature, and to quantify the relationships shown in fig. 13.1 further, two things are now required: first, fieldwork that examines the causes and consequences of variation in antipredator behavior (links 1–14 in fig. 13.1), and second, models that predict the consequences of class differences in fecundity and survival for population dynamics (links 15 and 16 in fig. 13.1).

Adjustments in investment in predator avoidance, in response to predator density and other factors, influence predation rates and consequently the predator population growth rate. Theoretically this interaction can dampen oscillations in predator and prey densities (Levins 1975; Ives and Dobson 1987). Predator-prey population dynamics, in turn, influence the strength of selection on antipredator behavior (appendix 13.1; Abrams 1990).

The model of Ives and Dobson (1987) illustrates how the costs and benefits of antipredator behavior influence predator-prey dynamics, but assumes that all individuals adopt identical antipredator behavior. When individual investment in defensive behavior varies, as shown in this chapter, individual fitness is no longer equivalent to the per capita growth rate of the population. The implications for predator-prey systems of the age, sex, reproductive status, and condition effects discussed here therefore remain to be formally explored. In spite of this, Ives and Dobson's conclusions suggest that where predation is implicated in the regulation of the Serengeti's ungulate populations (Fryxell, Greever, and Sinclair 1988; Sinclair 1989, Sinclair, chap. 9; Skogland 1991), antipredator behavior will play a role in the regulatory process.

ACKNOWLEDGMENTS

We thank the Tanzania Commission for Science and Technology, David Babu, Director of Tanzania National Parks, Bernard Maregesi, Chief Warden, and Professor Karim Hirji and Dr. George Sabuni, Serengeti Wildlife Research Institute, for permission to carry out our studies in the Serengeti. C. F.'s research was funded by a studentship from the Science and Engineering Research Council, and the chapter was written while in receipt of a postdoctoral research fellowship from New Hall, Cambridge. J. L.'s research, too recent to be included here, was funded by the University of Newcastle upon Tyne Research Committee, the Association for the Study of Animal Behaviour, and the British Ecological Society, and was supported logistically by Markus Borner, Tim Caro, Marion East, and Heribert Hofer. We are grateful to Andrew Balmford, Tim Caro, Robin Dunbar, James Deutsch, Karen Laurenson, Martyn Murray, and David Scheel for providing unpublished manuscripts and data at short notice, and to Stuart Laws for preparing figure 13.1. We are particularly grateful to Tim Caro for a number of discussions and for allowing us to present data from his recent book. Finally, we thank Tim Caro, the editors, and anonymous reviewers for their helpful comments on an earlier draft of the chapter.

APPENDIX 13.1 THE STRENGTH OF SELECTION ON ANTIPREDATOR BEHAVIOR

A knowledge of prey mortality due to predation and the benefit of antipredator behavior makes it possible to compare the relative strengths of selection acting on a particular defensive behavior as a result of interaction with different predators. We describe the general method and give an example of its application.

The strength of selection, W, is proportional to:

(Rate of capture/individual prey/unit time) × (the amount by which unit investment in defensive behavior reduces the probability of capture/attack).

This assumes, very simply, that each unit of investment is equally beneficial; more realistically, benefit will be a diminishing returns function of investment. A more complex model is presented by Abrams (1990). This simplification does not affect the result of the example described below. Symbolically,

$$W \text{ is proportional to } (P/Q)dS \tag{13.1}$$

where P = number of prey captured by a predator species/unit time, Q = prey population size, and dS = reduction in the probability of capture/attack resulting from unit investment in defense.

P is the product of three components:

$$P = NAS \tag{13.2}$$

where N = predator population size, A = rate of attack/predator individual/unit time, and S = probability of capture/attack.

The strength of selection on antipredator behavior is therefore a function of predator and prey population sizes, the rate of predator attacks, capture success rate, and the efficiency of the antipredator behavior in reducing capture success.

This is the case for a secondary defense: an antipredator strategy that reduces predation risk once an attack has been launched. For primary defenses, which reduce the probability of an attack occurring, dS is replaced by dA, since successful defense reduces A rather than S.

Comparing the strength of selection on the defensive behavior of a single prey species against two predator species, i and j, gives:

$$W_i/W_j = (P_i \times dS_i)/(P_j \times dS_j). \tag{13.3}$$

While values of P are available for some predator-prey systems, dS is difficult to measure. In the following example we have only the inequality $dS_i > dS_j$.

We now compare the relative strength of selection on vigilant behavior in Thomson's gazelle against cheetahs and against wild dogs. From table 13.3, P for cheetah = 5,134 gazelles per year, and for wild dog = 402 gazelles per year. The probability of capture/attack for cheetah decreases with the distance between predator and prey at the start of the attack, while it is independent of predator-prey distance for wild dog (see above section on ecological and social influences on vigilance); this is what might be expected for a stalking and a coursing predator, respectively. Consequently vigilance is more beneficial against cheetah than against wild dog ($dS_{\text{cheetah}} > dS_{\text{wild dog}}$) since, against the former predator, vigilance can reduce predation risk by facilitating detection of the predator while it is still at a distance and its probability of capture is low. Early detection of a wild dog, on the other hand, does not increase the probability of escape compared with later detection.

Thus $P \times dS_{\text{cheetah}} > P \times dS_{\text{wild dog}}$, and the strength of selection maintaining vigilance is greater as a result of encounters with the former predator. Even if cheetah and wild dog population sizes were equal, the same conclusion would hold, since the number of prey killed/predator/year (= AS) is 3.6 times higher for cheetah than for wild dog (calculated from table 13.3). The conclusion would also hold even if $P_{\text{cheetah}} = P_{\text{wild dog}}$ because of the inequality in dS, which is a consequence of the different hunting methods of stalking and coursing predators.

APPENDIX 13.2 EFFECT OF HIDING ON FAWN SURVIVAL AND RECRUITMENT: AN EXAMPLE

Hiding has been shown to reduce considerably the risk of predation from cheetahs, one of the most important predators of gazelle fawns in the Serengeti. Here we calculate the effect of hiding in terms of increasing recruitment, comparing it with a hypothetical situation in which no hiding occurs. Unless stated otherwise, data are from FitzGibbon 1990a.

Number of fawns killed by cheetahs while hidden = 14
Number of fawns killed by cheetahs while active = 45
Total = 59

Fawns spend an average of 39.6% of their time active, so in the absence of hiding, (100/39.6) × 45 = 114 fawns would have been killed—that is, 1.9 times (114/59) as many as with hiding.

Number of adult gazelles killed/cheetah/year = 24.4 (calculated from table 13.3).

Cheetahs eat 1.4 times as many fawns as they do adults (FitzGibbon and Fanshawe 1989), so number of fawns killed/cheetah/year = 34.2.

Number of cheetahs in ecosystem = approximately 210 (table 13.3).

Total number of fawns killed by cheetahs per year = 210 × 34.2 = 7,182.

Therefore without hiding, cheetahs would kill 1.9 x 7,182 = 13,646 fawns, an increase of 6,464.

Assuming each female has one fawn per year, and that the number of females is 265,262 (population size = 572,920 [Dublin et al. 1990], and females constitute 46.3% of the population [Laurenson 1992]), the number of fawns produced each year = 265,262.

Assuming cheetahs are the only predators of gazelle fawns, the number of fawns surviving in the absence of hiding behavior would be 265,262 − 13,646 = 251,616.

Thus hiding results in an increase in fawn survivorship/year of 6,464/251,616 = 2.6%, from decreased predation by cheetahs alone.

REFERENCES

Abrams, P. 1990. The evolution of anti-predator traits in response to evolutionary change in predators. *Oikos* 59:147–56.

———. 1992. Predators that benefit prey and prey that harm predators: Unusual effects of interacting foraging adaptations. *Am. Nat.* 140:573–600.

Balmford, A., Rosser, A. M., and Albon, S. D. 1992. Correlates of female choice in resource-defending antelope. *Behav. Ecol. Sociobiol.* 31:107–14.

Borner, M., FitzGibbon, C. D., Borner, M., Caro, T. M., Lindsay, W. K., Collins, D. A., and Holt, M. E. 1987. The decline in the Serengeti Thomson's gazelle population. *Oecologia* 73:32–40.

Bradley, R. M. 1977. Aspects of the ecology of Thomson's gazelles in the Serengeti National Park, Tanzania. Ph.D. thesis, Texas A&M University.

Campbell, K. L. I., Kajuni, A. R., Huish, S. A., and Mng'ong'o, G. B. 1990. *Serengeti ecological monitoring programme.* Serengeti Wildlife Research Centre, Biennial Report, 1988–89.

Caro, T. M. 1986a. The functions of stotting: A review of the hypotheses. *Anim. Behav.* 34:649–62.

———. 1986b. The functions of stotting in Thomson's gazelles: Some tests of the predictions. *Anim. Behav.* 34:663–84.

———. 1994a. *Cheetahs of the Serengeti plains: Group living in an asocial species.* Chicago: University of Chicago Press.

———. 1994b. Ungulate antipredator behaviour: Preliminary and comparative data from African bovids. *Behaviour* 128:189–228.

Caro, T. M., and FitzGibbon, C. D. 1992. Large carnivores and their prey: The quick and the dead. In *Natural enemies: The population biology of*

predators, parasites, and diseases, ed. M. J. Crawley, 117–42. Oxford: Blackwell.
Deutsch, J. C., and Weeks, P. 1992. Uganda kob prefer high-visibility leks and territories. *Behav. Ecol.* 3:223–33.
Dublin H. T., Sinclair, A. R. E., Boutin, S., Anderson E., Jago, M., and Arcese P. 1990. Does competition regulate ungulate populations? Further evidence from Serengeti, Tanzania. *Oecologia* 82:283–88.
Dunbar, R. I. M., and Dunbar, E. P. 1980. The pairbond in klipspringer. *Anim. Behav.* 28:219–29.
Estes, R. D., and Goddard, J. 1967. Prey selection and hunting behavior of the African wild dog. *J. Wildl. Mgmt.* 31:52–70.
Fanshawe, J. H., and FitzGibbon, C. D. 1993. Factors influencing the hunting success of a wild dog pack. *Anim. Behav.* 45:479–90.
FitzGibbon, C. D. 1988. The antipredator behaviour of Thomson's gazelles. Ph.D. thesis, University of Cambridge.
———. 1989. A cost to individuals with reduced vigilance in groups of Thomson's gazelles hunted by cheetahs. *Anim. Behav.* 37:508–10.
———. 1990a. Antipredator strategies of immature Thomson's gazelles: Hiding and the prone response. *Anim. Behav.* 40:846–55.
———. 1990b. Mixed-species grouping in Thomson's and Grant's gazelles: The antipredator benefits. *Anim. Behav.* 39:1116–26.
———. 1990c. Why do hunting cheetahs prefer male gazelles? *Anim. Behav.* 40:837–45.
———. 1994. Anti-predator strategies of female Thomson's gazelles with hidden fawns. *J. Mammal.* 74:758–82.
FitzGibbon, C. D., and Fanshawe, J. H. 1988. Stotting in Thomson's gazelles: An honest signal of condition. *Behav. Ecol. Sociobiol.* 23:69–74.
———. 1989. The condition and age of Thomson's gazelles killed by cheetahs and wild dogs. *J. Zool.* (Lond.) 218:99–107.
Fryxell, J. M., Greever, J., and Sinclair, A. R. E. 1988. Why are migratory ungulates so abundant? *Am. Nat.* 131:781–98.
Fuller, T. K., and Kat, P. W. 1990. Movements, activity, and prey relationships of African wild dogs *(Lycaon pictus)* near Aitong, southwestern Kenya. *Afr. J. Ecol.* 28:330–50.
Gosling, L. M., and Petrie, M. 1990. Lekking in topi: A consequence of satellite behaviour by small males at hotspots. *Anim. Behav.* 40:272–87.
Greene, C. H. 1986. Patterns of prey selection: Implications for predator foraging tactics. *Am. Nat.* 128:824–39.
Hamilton, W. D. 1971. Geometry for the selfish herd. *J. theor. Biol.* 31:295–311.
Hirth, D. H., and McCullough, D. R. 1977. Evolution of alarm signals in ungulates with special reference to white-tailed deer. *Am. Nat.* 111:31–42.
Ives, A. R., and Dobson, A. P. 1987. Antipredator behaviour and the population dynamics of simple predator-prey systems. *Am. Nat.* 130:431–47.
Jarman, P. J. 1974. The social organisation of antelope in relation to their ecology. *Behaviour* 48:215–67.
Kruuk, H. 1972. *The spotted hyena: A study of predation and social behavior.* Chicago: University of Chicago Press.
———. 1986. Interactions between felidae and their prey species: A review. In

Cats of the world: Biology, conservation, and management, ed. D. S. Miller and D. D. Everett. Washington, D.C.: National Wildlife Federation.

LaGory, K. E. 1986. Habitat, group size, and the behaviour of white-tailed deer. *Behaviour* 98:168–79.

Laurenson, K. 1992. Reproductive strategies in wild female cheetahs. Ph.D. dissertation, University of Cambridge.

Lazarus, J., and Symonds, M. 1992. Contrasting effects of protective and obstructive cover on avian vigilance. *Anim. Behav.* 43:519–21.

Leuthold, W. 1970. Observations on the social organization of impala *(Aepyceros melampus). Z. Tierpsychol.* 27:693–721.

———. 1977. *African ungulates: A comparative review of their ethology and behavioral ecology.* Berlin: Springer-Verlag.

Leuthold, W., and Leuthold, B. M. 1975. Patterns of social grouping in ungulates of Tsavo National Park, Kenya. *J. Zool.* (Lond.) 175:405–20.

Levins, R. 1975. Evolution in communities near equilibrium. In *Ecology and evolution of communities,* ed. M. L. Cody and J. M. Diamond, 16–50. Cambridge, Mass.: Harvard University Press.

Lima, S. L. 1987. Vigilance while feeding and its relation to the risk of predation. *J. theor. Biol.* 124:303–16.

Lima, S. L., and Dill, L. M. 1990. Behavioral decisions made under the risk of predation: A review and prospectus. *Can. J. Zool.* 68:619–40.

Maddock, L. 1979. The "migration" and grazing succession. In *Serengeti: Dynamics of an ecosystem,* ed. A. R. E. Sinclair and M. Norton-Griffiths, 104–29. Chicago: University of Chicago Press.

Prins, H. H. T. 1989. Condition changes and choice of social environment in African buffalo bulls. *Behaviour* 108:298–324.

Prins, H. H. T., and Iason, G. R. 1989. Dangerous lions and nonchalant buffalo. *Behaviour* 108:262–96.

Rowe-Rowe, D. T. 1974. Flight behaviour and flight distances of blesbok. *Z. Tierpsychol.* 34:208–11.

Schaller, G. B. 1972. *The Serengeti lion: A study of predator-prey relations.* Chicago: University of Chicago Press.

Scheel, D. 1992. Foraging behavior and predator avoidance: Lions and their prey in the Serengeti. Ph.D. dissertation, University of Minnesota.

Sibly, R., and Smith, R. H., eds. 1985. *Behavioural ecology: Ecological consequences of adaptive behaviour.* Oxford: Blackwell.

Sinclair, A. R. E. 1977. *The African buffalo: A study of resource limitation of populations.* Chicago: University of Chicago Press.

———. 1985. Does interspecific competition or predation shape the African ungulate community? *J. Anim. Ecol.* 54:899–918.

———. 1989. Population regulation in mammals. In *Ecological concepts,* ed. J. M. Cherrett, 197–241. Oxford: Blackwell.

Skogland, T. 1991. What are the effects of predators on large ungulate populations? *Oikos* 61:401–11.

Treisman, M. 1975a. Predation and the evolution of gregariousness. I. Models for concealment and evasion. *Anim. Behav.* 23:779–800.

———. 1975b. Predation and the evolution of gregariousness. II. An economic model for predator-prey interaction. *Anim. Behav.* 23:801–25.

Underwood, R. 1982. Vigilance behaviour in grazing African antelopes. *Behaviour* 79:81–108.

Van Orsdol, K. G. 1984. Foraging behaviour and hunting success of lions in Queen Elizabeth National Park, Uganda. *Afr. J. Ecol.* 22:79–99.

Walther, F. R. 1969. Flight behavior and avoidance of predators in Thomson's gazelle (*Gazella thomsoni* Guenther 1884). *Behaviour* 34:184–221.

———. 1973. On age class recognition and individual identification of Thomson's gazelle in the field. *J. S. Afr. Wildl. Mgmt. Assoc.* 2:9–15.

———. 1978. Quantitative and functional variations of certain behaviour patterns in male Thomson's gazelle of different social status. *Behaviour* 65:212–40.

IV Predator Demography and Behavior

FOURTEEN

Variation in Predation by Lions: Tracking a Movable Feast

D. Scheel and C. Packer

The Serengeti ecosystem is characterized by the annual migration of wildebeest, zebra, and gazelle. The Serengeti also supports sizable populations of resident ungulates. Even though the migratory species are their most frequent prey, Serengeti lions are territorial except during periods of extreme hardship (Packer, Scheel, and Pusey 1990). Lions must therefore endure wide fluctuations in the local densities of certain prey species, relying on the stable abundance of resident species during the lean season. The influence of local prey abundance can be measured by the predators' functional response (e.g., Holling 1959; Hilborn and Sinclair 1979) and by economic models from foraging theory (e.g., Stephens and Krebs 1986). Foraging theory successfully predicts the prey preferences of hunting lions (Scheel 1993). By preferring wildebeest and zebra during the migration and specializing on warthog and buffalo when the migrants are scarce, lions appear to be risk-sensitive foragers that maximize food intake rate.

The precise timing and pattern of the Serengeti migration is complex and erratic. Thus, local prey densities can vary dramatically on a weekly or even daily basis, and no two years are exactly the same. In addition, the sizes of the Serengeti ungulate populations have changed markedly over the past quarter century. In this chapter, we show that lion predation patterns vary not only with short-term changes in local prey density but also with long-term changes in herbivore population sizes.

METHODS

Lions and Their Habitats

Lions in a 2,000 km^2 area of southeastern Serengeti National Park have been studied continuously since 1966 (fig.14.1; Schaller 1972; Bertram 1979; Hanby and Bygott 1979; Packer et al. 1988; Packer, Scheel, and Pusey 1990). Forty-four different prides have occupied this area over the

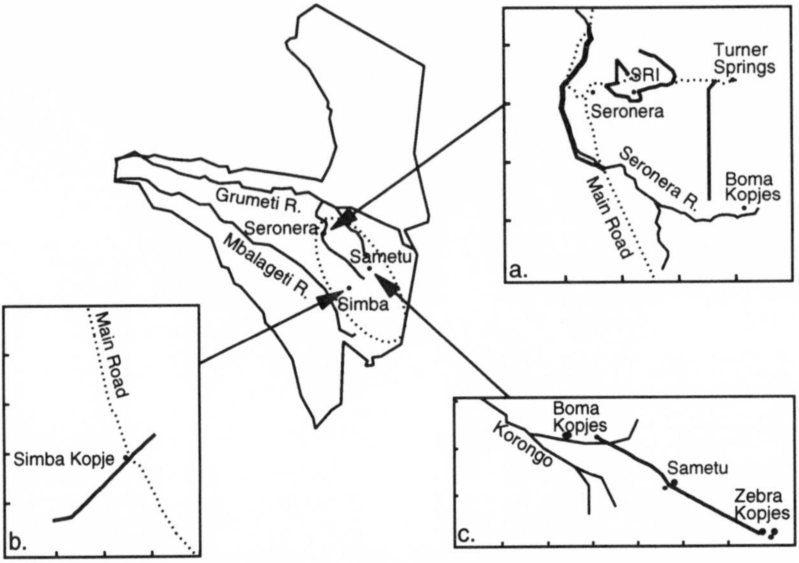

Figure 14.1 Location of transects in: (*a*) the Seronera woodlands and (*b* and *c*) the central plains. Ticks along the axes mark 5 km intervals. Dotted oval in the central map indicates the approximate limits of the long-term records.

past 25 years, and the study population currently comprises about 200 individuals in twenty prides. "Woodlands" prides live in habitat dominated by *Acacia, Commiphora,* and *Balanites* trees in the area located between Nyaraswiga Hill and Turner's Springs. "Plains" prides range in the open grasslands of the central plains south of the Seronera and Ngare Nanyuki Rivers. "Edge" prides live along the woodlands/plains boundary.

Short-Term Data

Hunting Observations. Data in this chapter are restricted to female lions, since they are the principal hunters (Schaller 1972; Scheel and Packer 1991). Hunting activities of radio-collared females and their companions were recorded during 96-hour watches just before or after each full moon. Lions were located by radiotelemetry and followed continuously for 96 hours. Night observations were made with light-intensifying goggles and 8 × 35 binoculars. Between September 1984 and December 1987, 198 hunts were recorded in 3,500 hours of observation of prides from all three habitats (Scheel and Packer 1991; Scheel 1993). Lion hunting frequency is measured as the number of hunts per day, whether or not the hunt was successful. Hunts are defined as movement toward potential prey by at least one lion exhibiting a typical stalking stance (see Scheel and Packer 1991).

Prey Density. Each month between July 1986 and December 1987, herbivores were censused along five fixed transects (fig. 14.1) varying from 11 to 21 km in length. All animals within 500 m of the transect were recorded, except when visibility was limited by brush or terrain, in which case the sampling area was reduced accordingly. Three transects were located in woodland habitat, two in the plains. One woodlands transect followed the course of the Seronera River.

In addition, all herbivores within 1,000 m of lions were censused hourly during the 96-hour follows. Prey censuses could be performed during moonlit nights, but no counts were attempted on dark nights. Prey density is the number of individuals recorded each day within 1,000 m of the lions. Analyses of lion hunting frequency on each prey species are restricted to only those observation periods when at least one animal of that species was present.

Long-Term Data

Lion Sightings and Carcass Records. Most data have been collected from opportunistic "sightings" between June 1966 and September 1991. More systematic observations date from 1984, when females in a dozen prides were fitted with radio collars. All sightings include the location of each lion group and details of each prey item (species, age/sex class, and, where known, whether the carcass was obtained by predation or scavenging). Only one sighting is included from each pride each day, and we exclude all sightings in which prides had moved outside their typical habitat ($N = 51$ of 9,436 sightings and 22 of 1,481 carcasses). Scavenged carcasses are excluded from all analyses. Quantity of meat available from each carcass is estimated as in Packer, Scheel, and Pusey (1990).

Rainfall. Monthly rainfall totals are available from gauges maintained by the Serengeti Ecological Monitoring Programme. Representative gauges from the woodlands ($n = 6$) and plains ($n = 10$) provide data for at least 200 months between June 1966 and September 1991. Rainfall is averaged across all gauges within each habitat each month, and "seasonal rainfall" is the total of these averages for the entire season. The wet season runs from November to the following May, the dry season from June to October (Hilborn et al., chap. 29; Sinclair 1979b). Note that seasonal rainfall is correlated between plains and woodlands (fig. 14.2).

Prey Population Sizes. Ungulate population sizes are estimated from published censuses (wildebeest and zebra: see Campbell 1989; Thomson's gazelle: Borner et al. 1987; Dublin et al. 1990; buffalo: Sinclair 1977; Campbell 1989). We estimate population size between censuses by linear interpolation and assume that populations have remained constant in the

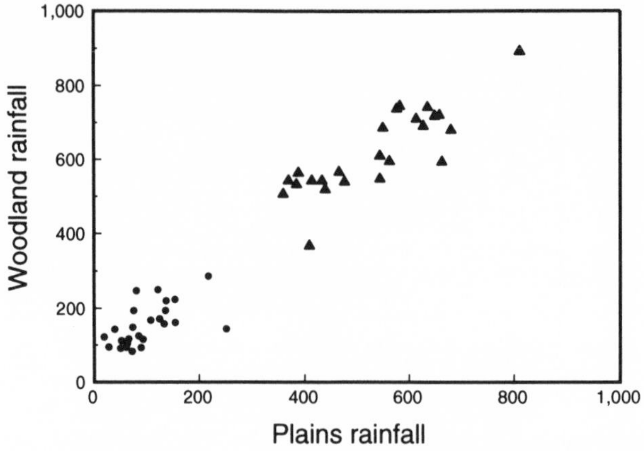

Figure 14.2 Correlations of seasonal rainfall between plains and woodlands habitats, 1966–1991. Dry season, circles (Spearman rank correlation = .65, n = 25 years); wet season, triangles (Spearman = .81).

years following the most recent census. Because of conflicting estimates for the population size of Thomson's gazelle in the early 1980s, we estimate the population from 1983 to 1988 to have been the average of figures presented by Borner et al. (1987) and Dublin et al. (1990).

The Serengeti buffalo population decreased significantly between 1975 and 1986, but the decrease was limited to the far northern and western sectors of the Serengeti (Dublin et al. 1990). Because these regions are well outside the lion study area, and the buffalo population in the southeastern Serengeti has not been affected by high levels of poaching (Campbell 1989), we have excluded this decrease from our population estimates. Note that the wildebeest and buffalo population sizes have been closely correlated over the past 25 years (fig. 14.3).

Statistical Analysis

To analyze hunting frequency in relation to local prey density, we use stepwise linear regression at a significance level of .01. All data collected during the same 96-hour watch are treated as a single independent point (n = 36).

Seasonal differences across habitats in both transect prey density and carcass frequency are analyzed by ANOVAs at a significance level of .05 (SYSTAT: Wilkinson 1988). Each census of the same transect is treated as an independent sample (n = 103). For seasonal variation in carcass frequency, each pride is treated as a single sample, data from the appropriate season are lumped across years, and the dependent variable is the proportion of sightings with carcasses of each species. Each species is considered separately, and the proportions are transformed by an arc-

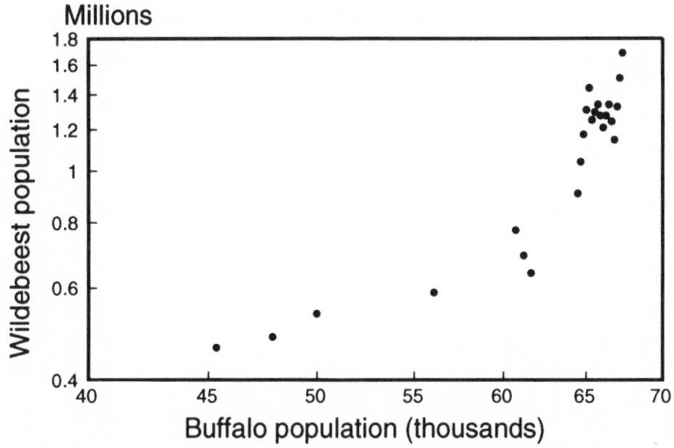

Figure 14.3 Correlation between wildebeest and buffalo populations in the Serengeti (Spearman = .86).

sine—square root transformation. The categorical variables are season (wet or dry) and habitat (woodland, edge, or plains).

Changes in carcass frequency across years (1966–1991) are analyzed with logistic regression models, and each lion sighting is treated as an independent point. The dependent variable is the presence or absence of a carcass at each lion sighting, and carcasses that persisted longer than one day are counted only once. Analyses are separated by species, season, and habitat, and regressions include sightings from all prides in a given habitat. We test six independent variables: the population size of the prey species (when available), size of the wildebeest population (for all species), size of the buffalo population, the preceding season's rainfall, the current season's rainfall, and the proportion of sightings at which a wildebeest carcass was found (this variable is not included in the wildebeest analysis). Population data are entered into all analyses as log(population size), and independent variables are removed from the logistic models in a reverse stepwise fashion. Because six regressions (two seasons by three habitats) are performed on each species, the significance level is set at .01 to reduce the incidence of spurious correlations.

RESULTS AND DISCUSSION

Between 1966 and 1991, female lions were observed feeding from 1,459 carcasses. Seven species accounted for over 90% of the total (both in numbers of carcasses and in kilograms of meat): wildebeest, zebra, Thomson's gazelle, buffalo, warthog, kongoni (hartebeest), and topi (fig. 14.4). Because of their adjacent rank in the lions' diet, relatively small sample size, and similar body size, we have combined topi and kongoni into a single prey "type" for the following analyses. Note that by including op-

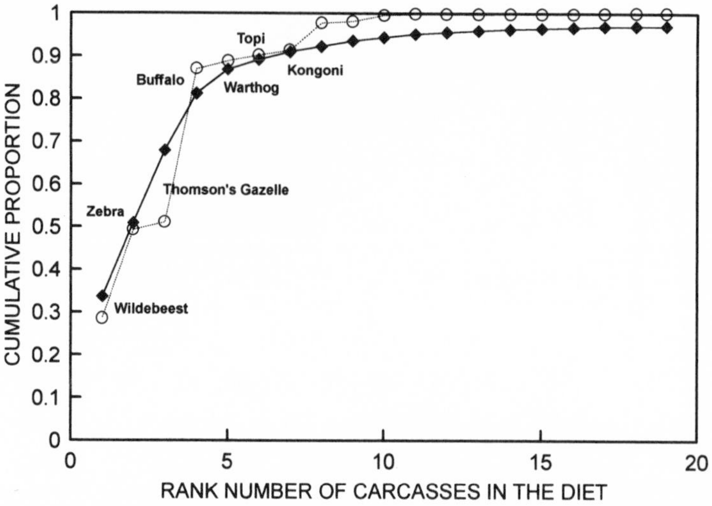

Figure 14.4 Proportion of each prey species in the lions' diet during 1966–1991, ranked by number of carcasses (triangles) and by kilograms of meat (circles). Note that although buffalo ranked only third in the number of carcasses, buffalo make the greatest contribution to the lions' diet in terms of kilograms of meat.

portunistic observations we underestimate the contribution of small prey species because small prey items are often consumed as soon as they are captured (see Bertram 1979). However, direct observations suggest that such small prey are rare and, in terms of biomass, make an insignificant contribution to the lions' total food intake (Schaller 1972; Packer, Scheel, and Pusey 1990; Hanby, Bygott, and Packer, chap. 15).

Hunting Frequencies

Local densities of the six major prey types varied dramatically from one 96-hour watch to the next, and the lions hunted wildebeest, warthog, and Thomson's gazelle significantly more often when those species were most abundant (fig. 14.5). Hunting rates for zebra and topi/kongoni show similar trends, but these are not statistically significant. In contrast, lions showed no tendency to vary their hunting frequency according to the local density of buffalo. Only large prides attempt to capture buffalo, and lions prefer to attack solitary bulls rather than herds (Packer, Scheel, and Pusey 1990; Scheel 1993), presumably because buffalo herds actively defend themselves and can even kill a lion (Packer 1986).

Prey Density and Carcass Records

Precise records of local prey density do not exist for each of the past 25 years. However, vegetation growth and hence migratory movements depend on rainfall (McNaughton 1979; Maddock 1979; Sinclair 1979a);

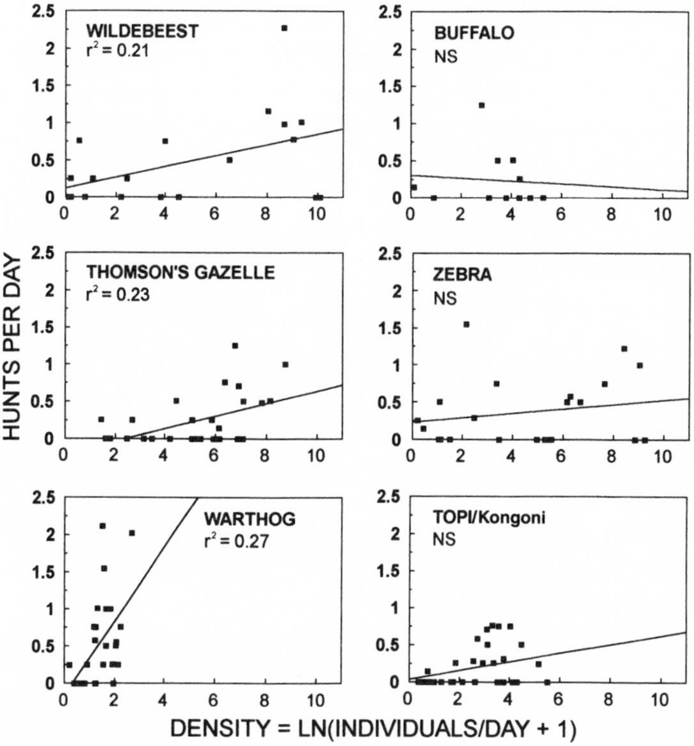

Figure 14.5 Frequency of lion hunts on each prey species plotted against the density of that species over a 4-day period. The regressions for wildebeest ($T = 2.8$), Thomson's gazelle ($T = 2.8$), and warthog ($T = 3.0$) are all significant ($P < .01$).

thus wildebeest and zebra are abundant in the southeastern Serengeti only during the wet season (Maddock 1979). In the following sections, we confirm the overall association between rainfall and local prey density (both between and within seasons), then use variation in rainfall and in the population size of each prey species as correlates of local prey abundance. These analyses assume that different conditions (e.g., rainfall, season, etc.) do not alter the probability of sighting lions with a carcass of each species, and thus that the proportion of sightings with carcasses reflects the underlying predation rate. Note, however, that this measure is too coarse to distinguish whether variation in predation rates results exclusively from changes in local prey density or from changes in lion preference.

Seasonal Variation. Across seasons, the transect data from 1986–1987 and the diet data from 1966–1991 show a similar pattern (table 14.1). Wildebeest were more locally abundant and more commonly found as

Table 14.1 Variation in local prey density across seasons.

Prey species	Prey density along transects (1986–1987)	Carcasses at lion sightings (1966–1991)
Wildebeest	Wet > Dry*	Wet > Dry***
Zebra	NS	Wet > Dry*
Thomson's gazelle	Dry > Wet**	Dry > Wet**
Buffalo	NS	NS
Topi/Kongoni	NS	NS
Warthog	Dry > Wet***	Dry > Wet***

$^*P \le .05$, $^{**}P \le .01$, $^{***}P \le .001$.

carcasses in the wet season, whereas Thomson's gazelle and warthog were more common in the dry season. While zebra densities did not vary significantly across seasons, zebra carcasses were more common in the wet season. There were no significant seasonal differences in the transect density or carcass frequency of buffalo or topi/hartebeest.

The prey censuses from the 96-hour watches indicate that prey abundance on the plains varies with rainfall within the same season. During the dry season, prey density correlates with average monthly rainfall from the preceding month. Wildebeest density was significantly higher following wet months (linear regression, $T = 23.4$, $r^2 = .99$, $n = 5$, $p < .01$), whereas the density of Thomson's gazelle declined following wet months ($T = -5.0$, $r^2 = .85$, $n = 5$, $p < .05$). Trends in the other prey species are not significant, but rainfall data are available for only a subset of the 96-hour watches. Too few data are available to perform a similar analysis in the woodlands.

Variation across Habitats. Only topi/kongoni density varied significantly across transects in 1986–1987 (table 14.2), and only these species were more common in the woodlands than on the plains (fig. 14.6). Over the past 25 years, the predation rate on several species has varied significantly across habitats. Compared with edge and plains lions, the woodlands lions were found more frequently with buffalo and warthog carcasses during the wet season and with buffalo carcasses during the dry season (table 14.2, fig. 14.6). In the dry season, edge lions were found more frequently with warthog carcasses.

Note that the transects were censused only over a 2-year period and were located primarily in the southern woodlands and northern plains, while the lion sightings extended over a far longer time period and a broader area of woodland and plains habitat (see fig. 14.1). Thus differences in predation rates (carcasses) that were not mirrored in the transect prey densities may still reflect substantial differences in local prey abundance: all other surveys have shown that buffalo and warthog are more common in the woodlands than on the plains (Jarman and Sinclair 1979; Hanby and Bygott 1979; Campbell 1989).

Table 14.2 Variation in local prey density across habitats.

Prey species	Prey density along transects (1986–1987)	Carcasses at lion sightings (1966–1991) Wet season	Dry season
Wildebeest	NS	NS	NS
Zebra	NS	NS	NS
Thomson's gazelle	NS	NS	NS
Buffalo	NS	W > E > P***	W > E > P***
Topi/Kongoni	W > P***	NS	NS
Warthog	NS	W > E and P**	E > W and P**

Note: W, woodlands; E, edge; P, plains. See table 14.1 for significance levels.

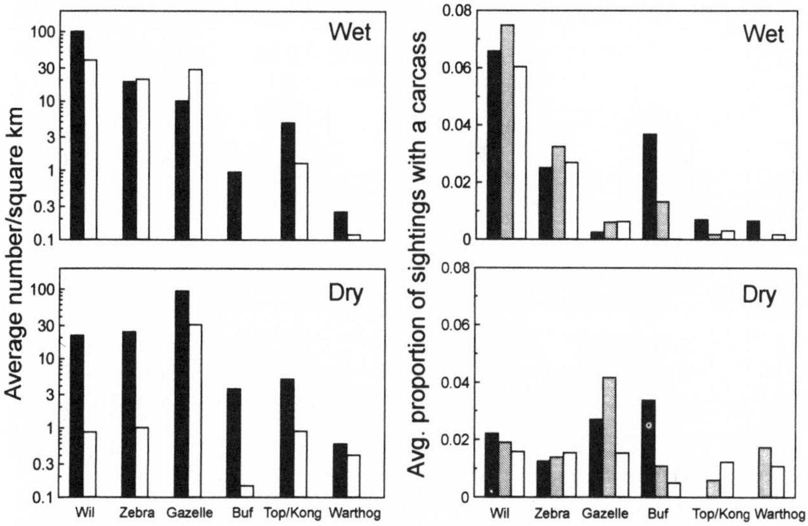

Figure 14.6 Seasonal variation in herbivore density along transects (left) and in proportion of sightings with carcasses (right) in the wet season (top) and dry season (bottom). Woodlands, solid bars; edge, hatched bars; plains, open bars. See tables 14.1 and 14.2 for statistics.

Rainfall Variation across Years. Because of the extreme variation in rainfall from one year to the next (Norton-Griffiths, Herlocker, and Pennycuick 1975; Hilborn et al., chap. 29), local prey density during a given season can vary strikingly across years (see above; Maddock 1979; Packer et al. 1988).

Rainfall during the dry season attracts wildebeest into our study area (see above; Maddock 1979; McNaughton 1979), and thus dry season rainfall is positively correlated with the incidence of wildebeest carcasses in each habitat (table 14.3, fig. 14.7). High wet season rainfall on the plains delays the wildebeest migration through the Seronera region until the beginning of the dry season (Maddock 1979), thus increasing dry season predation rates (fig. 14.7). Finally, higher wet season rainfall in the

Table 14.3 Correlations of four rainfall variables (woodland/plains wet/dry season totals) with the presence of carcasses at lion sightings during the dry season.

Prey species	Habitat		
	Woodland	Edge	Plains
Wildebeest	Woodland dry +***	Woodland dry +*** Plains wet +**	Plains dry +**
Zebra	NS	NS	Woodland wet −***
Thomson's gazelle	NS	NS	NS
Buffalo	NS	NS	n.d.
Topi/Kongoni	NS	n.d.	NS
Warthog	NS	NS	NS

Note: Columns indicate habitat in which lions were found with carcasses of each prey species. Text in cells indicates habitat and season for which rainfall was significantly correlated with the frequency of carcasses. Sign indicates direction of correlation. See table 14.1 for significance levels. n.d., insufficient data.

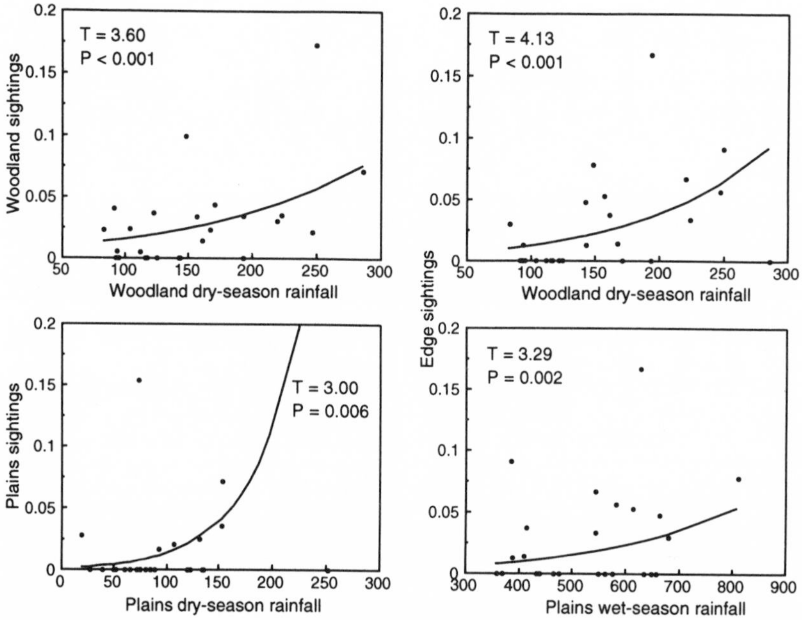

Figure 14.7 Significant correlations between rainfall and dry season sightings of lions with wildebeest carcasses (see table 14.3). Each point represents the proportion of lion sightings at which a carcass was found in a given season in that habitat. Sample size per point varies from 5 to 300. Statistics are from logistic regressions containing all significant predictors indicated in tables 14.3–14.6. Solid lines indicate values predicted by the logistic models.

woodlands decreases dry season predation on zebra on the plains (table 14.3). This pattern may result from the tendency of zebra to leave the plains sooner in years of heavy wet season rainfall (Maddock 1979).

Correlations between rainfall and wet season predation rates (table 14.4) are also consistent with patterns of prey movement. Following dry seasons with above-average rainfall, wet season predation rates decline for wildebeest, topi/kongoni, zebra, and Thomson's gazelle (fig. 14.8). Migrants that have been attracted to the southeastern Serengeti by high dry season rainfall move out to the eastern plains at the onset of the wet season (also see Maddock 1979; McNaughton 1979), thus reducing predation rates in our study area for the remainder of the wet season.

Wildebeest predation is significantly affected by rainfall within the wet season (table 14.4). The wildebeest rarely return to the woodlands during heavy wet seasons, thus remaining out of reach of the woodlands lions.

Ungulate Population Sizes and Changes in Vegetation. Testing for correlations between predation rates and prey population sizes is complicated by the similar recoveries in the wildebeest and buffalo population sizes following the rinderpest epizootic (Sinclair 1979a; Campbell 1989; see fig. 14.2). Further, the Serengeti landscape has changed markedly over the past 25 years. The region surrounding the Seronera River was kept clear of brush by park policy until 1969 (G. Schaller, B. C. R. Bertram, pers. comm.). Subsequently, reduced levels of burning led to a continuous increase in brushy vegetation in the woodlands and edge habitats (Sinclair, chap. 5). Because of covariation between these variables, the following analyses must be interpreted with caution.

Lion predation rates are correlated with the population sizes of several Serengeti ungulates (tables 14.5 and 14.6; note that insufficient data are available to include warthog or topi/kongoni population sizes in this analysis). Woodland predation on buffalo increased with the buffalo population (fig. 14.9), and predation rates on Thomson's gazelle varied with the gazelle population in two habitats. However, predation rates on wildebeest, zebra, and gazelle all declined with increasing populations of either wildebeest or buffalo (fig. 14.9). Because of colinearity between the wildebeest and buffalo population sizes, we cannot reliably distinguish which species has the greater effect in these last three cases. However, because lions have *reduced* their predation rates on wildebeest as the wildebeest/buffalo populations *increased,* the simplest explanation would be that lions have replaced wildebeest, zebra, and gazelle with greater numbers of buffalo. Alternatively, these three species may have spent less time in the southeastern Serengeti as the buffalo population, wildebeest population, or woody vegetation increased.

Table 14.4 Correlations of four rainfall variables (woodland/plains wet/dry season totals) with the presence of carcasses at lion sightings during the wet season.

	Habitat		
Prey species	Woodland	Edge	Plains
Wildebeest	Woodland wet −*** Plains dry −***	NS	NS
Zebra	NS	Plains dry −***	NS
Thomson's gazelle	NS	NS	Woodland dry −**
Buffalo	NS	NS	n.d.
Topi/Kongoni	Plains dry −**	n.d.	n.d.
Warthog	NS	n.d.	n.d.

Note: Columns indicate habitat in which lions were found with carcasses of each prey species. Text in cells indicates habitat and season for which rainfall was significantly correlated with the frequency of carcasses. Sign indicates direction of correlation. See table 14.1 for significance levels. n.d., insufficient data.

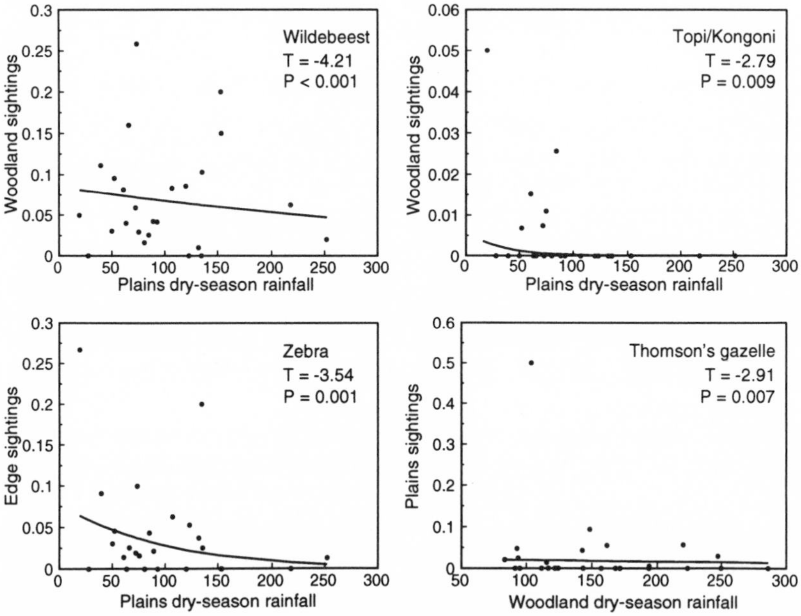

Figure 14.8 Significant correlations between dry season rainfall and sightings in the following wet season of lions with carcasses of each species (see table 14.4 for predictors). Details as in figure 14.7.

CONCLUSIONS

Lion predation on each prey species changes between seasons, across habitats, and from year to year. Most of this variation can be attributed to the annual migration of wildebeest, zebra, and gazelle. Prey movements throughout the year are driven by rainfall, and rainfall varies from one year to the next. Lions hunt several prey species in proportion to their

Table 14.5 Correlations of prey population sizes with the presence of carcasses at lion sightings during the dry season.

	Habitat		
Prey species	Woodland	Edge	Plains
Wildebeest	NS	NS	Buffalo or wildebeest −**
Zebra	NS	Buffalo or wildebeest −***	NS
Thomson's gazelle	Buffalo or wildebeest −***	Thomson's gazelle +***	NS
Buffalo	Buffalo only +**	NS	n.d.
Topi/Kongoni	NS	n.d.	NS
Warthog	NS	NS	NS

Note: Columns indicate habitat in which lions were found with carcasses of each prey species. Cells indicate Serengeti-wide population sizes (prey species were considered only when sufficient data were available) that were significantly correlated with the presence of carcasses. Sign indicates direction of correlation. See table 14.1 for significance levels. n.d., insufficient data.

Table 14.6 Correlations of prey population sizes with the presence of carcasses at lion sightings during the wet season.

	Habitat		
Prey species	Woodland	Edge	Plains
Wildebeest	Buffalo only −**	Buffalo or wildebeest −**	NS
Zebra	NS	Buffalo or wildebeest −**	NS
Thomson's gazelle	NS	NS	Thomson's gazelle +***
Buffalo	NS	NS	n.d.
Topi/Kongoni	NS	n.d.	n.d.
Warthog	NS	n.d.	n.d.

Note: Columns indicate habitat in which lions were found with carcasses of each prey species. Cells indicate Serengeti-wide population sizes (prey species were considered only when sufficient data were available) that were significantly correlated with the presence of carcasses. Sign indicates direction of correlation. See table 14.1 for significance levels. n.d., insufficient data.

local density (fig. 14.5), and that density is constantly changing (fig. 14.6, tables 14.1 and 14.2).

Predation on buffalo is consistently greater in the woodlands than on the plains (fig. 14.6). Thus, the woodlands lions may be buffered against seasonal changes in prey density by greater access to prey during those seasons when migrant species are locally scarce (see also Hanby, Bygott, and Packer, chap. 15).

Predation on wildebeest is highest when dry season rainfall attracts the herds to the study area (table 14.3) and lowest when wet season rainfall is high (table 14.4). However, wet season predation on migratory or semimigratory species declines following a rainy dry season (table 14.4), and wet season predation on resident species does not increase in compensation. Lion reproductive rates are highest in years with rainy dry sea-

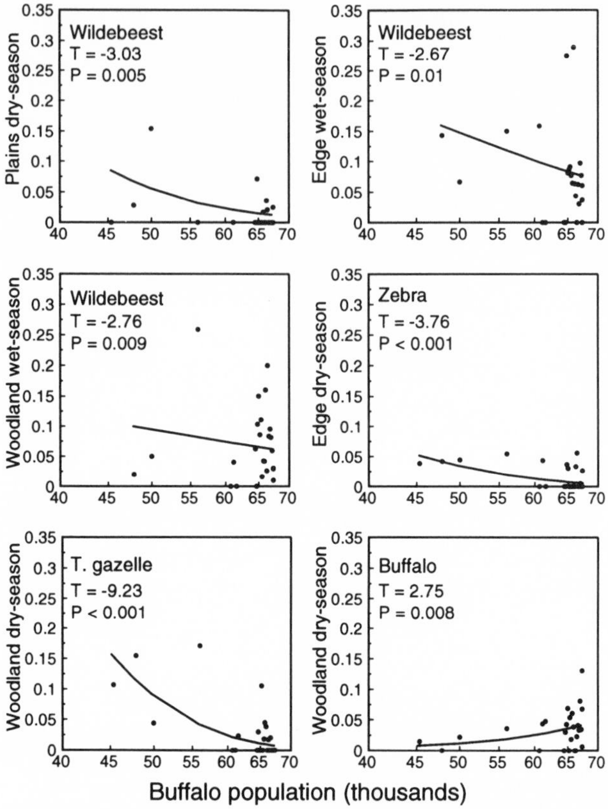

Figure 14.9 Significant correlations between the overall population size of Serengeti buffalo and lion predation rates on different prey species (see tables 14.5 and 14.6). Details as in figure 14.7.

sons or dry wet seasons (Packer et al. 1988), years when the wildebeest remain in the study area for more months of the year.

Predation on Thomson's gazelle varies with the size of the gazelle population. Predation rates on wildebeest, zebra and gazelle were higher in the 1960s (Schaller 1972) than in either the 1970s (see Hanby, Bygott, and Packer, chap. 15) or the 1980s (Packer, Scheel, and Pusey 1990). These declines have coincided with the recovery of the wildebeest and buffalo populations from the rinderpest epizootic. Predation on buffalo has increased over the same time, suggesting that the size of the buffalo population has had the most significant long-term effect (tables 14.5 and 14.6).

Our findings not only confirm the importance of the migratory prey to the lions, but also highlight the extent to which lions rely on buffalo in the absence of the migratory species. While buffalo in our part of the

Serengeti have been well protected over the past 25 years, buffalo numbers have declined drastically in the northern part of the park. This decline could well have altered the diet or population size of the northern lions.

ACKNOWLEDGMENTS

We thank George Schaller, Brian Bertram, Jeannette Hanby, and David Bygott for access to their long-term records; John Fryxell and an anonymous reviewer for comments on an earlier draft of the manuscript; David S. Babu, Director of Tanzania National Parks, George Sabuni, Coordinator of the Serengeti Wildlife Research Institute, and the Tanzanian National Scientific Research Council for permission and facilities; and Anne Pusey, Steve Scheel, Jon Grinnell, Barbie Allen, Marcus Borner, Tony Collins, John Fanshawe, Rob Heinsohn, Larry Herbst, Bruce Davidson, Karen McComb, and Charlie Trout for assistance with fieldwork. This research was supported by NSF grants BSR 8406935, 8507087, 8807702, and 9107397 to C. P. and Anne Pusey and by the Dayton Natural History Fund of the Bell Museum of Natural History to D. S.

REFERENCES

Bertram, B. C. R. 1979. Serengeti predators and their social systems. In *Serengeti: Dynamics of an ecosystem,* ed. A. R. E. Sinclair and M. Norton-Griffiths, 221–48. Chicago: University of Chicago Press.

Borner, M., FitzGibbon, C. D., Borner, M., Caro, T. M., Lindsay, W. K., Collins, D. A., and Holt, M. E. 1987. The decline of the Serengeti Thomson's gazelle population. *Oecologia* 73:32–40.

Campbell, K. L. I. 1989. *Serengeti Ecological Monitoring Programme: Programme report, September 1989.* Serengeti Wildlife Research Centre.

Dublin, H. T., Sinclair, A. R. E., Boutin, S., Anderson, E., Jago, M., and Arcese, P. 1990. Does competition regulate ungulate populations? Further evidence from Serengeti, Tanzania. *Oecologia* 82:283–88.

Hanby, J. P., and Bygott, J. D. 1979. Population changes in lions and other predators. In *Serengeti: Dynamics of an ecosystem,* ed. A. R. E. Sinclair and M. Norton-Griffiths, 249–62. Chicago: University of Chicago Press.

Hilborn, R., and Sinclair, A. R. E. 1979. A simulation of the wildebeest populations, other ungulates, and their predators. In *Serengeti: Dynamics of an ecosystem,* ed. A. R. E. Sinclair and M. Norton-Griffiths, 287–309. Chicago: University of Chicago Press.

Holling, C. S. 1959. Some characteristics of simple types of predation and parasitism. *Can. Entomol.* 41:385–98.

Jarman, P., and Sinclair, A. R. E. 1979. Feeding strategy and the pattern of resource partitioning in ungulates. In *Serengeti: Dynamics of an ecosystem,* ed.

A. R. E. Sinclair and M. Norton-Griffiths, 130–63. Chicago: University of Chicago Press.

Maddock, L. 1979. The "migration" and grazing succession. In *Serengeti: Dynamics of an ecosystem,* ed. A. R. E. Sinclair and M. Norton-Griffiths, 104–29. Chicago: University of Chicago Press.

McNaughton, S. J. 1979. Grassland-herbivore dynamics. In *Serengeti: Dynamics of an ecosystem,* ed. A. R. E. Sinclair and M. Norton-Griffiths, 46–81. Chicago: University of Chicago Press.

Norton-Griffiths, M., Herlocker, D., and Pennycuick, L. 1975. The patterns of rainfall in the Serengeti ecosystem, Tanzania. *E. Afr. Wildl. J.* 13:347–74.

Packer, C. 1986. The ecology of sociality in felids. In *Ecological aspects of social evolution: Birds and mammals,* ed. D. I. Rubenstein and R. W. Wrangham, 429–51. Princeton, N.J.: Princeton University Press.

Packer, C., Herbst, L., Pusey, A. E., Bygott, J. D., Hanby, J. P., Cairns, S. J., and Borgerhoff Mulder, M. 1988. Reproductive success of lions. In *Reproductive success: Studies of individual variation in contrasting breeding systems,* ed. T. H. Clutton-Brock, 363–83. Chicago: University of Chicago Press.

Packer, C., Scheel D., and Pusey, A. E. 1990. Why lions form groups: Food is not enough. *Am. Nat.* 136:1–19.

Schaller, G. B. 1972. *The Serengeti Lion: A Study of Predator-Prey Relations.* Chicago: University of Chicago Press.

Scheel, D. 1993. Profitability, encounter rates and prey choice of African lions. *Behav. Ecol.* 4:90–97.

Scheel, D., and Packer, C. 1991. Group hunting behavior of lions: A search for cooperation. *Anim. Behav.* 41:697–709.

Sinclair, A. R. E. 1977. *The African buffalo: A study of resource limitation of populations.* Chicago: University of Chicago Press.

Sinclair, A. R. E. 1979a. The eruption of the ruminants. In *Serengeti: Dynamics of an ecosystem,* ed. A. R. E. Sinclair and M. Norton-Griffiths, 82–103. Chicago: University of Chicago Press.

Sinclair, A. R. E. 1979b. The Serengeti environment. In *Serengeti: Dynamics of an ecosystem,* ed. A. R. E. Sinclair and M. Norton-Griffiths, 31–45. Chicago: University of Chicago Press.

Stephens, D. W., and Krebs, J. R. 1986. *Foraging Theory.* Princeton, N.J.: Princeton University Press.

Wilkinson, L. 1988. *SYSTAT: The system for statistics.* Evanston, Ill.: SYSTAT, Inc.

FIFTEEN

Ecology, Demography, and Behavior of Lions in Two Contrasting Habitats: Ngorongoro Crater and the Serengeti Plains

J. P. Hanby, J. D. Bygott, and C. Packer

African lions have been studied in a wide variety of habitats. Previous comparisons between populations have primarily focused on the effects of prey availability on broad measures of lion ecology and demography (see Van Orsdol, Hanby, and Bygott 1985). These studies have shown that population density, cub survival, and dispersal rates of subadults are all highest where prey is most abundant (Van Orsdol, Hanby, and Bygott 1985; Hanby and Bygott 1987; Pusey and Packer 1987; Packer et al. 1988). In this chapter we present a detailed comparison of two groups of lions living in adjacent areas that differ strikingly in prey availability. We document the seasonal food intake rates in both habitats and their effects on demographic parameters. We also examine how ecological factors influence lion foraging, ranging, parental behavior, and social behavior.

Study Sites

The Ngorongoro Crater is an extinct volcanic caldera located at the western edge of the Gregory Rift in northern Tanzania. The Crater Highlands are the source of the volcanic soil that formed the Serengeti plains immediately to the west (Sinclair and Norton-Griffiths 1979). The Highlands also act as a barrier to the moisture in the prevailing winds off the Indian Ocean. Consequently, the Crater is flanked by dense forest to the north, east, and south but by arid land to the west. The 250 km^2 Crater is thus a natural island of savanna habitat: the Crater floor is primarily open grassland, and the combination of rich soil, plentiful rainfall, and wet season flooding sustains a remarkable abundance of nonmigratory plains herbivores. The surrounding areas support far lower densities of these species over most of the year.

The Serengeti plains are open rolling grasslands to the west of the Ngorongoro Crater. Our study area includes a 1,700 km^2 section of the plains from the Seronera River near the center of Serengeti National Park to the southern and eastern park boundary. Large herbivores dominate the 25,000 km^2 Serengeti ecosystem, and migratory wildebeest, zebra, and gazelle move over the entire area. These herds are common on the plains from late November until late May each year, then move to the north and west, where they remain for most of the dry season. The Crater Highlands create a rain shadow over the plains, and rainfall is more strictly seasonal on the plains than elsewhere in the Serengeti (Sinclair and Norton-Griffiths 1979).

Lion Social Organization

Lions live in stable social groups ("prides") that typically contain 2–9 adult females (range: 1–16), their dependent young, and a coalition of 2–6 adult males (range 1–9) that has entered the pride from elsewhere (Packer et al. 1988). Prides are territorial and often occupy the same range for generations. Births tend to be synchronous within a pride (Bertram 1975); cubs born less than 1 year apart constitute a "cohort." Prides are maintained by the recruitment of daughters, although some cohorts of young females leave their natal pride to form a new pride nearby. All young males eventually leave their natal pride (Hanby and Bygott 1987; Pusey and Packer 1987), and breeding males generally remain in the same pride for only 2 or 3 years (Bygott, Bertram, and Hanby 1979; Pusey and Packer 1994).

METHODS

Lions in the two areas have been studied continuously since the 1960s (Schaller 1972; Bertram 1975; Elliott and Cowan 1978; Hanby and Bygott 1979; Packer et al. 1988, 1991). All animals are individually recognized by natural markings (Packer and Pusey 1993). Most of the data in this chapter were collected by J. P. H. and J. D. B. between 1974 and 1978. Ecological conditions have varied dramatically on the Serengeti plains over the past 25 years (Sinclair and Norton-Griffiths 1979); by focusing on a narrow time span we aim to present a comprehensive picture of these lions under a specific set of circumstances. However, we also include data collected between 1966 and 1990 for certain analyses and also refer to long-term data that have been presented elsewhere.

Demographic and ranging data in this chapter were collected on eight prides on the Serengeti plains and five in the Crater between September 1974 and February 1978. Between June 1976 and July 1977, intensive behavioral observations were made on one representative pride from each habitat. These prides were closely matched in size and composition (table 15.1) and were representative of the prides in their respective habitats.

Table 15.1 Composition of main study prides.

		Average		Total number of					
				Adults		Subadults		Cubs	
Habitat	Pride name	Pride size	FEQ[a]	Males	Females	Males	Females	1–2 yrs.	< 1 yr.
Crater	Gorigor	20	15	1+2[b]	6	0	2	12	0–6
Plains	Sametu	19	15	1	8	1	2	5	3–9

[a]FEQ = female equivalents = total weight of lions/weight of average adult female.
[b]One male from a coalition of three associated regularly with the Gorigor pride; the other two males associated primarily with a neighboring pride.

Each pride was observed continuously for 4 consecutive days around the full moon during alternate months. These "4-day follows" were made by remaining with those animals that lagged farthest behind while the pride hunted or traveled. No night vision equipment or radiotelemetry was employed. Observations were made on the largest subgroup if the pride split up temporarily. To compensate for slight differences in pride composition, several results are presented in terms of "female equivalents" (FEQ: Bertram 1973), which are simply the combined weight of all pride members divided by the weight of a typical adult female. For example, adult males are about 50% heavier than females, so one male contributes 1.5 female equivalents to the pride total.

The Serengeti pride was also followed in September 1977, but this observation period is included only in certain analyses. The pride remained in much smaller subgroups during this time, so these observations could not be compared with the remaining data. Group size has an important effect on individual food intake rate during the dry season in the Serengeti (Packer, Scheel, and Pusey 1990).

During the 4-day follows, the activity of each lion was recorded at 15-minute intervals and behaviors were recorded by event sampling (Altmann 1974). The quantity of meat eaten by each foraging group was estimated from the age, sex, and species of prey and the proportion of the carcass that was actually consumed by the lions (see Ledger 1968; Sachs 1967). The distance traveled by each pride was measured with a car odometer. Statistical analyses of all data from the 4-day follows provide a single average from each follow, and each follow is considered to be statistically independent.

Herbivore biomass was calculated following the method of Coe, Cumming, and Phillipson (1976). Biomass was measured at six locations in the Serengeti study area each month from April 1975 to October 1977, and at three locations in the Crater pride range on six occasions between January 1974 and February 1978.

Ground transect counts of hyenas were performed in the Crater and Serengeti in 1977. A total of thirty-nine fixed-width transects, spaced 2.5

km apart, were driven over 3,000 km² of the eastern Serengeti plains in May 1977 (for details see Serengeti Research Institute 1977). Twenty-four fixed-width transects, spaced 0.7 km apart, were driven over the entire Crater floor in August 1977.

RESULTS AND DISCUSSION

Prey Abundance

Serengeti Plains. The annual migration of large herbivores in the Serengeti results in marked seasonal changes in herbivore biomass on the plains. Attracted by short grass on nutrient-rich soil, large herds of wildebeest, zebra, and gazelle remain on the plains during the wet season (Sinclair and Norton-Griffiths 1979; McNaughton 1990). The average herbivore biomass during the wet season in this study was over 20,000 kg/km², and was composed mostly of migratory species (table 15.2). However, because these species move around the plains in response to local rainfall, prey availability during the wet season was highly variable within the Sametu pride range.

At the beginning of the dry season, the herds move to the Serengeti woodlands. The average dry season biomass on the plains during this study dropped to less than 1,000 kg/km² (table 15.2). Most of the dry season biomass consisted of Grant's and Thomson's gazelle; the remainder included topi, kongoni, warthog, ostrich, and the occasional wildebeest, eland, giraffe, and oryx.

Ngorongoro Crater. Over the entire Crater floor, prey biomass is almost constant throughout the year: the prey biomass on the Crater floor averaged 11,693 kg/km² in the wet season versus 12,000 kg/km² in the dry season. The large herbivores do show seasonal movements across the Crater floor, but these are much less pronounced than in the Serengeti ecosystem.

Several permanent streams flow through the Gorigor pride range, and

Table 15.2 Herbivore biomass on the Serengeti plains and in the Ngorongoro Crater.

		Biomass (kg/km²)			% of biomass			
Habitat	Season	Mean	SD	No. of surveys	Wildebeest	Zebra	Gazelle	Other
Plains	Wet	20,167	18,627	19	41%	13%	43%	3%
	Dry	970	1,326	18	4%	3%	82%	11%
Crater	Wet	8,400	2,617	3	31%	49%	10%	10%
	Dry	15,660	932	3	44%	48%	5%	3%

Note: On the plains prey biomass is significantly higher in the wet season than the dry season ($P < .001$), but prey abundance is also more variable in the wet season than in the dry ($P < .001$). Within the Crater, biomass is slightly higher in the dry season ($P = .10$). Across habitats, prey biomass is lowest in the Serengeti dry season ($P < .01$) and most variable in the Serengeti wet season ($P < .01$).

thus prey biomass on this part of the Crater floor was nearly twice as high in the dry season as in the wet season. Even in the "low" season, however, prey biomass was over eight times higher in the Gorigor pride range than on the Serengeti plains (table 15.2). During both seasons, at least 80% of the biomass in the Gorigor range was composed of wildebeest and zebra. Buffalo, eland, kongoni, and gazelle made up most of the remainder.

Hunting and Food Intake Rates

The seasonal patterns of food intake rates differed between the two habitats (table 15.3). In the Crater, the individual food intake rate (kg/FEQ/day) was virtually constant over the year. However, food intake rates in the Serengeti plains pride approached those of the Crater pride only when the migratory herds were abundant on the plains. During the "poor season" on the plains, the individual food intake rate was lower than during either season in the Crater ($U = 0$, $n_1 = 3$, $n_2 = 6$, $P < .05$, two-tailed, Mann-Whitney U test).

In both habitats, 15 of 17 kills (88%) were made at night. Most of the prey captured were middle-sized herbivores (adult weight = 100–200 kg: topi, wildebeest, kongoni, and zebra): 11 of 16 (69%) in the Crater; 13 of 17 (76%) on the plains. Note that during the dry season on the Serengeti plains most of the herbivore biomass consisted of gazelle (see table 15.2). Although gazelle form a large part of the lions' diet in the Serengeti woodlands (see Scheel and Packer, chap. 14), they are only rarely captured on the plains (table 15.3). Thus the plains lions suffered not only an overall loss in prey biomass in the dry season, but also a virtual absence of preferred prey.

Both prides acquired a similar proportion of carcasses by scavenging: 5 of 22 (22%) carcasses in the Crater versus 3 of 20 (15%) on the plains. However, there was a striking contrast between the two prides in the proportion of meat acquired by scavenging: 21% in the Crater versus 1% on the plains. At least three and possibly four of the scavenged carcasses in the Crater were acquired from spotted hyenas (the fifth was scavenged from other lions); whereas only one Serengeti meal was taken from hyenas, and the other two were stolen from jackals and vultures respectively.

Lions and hyenas show similar prey preferences (Kruuk 1972; Schaller 1972). Data collected in 1977 show that hyenas were considerably more abundant in the Crater than on the Serengeti plains. Transect counts of spotted hyenas in the two habitats provided estimates (with 95% confidence limits) of 451 ± 176 hyenas (1.8/km^2) in the 250 km^2 Crater versus 3,393 ± 814 (1.1/km^2) in the wet season and 852 ± 409 (0.3/km^2) in the dry season on the 3,000 km^2 Serengeti plains. Because of the closer proximity of the two carnivore species in the Crater, it is not

Table 15.3 Hunting and feeding during 4-day follows.

Date	Seasonal prey avail.	Total number of hunts	Prey killed	Meal scavenged	Group food intake (kg)	Lion group size (FEQ[a])	Kg food/FEQ[a]/day
CRATER							
June 1976	Good	14	3 wildebeest 1 reedbuck 1 unknown		399	12.6	7.9
August 1976	Good	11	1 yearling wildebeest 1 zebra		203	12.1	4.2
October 1976	Good	7	2 zebra 1 reedbuck	1 zebra	442	14.1	7.8
	Average = 11 ± 3				*348 ± 104*	*12.9 ± 0.8*	*6.6 ± 1.7*
December 1976	Poor	12	4 young zebra		464	14.7	7.9
February 1977	Poor	11	1 Grant's gazelle 1 goose	2 wildebeest 1 zebra	221	15.2	3.6
April 1977	Poor	3	1 eland	1 eland	480	15.3	7.8
	Average = 9 ± 4				*388 ± 118*	*15.0 ± 0.3*	*6.4 ± 2.0*

PLAINS							
January 1977	Good	9	1 topi 3 wildebeest 1 young Grant's gazelle	1 Thomson's gazelle	471	13.8	8.5
March 1977	Good	21	2 young zebra 1 hare		141	12.9	2.7
May 1977	Good	18	2 young zebra 1 wildebeest		273	13.1	5.2
	Average = 16 ± 5				*295 ± 135*	*13.2 ± 0.4*	*5.5 ± 2.4*
September 1976	Poor	14	1 topi 1 warthog 1 young kongoni	1 hare 1 unknown	162	13.0	3.1
November 1976	Poor	12	1 topi 1 warthog		126	12.7	2.5
July 1977	Poor	13	1 wildebeest		133	13.3	2.5
	Average = 13 ± 1				*140 ± 16*	*13.0 ± 0.2*	*2.7 ± 0.3*

[a]FEQ = female equivalent = total weight of lions/weight of average adult female.

surprising that the Crater lions were able to acquire a larger proportion of meat by scavenging.

Thus the Crater lions achieved higher rates of food intake throughout the year than did the plains lions during the dry season, and a considerable proportion of the meat consumed by Crater lions was obtained by scavenging. In both habitats, lions hunted mostly at night, and the two prides focused on prey of similar size.

Lion Density and Reproduction

Over a variety of measures of prey availability, "poor season" prey biomass correlates most closely with lion density across habitats (Van Orsdol, Hanby, and Bygott 1985). From our data, it is clear that lions on the plains had a difficult time acquiring adequate food during the poor season. Schaller (1972) and Packer, Scheel, and Pusey (1990) estimate that a female lion requires 5.0–8.5 kg of meat each day. Lions in the plains pride were unable to achieve these levels during any of the poor season 4-day follows (see table 15.3).

Lion population density increased on the Serengeti plains between Schaller's (1972) study in the late 1960s and the mid-1970s (Hanby and Bygott 1979; also see below). This increase may have resulted from improved prey availability during the poor season. Schaller's only dry season prey count on the plains indicated a prey biomass of only 131 kg/km^2. Poor season prey abundance subsequently increased for two reasons (Sinclair and Norton-Griffiths 1979; Hanby and Bygott 1979; Packer et al. 1988). First, the Serengeti herbivore populations increased continuously from the early 1960s through the 1970s after the eradication of rinderpest. Second, favorable rainfall patterns throughout the 1970s resulted in a higher population of resident species and a more continuous presence of migratory species on the plains.

Over the course of our study in the mid-1970s, an average of 119 lions were resident in eight prides in the 1,700 km^2 study area on the Serengeti plains. An additional 50 or so nomadic lions were also present each year when the migratory herds were on the plains, making the total density on the plains 0.1 lions/km^2. On the 250 km^2 Crater floor, an average of 97 lions were resident in five prides with an additional 5 nomads present (0.4 lions/km^2). The Crater population, however, contained a somewhat higher proportion of immatures (61% vs. 48% on the plains, see below). Thus the biomass of lions in the Crater was about 3.3 times higher per km^2 than on the plains.

Interbirth interval and age at first reproduction are similar in the two habitats (Packer et al. 1988). During this study there was no difference in litter size between the two populations (fig. 15.1). In contrast, survival of cubs to 3 years of age was over twice as high in the Crater as on the

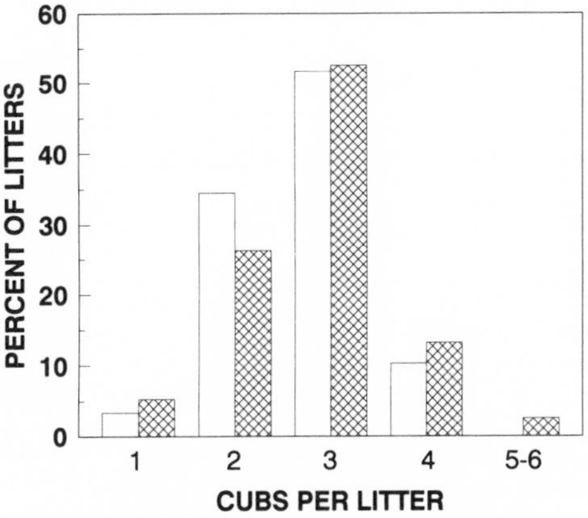

Figure 15.1 Distribution of litter sizes in the Crater (open bars; $N = 28$, mean = 2.8) and the plains (hatched bars; $N = 38$, mean = 2.7) for cubs born between 1974 and 1977.

Serengeti plains, and most mortality on the plains was concentrated in the first 12 months (fig. 15.2). Although female reproductive rates are consistently higher in the Crater (Packer et al. 1988), pride size there is similar to that of Serengeti plains prides (fig. 15.3) because of higher dispersal rates by subadults (Hanby and Bygott 1987; Pusey and Packer 1987) and higher mortality of adult females in the Crater (Packer et al. 1988).

Ranging, Denning, and Territoriality

Over a 3-year period, the mean range size of the five Crater prides was 45 km^2 compared with 200 km^2 for the six most frequently observed plains prides. The Crater pride ranges showed very little overlap (also see Elliott and Cowan 1978). In contrast, the Serengeti prides shifted their ranges each year with the arrival of the migration (see also Schaller 1972), with the result that there was considerable overlap between prides in their annual ranges. A detailed analysis of long-term ranging patterns in the two habitats will be presented elsewhere.

Data from the 4-day follows reveal differences in the daily ranging patterns of the two study prides (fig. 15.4). The average distance traveled over a 24-hour period was only about two-thirds as far for the Crater pride as for the plains pride. Although it might seem that this difference was due solely to the greater abundance of prey in the Crater, distance traveled *within* each habitat did not vary with prey abundance: Crater lions traveled only slightly (but not significantly) farther in the poor season than in the good season, and the plains pride traveled similar distances over both seasons.

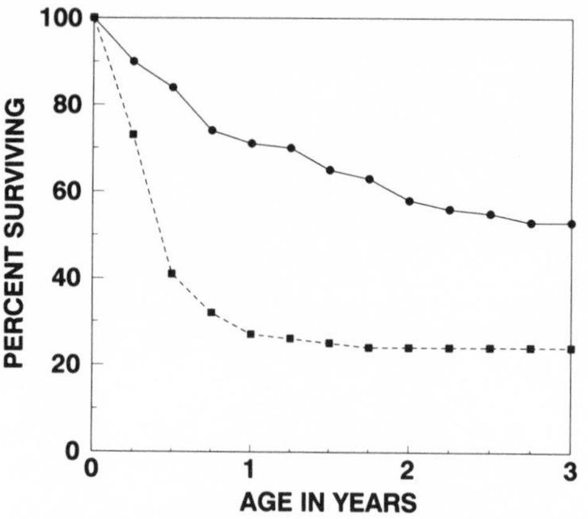

Figure 15.2 Cub survival for cubs born in the Crater (solid line and circles, $N = 109$ cubs) and the plains (dashed line and squares, $N = 142$ cubs) between 1974 and 1977. Treating each litter as statistically independent, a significantly higher proportion of Crater cubs survive to 1 year of age than do cubs on the plains (chi-square = 17.77, $N = 101$ litters, $P < .01$).

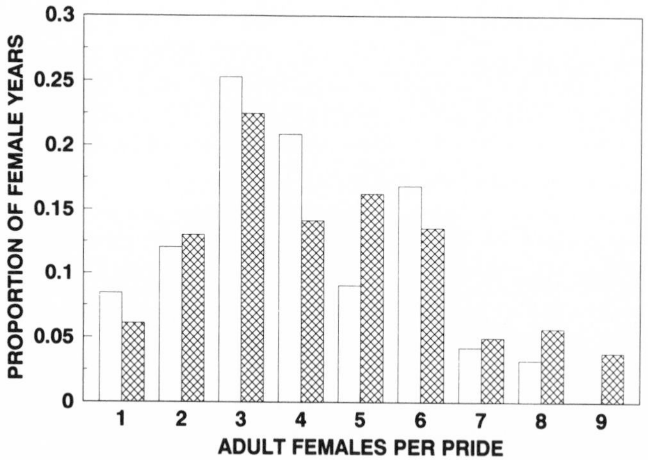

Figure 15.3 Proportion of females that have resided in prides of different sizes in the Ngorongoro Crater between 1963 and 1990 (open bars, $N = 498$ female years) and on the Serengeti plains between 1966 and 1990 (hatched bars, $N = 707$ female years). The distribution of pride sizes is measured in terms of the sizes experienced by individual females (see Pusey and Packer 1987). The proportion of females in each pride size is the proportion of "female years" in which prides of that size existed within each study area. Thus a pride that comprised two females for 5 years and three females for 2 years would contribute $2 \times 5 = 10$ female years to the pride size of two and 6 female years to the size of three.

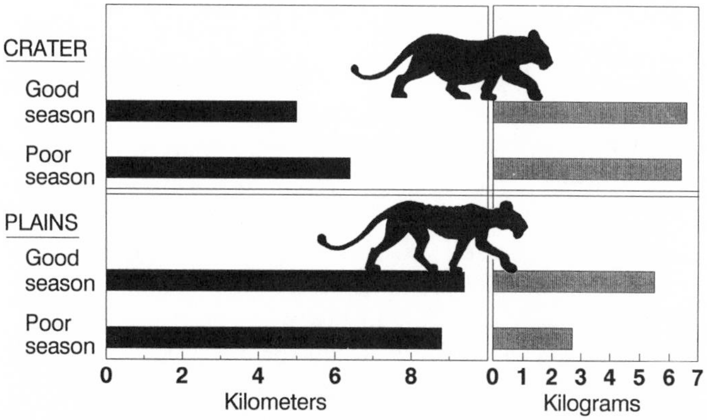

Figure 15.4 Average distance lions traveled per day in each habitat (solid bars) and their average food intake rates (shaded bars) in different seasons. The Serengeti lions traveled significantly farther than the Crater lions ($U = 4$, $n_1 = 6$ 4-day follows in the Crater, $n_2 = 7$ follows in the Serengeti, $P < .02$), but there was no significant effect of season within or between the two habitats. Food intake rate varied significantly between seasons in the Serengeti, but not in the Crater (see table 15.3).

Lions drink after every meal, and females keep their cubs hidden in dense brush until they are 2 months old. The distribution of water and cover probably had an important influence on the contrasting movements of the two prides. The Crater pride's range included open grassland, *Acacia* forest, lakeshore, freshwater streams, bushy hillside, and brush-covered mounds. Thus adequate cover was available over most of the pride's range, and no part of the range was more than 2 km from fresh water at any time of year.

The plains pride's range was by contrast featureless and harsh. Rocky outcrops (kopjes) and alkaline marshes provided cover, shade, or den sites in only four locations. One female traveled 20 km in 12 hours between her cubs' den and the remainder of her pride. In the wet season, water could be found in rock clefts, water holes, and low-lying areas on the plains. But in the driest months, water was available only in two marshes located 10 km apart.

The long distances traveled by the plains pride probably contributed to the higher mortality of its cubs (see fig. 15.2). Once cubs are old enough to leave their dens, they join the other cubs in the pride to form a stable "crèche," thus spending more time with their mothers (Packer, Scheel, and Pusey 1990). By this age, cubs feed on meat and must be able to range widely. Mothers give cubs access to kills, no matter how thin they are themselves, but they do not carry their cubs to kills, nor do they carry kills to their cubs. When prey is scarce on the plains, weakened cubs are often unable to keep up and may be abandoned (Packer and Pusey 1984). Most cub mortality on the plains occurs in the dry season, whereas there is no seasonality in cub mortality in the Crater (Packer et al. 1988). Thus, although the Serengeti pride traveled as far in the wet season as in the dry season, the cubs were less able to cope with long-distance travel when food was scarce and water was widely scattered.

Lions are intolerant of same-sex intruders within their territories. Although ranges in the Serengeti often overlap, members of adjacent prides usually remain several kilometers apart. The *annual* overlap results from *seasonal* shifts in ranges due to the changing distribution of prey. Interpride encounters usually end in aggression, with the larger group chasing the smaller (Packer, Scheel, and Pusey 1990). Roaring and marking play conspicuous roles in the territorial behavior of lions, although these behaviors also function in long-range communication between pridemates (Schaller 1972; Rudnai 1974; McComb et al. 1993; Grinnell, Packer, and Pusey 1995).

Males roared and marked more frequently than females, and both behaviors were broadly correlated with prey availability (table 15.4). Females in both habitats roared more in the good season than in the poor season, and males marked more in the good season.

Although there was no difference in the mean frequency of roaring between the two habitats, the Serengeti lions showed significantly higher variance in roaring behavior than the Crater lions (table 15.4). This may have been due to the fact that Serengeti lions frequently readjust their territorial boundaries over the course of a year and are often widely separated from their neighbors. The need to announce their location to their rivals may therefore be more intermittent than in the Crater.

Overall, these results suggest that roaring and marking within each habitat are most frequent when prey is most abundant. These are the conditions in which intruder pressure from nomadic lions is highest (Schaller 1972).

Activity Patterns and Social Behavior

Lions in the two habitats showed striking similarities in their activity patterns (fig. 15.5). In spite of very different physical habitats and prey abundance, lions in both habitats spent about 80% of their time sleeping, lying down, or sitting.

A detailed analysis of social behavior suggests that prey availability may affect frequency of social behavior in both habitats. Within each habitat, the frequency of social interactions was higher in the good season (fig. 15.6). The seasonal decrease in social interaction on the plains could be due to the energetic consequences of lower food intake when prey are scarce. However, a similar seasonal decrease was found in the Crater even though food intake rates there do not vary over the year (see table 15.3). This finding suggests that social interaction is correlated with changes in prey availability per se rather than with food intake. Lions may need to

Table 15.4 Frequency of roaring and marking (event/individual/hour).

Sex	Season	Habitat: Plains	Habitat: Crater
Roaring			
Males	Good	0.76 ± 0.01 (N = 2)	0.12 ± 0.07 (N = 3)
	Poor	0.26 ± 0.23 (N = 2)	0.08 ± 0.03 (N = 3)
Females	Good	0.27 ± 0.22 (N = 3)	0.08 ± 0.04 (N = 3)
	Poor	0 (N = 4)	0.04 ± 0.02 (N = 3)
Marking			
Males	Good	0.29 ± 0.24	0.18 ± 0.09
	Poor	0	0.03 ± 0.02
Females	Good	0	0.01 ± 0.01
	Poor	0.10 ± 0.17	0

Note: Across all conditions, males roar and mark significantly more often than females (roaring: $T = 1$, $n = 10$ 4-day follows when both males and females were present, $P < .01$, Wilcoxon signed ranks test; marking: $T = 3$, $P < .01$). Across habitats, females roar more during the "good" season than during the "poor" season ($U = 6.5$, $n_1 = 6$, $n_2 = 7$, $P < .05$); and males mark more during the "good" season ($U = 0$, $n_1 = 5$, $n_2 = 5$, $P < .01$). Differences in the variance in roaring frequency between the two habitats are significant for both males and females ($P < .01$). *N*, number of 4-day follows; sample sizes for marking behavior are the same as for roaring.

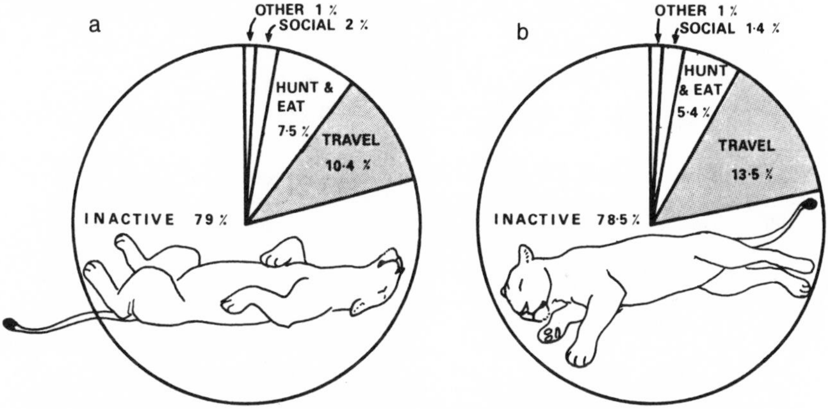

Figure 15.5 Pie charts of lion activity during the 4-day follows. A = Crater, B = plains. Data include all seven 4-day follows on the plains pride and are not separated by season because activity patterns did not vary seasonally (except for social behavior; see text and fig. 15.6). "Inactive" includes sleeping, lying, and sitting. "Other" includes marking, patrolling, roaring, self-grooming, and defecation/urination.

take more care to remain hidden from their prey when food is scarce and therefore may engage in lower levels of conspicuous activity.

CONCLUSION

Lions on the Serengeti plains clearly lead a harder life than their Crater counterparts do: food supplies are more ephemeral, water is scarcer, and denning sites are widely scattered. Cub mortality on the plains is highly seasonal and far higher than in the Crater. Moreover, most of our data

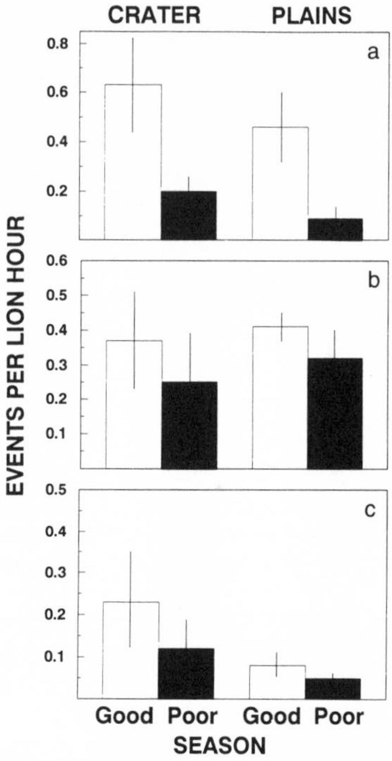

Figure 15.6 Frequency of social behavior during 4-day follows. Data include all seven 4-day follows on the plains pride. Play (*a*) includes mock fighting and hunting behavior. Affiliative behavior (*b*) involves licking, grooming, and head rubbing. Hostile behavior (*c*) involves growling, snarling, biting, swatting, and chasing another lion. None of the differences within or between habitats are statistically significant; however, lions in both habitats engaged in significantly more play in the "good" season than in the "poor" season ($U = 5.5, n_1 = 6, n_2 = 7, P < .025$). Data from the Crater are based on 4,056 lion hours; from the plains, on 6,610 lion hours.

were collected when conditions were unusually benign for the plains lions: successive years with favorable rainfall coupled with a continuous increase in herbivore populations allowed the plains lion population to increase (Hanby and Bygott 1979). These conditions persisted until 1979. Thereafter rainfall returned to a more typical pattern, herbivore populations leveled off, and the plains prides had to rely on even lower levels of prey availability or leave the plains on temporary forays to the woodlands (Packer, Scheel, and Pusey 1990). Cub mortality on the plains subsequently increased (Packer et al. 1988), and the number of adult females resident on the plains reached a peak in the early 1980s (fig. 15.7).

Many of the plains prides originate from Serengeti prides that live along the woodlands/plains boundary. Dispersing groups of young females leave the more crowded parts of the Serengeti to settle in the relatively vacant plains, thus avoiding the more intense intergroup competi-

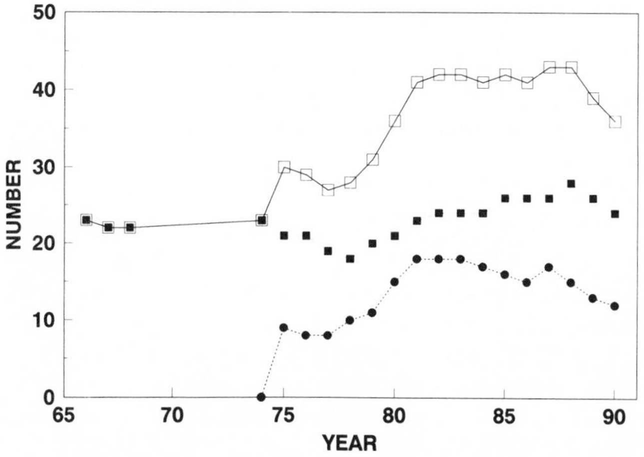

Figure 15.7. Number of adult females (≥ 4 years old) that resided in the eastern part of the Serengeti plains between 1966 and 1990. "Original prides" (solid squares, dotted line) are prides first identified and censused by George Schaller in 1966; "immigrant prides" (solid circles, dashed line) are those that were born in the Serengeti woodlands but dispersed onto the plains. The total number of females living on the plains (open squares, solid line) increased in the late 1970s following the immigration of excess females from the woodlands.

tion in the woodlands (Pusey and Packer 1987; C. Packer et al., unpub.). In fact, most of the increase in resident females on the plains over the past two decades is due to immigration from the woodlands (fig. 15.7). Thus the plains can be viewed as a sink for excess lions from the woodlands.

Like the Serengeti woodlands, the Crater is a net exporter of lions. Crater lions gain access to adequate food each year, and reproductive rates are consistently higher than on the plains (Packer et al. 1988). Competition for space in the Crater is intense: the reproductive rate of the Crater lions is sufficiently high to permit rapid population growth, yet the number of prides on the Crater floor has remained relatively constant (Packer et al. 1991), and pride size in the Crater is comparable to that on the Serengeti plains (see fig. 15.3). A large proportion of subadults born in the Crater disperse and settle in areas (both in the Crater Highlands and in the Serengeti) with far lower prey availability than on the Crater floor (Hanby and Bygott 1987; Pusey and Packer 1987).

Thus the continued presence of lions in suboptimal habitats may be guaranteed only if excess animals from superior habitats are able to travel freely into them. Where large lion populations have been subdivided by the creation of small parks or reserves surrounded by agricultural areas or human habitation, it may be necessary to create corridors between reserves to maintain viable lion populations within each reserve.

Of special relevance to the Serengeti is the fact that most poaching activity occurs in the northern and western woodlands (see Arcese, Hando, and Campbell, chap. 24; Campbell and Hofer, chap. 25). Although the woodlands lions may be less conspicuous than their photogenic plains counterparts, their numbers are far more important to the long-term viability of the Serengeti lion population.

ACKNOWLEDGMENTS

We thank Tanzania National Parks, Ngorongoro Conservation Area Authority, and the Tanzanian Scientific Research Council for permission to conduct this research. J. P. H. and J. D. B. are grateful to the Science Research Council (U.K.) and the New York Zoological Society for funding the fieldwork, and to the Serengeti Research Institute for participating in the hyena censuses. C. P. gratefully acknowledges the National Geographic Society, NSF grants 8507087 and 8807702, and the J. S. Guggenheim Foundation.

REFERENCES

Altmann, J. 1974. Observational study of behavior: Sampling methods. *Behaviour* 49:227–67.

Bertram, B. C. R. 1973. Lion population regulation. *E. Afr. Wildl. J.* 11:215–25.

———. 1975. Social factors influencing reproduction in wild lions. *J. Zool.* (Lond.) 177:463–82.

Bygott, J. D., Bertram, B. C. R., and Hanby, J. P. 1979. Male lions in large coalitions gain reproductive advantages. *Nature* 282:839–41.

Coe, M. J., Cumming, D. H., and Phillipson, J. 1976. Biomass and production of large African herbivores in relation to rainfall and primary production. *Oecologia* 22:341–54.

Elliott, J. P., and Cowan, I. McT. 1978. Territoriality, density and prey of the lion in Ngorongoro Crater, Tanzania. *Can. J. Zool.* (Lond.) 56:1726–34.

Grinnell, J., Packer, C., and Pusey, A. E. 1995. Cooperation in male lions: Kinship, reciprocity or mutualism? *Animal Behaviour.* 49:95–105.

Hanby, J. P., and Bygott, J. D. 1979. Population changes in lions and other predators. In *Serengeti: Dynamics of an ecosystem,* ed. A. R. E. Sinclair and M. Norton-Griffiths, 249–62. Chicago: University of Chicago Press.

———. 1987. Emigration of subadult lions. *Animal Behaviour* 35:161–69.

Kruuk, H. 1972. *The spotted hyena: A study of predation and social behavior.* Chicago: University of Chicago Press.

Ledger, H. P. 1968. Body composition as a basis for a comparative study of some East African mammals. *Symp. Zool. Soc. Lond.* 21:289–310.

McComb, K., Pusey, A. E., Packer, C., and Grinnell, J. 1993. Female lions can identify potentially infanticidal males from their roars. *Proc. R. Soc. Lond. B* 252:59–64.

McNaughton, S. 1990. Mineral nutrition and seasonal movements of African migratory ungulates. *Nature* 345:613–15.

Packer, C., Herbst, L., Pusey, A. E., Bygott, J. D., Hanby, J. P., Cairns, S. J., and Borgerhoff-Mulder, M. 1988. Reproductive success of lions. In *Reproductive success: Studies of individual variation in contrasting breeding systems,* ed. T. H. Clutton-Brock, 363–83. Chicago: University of Chicago Press.

Packer, C., and Pusey, A. E. 1984. Infanticide in carnivores. In *Infanticide: Comparative and evolutionary perspectives,* ed. G. Hausfater and S. B. Hrdy, 31–42. Hawthorne, N.Y.: Aldine.

———. 1993. Should a lion change its spots? *Nature* 362:595.

Packer, C., Pusey, A. E., Rowley, H., Gilbert, D. A., Martenson, J., and O'Brien, S. J. 1991. Case study of a population bottleneck: Lions of the Ngorongoro Crater. *Conserv. Biol.* 5:219–30.

Packer, C., Scheel, D., and Pusey, A. E. 1990. Why lions form groups: Food is not enough. *Am. Nat.* 136:1–19.

Pusey, A. E., and Packer, C. 1987. The evolution of sex-biased dispersal in lions. *Behaviour* 101:275–310.

———. 1994. Infanticide in lions. In *Protection and abuse of young in animals and man,* ed. S. Parmigiani, B. Svare, and F. vom Saal, 277–99. London: Harwood Academic.

Rudnai, J. 1974. *The social life of the lion.* St. Leonardsgate, U.K.: Medical and Technical Publishing.

Sachs, R. 1967. Liveweights and measurements of Serengeti game animals. *East Afr. Wildl. J.* 5:24–36.

Schaller, G. B. 1972. *The Serengeti lion: A study of predator-prey relations.* Chicago: University of Chicago Press.

Serengeti Research Institute. 1977. *Census of predators and other animals on the Serengeti plains, May 1977.* Serengeti National Park Report no. 52.

Sinclair, A. R. E., and Norton-Griffiths, M. 1979. *Serengeti: Dynamics of an ecosystem.* Chicago: University of Chicago Press.

Van Orsdol, K. G., Hanby, J. P., and Bygott, J. D. 1985. Ecological correlates of lion social organization. *J. Zool.* (Lond.) 206:97–112.

SIXTEEN

Population Dynamics, Population Size, and the Commuting System of Serengeti Spotted Hyenas

Heribert Hofer and Marion East

The spotted hyena is the most numerous of the large predators in the Serengeti (Hanby and Bygott 1979). The first research on the ecology of Serengeti spotted hyenas was conducted by Hans Kruuk between 1964 and 1968. Kruuk (1972) showed that hyenas are important predators that prefer to prey on migratory herbivores, especially wildebeest. He thought that Serengeti hyenas were seminomadic and formed short-term unstable groups in temporary territories, in contrast to those in the Ngorongoro Crater, where he discovered large, stable hyena clans, vigorously defending exclusive group territories (Kruuk 1966, 1972). In 1987 we established a long-term research program on the behavioral ecology of Serengeti spotted hyenas. Using individual recognition and radiotelemetry, methods not extensively used by Kruuk (1972) during his pioneering study, we have demonstrated that Serengeti hyenas do live in large, stable clans that defend exclusive group territories (Hofer and East 1993a). In contrast to Ngorongoro hyenas, however, they combine a residential existence with frequent short-term (several days), long-distance (40–80 km) foraging trips to the nearest migratory herds. Thus individuals regularly forage far beyond the boundaries of their group territory (Hofer and East 1993b). We call this combination of a residential existence with extended foraging trips a commuting system, following Kruuk (1966, 1972). The commuting system and the associated large population size make the spotted hyena a keystone predator (Paine 1969) in the Serengeti ecosystem and thus a species of particular interest to both managers and ecologists.

In this chapter we summarize our knowledge of the commuting system and apply it to three aspects of the population ecology of hyenas. First, previous assessments of changes in the hyena population were unsatisfactory because the commuting system prevented a straightforward interpretation of population census results. Here, we develop a method to estimate population size and convert previous census results into comparable population estimates. Second, we present data on demography and population dynamics and discuss these with reference to natural and human-imposed mortality. Third, the Serengeti spotted hyenas exist in larger groups and at higher densities than would be predicted from the biomass of resident prey within clan ranges. This is possible because commuting trips permit a year-round, population-wide exploitation of migratory herbivores (Hofer and East 1993a). We evaluate how the discovery of the commuting system changes our understanding of the relationship between hyenas, resident herbivores, and migratory herbivores, and we discuss whether the life history traits and spacing behavior of carnivores permit migratory herbivores in African ecosystems to escape predation (Fryxell, Greever, and Sinclair 1988).

SOCIAL ORGANIZATION AND FORAGING

Here we summarize the key features of the commuting system; for more details see Hofer and East (1993a,b,c). Our ongoing research is based on more than 500 individually known spotted hyenas, which we recognize by their distinct spot patterns, scars, and ear notches. The social organization of Serengeti clans we studied was similar to that of a large clan in the Masai Mara (Frank 1986a,b) and clans in other areas (Henschel and Skinner 1987; Cooper 1989), with intrasexual hierarchies in both females and males, and females dominating males (East and Hofer 1991a,b; Hofer and East 1993a). All clans were large, stable social groups with a mean group size of 45 adults and subadults (n = 7 clans). Daughters remained with their natal clan throughout life, while young males typically dispersed after they reached sexual maturity at 2–2.5 years. Adult males either joined study clans for short periods of less than 3 months (transient males) or became clan members (resident immigrant males). The territories of the majority of our study clans were situated in the central study area at the plains/woodlands boundary in the center of Serengeti National Park. A second study area contained clans living on the short-grass plains in the southeastern part of the ecosystem. All study clans defended a territory by scent marking, vocal displays (whooping, East and Hofer 1991b), boundary patrols, aggressive expulsion of nonresidents, and disputes with neighboring clans. Female clan members reared their cubs at a communal den inside the territory. In the central study area clan territories of approximately 56 km^2 formed a continuous mosaic. The density of clan members in this area was 0.81 adults and subadults per km^2 (Hofer and East 1993a).

Clan territories experienced fluctuations in prey abundance of several orders of magnitude due to the migratory behavior of some herbivores.

Prey abundance inside a clan territory was classified as: (1) *low* when migratory gazelles (Thomson's gazelle and Grant's gazelle) and the large migratory herds of wildebeest and zebra were absent, and resident herbivores were the only prey category present; (2) *medium* when herds of migratory gazelles were present in addition to resident herbivores, but the large migratory herds of wildebeest and zebra were absent; (3) *high* when large migratory herds of wildebeest and zebra were present (table 16.1). The large migratory herds passed through the territories in the central study area at least twice per year but were never present for long periods, so these clan territories held migratory herbivores for only 26% of the year (table 16.1).

Wildebeest were the prey most frequently killed inside clan territories (54%, n = 80), followed by Thomson's gazelles (15%: Hofer and East 1993a). All clan members fed inside their territories during periods of high prey abundance. During periods of low and medium prey abundance, however, clan members regularly moved long distances from their territories to feed in areas containing large concentrations of migratory

Table 16.1 Patterns of prey abundance in the Isiaka clan territory (the main study clan) during the period 1988–1990, and the response by spotted hyenas in terms of commuting.

	Level of prey abundance			
	Low	Medium	High	*P* value
Herbivore densities (animals/km²)				
Resident herbivores[a]	3.8 ± 0.8	3.1 ± 0.5	3.3 ± 0.9	NS
Gazelles[b]	3.1 ± 2.1	23.9 ± 6.1	16.0 ± 8.2	<.025
Wildebeest and zebra	0.3 ± 0.2	4.1 ± 1.8	219.1 ± 100.2	<.001
Total	7.2	31.0	238.5	
Mean number of days of each level of prey abundance[c]				
Wet season: January–May[d]	97	27	30	
Dry season: June–December[d]	38	97	63	
Total[d]	135 (38%)	124 (35%)	93 (26%)	
Mean duration of period of equal prey abundance (days)	54.0 ± 13.6	35.1 ± 6.3	26.4 ± 4.9	.051
Commuting				
Percentage of clan members foraging by commuting	86%–100%	49%–58%	0%	<.001
Percentage of clan members foraging in the clan's territory	0%–14%	42%–51%	100%	<.001
Percentage of clan members temporarily present in clan territory	22%	56%	100%	<.001

Source; Hofer and East (1993a,b,c), this study. Values are means (± SEM where available). *P* values are for Kruskal-Wallis tests on differences between levels of prey abundance, except for commuting section, where they are for *G* tests.

[a]Buffalo, kongoni, topi, warthog, giraffe, impala, ostrich.

[b]Thomson's gazelle, Grant's gazelle

[c]Association between level of prey abundance and season $G_1 = 68.15$, $P < .0001$.

[d]Means of 2 study years

herbivores, as we demonstrated by means of aerial and ground tracking of radio-collared individuals (Hofer and East 1993b). These commuting trips lasted on average 3–4 days for denning females (females with dependent cubs at the communal den in the clan's territory) and 6–10 days for non-denning females and males (Hofer and East 1993c). During the wet season individuals from clans in the central study area commuted on average 46.9 ± 11.4 km (± SD, n = 39, Hofer and East 1993b) to areas on the short-grass plains that contained migratory herds (fig. 16.1). In the dry season, radio-collared individuals from both plains clans and the central study clans were located in the vicinity of the national park boundary in the west and north of the Serengeti (fig. 16.1). Individuals from plains clans commuted three times the distance (mean 61.1 ± 10.6 km, n = 7) of individuals from the central study clans during the dry season (mean 20.7 ± 10.9 km, n = 17, Hofer and East 1993b). Observations during two study years indicated that denning females undertook on average 42–51 commuting trips annually, equivalent to a total minimum travel distance of 2,900–3,700 km, while non-denning females and males undertook 15–18 trips, traveling at least 1,000–1,300 km per year (Hofer and East 1993c).

Clans experienced a constant traffic of commuters in transit between the clan territory and the migratory herds. Numbers of commuters *foraging* inside a clan's territory increased dramatically when that territory contained large migratory herds. An analysis of the observed distribution of commuter locations showed that commuters (1) minimized interference from resident clans by avoiding their denning areas and (2) selected areas known to have migratory herds present throughout much of the wet or dry season (Hofer and East 1993b).

TERRITORY CARRYING CAPACITY AND COMMUTING THRESHOLD

If commuting is a response of clan members to insufficient prey abundance in the clan territory, then the proportion of clan members commuting, as revealed by aerial tracking of radio-collared individuals, can provide us with information on how clan members assess temporal changes in the carrying capacity of the clan's territory. Records of the proportion of clan members commuting are subject to a number of sampling biases, however, whose magnitude depends on the efficacy of the tracking and the actual duration of commuting trips, which vary on average between 3 and 10 days depending on the sex and reproductive status of clan members (see above). Using corrections for these biases developed elsewhere (Hofer and East 1993c), we calculated the *actual* proportion of clan members that foraged by commuting at each level of prey abundance (table

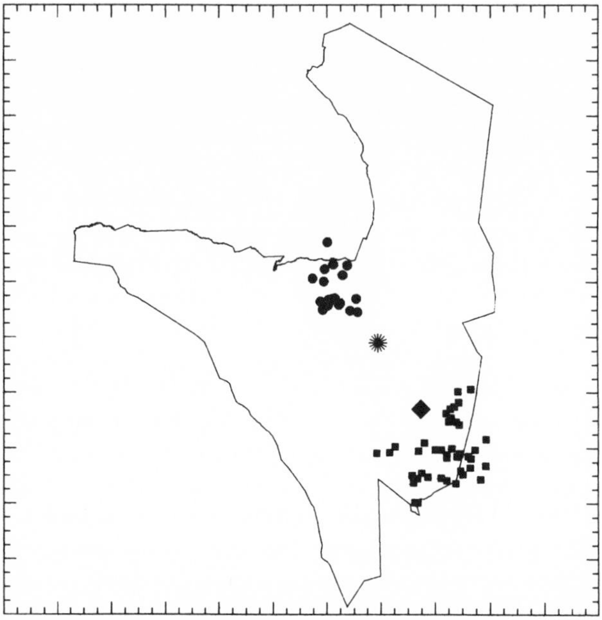

Figure 16.1 Commuting destinations during the wet (squares) and during the dry (circles) season of radio-collared spotted hyenas from the central (mean location denoted by asterisk) and plains (mean location denoted by diamond) study areas. Each symbol represents one commuting trip. One individual may be represented by several trips.

16.1). As commuters return to the clan's territory between commuting trips, the mean percentage of clan members actually temporarily present in the clan's territory (based on aerial tracking records, continuous watches at the communal den, and other sources: Hofer and East 1993c) is slightly higher than the percentage of clan members foraging at home (table 16.1). At low prey abundance less than a quarter, and at medium prey abundance only half, of the clan was present in the clan's territory at any one moment.

The *observed* proportion of clan members commuting increased as the biomass and density of migratory prey in a clan's territory declined (Hofer and East (1993b). Using the corrections for sampling biases introduced above, it is possible to establish a quantitative relationship between the *actual* proportion of clan members commuting and parameters of territory prey presence for each study year and for different commuting trip durations. This calculation provides an estimate of the density and biomass above which the first clan members switched from commuting to foraging inside their clan territory (i.e., when the proportion of clan members commuting dropped below 1), and above which all clan members foraged inside their clan territory (i.e., when the proportion of clan members commuting approached 0). Figure 16.2 displays the results as the domain between the minimum and maximum values calculated for two study years and the range of average commuting trip durations observed. Our estimates suggest that the first clan members switched to foraging inside their clan territory at densities between 0.9 and 5.3 animals/km^2 (750–7,400 kg biomass), a density of migratory animals approximately equivalent to the average density of resident herbivores in the territory (table 16.1). All clan members foraged inside their clan territory at herbivore densities above 340 animals/km^2 (1,360,000 kg biomass).

POPULATION SIZE

The population size of spotted hyenas on the Serengeti plains was estimated for 1966–1968 by Kruuk (1972) based on resightings of marked individuals, and for the wet and dry seasons of 1977 and the wet season of 1986 by the Serengeti Research Institute (1977a,b) and Campbell and Borner (1986) based on transect counts. SRI (1977a,b) and Hanby and Bygott (1979) were unable to compare Kruuk's estimate for 1966–1968 with the 1977 census because census area, field methods, and data analysis techniques differed and the implications of the commuting system for the interpretation of different censusing techniques were not fully appreciated. Appendix 16.1 discusses the difficulties caused by the commuting system for estimating population size and develops a censusing protocol that circumvents these problems. Fundamental to this protocol is the definition of a biologically meaningful segment of the total ecosystem population of hyenas that can be reasonably censused. One such segment comprises all spotted hyenas that commute to or live on the calving grounds of the large migratory herds in the wet season. We call this segment the "source population" of previous censuses. In appendix 16.2 we explain how we converted the results of previous censuses to compatible estimates of the size of the source population.

Previous studies hypothesized (SRI 1977a,b; Hanby and Bygott 1979; Frame 1986) that the source population ought to have increased from 1966–1968 to 1977 because the number of wildebeest substantially increased during this period (Dublin et al. 1990). The total source population did increase significantly between 1966–1968 and 1977 ($t = 2.44$, $P < .007$, one-tailed, table 16.2). However, in contrast to the hypotheses of SRI (1977a,b) and Frame (1986), the number of short-grass plains residents did not change significantly during the same period ($t = 0.43$, NS). Although the number of wildebeest has been constant since 1977 (Dublin et al. 1990), the hyena population should have increased since 1977, as predator populations usually lag behind their main prey (e.g., Crawley 1992). The increase in the source population between 1977 and 1986 was

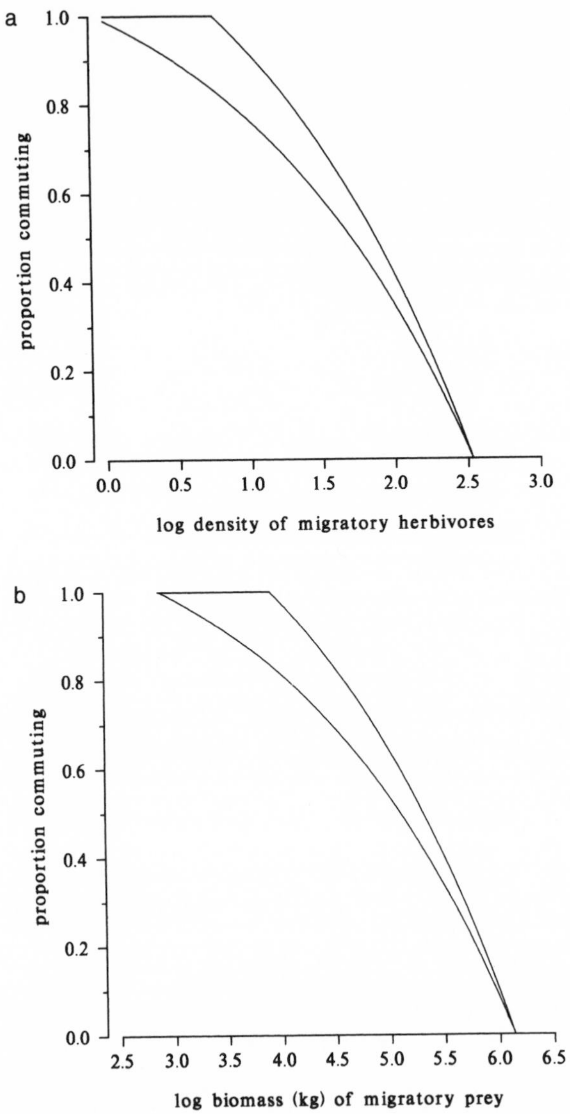

Figure 16.2 The proportion of members of the main study clan (Isiaka Clan) commuting in relation to the logarithm of (*a*) the density and (*b*) the biomass of migratory herbivores (gazelles, wildebeest, zebra) present in their territory. Based on empirical regressions and correction factors derived for aerial tracking data (see text), the enclosed areas encompass the range of values for different study years and commuting trip durations.

Table 16.2 Size of the source population of spotted hyenas that are resident on or commute during the wet season to the Serengeti plains.

Parameter	Symbol	Calculation	Census date 1966–1968	1977	1986
Dry season census count	N_D		368 ± 63[a]	500 ± 163	—
Total number of plains residents	$N_P =$	N_D/p_D[b]	658 ± 113[a]	893	—
Wet season census count	N_W		—	2,775 ± 388	4,263 ± 699
Total number of commuters	$N_C =$	$N_W - N_P$	—	1,882	3,370[d]
Total number of hyenas represented by wet season commuters	$N_R =$	$N_C/(1 - p_W)$[c]	—	2,413	4,321[d]
Total source population	$N_S =$	$N_R + N_P$	2207 ± 120	3,306 ± 432	5,214 ± 828[d]

Note: Calculations (means ± SE where applicable), based on data from table 16.A2, are described in detail in appendix 16.2.

[a]Block I of 1977 census only.

[b]p_D set as 0.56 (the proportion of clan members temporarily inside their clan territory during medium prey abundance, table 16.1)

[c]p_W set as 0.22 (the proportion of clan members temporarily inside their clan territory during low prey abundance, table 16.1)

[d]Assuming the total number of plains residents to equal the number in 1977 (see appendix 16.2 for discussion).

significant ($t = 2.04$, $P < .025$, one-tailed). Our detailed demographic records of clans in the central study area (part of the source population) between 1987 and 1992 suggest that the population has remained constant or declined slightly since 1986 (see below), so that the 1986 estimate is a maximum estimate of the size of the source population in 1992.

We can estimate the total number of clans in the current source population by dividing total population size by the average group size (45 adults and subadults, Hofer and East 1993a). Similarly, the total area occupied by clan territories in the source population is the product of the number of clans and average territory size (56 km^2, Hofer and East 1993a). We used computer simulations (appendix 16.2) in which the key parameters were varied to (1) obtain robust estimates of the dependent variables and (2) explore the importance of variation in key input parameters. The result of 10,000 simulations produced an estimate of 123 ± 37 (± SD) clans in the source population with territories covering a total of 6,900 ± 2,400 km^2. This is equivalent to 53 ± 19% of the area of Serengeti National Park or 29 ± 10% of the ecosystem. If the area of the short-grass plains covered by the 1986 census (2,364 km^2) is added, then the total area occupied measures 8,100 km^2 (63% of the area of Serengeti National Park, or 34% of the ecosystem). A sensitivity analysis (appendix 16.2) demonstrated that population density (group size divided by size of clan territory) and the proportion of clan members temporarily at home were important parameters, whereas the number of plains residents was not.

These estimates do not consider the possibility of differences in clan territory size between habitats, and they ignore dispersing, nomadic males. There is evidence that group sizes of clans in the woodlands in the north may be smaller than in the central study area from which our estimates are derived (Hofer and East 1993a). Also, human interference may have depopulated parts of the central, northern, and western areas of the Serengeti (Hofer, East, and Campbell 1993); these may not therefore be covered by a continuous mosaic of territories as is the central study area. Both factors suggest that the estimates of the percentage of the Serengeti covered by the source population are minimum estimates. Hence, the source population probably constitutes the majority of the total hyena population of the Tanzanian portion of the ecosystem, and the short-grass plains contain a sizable portion of the entire Serengeti population during the wet season.

Because of dense woodlands and persecution by humans (see below), woodland hyenas in the north and the west probably live at lower densities than do those in the censused population. In accordance with Kruuk (1972), our informed guess is that the remaining area probably contains an additional 2,000–2,500 spotted hyenas, giving a total of 7,200–7,700 hyenas for the Tanzanian part of the ecosystem.

ADULT MORTALITY

Kruuk (1972) thought that the Serengeti population was stable during the 1960s. He further concluded from the age distribution of Serengeti hyenas that their mortality rate must be lower than that of the hyenas in the Ngorongoro Crater. We estimated adult mortality rates (expressed as annual rates) of clan members from the central study area separately for each dry and wet season during the period 1987–1991 (Hofer, East, and Campbell 1993). Mean wet season mortality was 6.2 ± 2.8% (mean ± SEM) for females and 9.7 ± 6.8% for immigrant resident males. Mean dry season mortality rates for this period, however, at 19.8 ± 5% (females) and 22.7 ± 4.5% (males), were 2.5–3 times higher than wet season mortality rates, giving an average annual rate of 13 ± 3.4% for females, 15.6 ± 4.5% for males, and an overall mortality rate of 14.3 ± 2.7%.

Three factors argue against an increase in *natural* mortality from wet season to dry season. First, mean commuting distances in the dry season are less than half of wet season distances, so the increase cannot be due to an increase in commuting effort. Second, herds of migratory gazelles are present inside the territories throughout much of the dry season (table 16.1), and thus prey abundance in the territories is higher than during the wet season; hence food stress cannot be responsible for the increased mortality. Third, it is unlikely that lack of water is a source of mortality because (1) dry season commuters travel through a region with several permanent rivers, and (2) if water is a problem, then dry season rainfall should be negatively correlated with dry season mortality. Instead, we found that dry season mortality *increased* exponentially with dry season rainfall (Hofer, East, and Campbell 1993). This is the expected pattern if mortality is caused by the killing (mostly incidental snaring) of hyenas by game meat hunters who snare the migratory herds at the dry season commuting destinations (Hofer, East, and Campbell 1993). (There is no game meat hunting on the short-grass plains during the wet season). Because the movements of migratory herds depend on dry season rainfall, we hypothesized that as dry season rainfall increases in a particular area, migratory herds are attracted to and utilize that area, prompting an increase in game meat hunting effort, and that accordingly more hyenas are inadvertently (through snares set for wildebeest) or intentionally (through poisoning around hunter camps) killed. Direct evidence from radio-collared individuals confirmed that commuting hyenas are killed by game meat hunters (Hofer, East, and Campbell 1993). Game meat hunters have killed commuting hyenas since the early 1960s (Kruuk 1972; Schaller 1972), but apparently only since the mid-1970s has game meat hunting expanded rapidly as more people moved within walking distance of the boundaries of protected areas north and west of the Serengeti (Campbell and Hofer, chap. 25).

Other sources of mortality are (1) road accidents, as hyenas prefer to commute along major access roads and frequently lie in puddles on the verge of tracks or stay in culverts (Geigy and Boreham 1976); (2) predation by lions, usually at kills contested by both species (Kruuk 1972; Schaller 1972); and (3) violent encounters between hyenas at kills or in clan wars (van Lawick and van Lawick-Goodall 1970; Kruuk 1972; Henschel and Skinner 1991). Although hyenas act as hosts to trypanosomes, a role as a transmission agent has been discounted (Baker 1968; Geigy, Mwambu, and Kauffmann 1971; Bertram 1973; Geigy and Kauffmann 1973; Geigy et al. 1973; Beglinger, Kauffmann, and Müller 1976). Serengeti adult hyenas have been found with antibodies against rabies, canine herpes, canine distemper, canine brucella, canine adenovirus, canine parvovirus, feline calysi, leptospirosis (Hofer and East, unpub.) as well as bovine brucella, rinderpest, and anaplasmosis (Sachs and Staak 1966), but there is no evidence that they succumb to these pathogens.

REPRODUCTIVE EFFORT AND JUVENILE SURVIVAL

Spotted hyena cubs are born in litters of one or two (rarely three: Frank, Glickman, and Licht 1991) in a private birth den and transferred at a mean age of 14 days to the communal den (East, Hofer, and Türk 1989). Serengeti cubs are stationed at the communal den for a median of 12 months (Hofer and East 1993c) and are not weaned before the age of 12–18 months (H. Hofer and M. East, unpub.). The milk of spotted hyena females in the Serengeti has the highest protein content (mean 14.9%) recorded for any fissiped carnivore, a fat content (mean 14.1%) exceeded only by that of palearctic bears and sea otters *(Enhydra lutris),* and a higher gross energy density (mean 9.70 kJ/g) than that of most fissiped carnivores, including all species in the Serengeti (H. Hofer and M. East, unpub.). The spotted hyena's long period of cub dependence, prolonged lactation period, and highly nutritious milk are probably part of a suite of coevolved traits that have led to the highest maternal energy output of any fissiped carnivore (Oftedal and Gittleman 1989). High parental investment of this kind is predicted by some models that link elevated parental investment with extensive environmental variability (McGinley, Temme, and Geber 1987; Schultz 1991). We would therefore expect cub survival to be high compared with that of other carnivores in the Serengeti.

We recorded survival of cubs and subadults in five clans for a total of 218 cubs. We assessed cub survival for the following periods: (1) the first 30 days (n = 51), when cubs are transferred from the birth den to the communal den and the mother is in continuous attendance (Hofer and East 1993c); (2) 31–90 days (n = 86), when cubs do not emerge from the den unless called out by their mother; (3) 91–365 days (n = 86), when cubs emerge from the den by themselves, increasingly venture farther from the den but are still stationed there, and begin to take solids but are still dependent on milk; (4) 365 days to weaning (n = 66), when cubs, while still being suckled, accompany their mothers on commuting trips and increase their intake of solids; (5) weaning to 24 months (n = 52), when cubs become independent; (6) 24–36 months (n = 18), when female subadults become sexually mature; males disperse at this age and were not considered. Weaning was defined as the time when the cub was observed to suckle for the last time. It was possible to determine cub age on the basis of pelage, size, locomotory abilities, and behavioral development, and in one case because the birth of a cub was witnessed (Pournelle 1965; Golding 1969; Kruuk 1972; East, Hofer, and Türk 1989). Cubs were included in a sample for a given period if (1) they were alive and known to us at an age younger than the beginning of the period, and (2) sufficient time had elapsed that we could have witnessed their survival to the end of the period. Cubs were included in the evaluation of the first period if the age at which they were first observed was less than 30 days. Cub mortality during the first 30 days was probably underestimated because cubs first emerged at approximately 7 days, and stillbirths and early mortality would have gone unnoticed. Furthermore, most cubs were first sighted at the communal den, and thus some mortality in the birth den due to siblicide (Frank, Glickman, and Licht 1991; Hofer and East 1992) would have been missed.

We observed a dramatic decline in mortality after the first 3 months, and a more shallow decline during later periods (fig. 16.3a; approximate 95% confidence limits calculated according to Snedecor and Cochran

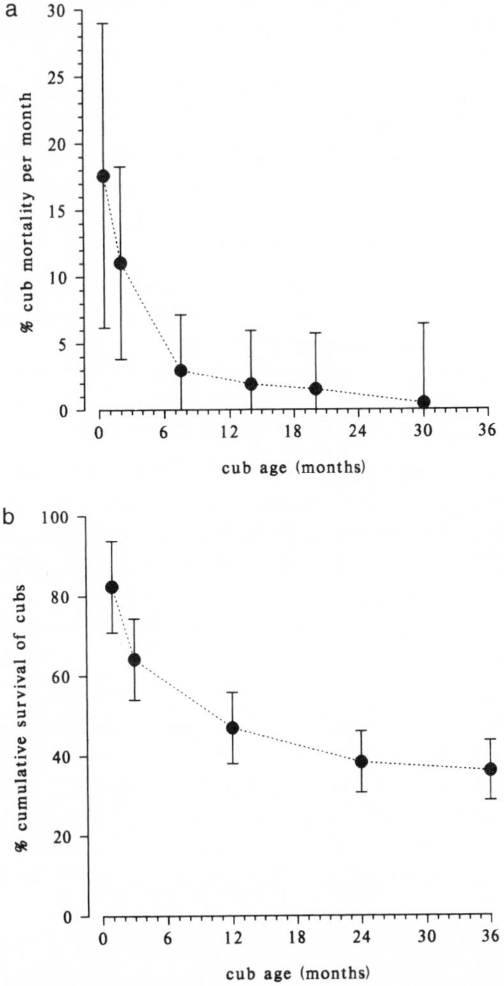

Figure 16.3 (*a*) Percent cub mortality per month (data plotted at the center of each age period: 0–1, 1–3, 3–12, 12–weaning, weaning–24, 24–36 months). (*b*) Cumulative cub survival (data plotted at the end of each period: 0–1, 1–3, 3–12, 12–24, 24–36 months). Estimates plus approximate 95% confidence limits are given. 0–24 months, both sexes combined; 24–36 months, females only.

1980). Mortality did not increase after weaning. Since Serengeti hyena cubs' transition to independence is gradual, mortality is unlikely to increase at that time, as it does in other species in which young animals still need to develop predatory skills after abrupt transitions to independence (e.g., cats, *Felis catus:* Tan and Counsilman 1985; pipistrelle bats, *Pipistrellus pipistrellus:* Racey and Swift 1985). Cumulative cub survival was high, with 47.0% surviving to 1 year, 38.4% surviving to 2 years, and 36.2% surviving to 3 years (fig. 16.3b). Mortality of female subadults during the period 24–36 months at 5.6% was slightly lower than adult female mortality during the wet season (see above).

Sources of cub mortality were diverse. Intraspecific social factors in the form of observed and presumed infanticide by adult clan members (H. Hofer and M. East, unpub.), and observed and presumed siblicide (see Frank, Glickman, and Licht 1991) were the most important known source of mortality (table 16.3). Siblicide usually took the form of the dominant cub of a twin litter consistently restricting the subordinate cub's access to the mother and the subordinate subsequently starving to death, even though the mother was regularly present at the communal den. In contrast, cubs listed as starved in table 16.3 starved to death because the mother failed to return to the communal den. In all but two cases the mothers were known to have died, the majority of them killed by game meat hunters during the dry season. Observed and presumed predation occurred at the communal den, mostly due to attacks by coalitions of male lions, and while mothers transferred cubs between dens (East, Hofer, and Türk 1989; Hofer and East 1993a). Several cubs died when communal dens collapsed after heavy rainstorms. In late 1993/early 1994 several cubs below 6 months of age died due to canine distemper (Hofer and East, unpub.). There was a significant association between the source of mortality and the age at death ($F_{3,40} = 3.57$, $P = .02$; table 16.3). The mean ages in table 16.3 must be viewed with caution because the source of mortality of cubs dying early is more likely to be missed, and hence ages at death in cases with known sources of mortality were biased toward older ages (median age at death of all cubs 135 days, $n = 88$; of cubs with known source of mortality 174 days, $n = 44$).

RECRUITMENT AND POPULATION DYNAMICS

We estimated rates of recruitment into the adult population as the number of males attaining 2 years of age and the number of females attaining 3 years of age during each dry and wet season as a percentage of the number of adult females and immigrant resident males present in the natal clan at the beginning of each season. For females, these rates are equivalent to the recruitment rate into the *breeding population;* for males,

Table 16.3 Sources of cub mortality and age at death in Serengeti spotted hyenas.

	Cubs<365 days			All cubs			
		% mortality			% mortality		
Source of mortality	*N*	All deaths	Known sources	*N*	All deaths	Known sources	Age at death (days±SEM)
Intraspecific/social	16	21.6	42.1	16	18.2	35.6	151±26
Interspecific/predation	8	10.8	21.1	9	10.2	20.0	241±63
Starvation	8	10.8	21.1	14	15.9	31.1	287±45
Environment/disease	6	8.1	15.8	6	6.8	13.3	88±15
No information	36	48.7	—	43	48.9	—	—
Total	74	100	100.1	88	100	100	—

these rates ignore mortality during dispersal. The results are shown in table 16.4. For 2-year-olds, annual recruitment rates did not vary significantly between males and females (n = 4 years for each sex, Mann-Whitney U = 5, NS), although the value for females appears to be lower. Recruitment rates did not vary between wet and dry seasons for females (3-year-olds, Mann-Whitney U = 21.5, n_1 = 6, n_2 = 6, NS) but were significantly higher during the wet season for males (Mann-Whitney U = 3, n_1 = 6, n_2 = 5, P = .028, two-tailed). As there was no difference between the number of cubs of either sex born during wet and dry seasons (males: Mann-Whitney U = 45.5, n_1 = 9 seasons, n_2 = 8 seasons, NS; females: U = 38.5, n_1 = 9, n_2 = 8, NS), the seasonal variation in male recruitment rates may be due to differential mortality after weaning during the wet season prior to the recruitment dry season. Prey abundance during the wet season was low (table 16.1), and individuals commuted longer distances during that time. Competition for food and potentially lethal encounters with lions may be higher on the short-grass plains because the olfactory, auditory, or visual location of kills is easier there than at the dry season commuting destinations in wooded areas. There was no evidence that the distribution of birth dates was seasonal, as cubs were born throughout the year (fig. 16.4).

We compared annual rates of recruitment with annual rates of adult mortality for females to assess the degree and direction of change in the size of the female breeding population in the central study area. The mean population change between 1987 and 1991 was 10.6%–13.0%—an annual decline of 2.4%. As 68.7% of dry season mortality was estimated to be caused by game meat hunters, we also calculated the potential rate of increase of the hyena population by excluding mortality due to game meat hunters. Assuming "natural" mortality to be equivalent to wet season mortality (for a justification of this assumption see discussion above and in Hofer, East, and Campbell 1993), the rate of increase, corrected

Table 16.4 Rates of recruitment (expressed as annual rates) of 2-year-old males and 2-year-old and 3-year-old females into the adult population of their natal clan.

Season	Males	Females 2-year-old	Females 3-year-old
Wet	0.467±0.117 (6)	0.210±0.085 (6)	0.085±0.043 (6)
Dry	0.141±0.071 (5)	0.142±0.037 (6)	0.128±0.046 (6)
P value[a]	0.028	NS	NS
Total	0.318±0.085 (11)	0.176±0.045 (12)	0.106±0.031 (12)
Annual	0.238±0.078 (4)	0.147±0.036 (4)	0.099±0.046 (4)

Note: Means ± S.E.M.; sample sizes (number of seasons or years) in parentheses.
[a]Intrasexual comparisons between seasons, Mann-Whitney *U* test.

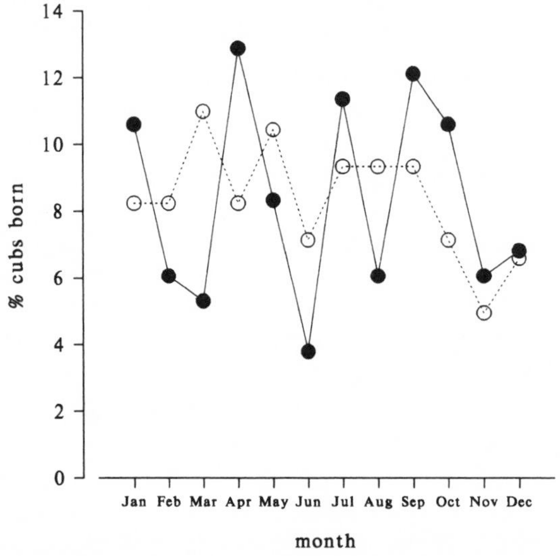

Figure 16.4 Monthly cub production of Serengeti spotted hyenas in 1964–1968 (open circles; data from Kruuk 1972, appendix D.4, $n = 182$) and 1987–1991 (solid circles, this study, $n = 264$). The two distributions are similar (Kolmogornov-Smirnov test, $d_{max} = 0.333$, NS).

for cub losses due to snaring of mothers, would be 10.8%–6.2%, or an increase of 4.6%. These estimates suggest that game meat hunters reduced the annual rate of population increase by as much as 7%.

The effect of human-induced mortality was also evident from a shift in the age distribution of dead animals. The age distribution of individuals that died between 1987 and 1991 was significantly different from the age distribution of skulls of Serengeti hyenas collected by Kruuk (1972) in 1966–1969 (Hofer, East, and Campbell 1993), a time when presumably the impact of game meat hunters was below its current level. Between 1987 and 1991 relatively more medium-aged individuals died than were represented among skulls in the 1960s.

PREDATOR-PREY DYNAMICS

Fryxell, Greever, and Sinclair (1988) developed a model to show that populations of migratory ungulates can escape regulation by predators but that resident ungulates will usually be regulated by predators. Their model assumed that predator populations were limited by resident prey because (1) migratory ungulates were thought to "escape" predation by removing themselves from predators and (2) predators were thought to be restricted in their movements due to the immobility of carnivore offspring during the first few months of life and their long nutritional dependence (Mech 1970; Schaller 1972; Sunquist and Sunquist 1989). In this section we use data from large carnivores in African ecosystems dominated by migratory ungulates to assess prey preferences and test the two underlying arguments of Fryxell et al.'s assumption. We ask: (1) Is the space use of predators restricted, permitting migratory ungulates to remove themselves from predators? (2) Does parental care restrict predator movements? (3) Are spotted hyenas limited by resident prey? (4) Do resident ungulates suffer heavy predation pressure—that is, is the proportion of any resident species in the predator's diet higher than its proportional availability in the environment? (5) Do migratory ungulates suffer reduced predation pressure—that is, is the proportion of any migratory species in the predator's diet lower than its proportional availability in the environment?

Movements

Table 16.5 summarizes data on movements from radiotelemetry studies and life history parameters for the five large African predators: African wild dog, cheetah, leopard, spotted hyena, and lion. All the predator populations have a nomadic segment, typically males, that usually moves with migratory herds. Among the resident segment of predator populations, movements and range sizes in the different ecosystems reflect the temporal and spatial scale of the movements of migratory herds and the availability of alternative prey. In social predators, however, the maintenance of group ranges may involve factors other than those directly associated with feeding ecology, and species-specific social factors may be in conflict with and override the ability of predators to exploit migratory prey (e.g., infanticide and territory defense in Serengeti male and female lions: Packer and Pusey 1983; Packer et al. 1988; mating territories in Serengeti male cheetahs: Caro and Collins 1987a,b). In many cases, therefore, migratory ungulates do not move out of reach of their predators, and in those cases in which they do it is not because predators are unable to follow them (perhaps with the exception of leopards), but because social factors prevent the predators from doing so. Consequently, if the importance of social

Table 16.5 Life history traits, ranging patterns, and feeding ecology of carnivores in three African ecosystems with substantial populations of migratory herbivores.

Species	Ecosystem[a]	Spatial organization[b]	Movements
African wild dog (20 kg)	Serengeti	Large group ranges or nomadic (1,500–2,000 km²)	Coincide with the movements of migratory herds
Cheetah (38 kg)	Serengeti	Large ranges (800 km²); nomadic or mating territories	Following movements of principal prey
	Southern Kalahari	Large ranges (>300 km²)	Opportunistic?
Leopard (38 kg)	Serengeti	F: small ranges (14–60 km²); M: small ranges (ca. 60km²)	Restricted to range; needs vegetation cover
	Southern Kalahari	Large ranges (200–400 km²)	Opportunistic
Spotted hyena (55–70kg)	Serengeti	Defended clan territories (56 km²)	Commuting system to migratory herds
	Savuti (Okavango)	Defended clan ranges (>100 km²)	Occasional visits to neighboring ranges
	Southern Kalahari	Defended clan territories or nomadic (553–1,776 km²)	Movements restricted to territory
Lion (135 kg)	Serengeti	F: pride territory or nomadic; M: mating territory or nomadic (both 20–100 km²)	Extensive movements coinciding with movements of migratory prey outside pride territory possible; plains prides show seasonal range shifts to woods
	Savuti	Large ranges	Range expansion coincides with movements of prey
	Central Kalahari	Pride range or nomadic (700–3,900 km²)	Seasonal range expansion coincides with movements of prey
	Southern Kalahari	Pride range or nomadic	?

[a]Savuti: dry and wet season migrations of zebra and tsessebe, high density of resident herbivores; Kalahari: infrequent wildebeest migration, low density of resident herbivores.
[b]F, females; M, males.

factors in a predator species varies between ecosystems, then its movement patterns and effect on prey might change accordingly.

Life History Parameters

We recognized three stages of cub development (table 16.5): (1) the age at which the period of restricted cub movements terminates—this is usually the end of the denning period or period in a lair, when cubs can follow adults; (2) weaning age; and (3) the age of independence, when individuals are able to hunt by themselves and are nutritionally independent of their mothers or other caretakers. In all predators nutritional de-

Table 16.5 (continued)

Principal prey[c]	Life history stages[d] 1	2	3	Source[e]
Thomson's gazelle,[M] wildebeest,[M] zebra[M]	2.5–3	2–3	14	1
Thomson's gazelle[M]	1.5	3	13–17	2
Springbok[M], steinbok[R]	—	—	—	3
Thomson's gazelle,[M] reedbuck[R]	2	3	13–20	4
Springbok,[M] kongoni[M]	—	—	—	5
Wildebeest,[M] Thomson's gazelle,[M] zebra[M]	12	12–18	>18	6
Zebra,[M] impala[R]	—	—	—	7
Gemsbok,[R] wildebeest[M]	6–10	15	18	8
Wildebeest,[M] Thomson's gazelle,[M] zebra[M]	12	12–18	>18	6
Warthog,[R] tsessebe[M]	—	—	—	10
Gemsbok,[M] springbok[M]	—	—	—	11
Wildebeest,[M] Gemsbok,[R] springbok[M]	—	—	—	12

[c]Principal prey in kill statistics ordered by frequency of occurrence. Prey species: gemsbok (*Oryx gazella*), springbok (*Antidorcas marsupialis*), tsessebe (*Damaliscus lunatus*), for others see Appendix A. M, migratory; R, resident.

[d]1, end of period of restricted cub movements; 2, weaning age; 3, age of independence (all in months).

[e]1: Kruuk and Turner 1967; Frame et al. 1979; Malcolm and Marten 1982; 2: Frame 1984; Caro and Collins 1987b; Durant et al. 1988; Caro 1989; 3: Labuschagne 1979; Mills 1984; 4: Kruuk and Turner 1967; Schaller 1972; Kingdon 1977; Bertram 1978, 1982; 5: Bothma and Le Riche 1984; Mills 1984; 6: Kruuk 1972; Hofer and East 1993c, unpubl.; 7: Cooper 1989, 1990; 8: Mills 1990; 9: Kruuk and Turner 1967; Schaller 1972; Bertram 1975; Packer et al. 1988; C. Packer, pers. comm.; 10: Viljoen 1993; 11: Owens and Owens 1984; 12: Eloff 1984; Mills 1984.

pendence extends well beyond the period of restricted cub movements, and each of these traits shows little allometric dependence. Life history traits of these carnivores must have evolved along two diverging paths: in all species except spotted hyenas, the period of restricted movement is short (ca. 2 months) and within the time limits imposed by the seasonal movements of migratory herbivores. In spotted hyenas, this period extends well beyond these limits, and weaning age is much closer to the age of nutritional independence. Thus, in this sample, life history traits do not constrain the exploitation of migratory prey by predators, with the exception of spotted hyenas.

Are Serengeti Hyenas Limited by Resident Prey?

Spotted hyena populations with clan sizes similar to that found in the Serengeti (Ngorongoro Crater, 54: Kruuk 1972; Masai Mara, 52: Frank 1986a; Savuti, 42: Cooper 1989) live in areas with densities of resident herbivores that exceed those in the Serengeti by a factor of five or larger (Hofer and East 1993a). Spotted hyena populations that live in areas with densities of resident herbivores similar to that of the Serengeti, however, live in clans of much smaller sizes (southern Kalahari, 8: Mills 1990; Kruger, 11: Henschel and Skinner 1991; Etosha, 21: Gasaway et al. 1989). This indicates that the ability of Serengeti spotted hyenas to exploit migratory herbivores throughout the year by commuting permits clan sizes in the Serengeti to exceed the carrying capacity of the territory (Hofer and East 1993a). Spotted hyenas in the Serengeti are therefore not limited by resident prey.

Prey Preferences

The effect of predators on resident and migratory ungulates depends on the functional response and prey choice of the predators. Fryxell, Greever, and Sinclair (1988) assumed a type III functional response and modeled prey choice according to the rule, "exploit the migratory prey when it is the most abundant prey species; if it declines below the abundance of alternative prey, switch to the alternative prey." This rule gives migratory species a reprieve from predation when they are less abundant than resident prey. Data on the two to three most important prey species in the diets (kills) of predators are summarized in table 16.5, while the results of statistical tests on prey preference, expressed as a significant deviation from the null model of a predator killing a species in proportion to its availability in the environment, are listed in table 16.6. Weaker predation pressure on migratory ungulates than assumed by the model occurs if there is positive preference for resident ungulates if they are not the most common prey, or negative preference for migratory ungulates if they are the most abundant prey. Stronger predation pres-

Table 16.6 Feeding preferences of carnivores in three African ecosystems with substantial populations of migratory herbivores.

		Evidence in favor of "prey escape" hypothesis: Resident prey taken more often than expected; migratory prey taken less often than expected				Evidence against "prey escape" hypothesis: Resident prey taken less often than expected; migra tory prey taken more often than expected		
Ecosystem		Predator	Prey	G test *P* value		Predator	Prey	G test *P* value
Kalahari	R	Hyena	*Gemsbok*	< .0001				
		Lion	*Gemsbok*	< .005				
	M	Hyena	*Wildebeest*	< .0001	M	Cheetah	Springbok	< .0001
		Leopard	*Wildebeest*	< .002				
		Cheetah	*Wildebeest*	< .0001				
		Lion	Eland	.043				
		Cheetah	Kongoni	.033				
Savuti	R	*Hyena*	Warthog[a]	< .0001	R	*Hyena*	*Impala*[b]	.009
		Lion	Warthog[a]	< .0001	M	*Hyena*	*Zebra*[b]	.001
Serengeti	R	Wild dog	Warthog	< .0001	R	*Hyena*	*Impala*	< .0001
		Lion	Reedbuck	< .009		Lion	*Impala*	< .0001
		Leopard	Reedbuck	< .0001		Cheetah	*Impala*	< .0001
						Leopard	*Impala*	< .0001
						Wild dog	*Impala*	< .0001
						Cheetah	Topi	.043
						Lion	Kongoni	< .0001
						Cheetah	Kongoni	.026
						Leopard	Warthog	< .03
	M	Cheetah	*Wildebeest*	< .0001	M	*Hyena*	*Wildebeest*	< .0001
		Leopard	*Wildebeest*	< .0001		*Hyena*	Th. gazelle	.04
		Hyena	Zebra	< .0001		Lion	Th. gazelle	< .0001
		Lion	Zebra	< .002		Cheetah	Th. gazelle	< .0001
		Cheetah	Zebra	< .0001		Leopard	Th. gazelle	< .0001
		Leopard	Zebra	< .0001		Wild dog	Th. gazelle	< .0001
		Wild dog	Zebra	< .0001		Leopard	Gr. gazelle	< .0001
						Wild dog	Gr. gazelle	< .0001

Sources: Mills (1990: tables 2.14, 2.18, 2.21, 2.22, appendix B); Kruuk (1972: tables 11, 19); Schaller (1972: tables 36, 63, 66); Cooper (1990); Viljoen (1993); Craig (in press).
Note: Expected proportion in diet (kills only) based on absolute population size of prey populations. Most abundant carnivore, resident prey, and migratory prey per ecosystem are given in italics. R, resident prey; M, migratory prey.
[a]The aerial census results for warthog are thought to grossly underestimate the actual warthog population, inflating significance of test (Bart Vandepitte, pers. comm.)
[b]Calculated by excluding warthog from total prey population.

sure on migratory ungulates than assumed by the model occurs if there is positive preference for less abundant migratory ungulates, or negative preference for resident ungulates if they are the most common prey. Stronger predation pressure on migratory ungulates than assumed by the model *might* occur if there is negative preference for resident ungulates if they are not the most abundant ungulate, as this depends on the slope of the functional response (right section of table 16.6).

The most abundant herbivores in the Serengeti and southern Kalahari are migratory wildebeest, while in Savuti the most common herbivores are resident impala (Craig, in press). Migratory ungulates are the most important prey (with two exceptions), although not necessarily the most common species (cf. table 16.5 with table 16.6). Prey preferences of predators in the southern Kalahari fit the assumptions of Fryxell et al.'s model well, and the model's conclusions can be applied to this system. In an independent qualitative assessment of predator-prey dynamics in the southern Kalahari, Mills (1990) concluded that the predator community was probably limited by resident ungulates. On the other hand, prey choice by spotted hyenas in Savuti clearly violated the assumptions of the model. Evidence from the Serengeti is mixed. All predators in the studies considered had a negative preference for impala, the most common resident ungulate, and a positive preference for Thomson's gazelle, then the third most abundant migratory ungulate. Zebra, the second most abundant ungulate, were taken less often than expected. Impala may have been underrepresented as their main habitat, woodland, was probably underrepresented in the studies. The results suggest that several migratory and resident ungulates may interact in complex ways with predators, which results in considerable variation in the population sizes of different species of migratory and resident ungulates.

Discussion

The ranging patterns of predators can be modified in many cases to cope with the movements of migratory prey, and life history traits should not prevent any carnivore except the spotted hyena from exploiting migratory herbivores. The commuting system in the Serengeti, however, permits the spotted hyena to combine a long denning period with the exploitation of migratory ungulates throughout the year. In particular, the plasticity of the response of territory owners to intruders may be viewed as an adaptive trait of the Serengeti population designed to facilitate the operation of the commuting system (see Hofer and East 1993b). For Serengeti spotted hyenas we can therefore reject Fryxell et al.'s assumption that they are limited by resident prey. For other predators, the two arguments used to justify this assumption are not substantiated. Cub immobility does not restrict predators, and in many cases migratory ungulates are unable to move out of reach of their predators. If predators other than Serengeti spotted hyenas are limited by resident prey, it is usually not due to these two factors; however, it may be related to social factors and patterns of prey choice caused by factors such as hunting specializations. The data on prey preferences reveal a complex pattern within ecosystems and considerable variation in the same predator species between ecosystems, restricting the usefulness of general models with simple prey preference rules. The data on prey preferences also suggest that it cannot be assumed that resident ungulates experience heavy predation while migratory ungulates do not.

Fryxell et al.'s conclusion may be restated as the hypothesis that the natural rate of increase of predators restricts them to population sizes insufficient to exert a regulatory effect on migratory ungulates. A regulatory effect was defined by Fryxell, Greever, and Sinclair (1988) as any density-dependent effect that tends to stabilize population numbers over time. Ideally this hypothesis would be tested by comparing rates of natural increase of predators, resident prey, and migratory prey. We can test it indirectly by asking whether the reproductive rate of spotted hyenas in the Serengeti, a system dominated by migratory ungulates, is lower than that in Ngorongoro, a system dominated by resident ungulates. We compared data on (1) interbirth intervals of females that had raised cubs successfully and (2) recruitment rates from our Serengeti study with data provided by Kruuk (1972) for Ngorongoro. The mean interbirth interval in the Serengeti of 21.7 ± 0.85 months (± SEM, $n = 21$; data from our main study clan collected over 6 years) was not significantly different from the mean of 18.8 ± 2.3 months ($n = 6$) recorded for Ngorongoro (Mann-Whitney $U = 87$, NS). Similarly, the average annual recruitment rate of 9.9 ± 4.6% (table 16.4) for the Serengeti was not different from the average rate of 13.5 ± 1.3% ($n = 2$) for Ngorongoro. Because sample sizes are small, the results must be viewed with caution, but they suggest that rates of increase of predators in systems dominated by migratory ungulates are not necessarily smaller than in those dominated by resident ungulates.

CONCLUSIONS

The two most important factors currently affecting the Serengeti spotted hyena population are (1) the removal of 8% of breeding females per year through snaring and poisoning by game meat hunters and (2) social factors (infanticide and siblicide). The enormous increase in the wildebeest population has permitted hyena numbers to more than double between 1968 and 1986. At the same time recruitment rose from 5.2% to 9.9%. One explanation for this increase might be that the increase in the population of migratory herbivores permitted a higher net energy intake per commuting trip, shorten those trips, and improved the chances of survival of young. The historical perspective suggests that the key factor limiting the hyena population was food supply, but unlike many other carnivores and birds (Newton 1979; Lindström 1986; Brown 1987; Moehlman 1989), spotted hyenas are not limited by the abundance of resident prey. Two previously championed arguments (Bergerud 1988; Fryxell and Sinclair 1988; Fryxell, Greever, and Sinclair 1988) suggesting that predators ought to be limited by resident ungulates were not well supported by our data: migratory herbivores do not necessarily escape predators, nor does cub immobility restrict predator movements. Our study showed that interactions between people and wildlife at the periphery may affect wildlife *throughout* a protected area, and that detailed knowledge of the social and spatial organization of a species is important for accurate assessment by park management of the impact of people on wildlife.

ACKNOWLEDGMENTS

We are grateful to the Tanzania Commission of Science and Technology for permission to conduct the study; the Directors General of the Serengeti Wildlife Research Institute and Tanzania National Parks, and the Conservator of the Ngorongoro Conservation Area Authority for cooperation and support; the Fritz-Thyssen-Stiftung, the Stifterverband der deutschen Wissenschaft, and the Max-Planck-Gesellschaft for financial assistance; and J. vom Baur, R. Burrows, D. Bygott, K. L. I. Campbell, J. Corlett, W. Golla, G. von Hegel, P. Heinecke, S. Huish, M. Jago, R. Klein, B. Knauer, P. D. Moehlman, W. Mohren, D. Schmidl, U. Seibt, A. Türk, H. Wiesner, and W. Wickler for information, assistance, and advice. We are particularly grateful to C. Trout for his enthusiastic, professional, and accurate aerial radio tracking of the hyenas. Stimulating discussions with H. Kruuk and G. W. Frame and comments by P. Arcese, S. Durant, L. G. Frank, H. Kruuk, W. Wickler, and an anonymous referee improved the chapter.

APPENDIX 16.1 CENSUSING COMMUTING HYENAS

In standard population censuses, the area occupied by a population is clearly defined, the animals can easily be observed, and the sample area is censused assuming that the result is representative for the total area. Spotted hyenas are difficult to census because they often spend the day in dense vegetation or half-submerged in holes. Furthermore, our study showed that (1) any given area contains a mixture of territory residents and commuters; (2) the ratio of residents to commuters cannot be derived from census results; (3) the ratio of residents to commuters varies between different areas; and (4) the ratio of residents to commuters in an area varies between seasons. Because commuters may have traveled a considerable distance to their current location, the geographic area that incorporates all territories of the censused residents and commuters (the "source area") is likely to extend beyond the censused area. We therefore distinguished the census result (the "census estimate") from the estimate of population size for the population that lives in the source area, which we call the "source population" of the census. This source population, formally defined as the segment of the Serengeti hyena population that commutes to or lives on the calving grounds of the migratory herds, contains the majority of hyenas in the ecosystem.

Censusing Protocol

1. Define as censusing area precisely those parts of the short-grass plains that are occupied by the large migratory herds of wildebeest and zebra during the wet season. The wet season distribution of the large migratory herds is more clearly defined than the dry season distribution, and the habitat is ideal for censusing hyenas. During the wet season, all spotted hyena territorial residents on the plains would forage inside their clan territory (because each clan territory contains migratory herds) and all commuters on the plains would come from territories located *outside* the censusing area. Conduct a first census during the wet season—for example, by driving transects.

2. Conduct a second census in the same censusing area during the following dry season. During the dry season the area would contain short-grass plains residents temporarily present inside their clan territory, and some plains residents in transit to or from commuting destinations, but *no* individual that had commuted to the plains during the wet season. Without the dry season census, population estimates are not possible.

3. Estimate the proportion of clan members commuting and, conversely, the proportion that are temporarily present in the clan territory during the dry (p_D) and wet (p_W) season (i.e., take estimates from table 16.1).

Computation of Estimate.
The total source population N_S is

$$N_S = (p_D \times N_W - p_W \times N_D)/[p_D \times (1 - p_W)],$$

where N_W and N_D are the census estimates from the wet and dry seasons and p_W and p_D are 0.22 and 0.56 (table 16.1). The census estimates are the estimates derived from methods such as transect counts and calculated in the usual fashion (Norton-Griffiths 1978).

Derivation

1. The total number N_P of short-grass plains residents. During the dry season census, some of the plains residents are at commuting destinations outside the short-grass plains; therefore the total number of plains residents N_P is larger than N_H, the number of short-grass plains residents present inside their clan territory at the point of censusing ($N_H = N_P \times p_D$). The dry season census estimate N_D is also larger than N_H because it includes N_T, the number of plains residents just in transit to or from commuting destinations. This number is unknown but is unlikely to be large because commuters travel chiefly at dawn, dusk, or during the night and are therefore less likely to be counted during a census. If we ignore N_T, the resulting error would be small, and we could approximate N_H by N_D, which gives us

$$N_P = N_D/p_D.$$

2. The total number N_R of hyenas represented by commuters on the short-grass plains during the wet season. The number of commuters N_C during the wet season is the difference between the census estimate for the wet season N_W and the number of plains residents, $N_C = N_W - N_P$. Some members of the clans from

which the wet season commuters originate are at home (N_O) and therefore occur outside the censusing area during the wet season census. Thus, the total number N_R of hyenas *represented* by the commuters during the wet season census is larger than N_C, the number of commuters, that is, $N_R = N_O + N_C$. With $N_O = N_R \times p_W$ (following the same line of argument as for $N_H = N_P \times p_D$) we get

$$N_R = N_C/(1 - p_W).$$

3. The total source population N_S. The total source population N_S is the number of hyenas represented by commuters on the short-grass plains during the wet season plus the short-grass plains residents, $N_S = N_R + N_P$. If N_R and N_P are replaced by the equivalent expressions from above, we arrive at a formula that requires only the census estimates for the wet and dry season and two constants:

$$N_S = (p_D \times N_W - p_W \times N_D)/[p_D \times (1 - p_W)].$$

APPENDIX 16.2 CONVERTING PREVIOUS CENSUS ESTIMATES

Census estimates and estimates of the size of the source population (see appendix 16.1) were calculated for the censuses of 1966–1968, 1977, and 1986 as follows. We compared maps of the censusing areas of the 1966–1968, 1977, and 1986 counts and assigned Kruuk's (1972, his fig. 11) resighting areas nos. 5, 8, 11, 12, 13, 14, 15, 16, two-thirds of 4, and half of 10 to Block I of 1977; nos. 2, 3, 6, 7, 9, one-third of 1, one-third of 4, and half of 10 to Block III of 1977; and excluded areas 17, 18, 19, 20, 21, 22 and two-thirds of 1. Block II of 1977 was not covered by Kruuk's data. Block I of the 1986 census was identical with Blocks I and II of the 1977 census; Block II of the 1986 census was congruent with Block III of the 1977 census. Details of the number and size of areas censused in 1977 and 1986 are given in table 16.A1.

The overall census estimates for the 1966–1968 period were recalculated from data in Kruuk (1972, table D.3) using Bailey's binomial model of the Lincoln index, the most appropriate model for a situation of random sampling with replacement (Seber 1982). This calculation produced estimates slightly different from those listed in Kruuk (1972). The new estimates are listed in table 16.A2. We calculated the variance of the Lincoln index estimates from the formula given by Seber (1982, 61).

Do the Lincoln index estimates provide a reasonable approximation of the size of the source population as defined in appendix 16.1? This would be the case if Kruuk's study area was equivalent to the area occupied by the source population—that is, if it included the complete set of (1) clan territories from which individuals originated when they were marked by Kruuk during a commuting trip and (2) commuting destinations of those individuals marked inside their clan territories. The following evidence suggests that this is the case:

1. If a substantial number of marked individuals commuted to destinations or originated in clan territories outside Kruuk's resighting areas, then the Lincoln index estimates based on dry season and wet season resightings should differ markedly. This was not the case (Kruuk 1972, 301–2; our table 16.A2). We have chosen the estimate based on wet season sightings (tables 16.A2 and 16.2) as an estimate of the source population, as it makes the statistical assessment of the significance of changes in the source population conservative.

Table 16.A1 Details of strip sample transect censuses of predators on the Serengeti plains in 1977 and 1986.

Census	Season	Area	1977 Block	Size of area (km²) Total	Sampled	Transects N	km	Baseline (km)	# sample units in population
1977	Wet	Short-grass plains	I + II	2077.5	166.2	25	831.0	62.9	313
		Long-grass plains	III	627.5	50.2	14	251.0	35.2	175
		Total		2705.0	216.4	39	1082.0	98.1	488
1977	Dry	Short-grass plains	I + II	2072.5	165.8	25	829.0	62.9	313
		Long-grass plains	III	617.5	49.4	14	247.0	35.2	175
		Total		2690.0	215.2	39	1076.0	98.1	488
1986	Wet	Short-grass plains	I + II	2363.5	189.1	24	945.4	60.2	300
		Long-grass plains	III	596.5	47.7	8	238.6	20.2	101
		Total		2960.0	236.8	32	1184.0	80.4	401

Source: For 1977 censuses recalculated from original data sheets except for transect numbers and lengths (Serengeti Research Institute 1977a,b); data for 1986 census from Campbell and Borner 1986.
Note: Sampling intensity 8% in all cases.

Table 16.A2 Census estimates (mean ± SE) of spotted hyenas on the Serengeti plains, 1966–1986.

Area		1977 Block	Census estimate 1966–1968	1977	1986[a]
Dry season					
Short-grass plains	SNP[b]	I	368 ± 63	450 ± 161	—
Short-grass plains	NCA[c]	II	—	50 ± 23	—
Long-grass plains	SNP[d]	III	—	263 ± 109	—
Total			1987 ± 153	763 ± 196	—
Wet season					
Short-grass plains	SNP[b]	I	—	2688 ± 387	} 4263 ± 699
Short-grass plains	NCA[c]	II	—	88 ± 22	}
Long-grass plains	SNP[d]	III	—	263 ± 87	250 ± 159
Total			2207 ± 124	3038 ± 397	4513 ± 717

Sources: Data for 1966–1968 calculated from Kruuk (1972) as described in appendix 16.2; data for 1977 recalculated from the original data as described in appendix 16.2; data for 1986 from Campbell and Borner (1986).
[a]Block I of 1986 was identical with Blocks I + II of 1977, and Block II of 1986 was equivalent to Block III of 1977.
[b]Mostly inside Serengeti National Park.
[c]Mostly inside the Ngorongoro Conservation Area.
[d]Inside Serengeti National Park.

2. Kruuk (1972) showed that his resighting areas encompassed dry season and wet season commuting destinations of marked individuals as well as their clan territories. For instance, the seasonal variation in resightings of individuals marked on the Musabi plains (Kruuk 1972, 48, fig. 12) indicates that they commuted to the Musabi plains during the dry season, originated from clan ranges south of Mukoma Hill and near the Moru track, and commuted to the short-grass plains during the wet season. Similarly, Kruuk's figures 13–16, read in reverse, suggest that the clan ranges of commuters marked in the wet season as well as the dry season commuting destinations of short-grass plains residents were principally inside his resighting areas.

3. Kruuk (1972) located dens on the short-grass plains during the dry sea-

Table 16.A3 Results of predator censuses on Serengeti plains during the wet and dry season 1977 and the wet season 1986.

Area	Census	Golden Jackals Est	CL	%	Den	Silverbacked Jackals Est	CL	%	Den
short-grass plains	Dry 1977	1125	360	32	0.54	63	72	115	0.03
	Wet 1977	1650	659	40	0.79	425	218	51	0.20
	Wet 1986	1575	378	24	0.67	50	68	136	0.02
			NS			<0	.001		
long-grass plains	Dry 1977	13	25	202	0.02	125	103	82	0.20
	Wet 1977	0	0	—	0	13	25	198	0.02
	Wet 1986	13	29	232	0.02	0	0	—	0
			NS				NS		
total	Dry 1977	1138	342	30	0.42	188	116	62	0.07
	Wet 1977	1650	626	38	0.61	438	208	48	0.16
	Wet 1986	1588	359	23	0.54	50	65	129	0.02
			NS			<0	.001		

Note: Est = Census estimate; *CL* = 95% Confidence Limits; % = 95% Confidence Limits as % of estimate; *Den* = density in individuals/km² and two-tailed *P* values for significance of difference in census estimate between wet season 1977 and 1986.

son, which suggests that year-round residence on the short-grass plains occurred in the 1960s.

Based on Kruuk's (1972, his figures 12–16) and our evidence (see fig. 16.1), the short-grass plains (i.e., Blocks I and II of the 1977 census) are not commuting destinations during the dry season. Resightings of marked individuals in these areas during the dry season therefore indicate strongly that these are observations of plains residents *at home* (appendix 16.1). If we assume that the probability of resighting a marked plains resident at home equals that of resighting an unmarked resident, we arrive at a census estimate of 380 plains residents in Block I (1977) and a total number of plains residents of 679 (see table 16.2).

When we checked the original data sheets of the 1977 censuses, we discovered that the population size estimates published in the original analyses of the 1977 censuses and used by Hanby and Bygott (1979) and in subsequent studies were incorrectly derived. The derivation used Jolly's method 2 for unequal-sized sampling units (Jolly 1969; Norton-Griffiths 1978), a method sensitive to the estimate of the total area censused. The size of the total area censused was calculated by drawing known-area grids measuring 100 km² over the surface of a map with the *designated* census blocks, instead of calculating it directly from the actual transects driven (Serengeti Research Institute 1977a, G. W. Frame, pers. comm.) This caused an overestimate of the total area censused by 11%–12% (slightly different values for different blocks), and population size estimates were thus inflated by 11%–12%. The correct parameters of the 1977 censuses recalculated from the original transect data are listed in table 16.A1, and the corrected census estimates for the other carnivore species, with 95% confidence limits and 95% confidence limits expressed as a percentage of the population size estimate, are listed in table 16.A2.

The drastic decline in the 1977 census estimate from wet season (commuters

Table 16.A3 (continued)

Lions Est	CL	%	Den	Cheetahs Est	CL	%	Den	Bat-eared Foxes Est	CL	%	Den
13	25	199	0.01	13	25	200	0.01	25	50	202	0.01
350	238	68	0.17	225	187	83	0.11	388	275	71	0.19
188	168	90	0.08	63	71	114	0.03	0	0	—	0.00
	NS				<.1			<0	.005		
75	107	142	0.12	63	69	111	0.10	100	104	104	0.16
238	275	116	0.38	0	0	—	0.00	0	0	—	0.00
63	118	188	0.10	0	0	—	0.00	63	145	232	0.10
	NS				—				NS		
88	100	114	0.03	75	67	89	0.03	125	106	85	0.05
588	337	57	0.22	225	178	79	0.08	388	261	67	0.14
250	187	75	0.08	63	68	108	0.02	63	145	232	0.02
	<0.09				<.1				<0.03		

and plains residents) to dry season (plains residents only) provides strong evidence for the commuting system being in full operation during this period. Estimates and variances were calculated following appendix 16.1 and are listed in table 16.2.

Because there was no dry season count during the 1986 census, we had to estimate the size of the plains population for 1986. The most conservative assumption is that the plains population did not increase from 1977 to 1986 (there is no evidence to suggest that it decreased); this assumption provides a conservative estimate of the size of the source population. The 1977 figure of 893 estimated plains residents is equivalent to 20 clans, assuming that the average clan size of 45 adults and subadults recorded between 1987 and 1992 (Hofer and East 1993a) is appropriate. Dry season surveys of the short-grass plains (H. Hofer and M. East, unpub.) suggest that this is a reasonable estimate of the total number of clans today. The census estimates indicate that while the total source population grew from 1966–1968 to 1986, the proportion of plains residents in the source population declined (31% in 1966–1968, 24% in 1977). Using the 1977 number of plains residents for 1986 would imply that the plains residents constituted only 15% of the source population in 1986.

In order to assess how strongly the precise value of the number of plains residents and other parameters affected the estimate of the source population, we varied the key parameters in a computer simulation. We took the mean of 10,000 simulations in which the proportion of commuters temporarily present in a clan territory (0.22, see table 16.1), the clan size (45), the size of the clan territory (56 km^2), the number of hyenas censused on the short-grass plains in 1986 (4,263), and the number of plains residents (500/0.56) were drawn as normally distributed variates independently from each other by a random number generator (GASDEV: Press et al. 1986) with a variance set equal to the variance in the census count of 1986. A subsequent sensitivity analysis indicated that variation in the number of plains residents had little effect on the estimate of the size of the source population.

REFERENCES

Baker, J. R. 1968. Trypanosomes of wild mammals in the neighbourhood of the Serengeti National Park. *Symp. Zool. Soc. Lond.* 24:147–58.

Beglinger, R., Kauffmann, M., and Müller, R. 1976. Culverts and trypanosome transmission in the Serengeti National Park (Tanzania). Part II. Immobilization of animals and isolation of trypanosomes. *Acta Trop.* 33:68–73.

Bergerud, A. T. 1988. Caribou, wolves, and man. *Trends Ecol. Evol.* 3:68–72.

Bertram, B. C. R. 1973. Sleeping sickness survey in the Serengeti Area (Tanzania) 1971. Part III. Discussion of the relevance of the trypanosome survey to the biology of large mammals in the Serengeti. *Acta Trop.* 30:36–48.

———. 1975. Social factors influencing reproduction in wild lions. *J. Zool.* (Lond.) 177:463–82.

———. 1978. *Pride of lions.* London: Dent.

———. 1982. Leopard ecology as studied by radio tracking. *Symp. Zool. Soc. Lond.* 49:341–52.

Bothma, J. D. P., and Le Riche, E. A. N. 1984. Aspects of the ecology and the behaviour of the leopard *Panthera pardus* in the Kalahari desert. *Koedoe* (suppl.) 27:259–79.

Brown, J. L. 1987. *Helping and communal breeding in birds: Ecology and evolution.* Princeton, N. J. : Princeton University Press.

Campbell, K. L. I., and Borner, M. 1986. *Census of predators on the Serengeti plains May 1986.* Serengeti Ecological Monitoring Programme report SEMP-86-2.

Caro, T. M. 1989. Determinants of asociality in felids. In *Comparative socioecology: The behavioural ecology of humans and other mammals,* ed. V. Standen and R. A. Foley, 41–74. Oxford: Blackwell Scientific Publications.

Caro, T. M., and Collins, D. A. 1987a. Ecological characteristics of territories of male cheetahs *(Acinonyx jubatus). J. Zool* (Lond). 211:89–105.

———. 1987b. Male cheetah social organization and territoriality. *Ethology* 74:52–64.

Cooper, S. M. 1989. Clan sizes of spotted hyaenas in the Savuti Region of the Chobe National Park, Botswana. *Botswana Notes Rec.* 21:121–33.

———. 1990. The hunting behavior of spotted hyaenas *(Crocuta crocuta)* in a region containing both sedentary and migratory populations of herbivores. *Afr. J. Ecol.* 28:131–41.

Craig, C. In press. *Wildlife populations in Botswana.* Sarowe, Botswana: Department of Wildlife & National Parks.

Crawley, M. J. 1992. Population dynamics of natural enemies and their prey. In *Natural enemies: The population biology of predators, parasites, and diseases,* ed. M. J. Crawley, 40–89. Oxford: Blackwell Scientific Publications.

Dublin, H. T., Sinclair, A. R. E., Boutin, S., Anderson, E., Jago, M., and Arcese, P. 1990. Does competition regulate ungulate populations? Further evidence from Serengeti, Tanzania. *Oecologia* 82:283–88.

Durant, S. M., Caro, T. M., Collins, D. A., Alawi, R. M., and FitzGibbon, C. D. 1988. Migration patterns of Thomson's gazelles and cheetahs on the Serengeti plains. *Afr. J. Ecol.* 26:257–68.

East, M. L., and Hofer, H. 1991a. Loud-calling in a female-dominated mammalian society: I. Structure and composition of whooping bouts of spotted hyaenas, *Crocuta crocuta. Anim. Behav.* 42:637–49.

———. 1991b. Loud-calling in a female-dominated mammalian society: II. Behavioural contexts and functions of whooping of spotted hyaenas, *Crocuta crocuta. Anim. Behav.* 42:651–69.

East, M. L., Hofer, H., and Türk, A. 1989. Functions of birth dens in spotted hyaenas (*Crocuta crocuta*). *J. Zool.* (Lond.) 219:690–97.

Eloff, F. C. 1984. Food ecology of the Kalahari lion *Panthera leo vernayi. Koedoe* (suppl.) 27:249–58.

Frame, G. W. 1984. Cheetah. In *The encyclopedia of mammals,* vol. 1, ed. D. W. Macdonald, 40–43. London: Allen & Unwin.

———. 1986. *Carnivore competition and resource use in the Serengeti ecosystem of Tanzania.* Ph.D. thesis, Utah State University, Logan.

Frame, L. H., Malcolm, J. R., Frame, G. W., and van Lawick, H. 1979. Social organization of African wild dogs *(Lycaon pictus)* on the Serengeti Plains, Tanzania. *Z. Tierpsychol.* 50:225–49.

Frank, L. G. 1986a. Social organisation of the spotted hyaena *(Crocuta crocuta).* I. Demography. *Anim. Behav.* 35:1500–1509.

———. 1986b. Social organisation of the spotted hyaena *(Crocuta crocuta).* II. Dominance and reproduction. *Anim. Behav.* 35:1510–27.

Frank, L. G., Glickman, S. E., and Licht, P. 1991. Fatal sibling aggression, precocial development, and androgens in neonatal spotted hyenas. *Science* 252:702–4.

Fryxell, J. M., Greever, J., and Sinclair, A. R. E. 1988. Why are migratory ungulates so abundant? *Am. Nat.* 131:781–98.

Fryxell, J. M., and Sinclair, A. R. E. 1988. Causes and consequences of migration by large herbivores. *Trends Ecol. Evol.* 3:237–41.

Gasaway, W. C., Mossestad, K. T., and Stander, P. E. 1989. Demography of spotted hyaenas in an arid savanna, Etosha National Park, South West Africa/Namibia. *Madoqua* 16:121–27.

Geigy, R., and Boreham, P. F. L. 1976. Culverts and trypanosome transmission in the Serengeti National Park (Tanzania). Part I. Survey of the culverts. *Acta Trop.* 33:57–67.

Geigy, R., and Kauffmann, M. 1973. Sleeping sickness survey in the Serengeti Area (Tanzania) 1971. Part I. Examination of large mammals for trypanosomes. *Acta Trop.* 30:12–23.

Geigy, R., Kauffmann, M., Mayende, J. S. P., Mwambu, P. M., and Onyango, R. J. 1973. Isolation of *Trypanosoma (Trypanozoon) rhodesiense* from game and domestic animals in Musoma district, Tanzania. *Acta Trop.* 30:49–56.

Geigy, R., Mwambu, P. M., and Kauffmann, M. 1971. Sleeping sickness survey in Musoma District, Tanzania. Part IV. Examination of wild mammals as a potential reservoir for *T. rhodesiense. Acta Trop.* 28:211–20.

Golding, R. R. 1969. Birth and development of spotted hyaenas *Crocuta crocuta* at the University of Ibadan Zoo, Nigeria. *Int. Zoo Yearb.* 9:93–95.

Hanby, J. P., and Bygott, J. D. 1979. Population changes in lions and other predators. In *Serengeti: Dynamics of an ecosystem,* ed. A. R. E. Sinclair and M. Norton-Griffiths, 249–62. Chicago: University of Chicago Press.

Henschel, J. R., and Skinner, J. D. 1987. Social relationships and dispersal patterns in a clan of spotted hyaenas, *Crocuta crocuta* in the Kruger National Park. *S. Afr. J. Zool.* 22:18–24.

———. 1991. Territorial behaviour by a clan of spotted hyaenas *Crocuta crocuta. Ethology,* 88:223–35.

Hofer, H., and East, M. L. 1992. Siblicide of spotted hyaenas is a consequence of allocation of maternal effort and a cause of skewed litter sex ratios. *Abstracts Fourth Int. Behav. Ecol. Congr.,* Princeton, N. J.

———. 1993a. The commuting system of Serengeti spotted hyaenas: How a predator copes with migratory prey. I. Social organization. *Anim. Behav.* 46:547–57.

———. 1993b. The commuting system of Serengeti spotted hyaenas: How a predator copes with migratory prey. II. Intrusion pressure and commuters' space use. *Anim. Behav.* 46:559–74.

———. 1993c. The commuting system of Serengeti spotted hyaenas: How a predator copes with migratory prey. III. Attendance and maternal care. *Anim. Behav.* 46:575–89.

Hofer, H., East, M. L., and Campbell, K. L. I. 1993. Snaring, commuting hyaenas, and migratory herbivores: Humans as predators in the Serengeti. *Symp. Zool. Soc. Lond.* 65:347–66.

Jolly, G. M. 1969. Sampling methods for aerial census of wildlife populations. *E. Afr. Agric. For. J.* 34:46–49.

Kingdon, J. 1977. *East African mammals: An atlas of evolution in East Africa.* vol. 3A: Carnivores. London: Academic Press.

Kruuk, H. 1966. Clan-system and feeding habits of spotted hyaenas (*Crocuta crocuta* Erxleben). *Nature* 209:1257–58.

———. *The spotted hyena: A study of predation and social behavior.* Chicago: University of Chicago Press.

Kruuk, H., and Turner, M. 1967. Comparative notes on predation by lion, leopard, cheetah, and wild dog in the Serengeti area, East Africa. *Mammalia* 31:1–27.

Labuschagne, W. 1979. *'n Bio-ekologiese en gedragstudie van die jagluiperd* Acinonyx jubatus jubatus *(Schreber, 1776).* M.Sc. thesis, University of Pretoria, South Africa.

Lindström, E. 1986. Territory inheritance in the evolution of group-living in carnivores. *Anim. Behav.* 34:1825–35.

Malcolm, J., and Marten, K. 1982. Natural selection and the communal rearing of pups in African wild dogs *(Lycaon pictus). Behav. Ecol. Sociobiol.* 10:1–13.

McGinley, M. A., Temme, D. H., and Geber, M. A. 1987. Parental investment in offspring in variable environments—theoretical and empirical considerations. *Am. Nat.* 130:370–98.

Mech, L. D. 1970. *The wolf: The natural history of an endangered species.* Minnesota: University of Minnesota Press.

Mills, M. G. L. 1984. Prey selection and feeding habits of the large carnivores in the southern Kalahari. *Koedoe* (suppl.) 27:281–94.

———. 1990. *Kalahari hyaenas: The comparative behavioural ecology of two species.* London: Unwin Hyman.

Moehlman, P. D. 1989. Intraspecific variation in canid social systems. In *Carnivore behavior, ecology, and evolution,* ed. J. L. Gittleman, 143–63. London: Chapman & Hall.

Newton, I. 1979. *Population ecology of raptors.* Calton: T & A D Poyser.
Norton-Griffiths, M. 1978. *Counting animals.* 2d ed. Handbook no. 1. Techniques in African Wildlife Ecology. Nairobi: African Wildlife Foundation.
Oftedal, O. T., and Gittleman, J. G. 1989. Patterns of energy output during reproduction in carnivores. In *Carnivore behavior, ecology, and evolution,* ed. J. G. Gittleman, 355–78. London: Chapman & Hall.
Owens, M., and Owens, D. 1984. *Cry of the Kalahari.* Boston: Houghton Mifflin.
Packer, C., Herbst, L., Pusey, A. E., Bygott, D., Hanby, J. P., Cairns, S. J., and Borgerhoff Mulder, M. 1988. Reproductive success of lions. In *Reproductive success: Studies of individual variation in contrasting breeding systems,* ed. T. H. Clutton-Brock, 363–83. Chicago: University of Chicago Press.
Packer, C., and Pusey, A. E. 1983. Adaptations of female lions to infanticide by incoming males. *Am. Nat.* 121:716–28.
Paine, R. T. 1969. A note on trophic complexity and community stability. *Am. Nat.* 103:91–93.
Pournelle, G. 1965. Observations on birth and early development of the spotted hyaena. *J. Mammal.* 46:503.
Press, W. H., Flannery, B. P., Teukolsky, S. A., and Vetterling, W. T. 1986. *Numerical recipes: The art of scientific computing.* Cambridge: Cambridge University Press.
Racey, P. A., and Swift, S. M. 1985. Feeding ecology of *Pipistrellus pipistrellus* (Chiroptera: Vespertilionidae) during pregnancy and lactation. I. Foraging behaviour. *J. Anim. Ecol.* 54:205–15.
Sachs, R., and Staak, C. 1966. Evidence of brucellosis in antelopes of the Serengeti. *Vet. Rec.* 79:857–58.
Schaller, G. B. 1972. *The Serengeti lion: A study of predator-prey relations.* Chicago: University of Chicago Press.
Schultz, D. L. 1991. Parental investment in temporally varying environments. *Evol. Ecol.* 5:415–27.
Seber, G. A. F. 1982. *The estimation of animal abundance and related parameters.* 2d ed. London: Charles Griffin.
Serengeti Research Institute. 1977a. *Census of predators and other animals on the Serengeti plains, May 1977.* Serengeti National Park report 52.
———. 1977b. *Census of predators and other animals on the Serengeti plains, October 1977.* Serengeti National Park report 73.
Snedecor, G. W., and Cochran, W. G. 1980. *Statistical methods.* 7th ed. Ames: Iowa State University Press.
Sunquist, M. E., and Sunquist, F. C. 1989. Ecological constraints on predation by large felids. In *Carnivore behavior, ecology, and evolution,* ed. J. L. Gittleman, 283–301. London: Chapman & Hall.
Tan, P. L., and Counsilman, J. J. 1985. The influence of weaning on prey-catching behaviour in kittens. *Z. Tierpsych.* 70:148–64.
van Lawick H., and van Lawick-Goodall, J. 1970. *Innocent killers.* London: Collins.
Viljoen, P. C. 1993. The effects of changes in prey availability on lion predation in a large, natural ecosystem in northern Botswana. *Symp. Zool. Soc. Lond.* 65:193–213.

SEVENTEEN

Dominance, Demography, and Reproductive Success of Female Spotted Hyenas

Laurence G. Frank, Kay E. Holekamp,
and Laura Smale

The social organization of a species may have profound effects on many aspects of its population ecology. The social rank of an individual may influence its access to food or cover, affecting its exposure to mortality from malnutrition or predation. Rate of reproduction and quality of offspring may be affected by socially mediated access to food or mates, or by aggression-related stress. In many species, variance in male mating success is related to rank and results in strong sexual selection, producing sexual dimorphism in size, ornamentation, or weaponry. The stress and injury associated with competition among males may result in their having significantly abbreviated life spans compared with those of females. In some species, socially mediated reproductive suppression may limit successful reproduction to high-ranking individuals. The sex ratio of a female's offspring may be influenced by interacting factors including social rank, the nature of the mating system, sex-specific dispersal or helping behavior, and resource abundance.

Due to their long life spans, relatively few large mammal species have been studied in the wild over time periods long enough to yield reliable measures of reproductive success and the factors that affect it. In this chapter, we summarize the reproductive performance of female spotted hyenas in a social group of 70–80 individuals in the Masai Mara National Reserve, Kenya. Hyenas are long-lived animals; the 13-year period covered by this study approximates the maximum female reproductive life span. Because dominance rank appears to be a central organizing feature of hyena social structure, our emphasis is on the influence of female rank on reproductive success.

Four years into our study, Frank (1986a,b) found that there appeared to be no relationship between the social rank of females and their reproductive success. At that time we defined reproductive success as the number of offspring surviving to 1 year of age. However, significant juvenile mortality occurs between the age of 1 year, when juveniles begin to become independent of their mothers, and the age of reproductive maturity. We are now able to reexamine the question of social effects on reproduction and survival with a much larger data set. In these comparisons we use offspring's attainment of sexual maturity, rather than survival to 1 year, as the primary measure of a female's reproductive success. Moreover, we can now compare the relative contributions of the original 1979 females to the current population of breeding females to assess individual variance in reproductive success. In contrast to the earlier results, these analyses show that social rank has a major effect on female reproductive success.

This chapter deals only with female reproductive success because mating is rarely seen and does not necessarily result in successful pregnancies. The mating system appears to be highly polygynous. In the earlier years, only the alpha male was observed to mate. In subsequent years, while most observed matings have been performed by the alpha male, we have also seen a small number of matings by mid-ranking males, and one instance of two transient males mating with a clan female. Biochemical studies of paternity may thus provide the most accurate measure of male reproductive success.

Spotted Hyena Social Organization

The spotted hyena is the most abundant large carnivore in sub-Saharan Africa, occurring in a wide range of habitats from desert to montane forest. In the prey-rich savannas of East Africa, hyena clans number up to 80 adults (Kruuk 1972). Their primary prey in the Mara are wildebeest, zebra, topi, and Thomson's gazelles. When a kill has been made, feeding is exceptionally competitive (Kruuk 1972); in the scramble to feed, high-ranking females and their offspring have a significant advantage (Frank 1986b).

Because females remain in their natal clan for life, the social group comprises several matrilines (Frank 1986a,b; Henschel and Skinner 1987; Mills 1990). All studies have found a stable linear social hierarchy among females, and because juveniles acquire their mother's rank (Frank 1986b; Holekamp and Smale 1993; Smale, Frank, and Holekamp 1993), the relative rankings of the clan matrilines remain stable over generations. Males emigrate from their natal clans following puberty, joining new clans where they gradually rise in the male dominance hierarchy. Adult females are strongly dominant over immigrant adult males (Kruuk 1972).

Female aggressiveness and dominance is associated with a syndrome of masculinization that includes a body size larger than that of the male, elevated levels of androgens, and dramatic masculinization of the external genitalia (Harrison Matthews 1939; Lindeque, Skinner, and Millar 1986; Frank, Glickman, and Zabel 1989; Glickman et al. 1992; Licht et al. 1992). The struggle for social rank commences at birth: spotted hyenas are born in a precocial state of development and fight violently during the first days of life (Frank, Glickman, and Licht 1991). Spotted hyenas normally give birth to twins; the neonatal fighting is thought to lead to the death of one sib when both are the same sex, accounting for many of the singleton litters observed.

METHODS

The data presented here represent the reproductive history of the Talek Clan from January 1979 through January 1992. The study area comprises the 60 km^2 home range of the clan in the north central Masai Mara National Reserve, Kenya (Frank 1986a); the home range boundaries have remained largely unchanged since 1979.

L. G. F. was present in the study area, with short gaps, until January 1983 and again after a 15-month gap in May—December 1984. He then monitored the population for 2–3 months per year through 1987, until continuous study by K. E. H. and L. S. commenced in May 1988. Most observations were made by daylight, between 0600–0900 and 1700–1930 hours. Some nocturnal observations were made with night vision goggles and 7x50 binoculars.

In the first years of the study, many animals were darted and ear-notched, but subsequent individual recognition was based on ID photographs of spot patterns. Identification of mother-offspring and sib pairs was based on nursing associations. In a number of cases, we could infer relationships between females that were already adults or subadults in 1979 based on frequent associations, strong affiliative behavior, and adjacent ranks in the clan hierarchy (Frank 1986b).

Sex was determined through methods described in Frank, Glickman, and Powch (1990). The discovery of the phallic glans difference in 1987 allowed us to sex some previously unsexed cubs using old ID photographs showing them with erections.

Demographic Categories

Females were considered to be adult at 3 years of age, the mean age of first reproduction (Frank 1986a). Juveniles were called cubs until 1 year of age, and subadults between 1 and 3 years. Resident males were immigrants that remained in the clan range for over 6 months. Annual clan composition was tabulated by counting the animals of each category present on 1 January of each year; thus, cubs that were born and died within a calendar year and immigrant males that remained in the clan briefly are not represented by this censusing convention. We have excluded data for the first year (ending 1 January 1980) because we were not confident that we recognized all clan members at that point.

Survival estimates for the first year of life are based on the 59 cubs brought to the communal den during the period 1 June 1988 through 1 February 1991, when intensive observations ensured that no early deaths were missed. Cubs that disappeared before sex was determined were assigned equally as males and females. Since loss in the natal den due to siblicidal aggression may be substantial (Frank, Glickman, and Licht 1991), data from captive hyenas (S. E. Glickman and L. G. Frank, unpub.) were used to adjust for offspring that died before they were brought to the communal den. In captivity, first-time breeders produced 78% singletons, while parous females produced 14% singleton litters, 64% twins, and 23% triplets (N = 9 first litters, 22 subsequent litters). To adjust the 59 wild cubs for death in the natal den, therefore, 86% of singletons produced by parous females were assumed to have originated as twins, and 23% of the twins to have originated as triplets. Clearly, data from captivity must be used with caution, but Henschel (1986) reported eight pairs of twins out of nine fetal litters from hyenas shot in Kruger National Park. Disappearance rates for animals older than 1 year are based on all known individuals born in the Talek Clan that have disappeared since the beginning of the study.

Males and females attain sexual maturity at approximately 2 and 3 years respectively (Harrison Matthews 1939; Frank 1986a). Moreover, males begin to disperse at about 2 years of age, and in practice it has usually been impossible to distinguish between death and dispersal. While females have been known to change clans in southern Africa (Tilson and Henschel 1986; Mills 1990), female immigration has not been recorded in East Africa (this study; Hofer and East 1993a), where female dispersal has been seen only during clan fissioning (Mills 1990; Holekamp et al. 1993). Therefore, disappearance of a female was assumed to indicate death, unless she was known to have participated in fissioning. Thus, we defined survival to sexual maturity as survival to 2 years for males and 3 years for females. Note that this convention, a consequence of male dispersal, does not permit direct comparison of survival rates of male and female juveniles after 2 years of age.

Date of disappearance was recorded as the month in which an animal was last seen. For animals that disappeared in the years 1983–1988, date of disappearance was assigned as midway through the interval of observer absence, resulting in the addition of several months to the animals' known age at last sighting.

Social Categories

Social ranks were determined based on the outcomes of agonistic interactions (Frank 1986b; Holekamp and Smale 1990); we did not see circular relationships. It was apparent that differences in various reproductive measures distinguished the highest-ranking females from all the rest. Because of this dichotomy, correlation statistics obscure the relationship between rank and reproductive success. Therefore, in most analyses of reproductive parameters, individuals were assigned to one of four rank levels, based on the highest rank achieved by each individual over the course of the study. Level One was defined as ranks one (alpha) and two; in practice, all individuals attaining rank two were daughters of alpha females, except for female KB, who was the mother of an alpha. All Level One females were members of the 04 matriline, which has been at the top of the clan hierarchy since the study began (Frank 1986b). Since the maximum number of breeding females in the clan was 23, Levels Two, Three, and Four were defined by dividing the remaining 21 possible ranks into equal numbers: 3–9, 10–16, and 17–23, respectively. Females born to mothers that had lost earlier alpha status were thus ranked in Level Two.

Reproduction

Because many females were represented by only partial reproductive life spans (animals that were already adults in 1979, and those born since then that are still alive), all measures of female reproductive success are based on annual rates of cub production. All analyses of reproductive success are based on 36 females for which there are data on at least 3 years of reproductive life. These data were analyzed via two-way ANOVA, with sex of offspring and rank level of mothers as factors. Statistical analyses were done on a microcomputer using SYSTAT (Wilkinson 1990). Analysis of age at first reproduction omitted females whose first litters appeared at ages greater than 4.5 years, because we may have missed litters that were lost at the natal den stage. Even with this correction, these data reflect production of the first successful litter, rather than physiological ability to breed.

Ages of cubs at communal dens were determined from their size and pelage when first encountered. In 1983–1988, when there were gaps in observer presence, month of birth frequently had to be estimated. In the analysis of interlitter intervals ($n = 83$), the beginning and end of an interval were accurately known to within 1 month in 53 cases and estimated to within 2 months in 30 cases. All rank levels and litter compositions were included in the analysis of estimated intervals, and there is no reason to expect a systematic bias among individuals or rank levels.

Interlitter interval is a measure of parental investment. Because a female may become pregnant soon after a litter is lost, an interval was excluded if the litter beginning the interval survived for less than 6 months ($n = 30$ out of 167 litters). This convention avoids confounding short intervals due to early mortality with those due to rapid successful reproduction. Intervals longer than 24 months were omitted, as these were probably due to loss of a litter before we detected it. Twin litters that were reduced to singletons after being brought to the communal den were also omitted.

Length of interbirth interval as a function of maternal rank level and the composition of the preceding litter (singleton male, singleton female, twin) was analyzed using two-way ANOVA. Each female contributed one data point for each litter type; for females that produced more than one litter of a given composition, the mean interval following that litter type was used in the analysis. Similarly, in analyses of litter size, a mean was computed for each female.

RESULTS

Clan Size

The territory boundaries of the Talek Clan remained relatively unchanged over the period of study. Numbers of adult females in the Talek Clan remained remarkably stable over the first 10 years of the study (fig. 17.1), fluctuating between 20 and 23 individuals ($\bar{X} \pm SE = 21.44 \pm 0.44$). The number of resident adult males varied between 14 and 20 but was normally less than the number of adult females ($\bar{X} = 14.33 \pm 1.01$), with a ratio of 1.49 females per male. The number of cubs present ($\bar{X} = 17.25 \pm 2.65$) varied widely, but was strongly correlated with the number of subadults ($\bar{X} = 22 \pm 3.12$) in the following year ($r = .83$, $n = 6$, $p < 0.02$), indicating that this variation reflected annual differences in clan fecundity or early cub survival.

A major change in the number of adult females occurred in 1990, when seven mid- and low-ranking females from three matrilines left the Talek Clan to form a new social group (Holekamp et al. 1993). They settled across the Talek River in an area where a neighboring clan had apparently been killed off by poisoning.

Survivorship

Data on 194 individuals born in the Talek Clan indicate that while survival rates of male and female juveniles were initially indistinguishable, after the age of 2 years different patterns of disappearance prevailed (fig. 17.2). Rate of male disappearance rose rapidly as males reached the age of dispersal, and all had left the clan by the age of 6 years. Females, however, spent their lifetimes in the clan; 1 female born in 1977 and 1 born in 1979 were still alive in early 1995. Male disappearance includes both death (particularly before the age of 2 years) and dispersal, whereas fe-

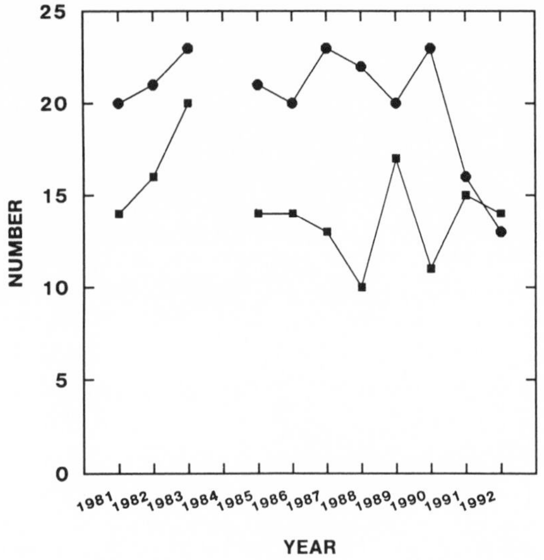

Figure 17.1 Population of the Talek Clan, 1981–1992, as of 1 January of each year. No data were available for 1984. The number of females was stable until the clan fissioned in 1990, apparently as a response to the poisoning of a neighboring clan outside the reserve. Circles indicate females; squares indicate males.

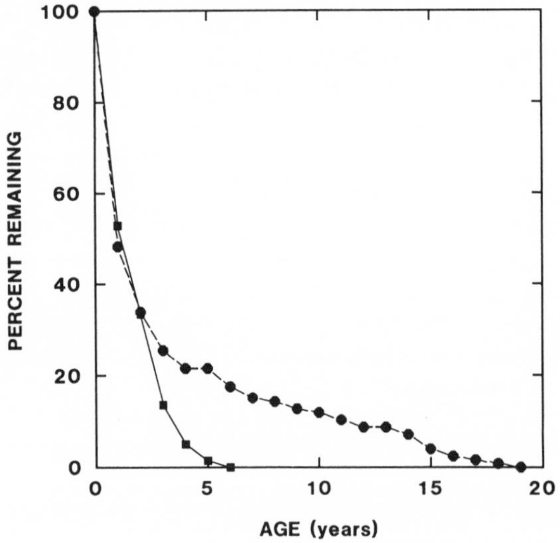

Figure 17.2 "Survivorship curve" based on 194 individuals born in the Talek Clan. See text for details. For females, disappearance corresponds to death. Males, however, start to disperse at 2 years of age, and in practice dispersal is rarely distinguishable from death. Circles indicate females; squares indicate males.

male disappearance is usually due only to death.

Mortality Factors

Juveniles. At least five litters born in October–November 1982 disappeared during torrential November rains that flooded the communal den. Thus, in some years, early cub survival may have been affected when heavy rains occurred during the primary birth period, November–January (Frank 1986a).

Six cubs older than 1 month, the age at which siblicidal aggression largely ceases (Frank, Glickman, and Licht 1991), were found dead at communal dens, showing tooth punctures and internal trauma. While these may have resulted from lion predation, the observation of infanticide in the wild (Henschel and Skinner 1991; H. Hofer and M. L. East, pers. comm.) and of infanticide by adult females in captivity (S. E. Glickman and L. G. Frank, unpub.) suggests that infanticide by adults may in some circumstances be responsible for cub deaths.

Adults. The primary source of mortality for older animals in the Mara Reserve was predation by lions. Of the fresh carcasses discovered (n = 16), seven adults, two cubs, and two subadults showed clear signs of lion predation (deep puncture marks on opposite sides of the head, neck, and thorax). Male lions were seen to stalk and attack hyenas on several occasions, and different coalitions of male lions seemed differentially likely to do so: three of the four cases of lion-caused mortality that were documented in 1979–1982 occurred after a male takeover of the local lion pride at the end of 1980. Similarly, S. Cooper, D. Joubert, and B. Joubert (pers. comm.) describe a difference between two male coalition partners in Savuti National Park, Botswana, in their tendencies to attack hyenas.

Aside from predation by lions, one cub, one subadult, and two females from another clan were known to have been killed by herdsmen's spears, and one adult male apparently was shot in a gun battle during a cattle raid. Immediately outside the study area, at least eight adult hyenas (none from the Talek Clan) have been killed near Masai villages, and many roadkills have been found between the Mara Reserve and the town of Narok.

Poisoning. We know of at least four incidents of mass poisonings north of the Mara Reserve during the course of our study. These reportedly occurred following hyena predation on livestock, when local people laced the cattle remains with pesticide. Although we have never been able to count hyena deaths directly following one of these incidents, M. Rainey (pers. comm.) found the carcasses of most members of a large clan in the Musiara region in 1984, and a similar event apparently occurred in the heavily populated Talek area in 1990 (Holekamp et al. 1993). Outside protected areas, human-caused mortality is probably more important than any other source, and appears to be increasing in southern Narok District.

Female Reproductive Success

Age at First Reproduction. Females of higher social rank had significantly earlier ages of first reproduction than lower-ranking females (r_s = 0.66, $p < .01$; fig. 17.3). Age of first breeding ranged from 2.5–4.3 years; ($\bar{X}$ = 3.44 ± 0.13, n = 19). The gestation period is approximately 110 days (Kruuk 1972). In captivity, with unlimited food, age at first reproduction has been as low as 21 months (S. E. Glickman and L. G. Frank, unpub.).

Life Expectancy and Social Rank. Thirty-two females that attained reproductive age died in the course of the study. Among these, there did not appear to be any relationship between female rank and life expectancy (r_s= 0.193, NS).

Losses of Entire Litters. There was no tendency for females of different rank levels to experience different rates of loss of whole litters, with loss defined as disappearance of a litter before reaching 6 months of age

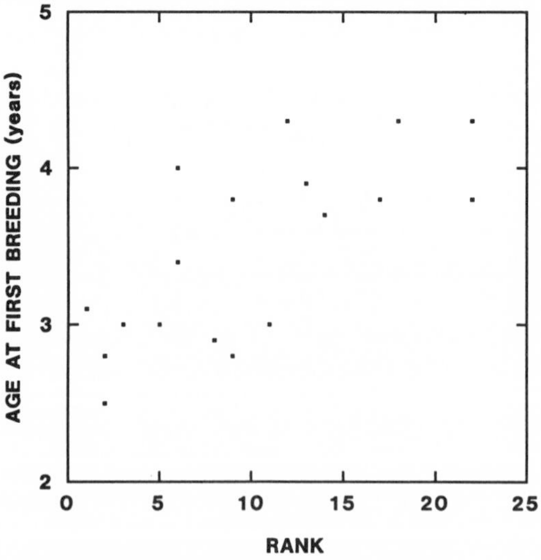

Figure 17.3 Age at first reproduction plotted against social rank. R_S = 0.66, n = 19, $p < .01$. Note that the relationship appears to reach an asymptote.

($F_{3,35}$ = 0.54, NS). However, some very long interbirth intervals were probably due to undetected loss.

Lifetime Offspring Sex Ratio. Higher-ranking females tended to produce an excess of sons over daughters, a pattern not seen in lower-ranking females (fig. 17.4). The lifetime sex ratio of offspring in the top half of the female hierarchy was significantly higher ($\bar{X}$ = 0.67 ± 0.06) than in the lower half ($\bar{X}$ = 0.50 ± 0.06, t = 2.015, df = 1, p = .05).

Interbirth Interval. Lower-ranking females had slightly longer interbirth intervals than Level One females regardless of litter sex composition ($F_{3,44}$ = 2.67, $p < .06$, fig. 17.5). There was no main effect of litter composition on the intervals (NS), but there was a strong interaction between rank level and litter composition ($F_{6,44}$ = 3.82, $p < .004$).

For Level One females, interbirth intervals were longest following singleton females, intermediate following twins, and shortest following singleton males ($F_{2,10}$ = 8.58, $p < .007$). The difference between means for singleton males and singleton females was 6.7 months. For females of Levels Two through Four, there were no effects of litter composition on subsequent intervals, except that Level Three females had longer intervals following twins than following singletons ($F_{3,17}$ = 7.09, $p < .003$).

The effect of rank on the length of the interval following singleton males ($F_{3,14}$ = 10.04, $p < .001$) was due exclusively to the difference between Level One and lower-level females. Rank level was unrelated to

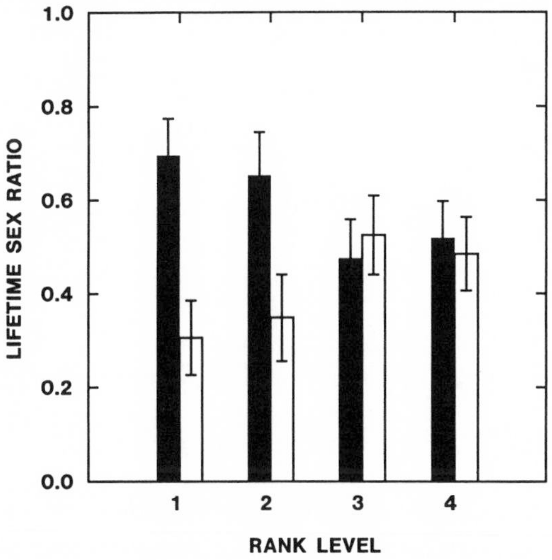

Figure 17.4 Sex ratio of cubs of known sex born to females of different rank levels. Solid bars indicate males; open bars indicate females.

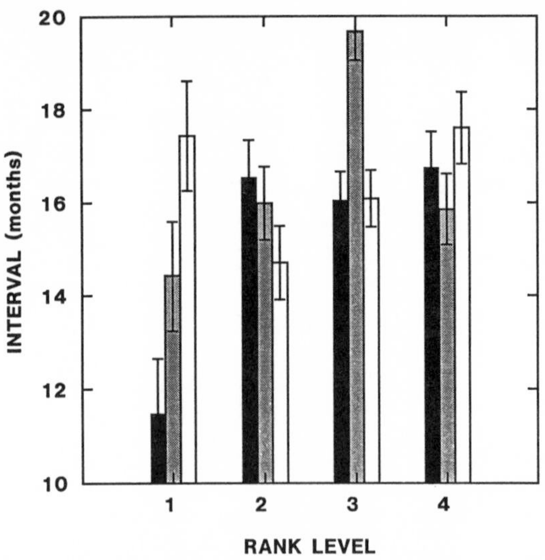

Figure 17.5 Interbirth intervals of females of different rank levels, according to composition of litters. Solid bars indicate singleton males; open bars indicate singleton females; shaded bars indicate twin litters.

interbirth interval length following female singletons. Although the sample size was small, Level One females had the longest mean intervals recorded following female singleton litters. Thus, high-ranking females have a tendency to produce singleton sons at a high rate and singleton daughters at a low rate.

Recruitment of Offspring. There was a strong effect of rank on rate of production of surviving offspring ($F_{3,32} = 6.12$, $p = .002$; fig. 17.6.) Because males and females attain reproductive status at different ages, the two sexes cannot be directly compared. However, Level One females produced surviving offspring of both sexes at rates more than 2.5 times greater than the rates of lower-ranking females (sons: $F_{1,32} = 12.61$, $p = .001$; daughters: $F_{1,32} = 6.70$, $p = .014$).

Recruitment of Breeding Female Descendants over a Decade. The cumulative effect of their more successful recruitment of daughters was reflected in the number of breeding descendants left by the original Level One females after 10 years. Between 1980 and 1990, the representation of the top-ranked matriline (Level One) increased from four to six adult females, while that of other matrilines remained stable (Level Two) or fell (Levels Three and Four) (fig. 17.7). Of the nine original females in the bottom rank level, two are still alive, but only one breeding female descendant remains.

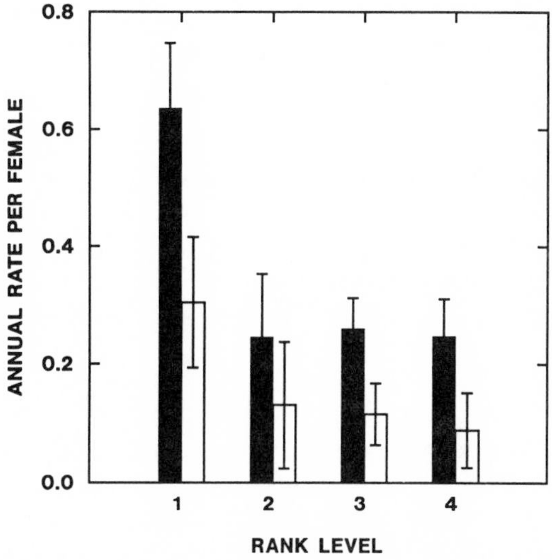

Figure 17.6 Rate at which females of different rank levels produced offspring of both sexes that reached the age of sexual maturity (2 years for males, 3 years for females). Solid bars indicate sons; open bars indicate daughters.

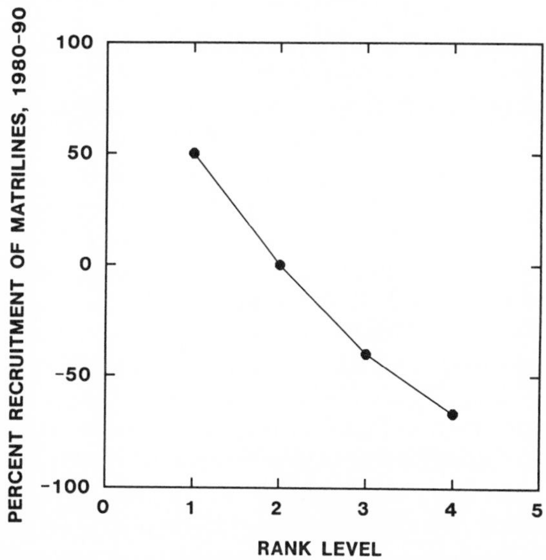

Figure 17.7 Recruitment rate of matrilines of different rank levels over the 10-year period 1980–1990. The descendants of Level One females (the top-ranked matriline) have increased by 50%, while matrilines constituting the lower rank levels have either remained stable or declined in membership.

Descendants of the original mid-ranking females tended to rank lower than their mothers because higher-ranking females produced more surviving daughters, which joined the female hierarchy just below their mothers. The declining rank of these matrilines was reflected in declining reproductive success.

In summary, while high-ranking females do not appear to have a greater life expectancy once they reach reproductive age, they start reproducing at an earlier age and produce more surviving offspring. The top-ranking matriline increased its representation in the clan over the 10-year interval while all others either remained stable or declined.

DISCUSSION

Breeding systems among the Carnivora range from solitary to highly cooperative. Most species lie at the solitary end of the spectrum: individuals of both sexes have discrete home ranges, males join females only to mate, and females raise their offspring alone. In these species, reproductive consequences of social competition are manifested largely in the ability of males to acquire a home range encompassing those of several females, and the ability of females to acquire a home range of sufficiently high quality to allow them to successfully raise offspring (Sandell 1989).

Whereas most families among the Carnivora contain species that form stable social groups, only among the Hyaenidae and the Canidae are most species social (Gittleman 1989). Monogamy is the norm among the Canidae, with several species forming packs through the retention of nonbreeding offspring. These nonbreeders perform a helping function in raising the current litter through food provisioning and possibly territorial defense against conspecifics or defense of kills against competitors. Sociality in some viverrids also confers antipredator benefits (Rood 1986). In pack-forming canids and viverrids, the alpha male and female suppress reproduction in subordinate group members (Creel et al. 1992; Malcolm and Marten 1982; Packard, Mech, and Seal 1983).

The lion, the only social felid, lies at the opposite extreme from reproductive suppression. There is little evidence of stable rank relations in either sex (Packer et al. 1988), and all adults share in reproduction, the females more equally than the males (Packer et al. 1991).

Thus, the spotted hyena occupies a middle position in the range from social suppression of reproduction to equal reproductive success. While all females produce litters, high social rank affects reproductive success through younger age at first breeding, shorter interlitter intervals, and improved survival of offspring. In virtually all of our analyses, the only significant differences in reproductive performance were between the highest rank level, alpha females and their daughters (all members of the

04 matriline), and all other rank levels. These high-ranking females thus contributed a disproportionate share of both surviving males and reproducing adult females to subsequent generations. Mills (1990) found a similar relationship between social rank and reproductive success in the smaller clans of the Kalahari.

Most long-term studies of reproductive success in mammals have concerned primate species, both wild and captive (reviewed by Silk 1986). Results have been varied and controversial. In some populations, findings were quite similar to those described here for spotted hyenas: high rank was associated with earlier reproduction, shorter interbirth intervals, improved survival, or biased sex ratios. Other studies have not found these patterns, but overall there appears to be a selective advantage to high social rank in primates, though it may rarely be as large as appears to be the case in spotted hyenas.

The proximate mechanisms favoring reproductive performance in high-ranking females are not clear. However, an important factor is probably their enhanced access to food in the exceptionally competitive melee that characterizes spotted hyena feeding (Kruuk 1972; Tilson and Hamilton 1984; Hamilton, Tilson, and Frank 1986; Frank 1986b; Holekamp and Smale 1990). It seems likely that the slower cub growth rates and prolonged interbirth intervals seen in lower-ranking females are largely due to their decreased access to food. However, it is possible that more subtle, or at least less observable, factors are also involved. The social stress of lower rank (Holekamp et al. 1993) may compromise females in ways that also affect reproductive physiology.

In a variety of cercopithecine primates, which have a social system similar to that of the spotted hyena but lacking female dominance, high-ranking females heavily emphasize production of female offspring, both at conception and through differential survival, while lower-ranking females produce more sons (van Schaik and Hrdy 1991). In these species, high-ranking mothers are able to influence the rank of their daughters, but male dispersal may reduce their influence on the mating success of their sons. The local resource competition hypothesis (Clark 1978; Silk 1983) suggests that in female-philopatric societies, high-ranking females may suppress production of daughters by lower-ranking females to reduce competition for food or other resources.

In spite of the many similarities between spotted hyena and cercopithecine social systems, high-ranking female hyenas emphasize production of sons. What factors account for this marked difference in the relationship between female rank and sex ratio of offspring? The high relative survival of male offspring of top-ranking females is in accord with the predictions of Trivers and Willard (1973) that good-quality or high-ranking females should emphasize production of sons in a polygynous mating system. In spotted hyenas, some of these females seem to specialize in singleton sons; singletons of both sexes are known to grow faster than members of twin litters (Hofer and East 1993b).

Given the polygynous mating system, Frank (1986b) speculated that these "alpha sons," maturing with both the nutritional and behavioral advantages of high maternal rank, might be expected to have an advantage over lower-ranking sons when joining a new clan and rising in the immigrant male hierarchy. The overproduction of sons by high-ranking females in a population at carrying capacity lends support to the suggestion that high-ranking sons may have a significant mating advantage.

Bias in offspring sex ratio may not be limited to high-ranking females. Recent data from both the Mara and Serengeti studies suggest that females may be able to adjust their offspring sex ratios in response to demographic changes within the population. Hofer and East (1992, chap. 16) report that the incidental snaring of hyenas in the dry season range of the Serengeti population contributes to annual mortality of adult females in excess of 13%, rising as high as 40% in one clan in one year. In the seven clans constituting their study population, they report an excess of female offspring, opposite to the bias seen in the long-term Talek data. Since 1991 the adult female component of the Talek Clan has been lower than seen in the previous 12 years (13 vs. 20+) due to clan fission in response to poisoning campaigns in the ranges of neighboring clans (Holekamp et al. 1993). At that time, high-ranking females started producing a significant excess of female offspring, and at least six pairs of female-female twins have survived (L. G. Frank, K. E. Holekamp, and L. Smale, unpub.). Prior to 1991, no female-female twins have survived siblicidal fighting in the natal den. These observations suggest that females may affect the sex ratio of their offspring in accordance with local ecological or demographic variables. The mechanism allowing females to bias the sex ratio of their litters is not known, but may involve an ability to influence the outcome of neonatal siblicidal aggression (Frank, Glickman, and Licht 1991; Frank 1992) or an unidentified prenatal mechanism.

The top-ranked matriline increased its representation in the population by 50% over a 10-year period. If this rate of increase is typical, and assuming a stable adult female population at the mean of 21–22 individuals, the entire adult female component of the clan could be descended from the one matriline after about 45 years, or fifteen generations. High variance in reproductive success results in a lowering of the effective population size (N_e) compared with that of a population in which more members contribute to subsequent generations (Lande and Barrowclough 1987). While low N_e has potentially negative consequences for the genetic diversity of the population (Caro and Durant, chap. 21), it may contribute to the rapid spread of mutations. In their monograph on hyena evolution,

Werdelin and Solounias (1991) point out that the apparent ability of high-ranking females to differentially produce male offspring may confer significant evolutionary flexibility on the species, as successful males appear to produce a large number of offspring. We do not yet have data on the reproductive success of sons of the different maternal rank levels, but it is now clear that high-ranking females produce a disproportionate share of surviving female descendants. If successful reproduction is largely the province of a relatively few high-ranking individuals, adaptations arising in these matrilines may be less likely to be lost through genetic drift than those in lower-ranking, less successful individuals. The strong selective advantage of high rank may thus help to explain the evolution of the spotted hyena's highly unlikely female reproductive anatomy and physiology (Glickman et al. 1992).

It seems counterintuitive that singletons, regardless of sex, should be more "expensive" than twin litters for a female to produce. However, it appears that top-ranked female hyenas take longer to recover from the production of a daughter than from twin litters, while intervals following singleton sons are very short. The number of singleton daughters contributing to this data set is small, and this effect needs to be verified with more litters. If interbirth interval can be taken as a measure of reproductive effort, singleton daughters born in high-ranking matrilines seem to require greater maternal investment than other litter types (see fig. 17.5), but singleton sons are produced at a higher rate (see fig. 17.6). High-ranking females may balance production of a small number of "expensive" daughters against a larger number of "cheaper" sons (Charnov 1982).

Infanticide has been inferred or witnessed several times in the wild, performed either by members of neighboring clans (Mills 1990; Henschel and Skinner 1991) or by higher-ranking females within the same clan (H. Hofer and M. L. East, pers. comm.). Several incidents of infanticide in captive hyenas confirm that females of the same social group may be a lethal threat to cubs under 3 months of age (S. E. Glickman and L. G. Frank, unpub.). Indirect evidence that infanticide is a significant threat may be found in the reluctance of small cubs to exit the burrow in the presence of adults, and the care they seem to show in ascertaining the identity of their mother before exiting. Some of the longer interbirth intervals are almost certainly due to early loss of a litter, and predation by other hyenas might be a significant cause of such losses.

Finally, chance effects of mortality are probably major factors in the reproductive success of all females, as has been documented in vervet monkeys (Cheney et al. 1988). We have recently reported a high incidence of birth complications in captive primiparous females (Frank and Glickman 1994) that may help to explain the high mortality rate of young adult females (Glickman et al. 1992). Loss of postpubertal daughters has had a severe impact on the size and survival of several free-living matrilines. Alpha female 03, for instance, lost three daughters soon after they reached reproductive age. While high social rank improves the probability of juvenile survival, even alpha females may lose a significant proportion of their offspring to predation or other random events.

The apparent increase in mass poisoning of predators outside the reserve is extremely worrisome. It is not clear why this is occurring, but the population of Masai, and hence cattle, adjoining the Talek Clan range is growing very rapidly. Constant large-scale incursions of cattle into the reserve increase the opportunities for hyenas to kill cattle, particularly since thousands of cattle are frequently present in the reserve at night. This situation leads not only to antipredator efforts by the livestock owners, but also directly affects the predators' wild ungulate prey. Cattle have a massive impact on the grassland, reducing forage for wildlife. Furthermore, the mere presence of cattle and herders, not to mention direct harassment of wildlife (e.g., driving off herds of wildebeest), also discourages wildlife use of the area.

The consequences of mass killing are clearly devastating for a slowly reproducing animal like the spotted hyena. Our results, along with those of Hofer and East (chap. 16), suggest that human-caused mortality is increasing and is a significant factor affecting hyena populations. Previous studies of recolonization after local extinction in Kruger National Park (Henschel 1986) showed that hyena numbers recover very slowly, while lions recolonize rapidly. Henschel suggested that slow recolonization was due to the extreme philopatry typical of female spotted hyenas, but the recent Mara experience (Holekamp et al. 1993) suggests that if hyena density is high, the range of a decimated clan may be reoccupied rapidly by lower-ranking females from neighboring groups. If poisoning occurs on a wide scale, however, population recovery will be very slow indeed.

CONCLUSION

A single clan of spotted hyenas in the Masai Mara National Reserve, Kenya, has been studied since 1979. Whereas the numbers of juveniles and immigrant adult males varied annually, the number of breeding females remained stable until the clan fissioned along matrilineal lines. Predation by lions is the primary cause of mortality at all ages after infancy, but human-caused mortality seems to be increasing outside the reserve. Number of offspring reaching reproductive age was used as a measure of reproductive success. Although all females bred, those in the top-ranking matriline were 2.5 times more successful than lower-ranking individuals. The correlation between female rank and reproductive success occurred because high-ranking females commence reproduction at an earlier age, have shorter interbirth intervals, and their offspring are more likely to survive to reproductive age than are those of lower-ranking females. At the observed recruitment rates, the top-ranked matriline will make up the entire adult female population of the clan in about fifteen generations, suggesting a high degree of relatedness. As long as the number of adult females was stable, there was a tendency for high-ranking females to produce an excess of sons, in accord with the Trivers-Willard hypothesis. There are indications that a population below carrying capacity may overproduce daughters, however, suggesting that offspring sex ratio may be affected by ecological or demographic variables.

ACKNOWLEDGMENTS

We thank the Office of the President, the Department of Wildlife Conservation and Management of the Republic of Kenya, and the Narok County Council for permission to carry out this research. Our work has been supported by grants BNS 7803614, BNS 8706939, BNS 9121461, and IBN 9296051 from the National Science Foundation, grant MH39917 from the National Institute of Mental Health, and grants from the National Geographic Society, the H. F. Guggenheim Foundation, the Center for Field Research, the National Wildlife Federation, the American Association of University Women, and the Charles A. Lindbergh Fund. S. Yoerg provided statistical expertise. T. Caro, M. East, H. Hofer, W. Koenig, and F. A. Pitelka made helpful comments on an earlier draft of the chapter.

REFERENCES

Charnov, E. 1982. *The theory of sex allocation.* Princeton, N.J.: Princeton University Press.

Cheney, D. L., Seyfarth, R. M., Andelman, S. J., and Lee, P. C. 1988. Reproductive success in vervet monkeys. In *Reproductive success: Studies of individual variation in contrasting breeding systems,* ed. T. H. Clutton-Brock, 384–402. Chicago: University of Chicago Press.

Clark, A. B. 1978. Sex ratio and local resource competition in a prosimian primate. *Science* 201:163–65.

Creel, S., Creel, N., Wildt, D. E., and Montfort, S. L. 1992. Behavioural and endocrine mechanisms of reproductive suppression in Serengeti dwarf mongooses. *Anim. Behav.* 43:231–45.

Frank, L. G. 1986a. Social organization of the spotted hyena *(Crocuta crocuta).* I. Demography. *Anim. Behav.* 34:1500–1509.

———. 1986b. Social organization of the spotted hyena *(Crocuta crocuta).* II. Dominance and reproduction. *Anim. Behav.* 34:1510–27.

———. 1992. Female spotted hyenas may use neonatal siblicide to bias sex ratios. Paper presented at the Fourth International Behavioral Ecology Congress, Princeton, N. J., 17–22 August 1992.

Frank, L. G., and Glickman, S. E. 1994. Giving birth through a penile clitoris: Parturition and dystocia in the spotted hyena *(Crocuta crocuta). J. Zool.* (Lond.) 234:659–65.

Frank, L. G., Glickman, S. E., and Licht, P. 1991. Fatal sibling aggression, precocial development, and androgens in neonatal spotted hyenas. *Science* 252:702–4.

Frank, L. G., Glickman, S. E., and Powch, I. 1990. Sexual dimorphism in the spotted hyena *(Crocuta crocuta). J. Zool.* (Lond.) 221:308–13.

Frank, L. G., Glickman, S. E., and Zabel, C. J. 1989. Ontogeny of female dominance in the spotted hyena: Perspectives from nature and captivity. In *The biology of large African mammals in their environment,* ed. P. A. Jewell and G. M. O. Maloiy, 127–46. Zool. Soc. London Symp. no. 61. Oxford: Clarendon Press.

Gittleman, J .L. 1989. Carnivore group living: Comparative trends. In *Carnivore behavior, ecology, and evolution,* ed. J. L. Gittleman, 183–207. Ithaca, N.Y.: Cornell University Press.

Glickman, S. E., Frank, L. G., Holekamp, K. E., Smale, L., and Licht, P. 1993. Costs and benefits of "androgenization" in the female spotted hyena: The natural selection of physiological mechanisms. In *Perspectives in ethology,* eds. P. P. G. Bateson, P. Klopfer, and N. Thompson, 87–117. New York: Plenum Press.

Glickman, S. E., Frank, L. G., Pavgi, S., and Licht, P. 1992. Hormonal correlates of "masculinization" in the female spotted hyena *(Crocuta crocuta).* I. Infancy through sexual maturity. *J. Reprod. Fert.* 95:451–62.

Hamilton, W. J., III, Tilson, R. L., and Frank, L. G. 1986. Sexual monomorphism in Spotted Hyenas, *Crocuta crocuta. Ethology* 71:63–73

Harrison Matthews, L. 1939. Reproduction of the spotted hyena (*Crocuta crocuta* Erxleben). *Phil. Trans. R. Soc. Lond. B.* 230:1–78.

Henschel, J. R. 1986. *The socio-ecology of a spotted hyena Crocuta crocuta clan in the Kruger National Park.* D.Sc. thesis, University of Pretoria.

Henschel J. R., and Skinner, J. D. 1987. Social relationships and dispersal patterns in a clan of spotted hyenas *Crocuta crocuta* in the Kruger National Park. *S. Afr. J. Zool.* 22:18–24.

———. 1991. Parturition and early maternal care of spotted hyenas *Crocuta crocuta:* A case report. *J. Zool.* (Lond.) 222:702–4.

Hofer, H., and East, M. L. 1992. Snares, commuting hyenas, and migratory herbivores: Humans as predators in the Serengeti. *Symp. Zool. Soc. Lond.* 65:347–66.

———. 1993a. The commuting system of Serengeti spotted hyenas: How a predator copes with migratory prey. I. Social organization. *Anim. Behav.* 46:547–57.

———. 1993b. The commuting system of Serengeti spotted hyaenas: How a predator copes with migratory prey. III. Attendance and maternal care. *Anim. Behav.* 46:575–89.

Holekamp, K. E., Ogutu, J. O., Frank, L. G., Dublin, H. T., and Smale, L. 1993. Fission of a spotted hyena clan: Consequences of prolonged female absenteeism and causes of female emigration. *Ethology* 93:285–99.

Holekamp, K. E., and Smale, L. 1990. Provisioning and food-sharing by lactating spotted hyenas *(Crocuta crocuta)*. *Ethology* 86:191–202.

———. 1993. Ontogeny of dominance in free-living spotted hyenas: II. Juvenile rank relations with other juveniles. *Anim. Behav.* 46:451–66.

Kruuk, H. 1972. *The spotted hyena: A study of predation and social behavior.* Chicago: University of Chicago Press.

Lande, R., and Barrowclough, G. F. 1987. Effective population size, genetic variation, and their use in population management. In *Viable populations for conservation,* ed. M. E. Soulé, 87–123. Cambridge: Cambridge University Press.

Licht, P., Frank, L. G., Pavgi, S., Yalcinkaya, T. M., Siiteri, P. K., and Glickman, S. E. 1992. Hormonal correlates of "masculinization" in the female spotted hyena *(Crocuta crocuta)*: II. Maternal and fetal steroids. *J. Reprod. Fert.* 95:463–74.

Lindeque, M., Skinner, J. D., and Millar, R. P. 1986. Adrenal and gonadal contribution to circulating androgens in spotted hyenas *(Crocuta crocuta)* as revealed by LHRH, HCG, and ACTH stimulation. *J. Reprod. Fert.* 78:211–17.

Malcolm, J. R., and Marten, K. 1982. Natural selection and the communal rearing of pups in the African wild dog *(Lycaon pictus)*. *Behav. Ecol. Sociobiol.* 10:1–13.

Mills, M. G. L. 1990. *Kalahari hyenas: The comparative behavioural ecology of two species.* London: Unwin Hyman.

Packard, J. M., Mech, L. D., and Seal, U.S. 1983. Social influences on reproduction in wolves. In *Wolves in Canada and Alaska,* ed. L. Carbyn, 78–85. Canadian Wildlife Service report no. 88. Ottawa: Canadian Wildlife Service.

Packer, C., Gilbert, D. A., Pusey, A. E., and O'Brien, S. J. 1991. A molecular genetic analysis of kinship and cooperation in African lions. *Nature* 351:562–65.

Packer, C., Herbst, L., Pusey, A. E., Bygott, J. D., Hanby, J. P., Cairns, S. J., and Borgerhoff Mulder, M. 1988. Reproductive success of lions. In *Reproductive success: Studies of individual variation in contrasting breeding systems,* ed. T. H. Clutton-Brock, 363–83. Chicago: University of Chicago Press.

Rood, J. P. 1986. Ecology and social evolution of the mongooses. In *Ecological aspects of social evolution,* eds. D. I. Rubenstein and W. R. Wrangham, 131–52. Princeton, N.J.: Princeton University Press.

Sandell, M. 1989. The mating tactics and spacing patterns of solitary carnivores. In *Carnivore behavior, ecology, and evolution,* ed. J. L. Gittleman, 164–82. Ithaca, N.Y.: Cornell University Press.

Silk, J. B. 1983. Local resource competition and facultative adjustment of sex ratios in relation to competitive activities. *Am. Nat.* 12:56–66.

———. 1986. Social behavior in evolutionary perspective. In *Primate societies,* ed. B. B. Smuts, D. L. Cheney, R. M. Seyfarth, R. W. Wrangham, and T. T. Struhsaker, 318–29. Chicago: University of Chicago Press.

Smale, L., Frank, L. G., and Holekamp, K. E. 1993. Ontogeny of dominance in free-living spotted hyenas: Juvenile rank relations with adult females and immigrant males. *Anim. Behav.* 46:467–77.

Tilson, R. L., and Hamilton, W. J. III. 1984. Social dominance and feeding patterns of spotted hyenas. *Anim. Behav.* 32:715–24.

Tilson, R. L., and Henschel, J. R. 1986. Spatial arrangement of spotted hyena groups in a desert environment, Namibia. *Afr. J. Ecol.* 24:173–80.

Trivers, R. L., and Willard, D. 1973. Natural selection of parental ability to vary the sex ratio of offspring. *Science* 179:90–92.

van Schaik, C. P., and Hrdy, S. B. 1991. Intensity of local resource competition shapes the relationship between maternal rank and sex ratio in cercopithecine primates. *Am. Nat.* 138:1555–62.

Werdelin, L., and Solounias, N. 1991. The Hyaenidae: Taxonomy, systematics and evolution. *Fossils and Strata* 30:1–104.

Wilkinson, L. 1990. *SYSTAT: The system for statistics.* Evanston, Ill.: SYSTAT, Inc.

EIGHTEEN

Implications of High Offspring Mortality for Cheetah Population Dynamics

M. Karen Laurenson

The modern cheetah once roamed throughout Africa, the Middle East, and India. A sharp decrease in both the range and density of this species over the last 100 years has left the largest remaining populations in eastern and southern Africa and only a small relict population in Iran (Myers 1975). Although it is often assumed that large carnivores will continue to exist in Africa's network of protected areas, cheetahs may face additional problems because they live at low density compared with other large carnivores. Species living at low densities rarely exist locally in large numbers and, as a consequence, may have difficulty in maintaining a minimum viable population size and genetic diversity (Franklin 1980; Gilpin and Soulé 1986). Small populations are also more vulnerable to demographic and environmental stochasticity (Simberloff 1986).

The reasons that cheetahs live at low density are not well understood. In general, carnivore numbers are thought to be limited by the size of the prey populations on which they depend (Bertram 1975; Brand and Keith 1979; Fuller 1989). Previous reports, however, suggest that juvenile mortality may be high in cheetahs and that predation by other large carnivores may account for a substantial proportion of this mortality (Schaller 1972; Frame and Frame 1981). The concept that predators themselves may be limited by other predators is unusual although there are no theoretical reasons to oppose it.

In this chapter I discuss recent findings on the causes of juvenile mortality in cheetahs (Laurenson 1994) and present preliminary data suggesting that predation on cheetah cubs is an important factor affecting the Serengeti cheetah population, with the hope that this will stimulate additional study. First, I consider the relative importance of factors affecting cheetah fecundity and mortality, and second, I present model simulations that demonstrate the effect of variation in fecundity and mortality factors on female lifetime reproductive success and cub recruitment rates. In addition, I discuss the implications for cheetahs of recent changes in carnivore numbers in the Serengeti-Mara ecosystem. Finally, I consider the relevance of these findings for cheetah populations elsewhere in Africa and their application to future conservation.

METHODS

A study of the cheetahs of the Serengeti plains and woodland edge has been carried out since 1980 (Caro 1994). As part of this long-term study, I collected data on cub mortality and reproduction from 1987 to 1990 by closely monitoring 20 radio-collared females and pinpointing the time when they gave birth. After locating lairs, I counted the cubs and estimated their ages. They were then checked weekly until they died or left the lair. A 5-day period of intensive observation was conducted when the cubs were approximately 4 weeks old (detailed methods are described in Laurenson, Caro, and Borner 1992; Laurenson 1993). Extensive analyses could find no effect of observations or handling on cub mortality (Laurenson and Caro 1994).

Determination of Cause of Cub Mortality

In some cases I witnessed cub deaths, and in other cases in which cubs disappeared between my visits to the lair, circumstantial evidence, such as cub remains or maternal behavior, allowed causes of cub death to be inferred (see Laurenson 1994 for methods). Only cases in which I knew or was almost certain of the cause of death were included in the analyses (table 18.1). Data from parallel studies and other observers in the ecosystem were used to assess the relative importance of different species as predators of cheetah cubs.

Cheetah Cub Recruitment

A simple equation was used to estimate cub recruitment because it was more convenient to work with the available reproductive and mortality parameter values than to use Lotka's equation or a Leslie matrix formulation (Begon, Harper, and Townsend 1991). The number of female cubs that a female cheetah could raise in her lifetime was estimated using the following formula, which takes account of the high litter mortality rate:

$$\text{Female cubs raised} = (1 - m) \times n \times e/(a + bc),$$

where a = number of months that each successful litter takes to reach independence (time taken to conceive after litter lost + gestation + months with mother); b = number of months spent by mother on each litter that is lost (conception time + gestation + life of litter); c = number of dead litters for every one that survives $[(1 - d)/d]$, where d is the proportion of surviving litters; e = reproductive life span of females in

Table 18.1 Percentage of cheetah cubs in this study that died due to each cause of mortality between birth and independence.

	Percentage of cub mortality		
Cause	In lair	After emergence	Total (birth to independence)
Predation	67.8	90.0	73.0
Abandonment	9.9		7.7
Fire	9.9		7.7
Exposure	7.4		5.7
Possibly inviable cubs	4.9		3.8
Other		10.0	2.2

Source: Laurenson 1994.
Note: Only cases in which the cause of death was definitely or almost certainly known (n = 40.5) were used in estimating the proportion of mortality attributable to each cause in the lair. A litter whose size was unknown was assigned the average litter size of 3.5 cubs. Of all cub deaths (n = 115), 77.4% occurred before cubs emerged from the lair, whereas 22.6% occurred after emergence. "Other" causes of mortality were assigned for 10% of cubs dying after emergence.

months; n = average number of female cubs in litters that reach independence; and m = proportion of adolescents that die. Parameter estimates were obtained from this study or from the long-term study of this cheetah population (Caro 1994).

RESULTS

Factors Affecting Cheetah Abundance

Juvenile Mortality. Juvenile mortality was found to be extremely high for cheetahs on the Serengeti plains, with approximately 72.2% of litters dying before they emerged from the lair at 8 weeks of age. An average of 83.3% of cubs alive at emergence died by adolescence at 14 months of age; thus newborn cheetah cubs were estimated to have only a 4.8% chance of reaching independence. Predation was the major source of mortality, accounting for 73.0% of cub deaths overall (see table 18.1). Lions were the primary predators of cubs in the lair, whereas spotted hyenas and lions took approximately equal proportions of emergent cubs (table 18.2).

Other causes of mortality were of relatively little importance, but some cubs died of starvation when abandoned by their mothers (7.7%) and others died as a result of unpredictable events such as fire (7.7%) or exposure (5.7%). The availability of prey and the difficulty of obtaining sufficient food may play a role in the probability of abandonment. Fewer Thomson's gazelles were counted around the lairs of litters that were definitely abandoned than around lairs from which litters emerged (n = 2, 10 respectively; medians: 2, 555; Mann-Whitney U test, $U = 0$, $p < .05$).

Table 18.2 Percentage of cheetah cub deaths from predation, by predator species, in studies in the Serengeti-Mara ecosystem.

Predator	Percentage of cubs killed by predator: In lair (n = 53.5)	After emergence (n = 12)	Birth to independence (n = 65.5)
Lion	88.8	33.3	78.6
Spotted hyena	5.6	41.7	12.2
Leopard		8.3	1.5
Cheetah	3.7		3.1
Masai dogs		16.7	3.1
Raptors	1.9		1.5

Source: From Laurenson 1994 (tables 5 and 6). Data from this study, Burney 1980; Frame and Frame, 1981; Ammann and Ammann, 1984; Caro et al. 1987; G. Domb and M. Smits van Oyen, pers. comm.; D. Richards, pers. comm.
Note: Litters of unknown size were assigned the average litter size of 3.5 cubs.

Adult Mortality. The causes of adult mortality and their relative importance are difficult to discern. Nevertheless, approximately 50% of young males that die are known to be killed as a result of intraspecific fights (Caro 1994). Predation by lions, spotted hyenas, and leopards undoubtedly occurs but may be secondary to other problems such as disease, starvation, or injuries; two unsuccessful predation attempts on healthy individuals were observed in this study. Serengeti cheetahs have been exposed to a variety of feline diseases (Heeney et al. 1990; Evermann et al. 1993), and some adult deaths may be attributable to disease (S. Gascoyne, pers. comm., A. Cunningham, pers. comm.), although sarcoptic mange infection, the most common overt health problem (pers. obs.), may be secondary to stress (Caro, FitzGibbon, and Holt 1989) or other causes of ill health. Starvation is most likely to occur in adolescents, particularly if few gazelle fawns or hares are available, as they depend heavily on these prey items while perfecting their hunting techniques (Caro 1994). Death by snaring is rare, but has occurred in this ecosystem (K. Campbell, pers. comm.).

Adult mortality rates are also difficult to quantify for wild cheetahs, but data from radio-collared females in this study were used to estimate mortality rates. Dividing the number of females in each of three age classes (adolescent; < 3 years; prime, 3–9 years; old, > 9 years) that died while radio-collared by the total number of years they were radio-collared gave the rate at which females died each year. Adolescent females (n = 6) died at a rate of 0.153 per year, prime females (n = 18) at 0.227 per year, and old females (n = 2) at 0.55 per year. The mean life expectancy of females reaching 3 years of age was therefore 3.9 years.

Female Fecundity. Female fecundity is affected by the age at first and last successful breeding attempt, litter size, and the interval between births. In many species these are primarily influenced by nutrition and food availability (Sadleir 1969; Mitchell 1973; Rattray 1977). There was some evidence that female cheetahs were more fertile in the wet season, when Thomson's gazelle fawns are abundant. More litters were conceived during the wet than during the dry season, and females that lost litters in the dry season took longer to conceive successfully again than those losing litters in a wet month (Laurenson, Caro, and Borner 1992). In addition, there was a nonsignificant trend for litters conceived in the wet season to be larger than those conceived in the dry season (Laurenson et al. 1992). Thus, although there is little information on the factors affecting the age of first and last breeding in female cheetahs, there is some evidence that nutritional factors and prey availability may affect reproductive rates in this species.

Cub Recruitment Rates

The effects of variations in fecundity and mortality on the number of female cubs produced in the lifetime of an adult female were simulated using the equation given above (table 18.3). Overall, factors affecting juvenile and adult survival had a greater simulated effect on cub recruitment than those affecting fecundity. Variations in offspring survival, through its effect on both parameters *d* and *n*, were substantial, yielding ranges of 0.35–1.86 and 0.37–1.32 respectively in the number of female cubs raised. Although these parameters should not be treated independently, as they positively covary, it was impossible to quantify this relationship. Thus changes in offspring survival would cause a greater alteration in the number of cubs raised than simulated here, where these parameters were only varied independently.

The number of cubs raised also varied considerably in response to changes in reproductive life span, although this was predominantly due to the wide range of the age at death (42–144 months), rather than that of the age at first reproduction (24–38 months). Variations in *a* and *b* (length of time that mothers provided care for their offspring and time taken to return to estrus after lost litters) had relatively little effect on lifetime reproductive success.

The effect of demographic conditions on cheetah population dynamics was also examined. The combinations of parameter values yielding net recruitment rates of 1 were calculated and a three-dimensional surface drawn (fig. 18.1), allowing values of *d, n,* and *e* to vary between 0–1, 0–144, and 0–2 respectively. The parameters *a, b,* and *m* were treated as constants (with values 21.7, 4.9, and 0.15 respectively) as they had relatively little effect on reproductive rates. Points lying in the region above the surface represent demographic conditions resulting in an expansion of the cheetah population, whereas points lying below the surface represent conditions under which the population will decline.

Table 18.3 The range of the number of female cubs that could be produced by a female cheetah during her lifespan (r) depending on the values of variables affecting fecundity and offspring survival.

	Range of parameter values			No. female cubs raised in lifetime		
Parameter	Min	Max	Mean	Min	Max	% variation from mean
a	17.3	24.65	21.7	0.56	0.76	+/−15.2
b	4.4	5.8	4.9			
d	0.05	1	0.24	0.35	1.86	+181.8, −47
n	0.5	1.8	0.9	0.37	1.32	+100, −43.9
e	12	120	52.8	0.15	1.50	+127.3, −77.2
m	0	0.35	0.15	0.54	0.77	+16.7, −18.2

Note: When one parameter was varied, others were held at their mean value. Minimum and maximum values of *a* and *b* could be varied only together. If mean parameter values were used, 0.66 female cubs were raised in the lifetime of a female cheetah.

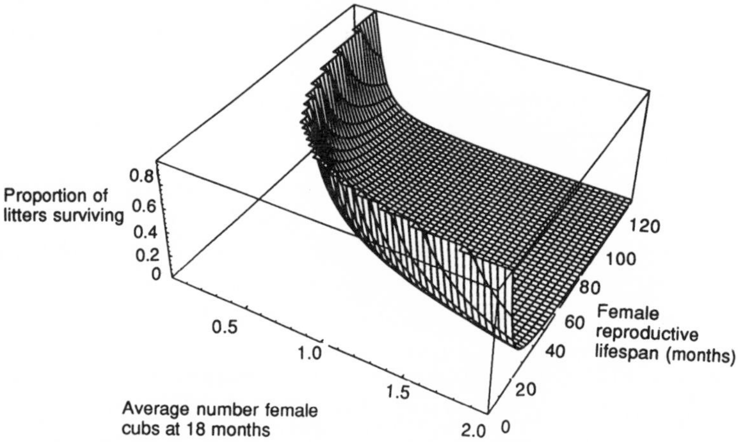

Figure 18.1 The relationship between the number of female cubs per litter at independence, female reproductive life span, and proportion of litters born that survive that give a stable cheetah population. Points lying above the surface represent combinations of values resulting in an expanding population, points below represent values associated with population decline.

Using mean values for these demographic parameters calculated from individuals in this study (from Laurenson, Caro, and Borner 1992) only 0.66 female cubs will be raised by each adult female. This suggests that cub recruitment is not presently sufficient to maintain the cheetah population on the Serengeti plains.

The Influence of Changes in Carnivore Numbers on the Cheetah Population

Natural changes in predator numbers have recently occurred in the Serengeti-Mara ecosystem due to changes in prey numbers. The increase in the population of wildebeest during the 1960s and 1970s due to the control of rinderpest (Sinclair 1979) to its present level of approximately 1.4 million (Campbell 1989), combined with a series of years of favorable rainfall, has led to an increase in the lion population (Hanby and Bygott 1979). In particular, lion numbers on the plains have increased from approximately 25 in the 1960s (Schaller 1972) to about 80 in the 1970s, and possibly 250 recently (Hanby, Bygott, and Packer, chap. 15). The total number of spotted hyenas has probably also increased since the mid-1960s to current estimates of 7,200–7,700 in the ecosystem, with a core population on the plains of some 5,200 (Hofer and East, chap. 16). If predation by these carnivores is important in limiting the cheetah population, then they may have had an increasing effect recently.

Although there are no long-term census data for cheetahs between the 1960s and 1990s, some demographic parameters are available from George Frame's study in 1974–1976 (Frame 1976). The average litter size of cubs aged 8–18 months in the study area decreased significantly between the mid-1970s and the late 1980s (fig. 18.2). If adolescent mortality did not change over the same period, recruitment of cheetahs into the adult population probably declined.

Decreased litter size at independence could be explained by a reduction in the number of cubs born or by an increase in partial litter mortality after birth. Reduced litter size at birth is less likely because the average litter size of cubs less than 4 weeks old did not change significantly between 1969–1976 and 1987–1990 (fig. 18.2) and because both figures are comparable to captive litter sizes (Marker and O'Brien 1989). In consequence, the observed decline in the litter size of grown cubs is probably due to increased mortality rates after birth. Partial litter mortality before emergence from the lair was rare when predation of cubs was witnessed (11.8%, n = 17 predation events) but common (85.7%, n = 7) post-emergence when predation of cubs was witnessed (Laurenson 1994).

Litter size at independence may be a useful indicator of the level of predation pressure on cheetah cubs. Support for this relationship comes from Namibia, where the average size of nine litters of 10-month-old cubs was 4.0 on ranchland where lions and hyenas had been eliminated (McVittie 1979). This discrepancy is highly unlikely to be due to latitude or ecological conditions because litter sizes in the nearby Etosha National Park are no greater than on the Serengeti plains (P. Stander, pers. comm.). Thus an increase in the rate of predation on cubs 2–18 months old is the most likely explanation for the decrease in observed litter size at independence over the last 15 years on the Serengeti plains.

Do Large Carnivores Influence Cheetah Populations Elsewhere?

Cheetahs live at low densities throughout their range in a wide variety of habitats and ecological conditions (Myers 1975; Stander 1991). To assess

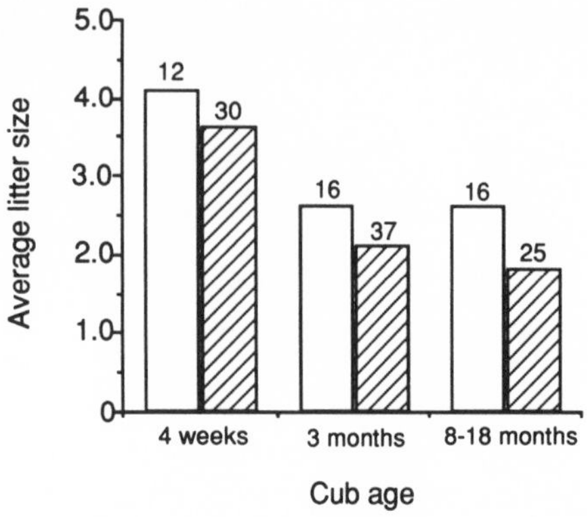

Figure 18.2 The average size of cheetah litters on the Serengeti plains at different cub ages during the mid-1970s (Frame 1976; open bars) and late 1980s (this study; hatched bars). Significant differences in litter size occur only between the oldest age class of cubs (Mann-Whitney U test, $U = 581, p < .001$).

whether other carnivores have an effect on cheetahs elsewhere in Africa, the relationship between cheetah, prey, and predator biomass was examined using data collated by Stander (1991) from nine protected areas in eastern and southern Africa, but including updated data from the Ngorongoro Conservation Area and Serengeti National Park (Hanby, Bygott, and Packer, chap. 15; Hofer and East, chap. 16). The model that best explained cheetah biomass included the variables medium-sized prey biomass (i.e., prey that were in the size range 15–60 kg—e.g., Thomson's gazelles, Grant's gazelles, impala) and lion biomass ($r^2 = .82$, df = 6, $p < .01$). Prey biomass had a positive relationship with cheetah biomass ($t = 5.23, p < .01$; fig. 18.3a), whereas lion biomass had a negative effect on cheetah biomass ($t = -2.69, p < .04$; fig. 18.3b). The combined biomass of lions and spotted hyenas also had a significant negative effect on cheetah biomass, taking into account the effects of prey biomass ($r^2 = .79$, df = 6, $p < .01$).

These results suggest a negative effect of large predators on the size of cheetah populations. Nevertheless, it should be noticed, first, that population estimates of carnivores are often unreliable and that the total number of areas censused is few. Second, the result was primarily driven by the inclusion of the Ngorongoro Crater and Serengeti data points (regression statistics for variables in the model excluding Ngorongoro: prey biomass, $t = 4.55$, $p = .04$; predator biomass, $t = -1.37$, NS; lion, $t = -1.07$, NS; spotted hyena, $t = -1.04$, NS). As few data are available, the analysis does not take into account possible effects of differences in ecological conditions, such as prey distribution, migration, or habitat type, which might explain much of the variance in cheetah biomass.

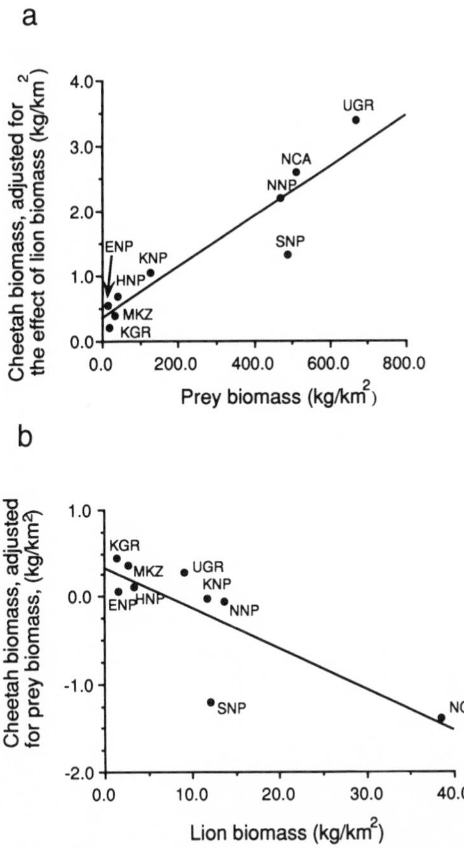

Figure 18.3 (*a*) The relationship between cheetah biomass and prey biomass, across nine African protected areas, taking into account the effect of lion biomass, such that $y = 0.34 + 0.004x$. (*b*) The relationship between cheetah biomass and lion biomass across nine African protected areas, taking into account the effect of prey biomass, such that $y = 0.34 - 0.046x$. ENP, Etosha National Park, Namibia; HNP, Hwange National Park, Zimbabwe; KGR, Kalahari Gemsbok National Park, R.S.A; KNP, Kruger National Park, R.S.A; MKZ, Mkomazi Game Reserve, Tanzania; NCA, Ngorongoro Conservation Area, Tanzania; NNP, Nairobi National Park, Kenya; SNP, Serengeti National Park, Tanzania; UGR, Umfolozi Game Reserve, R.S.A.

DISCUSSION

Mortality, Fecundity, and the Population Dynamics of Serengeti Cheetahs

It seems likely that cub mortality is a major factor affecting the size of the cheetah population on the Serengeti plains. When compared with demographic patterns of other large mammals (Caughley 1966; Loudon 1985), cheetah cub mortality in this study was extremely high, with 95.2% of cubs dying before reaching independence. Cheetah cub mortality is also high compared with former estimates (30%–60%) in East Africa (McLaughlin 1970; Schaller 1972; Frame and Frame 1981; Burney 1980) due to a complete underestimation of the scale of mortality in the lair. Predation by other carnivores has been commonly cited as the most important source of cub mortality in cheetahs, but there has been little evidence to substantiate these claims (Schaller 1972; Eaton 1974; Myers 1975). Results of this study indicate that predation, mainly by lions, is indeed the major cause of cub mortality, accounting for approximately 73% of mortality between birth and independence. Abandonment, which has not previously been reported as a cause of mortality in this species, fire, and bad weather each accounted for a smaller proportion of mortality.

The magnitude of predation on cheetah cubs (see table 18.1) suggests that other large predators may be affecting cheetah population dynamics in the Serengeti. This conclusion is supported by the observed decline in cheetah litter size at independence since the 1970s, which has coincided with a rise in lion and spotted hyena numbers on the Serengeti plains. Assuming that mortality from other sources has continued at a constant rate, the decrease in litter size at independence could be explained by additional mortality from predation on cheetah cubs after emergence. The increase in lion numbers may, however, have had a disproportionate effect on cheetahs above that of a simple increase in pride size, because it has led to an expansion of the lions' range. Lion prides now inhabit a large area of the plains, rather than just the woodland edge, year-round (Hanby, Bygott, and Packer, chap. 15). Lair sites used by female cheetahs on the short- and medium-grass plains (Laurenson 1993), probably relatively free of predation previously, may now be suffering substantially higher predation rates. The recent reduction of the lion population due to an epizootic of canine distemper will serve as a further natural experiment on the relationship between cheetah cub mortality and population dynamics.

Other factors, however, such as food availability, disease, parasites, and social structure are known to be important in limiting the size of vertebrate populations (Sinclair 1989), with the availability of prey being the major factor limiting many predator populations (lions: Bertram 1975; lynx, *Lynx canadensis:* Brand and Keith 1979; wolves, *Canis lupus:* Fuller 1989). Food limitation could act through decreasing reproductive rates, such as lowered conception rates or litter sizes, or alternatively by increasing mortality from starvation. There is, however, little evidence that cheetah numbers in the Serengeti-Mara ecosystem are currently determined by prey abundance. Conception rates and litter sizes are similar to those in captivity, and cub growth rates are also equivalent (Laurenson, in press). Wild cheetahs were also in comparable physical condition to captive cheetahs according to physical, hematological, biochemical, and hormonal measures (Caro et al. 1987; Laurenson 1992). Furthermore, the biomass of prey weighing 15–60 kg in Serengeti is greater than that required to support an equivalent biomass of cheetahs elsewhere in Africa (fig. 18.3), and the availability of prey, such as hares, weighing less than 15 kg is not even included in this calculation. These prey species can be an important component of the cheetah's diet, particularly for adolescents or when other prey are scarce (Caro 1994). Finally, cub abandonment accounted for only 9.5% of juvenile mortality in this study and was related to local scarcity of migrating Thomson's gazelles. In summary, although nutritional status may cause some variations in fertility, such as lowered conception rates in the dry season (Laurenson, Caro, and Borner 1992), food availability probably has little effect on cheetah numbers in this ecosystem at present.

In the simulations, the effect of variations in fecundity and mortality rates also suggested that offspring survival has a powerful influence on cub recruitment through its effect on the number of litters that do not survive. Although partial litter mortality also affected recruitment, it produced less variation in total cub production, primarily because initial litter size was determined by female fecundity. Differences in fecundity, such as the time taken to return to estrus or the age at first reproduction, appeared to have little effect on the potential number of cubs produced compared with increasing cub survival by avoiding predators.

In summary, although there is little information on the factors affecting adult mortality in females, these findings provide tentative evidence that offspring mortality, particularly from lion predation, may have a critical effect on the size of the Serengeti cheetah population. Interactions between cheetahs and other predators are potentially important for the population dynamics of cheetahs in the Serengeti-Mara ecosystem and deserve further scrutiny.

Influence of Major Carnivores on Cheetah Populations in Africa

Initial analysis of the relationship of the biomass of cheetahs, their prey, and their predators suggested that cheetah biomass across nine protected areas in Africa was primarily determined by the biomass of prey available to cheetahs. In most protected areas, therefore, their low density can be explained primarily by the low density of the prey species on which they depend. Nevertheless, the biomass of lions and spotted hyenas combined, or of lions alone, also had a significant negative effect on cheetah biomass in this analysis. This result, however, needs further validation, and furthermore, makes no distinction between the effect of predators on cheetah numbers through competition for resources and the more direct effect of cub mortality observed in the Serengeti. In addition, the relationship between lions and cheetah density was driven by the high biomass of lions and hyenas in the Ngorongoro Crater and a scarcity of cheetahs in the Serengeti ecosystem; differences may exist in the cheetah-predator relationship in these areas. For example, the migratory system in the Serengeti may depress the biomass of cheetahs that can be supported, as may the exceptional productivity of the Ngorongoro Crater, which gives rise to extremely high lion and spotted hyena densities and thus results in complete exclusion of smaller predators such as cheetahs or wild dogs.

Despite this caveat, there is a suggestion that other predators may depress cheetah density even when predator density is not exceptionally high. On ranchland in Namibia, other predators have been largely eliminated. Under these conditions of release from predation pressure cheetahs appear to flourish, and the litter size at independence is extremely high at 4.0 (McVittie 1979). Thus large predators can potentially have an effect on cheetah populations in a variety of ecological conditions. The prospective analyses presented here and the questions they raise clearly point out that closer scrutiny of the cheetah-predator relationship is warranted.

Implications for Conservation

The suggestion that other large predators have a detrimental effect on cheetah population size is important from a conservation perspective. Although protected areas such as national parks are often considered to be a universal panacea for species survival, this may not be the case for cheetahs because of the protection afforded to other large predators. The elimination of these large predators in these areas is not, however, a desirable or realistic management option. Nevertheless there are areas where these predators exist at very low numbers or have been eliminated by humans. Some pastoralists and ranchers tolerate cheetahs to a greater extent than lions or hyenas, and cheetahs seem to prosper in these areas (McVittie 1979; Burney 1980). As cheetahs may have difficulty in reaching large numbers in isolated protected areas, it is perhaps in these multiple land use areas that conservation efforts should be concentrated to find ways in which continuing conflict between cheetahs and humans can be minimized (Laurenson et al. 1992).

ACKNOWLEDGMENTS

I am grateful to the Government of Tanzania, Tanzania National Parks, and the Serengeti Wildlife Research Institute for permission to conduct this research, and in particular, to David Babu, Karim Hirji, Hassan Nkya, and Asukile Kajuni for their assistance. My thanks also to Barbie Allen, Markus Borner, Peter Hetz, Marianne Kuitert, and Charlie and Lyn Trout, who helped me enormously while I was in Serengeti, and to Tim Caro for his support at all stages of this project. Steve Albon, Peter

Arcese, Tim Caro, Bryan Grenfell, Philip Stander, and an anonymous reviewer provided constructive comments on the manuscript. This work was supported by the Frankfurt Zoological Society, the Leverhulme Trust, The Messerli Foundation, the National Geographic Society, and the Cambridge Philosophical Society.

REFERENCES

Ammann, K., and Ammann, K. 1984. *Cheetah.* Nairobi: Camerapix Pub. Int.

Begon, M., Harper, J. L., and Townsend, C. R. 1991. *Ecology: Individuals, populations and communities.* 2d ed. Oxford: Blackwell Scientific Publications.

Bertram, D. 1975. Social factors influencing reproduction in wild lions. *J. Zool.* (Lond.) 177:463–82.

Bothma, J. du P., and Le Riche, E. A. N. 1986. The influence of increasing hunger on the hunting behaviour of southern Kalahari leopards. *J. Arid Environ.* 18:79–84.

Brand, C. J., and Keith, L. B. 1979. Lynx demography during a snowshoe hare decline in Alberta. *J. Wildl. Mgmt.* 43 (4): 827–49.

Burney, D. A. 1980. The effects of human activities on cheetahs *(Acinonyx jubatus schr.)* in the Mara region of Kenya. M.Sc. thesis, University of Nairobi.

Campbell, K. L. I. Serengeti Ecological Monitoring Programme. In *S. W. R. C. Biennial Report,* 1988–89, 5–13.

Caro, T. M. 1994. *Cheetahs of the Serengeti plains: Group living in an asocial species.* Chicago: University of Chicago Press.

Caro, T. M., FitzGibbon, C. M., and Holt, M. E. 1989. Physiological costs of behavioural strategies of male cheetahs. *Anim. Behav.* 38:309–17.

Caro, T. M., Holt, M. E., FitzGibbon, C. D., Bush, M., Hawkey, C. M., and Kock, R. A. 1987. Health of adult free-living cheetahs. *J. Zool.* (Lond.) 212:573–84.

Caughley, G. 1966. Mortality patterns in mammals. *Ecology* 47:906–18.

Eaton, R. 1974. *The cheetah: The biology, ecology, and behaviour of an endangered species.* New York: Van Nostrand Reinhold.

Evermann, J., Laurenson, M. K., Caro, T. M., and Mc Kiernan, A. 1993. Infectious disease surveillance in captive and free-living cheetahs: An integral part of the species survival plan. *Zoo Biol.* 12:125–33.

Frame, G. W. 1976. Cheetah ecology and behaviour. In *S. R. I. Annual report,* 1975–76, 74–87.

Frame, G. W., and Frame, L. H. 1981. *Swift and enduring: Cheetahs and wild dogs of the Serengeti.* New York: E. P. Dutton.

Franklin, I. A. 1980. Evolutionary change in small populations. In *Conservation biology: An evolutionary-ecological perspective,* ed. M. E. Soulé and B. A. Wilcox, 135–49. Sunderland, Mass.: Sinauer Associates.

Fuller, T. K. 1989. *Population dynamics of wolves in north-central Minnesota.* Wildlife Monographs no. 105.

Gilpin, M. E., and Soulé, M. E. 1986. Minimum viable populations: Processes of species extinction. In *Conservation biology: The science of scarcity and diversity,* ed. M. E. Soulé, 19–34. Sunderland, Mass.: Sinauer Associates.

Hanby, J. P., and Bygott, D. 1979. Population changes in lions and other predators. In *Serengeti: Dynamics of an ecosystem,* ed. A. R. E. Sinclair and M. Norton-Griffiths, 249–62. Chicago: University of Chicago Press.

Heeney, J. L., Evermann, J. F., McKeiran, A. J., Marker-Kraus, L., Bush, M., Wildt, D. E., Meltzer, D. G., Colly, L., Lukas, J., Mantan, J., Caro, T., and O'Brien, S. J. 1990. Prevalence and implications of feline coronavirus infections of captive and free-ranging cheetahs *(Acinonyx jubatus). J. Virol.* 64:1964–72.

Laurenson, M. K. 1992. Reproductive strategies in wild female cheetahs. Ph.D. thesis, University of Cambridge.

———. 1993. Early maternal behavior of cheetahs in the wild: Implications for captive husbandry. *Zoo Biol.* 12:31–45.

———. In press. Cub growth and maternal care in cheetahs. *Behav. Ecol.*

———. 1994. High juvenile mortality in cheetahs and its consequences for maternal care. *J. Zool.* (Lond.) 234:387–408.

Laurenson, M. K., and Caro, T. M. 1994. Monitoring the effects of non-trivial handling in free-living cheetahs. *Anim Behav.* 47:547–57.

Laurenson, M. K., Caro, T. M., and Borner, M. 1992. Patterns of female reproduction in wild cheetahs. *Natl. Geogr. Res. Explor.* 8(1):64–75.

Loudon, A. S. I. 1985. Lactation and neonatal survival in mammals. *Symp. Zool. Soc. Lond.* 54:183–207.

Marker, L., and O'Brien, S. J. 1989. Captive breeding of the cheetah *(Acinonyx jubatus)* in North American zoos (1871–1986). *Zoo Biol.* 8:3–16.

McLaughlin, R. T. 1970. Aspects of the biology of cheetahs (*Acinonyx jubatus* (Schreber) in Nairobi National Park. M.Sc. thesis, University of Nairobi.

McVittie, R. 1979. Changes in the social behaviour of South West African cheetah. *Madoqua* 2(3):171–89.

Mitchell, B. 1973. The reproductive performance of wild Scottish red deer *Cervus elaphus. J. Reprod. Fert.* 19:271–85.

Myers, N. 1975. *The status of the cheetah in Africa south of the Sahara.* Morges, Switzerland: IUCN.

Rattray, P. V. 1977. Nutrition and reproductive efficiency. In *Reproduction in domestic mammals,* eds. H. H. Cole and P. T. Cupps, 553–75. London: Academic Press.

Sadleir, R. M. F. S. 1969. *The ecology of reproduction in wild and domestic animals.* London: Methuen.

Schaller, G. B. 1972. *The Serengeti lion: A study of predator-prey relations.* Chicago: University of Chicago Press.

Simberloff, D. S. 1986. The proximate causes of extinction. In *Patterns and processes in the history of life,* eds. D. M. Raup and D. Jablonski, 259–76. Berlin: Springer-Verlag.

Sinclair, A. R. E. 1979. The Serengeti environment. In *Serengeti: Dynamics of an ecosystem,* ed. A. R. E. Sinclair and M. Norton-Griffiths, 31–45. Chicago: University of Chicago Press.

———. 1989. Population regulation in animals. In *Ecological concepts,* ed. J. M. Cherret, 197–241. Oxford: Blackwell.

Stander, P. E. 1991. Aspects of the ecology and scientific management of large carnivores in sub-Saharan Africa. M.Phil. thesis, University of Cambridge.

NINETEEN

Demographic Changes and Social Consequences in Wild Dogs, 1964–1992

Roger Burrows

The African wild dog or hunting dog is one of the most endangered large carnivores (Ginsberg and Macdonald 1990) and is the object of active conservation measures in both the Tanzanian and Kenyan sectors of the Serengeti ecosystem. During the last 25 years the status of the wild dog has changed from that of "vermin," suffering remorseless persecution by conservationists (Bere 1956) as well as hunters (Maugham 1914), to one of the top attractions for visitors to the Serengeti-Mara ecosystem. A succession of filmmakers, photographers, naturalists, and scientists who since 1964 have focused on wild dogs have contributed to this change in attitude.

The periods of scientific research on wild dogs can be conveniently divided into two eras: pre- and post-1980, with a cessation between 1978 and 1984. Between April 1967 and January 1978 several wild dog packs were studied mainly on the short and intermediate grasslands of the Serengeti plains (Sinclair 1979); this work was normally restricted to a maximum study area of approximately 5,200 km^2 (Malcolm 1979). Most of the original information on the hunting behavior and social life of the species was obtained during the short denning period of 2–3 months during the wet season, when packs were confined to a small core area of 10–50 km^2 around the natal den (Estes and Goddard 1967; van Lawick and van Lawick-Goodall 1970; Schaller 1972; Frame et al. 1979; Malcolm and Marten 1982). Only limited use was made of radiotelemetry (Frame and Frame 1981).

Following the cessation of scientific research on wild dogs in 1978, monitoring of the wild dog population in Serengeti National Park (SNP) and Ngorongoro Conservation Area (NCA) was reestablished in 1985 with the Hunting Dog Project (HDP) (Fanshawe 1988), part of the existing Serengeti Ecological Monitoring Programme (SEMP). In 1989 the HDP became a joint Frankfurt Zoological Society and World Wide Fund for Nature project and became known as the Serengeti Wild Dog Project

(WDP). In 1987 the Loita Wild Dog Project (LWDP) (Scott 1991) began in Kenya (Mara/Loita regions). In the Tanzanian sector of the ecosystem the wild dog study packs discovered and monitored post-1985 were either resident on the Serengeti plains of SNP/NCA (i.e., the pre-1980 study area) or in the western corridor of SNP (with the exception of occasional observations of a pack whose home range straddled the international border) (see figure 19.1. for pack names and locations). The wild dog population resident in the ecosystem outside of the study areas is largely unknown. One of the main aims of both projects was to discover the reasons for the decline in the wild dog population in areas where wild dogs were officially protected, concentrating particularly on the roles of disease and interspecific competition from spotted hyenas. Beginning in October 1989, the present study on the ecology, behavior, and demography of wild dogs, based on intensive ground observations in SNP/NCA, complemented the aerial monitoring undertaken by WDP (Burrows 1990, 1993). I collate these data with those of the other studies over the past 25 years.

METHODS

Radiotelemetry has been the main technique used in SNP/NCA from 1985 and in Mara/Loita from 1987 (Scott 1991; Ginsberg et al. 1995). At least one dog from each known pack was radio-collared for tracking by aircraft (collar weight 250–550 g, approximately 1%–2.5% of adult dog body weight [22 kg]). A total of 22 individuals were radio-collared in SNP/NCA between 1985 and 1991 (table 19.1): 64% of the 14 collared males were more than 24 months old when collared; the mean age at collaring of 5 others was 16.4 months (3.6 SE). Of 8 collared females, 2 were more than 24 months old, and 6 others had a mean age of 17 months (6.4 SE) when collared. Serum samples were normally collected at the time of collaring to screen for pathogens and evidence of exposure to disease. A similar program was carried out in the Mara/Loita region (Scott 1991).

Observations and photographs of wild dogs throughout the ecosystem were solicited from tourists (Fuller et al. 1992; Burrows 1990) and record cards of individuals were compiled. In 1989, to counter a perceived threat to the wild dog population in the ecosystem from rabies in domestic dogs, an anti-rabies vaccination campaign was initiated by LWDP in Mara/Loita (Kat 1990). In Tanzania, the only two known packs in SNP/NCA, the Salei and Ndoha, were vaccinated in September 1990 using an inactivated vaccine (Gascoyne et al. 1993). Thirty-four dogs (24 adults and yearlings and 10 pups), forming 81% of the study population in SNP/NCA, were vaccinated in September 1990 (excluding four pups of the Ndoha pack). They were either hand-vaccinated after immobilization

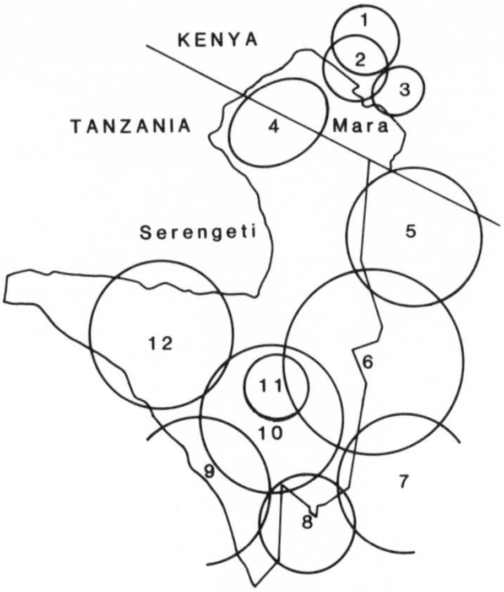

Figure 19.1 Approximate locations of wild dog packs. Year when pack was first reported and year of last confirmed report/demise. 1, Aitong (1985–1989); 2, Intrepids (1988–1990); 3, Ole Sere (1990); 4, Border Rovers/Triangle (1988–1991?); 5, *Loliondo (1989–1992); 6, Genghis (1970–1979), *Semetu (1974–1975?), Pedallers (1986), Barafu (1989), Mountain (1989–1990), New Barafu (1991); 7, Lemuta (1990); 8, Genghis (1967–1969), Pimpernel (1969–1971), Planes (1970–1973), *Spitfire (1974–1977?), Ndutu (1989–1990), Trail Blazers (1991); 9, *Herod (1969–1970), *Cassidy (1971?), Simba (1973–1975?), Kühme (1973–1977?), Falcon (1977?), *Moru Track (1990–1991?); 10, Kühme (1964–1973), Plains (1974–1985?), Naabi (1985–1988), Salei (1988–1991); 11, Seronera (1971–1977), M and S (1991); 12, *Kirawira (1972–1977?), Ndoha (1987–1991).
Note: * = nonstudy packs; ? = fate unknown.

Table 19.1 Study pack handling and longevity (SNP/NCA only).

	1970–1978	1985–1991
Study packs	14	12
Radio-collared packs	2	12
Radio-collared dogs	3	22
Vaccinated dogs	0	34[a]
Total all handled dogs	3	52
Mean pack life in years (*n,* SE)[b]	5.6 (7, 0.2)	2.7 (11, 0.1)

Source: From Burrows et al., 1994. Apart from this study, references for 1970–1978 are Malcolm (1979), Malcolm and Marten (1982), and Frame and Frame (1981), and for 1985–1991, Fanshawe (1988) and Scott (1991).
[a]Including 4 dogs also radio-collared.
[b]8 complete packs died out between 1985 and 1991.

(n = 4) or "darted" (n = 30) using an air-powered gun firing a pressurized, vaccine-charged, hypodermic syringe. Those dogs that were vaccinated and/or radio-collared will be referred to as "handled." All the packs studied post-1980 (n = 12 SNP/NCA, n = 3 Mara/Loita) contained handled dogs (table 19.1). From mid-1990 the Moru Track pack entered the study area sporadically and was reported by tourists. Because this pack was not observed by researchers it never became a study pack.

This review of changes in the demography of wild dogs concentrates on the evidence obtained from the subpopulation in the Serengeti plains study area (approximately 5,200 km^2) in SNP/NCA, the only population monitored in both research periods. Post-1985, the study area was expanded to include the western corridor with occasional observations on a pack discovered in the north of SNP. Wild dogs in the new study areas, those non-study (unhandled) dogs discovered sporadically in those areas, and dogs in the areas adjacent to the study areas were considered the non-plains population (fig. 19.2). Whenever possible, results and data from the pre-1980 period have been compared with those of the post-1980 period. The small wild dog population monitored in the Serengeti ecosystem since 1985 makes statistical treatment of the data very difficult.

Pack life span is defined as the period of continuity through the male kin line until the date on which a well-observed study pack was last seen in its home range or the date on which the last pack member died in the pack's home range. Individual longevity is defined as the period from birth until the date of the last verified sighting. A sighting was considered verified if individual identification was confirmed by the observer or the sighting was supported by a dated photograph from which identification was possible using a reference collection of photographs.

RESULTS

Distribution

The distribution of wild dogs throughout the ecosystem is patchy (fig. 19.1) and coincides with the occurrence of grassland plains and open woodland. The existence of woodland breeding packs of wild dogs in Serengeti (Frame 1986; Malcolm 1979) has not been substantiated. Only 2 of 160 (1.25%) aerial locations of radio-collared dogs over a 6-year period were in woodland. Certain plains areas were regularly used for denning or regularly visited by wild dogs, while other areas, particularly those in woodland, were rarely visited. The Naabi Hill area was continuously occupied between 1964 and 1991, and I have traced through photographic material a kinship line from the first pack of wild dogs studied in this area (Kühme 1965) through 1991. At least one male emigrated from the Plains pack in the Naabi area, probably in 1985, to the western corridor and between 1987 and 1990 was observed as the alpha male of the Ndoha pack.

Home Range

Home range was defined as the area regularly used by the pack during the denning season with the addition of other areas visited at other times. Estimates of home range sizes depend on the method of calculation and the timing of samples (Fuller et al. 1992). These varied from 50–260 km^2

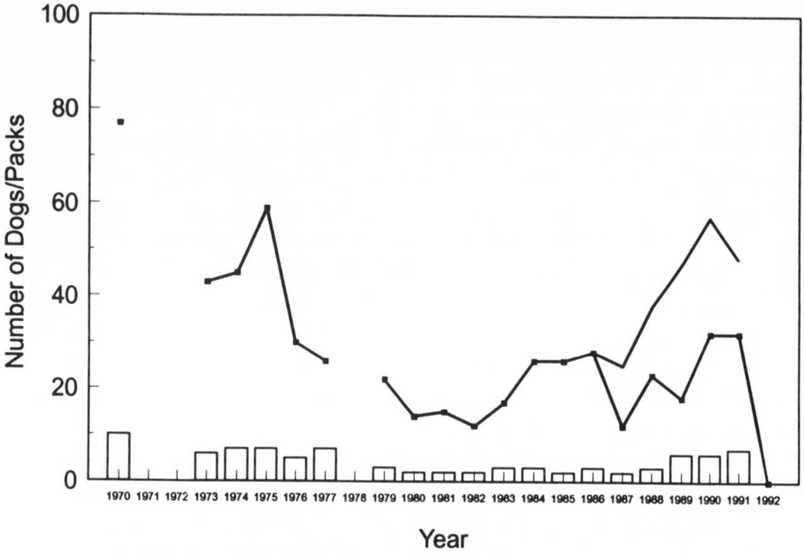

Figure 19.2 Numbers of dogs and packs in the study areas, 1970–1992. Solid lines with squares: number of yearlings/adult dogs in study packs/groups in the Serengeti plains (SNP/NCA) study area only, 1970–1992. No data for 1971, 1972, and 1978. Data for 1979–1984 minimal values based on casual reports and photographs. Solid lines without squares: total number of yearling/adult dogs in Tanzanian sector of the ecosystem (i.e., those in the Serengeti plains, western corridor, and international border study areas but also including non-study packs/groups 1985–1992. Histograms: total number of study packs in Serengeti plains study area 1970–1977; minimum number of packs in same study area 1979–1984; and all packs in the Tanzanian sector of the ecosystem 1985–1991.

in the denning period to 1,500–2,000 km^2 in the dry season. Home ranges were not exclusive, and considerable overlap occurred in the ranges of pre-1980s packs (G. W. Frame and L. H. Frame 1976). The post-1980 home ranges of two SNP packs had a limited overlap of about 400 km^2 between December 1987 and March 1988 and between October and December 1990, in each case just prior to the emigration of the "yearlings" (12–24 months old) (Burrows 1991a,b).

Population Trends

A "well-documented" decline in the wild dog population of the Serengeti plains from 1970 to 1978 was reported (Malcolm 1979). Between 1970 and 1973 three packs were shot (L. H. Frame and G. W. Frame 1976; Malcolm 1979), and between September 1970 and October 1973 no new wild dogs were found in the study area (Frame et al. 1979). Although shooting of wild dogs had ended by 1974, between 1974 and 1978 there was an overall decline in the Serengeti study population, and in mean adult pack size, but not in the number of packs on the plains (fig. 19.2)

(Frame et al. 1979). However, the number of study individuals fluctuated rapidly after 1985. Data for 1979–1984, and for the non-plains population in 1985–1991, come from casual records and photographs only and are therefore minimal values.

Three study packs died out in the Mara/Loita between 1989 and 1991, while in SNP/NCA one study pack also died out in each of the years 1986, 1988 and 1989, and two in 1990. Between January and July 1991, all five SNP/NCA study packs disappeared (the original two packs had split by January 1991) (fig. 19.3) (Burrows 1993). Subsequently there have been no confirmed sightings of any member of any study pack. However, unhandled packs/groups still exist in the ecosystem and surrounding regions (Burrows, Hofer, and East 1994).

Pack Formation and Composition

Ninety-four percent of 21 packs formed in the ecosystem between 1969 and 1991 originated when all-male groups of dogs (group size range 1–9 individuals; seven groups contained related dogs of different cohorts) joined groups of females (range 1–6 individuals) to which they were not closely related. In three cases, however, new packs were formed when subordinate pairs split from the main pack (pack fission) and bred within the home range of the original pack. One pack was formed when five brothers joined their lone aunt. The number of adults in 18 newly formed packs ranged from 2 to 15 (mean 5.6), but packs rarely had more than 12 mature adults (Fuller et al. 1992). After 1985 the number of adult dogs in packs fluctuated due to emigration, and some packs both before and after 1980 were reduced to a single breeding pair with offspring. In four cases post-1980, and in at least three cases pre-1980 (van Lawick and van Lawick-Goodall 1970), adults joined existing packs. Some, but not all, of these dogs were related to existing pack members.

Over both periods of study, the sex ratio of founder members (primary ratio) in 18 new packs was 46% female in the first year, 35% (n = 9 packs) in the second year, 30% (n = 7 packs) in the third, and 23% (n = 4) in the fourth. However, the sex ratio of adults in the pack did not differ significantly from a 50:50 ratio (table 19.2).

Reproduction

Although breeding in a pack was mainly the prerogative of the alpha (dominant) female, at least 14 beta (subordinate) females succeeded in producing pups, accounting for 25% of 57 breeding attempts recorded between 1964 and 1991. In eight of ten packs in which dominance status was known, the beta females gave birth after the alphas (Burrows 1992b). The interval between the births ranged from a few days to 2 months. In SNP/NCA post-1980, 23% of 21 litters were born to beta females, and

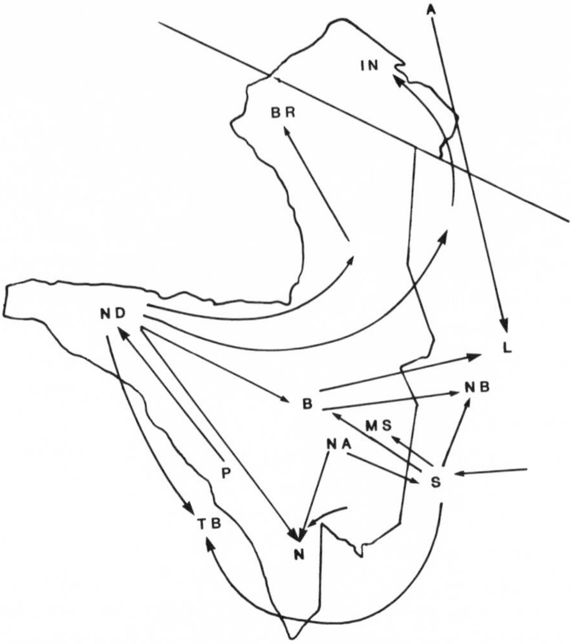

Figure 19.3 General direction of dispersal movements during emigration from packs during 1985–1991. A, Aitong; IN, Intrepids; BR, Border Rovers/Triangle; L, Lemuta; NB, New Barafu; B, Barafu; S, Salei; N, Ndutu; TB, Trail Blazers; P, Plains; NA, Naabi; ND, Ndoha; MS, M and S.

Table 19.2 Sex ratio of all adult wild dogs in all packs.

Period	% female	*n*	Location	References[a]
1966–1969	42	100	SNP/NCA	1
1970–1978	36	151	SNP/NCA	2
1970–1978	32	37	SNP/NCA	3
1985–1991	52	68	SNP/NCA	4,5
1988–1991	42	26	Mara	6

Note: None of the ratios differs significantly from equality.
[a]1, Schaller 1972; 2, Malcolm 1979; 3, Malcolm and Marten 1982; 4, Scott 1991; 5, this study; 6, Fuller et al. 1992.

in the Loita region the Aitong pack produced beta litters in 3 successive years from 1986 to 1988 (Ammann 1987; Scott 1991). In six of eight cases in which more than one breeding female was present, more alpha pups survived to 3 months than beta pups (van Lawick 1973; Frame et al. 1979; Ammann 1987; Scott 1991; Burrows 1993). The alpha female invariably took over the beta female's pups (van Lawick 1973; Burrows 1993), resulting in combined litters. This may explain some of the very large litters reported by van Lawick and van Lawick-Goodall (1970).

When the beta female left the Salei pack (secondary emigration) in 1990, her pups remained with the alpha female (Burrows 1993).

Denning in wild dogs often coincided with the movements of migratory wildebeest and Thomson's gazelle (Fuller et al. 1992). Litter size (range 3–16) and composition were determined when pups first appeared above ground at 3–4 weeks, or when the female was observed moving her litter prior to this. The sex ratio of pups at about 3–4 weeks showed no significant difference from an expected 50:50 sex ratio (table 19.3).

Between 1985 and 1990, if losses of whole litters from packs that died out are excluded, some pups from all litters that left the den survived to 1 year of age. This high pup survival in both study areas contrasts with the situation before 1980, when high pup mortality occurred even after the pups left the den (Malcolm 1979). In SNP/NCA, of 9 litters lost pre-1980, 7 were born in the second half of the wet season (Frame et al. 1979). Between 1970 and 1991 some pups survived to adulthood from 5 beta litters in the ecosystem (Malcolm 1979; Scott 1991; Burrows 1993).

Role and Nature of Helpers

Helpers in a pack were those adults other than the breeding pair. Before 1980 yearlings were excluded from consideration as helpers due to uncertainty about their role in raising young. In that period there was a positive but nonsignificant correlation between the number of adult helpers and the number of pups surviving to 1 year (Malcolm and Marten 1982). After 1985 a positive correlation was again found between the number of helpers and pup survival, but only if yearlings were included as helpers ($r = .62$, $p < .05$). Indeed, the strongest correlation was between pup survival and the number of yearlings ($r = .71$, $p < .05$). No correlation was found if only post-yearling adults were considered as helpers.

Recruitment

Before 1980, male recruitment into the natal pack was common (Malcolm and Marten 1982), but after 1980 this did not occur in the Mara/Loita (Burrows 1991b; Fuller et al. 1992) or SNP/NCA populations. No young females were recruited into their natal packs pre- or post-1980 in either study area. However, at least two sisters and two lone females joined existing packs containing a breeding female.

Dispersal

Before 1980, natal emigration was reported for all yearling females, but there were only two confirmed cases in males (L. H. Frame and G. W. Frame 1976; Frame et al. 1979). After 1985 all yearlings of both sexes emigrated in both the Serengeti and Mara populations (Burrows 1991a,b; Fuller et al. 1992), coincident with high pup survival in the packs. In

Table 19.3 Sex ratio in wild dog pups at 3–4 weeks.

Period	% female	*n*	Location	References[a]
1966–1969	41	37	SNP/NCA	1
1970–1978	41	96	SNP/NCA	2
1985–1989	52	135	SNP/NCA/Mara	3
1988–1990	51	53	Mara	4
1990–1991	42	19	SNP/NCA	5

Note: None of the ratios differs significantly from equality.
[a]1, Schaller 1972; 2, Frame et al. 1979; 3, Scott 1991; 4, Fuller et al. 1992; 5, this study.

seven of eight cases of yearling emigration between 1977 and 1991, siblings (5–12 months old) were left behind (Malcolm and Marten 1982; R. Burrows, unpub.). There was a tendency for young females to undertake natal emigration prior to their male littermates: the mean age of 37 emigrating males was 24.3 months, and of 38 females was 21.4 months. There was no significant sex bias in the yearlings present in the packs (table 19.4). Before 1980, emigrant groups of females had only to travel to an adjoining home range (L. H. Frame and G. W. Frame 1976) to find an established all-male group. Seven such groups of available males were recorded in the period 1966–1975, and some existed for several months (mean 9.5 months) before being joined by females (Malcolm 1979). After 1980, single-sex groups of young dogs and single females roamed nomadically up to 100 km from their natal area for up to 2 years before finding mates, and no all-male groups occupying a home range were reported. Between 1985 and 1991 at least 30 dogs (16 females), including 8 radio-collared individuals, emigrated from their natal packs in SNP/NCA, and at least 83% of these formed or joined new packs (fig. 19.3). The mean period between emigration and death was 19.08 months (n = 25, 14.3 SD), with no mortality of radio-collared dogs during emigration.

Secondary emigration occurred when dogs left the pack they had joined after leaving their natal pack and were observed in other groups or alone. In nine cases before 1980, subordinate sisters of alpha females were involved in secondary emigration. Two alpha females also undertook secondary emigration, in one of these cases accompanied by three daughters (Malcolm and Marten 1982). Secondary emigration by males was not recorded pre-1980, but occurred in two packs in 1990.

Three cases of pack fission, when at least one male and an unrelated female emigrated together to form a new pack, were verified pre-1980 (Malcolm and Marten 1982). In late 1990 the Salei pack (20 dogs) underwent fission followed by the dispersal of two single-sex yearling groups, resulting in the formation of three new packs (Burrows 1993).

Wild dog immigrants to the Serengeti plains study area of SNP/NCA were rarely observed in the pre-1980 period (Frame et al. 1979). By contrast, between 1985 and 1991 a minimum of 50 dogs in 16 single-sex

Table 19.4 Sex ratio of yearling wild dogs.

Period	% female	*n*	Location	References[a]
1966–1969	45	53	SNP/NCA	1
1970–1978	46	16	SNP/NCA	2
1987–1990	43	35	Mara	3
1985–1991	48	87	SNP/NCA	4, 5, 6

Note: None of the ratios differs significantly from equality.
[a]1, Schaller 1972; 2, J. Malcolm, pers. comm.; 3, Fuller et al. 1992; 4, Fanshawe 1988; 5, Lelo 1990; 6, this study.

groups, including two groups of females whose natal pack was the Aitong in Kenya, and one new pack sporadically entered the same study area.

Two unknown lone females were the only wild dogs confirmed in the SNP/NCA plains and western corridor study areas in 1992 (fig. 19.2). From May 1993 five unknown young females were resident in SNP for approximately 3 months (S. Huish and K. Campbell, pers. comm.; Burrows, Hofer, and East 1994).

Mortality and Disease

From 1969 to 1971, 12 of 64, and between 1974 and 1977, 13 of 79 known adults died, giving an annual survival of 0.825 (J. Malcolm, pers. comm.). Before 1980, at least four entire packs were lost due to shooting, three packs in SNP and one pack of 45 in the Ngorongoro Crater in 1951 (Frame 1977). Since 1985 at least 7 dogs have been killed by vehicles within SNP/NCA, but the only other known deaths, other than those involving whole packs, were of two adults. Evidence from observations of young pups being moved between dens suggests that high pup mortality takes place during the first 4 weeks of life. Disease and predation by spotted hyenas have been suggested as possible causes (Frame et al. 1979; Frame and Frame 1981).

An impermeable calcium carbonate hardpan exists approximately 1 m below the surface of the short-grass plains (Sinclair 1979). This hardpan, acting as a subsoil drainage surface, forms the floor on which the pups lie within the den. Thus dens on the short-grass plains are often damp in the wet season (R. Burrows, unpub.). Hypothermia or drowning due to den flooding may be responsible for the deaths of some pups (Schaller 1972; R. Burrows, unpub.). A possible case of infanticide was reported pre-1980 (van Lawick 1973). Between 1985 and 1991 two litters died, again possibly due to infanticide by the alpha female, but the evidence for this is circumstantial (Scott 1991). Another nine litters probably died from disease. Litter mortality was high within the first year both pre- and post-1980 (Burrows, Hofer, and East 1994), but in the latter period, in 81% of cases in which whole litters died, the pack also died.

Each pack studied during 1964–1991 had a continuous ancestry through the original founding alpha male and before 1980 from males recruited into the natal pack (Malcolm 1979). Between 1985 and 1991, when a pack died out, it was often replaced by another group of unrelated or distantly related dogs that formed a new pack and used a similar home range (fig. 19.1) (Burrows, Hofer, and East 1994). Pack life span fell significantly ($p < .001$) between 1985 and 1991 (table 19.1) (Burrows et al., 1994). In SNP/NCA post-1985 and in the Mara/Loita regions post-1987, a new phenomenon of rapid total pack extinction was recorded, with dogs of all ages dying/disappearing within a few weeks. Of the 22 dogs radio-collared between 1985 and 1991, all had died or disappeared by July 1991. Fifty percent of these radio-collared dogs died within 10 months (n = 11, mean 5.72, 3.06 SD) and with them died eight (66%) of the packs studied.

The direct cause of disease-related mortality in wild dogs in the ecosystem is rarely known, but serum samples taken from wild dogs (Lelo 1990; Alexander et al. 1991) between 1985 and 1990 showed that some packs had been exposed to a variety of viral diseases, including parvovirus, coronovirus, adenovirus, and canine herpes. However, the effect of these diseases on the survival of wild dogs is unknown. Four SNP/NCA packs had been exposed to rabies prior to vaccination (Gascoyne, Laurenson, and Borner, 1993), but despite the spread of rabies into the region adjacent to the Serengeti in the late 1970s (Rweyemamu et al. 1973; Magambe 1985), rabies was not confirmed in wildlife in SNP until 1987, when it was found in bat-eared foxes (B. Maas, pers. comm.). In September 1989 rabies was confirmed for the first time in free-living wild dogs, in a pack that died out in the Masai Mara region of Kenya (Scott 1991). In August 1990 a study pack died in NCA, and again rabies was confirmed (Gascoyne et al. 1993). No rabies was reported or suspected in any other species at the time in either case. Rabies was again confirmed in samples from two study packs that died in late 1990 (Alexander, Richardson, and Kat 1992) when a "rabies related extirpation of the African Wild Dog in the Massai Mara, Kenya" took place (Alexander 1992).

In SNP/NCA between June 1986 and July 1991, the whole of the known study population of twelve packs (80 adults and 73 pups) died or disappeared (Fanshawe 1988; Scott 1991; Burrows 1992a; Burrows, Hofer, and East 1994). In all packs at least one member had been immobilized in connection with radio-collaring during the previous 10 months, and five packs contained vaccinated individuals.

Intraspecific Competition

Aggressive pack interactions (van Lawick 1976; Frame and Frame 1981) and serious fighting among individuals, particularly between females for dominance status and possession of pups, have been reported (van Lawick and van Lawick-Goodall 1970; van Lawick 1973; Scott 1991). In 1991 one of a pair of young females who joined the New Barafu pack attempted unsuccessfully to take over the alpha role from the incumbent, unrelated alpha female (Burrows 1993).

Interspecific Competition

Small groups of wild dogs lose kills to spotted hyenas, particularly in the dry season (Kruuk 1972; Frame and Frame 1981), and sometimes to lions (Schaller 1972; R. Burrows, unpub.). Lions capture and sometimes kill adult wild dogs and pups (Reich 1981; Scott 1991; R. Burrows, unpub.). There is some circumstantial but no direct evidence of predation by spotted hyenas on wild dog pups (Frame and Frame 1981). However, in 1991, a pair of wild dogs with no helpers succeeded in raising four pups to at least 3 months of age within 500 m of an active hyena den (R. Burrows, unpub.).

DISCUSSION

Population Trends

Prior to 1974 any sudden disappearance of a pack of wild dogs may have been due to shooting (Frame 1977). The known loss of three packs in this way, plus losses from packs due to individual mortality and dispersal, particularly of females, and lack of immigrants into the study area can explain the reported decline from twelve packs to seven between 1970 and 1974 (Frame et al. 1979), particularly as the number of packs in 1970 had been recently revised to ten (J. Malcolm, cited in Burrows 1993). The decline in the Serengeti wild dog population post-1970 was, as suggested by Frame (1977), more apparent than real (see fig. 19.2).

Reproduction

In wild dogs the alpha pair, particularly the female, influences the breeding of subordinates by overt aggression and agonistic behavior and so limits the pack's annual reproduction. Behavioral control of breeding by subordinates is also reported for wolves (Packard, Mech, and Seal 1983) and jackals (Moehlman 1987). Reproductive suppression in female wild dogs is possibly due to stress-induced hormonal changes leading to the inhibition of, or delay of, ovulation (van Heerden and Kuhn 1985). Support for this theory is provided by the fact that 80% of alpha females in the Serengeti-Mara ecosystem produced their pups before subordinates did.

The use of helpers is common in birds (Brown 1987), and occurs in some mammals such as red foxes *(Vulpes vulpes)* (Macdonald 1983) and black-backed and golden jackals (Moehlman 1979). In a new wild dog pack survival of the first litter will depend upon the number of adult founder members available as helpers to the alpha female. In subsequent years secondary emigration and pack fission may deprive her of these essential helpers, but by then there are typically yearlings in the pack.

Dispersal

High pup mortality and high alpha dog mortality (7 males and 5 females; data from Malcolm 1979) pre-1980 enabled young males to gain breeding opportunities within their natal packs, so male recruitment was high and emigration rare. With alpha mortality confined to packs that died out completely and high pup survival between 1985 and 1990, all yearling males sought breeding opportunities outside their natal packs. This resulted in no male recruitment and total emigration after 1980, a strategy always adopted by yearling females pre-1980 (Frame and Frame 1976; Burrows 1991b; Fuller et al. 1992).

The lack of immigrants pre-1980 in the Serengeti plains study area suggests that packs in the surrounding areas were, like the study packs, not breeding successfully. By contrast, the number of mainly yearling immigrant groups entering the Serengeti plains study area post-1984, even after the study pack deaths of 1991, demonstrates the continuing reproductive success of packs in the regions surrounding this study area.

Mortality and Disease

Changes in the pattern of mortality of the wild dog population between the 1970s and 1980s suggest that different causal agents were involved during the two periods. Between 1967 and 1973 high mortality occurred in some packs (Malcolm 1979), in part attributed to an unidentified "distemper-like disease" (Schaller 1972), which affected adults but was mainly associated with the deaths of pups or whole litters. The virus *Brucella abortus* was found in some dogs (Sachs, Staak, and Groocock 1968), and this may also have led to some litter deaths. From 1974 to 1977 no ailing dogs were seen (Malcolm 1979), and no whole-pack mortality from disease was reported pre-1980.

After 1980, apart from road deaths and one possible case of lion predation, all known adult and yearling mortality occurred when entire study packs died out within a few weeks. Rabies was confirmed in five carcasses (Kock and Gascoyne 1991), and vaccinated dogs died (P. Kat and J. Richardson cited in Macdonald et al. 1992). Rabies is virtually absent in other wildlife regions of Africa despite its presence in domestic animals in the surrounding areas (King 1993; Thomson and Meredith 1993).

The presence of rabies-neutralizing antibodies in the serum of some wild dogs sampled in SNP/NCA up to 2 years prior to the vaccination program (Gascoyne, Laurenson, and Borner 1993) suggested that dogs from four packs had probably been exposed to the rabies virus and survived.

The five handled packs that died or disappeared in SNP/NCA in 1991 all contained rabies-vaccinated individuals and some carried prevaccination antibodies (Gascoyne et al. 1993; Gascoyne, Laurenson, and Borner 1993; Gascoyne et al. 1994; Macdonald et al. 1992). As no tissue samples were collected in SNP/NCA in 1991, the cause of pack disappearance remains unknown, but rabies is suspected (Kock and Gascoyne 1991). This suggestion is supported by the evidence of previous pack exposure and the confirmed rabies-related deaths of wild dogs from three other packs in the ecosystem (Alexander, Richardson, and Kat 1992; Gascoyne et al. 1993). There is therefore good evidence of study pack deaths over a period of 3 years associated with rabies, which is the only disease proven to be associated with pack extinction.

There are similarities between the Mara and Serengeti packs in the way in which the packs were handled and the pattern of subsequent pack mortality. In both sectors of the ecosystem all seven packs that died or disappeared in 1991 did so within a few weeks, and all contained individuals vaccinated against rabies in the previous 12 months (Nicholls 1990; P. Kat and J. Richardson, cited in Macdonald et al. 1992; Gascoyne et al. 1993).

Sightings of unhandled wild dogs after June 1991 outside the study areas suggest that the population crash occurred mainly within the Serengeti and Mara study areas. Macdonald et al. (1992) suggest that the Serengeti dogs did not die from rabies in 1991, but from some other disease such as canine distemper (CDV). However, serum samples taken in the ecosystem post-1985 were negative and exposure to CDV very low in domestic dogs in adjacent areas (Lelo 1990; Alexander et al. 1991). Domestic dogs, a potential but unproven source of rabies and CDV, are largely absent from Serengeti National Park itself but are common in the surrounding areas (Alexander, Richardson, and Kat 1992).

Creel (1992) suggests that the Serengeti wild dog's "low population density and repeated bottlenecks" could be factors producing "the rollercoaster Serengeti population." However, the frequent appearance of immigrant groups post-1980 does not support a genetic bottleneck hypothesis for the 1991 crash, nor is this necessary to explain the decline in the population from 1970 to 1973 in Serengeti.

The significant decline in pack and individual longevity (Burrows, Hofer, and East 1994) post-1980 coincided with the period when extensive handling of packs took place (see table 19.1). This decline, combined with the deaths of 50% of the radio-collared dogs within 12 months of handling, and the disappearance of all the vaccinated dogs within 12 months of vaccination, suggests an association between reduced longevity and handling (Burrows, Hofer, and East 1994). After excluding the Mara vaccinations as handling, another study claims no similar effects (Ginsberg et al. 1995). However, no other wild dog population has the combination of prevaccination rabies antibody titers, postexposure vaccination, and confirmed rabies-related mortality (Burrows and East 1995).

The arrival of new dogs in the SNP/NCA study area post-June 1991 demonstrates that whatever caused the 1991 demise of the SNP/NCA study population did not eliminate the entire non-study population in surrounding areas. These observations led to the handling-stress-rabies hypothesis (Burrows 1992a; Burrows 1994), which suggests that stress resulting from handling for radio collaring and anti-rabies vaccinations may lead to the reactivation of latent rabies virus. This mechanism could account for the sporadic radio-collared pack deaths up to 1990. Alternatively, the vaccinations, by compromising natural immunity to rabies challenge in some individuals, may have led to the die-off of packs containing vaccinated dogs in 1991. The dangers of vaccinating free-living wild animals are set out by Hall and Harwood (1990). Death could result directly from rabies virus—induced disease or indirectly as a consequence of reduction in hunting ability prior to the development of full clinical symptoms, which would lead to death by starvation or secondary infection. Rabies virus could be quickly spread among pack members via oral social greeting (J. Richardson, cited in Scott 1991) or wounds.

Intraspecific Competition

It is likely that ritualized aggression, including play fighting in the first year of life, determines social rank and contributes to the apparently harmonious relationships among pack members. In seven cases the alpha male of a pack was displaced by a younger relative, but displaced alpha males remained within the pack (Malcolm 1979). The few records of competition between adult males leading to death (Frame and Frame 1981; Reich 1981) suggest that such an outcome is unusual.

There is little direct evidence of infanticide; any pup deaths are probably a direct or indirect consequence, due to starvation, or the alpha female's attempts to take over a subordinate's litter. There is, however, considerable evidence for aggression associated with breeding opportunities and access to pups leading to physical injury in pack females (van Lawick and van Lawick-Goodall 1970; van Lawick 1973; Scott 1991; Burrows 1993).

Interspecific Competition

Between 1969 and 1977 there was an increase in the lion population in Serengeti (Hanby and Bygott 1979), particularly on the plains, which is the preferred habitat of wild dogs. Wild dogs have not bred in the Ngorongoro Crater for 25 years (Packer et al. 1991) although prey abundance there is the highest in Africa (van Orsdol, Hanby, and Bygott 1985). In 1962 lion numbers in the Ngorongoro Crater fell dramatically from 60–75 individuals to about 10 following the outbreak of a plague of the bloodsucking fly *Stomoxys calcitrans* (Fosbrooke 1963). In 1966, when the lion population was still at less than a third of its previous numbers, a pack of wild dogs bred successfully in the Crater and raised pups (Estes and Goddard 1967). By 1969, the lion population had built up again to its former level (Schaller 1972), and since then wild dogs have not bred in the Ngorongoro Crater. This suggests that interference competition may exclude wild dogs from certain areas. This could explain the dogs' scarcity in both the Ngorongoro Crater and the Masai Mara National Reserve, where lion density is also high, as well as their continuing presence in areas with low lion densities, adjacent to human settlements and on Masai grazing land at the margins of the ecosystem.

A reported increase in the number of spotted hyenas resident on the plains during Serengeti wet seasons (Hanby and Bygott 1979) also coincided with the reported decline in Serengeti wild dogs in the 1970s (Frame et al. 1979), and hyenas were thus considered a threat to the survival of wild dogs on the plains (Frame 1986). Despite this decline, however, numbers of dogs built up again between 1984 and 1985 (T. Caro, cited in Frame 1986, and photographic evidence from P. Moehlman, D. Rechsteiner, and A. Earnshaw). Moreover, Hofer and East (chap. 16) show that the resident population of spotted hyenas on the plains remained constant throughout this period. When wild dogs last bred successfully in the Ngorongoro Crater in 1966 there was no evidence of a reduction in the hyena population there. Published evidence for hyena predation on wild dog pups is circumstantial (Frame and Frame 1981) and is not reported for the post-1985 research period. Prey stealing by spotted hyenas is common (Kruuk 1972; Schaller 1972; R. Burrows, unpub.). Food competition is considered a major cause of wild dog pup mortality (Frame et al. 1979) and a limiting factor on wild dog distribution in SNP/NCA (Frame 1986). Interspecific competition from hyenas is also thought to be a factor favoring the maintenance of large wild dog packs (Fanshawe and FitzGibbon 1993); this suggestion is supported by the lack of significant prey stealing by hyenas from large packs of wild dogs (Fuller and Kat 1990; Fanshawe and FitzGibbon 1993). However, any effects of hyenas on wild dog survival or distribution remain to be quantified.

CONCLUSIONS

The main conclusions drawn from this study are as follows:

1. The wild dogs in the Serengeti-Mara ecosystem form one breeding population and are not genetically isolated.

2. Changes in the pattern of mortality, particularly in pups, produced contrasting pack structures and dynamics before and after 1980.

3. The 1970s study population decline on the Serengeti plains continued after 1985 but began to level off. The extinction of all study packs in 1991 was unlikely to have been due to chance.

4. Until 1991, a few successful packs in each area of the ecosystem were able to maintain the population by emigration and pack fission.

5. High pup and adult mortality resulted in recruitment of male yearlings, producing highly skewed secondary sex ratios in packs.

6. Young males were recruited into their natal pack only if pup survival was low or one parent died. A male from the youngest cohort present typically became alpha male.

7. Low mortality in packs led to an expanding population, yearling emigration (except as in 6.), and rapid fluctuations in pack size.

8. Genetic relationships between members of the same sex in packs were variable.

9. The alpha pair monopolized reproduction. Beta pups were taken over by the alpha female, and they survived when food was not limiting.

10. Intraspecific aggression was not an important direct cause of mortality but could have led to stress and injury.

11. The suggested link between increased hyena numbers in the 1970s and the decline/extinction of the wild dog study populations is not supported by the data, but lions may exclude wild dogs from some areas.

12. Handling was correlated with reduced longevity in the Serengeti wild dog study population.

13. Shooting and disease in pups were probably the major limiting factors for wild dogs in the ecosystem before 1980 but disease, the only limiting factor identified after 1985, resulted in a new phenomenon, the loss of whole study packs.

14. Rabies, possibly handling-stress related, resulted in study pack deaths throughout the ecosystem after 1985. Other unidentified factors may have contributed to the study population decline. Hypotheses on the cause(s) of the extinction of all study packs throughout the ecosystem need further investigation. However, the value of the information gained from handling wild dogs in an effort to conserve them must be weighed against known risks to the species' survival.

ACKNOWLEDGMENTS

I wish to acknowledge the contributions of G. and L. Frame, J. Malcolm, and members and associates of HDP/SEMP: M. Borner, T. Caro, K. Campbell, J. Fanshawe, C. FitzGibbon, S. Gascoyne, S. Huish, K. Laurenson, S. Lelo, J. Malcolm, J. Scott, and C. and L. Trout; Loita Wild Dog Project members P. Kat and J. Richardson; scientists and other friends at SWRC and Ndutu; Serengeti National Park rangers and tourists for supplying dog sightings and photographs; J. Corlett, M. Deeble, B. Figenschou, P. Moehlman, H. van Lawick, and V. Stone for logistical support; and P. Chanin, M. East, H. Hofer, and J. Malcolm for scientific and statistical advice. This study was partly funded by Mr. and Mrs. N. Silverman via World Wildlife Fund, and I thank them for their support.

REFERENCES

Alexander, K. 1992. Reduction of biodiversity: The role of disease. *Tail Tips: A Newsletter from the University of Minnesota African Wild Dog Committee,* ed T. Nicholls, 3:4.

Alexander, K. A., Conrad, P. A., Gardener, I. A., Lerche, N., Parish, C., Appel, M. G., Levy, M. G., and Kat, P. W. 1991. Serologic survey for selected microbial pathogens in African wild dogs *(Lycaon pictus)* and sympatric domestic dogs *(Canis familiaris)* in Maasai Mara, Kenya. In *Proc. Am. Assoc. Zoo Vet.,* 242–43.

Alexander K. A., Richardson, J. D., and Kat, P. W. 1992. Disease and conservation of African wild dogs. *Swara* 15(6):13–14.

Ammann, K. 1987. Wild dogs in the Masai Mara. *Swara* 10(5):8–9.

Bere, R. M. 1956. The African wild dog. *Oryx* 3:180–82.

Brown, J. L., 1987. *Helping and communal breeding in birds: Ecology and evolution.* Princeton, N.J.: Princeton University Press.

Burrows, R. 1990. Wild dogs. *Conserv. Monitoring News,* 2:6–7. Arusha, Tanzania: Tanzania Wildlife Conservation Monitoring.

———. 1991a. Observations on the behaviour, ecology and conservation status of African wild dogs *(Lycaon pictus)* in Serengeti National Park. Seronera, Tanzania: Serengeti Wildlife Research Institute.

———. 1991b. Wild dogs of the Serengeti. Part 1. *Miombo* 7:4–5.

———. 1992a. Rabies in wild dogs. *Nature* 359:277.

———. 1992b. Wild dogs of the Serengeti. Part 2. *Miombo* 8:4–5.

———. 1993. Observations on the behaviour, ecology, and conservation status of African wild dog in SNP, October 1989—December 1992. Scientific report 1990–1992, 53–59. Serengeti Wildlife Research Centre.

———. 1994. Reply to Gascoyne et al. *J. Wildl. Dis.* 30:297–99.

Burrows, R., and East, M. L. 1995. Reply to Ginsberg et al. *Conserv. Biol.* (in press).

Burrows R., Hofer H., and East, M. L. 1994. Demography, extinction and intervention in a small population: The case of the Serengeti wild dogs. *Proc. R. Soc. Lond. B* 256:281–92.

Creel, S. 1992. Causes of wild dog deaths. *Nature* 360:633

Estes, R. D., and Goddard, J. 1967. Prey selection and hunting behavior of the African Wild Dog. *J. Wildl. Mgmt.* 31(1):52–70.

Fanshawe, J. H. 1988. Serengeti Hunting Dog Project. In *Serengeti Wildlife Research Centre, Annual Report 1986–87,* 37–40.

Fanshawe, J. H., and FitzGibbon, C. D. 1993. Factors influencing the hunting success of an African Wild Dog pack. *Anim. Behav.* 45:479–90.

Fosbrooke, H. 1963. The *Stomoxys* plague in Ngorongoro 1962. *E. Afr. Wildl. J.* 1:124–26.

Frame, G. W. 1986. Carnivore competition and resource use in the Serengeti ecosystem of Tanzania. Ph.D. thesis, Utah State University, Logan.

Frame, G. W., and Frame, L. H. 1976. Population study of cheetahs and wild dogs. In *Serengeti Research Institute, Annual Report 1974–1975,* 129–45.

———. 1981. *Swift and enduring.* New York: Dutton.

Frame, L. H. 1977. Wild dog ecology and behaviour. In *Serengeti Research Institute Annual Report 1975–1976,* 87–103.

Frame, L. H., and Frame, G. W. 1976. Female African Wild Dogs emigrate. *Nature* 263:227–29.

Frame, L. H., Malcolm, J. R., Frame, G. W., and van Lawick, H. 1979. Social organization of African wild dog *(Lycaon pictus)* on the Serengeti Plains, Tanzania, 1967–78. *Z. Tierpsychol.* 50:225–49.

Fuller, T. K., and Kat, P. W. 1990. Movements, activity, and prey relationships of African wild dogs *(Lycaon pictus)* near Aitong, southwestern Kenya. *Afr. J. Ecol.* 28:330–50.

Fuller, T. K., Kat, P. K., Bulger, J. B., Maddock, A. H., Ginsberg, J. R., Burrows, R., McNutt, J. W., and Mills, M. G. L. 1992. Population dynamics of African wild dogs. In *Wildlife 2001: Populations,* ed. D. R. McCullough and R. H. Barrett, 1125–39. London: Elsevier Science Publishers.

Gascoyne, S. C., Laurenson, M. K., Lelo, S., and Borner, M. 1993. Rabies in African wild dogs *(Lycaon pictus)* in the Serengeti region, Tanzania. *J. Wildl. Dis.* 29:396–402.

Gascoyne, S. C., Laurenson, M. K., and Borner, M. 1993. Rabies and African wild dogs, *Lycaon pictus.* In *Proc. Int. Conf. Epidemiology: Control and Prevention of Rabies in E. and S. Africa,* ed. A. King, 133–40. Fondation Marcel Mérieux.

Gascoyne, S. C., King, A. A., Laurenson, M. K., Borner, M., Schildger, B., and Barrat, J. 1993. Aspects of rabies infection and control in the conservation of the African wild dog *(Lycaon pictus)* in the Serengeti region, Tanzania. *Onderstepoort J. Vet. Res.* 60:415–20.

Ginsberg, J. R., and Macdonald, D. M. 1990. *Foxes, wolves, jackals and dogs: An action plan for the conservation of canids.* Gland, Switzerland: IUCN.

Ginsberg, J. R., Alexander, K. A., Creel, S., Kat, P. W., McNutt, J. W., and Mills, M. G. L. 1995. Handling and survivorship in the Wild Dog *(Lycaon pictus)* in five African ecosystems. *Conserv. Biol.* In press.

Hall, A., and Harwood, J. 1990. *The Intervet guidelines to vaccinating wildlife.* Sea Mammal Research Unit, Cambridge.

Hanby, J. P., and Bygott, J. D. 1979. Population changes in lions and other predators. In *Serengeti: Dynamics of an ecosystem,* ed. A. R. E. Sinclair and M. Norton-Griffiths, 249–62. Chicago: University of Chicago Press.

Kat, P. 1990. Tragedy strikes the Aitong pack. In *Tail Tips: A Newsletter from the University of Minnesota African Wild Dog Committee,* ed. T. Nicholls, 1:2–3.

King, A. A. 1993. African overview and antigenic variation. In *Proc. Int. Conf. Epidemiology: Control and Prevention of Rabies in E. and S. Africa,* ed. A. King, 57–68. Fondation Marcel Mérieux.

Kock, R., and Gascoyne, S. 1991. *Rabies in the wild dog* (Lycaon pictus) *in East Africa: Crisis management.* Health and Management of Free-Ranging Mammals Symposium, Centre National d'Etudes Vétérinaires et Alimentaires.

Kruuk, H. 1972. *The spotted hyena: A study of predation and social behavior.* Chicago: University of Chicago Press.

Kühme, W. 1965. Communal food distribution and division of labour in African Hunting Dogs. *Nature* 205:443–44.

Lelo, S. 1990. Wild Dog Project. In *Serengeti Wildlife Research Centre Biennial Report, 1988–1989,* ed. S. A. Huish and K. L. I. Campbell, 27–30. Seronera: Serengeti Wildlife Research Centre.

Macdonald, D. W. 1983. The ecology of carnivore social behaviour. *Nature* 30:379–84.

Macdonald, D. W., Artois, M., Aubert, M., Bishop, D. L., Ginsberg, J. R., King, A. A., Kock, N., and Perry, B. D. 1992. Causes of wild dog deaths. *Nature* 360:633–34.

Magambe, S. R. 1985. Epidemiology of rabies in the United Republic of Tanzania. In *Rabies in the Tropics,* ed. E. Kuwert, C. Merieux, H. Koprowski, and K. Bogel, 392–98. New York: Springer-Verlag.

Malcolm, J. R. 1979. Social organisation and communal rearing in African Wild Dogs. Ph.D. thesis, Harvard University.

Malcolm, J. R., and Marten, K. 1982. Natural selection and the communal rearing of pups in African Wild Dogs *(Lycaon pictus) Behav. Ecol. Sociobiol.* 10:1–13.

Maugham, R. C. F. 1914. *Wild Game in Zambesia.* London: John Murray.

Moehlman, P. D. 1979. Jackal helpers and pup survival. *Nature* 277:382–83.

———. 1987. Social organization in jackals. *Am. Sci.* 75:366–75.

Nicholls, T. 1990. Tracking the Intrepid pack. In *Tail Tips: A Newsletter from the University of Minnesota African Wild Dog Committee,* ed. T. Nicholls, 1:3–4.

Packard, J. M., Mech, L. D., and Seal, U.S. 1983. Social influences on reproduction in wolves. In *Wolves in Canada and Alaska,* ed. L. Carbyn, 78–85. Canadian Wildlife Service report no. 88. Ottawa: Canadian Wildlife Service.

Packer, C., Pusey, A. E., Rowley, H., Gilbert, D. A., Martenson, J., and O'Brien, S. J. 1991. Case study of a population bottleneck: Lions of the Ngorongoro Crater. *Conserv. Biol.* 5:219–30.

Reich, A. 1981. The behavior and ecology of the African Wild Dog *(Lycaon pictus)* in the Kruger National Park. Ph.D. thesis, Yale University.

Rweyemamu, M. M., Loretu, K., Jakob, H., and Gorton, E. 1973. Observations on rabies in Tanzania. *Bull. Epizoot. Dis. Afr.* 21:19–27.

Sachs, R., Staak, C., and Groocock, C. 1968. Serological investigation of brucellosis in game animals in Tanzania. *Bull. Epizoot. Dis. Afr.* 16:93–100.

Schaller, G. B. 1972. *The Serengeti lion: A study of predator-prey relations.* Chicago: University of Chicago Press.

Scott, J. 1991. *Painted wolves: Wild dogs of the Serengeti-Mara.* London: Hamish Hamilton.

Sinclair, A. R. E. 1979. The Serengeti environment. In *Serengeti: Dynamics of an ecosystem,* ed. A. R. E. Sinclair and M. Norton-Griffiths, 31–45. Chicago: University of Chicago Press.

Thomson, G., and Meredith, C. 1993. Wildlife rabies in southern Africa. In *Proc. Int. Conf. Epidemiology: Control and Prevention of Rabies in E. and S. Africa,* ed. A. King, 166–74. Fondation Marcel Mérieux.

van Heerden, J., and Kuhn, F. 1985. Reproduction in captive hunting dogs *(Lycaon pictus). S. Afr. J. Wildl. Res.* 15:80–84.

van Lawick, H. 1973. *Solo.* London: Collins.

———. 1976. *Savage paradise.* London: Collins.

van Lawick, H., and van Lawick-Goodall, J. 1970. *Innocent killers.* London: Collins.

van Orsdol, K. G., Hanby, J. P., and Bygott, J. D. 1985. Ecological correlates of lion social organization. *J. Zool.* (Lond.) 206:97–112.

TWENTY

Habitat Variation and Mongoose Demography

Peter M. Waser, Lee F. Elliott, Nancy M. Creel, and Scott R. Creel

The conspicuousness of large Serengeti carnivores makes it easy to overlook the astounding density and diversity of small carnivores: twenty species of canids, felids, hyenids, mustelids, and viverrids are predators of insects and small vertebrates within the park. In this chapter we describe the demography and population dynamics of the three most common diurnal viverrids, the dwarf mongoose *(Helogale parvula),* banded mongoose *(Mungos mungo),* and slender mongoose *(Herpestes sanguineus).* These three species, along with the nocturnal white-tailed mongoose *(Ichneumia albicauda),* have been the focus of the Serengeti mongoose project, initiated in 1974 by the late Jon Rood.

Concentrating on *Helogale* but discussing the other species where data are available, we briefly review social structure and describe age-specific patterns of survival and fecundity. We then document long-term trends in population size as well as annual and seasonal patterns of fecundity and mortality. Some, but not all, of these patterns can be predicted from rainfall data. Rainfall influences both the density of prey available to mongooses and (through its effect on grass height) their vulnerability to predators.

Helogale and *Mungos* are predators of invertebrates (Rood 1975, 1986); *Herpestes* is more partial to small vertebrates (Rood and Waser 1978). All three species are potentially vulnerable to a wide variety of larger predators, both mammalian and avian. We examine the relationships between an index of vulnerability to predators (grass height), two indices of food availability (sweep samples, which assay Orthoptera, and dung counts, which assay Coleoptera), and *Helogale* survival and fecundity. Finally, we characterize spatial variation in vegetation, termite mound density, and other sources of cover or den sites within the study area. We examine the relationship of these indicators of habitat quality to *Helogale* distribution and reproduction.

METHODS

Population Density, Fecundity, and Survival

Our demographic data come from individually marked study populations established by Jon Rood. Rood's observations, and those of the rest of us since 1987, have emphasized *Helogale*. Between 1974 and 1991, we recorded individual histories for 713 marked individuals, constituting most of the dwarf mongooses living within Rood's 5 by 5 km "Sangere" study area, which includes the SWRC and the Sangere River drainage to the north and east. Since 1977, all dwarf mongooses in this area, including unmarked animals, have been censused at the beginning of each birth season (November).

We used these censuses to determine age-specific fecundity and survival rates. Data on annual survival are available for sixteen cohorts, including nine cohorts (1974–1983) whose members have all died. In estimating annual survival rates, we assumed that immigration onto and emigration off of the Sangere study site were in balance, and estimated the annual survival of age class x in year t by counting the animals of age class $x + 1$ in year $t + 1$, including immigrants onto the site but excluding emigrants off the site, even if they were found in adjacent areas. Immigrants onto the study site were aged by tooth wear (Rood 1980). Our life tables are thus cohort life tables, with annual survival and fecundity averaged over multiple cohorts.

Since the Sangere population was surrounded by similar habitat and by similarly dense mongoose populations, our assumption that immigration balanced emigration seems appropriate. However, some immigrants (18%) could not be aged by tooth wear, and these immigrants are the source of a potential error in our survival estimates: the numbers of immigrants in older age classes, and thus the number of survivors reaching those age classes, was underestimated. Since unknown-aged immigrants constitute a minority of all immigrants, and immigrants constitute a minority of each age class, the magnitude of this error is small; we estimate it to be less than 5%.

We also used the annual censuses to calculate adult mortality and recruitment rates each year. Again, to determine survival rates we estimated the number of adult deaths as the number of adults that died or disappeared from Sangere groups (including those that emigrated off the study site) minus the number of adult immigrants into those same groups. We estimated the overall recruitment rate in year t by counting the number of juveniles that survived to the census in year $t + 1$, divided by the number of adult females in year t. Similarly, we estimated the recruitment rate for alpha females as the number of juveniles recruited in year $t + 1$ per alpha female in year t.

Annual measures of recruitment are influenced by the number of litters produced per year, by litter size, and by juvenile survival. During 7 years, an observer was present throughout the breeding season. For closely observed groups in these years, we estimated juvenile survival by counting all litters when they first emerged from the den and subsequently counting the number of yearlings recruited into the same groups. These counts also gave us measures of the number of litters per pack and of litter size. Finally, we calculated age-specific fecundities from these data by determining the mean numbers of offspring produced per group and breeding season as a function of the age of the alpha female, and multiplying these means by the age-specific probabilities of being dominant. For simplicity, we assumed that all reproduction is by the alpha female, although DNA analyses indicate that a small proportion of offspring have subordinate parents (Keane et al. 1994).

Monthly variation in per capita birth rate was investigated by determining, for each month, how many marked Sangere females were under observation over the entire study period and by counting the number of litters produced by those females. We counted all litters when they emerged and when they were weaned (at an age of approximately 3 months). These data also allowed us to look at monthly variation in litter size, number of litters per alpha female, number of young produced and weaned per alpha female, and juvenile mortality between first count and weaning. Sample sizes varied among months, as observers were more often present during the wet season.

We investigated monthly variation in adult per capita death rates by counting the number of deaths and disappearances observed in the marked population each month and dividing it by the number of animals that were under observation during that month. Deaths and disappearances that occurred in months when the observer was absent were apportioned equally to those months (for instance, if 11 animals disappeared over a 3-month period, 11/3 animals were assumed to have died each month). Because of this, we probably underestimated the magnitude of seasonal variation in mortality.

In addition to *Helogale,* Rood also began marking *Mungos* in 1974, but until 1980 he concentrated his work on this species in other parts of the Serengeti. Between 1980 and 1987, he marked 108 banded mongooses within the Sangere study area, representing a majority of the individuals in three packs, and he censused all individuals within the study area each November. Most groups were also counted in 1987–1990.

Since 1975, 75 *Herpestes* have been marked, and locations of these animals have been recorded whenever they were sighted. Most observations have been concentrated in a 1.5 km^2 subset of the Sangere study area surrounding the SWRC.

Weights. *Mungos* and *Herpestes* were weighed when trapped. *Helogale* were weighed by placing a digital platform scale on den sites and reading the weight when animals climbed on the scale (Creel and Creel 1991).

Environment

Rainfall. Data on monthly rainfall from daily records for the SWRC meteorological station were generously provided by Ken Campbell and Sally Huish of the Serengeti Ecological Monitoring Programme.

Temporal Environmental Variation. Between October 1989 and April 1991 we determined vegetation height monthly by measuring grass/scrub canopy height over a 1 m^2 area at twenty-five random locations within each of seven *Helogale* home ranges. Three of the seven home ranges, located north of SWRC, were classified as woodland and were characterized by approximately 89% tree and scrub and 8% grass cover (see table 20.3); the other four, south of SWRC, were classified as grassland (21% tree and scrub, 68% grass cover). Canopy height was measured by dropping a light wire frame strung with nylon mesh down a pole marked out in centimeters.

Between grass/scrub canopy height measurements, we walked a 50 m transect, counting all ungulate droppings in a 2 m wide corridor, which gave a measure of dropping densities over an area of 2,500 m^2 for each home range. A variety of efforts were made to sample soil coleopteran larval densities more directly, but we concluded that adequate sampling was unlikely without a backhoe.

A relative measure of the seasonal and habitat distribution of orthopteran biomass was determined by sweep samples performed at quarterly intervals in six of the seven home ranges, three in grassland and three in woodland. Seasons were distinguished as early dry season (May, June, July), late dry season (August, September, October), early wet season (November, December, January), and late wet season (February, March, April). In each home range, three 0.1 ha plots were thoroughly swept using a net with a 38 cm opening diameter. Approximately five hundred sweeps (side-to-side distance of about 2 m) were made in each plot. Folse (1978) determined that the number of acridid individuals per sweep was proportional to the number of individuals per square meter. Sampling was done on clear and windless days between the hours of 1000 and 1400 when temperatures ranged between 29°C and 35°C. Acridids clearly constituted the majority of the biomass in the sweep samples, as has been observed by others (Sinclair 1975; Folse 1978). Adult and nymphal acridid grasshoppers were categorized into 0.5 cm size classes, and numbers were recorded for each age-size class. Individuals from each age-size class were oven-dried at 70°C for 24 hours and weighed to the nearest

0.0001 g. Biomass was then calculated for each plot. During the late wet season sampling period, when vegetation was at its peak, vegetation height was measured directly at 25 points in each 0.1 ha insect plot by the methods described above.

Helogale Home Range Quality

Together with Markus Borner of the Serengeti Ecological Monitoring Programme, we made an aerial survey of our study site, taking photographs from 1,500 feet (measured by radar altimeter) with a vertically mounted Nikon F3 camera, an 18 mm Nikkor lens, and Ektachrome 5036 film. The resulting slides were projected at 10 times magnification, yielding a scale of approximately 1:2,500, onto a 12 by 18 grid with two randomly located dots within each grid cell. For the same six *Helogale* home ranges sampled for grass height and orthopteran density, we estimated home range size by counting grid cells on these projections. We used the dot grid to estimate the percentages of each home range made up of grass, scrub, trees with > 5 m canopy diameter, kopje (rock outcropping), and korongo (seasonal watercourse). The numbers of trees (> 5m diameter) and termitaries (primarily *Macrotermes*) were also counted, allowing a direct measure of density. Unused areas immediately adjacent to the six home ranges were also analyzed, allowing Wilcoxon matched-pairs signed-rank tests to compare habitat within and outside of home ranges.

RESULTS

Helogale

Social Structure. The dwarf mongoose population is a collection of stable social groups, ranging in size from 2 to 21 (mean size at the beginning of the breeding season = 8.9), among which both sexes can transfer as individuals. Emigration rates peaked at ages 1–2 for both sexes. A higher proportion of males than females emigrated, and they did so over a broader range of ages. Occasionally, dwarf mongooses dispersed in groups of two or three. Dispersers usually immigrated as subordinates, but transient males occasionally "took over" groups, evicting resident males, and occasionally founded new groups (Rood 1987, 1990; Waser, Creel, and Lucas 1994; Creel and Waser 1994).

Some groups failed during the study due to the death of all adult females followed by the departure of males; other small groups died out completely. New groups formed when solitary females remained resident in an area for a period of months and were joined by transient males. No cases of groups merging or splitting were observed. The distribution of Sangere group longevities was bimodal: of 41 groups followed during the study, 18 groups lasted only 1–2 years, while 6 lasted for the entire 16 years of the study.

As has been described in detail elsewhere, subordinate dwarf mongooses of both sexes are reproductively suppressed, and most reproduction is by the oldest, dominant individuals of each group. Subordinates helped to raise the young of dominants by guarding them from predators, by grooming and carrying them, by bringing them food, and on occasion, by lactating; this help significantly increased juvenile recruitment (Rasa 1973; Rood 1978, 1980, 1983b, 1990; Creel et al. 1991, 1992; Keane et al. 1994).

Growth, Survival, and Fecundity. Mean adult body mass was 354 g for dwarf mongooses. Full adult weight was attained at roughly 2 years (fig. 20.1a). At sexual maturity (1 year) dwarf mongooses were at 88% of asymptotic weight. *Helogale* were monomorphic in adult body mass and in growth profiles. Growth was negatively affected by litter size, and positively affected by the number of adult and yearling helpers within the pack, and by rainfall during the breeding season (Creel and Creel 1991).

Of 374 juveniles counted soon after emergence from the den, 155 (41%) survived to their first November. Marked juvenile males apparently disappeared at the same rate as females ($G = 1.66$, $N = 167$, $P = .20$). Juveniles remained in the den for their first few weeks of life and additional mortality may have occurred at this time.

Survival among yearlings and adults was much higher (table 20.1). Averaged across ages 1–14, mean male annual survival was 0.68, female 0.74. Maximum observed longevity was 10 years for males, 14 for females. Survivorship declined almost linearly with age, at nearly identical rates for both sexes (fig. 20.2).

Helogale fecundity schedules (table 20.1, figure 20.2) show that for both sexes, age-specific fecundity increased steadily and markedly from ages 1 to 10 (females: $b = 0.28$, $t = 3.83$, $P = .003$; males: $b = 0.39$, $t = 4.89$, $P = .002$). For males, this effect was entirely because a higher percentage of males bred in each successive age class (Creel and Waser 1991). Dominance is highly dependent on age for both sexes in dwarf mongooses (Rood 1980; Creel et al. 1992). Only 1% of yearling males attained dominant, breeding positions, although radioimmunoassay of urinary androgen metabolites indicated that yearlings are sexually mature (Creel et al. 1992). This percentage increased steadily with age up to 70% for 6-year-olds and remained constant thereafter. A very similar pattern held among females: 1% of yearlings attained breeding positions, with this percentage increasing to 71% for 6-year-olds and remaining steady to age 10. However, an additional effect operated among females only: annual reproductive success increased with experience as a breeder among females, but not among males (Creel and Waser 1991, 1994).

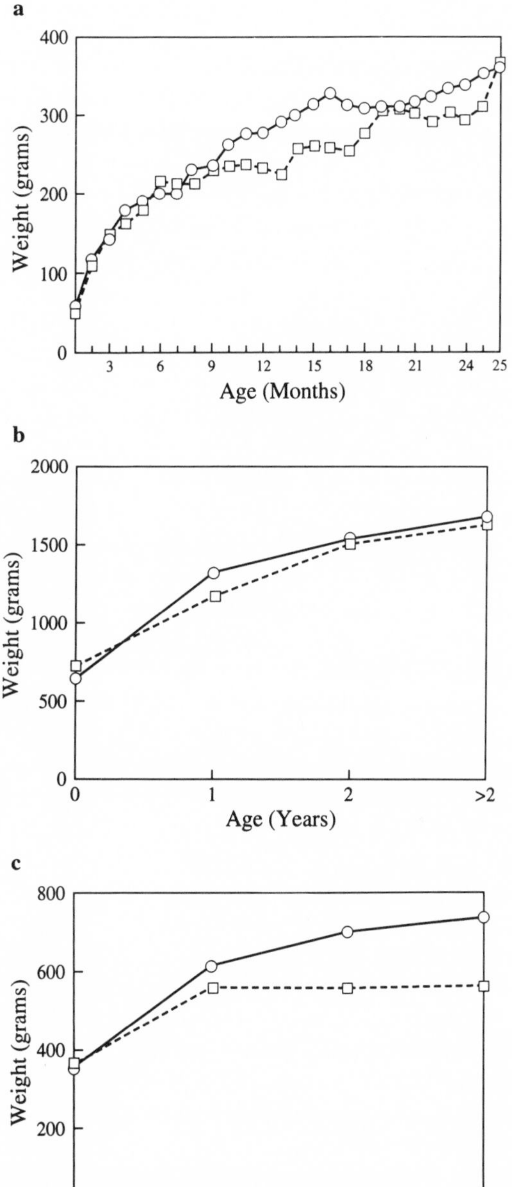

Figure 20.1 Mean weight versus age for (*a*) *Helogale,* (*b*) *Mungos,* and (*c*) *Herpestes.* Circles, males; squares, females. Pregnant females are excluded, and ages >2 for *Mungos* and >3 for *Helogale* are lumped because of small sample size. For *Herpestes,* few adults were of known age, so weight is broken down by categories of tooth wear: deciduous (<1 year); slight (adult dentition, molars and carnassial premolars unworn or slightly worn, 1–2 years); medium (probably 3–4 years); and heavy (probably >4 years).

Table 20.1 *Helogale* life table, based on nine cohorts whose members have all died.

	Males				Females			
x	n_x	l_x	s_x	m_x	n_x	l_x	s_x	m_x
0	444	1.000	0.410	0.00	402	1.000	0.410	0.00
1	182	0.410	0.593	0.00	165	0.410	0.800	0.00
2	102	0.243	0.814	0.09	134	0.328	0.769	0.21
3	77	0.198	0.766	0.72	101	0.252	0.723	0.39
4	51	0.152	0.667	1.70	63	0.182	0.778	0.95
5	30	0.101	0.633	1.46	40	0.142	0.600	1.32
6	19	0.064	0.789	2.13	18	0.085	0.667	1.48
7	14	0.051	0.571	1.93	11	0.057	0.545	2.45
8	8	0.029	0.500	1.83	9	0.031	0.667	3.78
9	3	0.014	0.667	2.44	6	0.021	0.667	2.56
10	1	0.010	0.000	4.44	3	0.014	0.333	4.07
11	0	0.000		0.00	2	0.005	1.000	3.76
12	0	0.000		0.00	2	0.005	0.500	3.00
13	0	0.000		0.00	1	0.002	1.000	2.00
14	0	0.000		0.00	1	0.002	0.000	0.00

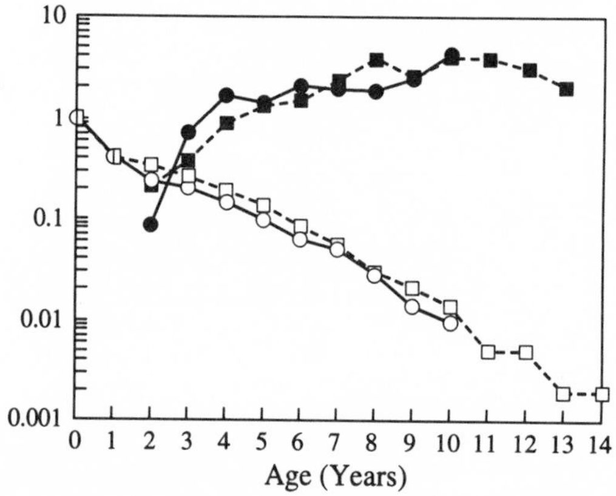

Figure 20.2 *Helogale* survivorship (l_x; open symbols) and fecundity (m_x; solid symbols). Note log scale. Circles, males; squares, females.

Net reproductive rate calculated from table 20.1 was 1.06 (based on females) or 0.94 (based on males). The lower R_0 for males may reflect unaged immigrants (see above). These values indicate a stable or slightly increasing *Helogale* population.

Mungos

Social Structure. Like the dwarf mongoose population, the Sangere banded mongoose population is a collection of social groups; mean group size was 15 at the beginning of the breeding season (range 6–21 on the Sangere site, 4–29 elsewhere). Like dwarf mongoose groups, banded mongoose groups were linked by both male and female transfers. Of 24 known-aged marked males, only 2 remained in their natal groups as 4-year-olds, while 13 were observed to emigrate or disappeared under circumstances suggestive of emigration (group takeovers, see below). Of 18 known-aged marked females, 4 remained as 4-year-olds and 1 was observed to emigrate. Thus males were more likely than females to transfer among groups (Rood 1975, 1982, 1986; J. P. Rood, unpub.).

Several banded mongoose males often transferred between groups as a unit. Males were sometimes observed to leave their groups and temporarily invade the ranges of neighboring packs before returning. In addition, transient groups of 1–4 males were commonly sighted, sometimes repeatedly over a period of up to 3 months. On three occasions marked groups were subject to apparent takeovers, involving the eviction of all pack males and simultaneous immigration of several unmarked animals.

Banded mongoose groups were more dynamic than dwarf mongoose groups in other ways. Groups not infrequently split into subgroups that foraged independently for periods of days. Groups were twice observed to split permanently, and on several occasions groups suddenly decreased markedly in size under circumstances that suggested a fission.

Within groups, females produced young synchronously, and in contrast to dwarf mongooses, reproductive suppression of females was incomplete or absent. Adults of both sexes guarded, carried, and fed young, and females with simultaneous litters apparently suckled each other's offspring; in some cases, predator defense by nonparents was literally lifesaving (Rood 1974, 1975, 1983a, 1986). Perhaps because of the small number of group-years available for analysis, group size did not significantly affect the number of yearlings recruited (b = 0.34 yearlings/adult, t_{12} = 1.13, P = .28).

Growth, Survival, and Fecundity. Banded mongooses reached adult weight by age 3 years. Males were significantly heavier than females as yearlings (t_{62} = 4.08, $P < .001$) but the difference was very slight by age 3 (males, 1,683 g; females, 1,637 g; t_{64} = 1.14, P = .26). At age 1, males were at 80% of their adult weight, females at 72% (fig. 20.1b).

Of 115 banded mongoose juveniles counted at 5–10 weeks of age, 46% survived to their first November. Survival rates of marked male and female juveniles were equal (G = 0.07, N = 48, P = .79). As with dwarf mongooses, additional mortality might have occurred prior to our first counts.

Based on 1980–1985 cohorts from the three most completely marked groups, yearling banded mongooses survived only slightly better than juveniles (table 20.2). Even adults apparently survived at rather low rates: adult males 2 years and older survived at an annual rate of 0.65, adult

Table 20.2 *Mungos* life table, based on marked animals in three groups.

	Males			Females			
x	n_x	l_x	s_x	n_x	l_x	s_x	m_x
0		1.00	0.46		1.00	0.46	0.00
1	27	0.46	0.48	18	0.46	0.44	0.00
2	12	0.22	0.83	6	0.20	0.67	1.24
3+	28	0.18	0.57	29	0.14	0.69	1.77

females at 0.69. Marked females were observed to live at least 8 years, males at least 6.

Among marked females, reproduction was first observed in 2-year-olds (even though captive yearlings are capable of reproduction: Rood 1975). Two-year-olds became pregnant at lower rates than older females; in cases of synchronous litters involving known-aged females, 8 of 12 2-year-olds became pregnant versus 22 of 23 females aged 3 years and older ($P = .04$, Fisher exact test). The size of the joint litter increased significantly with the number of synchronously pregnant females ($b = 1.79$ juveniles/female, $t_{13} = 2.32$, $P = .037$). It was not possible to apportion offspring among females that were synchronously pregnant; at first count, 15 litters involving 49 females produced 126 young, indicating a mean contribution of 2.6 young per female. Marked packs produced 1.4 litters per year (table 20.2).

Herpestes

Social Structure. With few exceptions, Sangere slender mongooses interacted as individuals. *Herpestes* juveniles generally dispersed during their first 6 months (Rood and Waser 1978). Of sixteen females trapped as juveniles, twelve disappeared from the study site after short periods of independent foraging within the maternal range; some of these probably dispersed off the site. One female dispersed within the study site, and three continued to forage as independent adults on their natal home ranges for up to 2 years. When their mothers died, these females inherited their home ranges.

While both sexes evidently dispersed substantial distances, males appeared to disperse earlier and farther. Males were only 25% as likely as females to be captured with their mothers, suggesting that they emigrated earlier. Only one young male remained on the study site as a yearling, and one was sighted as a 5-year-old more than 3 km from his natal range. Once established on an adult home range, however, males maintained stable, amicable relationships with other males, with which they shared home ranges for periods of up to 7 years (Rood 1989; Waser et al. 1994).

Growth, Survival, and Fecundity. By age 1, slender mongooses were close to adult weight and were considerably more dimorphic than the other two species (males: 713 g; females, 574 g; $t_{53} = 8.29$, $P < .001$) (fig. 20.1c).

During the study, 35 juveniles were detected and 22 young animals settled on the site (including the 5 that remained on their natal ranges). Assuming that the study site is neither a source nor a sink for dispersing juveniles, this suggests a juvenile survival rate of 0.63. Disappearance rates of marked males and females did not differ ($p = .66$, Fisher test, $N = 20$).

Once established as adults, male and female *Herpestes* had high annual survival rates: 0.82 for males (9 disappearances in 51 male-years) and 0.79 for females (8 disappearances in 39 female-years). Maximum observed longevity was 8 years for both sexes.

Herpestes females were capable of reproduction as yearlings. Females produced 1–2 litters of 1–3 young per year; 35 young were observed in 39 female-years, suggesting a mean of 0.90 young per female per year.

Population Density, Recruitment, and Mortality Patterns

Helogale, Mungos, and *Herpestes* all maintained high population densities on the Sangere study site (fig. 20.3). All three species apparently increased in number during the mid-1980s, decreasing again in the last few years of the study. Overall, there has been a trend toward increasing population size in *Helogale* and *Herpestes,* but neither a net increase nor a decrease for *Mungos.* This relative population stability is striking given the major changes in vegetation that have occurred during this same pe-

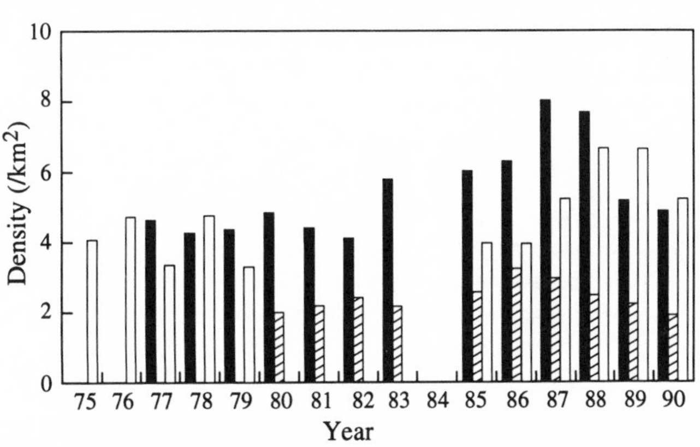

Figure 20.3 Annual trends, 1975–1990, in mongoose population density. Open bars, *Herpestes;* hatched bars, *Mungos;* solid bars, *Helogale.* Data are missing in years when observers did not census the total population.

riod: woody plant densities have increased throughout the study area and *Acacia-Commiphora* woodland has gradually spread south into areas that were grassland in the mid-1970s.

Surprisingly, patterns of *Helogale* fecundity across years appeared to be little influenced by rainfall (fig. 20.4). We regressed annual recruitment against annual rainfall (the same and the preceding year), seasonal rainfall (wet season, dry season, preceding wet season), and quarterly rainfall (during the same and the preceding year), as well as against rainfall over the wettest 3 months, the driest 3 months, and the driest 3 months of the preceding year. The only pattern that emerged is that recruitment was increased by high rainfall during the preceding dry season. The number of yearlings produced per alpha female was higher in years with higher rainfall in the preceding early dry season ($P = .01$, $R^2 = .45$), the preceding dry season overall ($P = .01$, $R^2 = .54$), or the driest 3 months of the preceding year ($P = .01$, $R^2 = .55$). The number of yearlings per female showed the same trends less strongly (preceding early dry season $P = .08$, $R^2 = .28$; preceding dry season overall $P = .06$, $R^2 = .32$).

Annual variation in recruitment reflects variation in litter size and in the number of litters produced. Perhaps because these parameters could be measured only during those years when an observer was present through the entire breeding season, no relationships emerged between them and annual, seasonal, or quarterly rainfall.

Similar results held for *Helogale* mortality. Juvenile and adult death rates varied less among years than did recruitment, and neither was

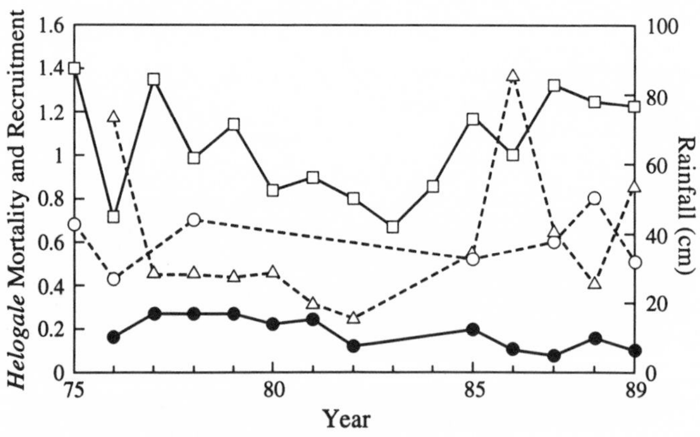

Figure 20.4 Annual trends, 1975–1989, in rainfall and in *Helogale* recruitment and mortality. Squares with solid lines, annual rainfall; solid circles with solid lines, adult mortality rates; triangles with dashed lines, recruitment rates (yearlings/female); open circles with dashed lines, juvenile mortality. Juvenile mortality data are available only for years during which observers were present throughout the breeding season.

strongly related to any of the annual measures of rainfall that we investigated (fig. 20.4).

Monthly Birth and Death Rates

All three species reproduced primarily during the rains, but seasonality was more pronounced in *Helogale* and *Mungos* than in *Herpestes* (fig. 20.5a–c).

In *Helogale,* birth rates were significantly associated with seasonal patterns of rainfall. Litters were born almost exclusively between the months of November and May, with occasional litters in October and June (fig. 20.5c). The annual peak of births was timed so that most litters were born during periods of substantial rainfall and with at least a month of good rainfall yet to come. This result is most clearly illustrated by regressions of mean monthly per female birth rate on mean monthly rainfall with various lag periods (fig. 20.6). Six of the twelve possible regressions were significant; birth rate was positively associated with rainfall during the birth month and the two subsequent months.

Litter size (at first count) and the number of litters produced per alpha female also fluctuated seasonally, peaking in the mid-wet season (fig. 20.5c). Seasonality in "litter size" may have been caused by high mortality between birth and first count during dry season months. Juvenile mortality after first count but before weaning increased rapidly as the beginning of the dry season approached (fig. 20.5d).

Adult *Helogale* death rates followed a similar seasonal pattern: monthly death rates doubled during the dry season (fig. 20.5d).

Temporal and Spatial Environmental Variation and *Helogale* Birth and Death Rates

Grass height, dung counts, and acridid biomass all showed striking seasonal variation, but only slight differences between woodland and grassland (fig. 20.7).

Grass height increased steadily through the end of the rains and then declined in the late dry season (fig. 20.7b). There was a significant correlation between vegetation height and rainfall 3 months previously ($r = .59$, $P = .02$). Analysis of variance comparing vegetation height between habitats and among seasons showed significant differences among seasons (lowest in the early wet season, $F = 19.27$, $P < .001$) and between habitats (lower in woodland, $F = 10.04$, $P = .002$).

Ungulate dung counts were also cyclical (fig. 20.7d), but the cycle was negatively correlated with that of grass height ($r = -.62$, $R^2 = .38$, $P < .001$) and no significant correlation with monthly rainfall was detected. Movement of ungulates south through the Sangere area had little effect

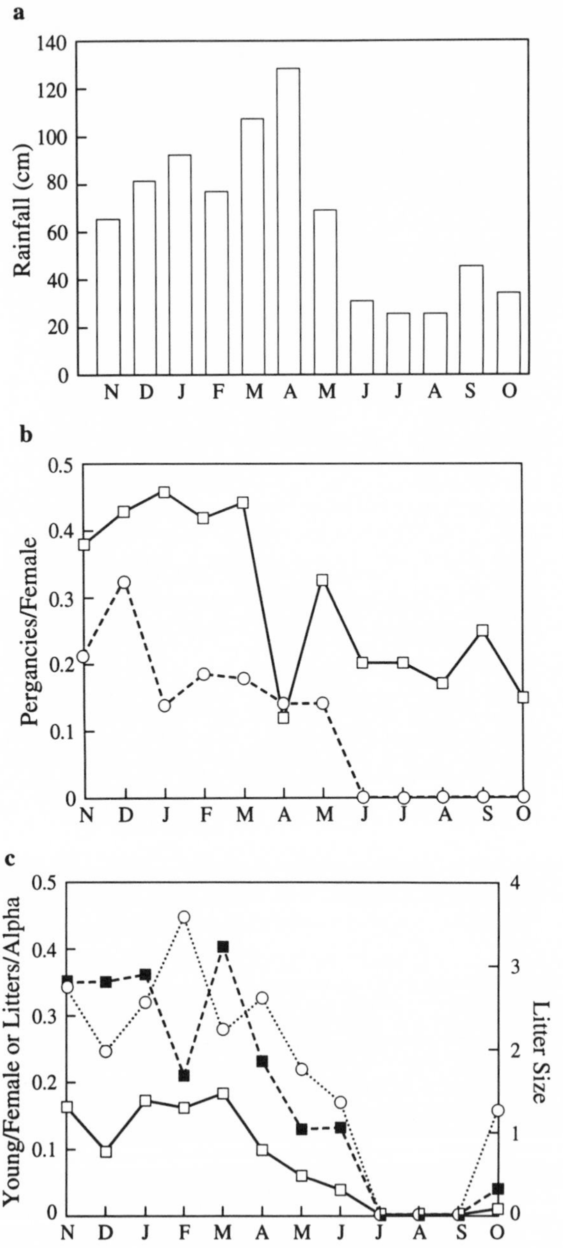
a
Rainfall (cm)
140
120
100
80
60
40
20
0
N D J F M A M J J A S O
b
Pergancies/Female
0.5
0.4
0.3
0.2
0.1
0
N D J F M A M J J A S O
c
Young/Female or Litters/Alpha
0.5
0.4
0.3
0.2
0.1
0
Litter Size
4
3
2
1
0
N D J F M A M J J A S O

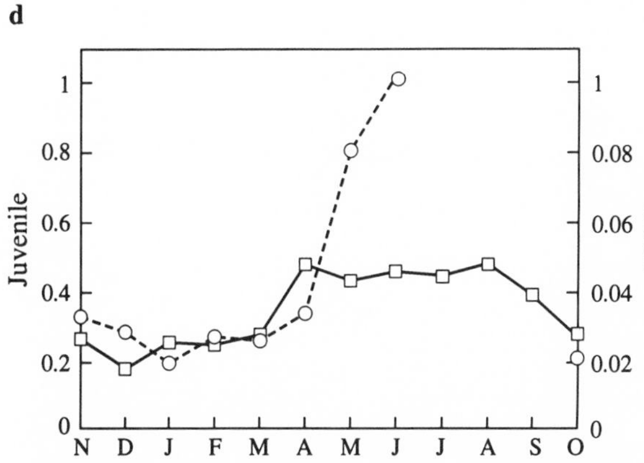

Figure 20.5 Seasonal trends, 1975–1990. (*a*) Rainfall. (*b*) *Herpestes* and *Mungos* reproduction. Circles with dashed lines, number of pregnancies per female *Mungos;* squares with solid lines, number of pregnancies per female *Herpestes.* (*c*) *Helogale* reproduction. Circles with dotted lines, litter size; solid squares with dashed lines, litters per alpha female; open squares with solid lines, young per adult female. Note that the pronounced dip in number of litters in February likely reflects the fact that most packs have produced a litter in December or January, so cannot produce another in February no matter what the rainfall. (*d*) *Helogale* mortality. Squares with solid lines, monthly per capita death rate for adults; circles with dashed lines, death rates between first count and 3 months for juveniles.

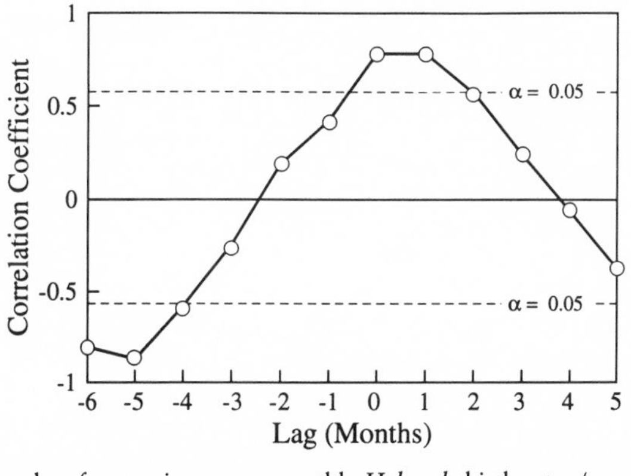

Figure 20.6 Results of regressing mean monthly *Helogale* birth rates (young/female) against mean monthly rainfall in the preceding and following 6 months. The significant positive correlations indicate that high birth rates occur during months with high rainfall in the same month and succeeding 2 months (and that low birth rates occur when it is dry, or dry months are in the offing).

on grass height because it occurred prior to most grass growth. When the migration north occurred in the early dry season, large numbers of ungulates produced a dramatic reduction in vegetation height and, coincidentally, high dung counts. ANOVA indicated a significant difference in dung counts among seasons (lowest in early dry season, $F = 13.02$, $P < .001$), but no difference between habitats.

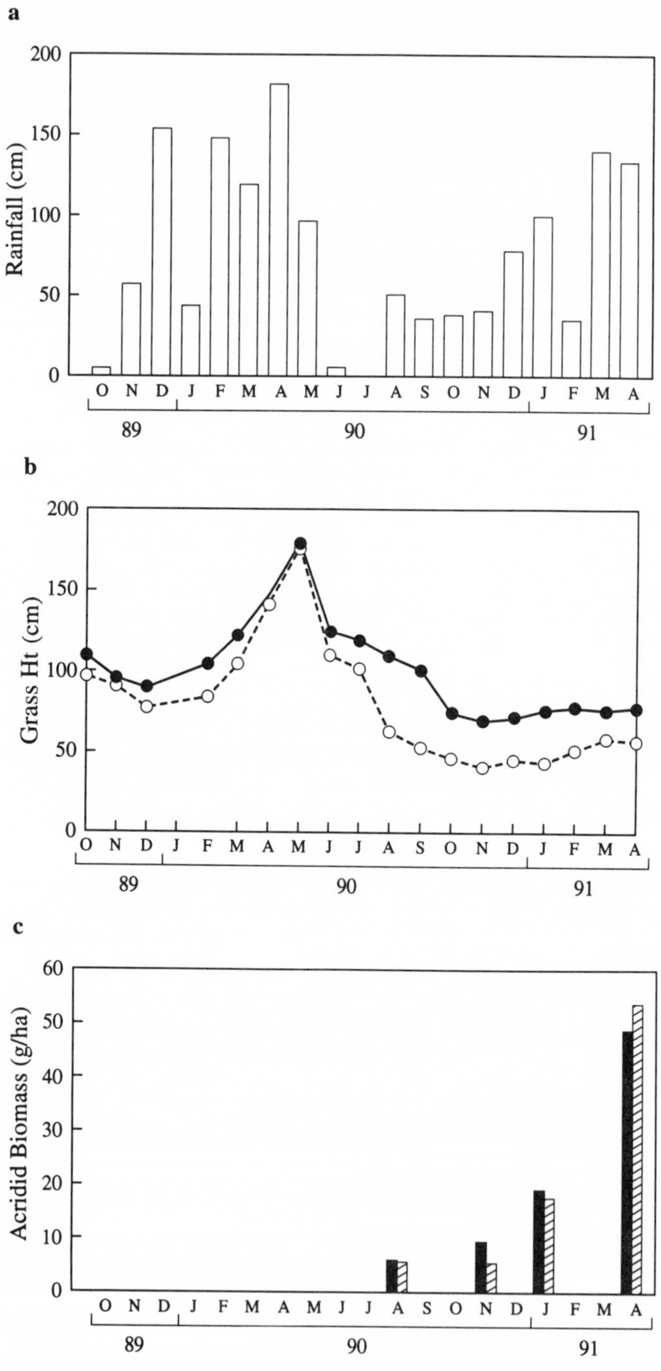
a
Rainfall (cm)
200
150
100
50
0
O N D J F M A M J J A S O N D J F M A
89
90
91
b
Grass Ht (cm)
200
150
100
50
0
O N D J F M A M J J A S O N D J F M A
89
90
91
c
Acridid Biomass (g/ha)
60
50
40
30
20
10
0
O N D J F M A M J J A S O N D J F M A
89
90
91

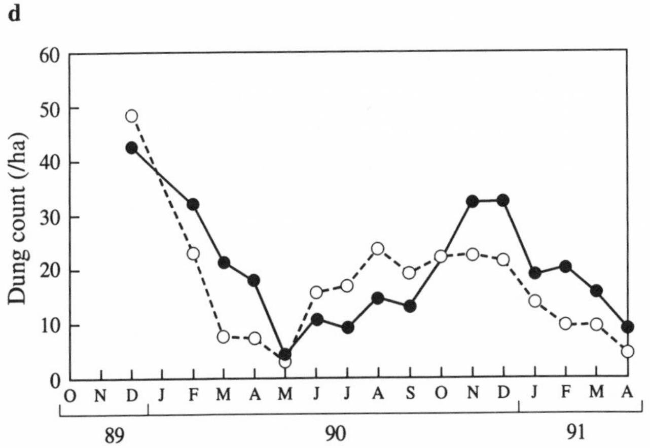

Figure 20.7 Seasonal trends on seven *Helogale* home ranges during 1989–1991. (*a*) Rainfall. (*b*) grass height. Open circles, home ranges in grassland; solid circles, home ranges in woodland. Grass height is an index of vulnerability to predators. (*c*) Acridid biomass. Solid bars, home ranges in grassland; hatched bars, home ranges in woodland. Note that acridid biomass was measured quarterly in 1990–1991 only. (*d*) Dung counts. Open circles, home ranges in grassland; solid circles, home ranges in woodland. Acridid biomass and dung counts (because they reflect beetle densities) are indices of food availability.

The late wet season peak in acridid biomass (fig. 20.7c) is consistent with previous reports (Dingle and Khamala 1972; Folse 1978; Sinclair 1975, 1978). Since data were collected quarterly, no monthly correlations with rainfall were possible. Nevertheless acridid biomass and vegetation height were significantly correlated ($r = .72$, $R^2 = .52$, $P = .001$) during the late wet season when vegetation height was measured directly on the acridid sampling plots, suggesting that rainfall effects on grass height may influence acridid biomass. ANOVA indicated a significant difference among seasons for acridid biomass ($F = 105.38$, $P < .001$), but no difference between habitats.

We tabulated *Helogale* birth and death rates on each of the six pack home ranges during each of the 20 months for which environmental variables were measured. We then summarized these rates by habitat and season. Again, we used ANOVA to examine birth and death rate differences among seasons and between habitats. We then used regressions to ask whether temporal differences in pack birth and death rates were predicted by differences in grass height, dung count, or acridid biomass.

Birth rate differed significantly among seasons ($F = 2.84$, $P = .04$) but not between habitats. Regression analysis revealed grass height as the only significant predictor of birth rate; the slope was negative ($r = -.34$, $P = .02$), reflecting the dramatic decrease in birth rate during the late wet season when vegetation remained high.

Analysis of the variance in mortality rates indicated no differences among seasons or between habitats. Of all regressions of mortality rates against environmental variables, only one was marginally significant: mortality rates were slightly higher in seasons with higher ungulate dropping counts (for habitat-season comparisons, $r = .58$, $P = .05$).

Helogale Home Range Quality

Dwarf mongoose home ranges averaged 27.4 ± 3.3 hectares. Grass and scrubby *Acacia* and *Commiphora* combined to form the majority of all ranges, with larger trees, kopjes, and korongos accounting for smaller areas, but home ranges exhibited extreme variation in most of the habitat parameters we measured (table 20.3).

Comparison of habitat parameter values within *Helogale* home ranges with the same parameters in adjacent areas not used by *Helogale* (table 20.3) suggests that the most important determinant of dwarf mongoose habitat preference is the density of potential dens. Dwarf mongooses den in termitaries, rock crevices, and occasionally hollow logs. The percentage of kopje within home ranges was always equal to or greater than that in matched areas unused by mongooses ($P = .02$, Wilcoxon); termitary density was higher in five of six home ranges than in "control" areas ($P = .06$, Wilcoxon). The mean density of termitaries was twice as high on home ranges as off.

While mean percentage of scrub and percentage of trees were slightly higher on home ranges than off, we found no significant effect of vegetation characteristics on habitat preferences. Some home ranges used by *Helogale* contained as high a percentage of grass, and as low a percentage

Table 20.3 Habitat parameters determined by examination of aerial photographs of six *Helogale* home ranges and adjacent unused areas.

Pack	% scrub	% grass	% trees	% kopje	% korongo	Trees (no./ha)	Termitaries (no./ha)
Helogale home ranges							
W1	8	88	3	1	0	1.53	4.64
T1	35	48	5	8	3	0.39	0.90
E1	6	67	5	14	9	2.19	1.39
I1	91	3	6	0	0	3.42	2.05
O	89	1	10	0	0	2.98	1.86
F	54	20	17	0	10	2.06	1.19
Mean	47.1	37.8	7.6	7.6	3.7	2.23	1.87
Unused areas							
W1	38	49	6	0	6	1.42	0.81
T1	39	48	6	0	7	0.43	0.34
E1	11	82	3	3	0	1.00	1.38
I1	27	53	8	0	12	2.52	0.61
O	29	65	6	0	0	2.56	0.68
F	48	48	4	0	0	3.44	1.60
Mean	32.0	57.5	5.5	0.5	4.2	1.71	0.81

of trees and scrub, as matched areas avoided by them ($P > .05$, Wilcoxon).

A parallel result was obtained using a larger-scale analysis based on strip counts of *Macrotermes* mounds in 1 km^2 areas by Jon and Josephine Rood. Within an 8 km^2 area used by *Helogale* groups that persisted throughout the entire study, the median number of termitaries was 175/km^2 (range 50–320); on a 19 km^2 area not used by *Helogale* or used only by small, short-lived groups, the median number of termitaries was only 10 km^2 (range 0–170; Mann-Whitney $U = 9.5$, $P = .002$).

Home Range Size, Habitat Parameters, and Reproductive Success

The above analyses describe the variables that affect where dwarf mongooses locate their home ranges. One can ask further whether variation among home ranges predicts reproductive success (mean number of surviving yearlings in each pack between 1987 and 1989) in the packs living on them. Forward stepwise multiple regression showed that of the ecological variables described above, only home range size had a significant effect on reproductive success within a pack ($R^2 = .85$, partial $b = 0.07 \pm 0.02$, $N = 6$, $t = 3.88$, $P = .02$). The next most influential variable was percentage of kopje (partial $b = 0.17 \pm 0.07$, $t = 2.22$, $P = .09$). Other variables exerted little effect.

Home range size is the only one of the ecological variables discussed above that might be affected by the pack itself. A pack might expand or contract its range from year to year, but cannot easily affect the percentage of its range covered by scrub, or the density of termitaries. A straightforward possibility is that home range size is directly affected by pack size. To determine whether pack size was by itself a sufficient predictor of reproductive success, we added pack size to the pool of ecological independent variables and used forward stepwise multiple regression to determine which variables best predicted reproductive success. Only pack size entered the stepwise regression, indicating that pack size was indeed a sufficient predictor of reproductive success ($b = 0.26 \pm 0.03$, $t = 7.74$, $P < .001$, $R^2 = .92$). None of the ecological variables entered the model using either forward or backward stepwise regression.

DISCUSSION

Mongoose social structure, in Serengeti as elsewhere, has been influenced by both diet and predation (Gorman 1979; Rood 1986; Waser and Waser 1985). Gregariousness is a characteristic of small, diurnal insectivores, like *Helogale* and *Mungos,* whose food is clumped or rapidly renewing; these prey characteristics are likely to reduce the costs of foraging in a group. At the same time, survival rates increase with group size because larger groups are more likely to detect and deter predators, as demonstrated in *Helogale* (Rasa 1983, 1987; Rood 1983a, 1990). In contrast, mongooses that prey on small vertebrates, like *Herpestes,* are largely solitary; such prey are not divisible among group members and generally occur in relatively closed microhabitats where the mongoose hunting them is itself relatively less vulnerable to predators.

The data summarized in this chapter suggest that these same factors, food availability and predation, also influence mongoose demography and distribution. Seasonal patterns of mortality and fecundity appear to reflect seasonal patterns of food availability; population density and distribution appear most strongly influenced by predation.

Demography

Annual survival rates of the three species are surprisingly high for such small animals. The observed longevity of the dwarf mongoose, in particular, is far above the average for carnivores of its size (Gittleman 1986). Annual survival rates of the three species are also surprisingly similar given the fourfold difference among species in body weight. Banded mongoose survival rates, which appear to be lower than those of the other two species, are probably underestimates: the proportion of animals marked was too low to count immigrants accurately and the number of groups monitored was too small to track emigrants. This problem was particularly severe for males, which dispersed more often. The apparently greater longevity of dwarf mongooses than banded and slender mongooses is misleading for similar reasons; many more dwarf mongooses were marked, and more were marked early in the study, so that the chance of seeing a mongoose survive into its second decade was much higher for *Helogale* than for *Mungos* or *Herpestes.*

All three species die at higher rates during their first year, but subsequent mortality rates are approximately age-independent. Females appear to survive at higher rates than males, perhaps reflecting male-biased dispersal, but differences are not marked, even in the highly dimorphic slender mongoose.

Age-specific fecundity patterns differ markedly among the three species; *Herpestes* females routinely reproduce as yearlings, while the gregarious species rarely do, and young *Helogale* females in groups with older females are profoundly suppressed reproductively, a phenomenon exhibited only slightly in *Mungos.* These differences suggest that dwarf mongoose reproduction should be buffered from some types of environmental variation: if half of all group members were to starve, the number of reproductively active females would change very little.

Comparative data on the demography of other mongoose populations are rare. Among banded mongooses in Queen Elizabeth Park,

Uganda, an area with year-round food availability, annual survival rates are very high (0.89: Rood 1975).

Seasonal Patterns

The periodicity of reproduction in small Serengeti carnivores is clearly shaped by rainfall. Perhaps surprisingly, dwarf mongoose birth rates are significantly associated with rainfall up to 2 months after, but not prior to, birth. Gestation in *Helogale* lasts 7 weeks (Rasa 1973), so that one might expect a significant association between birth rate and rainfall 2 months previously. However, the observed lag between monthly birth rates and rainfall makes sense considering the relative costs of gestation and lactation. The energetic costs of gestation are much lower than those of lactation (Creel and Creel 1991), which lasts approximately 2 months.

For dwarf mongooses, higher juvenile mortality among late-born litters selects against dry season reproductive attempts; the same trend has been reported in Serengeti white-tailed mongooses (Waser 1980) and presumably also shapes banded and slender mongoose breeding seasons. In Uganda, where rainfall is less seasonal and food is easily available year-round, *Mungos* breeding is also less seasonal (Rood 1975). Rainfall also shapes seasonal patterns of adult mortality, at least in dwarf mongooses.

Rainfall must influence seasonal patterns of reproduction and mortality through its effects on vegetation growth or prey density, but demonstrating this, and determining the relative importance of resources and predators, has proven surprisingly difficult. To some extent, the lack of a clear relationship between the measures of food abundance we used and mongoose survival and reproduction reflects the variety of prey taken by mongooses. Dwarf mongooses feed heavily on termites, which we did not sample; termites become more available after the rains begin (Sinclair 1978). Alates emerge then, and workers engage in mound building, thus becoming accessible as prey. Mongooses also feed heavily on adult scarabs; adult scarab activity is increased considerably in the wet season (R. Foster, pers. comm.) and dung counts may be insufficient to portray the availability of scarabs, let alone other beetles, to the mongooses. Even more important, scarabs may be present as pupae or diapausal adults below the ground, where they are difficult for human observers to find, yet the insects may be available to a digging mongoose. Dwarf and particularly banded mongooses dig for much of their food.

Several lines of evidence suggest that demographic seasonality reflects seasonality in food availability rather than in intensity of predation. Dwarf mongoose adult and juvenile mortality rates are low during the early wet season despite the absence of cover (figs. 20.6 and 20.7). Dwarf mongoose juvenile growth rates are increased by higher rainfall (Creel and Creel 1991), a result that is presumably driven by higher food levels.

The timing of dwarf mongoose births causes lactation to coincide with months of highest insect density, at least as indexed by acridid numbers (fig. 20.7c). Reproduction is less seasonal in the slender mongoose, a predator of small vertebrates, than in the insectivorous species, whose prey are more seasonal (fig. 20.5).

Predation effects on seasonal patterns of survival and reproduction cannot be ruled out, however, because of an expected interaction between predation risk and prey densities: high prey density allows adults to spend more time in vigilance, spend less time foraging, and perhaps forage closer together, thereby reducing their exposure to predators and increasing the likelihood that potential predators are detected.

Population Regulation

While seasonal patterns of births and deaths may be food-driven, annual variation in population density is not. The lack of an effect of annual rainfall on mongoose numbers may indicate that predation, rather than food limitation, regulates mongoose numbers. Mongooses are attacked by a suite of predators in Serengeti ranging from relatively small raptors (snake eagles, *Circateus* spp.) to large mammalian carnivores (wild dogs); for the smaller mongoose species, the list of predators includes the larger mongooses (Rood 1983a, 1990). With the exception of several deaths caused by a grass fire, all mongooses whose deaths were witnessed were taken by predators. Dwarf and banded mongooses sometimes lost entire litters under circumstances suggestive of predation. Among banded mongooses, presumably more vulnerable because of their use of more open habitat, larger groups in our sample did not raise more yearlings than smaller groups did, even though litter sizes at birth were higher. In large part, this reflected the high proportion of groups of all sizes that lost complete litters.

The importance of predation may vary with location: banded mongooses occur at low densities in the short-grass plains, where they may be food-limited during the dry season; at that time of year, *Mungos* subsists largely on insects found in buried ungulate dung, a legacy of the wet season migration. But differences in density also covary with predation risk: on the short-grass plains, *Mungos* are highly vulnerable to larger predators, themselves facing food shortages during the dry season. Similarly, the high densities of *Mungos* in Uganda might be caused by higher year-round food levels, but they also might reflect lower densities of large predators there (Rood 1975, 1982).

Disease, implicated as a factor in regulating other carnivore populations, has not been evident on the study site. The absence of disease outbreaks on the study site may not, however, be universal. Diseased white-tailed mongooses were observed in the western corridor prior to a local population crash in 1977–1978, and banded mongooses in poor condition and suffering extensive hair loss were also seen coincident with disease outbreaks in western corridor canids (H. Rood, pers. comm.).

Distribution and Population Density

The distribution of dwarf mongooses on the study site is clearly shaped by the distribution of den sites, particularly termitaries. Termite mounds are the focus of almost all daily activity, serving as a refuge from predators, a high point for antipredator vigilance, an underground den for dependent offspring that cannot forage with the group, and a safe base for social interaction throughout the day. Dwarf mongooses favor areas with high termitary density, and that tendency is adaptive, as indicated by the long-term persistence of groups in areas with high termitary density, presumably reflecting higher birth rates and individual survival in those areas.

Termites are also an important component of the dwarf mongoose diet, but this effect of termitaries on mongoose distribution is likely to be secondary. There is no suggestion that dwarf mongooses abandon areas with high termitary densities during periods when the termites themselves are not available as food. In the Selous, dwarf mongooses occur at high density in areas without termite mounds, substituting large cracks in black-cotton soil for termitaries as a source of shelter (S. Creel, unpub.).

In general, dwarf mongooses are absent from short-grass plains and have been characterized as woodland species. Surprisingly, on the Sangere site the differences in food availability between woodland and grassland areas are quite small. Predation, therefore, seems the most likely factor that usually confines dwarf mongooses to woodland. Increased vulnerability to predators in short-grass habitats might tip the balance against dwarf mongooses in competition with the larger banded mongooses (which also act as predators on dwarf mongoose young). On a local scale, it seems likely that the key resource necessary to sustain dwarf mongooses is not tall grass but a safe hole within reasonable reach, and that the predilection of dwarf mongooses for woodland is a secondary effect of the density of termitaries there. On the Sangere site, dwarf mongooses occupy extremely variable habitat, as long as it contains a sufficient density of termite mounds or kopjes for use as dens. Mongoose populations at Kirawira in the western corridor provide a test case: dwarf mongooses occur at high densities in short grass, but termitaries are numerous there.

Banded mongooses appear more flexible in their habitat requirements than *Helogale*. In Sangere, they tend to be more conspicuous than dwarf mongooses in open grassland, but they use woodland as well. Banded mongooses occur in even the most open areas of the Serengeti short-grass plains, where they den in rock outcrops or spring hare holes. Because of their larger size and their tendency to forage in tighter groups, banded mongooses may be slightly less vulnerable to predators than dwarf mongooses. Slender mongooses appear common wherever there is cover (Rood and Waser 1978).

Serengeti mongooses attain population densities that are spectacularly high. Per square kilometer, the Sangere area in 1990 contained 5.0 dwarf mongooses, 2.0 banded mongooses, and 5.3 slender mongooses (biomass densities thus reach 1.7 kg/km^2 for dwarf mongooses, 3.3 kg/km^2 for banded mongooses, and 3.4 kg/km^2 for slender mongooses). These numbers are matched by those of white-tailed mongooses in the western Serengeti (4.3/km^2: Waser 1980), and probably by mongooses elsewhere in eastern and southern Africa (S. Creel, unpub.; P. Cavallini, pers. comm.) Elsewhere, such densities are approached only by palm civets *(Nandinia binotata)* in Gabon, badgers *(Taxidea taxus)* in the northwestern United States, and ringtails *(Bassariscus astutus)* in the southwestern United States (data summarized by Sandell 1989). These densities are particularly striking given the mongooses' strictly carnivorous and insectivorous diets. Palm civets and ringtails are partially frugivorous; the highest density reported for a temperate mustelid with a diet similar to that of mongooses is 1.0/km^2 (stoats, *Mustela erminea:* Sandell 1989).

Rood (1987; J. P. Rood, unpub.) estimated *Helogale* and *Mungos* densities even higher than those reported here for Serengeti woodlands. He also estimated densities of 2 *Helogale*/km^2 in the park's western corridor, 0.5 *Mungos*/km^2 in the short-grass plains, and 1.5 *Herpestes*/km^2 in both the western corridor and woodlands. Taking the more conservative estimates of density in all cases, and assuming 17,800 km^2 of woodland, 2,500 km^2 in the western corridor, and 5,200 km^2 in the short-grass plains, we arrive at estimates of 94,000 dwarf mongooses, 43,000 banded mongooses, and 30,000 slender mongooses in the Serengeti. These numbers are orders of magnitude higher than the population sizes of other Serengeti carnivores.

The high densities and population sizes attained by Serengeti mongooses appear to be stable. Changes in vegetation, with whatever consequences these have had for termites, beetles, and small mammals (mongoose prey) or raptors and larger carnivores (mongoose predators), have not markedly influenced densities since 1975. Censuses indicate a steady or slowly increasing population size for dwarf and slender mongooses, as does the net reproductive rate calculated for dwarf mongoose females. Net reproductive rate calculated for male dwarf mongooses indicates a population that is not increasing, but the problem of unaged (mostly male) immigrants suggests that this figure is a slight underestimate.

Future mongoose population trends can be viewed with optimism, but a number of questions regarding mongoose population dynamics clearly remain unanswered. Understanding the extent to which food levels control mongoose fecundity and mortality—and the effect that mongooses have on insects that affect nutrient cycling through the rest of the Serengeti ecosystem (Sinclair 1975)—will require much more comprehensive sampling of insect density and phenology, particularly of beetles and termites. Similarly, studies of small mammal population dynamics must be much more extensive before we can determine whether the year-to-year fluctuations documented by Senzota (1982) are a consequence of predator-prey interactions. On the other hand, we know virtually nothing about the extent to which mongoose populations are coupled with those of their predators, particularly raptors. Such studies of vertebrate predator-prey interactions will not be simple, but ultimately they would have wide applicability, since on a global scale, predators the size of mongooses are much more widespread than predators the size of lions or wild dogs. Because of their high density and visibility, Serengeti small carnivores will continue to provide unique opportunities for those interested in the dynamics of vertebrate predator-prey interactions.

ACKNOWLEDGMENTS

We thank Hazel, Josephine, and the late Jon Rood for their friendship and generosity; Jon pioneered the field study of mongooses, began the Serengeti mongoose project, built the data base on which this study was founded, and made our participation in the project possible. The Tanzania National Scientific Research Council, the Tanzania National Parks, and the Serengeti National Park have made the Serengeti mongoose project possible, and it has been funded in part by the National Science Foundation and the National Geographic Society. Hazel and Josephine Rood, Nancy Link, Paolo Cavallini, Brian Keane, Priyanga Amarasekare, and Peter Arcese improved earlier versions of the chapter.

REFERENCES

Creel, S. R., and Creel, N. M. 1991. Energetics, reproductive suppression, and obligate communal breeding in carnivores. *Behav. Ecol. Sociobiol.* 28:263–70.

Creel, S. R., Creel, N. M., Wildt, D. E., and Monfort, S. L. 1992. Behavioural and endocrine mechanisms of reproductive suppression in Serengeti dwarf mongooses. *Anim. Behav.* 43:231–45.

Creel, S. R., Monfort, S. L., Wildt, D. E., and Waser, P. M. 1991. Spontaneous lactation is an adaptive result of pseudopregnancy. *Nature* 351:660–62.

Creel, S. R., and Waser, P. M. 1991. Failures of reproductive suppression in dwarf mongooses *(Helogale parvula)*: Accident or adaptation? *Behav. Ecol.* 2:7–15.

———. 1994. Inclusive fitness and reproductive strategies in dwarf mongooses. *Behav. Ecol.* 5:339–48.

Dingle, H., and Khamala, C. P. M. 1972. Seasonal changes in insect abundance and biomass in an East African grassland with reference to breeding and migration in birds. *Ardea* 59:216–21.

Folse, L. J. Jr. 1978. Avifauna-resource relationships on the Serengeti plains. Ph.D. dissertation, Texas A&M University.

Gittleman, J. L. 1986. Carnivore life history patterns: Allometric, phylogenetic, and ecological associations. *Am. Nat.* 127:744–71.

Gorman, M. L. 1979. Dispersion and foraging of the small Indian mongoose, *Herpestes auropunctatus,* relative to the evolution of social viverrids. *J. Zool.* (Lond.) 187:65–73.

Keane, B., Waser, P. M., Creel, S. R., Creel, N. M., Elliott, L. F., and Minchella, D. J. 1994. Subordinate reproduction in dwarf mongooses. *Anim. Behav.* 47:65–75.

Rasa, O. A. E. 1973. Intra-familial sexual repression in the dwarf mongoose, *Helogale parvula. Naturwissenschaften* 60:303–4.

———. 1983. Dwarf mongoose and hornbill mutualism in the Taru desert, Kenya. *Behav. Ecol. Sociobiol.* 12:181–90.

———. 1987. The dwarf mongoose, a study of behavior and social structure in relation to ecology in a small, social carnivore. *Adv. Stud. Behav.* 12:121–63.

Rood, J. P. 1974. Banded mongoose males guard young. *Nature* 248:176.

———. 1975. Population dynamics and food habits of the banded mongoose. *E. Afr. Wildl. J.* 13:89–111.

———. 1978. Dwarf mongoose helpers at the den. *Z. Tierpsychol.* 48:277–88.

———. 1980. Mating relationships and breeding suppression in the dwarf mongoose. *Anim. Behav.* 28:143–50.

———. 1982. Ecology and social organization of the banded and dwarf mongoose. *Natl. Geogr. Soc. Res. Rep.* 14:571–76.

———. 1983a. Banded mongoose rescues pack member from eagle. *Anim. Behav.* 31:1261–62.

———. 1983b. The social system of the dwarf mongoose. In *Recent advances in the study of mammalian behavior,* ed. J. Eisenberg and D. Kleiman, 454–88. Special publication no. 7. Shippensburg, Pa.: American Society of Mammalogists.

———. 1986. Ecology and social evolution in the mongooses. In *Ecological aspects of social evolution,* ed. D. Rubenstein and R. Wrangham, 131–52. Princeton, N.J.: Princeton University Press.

———. 1987. Dispersal and intergroup transfer in the dwarf mongoose. In *Mammalian dispersal patterns: The effects of social structure on population genetics,* ed. B. D. Chepko-Sade and Z. Halpin, 85–103. Chicago: University of Chicago Press.

———. 1989. Male associations in a solitary mongoose. *Anim. Behav.* 38:725–28.

———. 1990. Group size, survival, reproduction, and routes to breeding in dwarf mongooses. *Anim. Behav.* 39:566–72.

Rood, J. P., and Waser, P. M. 1978. The slender mongoose, *Herpestes sanguineus,* in the Serengeti. *Carnivore* 1:54–58.

Sandell, M. 1989. The mating tactics and spacing patterns of solitary carnivores. In *Carnivore behavior, ecology, and evolution,* ed. J. L. Gittleman, 164–83. Ithaca, N.Y.: Cornell University Press.

Senzota, R. B. M. 1982. The habitat and food habits of the grass rats *(Arvicanthus niloticus)* in the Serengeti National Park, Tanzania. *Afr. J. Ecol.* 20:241–52.

Sinclair, A. R. E. 1975. The resource limitation of trophic levels in tropical grassland ecosystems. *J. Anim. Ecol.* 44:497–520.

———. 1978. Factors affecting the food supply and breeding season of resident birds and movements of palaearctic migrants in a tropical African savannah. *Ibis* 120:480–97.

Waser, P. M. 1980. Small nocturnal carnivores: Ecological studies in the Serengeti. *Afr. J. Ecol.* 18:167–85.

Waser, P. M., Creel, S. R., and Lucas, J. R. 1994. Death and disappearance: Estimating mortality risks associated with dispersal and philopatry. *Behav. Ecol.* 5:135–41.

Waser, P. M., Keane, B., Creel, S. R., Elliott, L. F., and Minchella, D. J. 1994. Possible male coalitions in a solitary mongoose. *Anim. Behav.* 47:289–94.

Waser, P. M., and M. S. Waser. 1985. *Ichneumia albicauda* and the evolution of viverrid gregariousness. *Z. Tierpsychol.* 68:137–51.

V Conservation and Management

TWENTY-ONE

The Importance of Behavioral Ecology for Conservation Biology: Examples from Serengeti Carnivores

T. M. Caro and S. M. Durant

Biologists working in East Africa and elsewhere are usually assigned to one of two camps: those who conduct "research" and those who practice "conservation." Though camp members may view each other amicably, often as not researchers see conservation as uninteresting or as a second-rate discipline, while conservationists regard research as irrelevant or esoteric. Conservationists often ask research biologists, and those studying animal behavior in particular, the galling question of whether their years in the field amount to anything.

In the past, the answer to this question often amounted to "no," but it is increasingly clear that biological research has an important role to play in conservation. Growing concern about rates of species extinction and habitat loss has led to the formation of a new applied discipline called conservation biology (Ehrlich and Ehrlich 1981). Conservation biology addresses processes by which populations go extinct (research-oriented questions) and strategies for preventing extinctions (conservation-oriented questions), and thus provides many important bridges between entrenched camps. The discipline draws upon population biology, biogeography, community ecology, and genetics to meet its objectives (Simberloff 1988). For most biologists, then, the links between their work and conservation are now much more obvious than in the past. For behavioral ecologists, however, who formed the majority of research personnel in the Serengeti during the 1980s, difficulties remain in justifying their work on conservation grounds because their field has been largely ignored by conservation biology (but see Soulé 1983; Simberloff 1986).

In this chapter we demonstrate the crucial role played by behavioral research in conservation biology by exploring the links between behavioral ecological research conducted on Serengeti carnivores over the last 15 years and conservation science. Large carnivores have a special significance for conservation biology for four reasons. First, they live at lower densities than the species on which they feed and are hence more vulnerable to extinction. Second, because they are at the top of the trophic pyramid, their presence is dependent on many lower trophic levels remaining intact. Third, they may therefore be sensitive indicators of ecosystem perturbations, since changes in the reproduction and population size of a predator may be easier to monitor than those in prey or vegetation (Landres, Verner, and Thomas 1988). Fourth, large carnivores are "flagship species" (Western 1987) capable of attracting disproportionate attention and funding (e.g., Rabinowitz 1986). Though their import is acknowledged (Terborgh 1988), as yet little attempt has been made to translate knowledge of carnivore population dynamics into conservation theory or practice.

To date, research on Serengeti carnivores has lain squarely within the realm of behavioral ecology. The principal studies, on black-backed jackals, cheetahs, dwarf mongooses, lions, spotted hyenas, and wild dogs, have each monitored recognized individuals over long time periods and have been concerned primarily with understanding aspects of their diverse breeding systems. Based on papers in print and manuscripts made available to us at the time of writing, we have summarized the main achievements of Serengeti carnivore studies in table 21.1. It should be noted, however, that useful work has also been conducted on bat-eared foxes (Lamprecht 1979; Malcolm 1986), banded mongooses (Rood 1975; Waser et al., chap. 20), slender mongooses (Waser et al., chap. 20), white-tailed mongooses (Waser and Waser 1985), golden jackals (Moehlman 1983), aardwolves (Kruuk and Sands 1972), striped hyenas (Kruuk 1976), and leopards (Bertram 1982; Cavallo 1990).

In this chapter, we first outline the reasons that populations go extinct. Then, using examples from Serengeti carnivores, we show how knowledge of an animal's behavior can assist in all key facets of conservation biology, including management strategies. Finally, we briefly discuss how diverse conservation studies of endangered species outside Serengeti have profited from consideration of behavior, reinforcing the point that behavioral ecologists have an important role to play in a biological world rapidly becoming dominated by conservation issues.

CAUSE OF POPULATION EXTINCTIONS

Populations can go extinct as a result of deterministic processes such as sustained habitat destruction or overhunting. Populations may also succumb to chance or stochastic events of either a demographic or environmental nature. Demographic stochasticity, which results from individual variation in birth and death rates, occurs in all populations, but its effects

Table 21.1 Principal topics in studies of Serengeti carnivores.

Species	Topics	Selected references[a]
Black-backed jackal	Helpers at the den	Moehlman 1979
	MtDNA sequence divergence	Wayne et al. 1990
Cheetah	Reproductive strategy	Caro and Collins 1987; Caro, FitzGibbon, and Holt 1989
	Consequences of grouping	Caro 1994
	Interspecific predation	Laurenson 1994
	Parental care	Caro 1987; Laurenson, 1994
	Hunting behavior	FitzGibbon 1990
	Genetics and reproduction	O'Brien et al. 1987; Wildt, O'Brien, et al. 1987
Dwarf mongoose	Social organization	Rood 1978, 1980, 1990
	Kin selection	Creel and Waser, 1994
	Dispersal	Rood 1987
	Reproductive suppression	Creel et al. 1991, 1992
Lion	Reproductive strategies	Bygott, Bertram, and Hanby 1979; Packer and Pusey 1982; Packer, Gilbert, et al. 1991
	Infanticide	Bertram 1975; Packer and Pusey 1983
	Dispersal and philopatry	Hanby and Bygott 1987; Pusey and Packer 1987
	Consequences of grouping	Packer, Scheel, and Pusey 1990
	Hunting behavior	Schaller 1972; Scheel and Packer 1991; Scheel 1993
	Comparative genetics	Wildt, Bush, et al. 1987; Packer, Pusey, et al. 1991
Spotted hyena	Social organization	Frank 1986a, b; Hofer and East 1993a
	Siblicide	Frank, Glickman, and Light 1991
	Hunting behavior	Kruuk 1972
	Ranging behavior	Hofer, East, and Campbell 1993; Hofer and East, 1993b
	Vocalizations	East and Hofer 1991; Hofer and East, 1993c
	Parental care	
Wild dog	Social organization	Frame et al. 1979; Malcolm and Marten 1982
	Dispersal	Frame and Frame 1976
	Hunting behavior	Fanshawe and FitzGibbon 1993

[a]Based on published material and preprints sent to the authors, excluding chapters in this volume.

increase as population size declines (Goodman 1987; Durant and Harwood 1992). Similarly, environmental stochasticity, which results from external factors such as drought or disease, acts on all populations, but its effects remain substantial even in large populations. In general, populations outside protected areas are most likely to be subject to deterministic extinction processes, whereas those inside are expected to increase or remain stable but be more vulnerable to stochastic events.

If a population remains at low numbers for a sustained period, genetic problems may also arise. Small populations may be subject to inbreeding depression during initial years of population decline since there is an increased chance of deleterious recessives being expressed, which may be manifested in high infant mortality (Ralls, Brugger, and Ballou 1979; Templeton 1987). In addition, small populations may suffer a loss in genetic variance (Miller 1979; Gilpin and Soulé 1986). It has been hypothesized that a population with a low level of genetic diversity has less ability to respond to natural selection under changing environmental conditions and may thus be more susceptible to environmental stochasticity (Dobzhansky and Wallace 1953; Franklin 1980; Selander 1983).

Conservation biologists and geneticists relate the census population size to the number of individuals contributing genetic material to the next generation by using a theoretical quantity called the genetic effective population size. It employs the concept of the ideal population in which each individual has an equal chance of mating with every other individual (including itself). In reality, individual lifetime contributions to fitness in most populations are nonrandom because of phenotypic differences (Clutton-Brock 1988), and accurate estimates of the effective population size strongly depend upon detailed knowledge of individuals' contributions to the next generation (Crow and Kimura 1970).

THE INFLUENCE OF BREEDING SYSTEM ON EXTINCTION PROBABILITY

Grouping Patterns

The extent to which individuals are grouped together alters the way in which a population is subdivided. If populations are subject to strong independent environmental stochasticity, then population subdivision (i.e., its metapopulation structure) can promote chances of persistence through time because of the "spreading of risk" (den Boer 1968): for example, while a catastrophic event could wipe out one group, the odds are that others would survive. In addition, theory suggests that grouping may influence effective population size and genetic differentiation within populations, both of which affect the rate at which genetic diversity is lost from a population (Gilpin 1991); however, few empirical studies have tested these hypotheses.

Carnivore species show striking differences in group size and the extent to which they are social (table 21.2; see also Bertram 1979). For example, leopards of both sexes live alone as adults (Bertram 1982), whereas female cheetahs are solitary while males either live alone or in small groups of two or three individuals (Caro 1989). Lionesses live in groups ranging in size from two to eighteen females, and these groups are held by coalitions of between one and nine males (Packer et al. 1988;

Table 21.2 Behavioral ecology of some Serengeti carnivores.

Cheetah	Dwarf mongoose	Lion	Spotted hyena	Black-backed jackal	Wild dog
Grouping					
Primarily asocial	Large social groups	Large social groups	Large social groups	Small social groups	Large social groups
Mating system					
Polygynous and polyandrous	Monogamous with reproductive suppression	Polygynous and polyandrous	Polygynous	Monogamous with reproductive suppression	Polyandrous with reproductive suppression
Dispersal					
Male-biased with kin	Both sexes with kin	Male-biased with kin	Male-biased	Both sexes	Both sexes with kin[a]
Intraspecific interactions					
Infrequent	Intense	Common	Intense	Intense	Intense
Interspecific interactions[b]					
Rare	Rare	Rare	Common	Common	Common
Range size					
Some very large	Small	Medium	Very large	Small	Very large

[a]Frame and Frame 1976; Fuller et al. 1992
[b]With other carnivores.

Packer, Gilbert, et al. 1991). Knowledge of social structure contributes to predictions about population persistence and highlights the problems in ascribing different species a single group size value. In addition, studies that collect demographic and genetic information, such as those on lions (Packer, Gilbert et al. 1991), are in a good position to relate grouping patterns to genetic differentiation.

Mating Systems

In species that breed polygynously, many males fail to reproduce; this reduces the effective population size below the actual population size. In species that are reproductively suppressed, in which typically only one male and one female breed per group, these effects may be even more striking: here the number of breeding individuals becomes equivalent to the number of groups. Furthermore, under panmictic breeding conditions the variance in birth rate is inversely related to population size. In reproductively suppressed populations the variance in birth rate is inversely related to the proportion of breeding individuals, and is thus higher than in monogamous populations.

Reproductive suppression is characteristic of several carnivores (Creel and Creel 1990; table 21.2). For example, in Serengeti, three species show some form of reproductive suppression or delayed breeding in which nonreproductives help in different ways. In wild dog packs usually only the alpha female and male breed; subordinate males normally stay on to help their brother raise the litter, regurgitating food for the pups, while females disperse to form new packs (Frame et al. 1979; Fuller et al. 1992). Similarly, only the alpha pair of dwarf mongooses normally breeds, while subordinates guard, carry, and bring food to the alphas' offspring (Rood 1980); subordinate females can also provide milk for the alphas' young (Creel et al. 1991). Offspring of black-backed jackals from the previous year often remain on their natal territory, helping to rear their full siblings by bringing food back to them, and do not reproduce in their first year (Moehlman 1979). As a consequence, wild dog and dwarf mongoose populations are subject to higher levels of demographic stochasticity than, for example, those of leopards or banded mongooses. Reproductive suppression also reduces the number of individuals contributing to the gene pool, and therefore greatly increases the rate of loss of genetic diversity.

DNA fingerprinting techniques now used extensively in behavioral ecology can shed additional light on the number and identity of individuals contributing to the next generation. For example, Packer, Gilbert, et al. (1991) have shown that certain males in large coalitions of male lions do not father any offspring; this finding alters assessments of effective population size based on behavioral observations.

Dispersal

Effective population size is also influenced by the extent and costs of dispersal between groups (Lande and Barrowclough 1987; Rogers 1987). Dispersal is defined here as movement from group of origin to the first or subsequent breeding group (Chepko-Sade et al. 1987). Evidence from computer simulations shows that both dispersal and metapopulation structure affect population persistence, but the manner in which they do so depends on dispersal costs and the types of stochastic events involved (Durant 1991; Hansson 1991). If, for example, dispersal reduces survivorship substantially, perhaps as a result of predation, it will reduce population persistence in comparison to populations in which dispersal costs are low. In saturated habitats where territorial openings become available only as a result of death or ousting of residents, however, dispersal costs may have little influence on population persistence. Kin-structured migration, in which relatives disperse together, also influences effective population size, although its effects depend on interactions with group size and dispersal rates. In general, however, genetic differentiation between groups increases when kin migrate together.

Behavioral ecological studies sometimes obtain good data on these aspects of dispersal. For example, it is known that dwarf mongooses normally transfer between packs with overlapping home ranges, with median dispersal distances being 0.5 km for males and 1.0 km for females (Rood 1987). In regard to costs, dispersing lionesses breed at a later age in Serengeti than do nondispersers, whereas in Ngorongoro Crater they suffer greater mortality than do nondispersers (Pusey and Packer 1987). Finally, the proportion of dispersers that transfer alone or in groups and the degree of relatedness between dispersers may also be known. For instance, simultaneous primary transfer by littermates is commonplace in cheetahs, dwarf mongooses, lions, and wild dogs (table 21.2).

THE INFLUENCE OF INTRASPECIFIC BEHAVIOR ON POPULATION SIZE

Intraspecific behavior strongly influences the intrinsic rate of increase for populations. In spotted hyenas, offspring are born with fully erupted canines and incisors and attack their siblings at birth, often killing like-sexed littermates in the narrow burrow where the mother cannot intervene (Frank, Glickman, and Licht 1991). This results in a mortality rate of approximately 20% in the first few weeks of life. Among wild dogs, in those rare instances in which a subordinate female gives birth, the alpha female attempts to control access to the litter, which interferes with pup provisioning and results in litter loss (Frame et al. 1979).

On entering a pride for the first time, male lions often kill cubs sired by former male residents and this accounts for 27% of all cub mortality in the first year of life (Packer et al. 1988). Infanticide also occurs in leopards (Ilany 1990; Cavallo 1991). The conservation implications of this behavior are severe. If resident males are shot and new males replace them, small cubs will be killed and subadults evicted (Packer and Pusey 1983). Frequent replacements can actually halt recruitment altogether in both species. On the basis of this information derived from behavioral and ecological research, Packer (1990) recommended that lion and leopard hunting be stopped in the Loliondo and Ikorongo Game Controlled Areas. Behavioral data on habitat requirements and recruitment rates can therefore help to determine hunting quotas or the level of trade that can be sustained.

Social behavior can also have negative consequences for genetic diversity, the most obvious example being male-male competition over access to females. Indeed, in Ngorongoro Crater, large coalitions of male lions have prevented any immigration of males from outside since 1970. This, together with a population crash in the early 1960s (Fosbrooke 1963), has resulted in the current Crater population of 75–125 animals being descended from just 15 founders (Packer, Pusey et al. 1991). In effect, the population is genetically isolated for behavioral reasons. Simulations show that heterozygosity has been declining since the mid-1970s and that female reproductive performance may have suffered as a result (Packer, Pusey et al. 1991).

Further research is required to determine the range of population sizes over which these intraspecific behaviors are manifested, but such behaviors can potentially reduce recruitment rates even at low population sizes.

THE INFLUENCE OF INTERSPECIFIC BEHAVIOR ON POPULATION SIZE

Predation and Competition

Predation and competition are important determinants of population size. Although these processes are usually studied by ecologists, they are now under increasing scrutiny from behavioral ecologists because direct observation of individual predators minimizes bias in diet estimation (Caro and FitzGibbon 1992). For example, by studying individually recognized female cheetahs intensively, Laurenson (1994) found that cheetah cub mortality was extremely high, with only 5% of cubs reaching independence. This mortality stemmed primarily from predation by lions, which kill cheetah cubs both in their lair and soon after emergence.

Laurenson (chap. 18) has argued that the principal factor limiting cheetah population size in Serengeti and probably in other protected areas is cub mortality due to sympatric predators. Predation is likely to be a far more important factor affecting this species than is its reduced genetic variability (Caro and Laurenson 1994). Litter loss may depend on the density of other predators and perhaps the availability of safe lair sites. Under current conditions cheetahs may fare best in game reserves or multiple-use areas where other predators are hunted or harassed, but where prey densities remain relatively high (Laurenson, Caro, and Borner 1992).

Disease

Disease epizootics may exert a particularly strong effect on the probability of population extinction (Scott 1988; Thorne and Williams 1988). Behavioral studies are crucial to understanding the epidemiology of disease and in designing disease management programs such as vaccination schemes. The incidence of epizootics depends upon the contact rate between and within subpopulations and the number of susceptible individuals (Anderson and May 1979). Disease will therefore affect social species with a high rate of contact between individuals most strongly (table 21.2). If disease is transmitted between groups or individuals intraspecifically, outbreaks are likely to be correlated in time between groups. However, in low-density species or where intraspecific interactions are infrequent, interspecific transmission may assume greater importance (see also Dobson, chap. 23).

As a result of long-term monitoring, we know that disease has affected both the lion and wild dog populations in the Serengeti ecosystem. Lions in Ngorongoro Crater underwent a population crash in the early 1960s as a result of a plague of *Stomoxys* biting flies, which reduced the lion population from about 60–75 to 10–15 animals (Fosbrooke 1963; Packer, Pusey et al. 1991). Lions are also suffering a new outbreak of disease at the time of writing.

Wild dogs in Serengeti may also have declined as a result of disease. The population dropped from 110 adults in 1970 to 26 in 1977 as a result of high pup mortality (Malcolm 1979), and clinical signs of disease were observed on several occasions (Schaller 1972; Malcolm 1979). In 1990 rabies was confirmed as the cause of death of one dog, and clinical signs were seen in other pack members (Gascoyne et al. 1993). A study of seroprevalence to rabies antibody showed that over 40% of wild dogs sampled had been exposed to the virus (Gascoyne et al. 1993). Knowledge of behavioral interaction rates between different wild dog packs and between packs and domestic animals would shed light on the relative importance of intraspecific and interspecific disease transmission in this species.

SPECIES-SPECIFIC BEHAVIOR AND EXTINCTION PROBABILITY

The probability that a species will go extinct depends on numerous ecological, behavioral, and life history factors. For example, large species and those with large ranges occur at low population sizes and are more vulnerable to extinction (see Gilpin and Diamond 1980; Higgs and Usher 1980). In addition, some evidence suggests that risk-prone species have low rates of dispersal, slow rates of reproduction, or specialized diets (Terborgh 1974; Wilcox 1980; Fowler and MacMahon 1982). Other, poorly understood behavioral factors, such as willingness to cross open areas (Willis 1974) and susceptibility to nest predation (Sieving 1992) have also been implicated in local extinctions. The relative importance of behavioral factors in promoting extinction requires urgent investigation since these factors greatly affect a population's response to range fragmentation.

THE IMPORTANCE OF BEHAVIORAL STUDIES FOR MANAGEMENT OPTIONS

Ranging Behavior

Many protected areas in Africa have been delineated to take account of the movements of species they are trying to protect. For example, the present boundaries of the Serengeti National Park and Ngorongoro Conservation Area were formed only after wildebeest ranging patterns were known (Grzimek and Grzimek 1959; Turner 1987). If intact ecosystems are to be conserved, it is essential that they account for the ranging behavior of large carnivores (table 21.2).

Serengeti cheetahs and wild dogs have enormous home ranges because they follow the movements of migratory prey (Durant et al. 1988) (an average of 833 km^2 and 777 km^2 for female and nonterritorial male cheetahs respectively: Caro 1994; 1,500–2,000 km^2 for wild dog packs: Frame et al. 1979). Both species range outside the park. Fortunately, buffer zones afford protection to both species, but their population sizes would almost certainly be reduced if agriculture directly abutted the park. Elsewhere in areas of high prey density, wild dog ranges are smaller, at about 600 km^2 (Reich 1981). Nonetheless, even in such favorable circumstances Frame and Fanshawe (1991) estimate that a reserve 2,300 km^2 in area could support only six packs or 30 adults maximum. Very few reserves in Africa are large enough to contain the 200–300 individuals thought to be a rough minimum figure for long-term population persistence, disregarding genetic deterioration (East 1981; Fanshawe, Frame, and Ginsberg 1991). In comparison to one large reserve, several small reserves with no gene flow between them would be completely inadequate for wild dog protection (Diamond 1976; Terborgh 1976; Newmark 1987).

While spotted hyenas occur at higher densities than wild dogs or cheetahs, they also range over huge areas. Hyenas commute from a core clan range to the wildebeest migration in order to hunt, with the result that a large proportion of the hyena population may converge on one small area (Hofer, East, and Campbell 1993). If this area is subject to snaring, as occurs in the west of the park and in the Grumeti and Ikorongo Game Controlled Areas, the whole population is at risk. Indeed, Hofer, East, and Campbell (1993) calculate that over 10% of adult spotted hyenas on the plains-woodland border are killed in snares each year.

Carnivores also cross large areas while dispersing. For example, male lions occasionally move from the central Serengeti plains to Ngorongoro Crater or into the Loliondo Game Controlled Area, where they have been shot (Packer 1990). Reserve designers must take the movements and dispersal patterns of wide-ranging species into account, but can possibly take advantage of long dispersal distances to construct corridors leading to other protected areas (Johnsingh, Narendra Prasad, and Goyal 1990; but see Hobbs 1992).

Calculating Minimum Viable Populations

One of the most important parameters for conservation is the size below which a population ceases to be viable over the long term. A minimum viable population (MVP) can be defined in two ways. First, the demographic MVP is the population size able to persist with a particular probability over a specified number of years (Shaffer 1981). The genetic MVP, however, is the size of the population able to maintain a particular level of genetic diversity over a specified time period (Foose et al. 1986; Ralls and Ballou 1986; Soulé et al. 1986). In general, the genetic MVP is higher than the demographic MVP, depending on factors such as the species' behavior and the risks of extinction that management is willing to accept (Soulé et al. 1986). In Serengeti, only wild dogs and possibly cheetahs fall below generally accepted MVP levels (table 21.3).

The demographic MVP can be calculated only by estimating the probability distribution of the time to extinction. Some models rely on mean population growth rate and variance estimates but need a long-term demographic data set characteristic of behavioral ecological studies in order to be accurate (Durant 1991). The genetic MVP is generally calculated using detailed life history statistics. Demographic records therefore have an important role to play in calculating both sorts of MVP.

Monitoring Populations

Management requires regular monitoring of species in order to protect them effectively, and knowledge of behavior underlies many different population monitoring schemes. For example, ungulate censuses rely on knowing of the whereabouts of species at different times of year (Sinclair and Norton-Griffiths 1982). Similarly, in species that commute, population estimates will be greatly affected by when and where censuses are conducted (Hofer and East, chap. 16). Transects use assumptions about the distribution of group sizes and how different habitats are utilized, and will be affected if group sizes vary between habitats. Capture-mark-recapture techniques assume an equal chance of recapturing or resighting individuals, which in turn depends on their tameness following handling and their activity schedules. Conversely, behavioral observations are useful in assessing whether monitoring techniques, such as radio-collaring, have an effect on study animals (Laurenson and Caro 1994).

Interventions

In some circumstances interventions may be desirable when populations exhibit low densities and poor recruitment, or when sex ratios become skewed. For example, when rabies was implicated in the reduction of the Serengeti wild dog population to very low levels, two packs were vaccinated against rabies following a trial program with four wild dogs at the Frankfurt Zoo (Gascoyne et al. 1993).

Reintroductions, another form of intervention, can best be applied in three situations (Stanley Price 1989): first, when a localized extinction has occurred; second, to bolster an existing population that has declined to

Table 21.3 Approximate 1991 population sizes of some carnivores in the Tanzanian and Kenyan portions of Serengeti ecosystem combined.

Species	Estimated numbers of adults	Source
Cheetah	200–250	Authors' estimate
Leopard	800–1,000	Borner et al. 1987
Lion	2,800	Packer 1990
Banded mongoose	43,000	Waser et al., chap. 20
Dwarf mongoose	94,000	Waser et al., chap. 20
Slender mongoose	30,000	Waser et al., chap. 20
Black-backed jackal	6,300	Authors' estimate[a]
Spotted hyena	9,000	Hofer and East, chap. 16[b]
Wild dog	50	Burrows 1991

[a]Based on dividing 11,425 km^2 of woodlands with >5% canopy (Tanzania Wildlife Conservation Monitoring database) by average home range size (Fuller et al. 1989).
[b]An estimated 1,500 spotted hyenas living in the Masai Mara (Hilborn et al., chap. 29) were added onto the rounded 7,500 quoted for the Tanzania ecosystem (Hofer and East, chap. 16).

low levels; and third, to introduce new genetic material into an inbred population. Successful reintroductions critically depend upon accurate behavioral data, including knowledge of a species' activity cycle, diet, and social behavior. Stanley Price (1989) has outlined a number of behavioral factors that can facilitate reintroductions: being tolerant of a broad range of habitats and thus adaptable to new situations; having a wide range of foods available; being exploratory and hence able to move into new areas; and being amenable to behavioral manipulation.

Currently, reintroducing cheetahs into reserves in Russia and India is under discussion, since their numbers increase rapidly in the absence of lions and spotted hyenas (Anderson 1984). The cheetah has many of the traits highlighted by Stanley Price. For example, Adamson (1969) was able to release a captive cheetah successfully by gradually reducing the amount of food it received from her.

In summary, it is clear that behavioral ecology can contribute to many critical facets of conservation biology, such as predictions of population persistence, reserve design, and management. The most important facts it can contribute are size and number of social or geographic units, knowledge of the mating system and dispersal, interaction rates, ranging patterns, and species-specific behavior (table 21.4).

LINKS BETWEEN CONSERVATION AND BEHAVIOR OUTSIDE SERENGETI

In this chapter we have shown how knowledge of behavior and ecology enhances conservation using data from Serengeti carnivores. Since large carnivores hold such a prominent place in conservation biology, it could be argued that our examples are special cases. Yet an examination of

Table 21.4 Relationships between behavioral ecology and conservation biology.

Aspect of behavioral ecology	Relevant information	Principal conservation significance
Demographic records	Mean and variance in reproductive success	Calculating MVPs
Grouping patterns	Mean and range	Population persistence Genetic diversity Monitoring
Mating system	Polygyny/monogamy, reproductive suppression	Population persistence Genetic diversity
Dispersal	Rate, costs, and kin structure	Population persistence Genetic diversity Reserve size and corridors
Ranging patterns	Overlap and home range size	Reserve size Monitoring Interventions
Intraspecific interactions	Rate	Population persistence Disease transmission Interventions
Interspecific interactions	Rate	Population persistence Disease transmission
Species-specific behavior	Various	Population persistence Monitoring Interventions

other conservation studies and programs shows that behavioral ecology has important ramifications for a wide variety of conservation agendas.

First, dispersal behavior is increasingly being incorporated into extinction models to make them more precise (Chepko-Sade et al. 1987). Durant and Mace (1994) showed that monk seals became increasingly vulnerable to extinction as migration increased because they risked moving to uninhabited localities where breeding was impossible, but mountain gorillas became less vulnerable with increasing migration because females transferred to breeding groups. Indeed, dispersal and colonization of new areas may be enhanced by the presence of conspecifics, thereby altering metapopulation dynamics (Smith and Peacock 1990). Similarly, knowledge of dispersal distances helps to address the related issue of how habitat fragmentation affects population persistence, and is now being incorporated into plans for the recovery of the northern spotted owl (Murphy and Noon 1992).

Second, empirical conservation studies now attempt to collect data on life histories and the ecological and social factors affecting individuals.

As illustrations, Laurance (1991) found that rainforest mammals in northern Queensland, Australia, were more prone to extinction if they had low fecundity and high longevity, or if they had specialized diets; while Soulé et al. (1988) showed that the occurrence of chaparral-requiring birds was positively associated with the presence of coyotes, since the latter reduced the abundance of avian "mesopredators" such as gray foxes and domestic cats.

Third, conservation strategies are now beginning to account for behavior even in their initial stages. For example, female grouping patterns and mating preferences in African elephants are seen as critical in predicting the chances of subpopulation recovery following poaching (Dobson and Poole, in press).

Fourth, as with carnivores, ranging patterns of other species have helped to delineate the size and location of reserve boundaries. Based on measurements of the huge territories of rainforest raptors, Thiollay (1989) argued that the size of proposed national parks in French Guiana should be as large as 1–10 million hectares in order to encompass a sufficient number of breeding pairs.

Fifth, rehabilitation programs have relied extensively on behavioral insights to be successful. In attempting to increase the number of nest sites for the highly endangered Puerto Rican parrot, researchers found that the pearly-eyed thrasher was driving parrots away from nesting holes or breaking their eggs. By carefully determining the size and shape of artificial nest boxes preferred by each species and erecting both in close proximity, the researchers enabled each species to lay, since the aggressive thrashers drove intruding conspecifics away from the parrots' nests (Snyder and Taapken 1978). Similarly, the successful reintroduction of Arabian oryx into Oman rested heavily on advance knowledge of the ranging patterns, diet, grouping, and reproductive behavior of the species (Stanley Price 1989).

Finally, the effects of tourism can be determined in part from observing animals' responses to human disturbance. For instance, Burger and Gochfeld (1991) showed that distances at which birds were flushed by humans were shorter in residents that were regularly exposed to people than in migrants, indicating that habituation had occurred. More studies of this nature would be useful in East Africa, where national revenue depends so much on the presence and viability of mammal populations in the face of mass tourism (see Burney and Burney 1979).

Though space limits us to these few examples, it should be clear that the bridges between behavioral ecology and conservation biology are numerous, and are often pivotal to conservation programs. Indeed, the benefits of knowing species' habitat requirements for in situ conservation and their behavioral needs for ex situ conservation are self-evident. Currently, however, links between the disciplines are constructed by conservation biologists seeking to make their models more realistic, or their management plans more successful. We additionally need behavioral ecologists to give more weight to conservation concerns in the course of their research and to present data in a form more suitable for the purposes of conservation biology, since their findings can greatly assist in predicting imminent population extinctions.

ACKNOWLEDGMENTS

We are grateful to the Government of Tanzania for its hospitality to long-term research, and to Steve Albon, Peter Arcese, Andy Dobson, Paule Gros, Susan Harrison, Karen Laurenson, and Peter Waser for comments, and to Ken Campbell and Sally Huish for help with word processing. T. M. C. was supported by Hatch funds granted to the University of California while writing.

REFERENCES

Adamson, J. 1969. *The spotted sphinx*. London: Collins.

Anderson, J. L. 1984. A strategy for cheetah conservation in Africa. In *The extinction alternative,* ed. P. J. Mundy, 127–35. Johannesburg: Endangered Wildlife Trust.

Anderson, R. M., and May, R. M. 1979. Population biology of infectious diseases. *Nature* 280:361–67.

Bertram, B. C. R. 1975. Social factors influencing reproduction in wild lions. *J. Zool.* (Lond.) 177:463–82.

———. 1979. Serengeti predators and their social systems. In *Serengeti: Dynamics of an ecosystem,* ed. A. R. E. Sinclair and M. Norton-Griffiths, 221–48. Chicago: University of Chicago Press.

———. 1982. Leopard ecology as studied by radio tracking. *Symp. Zool. Soc. Lond.* 49:341–52.

Borner, M., FitzGibbon, C. D., Borner, M., Caro, T. M., Lindsay, W., Collins, D. A., and Holt, M. E. 1987. The decline of the Serengeti Thomson's gazelle population. *Oecologia* 73:32–40.

Burger, J., and Gochfeld, M. 1991. Human distance and birds: Tolerance and response distances of resident and migrant species in India. *Environ. Conserv.* 18:158–65.

Burney, D., and Burney, L. 1979. Cheetah and man, part two. *Swara* 2(3):28–32.

Burrows, R. 1991. Observations on the behaviour, ecology, and conservation status of African wild dogs *(Lycaon pictus)* in Serengeti National Park, Tanzania. Unpublished typescript, June 1991.

Bygott, J. D., Bertram, B. C. R., and Hanby, J. P. 1979. Male lions in large coalitions gain reproductive advantages. *Nature* 282:839–41.

Caro, T. M. 1987. Cheetah mothers' vigilance: Looking out for prey or for predators? *Behav. Ecol. Sociobiol.* 20:351–61.

———. 1989. Determinants of asociality in felids. In *Comparative socioecology: The behavioural ecology of humans and other mammals,* ed. V. Standen and R. A. Foley, 41–74. Special publication of the British Ecological Society no. 8. Oxford: Blackwell Scientific Publications.

———. 1994. *Cheetahs of the Serengeti plains: Group living in an asocial species.* Chicago: University of Chicago Press.

Caro, T. M., and Collins, D. A. 1987. Male social organization and territoriality. *Ethology* 74:52–64.

Caro, T. M., and FitzGibbon, C. D. 1992. Large carnivores and their prey: The quick and the dead. In *Natural enemies: The population biology of predators, parasites, and diseases,* ed. M. J. Crawley, 117–42. Oxford: Blackwell Scientific Publications.

Caro, T. M., FitzGibbon, C. D., and Holt, M. E. 1989. Physiological costs of behavioural strategies for male cheetahs. *Anim. Behav.* 38:309–17.

Caro, T. M., and Laurenson, M. K. 1994. Ecological and genetic factors in conservation: A cautionary tale. *Science* 263:485–86.

Cavallo, J. A. 1990. Cat in the human cradle. *Nat. Hist.* 2:52–61.

———. 1991. A study of leopard behavior and ecology in the Seronera Valley, Serengeti National Park, Tanzania. Unpublished typescript to SWRI, TANAPA, COETECH, and MWEKA, February 1991.

Chepko-Sade, B. D., Shields, W. M., Berger, J., Halpin, Z. T., Jones, W. T., Rogers, L. L., Rood, J. P., and Smith, A. T. 1987. The effects of dispersal and social structure on effective population size. In *Mammalian dispersal patterns: The effects of social structure on population genetics,* ed. B. D. Chepko-Sade and Z. T. Halpin, 287–321. Chicago: University of Chicago Press.

Clutton-Brock, T. H. 1988. *Reproductive success: Studies of individual variation in contrasting breeding systems.* Chicago: University of Chicago Press.

Creel, E. R., and Creel, N. M. 1990. Energetics, reproductive suppression, and obligate communal breeding in carnivores. *Behav. Ecol.* 28:263–70.

Creel, E., Creel, N., Wildt, D. E., and Montfort, E. L. 1992. Behavioural and endocrine mechanisms of reproductive suppression in Serengeti dwarf mongooses. *Anim. Behav.* 43:231–45.

Creel, E. R., Montfort, E. L., Wildt, D. E., and Waser, P. M. 1991. Spontaneous lactation is an adaptive result of pseudopregnancy. *Nature* 351:660–62.

Creel, E. R., and Waser, P. M. 1994. Inclusive fitness and reproductive strategies in dwarf mongooses. *Behav. Ecol.* 5:339–48.

Crow, J. F., and Kimura, M. 1970. *An introduction to population genetics theory.* New York: Harper and Row.

den Boer, P. J. 1968. Spreading of risk and stabilization of animal numbers. *Acta Biotheoret.* 18:165–94.

Diamond, J. M. 1976. Island biogeography theory and conservation: Strategy and limitations. *Science* 193:1027–29.

Dobson, A. P., and Poole, J. H. In press. Ivory poaching and the viability of African elephant populations. *Conserv. Biol.*

Dobzhansky, J. H., and Wallace, B. 1953. The genetics of homeostasis in Drosophila. *Proc. Natl. Acad. Sci. USA* 39:162–71.

Durant, S. M. 1991. Individual variation and dynamics of small populations: Implications for conservation and management. Ph.D. thesis, Cambridge University.

Durant, S. M., Caro, T. M., Collins, D. A., Alawi, R. M., and FitzGibbon, C. D. 1988. Migration patterns of Thomson's gazelles and cheetahs on the Serengeti plains. *Afr. J. Ecol.* 26:257–68.

Durant, S. M., and Harwood, J. 1992. Assessment of monitoring and management strategies for local populations of the Mediterranean monk seal, *Monachus monachus. Biol. Conserv.* 61:81–92.

Durant, S. M., and Mace, G. M. 1994. Species differences and population structure in population viability analysis. In *Creative Conservation: Interactive management of wild and captive animals,* eds. P. J. S. Olney, G. M. Mace and A. T. C. Feistner, 67–91. London: Chapman and Hall.

East, M. L., and Hofer, H. 1991. Loud calling in a female-dominated mammalian society: I. Structure and composition of whooping bouts of spotted hyenas, *Crocuta crocuta. Anim. Behav.* 42:637–49.

East, R. 1981. Species-area curves and populations of large mammals in African savanna reserves. *Biol. Conserv.* 21:111–26.

Ehrlich, P. R., and Ehrlich, A. H. 1981. *Extinction: The causes and consequences of the disappearance of species.* New York: Random House.

Fanshawe, J. H., and FitzGibbon, C. D. 1993. Factors influencing the hunting success of an African wild dog pack. *Anim. Behav.* 45:479–90.

Fanshawe, J. H., Frame, L. H., and Ginsberg, J. R. 1991. The wild dog—Africa's vanishing carnivore. *Oryx* 25:137–46.

FitzGibbon, C. D. 1990. Why do cheetahs prefer male gazelles? *Anim. Behav.* 40:837–45.

Foose, T. J., Lande, R., Flesness, N. R., Rabb, G., and Read, B. 1986. Propagation plans. *Zoo Biol.* 5:139–46.

Fosbrooke, H. 1963. The *Stomoxys* plague in Ngorongoro, 1962. *E. Afr. Wildl. J.* 1:124–26.

Fowler, C. W., and MacMahon, J. A. 1982. Selective extinction and speciation: Their influence on the structure and functioning of communities and ecosystems. *Am. Nat.* 119:480–98.

Frame, L. H., and Fanshawe, J. H. 1991. African wild dog *Lycaon pictus:* A survey of status and distribution 1985–1988. Unpublished report to IUCN, Morges, Switzerland.

Frame, L. H., and Frame, G. W. 1976. Female African wild dogs emigrate. *Nature* 263:227–29.

Frame, L. H., Malcolm, J. R., Frame, G. W., and van Lawick, H. 1979. Social organization of African wild dogs *(Lycaon pictus)* on the Serengeti plains, Tanzania 1967–1978. *Z. Tierpsychol.* 50:225–49.

Frank, L. G. 1986a. Social organization of the spotted hyaena *(Crocuta crocuta).* I. Demography. *Anim. Behav.* 34:1500–1509.

———. 1986b. Social organization of the spotted hyaena *(Crocuta crocuta).* II. Dominance and reproduction. *Anim. Behav.* 34:1510–27.

Frank, L. G., Glickman, E. E., and Licht, P. 1991. Fatal sibling aggression, precocial development, and androgens in neonatal spotted hyaenas. *Science* 252:702–4.

Franklin, I. R. 1980. Evolutionary change in small populations. In *Conservation biology: An evolutionary-ecological perspective,* ed. M. E. Soulé and B. A. Wilcox, 135–49. Sunderland, Mass: Sinauer Associates.

Fuller, T. K., Biknevicius, A. R., Kat, P. W., van Valkenburgh, B., and Wayne, R. K. 1989. The ecology of three sympatric jackal species in the Rift Valley of Kenya. *Afr. J. Ecol.* 27:313–23.

Fuller, T. K., Kat, P. W., Bulger, J. B., Maddock, A. H., Ginsberg, J. R., Burrows, R., Weldon McNutt, J., and Mills, M. G. L. 1992. Population dynamics of African wild dogs. In *Wildlife 2001: Populations,* ed. D. R. McCullough and R. H. Barrett, 1125–39. London: Elsevier Applied Science.

Gascoyne, E. C., Laurenson, M. K., Lelo, S., and Borner, M. 1993. Rabies in African wild dogs *(Lycaon pictus)* in the Serengeti region, Tanzania. *J. Wildl. Dis.* 29:396–402.

Gilpin, M. E. 1991. The genetic effective size of a metapopulation. *Biol. J. Linn. Soc.* 42:165–75.

Gilpin, M. E., and Diamond, J. M. 1980. Subdivision of nature reserves and the maintenance of species diversity. *Nature* 285:567–68.

Gilpin, M. E., and Soulé, M. E. 1986. Minimum viable populations: Processes of species extinction. In *Conservation biology: The science of scarcity and diversity,* ed. M. E. Soulé, 19–34. Sunderland, Mass.: Sinauer Associates.

Goodman, D. 1987. The demography of chance extinction. In *Viable populations for conservation,* ed. M. E. Soulé, 11–34. Cambridge: Cambridge University Press.

Grzimek, B., and Grzimek, M. 1959. *Serengeti shall not die.* Berlin: Ullstein, A. G.

Hanby, J. P., and Bygott, J. D. 1987. Emigration of subadult lions. *Anim. Behav.* 35:161–69.

Hansson, L. 1991. Dispersal and connectivity in metapopulations. In *Metapopulation dynamics: Empirical and theoretical investigations,* ed. M. Gilpin and I. Hanski, 89–103. London: Academic Press.

Higgs, A. J., and Usher, M. B. 1980. Should nature reserves be large or small? *Nature* 85:568–69.

Hobbs, R. J. 1992. The role of corridors in conservation: Solution or bandwagon? *Trends Ecol. Evol.* 7:389–92.

Hofer, H., and East, M. L. 1993a. The commuting system of spotted hyaenas: How a predator copes with migratory prey. I. Social organization. *Anim. Behav.* 46:547–57.

———. 1993b. The commuting system of spotted hyaenas: How a predator copes with migratory prey. II. Intrusion pressure and commuters' space use. *Anim. Behav.* 46:559–74.

———. 1993c. The commuting system of spotted hyaenas: How a predator copes with migratory prey. III. Attendance and maternal care. *Anim. Behav.* 46:575–89.

Hofer, H., East, M., and Campbell, K. L. I. 1993. Snares, commuting hyaenas, and migratory herbivores: Humans as predators in the Serengeti. *Symp. Zool. Soc. Lond.* 65:347–66.

Ilany, G. 1990. The spotted ambassadors of a vanishing world. *Israelal* May/June 1990, 18–24.

Johnsingh, A. J. T., Narendra Prasad, E., and Goyal, E. P. 1990. Conservation status of the Chila-Motichur corridor for elephant movement in Rajaji-Corbett national parks area, India. *Biol. Conserv.* 51:125–38.

Kruuk, H. 1972. *The spotted hyaena: A study of predation and social behavior.* Chicago: University of Chicago Press.

———. 1976. Feeding and social behaviour of the striped hyaena. *E. Afr. Wildl. Ecol.* 14:91–111.

Kruuk, H., and Sands, W. A. 1972. The aardwolf (*Proteles cristatus* Eparrman) 1783 as predator on termites. *E. Afr. Wildl. J.* 10:211–27.

Lamprecht, J. 1979. Field observations on the behaviour and social system of the bat-eared fox *Otocyon megalotis* Desmarest. *Z. Tierpsychol.* 49:260–84.

Lande, R., and Barrowclough, G. F. 1987. Effective population size, genetic variation and their use in population management. In *Viable populations for conservation,* ed. M. E. Soulé, 87–123. Cambridge: Cambridge University Press.

Landres, P. B., Verner, J., and Thomas, J. W. 1988. Ecological uses of vertebrate indicator species: A critique. *Conserv. Biol.* 2:316–28.

Laurance, W. F. 1991. Ecological correlates of extinction proneness in Australian tropical rain forest mammals. *Conserv. Biol.* 5:79–89.

Laurenson, M. K. 1994. High juvenile mortality in cheetahs (*Acinonyx jubatus*) and its consequences for maternal care. *J. Zool.* (Lond.) 234:387–408.

Laurenson, M. K., and Caro, T. M. 1994. Monitoring the effects of non-trivial handling in free-living cheetahs. *Anim. Behav.* 47:547–57.

Laurenson, M. K., Caro, T. M., and Borner, M. 1992. Patterns of female reproduction in wild cheetahs. *Natl. Geogr. Res. Explor.* 8(1):64–75.

Malcolm, J. R. 1979. Social organisation and communal rearing in African Wild Dogs. Ph.D. thesis, Harvard University.

———. 1986. Socio-ecology of bat-eared foxes *(Otocyon megalotis)*. *J. Zool.* (Lond.) 208:457–67.

Malcolm, J. R., and Marten, K. 1982. Natural selection and the communal rearing of pups in African wild dogs *(Lycaon pictus)*. *Behav. Ecol. Sociobiol.* 10:1–13.

Miller, R. I. 1979. Conserving the genetic integrity of faunal populations and communities. *Environ. Conserv.* 6:297–304.

Moehlman, P. D. 1979. Jackal helpers and pup survival. *Nature* 277:382–83.

———. 1983. Socioecology of silverbacked and golden jackals *(Canis mesomelas, Canis aureus)*. In *Recent advances in the study of mammalian behavior,* ed. J. F. Eisenberg and D. G. Kleiman, 423–38. Special Publication no. 7. Lawrence, Kans.: American Society of Mammalogists.

Murphy, D. D., and Noon, B. R. 1992. Integrating scientific methods with habitat conservation planning: Reserve design for northern spotted owls. *Ecol. Appl.* 2:3–17.

Newmark, W. D. 1987. A land-bridge island perspective on mammalian extinctions in western North American parks. *Nature* 325:430–32.

O'Brien, E. J., Wildt, D. E., Bush, M., Caro, T. M., FitzGibbon, C. D., Aggundey, I., and Leakey, R. E. 1987. East African cheetahs: Evidence for two population bottlenecks? *Proc. Natl. Acad. Sci. USA* 84:508–11.

Packer, C. 1990. Serengeti lion survey. Unpublished typescript to TANAPA, EWRI, MWEKA and the Game Department, November 1990.

Packer, C., Gilbert, D. A., Pusey, A. E., and O'Brien, E. J. 1991. A molecular genetic analysis of kinship and cooperation in African lions. *Nature* 351:562–65.

Packer, C., Herbst, L., Pusey, A. E., Bygott, J. D., Hanby, J. P., Cairns, E. J., and Borgerhoff Mulder, M. 1988. Reproductive success of lions. In *Reproductive success: Studies of individual variation in contrasting breeding systems,* ed. T. H. Clutton-Brock, 363–83. Chicago: University of Chicago Press.

Packer, C., and Pusey, A. E. 1982. Cooperation and competition in coalitions of male lions: Kin selection or game theory? *Nature* 296:740–42.

———. 1983. Adaptations of female lions to infanticide by incoming males. *Am. Nat.* 121:716–28.

Packer, C., Pusey, A. E., Rowley, H., Gilbert, D. A., Martenson, J., and O'Brien, E. J. 1991. Case study of a population bottleneck: Lions of the Ngorongoro Crater. *Conserv. Biol.* 5:219–30.

Packer, C., Scheel, D., and Pusey, A. E. 1990. Why lions form groups: Food is not enough. *Am. Nat.* 136:1–19.

Pusey, A. E., and Packer, C. 1987. The evolution of sex-biased dispersal in lions. *Behaviour* 101:275–310.

Rabinowitz, A. 1986. *Jaguar.* New York: Random House.

Ralls, K., and Ballou, J. 1986. Captive breeding programs for populations with a small number of founders. *Trends Ecol. Evol.* 1:19–22.

Ralls, K., Brugger, K., and Ballou, J. 1979. Inbreeding and juvenile mortality in small populations of ungulates. *Science* 206:1101–3.

Reich, A. 1981. The behavior and ecology of the African Wild Dog *(Lycaon pictus)* in the Kruger National Park. Ph.D. thesis, Yale University.

Rogers, A. R. 1987. A model of kin structured migration. *Evolution* 41:417–26.

Rood, J. P. 1975. Population dynamics and food habits of the banded mongoose. *E. Afr. Wildl. J.* 13:89–111.

———. 1978. Dwarf mongoose helpers at the den. *Z. Tierpsychol.* 48:277–87.

———. 1980. Mating relationships and breeding suppression in the dwarf mongoose. *Anim. Behav.* 28:143–50.

———. 1987. Dispersal and intergroup transfer in the dwarf mongoose. In *Mammalian dispersal patterns: The effects of social structure on population genetics,* ed. B. D. Chepko-Sade and Z. T. Halpin, 85–103. Chicago: University of Chicago Press.

———. 1990. Group size, survival, reproduction, and routes to breeding in dwarf mongooses. *Anim. Behav.* 39:566–72.

Schaller, G. B. 1972. *The Serengeti lion: A study in predator-prey relations.* Chicago: University of Chicago Press.

Scheel, D. 1993. Profitability, encounter rates, and the prey choice of African lions. *Behav. Ecol.* 4:90–97.

Scheel, D., and Packer, C. 1991. Group hunting behaviour in lions: A search for cooperation. *Anim. Behav.* 41:697–709.

Scott, M. E. 1988. The impact of infection and disease on animal populations: Implications for conservation biology. *Conserv. Biol.* 2:40–56.

Selander, R. K. 1983. Evolutionary consequences of inbreeding. In *Genetics and conservation,* ed. C. M. Schonewald-Cox, E. M. Chambers, B. MacBryde, and W. L. Thomas, 201–15. Menlo Park, Calif.: Benjamin/Cummings.

Shaffer, M. L. 1981. Minimum population sizes for species conservation. *BioScience* 31:131–34.

Sieving, K. E. 1992. Nest predation and differential insular extinction among selected forest birds of central Panama. *Ecology* 73:2310–28.

Simberloff, D. 1986. The proximate causes of extinction. In *Patterns and processes in the history of life*, ed. D. M. Raup and D. Jablonski, 259–76. Berlin: Springer-Verlag.

———. 1988. The contribution of population and community biology to conservation science. *Annu. Rev. Ecol. Syst.* 19:473–511.

Sinclair, A. R. E., and Norton-Griffiths, M. 1982. Does competition or facilitation regulate migrant ungulate populations in the Serengeti? A test of hypotheses. *Oecologia* 53:364–69.

Smith, A. T., and Peacock, M. M. 1990. Conspecific attraction and the determination of metapopulation colonization rates. *Conserv. Biol.* 4:320–23.

Snyder, N. F. R., and Taapken, I. D. 1978. Puerto Rican parrots and nest site predation by pearly-eyed thrashers. In *Endangered birds: Management techniques for preserving threatened species*, ed. S. A. Temple, 113–20. Madison: University of Wisconsin Press.

Soulé, M. E. 1983. What do we really know about extinction? In *Genetics and conservation*, ed. C. M. Schonewald-Cox, E. M. Chambers, B. MacBryde, and W. L. Thomas, 111–24. Menlo Park, Calif.: Benjamin/Cummings.

Soulé, M. E., Bolger, D. T., Alberts, A. C., Wright, J., Sorice, M., and Hill, E. 1988. Reconstructed dynamics of rapid extinctions of chaparral-requiring birds in urban habitat islands. *Conserv. Biol.* 2:75–92.

Soulé, M. E., Gilpin, M., Conway, W., and Foose, T. 1986. The millennium ark: How long a voyage, how many staterooms, how many passengers? *Zoo Biol.* 5:101–13.

Stanley Price, M. R. 1989. *Animal re-introductions: The Arabian oryx in Oman.* Cambridge: Cambridge University Press.

Templeton, A. R. 1987. Inferences on natural population structure from genetic studies on captive mammalian populations. In *Mammalian dispersal patterns: The effects of social structure on population genetics*, ed. B. D. Chepko-Sade and Z. T. Halpin, 257–72. Chicago: University of Chicago Press.

Terborgh, J. 1974. Preservation of natural diversity: The problem of extinction prone species. *BioScience* 24:715–22.

———. 1976. Island biogeography theory and conservation: Strategy and limitations. *Science* 193:1029–30.

———. 1988. The big things that run the world—a sequel to E. O. Wilson. *Conserv. Biol.* 2:402–3.

Thiollay, J. M. 1989. Area requirements for the conservation of rain forest raptors and game birds in French Guiana. *Conserv. Biol.* 3:128–37.

Thorne, E. T., and Williams, E. E. 1988. Disease and endangered species: The black-footed ferret as a recent example. *Conserv. Biol.* 2:66–74.

Turner, M. 1987. *My Serengeti years: The memoirs of an African game warden*, ed. B. Jackman. London: Elm Tree Books, Hamish Hamilton.

Waser, P. M., and Waser, M. E. 1985. *Ichneumia albicauda* and the evolution of gregariousness. *Z. Tierpsychol.* 68:137–51.

Wayne, R. K., Meyer, A., Lehman, N., van Valkenburgh, B., Kat, P. W., Fuller, T. K., Girman, D., and O'Brien, E. J. 1990. Large sequence divergence among mitochondrial DNA genotypes within populations of eastern African black-backed jackals. *Proc. Natl. Acad. Sci. USA* 87:1772–76.

Western, D. 1987. Africa's elephants and rhinos: Flagships in crisis. *Trends Ecol. Evol.* 2:343–46.

Wilcox, B. A. 1980. Insular ecology and conservation. In *Conservation biology: An evolutionary-ecological perspective,* ed. M. E. Soulé and B. A. Wilcox, 95–117. Sunderland, Mass.: Sinauer Associates.

Wildt, D. E., Bush, M., Goodrowe, K. L., Packer, C., Pusey, A. E., Brown, J. L., Joslin, P., and O'Brien, E. J. 1987. Reproductive and genetic consequences of founding isolated lion populations. *Nature* 329:328–31.

Wildt, D. E., O'Brien, E. J., Howard, J. G., Caro, T. M., Roelke, M. E., Brown, J. L., and Bush, M. 1987. Similarity in ejaculate-endocrine characteristics in captive versus free-ranging cheetahs of two subspecies. *Biol. Reprod.* 36:351–60.

Willis, E. O. 1974. Populations and local extinctions of birds on Barro Colorado Island, Panama. *Ecol. Monogr.* 44:153–69.

TWENTY-TWO

Population Structure of Wildebeest: Implications for Conservation

Nicholas Georgiadis

The Serengeti ecosystem is often defined as the area encompassed by the wildebeest migration (McNaughton and Campbell 1991), and this definition has provided a compelling justification for shaping and reshaping the boundaries of a vast protected area. If one goal of protected areas is to maintain genetically intact wildlife populations, however, it is still unclear whether this definition is ecologically appropriate for other species in the ecosystem, or even for wildebeest. For example, effective conservation plans rest on assumptions about the spatial limits of populations and their interactions with one another. The problems involved in making such assumptions are shared by all protected areas whose fate it is to become "islands in a sea of humanity" (Leader-Williams, Harrison, and Green 1990). Nevertheless, we are rarely certain of the extent to which protected areas are conserving "intact" populations and communities. Two possibilities arise: (1) protected areas may represent natural islands that were maintaining independently evolving lineages at their inception; or (2) they may contain populations now confined within unnaturally isolated parks that were once linked by dispersal and gene flow. I suggest that answers to such questions are essential for the long-term management of protected areas as functionally intact communities. Ecological processes that operate over large areas and long time spans may be drastically modified when habitat fragments are set aside for conservation and the intervening lands are developed (Pimm 1991; Saunders, Hobbs, and Margules 1991). In this chapter I explore the questions raised above for wildebeest within Serengeti and several other protected African savannas.

Studies of large herbivore populations have typically been confined to single species in one area over a few years. The monitoring of the size of the Serengeti wildebeest population over 30 years provides an exceptionally long and intensive example (McNaughton and Campbell 1991). On occasion, migrations have also been followed by radiotelemetry, and such techniques are useful for revealing present-day movement patterns.

However, these studies are typically limited to a few individuals followed over spans considerably less than an average life span, and thus cannot be expected to reveal long-distance dispersal events that may affect populations over ecological time. More important, radiotelemetry cannot reveal the extent of gene flow, which, over many generations, will extend beyond the maximum dispersal distance in most species, and which ultimately determines the degree of population isolation over evolutionary time (Slatkin 1987).

An accurate impression of the natural population structure of a species can be derived using genetic techniques. Moreover, by combining contemporary patterns of genetic variation with information about phylogenetic history, evolutionary processes that have been operating for millennia can be inferred (Slatkin and Maddison 1990). By understanding patterns of gene flow, we can assess the extent to which habitat fragmentation and impediments to dispersal might disrupt the natural structure of wildlife populations. Such an assessment is possible even after natural ranges have been fragmented by humans, so long as wildlife populations have not been reduced to very small sizes and individuals have not been intentionally translocated between regions.

For African mammals in general, information about population structure and gene flow is scarce. Several recent studies, however, have begun to uncover the evolutionary histories of cheetahs (O'Brien et al. 1985), rhinos (Ashley, Melnick, and Western 1990), lions (Gilbert et al. 1991 and references therein), and elephants (Georgiadis et al. 1994).

Wildebeest Biology

Wildebeest are bovids of the tribe Alcelaphini, of which two additional species are found in Serengeti: the topi *(Damaliscus korrigum)* and kongoni *(Alcelaphus buselaphus)*. The fossil record of the Alcelaphini is the best known of the African bovid tribes (Vrba 1979, 1984). They are members of a rapidly speciating clade consisting of four extant genera and seven species. The evidence for twenty-five extinct species suggests that none arose more than 10 million years ago. The splitting of *Connochaetes taurinus* from its only extant congener, *C. gnou,* probably occurred about 3 million years ago, while the divergence of *Connocheates* from its sister genera *Damaliscus* and *Alcelaphus* occurred 4–5 million years ago (Vrba 1979; Georgiadis et al. 1990).

At least five "subspecies" of blue wildebeest have been described in Africa, based on morphological criteria. Two of these occur in East Africa, with *C. t. albojubatus* being found east and *C. t. mearnsi* being found west of the eastern Rift Valley in Kenya and Tanzania respectively. Three other subspecies, *C. t. johnstoni, C. t. cooksoni,* and *C. t. taurinus,* are found in southern Tanzania, Zambia's Luangwa Valley, and southern

Africa respectively (Estes 1991). "Subspecies," however, is used here as a convenient descriptive term. It does not imply that the taxa have equivalent evolutionary or systematic status.

Northern Tanzania provides an ideal setting for the analysis of large mammal population structure because most populations there remain effectively intact. It is a region of profound geologic, climatic, and ecological diversity (Lamprey 1963; Sinclair and Norton-Griffiths 1979; McNaughton 1985; de Boer and Prins 1990). In particular, wildebeest are abundant, easy to sample, widely distributed, and are a dominant component of the Serengeti ecosystem.

Wildebeest are strict grazers, preferring short green grass in open habitats, and they must drink during the dry season. They are found in open savannas with moderate to high rainfall (e.g., Serengeti-Mara and Nairobi), in more arid areas where groundwater sustains grass growth (e.g., Amboseli and Manyara), or where rivers provide permanent water (e.g., Tarangire). Food preferences of wildebeest permit large herds to congregate in high densities; the vast herds that amass on the Serengeti plains in the wet season provide an extreme example (McNaughton 1984, 1985). In their search for forage and minerals, wildebeest can be exceptionally mobile, sometimes migrating hundreds of kilometers (Maddock 1979). Although the Serengeti is best known for the migratory population that traverses the ecosystem on a seasonal basis (fig. 22.1, population E), three small resident populations persist throughout the year at Loliondo (G), Kirawira (H), and Loita in the Masai Mara (I). Because no ecological barriers exist between these populations, seasonal contact between migratory and resident populations may occur. Thus, any genetic differences among these populations would result from behavioral rather than ecological or geographic isolation.

The wildebeest mating system is polygynous, and males are territorial during the rut (Estes 1991). Males attempt to gather and mate with females within their territories, but these are temporary associations. Males defend territories until they are defeated or females have moved on.

Marked geographic relief separates the Serengeti wildebeest from a population that resides for most of the year in the Ngorongoro Crater (F). The Ngorongoro wildebeest are remarkably tame and habituated, suggesting that individuals reside there for long periods. However, large numbers of them are also known to disperse from the Crater during the rains and to mingle with the Serengeti migrants in the Ol Balbal area (R. Estes, pers. comm.). While there appears to be no sex bias in dispersal in the Serengeti migrants, females tend to be more philopatric than males in resident populations (R. Estes, pers. comm.). The Crater wildebeest are morphologically similar to those in Serengeti, but they are apparently independent in terms of their population dynamics. The Ngorongoro popu-

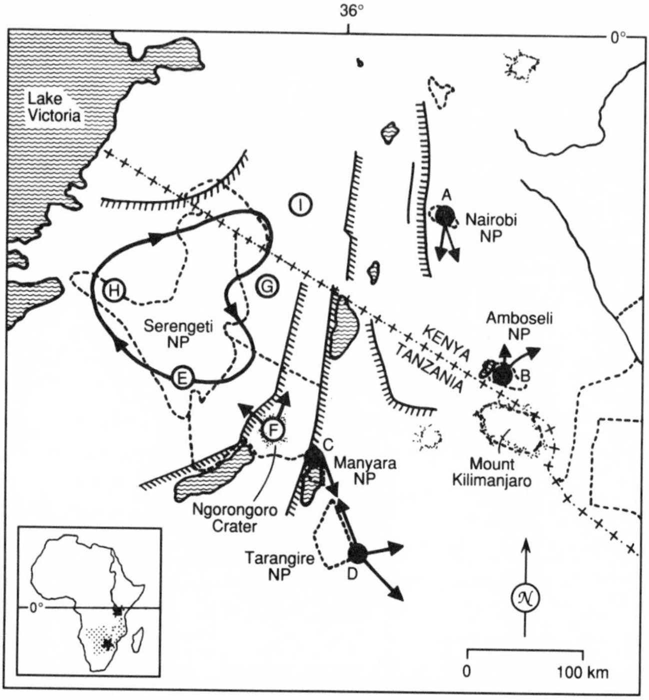

Figure 22.1 Distribution of the focal wildebeest populations in eastern Africa. Populations east of the Rift Valley (solid circles) were sampled at Nairobi (A), Manyara (C), Tarangire (D), and Monduli (immediately north of Manyara). Populations west of the Rift Valley (open circles) were sampled south of Naabi Hill (Serengeti migrants, E), and at Ngorongoro Crater (F), Loliondo (G), and Kirawira (H). Wildebeest at Amboseli (B) and Loita (I) were not sampled. Heavy arrowed lines mark directions and extents of principal wildebeest migrations. Faults associated with the Rift Valley are marked by feathered lines. Inset: The total range of blue wildebeest in Africa (stippled), the location of the study area in eastern Africa that is detailed in the main figure (solid square), and the location of the population sampled in Hwange, Zimbabwe (asterisk).

lation has been declining over the last decade, while Serengeti wildebeest have remained stable (Campbell and Borner, chap. 6; Runyoro et al., chap. 7).

Despite their vagility, wildebeest are selective of habitat, generally avoiding dense woodlands, and are therefore distributed discontinuously over their range. The Rift Valley is a particularly conspicuous geographic feature bisecting wildebeest distributions in eastern Africa, and it forms the boundary between two morphologically distinguishable groups of populations: *Connochaetes taurinus mearnsi* to the west and *C. t. albojubatus* to the east. The latter are larger and have a lighter grey pelage than the former. Populations east of the Rift Valley congregate around permanent water at Nairobi (fig. 22.1, population A), Amboseli (B), Manyara

(C), and Tarangire (D) during the dry season and emigrate to surrounding areas during the wet season.

I describe patterns of genetic subdivision among these populations, based on the analysis of mitochondrial DNA (mtDNA). Wildebeest in the Serengeti-Ngorongoro, the Tarangire-Manyara-Monduli, and the Nairobi ecosystems and a geographically distant population in Hwange (Zimbabwe; *C. t. taurinus*) were compared. The latter are morphologically distinguishable by their black beards, whereas wildebeest in eastern Africa are white-bearded.

METHODS

I used "skin biopsy darts" (Karesh, Smith, and Frazier-Taylor 1989) fired from a dart gun to obtain many tissue samples without immobilizing donors. Once recovered, samples were removed from the modified needle of the biopsy dart and preserved by immersion in liquid nitrogen. With this and other methods, samples were obtained from wildebeest in seven locations: Nairobi (from animals shot for meat); Serengeti migrants (from animals found in snares and using the biopsy dart); Loliondo residents (from animals shot by the Tanzania Wildlife Corporation); Kirawira residents (biopsy dart); Ngorongoro, Manyara, and Tarangire (biopsy dart), and Hwange in Zimbabwe (biopsy dart). Three dried skin samples were also collected from animals shot by professional hunters in Monduli. Samples from Tarangire, Monduli, and Manyara were combined to represent a single region, hereafter referred to as Tarangire.

DNA from all samples was extracted in Nairobi using standard protocols and taken under U.S. Department of Agriculture permit to Washington University in Saint Louis for analysis. Two DNA regions within the mitochondrial genome were amplified in each sample using the polymerase chain reaction and primers that were conserved among mammals and amphibians (e.g., the control region [1.8 kb] and the sequence between Glu and Leu tRNAs that spans the ND5/6 region [2.4 kb]). Fragments were digested with ten restriction enzymes (the control region segment by *Ase*I, *Dde*I, *Hha*I, *Hinc*II, *Hinf*I, *Mbo*I, *Mse*I, *Rsa*I, and *Ssp*I, the ND5/6 segment by *Hha*I and *Hinf*I) that had previously been found to yield useful genetic variation in wildebeest (pers. obs.). A total of 211 base pairs were screened for genetic variation within each individual using this method. Individuals sharing the same restriction fragment patterns over all enzymes were assumed to possess the same mitochondrial haplotype. Restriction fragments were separated in 1.5% agarose gels and directly visualized with ethidium bromide under ultraviolet light. This eliminated the need for Southern blot transfer and detection by radioactive probes. Restriction sites were mapped by partial digestion of ampli-

Table 22.1 Mitochondrial DNA haplotype frequencies in seven wildebeest populations in eastern and southern Africa.

		Haplotype number											
Country	Population[a]	1	2	3	4	5	6	7	8	9	10	11	12
Kenya	A. Nairobi	16	1	—	—	—	—	—	—	—	—	—	—
Tanzania	D. Tarangire	2	—	11	1	3	—	—	—	—	—	—	—
Tanzania	F. Ngorongoro	—	—	—	—	—	3	6	4	8	2	2	1
Tanzania	E. Serengeti Migrants	—	—	—	—	—	3	8	1	4	12	—	2
Tanzania	G. Loliondo Residents	—	—	—	—	—	1	3	4	4	2	—	1
Tanzania	H. Kirawira Residents	—	—	—	—	—	3	3	1	4	2	2	—
Zimbabwe	Hwange	—	—	—	—	—	—	—	—	—	—	—	—
Totals		18	1	11	1	3	10	20	10	20	18	4	4

[a]Letters identify the populations in fig. 22.1.

fied segments that were end-labeled with biotin. Fragments were separated in 2.0% agarose gels, blotted onto a nylon membrane, and the labeled fragments exposed using Photogene bioluminescence.

Population genetic subdivision was quantified using F_{st} (Weir and Cockerham 1987), which varies from 0 under panmixia to 1 under absolute subdivision caused by local fixation and a total lack of gene flow. Estimates of F_{st} were derived using the algorithm written by Alan Templeton and described in Davis et al. (1990), a method that also provides confidence intervals by resampling. Within-population genetic diversity (θ) was estimated using the method of Ewens, Spielman, and Harris (1981). Sequence divergence between pairs of haplotypes was estimated using the Jukes-Cantor method (Nei and Millar 1990) and a phenogram depicting relative sequence divergence between haplotypes was derived using the neighbor-joining method.

RESULTS

A total of 47 restriction sites were mapped, of which 28 were polymorphic within wildebeest. Twenty-five mitochondrial haplotypes were found among 144 individuals from seven sampling locations (table 22.1). Haplotypes were distributed among populations in a strongly nonrandom manner. There was no evidence for genetic subdivision within the four populations occupying the Serengeti ecosystem, including the Ngorongoro Crater population (F_{st} = 0.03, 95% CI: −0.14–0.15). Thus wildebeest in the entire Serengeti- Ngorongoro ecosystem likely constitute one randomly mating population. East of the Rift Valley, however, populations at Nairobi and Tarangire (including Manyara and Monduli) shared none of the haplotypes found in Serengeti. This finding strongly suggests a lack of recent maternal gene flow across the Rift Valley.

Markedly significant subdivision was also detected between popula-

Haplotype number													Sample size
13	14	15	16	17	18	19	20	21	22	23	24	25	
—	—	—	—	—	—	—	—	—	—	—	—	—	17
—	—	—	—	—	—	—	—	—	—	—	—	—	17
—	—	—	—	—	—	—	—	—	—	—	—	—	26
1	1	1	1	—	—	—	—	—	—	—	—	—	34
—	—	—	1	1	—	—	—	—	—	—	—	—	17
—	—	—	—	—	1	—	—	—	—	—	—	—	16
—	—	—	—	—	—	8	1	1	3	2	1	1	17
1	1	1	2	1	1	8	1	1	3	2	1	1	144

tions to the east of the Rift Valley. Only one haplotype was shared between Tarangire and Nairobi (table 22.1), but in strongly contrasting frequencies (F_{st} = 0.61, 95% CI: 0.34–0.79). Thus maternal gene flow is also limited or absent between these populations. There were no shared haplotypes between the populations in eastern Africa and those in Zimbabwe (table 22.1). While there were no fixed restriction site differences between populations in eastern Africa, two restriction site loci were fixed between wildebeest in eastern and southern Africa. This finding indicates long-term isolation and independent evolution between those regions.

Genetic diversity, θ, within populations varied by more than a factor of ten, from 0.00122 in Nairobi to 0.0158 in Hwange, with Tarangire (0.00474) and Ngorongoro-Serengeti (0.00824) exhibiting intermediate levels. An unrooted phenogram depicting degrees of relatedness among haplotypes (fig. 22.2) shows marked concordance between geographic location and sequence similarity. Populations in two regions (Hwange and Nairobi) appear to be monophyletic. The monophyletic grouping of haplotypes and low genetic diversity found in Nairobi, coupled with sequence similarity with a haplotype that is found only in Tarangire, suggest that the Nairobi population was founded by relatively few individuals from Tarangire, with no evidence of direct genetic contributions from Serengeti.

DISCUSSION

For a large-bodied, vagile species, wildebeest populations are surprisingly subdivided. An examination of the ecological and geographic context of this population genetic pattern does suggest that a causal mechanism for this subdivision exists. The lack of subdivision among the resident and migratory populations within the Serengeti ecosystem implies that social forces are not strong enough to affect population structure in wildebeest, at least with respect to female lineages in that ecosystem. This is true

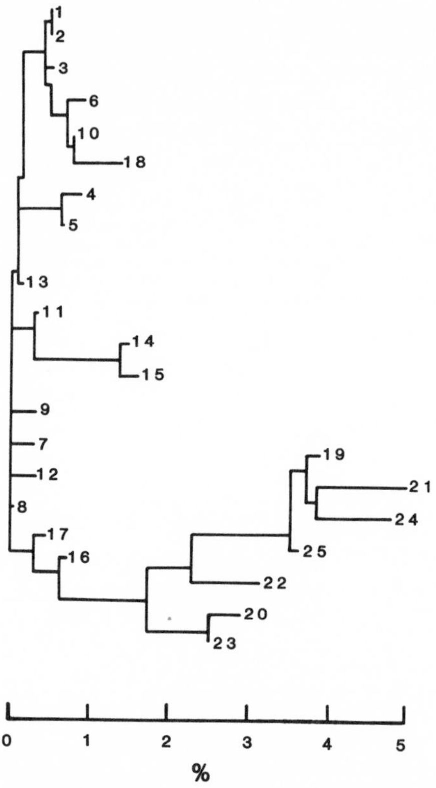

Figure 22.2 A neighbor-joining phenogram depicting degrees of DNA sequence divergence among 25 wildebeest mitochondrial haplotypes that were identified by restriction site analysis (numbers correspond to haplotypes in table 22.1). Sequence divergence is proportional to the sum of the horizontal branch lengths between each pair of haplotypes.

despite the fact that the Ngorongoro Crater and Serengeti are separated by marked relief. Thus, geographic relief alone is clearly insufficient to obstruct wildebeest dispersal. By contrast, lack of gene flow across the Rift Valley between Ngorongoro and Manyara, an area characterized not just by substantial relief but also by dense forest, suggests that habitat discontinuities do obstruct wildebeest dispersal. This forest band, now largely replaced by agriculture, is in places only 45 km wide. Although an individual wildebeest could traverse that distance in a day, the genetic evidence suggests that wildebeest (at least females) have not moved between Ngorongoro and Manyara for perhaps thousands of years. The preference of wildebeest for open habitats and their aversion to woody habitats with a closed canopy impose severe restrictions on their movements and gene flow. The same kind of mechanism has likely been limiting gene flow between Tarangire and Nairobi.

Savanna national parks are both characterized and justified by the large mammals that they conserve, if only for economic reasons, and most conservation policy decisions will be made in the interests of these conspicuous species. The overall result of this study shows that gene flow between wildebeest populations can be extreme or very limited, depending primarily on whether or not intervening habitats allow for wildebeest dispersal. These results imply that progressive isolation of protected areas by human development will not critically disrupt natural dispersal processes of wildebeest in most instances. Although wildebeest are capable of migrating much longer distances than those separating many protected areas in eastern Africa, their populations are nevertheless naturally confined to relatively small regions by habitat. The Serengeti-Ngorongoro, Tarangire-Manyara-Monduli, and Nairobi regions appear to be conserving wildebeest populations that are independent of each other in both demographic and evolutionary senses.

Within those regions, however, wildebeest disperse freely. For example, the population in Ngorongoro may appear to be demographically independent of the Serengeti wildebeest over ecological time, but has clearly been dependent on the Serengeti population over evolutionary time. Similarly, wildebeest commute between Tarangire and Manyara (Borner 1984; Mwalyosi 1991). The population in Manyara is susceptible to local extinction (by emigration due partly to fluctuations in the lake level, as has happened before; H. Prins, pers. comm.) and recolonization by individuals immigrating from the north or the east. If isolated from Tarangire, Manyara might not sustain a permanent wildebeest population, nor could wildebeest recolonize that area naturally. A protected corridor linking the wildlife communities in Manyara and Tarangire would increase the chances that wildebeest and other species will persist in Manyara. And if necessary, reintroduction by humans of wildebeest into Manyara should be done with individuals from east of the Rift Valley.

Although conclusions similar to my own would have been reached on the basis of morphological comparisons, morphology is not always a dependable index of population genetic subdivision. For example, the morphological similarity of wildebeest in Tarangire and Nairobi does not reveal the genetic subdivision detected by the mitochondrial data. Genetic comparisons thus provide more reliable (provided laboratory facilities are available) indices of population structure than do morphological comparisons, especially for living animals. For large mammals in particular, this study shows that the combination of biopsy darting and the polymerase chain reaction greatly facilitates the investigation of population genetic questions that are important for conservation, but that were previously too cumbersome and expensive to address.

Genetic analyses also reveal aspects of the evolutionary history of a species that are useful to conservation and management. For example, some wildebeest populations (e.g., Nairobi) have apparently been founded by relatively few individuals from a single source, followed by little genetic exchange with other populations (at least due to female dispersal). Under these conditions, rates of population differentiation can be increased by genetic drift when populations are small (Wade and McCauley 1988), and over evolutionary time by natural selection. Given that wildebeest will increasingly be translocated to new areas or reintroduced into areas where they have become locally extinct, genetic analyses also provide practical guidelines for conducting such manipulations in ways that are consistent with natural processes. As wildlife communities are confined within protected areas, their effective management will increasingly rely on this kind of information (Templeton 1991). Using these methods we can identify the species and communities that are most likely to be changed by fragmentation and plan accordingly. Moreover, by understanding how ecological and evolutionary processes affect the structure of wild populations, we can more effectively prevent them from being distorted by the very conservation measures that are meant to insure their continuity.

CONCLUSIONS

A long-term goal of conservation biology is to manage protected areas as communities that function as if they were intact. Given that the fate of most protected areas is to become islands in a sea of humanity, to what extent are they *natural* islands that are already maintaining independently evolving lineages and, by implication, dynamically independent populations? My results show that gene flow between wildebeest populations is either extreme or very limited, depending primarily on whether or not intervening habitats are suitable for wildebeest dispersal. Although wildebeest are capable of migrating much longer distances than those separating protected areas in eastern Africa, the Serengeti-Ngorongoro, Tarangire-Manyara-Monduli, and Nairobi ecosystems are all conserving wildebeest populations that are independent of each other in both demographic and evolutionary senses. Therefore, in most instances, progressive isolation of protected areas by human development will not critically disrupt dispersal processes in wildebeest that have been operating over evolutionary time. Similar conclusions may be true for species that are as habitat-specific as wildebeest, but they may not apply to habitat "generalist" species. The study shows that, for large mammals in particular, new genetic techniques greatly facilitate the solution of important conservation problems that were previously too cumbersome and expensive to address.

ACKNOWLEDGMENTS

I thank the Tanzania Commission for Science and Technology and Tanzania National Parks for permission to work in Tanzania, in particular D. Babu, B. Mwasaga, K. Hirji, and G. Sabuni. Advice from B. Maragesi, J. Hando, E. Kishe, and E. Lyanga, all of Tanzania National Parks, E. Chausi of Ngorongoro Conservation Authority, and R. Martin of Zimbabwe National Parks made sampling much easier. N. Johnson, L. Bischof, and S. and N. Creel all assisted with sampling. P. Kat donated ten samples from Nairobi. A. Hill and T. Corfield of Tanzania Game Trackers permitted me to cut small skin samples from hunters' trophies. The polymerase chain reaction primers were designed by J. Patton. A. Templeton gave me space in his laboratory. My parents were subjected to DNA extractions on their kitchen stove in Nairobi. I thank Peter Arcese for editorial comments. The work was initially supported by grants to S. McNaughton and P. Dunham. Later, I was funded by a grant to study elephant genetics from the Liz Claiborne-Art Ortenberg Foundation, through Wildlife Conservation International, a project launched by D. Western, and by the National Science Foundation.

REFERENCES

Ashley, M. V., Melnick, D., and Western, D. 1990. Conservation genetics of the black rhinoceros, *Diceros bicornis* I: Evidence from mitochondrial DNA of three populations. *Conserv. Biol.* 4:71–77.

Borner, M. 1984. The increasing isolation of Tarangire National Park. *Oryx* 19:91–96.

Davis, S., Strassman, J., Hughes, C., Pletscher, L., and Templeton, A. 1990. Population structure and kinship in Polistes, Hymenoptera, Vespidae: An analysis using ribosomal DNA and protein electrophoresis. *Evolution* 44:1242–53.

de Boer, W. F., and Prins, H. 1990. Large herbivores that thrive mightily but eat and drink as friends. *Oecologia* 82:264–74.

Estes, R. 1991. *The behavior guide to African mammals.* Berkeley: University of California Press.

Ewens, W. J., Spielman, R. S., and Harris, H. 1981. Estimation of genetic variation at the DNA level from restriction endonuclease data. *Proc. Natl. Acad. Sci. USA.* 78:3748–50.

Georgiadis, N. J., Bischof, L., Templeton, A. R., Patton, J., Karesh, W., and Western, D. 1994. Structure and history of African elephant populations: I. Eastern and Southern Africa. *J. Hered.* 85:100–104.

Georgiadis, N., Kat, P., Oketch, H., and Patton, J. 1990. Allozyme divergence within the Bovidae. *Evolution* 44:2135–49.

Gilbert, D., Packer, C., Pusey, A. E., Stephens, J. C., and O'Brien, S. J. 1991. Analytical DNA fingerprinting in lions: Parentage, genetic diversity, and kinship. *J. Hered.* 82:378–86.

Karesh, W., Smith, F., and Frazier-Taylor, H. 1989. A remote method for obtaining skin biopsy samples. *Conserv. Biol.* 1:261–62.

Lamprey, H. F. 1963. Ecological separation of the large mammal species in the Tarangire Game Reserve, Tanganyika. *E. Afr. Wildl. J.* 1:63–92.

Leader-Williams, N., Harrison, J., and Green, M. J. B. 1990. Designing protected areas to conserve natural resources. *Sci. Prog.* 74:189–204.

Maddock, L. 1979. The "migration" and grazing succession. In *Serengeti: Dynamics of an ecosystem,* ed. A. R. E. Sinclair and M. Norton-Griffiths, 104–29. Chicago: University of Chicago Press.

McNaughton, S. J. 1984. Grazing lawns: Animals in herds, plant form, and coevolution. *Am. Nat.* 124:863–86.

McNaughton S. J. 1985. Ecology of a grazing ecosystem: The Serengeti. *Ecol. Monogr.* 55:259–94.

McNaughton, S. J., and Campbell, K. 1991. Long-term ecological research in African ecosystems. In *Long-term ecological research,* ed. P. G. Risser, 173–89. Chichester: John Wiley and Sons.

Mwalyosi, R. B. B. 1991. Ecological evaluation for wildlife corridors and buffer zones for Lake Manyara National Park, Tanzania, and its immediate environment. *Biol. Conserv.* 57:171–86.

Nei, M., and Miller, J. C. 1990. A simple method for estimating average number of nucleotide substitutions within and between populations from restriction data. *Genetics* 70:639–51.

O'Brien, S. J., Roelke, M. E., Marker, L., Newman, A., Winkler, C. A., Meltzer, D., Colly, L., Evermann, J. F., Bush, M., and Wildt, D. E. 1985. Genetic basis for species vulnerability in the cheetah. *Science* 227:1428–34.

Pimm, S. 1991. *The balance of nature*? Chicago: University of Chicago Press.

Saunders, D., Hobbs, R., and Margules, C. 1991. Biological consequences of ecosystem fragmentation: A review. *Conserv. Biol.* 5:18–32.

Sinclair, A., and Norton-Griffiths, M., eds. 1979. *Serengeti: Dynamics of an ecosystem.* Chicago: University of Chicago Press.

Slatkin, M. 1987. Gene flow and the geographic structure of natural populations. *Science* 236:787–92.

Slatkin, M., and Maddison, W. P. 1990. Detecting isolation by distance using phylogenies of genes. *Genetics* 126:249–60.

Templeton, A. R. 1991. Genetics and conservation biology. In *Species conservation: A population biological approach,* ed. A. Seitz and V. Loeschcke, 15–29. Basel: Birkhauser Verlag.

Vrba, E. S. 1979. Phylogenetic analysis and classification of fossil and recent Alcelaphini Mammalia (Family Bovidae). *Biol. J. Linn. Soc.* 11:207–28.

Vrba, E. S. 1984. Evolutionary pattern and process in the sister group Alcelaphini-Aepycerotini, Mammalia: Bovidae. In *Living Fossils,* ed. N. Eldredge and S. M. Stanley, 62–79. New York: Springer.

Wade, M., and McCauley, D. 1988. Extinction and recolonization: Their effects on the genetic differentiation of local populations. *Evolution* 42:995–1005.

Weir, B. S., and Cockerham, C. C. 1984. Estimating *F*-statistics for the analysis of population structure. *Evolution* 34:1060–76.

TWENTY-THREE

The Ecology and Epidemiology of Rinderpest Virus in Serengeti and Ngorongoro Conservation Area

Andy Dobson

The popular perception of Serengeti is that of a predator-prey system where ecological interactions are dominated by large herbivores and the carnivores that prey upon them (Grzimek and Grzimek 1960; Schaller 1972; Kruuk 1972). This chapter argues that the population density of wildebeest and buffalo have been regulated for most of this century by rinderpest, a viral disease of wild and domestic ungulates. Removal of this pathogen through the vaccination of cattle has led to the eruption of wildebeest and an increase in buffalo numbers. This eruption in turn led to increases in the numbers of carnivores, particularly lions and hyenas. This chapter describes the ecology and epidemiology of rinderpest virus and its control in the Ngorongoro Conservation Area. The analysis suggests that maintaining a high level of rinderpest inoculation in the population of cattle that live in and around the park continues to be a crucial component of successful conservation in the whole Serengeti/Ngorongoro region.

Understanding the population biology of rinderpest in the Serengeti is a classic conservation problem. It involves resolving conflicts of interest between pastoralists and conservation authorities who have complementary interests in the health of domestic and wild animals. Although the virus has now been eradicated from the ecosystem, the possibility of its reintroduction from endemic sources outside the region poses a continual threat to the large herds of wildlife living in the Ngorongoro Conservation Area and Serengeti (Plowright 1982, 1985). At least two outbreaks have occurred in the last 10 years, and these have led to significant localized mortality in buffalo and other wildlife species (Rossiter et al. 1987; Prins and Weyerhauser 1987; Anderson et al. 1990). The main way to ensure the absence of this virus is to maintain high levels of immunity in the cattle population in the surrounding areas. Inoculating cattle in the

NCA and around Serengeti is expensive, and perception of the risk associated with a disease usually declines with increasing time since the last outbreak.

The importance of understanding the population biology of diseases in relation to conservation biology has received recent recognition (Dobson and May 1986; Dobson and Miller 1989; May 1988). Although the majority of recent developments in the population dynamics of infectious diseases have focused on the epidemiology of human pathogens (Anderson and May 1991), quantitative ecological studies on wildlife diseases are becoming more common (for examples, see Anderson and Trewhella 1985; Anderson et al. 1981). Similarly, a number of theoretical studies have examined the population dynamics of systems involving one pathogen and two host species (Holt and Pickering 1985; Dobson and May 1986; Begon et al. 1992). Understanding the population dynamics of such diseases is crucial in systems involving pathogens that infect both wildlife and domestic species, particularly in situations in which economic, health, and political factors are important. Determining the most cost-effective and least damaging ways to control pathogens will considerably enhance the ability of wildlife managers to control similar diseases in other situations in which they pose a major threat to the preservation of species diversity.

EPIDEMIOLOGICAL HISTORY OF RINDERPEST IN EAST AFRICA

Evolutionary Relationships of Rinderpest

Rinderpest is a member of the *Morbillivirus* genus in the order Paramyxoviridae. This genus contains four of the most important pathogens of humans and their domestic livestock: the others are canine distemper, measles, and Peste de Petite Ruminants. Recent work on the antigenic relationships of the four viruses suggests that they show a high degree of epitopic homology (Norrby et al. 1985; McCullogh et al. 1986). This work can be used to construct a phylogeny for the four species (fig. 23.1); the available evidence suggests that rinderpest virus is the ancestral root of this tree. The Bovidae are presumed to be ancestral hosts of the lineage. Most of the radiation of the species is likely to have occurred in India, the Far East, or Europe following the domestication of ungulates and canids (Plowright 1985; Clutton-Brock 1987). As ungulate species that live in the Sahara desert exist at very low population densities, they would be unlikely to support a continuous infection of rinderpest virus. The desert would thus act as an efficient barrier to the spread of rinderpest into the southern half of the African continent (Dobson 1988). The pandemic of rinderpest that occurred at the end of the last century was initiated by the

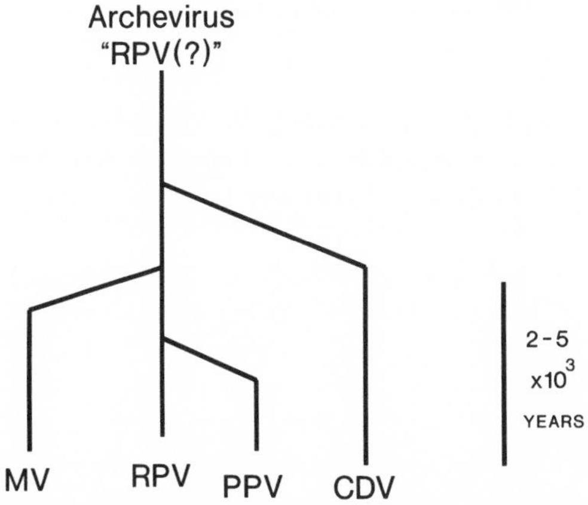

Figure 23.1 The evolutionary relationships of the *Morbillivirus* genus. The tree is drawn for the major species in the group: RPV, rinderpest virus; MV, measles virus; PPV, Peste des Petite Ruminants; CDV, canine distemper. The estimate of the split between MV and RPV is a rough estimate based on the available evidence for historical plagues whose etiological descriptions have similar diagnostic characteristics to present-day measles. (Adapted from McCullogh et al. 1986.)

accidental introduction of a few infected cattle into the horn of Africa. The epidemic caused massive mortality, confirming that sub-Saharan ungulate populations had no previous exposure to the pathogen (Plowright 1985).

History of Rinderpest Control in East Africa

Throughout the twentieth century rinderpest has caused disease outbreaks in cattle and a variety of wild ungulates throughout sub-Saharan Africa. Table 23.1 lists the main outbreaks of rinderpest in East Africa and the main species that were affected since its introduction in 1890 (Simon 1962). Outbreaks of the disease can be recognized by the pathology, which is characterized by the onset of a sudden high fever several days after infection; this persists for 2 to 3 days, after which time ulcers and erosions appear in and around the mouth. Infected animals then produce profuse nasal and ocular discharges, which may be accompanied by severe diarrhea. The animals quickly become emaciated, and a high proportion die from the more virulent strains of the pathogen (Scott 1964).

The first inoculation programs against rinderpest in East Africa were initiated in the early 1940s. By the 1950s wide-scale programs were employed throughout northern Tanzania (Branagan and Hammond 1965). These early vaccination programs used the Kenya goat-attenuated virus (KAG), which provided long-lasting immunity to animals vaccinated after the maternal immunity conferred by their colostral antibodies had

Table 23.1 The main recorded epidemics of rinderpest in East Africa and the species affected.

Year	Species affected
1890	Panzootic—most ungulate species
1897	Kongoni and kudu
1913–1921	Eland and giraffe, then buffalo, bushbuck, and reedbuck
1929	Buffalo, bushbuck, and warthog, then eland and waterbuck
1931	Buffalo, giraffe, and wildebeest
1937–1941	Buffalo, eland, and giraffe, then buffalo, eland, and kudu
1949	Cattle, eland, and then wildebeest
1960	Eland, kudu and warthog, buffalo, bushbuck and giraffe, impala, and oryx

Source: Adapted from Simon 1962.

waned (Plowright 1982; Scott 1964). Although mortality associated with vaccination was low in healthy cattle (<2% within 3 weeks of inoculation), vaccination mortality was significantly higher in calves (≅20%), particularly if they were suffering from protozoan infections such as coccidiosis, theileriasis, or trypanosomiasis (Branagan and Hammond 1965). This mortality discouraged owners from bringing calves for vaccination and produced a significant pool of susceptibles; outbreaks of rinderpest in cattle between 1940 and 1960 invariably affected calves and yearlings almost exclusively (Plowright 1982). This problem led to the development of culture-attenuated rinderpest vaccine that was innocuous to cattle of all ages, while still producing lifelong immunity (Plowright and Taylor 1967). More recent developments have produced a recombinant vaccine (Yilma et al. 1988). The use of recombinant vaccine has led to slight reductions in the cost of vaccine production, but these savings are offset by reports of rinderpest symptoms in immunocompromised human hosts in areas where this vaccine has been tested (P. Rossiter, pers. comm.).

The development of an entirely effective rinderpest vaccine led to a large-scale inoculation program (JP-15) for control of rinderpest in sub-Saharan Africa. The first three phases of JP-15 were launched in West Africa in 1962, and phase IV began in East Africa in 1968. The program attempted the mass immunization of 80×10^6 cattle in 22 countries. The key success of the program lay in reducing the proportion of young cattle that were susceptible in the population. In areas where vaccination was carried out efficiently, such as Ngorongoro, the proportion of susceptibles fell to about 45%. However, in areas where access made comprehensive coverage more difficult, the proportion of susceptibles remained at the 80%–90% level (fig. 23.2).

History of Rinderpest in Serengeti

Rinderpest was first rumored to be present in Serengeti in 1930, but its presence was not confirmed until 1933, when infected tissue was collected

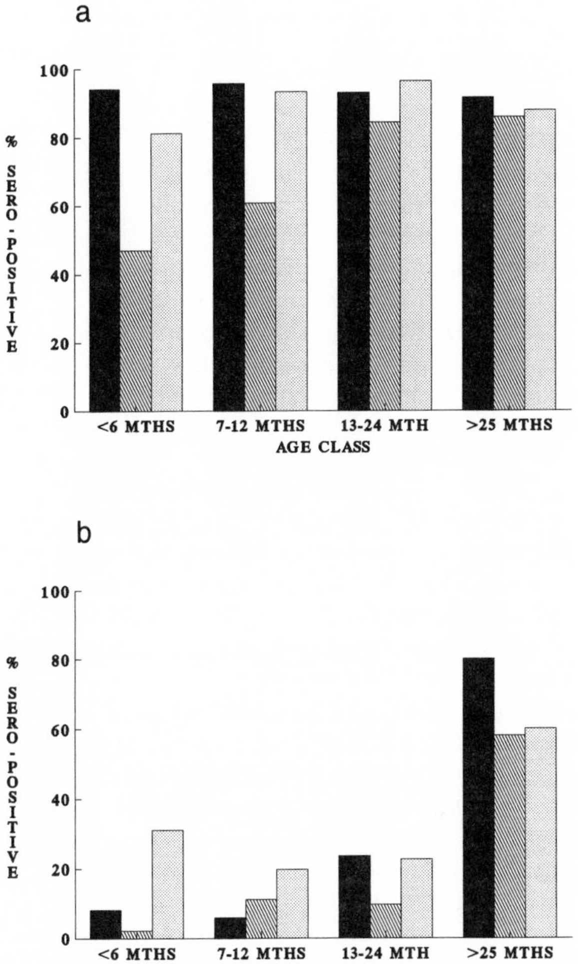

Figure 23.2 Serology profiles of cattle of different ages before and after vaccination in (*a*) Ngorongoro District in 1969–1970 and (*b*) Kajiado District in 1970. In Ngorongoro District serology samples were taken after vaccination in August 1969 (solid bars), and before vaccination in May/June and July 1970 (hatched and dotted bars). In Kajiado, three independent samples were taken January and January/February 1970. (Adapted from Plowright 1982.)

from a 2-year-old wildebeest near Ngorongoro Crater (Plowright and McCullough 1967). The available records for the region from 1946 through 1961 indicate that rinderpest was present in Serengeti wildebeest throughout this time, with outbreaks occurring at 1- to 2-year intervals in yearling and 2-year-old wildebeest as well as in buffalo, warthog, eland, and impala (Plowright and McCullough 1967).

Following the initiation of the vaccination scheme in cattle surrounding the national park in the late 1950s, wildebeest numbers in the Serengeti increased from about 300,000 to about 1.5 million (Sinclair, chap. 1). Increases also occurred in numbers of buffalo as well as in lions and hyenas, which are the main predators of those two species (Sinclair, chap. 1; Scheel and Packer, chap. 14; Hofer and East, chap. 16). These increases were matched by the initial disappearance of the virus from wildebeest, buffalo, and eland (fig. 23.3). The disappearance of rinderpest was matched by the decline in "yearling disease," which had previously produced levels of mortality in immature wildebeest that were estimated to be as high as 40% (Talbot and Talbot 1963).

Although there have been no major epidemics of rinderpest in the Serengeti/Ngorongoro region since vaccination schemes were modified to use cell-culture vaccine, there have been a number of outbreaks that have led to significant localized mortality in cattle, buffalo, and eland. In particular, an outbreak at Lobo in March 1982 killed many buffalo. This was followed by an outbreak in the Ngorongoro Crater area in June of the same year in which 2,000–4,000 buffalo and a number of giraffe, warthog, and eland died. Sera collected from both these areas indicated the presence of rinderpest antibody (Rossiter et al.1983). Subsequent serological surveys by Anderson et al. (1990) suggested that the disease persisted in the buffalo population for a number of years before fading out (fig. 23.4). However, it should be noted that these data, which were collected from animals of different ages over the course of 2–3 years, could also be interpreted as reflecting the continued presence of the virus in the buffalo population. The change in prevalence with age might simply reflect continued low-level exposure of animals to the virus. The absence of any pathological evidence would suggest that if the virus is present, it is a very mild strain. Although serum samples were taken from 94 wildebeest, no animals with positive titers for rinderpest antibodies were identified. Very low levels of prevalence (< 1%) were found in sheep and goats in the area.

Population Numbers in Ngorongoro Conservation Area

The different wildlife species living in Ngorongoro Crater have been censused fairly regularly by both the Ngorongoro Conservation Area Authority and Mweka Wildlife College since the early 1960s (Estes and Small 1981; Homewood and Rodgers 1991). Although buffalo numbers have increased significantly since rinderpest was eradicated, the numbers of most species have remained relatively constant, while numbers of wildebeest, eland, and rhinoceros have declined (fig. 23.5a). In contrast to the rest of the Serengeti, there is no evidence of an increase in the wildebeest population. The number of people living in the region has increased by 2%–3% each year, while the number of cattle has remained roughly

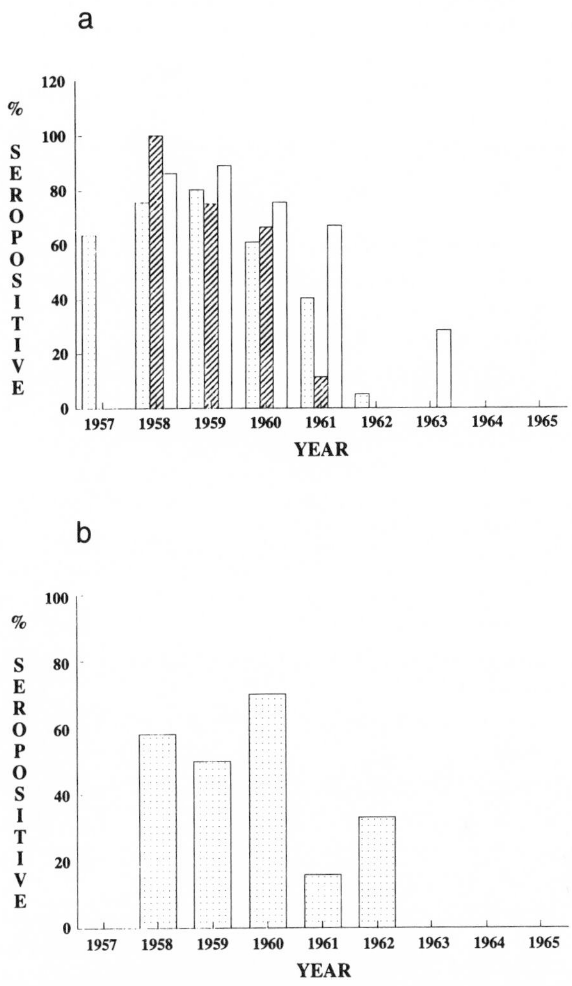

Figure 23.3 Serology profiles from (*a*) wildebeest and (*b*) eland in different parts of the Serengeti ecosystem following the onset of vaccination of cattle in the late 1950s. The wildebeest samples were collected in the northern Serengeti and Masai Mara (dotted bars), Ngorongoro Conservation Area (hatched bars), and the OlBal plains (open bars). The eland samples are all from Ngorongoro Conservation Area. (Data from Plowright and McCullough 1967; Taylor and Watson 1967.)

constant (fig. 23.5b). This has led to a decline in the ratio of cattle to humans and increasing reliance on sheep and goats as a source of protein (Homewood and Rodgers 1991). Nevertheless, cattle are still crucial to the Masai as a source of protein and are an intrinsic part of their culture. Ensuring that rinderpest remains absent from the area is not only important for wildlife, but crucial for the welfare of the Masai and their cattle.

a

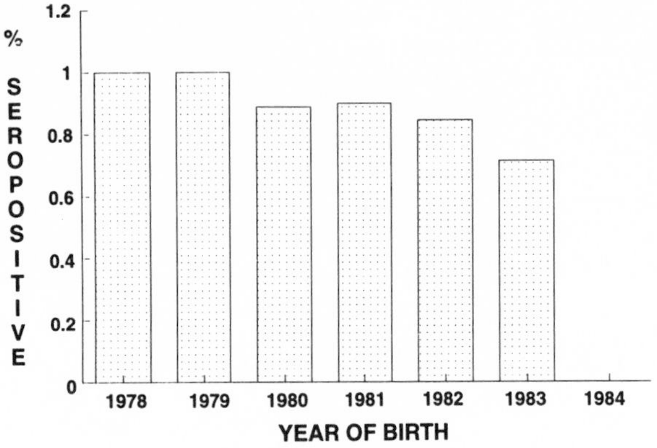

b

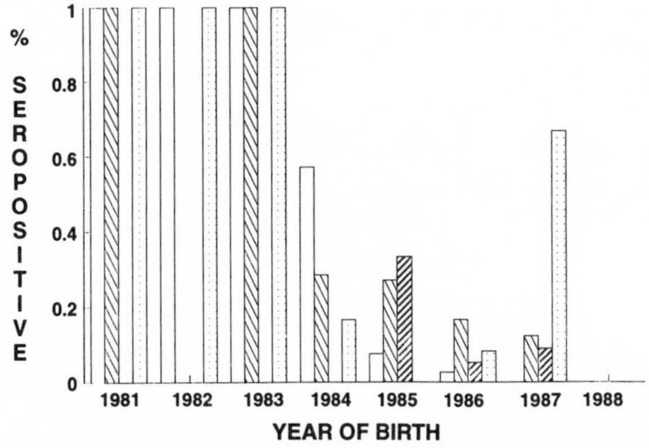

Figure 23.4 (*a*) Serology profiles of tissue samples collected from buffalo of different ages in Masai Mara in Kenya. (Adapted from Rossiter et al. 1987.) (*b*) Serology profiles of tissue samples collected from buffalo of different ages in three areas of Serengeti National Park. The northern samples (74, open bars) were taken around Kleins Camp; the western samples were taken around Musabi (85, light hatch) and also at Kirawira and the Ndabaka plains (34, open bars); samples were also taken around Seronera and Moru Kopjes (38, dark hatch). (Adapted from Anderson et al. 1990.) Some age classes were unavailable in some areas so some of the data are missing.

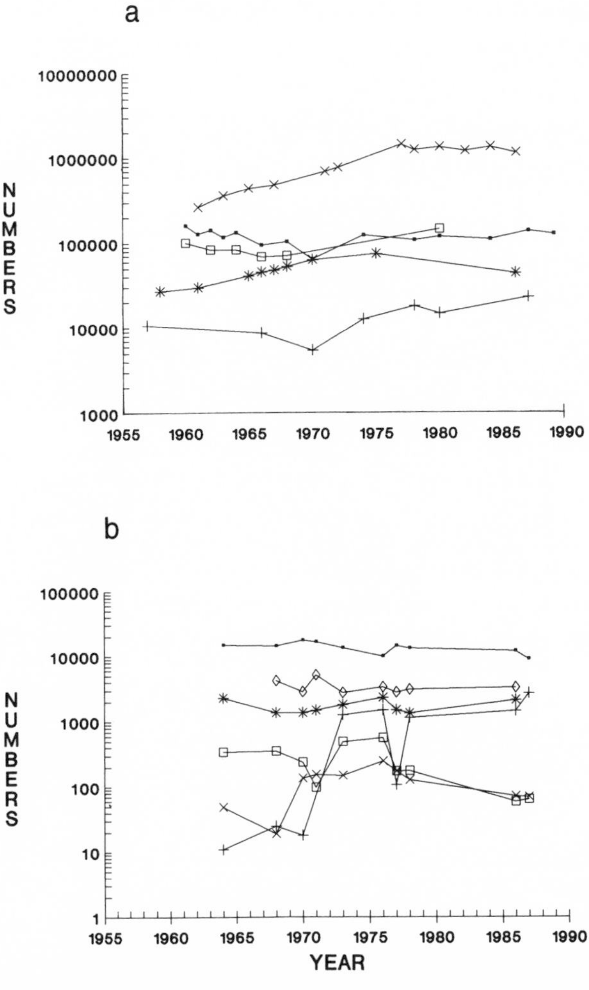

Figure 23.5 (*a*) Estimates of population size for the human population (plus signs), for cattle (small solid squares), and shoats (sheep and goats; large open squares), and for migratory wildebeest (X's) and buffalo (asterisks) in Ngorongoro District between 1955 and 1988. (Adapted from Homewood and Rodgers, 1991.) (*b*) Estimates of the principal ungulate populations in Ngorongoro Crater for the period 1966–1988. Data are provided for wildebeest (small solid squares), buffalo (plus signs), eland (large open squares), kongoni (X's), Grant's gazelle (asterisks), and Thomsons gazelle (diamonds). (Adapted from Estes and Small 1981; Homewood and Rodgers 1991.)

POPULATION BIOLOGY OF RINDERPEST

Single-Species Models

Knowledge of the dynamics of the pathogen in a single-species population is essential to our understanding of the dynamics of more complex, multispecies host-pathogen models. The dynamics of rinderpest may be modeled using standard SIR models (Anderson and May 1982, 1991). These models assume that the host population can be divided into three classes: susceptibles, S, infected, I, and recovered and resistant, R. Three coupled differential equations describe the rates of change of each of these classes of host:

$$dS/dt = a(S + I + R) - bS - \beta SI + \delta R \tag{23.1}$$
$$dI/dt = \beta SI - (\alpha + b + \nu) I \tag{23.2}$$
$$dR/dt = \nu I - (b + \delta) R \tag{23.3}$$

where a is the per capita population birth rate (assuming no effect of the pathogen on host fecundity), b is the per capita death rate of uninfected animals, β is the per capita transmission rate between infected and susceptible animals, δ is the rate at which resistant animals lose their immunity to reinfection ($1/\delta$ is the average time for which resistance lasts), α is the per capita increase in host mortality due to the pathogen, and ν is the recovery rate of infected animals ($1/\nu$ is the average duration of infectiousness). This basic model also assumes homogenous mixing between all animals in the population; this assumption will be modified later when we consider the spatial distribution of the different species that may act as hosts.

Most parameters of the basic SIR model for rinderpest can be estimated from population studies of wild ungulates or from veterinary records. Most wild and domestic ungulates live for 2 to 5 years, so b is of the order of 0.2 to 0.5. When infected with pathogenic strains of rinderpest virus, most hosts succumb within 10 days, so α is of the order of 30 to 50. If they survive, most animals cease to be infectious, so ν is of the order of 25. Estimating the rate of disease transmission, β, is always the hardest part of any epidemiological program. Derivation of two further fundamental parameters of the pathogen's epidemiology allows insights to be developed on how this might be undertaken. Equations (23.1–23.3) may be rearranged to obtain expressions for the basic reproductive rate of the pathogen, R_o

$$R_o = \beta N/(\alpha + b + \nu) \tag{23.4}$$

where N is the total population size ($S + I + R$), which in a population of susceptibles equals S; and the basic reproductive rate is the number of secondary infections that a single infectious individual would produce in a population of susceptibles (Anderson and May 1982). This expression must exceed unity if the pathogen is to persist in the population. The threshold number of hosts required to maintain an infection of the pathogen, N_T, is given by a simple rearrangement of this expression when $R_o = 1$;

$$N_T = (\alpha + b + \nu)/\beta \tag{23.5}$$

Even in the absence of any empirical data on the magnitude of the model's parameters, these equations tell us several important things about the dynamics of pathogens. First, there is an approximately inverse relationship between R_o and N_T; thus highly transmissible diseases need only small populations to sustain them. Second, parasites with increased virulence tend to have reduced basic reproductive rates and require large populations to sustain continuous infections. Plowright (1982) suggests that 100,000 susceptibles in a population of 500,000 cattle are sufficient to maintain the virus. This would suggest a preliminary estimate for β of 0.001. Analysis of age prevalence curves for cattle before and after the vaccination scheme was initiated suggest an average age of infection of about 230 days. These age-prevalence data may be used to estimate the "force of infection" for rinderpest (Anderson and May 1991), whence it is possible to obtain an estimate for β in the range 0.002 to 0.005 (B. Grenfell, pers. comm.). Substitution of this estimate for β into equation (23.5) suggests that between 20,000 and 50,000 susceptible cattle would be required for an outbreak to occur. However, these estimates of β may be inflated if significant amounts of transmission occur from wildlife back to cattle.

Examination of the system of equations at equilibrium allows us to derive an expression for the prevalence of infection:

$$I^*/N^* = r/\alpha. \tag{23.6}$$

As infected animals of all species tend to live for between 1 and 2 weeks once infected, and the intrinsic population growth rate of most ungulates is between 0.1 and 0.2, we would expect to see less than 0.5% of the animals infected *at any time* when the disease is endemic in a host population. Note that this expression also implies that the proportion of a population infected *declines* with the virulence of a pathogen.

The size of the equilibrium host population in the presence of the pathogen is given by

$$N^* = \alpha\,(\alpha + b + \nu)/\beta\{\alpha - r\,[1 + (\nu/b + \delta)]\} \tag{23.7}$$

This suggests that host population size decreases as pathogen transmission efficiency, β, increases. If rinderpest were restricted to a single species of host, we could eradicate the pathogen by maintaining the number of susceptible hosts below the threshold for disease transmission. In cattle this could be undertaken by vaccination. In the simplest case, the proportion of cattle, *p,* that would have to be resistant is given by Anderson and May (1982) as

$$p > 1 - 1/R_o \tag{23.8}$$

The above parameter estimates suggest that this would require between 57% and 91% of cattle to be vaccinated. However, this calculation assumes that cattle are the sole hosts for the pathogen. The crucial feature of rinderpest is that it is transmitted both within and between species. To examine the population dynamic consequences of interspecific transmission we need to extend our basic model framework and include both intra- and interspecific transmission rates.

INTRASPECIFIC AND INTERSPECIFIC PATHOGEN TRANSMISSION

Mixed-Species Models

The model framework may be readily extended to include a second species of host (Anderson and May 1986; Holt and Pickering 1985; Begon et al. 1992). The dynamics of infection are again described by a set of coupled differential equations for each population:

$$dS_i/dt = r_iS_i - \beta_{ii}S_iI_i + \beta_{ij}S_iI_j + a_iI_i + e_iR_i \tag{23.9}$$

$$dI_i/dt = \beta_{ii}S_iI_i + \beta_{ij}S_iI_j - d_iI_i \tag{23.10}$$

$$dR_i/dt = v_iI_i - (\delta - b_i)\,R_i \tag{23.11}$$

where $r_i = a_i - b_i$, $d_i = \alpha_i + b_i + v_i$, $e_i = a_i + v_i$, and β_{ij} and β_{ii} are the rates of inter- and intraspecific transmission respectively (Holt and Pickering 1985). These equations can again be examined at equilibrium to determine the influence of the various transmission and virulence rates on the numbers of hosts infected in each population. The number of hosts of each species infected at any time is now given by the expression

$$I_i^{**} = I_i^{*} - (\beta_{ij}/\beta_{ii})\, I_j^{*}/(1 - (\beta_{ij}\beta_{ji}/\beta_{ii}\beta_{jj})). \tag{23.12}$$

Here the I_i*'s are the number of hosts infected in each population in the absence of interspecific transmission, and I_i** is the number of hosts infected when interspecific transmission occurs. This result suggests that the ratio of between- to within-species transmission rates is as important in determining the numbers of animals infected as are the actual magnitudes of these rates. Where rates of interspecific transmission, β_{ij}, are low compared with rates of intraspecific transmission, β_{ii}, then an increase in the rate of interspecific transmission leads to a decline in the numbers of infected animals (see fig. 23.6). This is mainly because host population size is reduced due to more animals becoming infected and dying. However, as rates of interspecific transmission approach those of intraspecific transmission, the host population first experiences a rapid increase in the numbers infected, followed by a decline to extinction as all animals become infected and die. Similar effects occur if rates of interspecific transmission are intermediate but the population of the reservoir host is sufficiently large to ensure that some susceptible hosts are always infected. Establishment requires that within-species transmission be greater than between-species transmission.

Begon et al. (1992) provide a detailed analysis of the population dynamics of pathogens that infect two host species that are regulated at constant carrying capacities, K_1 and K_2, in the absence of the pathogen. Their analysis assumes that all animals infected by the pathogen either die or return to the susceptible pool (there are no immune or resistant individuals), but most of the major conclusions concerning pathogen establishment and host coexistence still apply to the dynamics of rinderpest. In particular, if only one of the species, H_1, has a population size sufficient to maintain the pathogen ($H_{1T} < K_1$), then persistence of the disease in both species requires either $\alpha_2 < r_2$ or $\beta_{22} < \beta_{21}$. In the case of rinderpest, species 1 would correspond to cattle, while species 2 would correspond to wildebeest or buffalo. The second condition is unlikely to be met for either wildebeest or cattle, as it requires that between-species transmission exceed within-species transmission. The first condition is dependent upon the virulence of the virus; if the host population growth rate, r_2, exceeds the virulence of the virus, α_2, then the pathogen will be present in both species, even if the wild species is below its individual threshold for establishment, H_{2T}. In contrast, where the virulence of the pathogen exceeds the growth rate of the host population ($\alpha_2 > r_2$), it is possible for the presence of the disease in cattle to drive an alternative wild host to extinction; this will occur when sufficient numbers of species 1 are infected that between-species transmission leads to a significant increase in the rate of infection of species 2 (formally $I_1 {}^* \beta_{21}/\beta_{22} > I_2$). This is most likely to occur when there are significant differences in the virulence and transmission efficiency of the virus in species 1 and in species 2 (with $\alpha_1 < \alpha_2$).

Estimation of Transmission Rates

The preceding analyses suggest that quantifying rates of transmission is crucial to increasing our understanding of the way the pathogen spreads, both within a single-species population and between different species. A number of different ecological and epidemiological factors are likely to be important in determining the relative rates of within- and between-species transmission. Population size is a major variable determining a population's ability to maintain a pathogen (Kermack and McKendrick 1927); diseases rapidly die out in species with small, scattered populations, but they are likely to remain present for a long time in species with large, aggregated populations (Anderson and May 1991). Population sizes will also be important in the establishment of pathogens that infect several species of host. Here we need to determine whether the presence of infected individuals of another species can allow a pathogen to become established in a population in which it would otherwise be unable to persist. Thus we will need to estimate not only rates of transmission between different individuals of different age, sex, or social status and other members of each individual species, but also transmission rates between different age and sex classes of different species.

This initially daunting task may be simplified if we can independently quantify ecological associations and interactions between individuals of the same and different species. Distances between animals are likely to be important in determining rates of transmission of respiratory diseases and can be fairly readily quantified in the field. As in other ecological processes, it is not only the mean distance between animals that will be of importance in determining transmission rates, but also the heterogeneity in these distances. In the case of intraspecific transmission, the observed population densities and levels of aggregation are determined both by the habitat in which the animal lives and its social system. Physiological factors are a second important component that determines the ability of each species to transmit a pathogen. Here a variety of factors, such as the quality of mucus and saliva produced and the rate at which an infected animal coughs and sneezes, all affect its ability to produce viable pathogen transmission stages.

If we assume that net transmission rate consists of a physiological and a spatial component, the different rates of transmission may be expressed in the simplest case as

$$\beta_{ii} = [\beta_c + \text{var}\ (\beta_c)]\ \beta_{p.} \qquad (23.13)$$

Here β_c represents the component of transmission due to the average distance between conspecifics, var (β_c) represents an increase in transmission rate due to the tendency of the species to aggregate into social groups and β_p represents the species-specific, physiological component of transmission. Notice that if it is possible to independently quantify the spatial components of transmission, $\beta_{c,}$ using an index of mean crowding such as the reciprocal of nearest neighbor distance, then the physiological component, $\beta_{p,}$ can be used to scale the net transmission rate. Because species with different social systems will exhibit different tendencies to aggregate it will be interesting to see how much of the variation in prevalence between species, is explained by differences in social behavior. Similarly, it should be possible to use these data to determine how differences in social behavior *within* a species are likely to affect expected levels of prevalence and compare these with observed serology data for different social classes of a single host population. The estimates can be rescaled using available serology profiles to produce expected patterns of prevalence in sectors of the population with different social or age characteristics.

Interspecific Transmission

Understanding the dynamics of rinderpest in a community of potential host species requires the construction of a WAIFW matrix ("Who Acquires Infection From Whom": Anderson and May 1985, 1991; Schenzle 1984). The matrix quantifies the rates of transmission between different classes of individuals in the pool of potential hosts. In the case of rinderpest, the average distance between members of different species and their average group size will be crucial in determining rates of interspecific disease transmission. Because the ratios of interspecific to intraspecific transmission are important in determining the prevalence of infection in different host species, it seems sensible to assume that the relative magnitudes of interspecific transmission rates scale inversely with some index of distance between individuals of different species. Figure 23.6 illustrates the reciprocal of average distances observed between the nearest individuals of the same and different species in one survey of Ngorongoro crater during August 1989; the data are taken from surveys of the spatial distribution of ungulates in the crater (A. Dobson, unpub.). These surveys suggest that species with different social systems exhibit different tendencies to aggregate, and that much of the variation in disease prevalence between species is explained by these differences in spatial distribution (A. Dobson, unpub.). Similarly, if the distances between members of different species are crucial in determining rates of intraspecific transmission, this analysis suggests that disease transmission rates within a species are likely to be much greater than rates between species (fig. 23.6). However, transmission may occur readily between some species (e.g., Thomson's and Grant's gazelles), but hardly at all between others (e.g., wildebeest and topi).

Vaccination Coverage in the Ngorongoro Region

Cattle in the Ngorongoro region are vaccinated annually with a live attenuated virus vaccine. During the dry season, cattle are driven by tribesmen to a local "crush," where as many as 4,000 individuals may be vaccinated in a single day. The records for different regional crushes are kept at the Office of the Conservator and at the regional veterinary center at Arusha. These data allow temporal and spatial variation in vaccination coverage to be monitored. Figure 23.7 illustrates the total number of adult and juvenile cattle inoculated in the entire NCA region for the years 1971–1990. Although coverage in the early years of the vaccination campaign was occasionally erratic, the level of coverage has increased steadily, so that by the late 1980s about 90,000 out of 130,000 cattle (about 70%) were protected at any one time. If Plowright's estimate that about 100,000 susceptible cattle are required to maintain the pathogen in the population is correct, then the present level of coverage should be suffi-

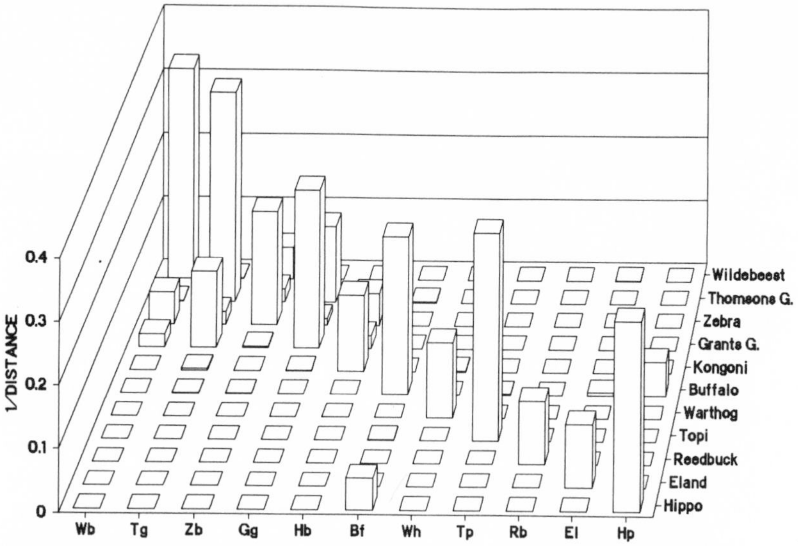

Figure 23.6 WAIFW ("who acquires infection from whom") matrix for transmission within and between the main ungulate species in Ngorongoro Crater. The estimates of transmission rates are the inverse of samples of nearest neighbor distance for each ungulate species in the Crater The estimates were obtained in August 1989 by driving around the crater and estimating the distance to the nearest conspecific and the nearest member of another species for between one and five animals in each social group of ungulates observed each day. These surveys have been repeated in the wet and dry season for each of 2 years. (Adapted from Dobson, unpub.)

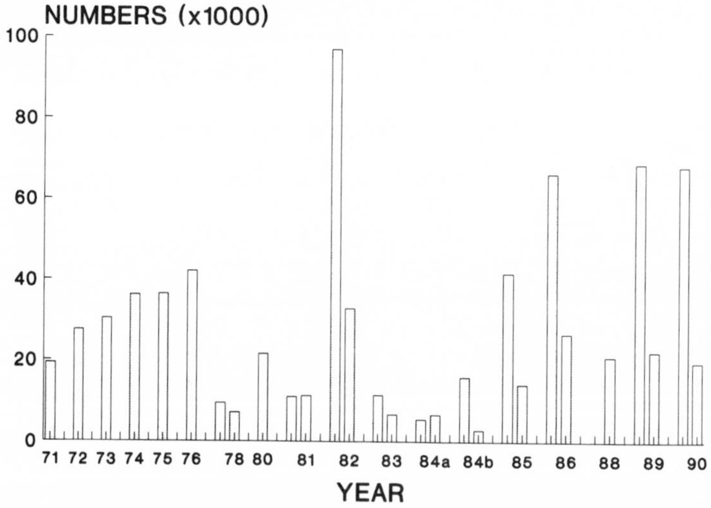

Figure 23.7 The total numbers of rinderpest vaccinations given to cattle in the Ngorongoro District between 1970 and 1990. The vaccination campaign lasts throughout the dry season, from July through September. The data illustrated are for total numbers of adult animals, which are branded immediately before vaccination (left bars after 1978), and calves, which are marked with an ear punch (right bars after 1978). (Adapted from Dobson, unpub.)

cient to prevent the disease from becoming reestablished. However, the estimate of transmission rate obtained above using serology data collected from cattle in the region suggests a threshold value more in the range of 20,000 to 50,000 animals. If we examine the estimated numbers of susceptible cattle in the Ngorongoro region over the last 20 years (fig. 23.8), we find that the numbers of susceptibles have exceeded this lower level at least three times.

The outbreak that occurred in 1982 spread into the wildlife and killed large numbers of buffalo. A larger epidemic was prevented by mounting a massive vaccination campaign, which may have inoculated about 99% of the cattle in the region. Although the cause of this outbreak has been traced to illegal cattle movements through the conservation area, the result underlies the importance of vaccination in containing the disease and in preventing future outbreaks. Furthermore, it should be noted that the outbreak occurred in NCA, where the vaccination coverage is very good (see fig. 23.2); where the vaccination coverage is lower, the potential for rinderpest persistence is higher. This may explain why both

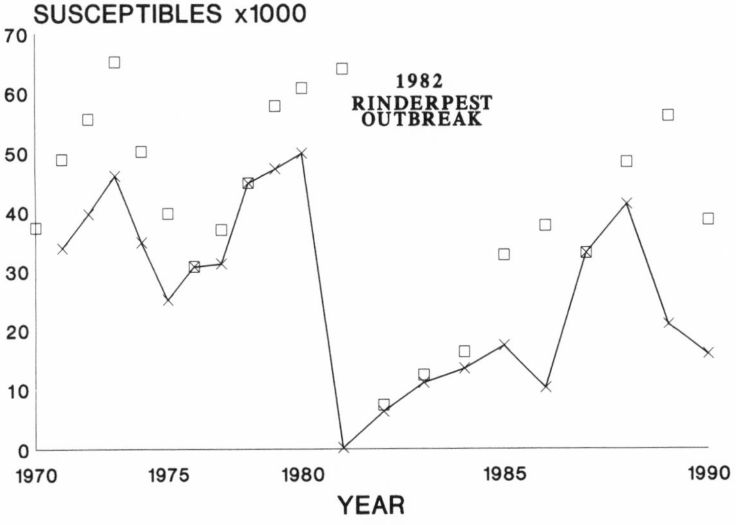

Figure 23.8 Estimates of the numbers of susceptible cattle in NCA during the period 1970–1990. The figure gives the numbers of susceptible individuals in the population, before (squares) and after (Xs) vaccination, in each year. As average survival and fecundity rates of cattle have been estimated for the region (Homewood and Rodgers 1991), it is possible to estimate the proportions of susceptible, resistant, and multiply innoculated animals in the host population. These estimates were obtained by assuming that a previous inoculation lasts for life, but that the number of cattle brought for inoculation includes animals that have been inoculated previously. Two estimates of the potential transmission threshold for rinderpest to become established are given in the text. The upper value at 100,000 susceptibles represents Plowright's estimate, the lower value at 50,000 is based on an estimate of transmission rate using the serology profile for cattle given in figure 23.2.

Rossiter et al. (1987) and Anderson et al. (1990) found evidence of rinderpest in buffalo from the northern part of Serengeti, where illegal cattle movements have been a common occurrence.

DISCUSSION AND CONCLUSIONS

The evidence presented in this chapter suggests that wildlife do not act as a long-term reservoir of rinderpest virus. Although no wildlife were ever vaccinated to protect them from infection, the wide-scale vaccination of cattle that was initiated in the 1950s ultimately led to the disappearance of the disease from wildlife. This resulted in large increases in the wildebeest and buffalo populations, and a six- to eightfold increase in wildebeest density. However, this increase *failed* to produce an outbreak of the pathogen. Although increases in buffalo numbers have also failed to sustain infections of rinderpest, outbreaks of rinderpest occurred in this species on a number of occasions in the Serengeti and Ngorongoro regions. Serological surveys suggest that these outbreaks involved mild strains of rinderpest virus that persist for a number of years before fading out. These findings would imply that wildebeest are well below the threshold density required for them to maintain rinderpest virus continuously, while buffalo may be fairly close to this critical density.

In the absence of the virus in cattle it is unlikely that populations of wild ungulates will maintain the pathogen, and rinderpest epidemiology in the region is likely to be characterized by occasional outbreaks in buffalo (and perhaps eland) whenever contact is made with infected cattle moved through the region. Rinderpest is therefore still a threat to the integrity of the Serengeti and Ngorongoro conservation areas. This is because its absence depends on an efficient vaccination program in cattle (Plowright 1985), but vaccination coverage is variable throughout the region and may be low in inaccessible areas (A. Dobson, unpub.). Furthermore, the annual vaccination program is expensive to operate; after road maintenance, it is the largest item in the parks' annual budgets. Although it is economically cheaper to vaccinate during the dry season, the cattle are on a lower nutritional plane at this time, and lower numbers are brought for in vaccination. Nevertheless, at times of prolonged drought, cattle are now grazed and watered in Ngorongoro Crater. Because droughts also lead to reduced attendance at vaccination crushes, this leads to a higher risk of contact between significant numbers of susceptible wild and domestic individuals. An outbreak that occurred during a drought period, when all species are on a low plane of nutrition and likely to be immunologically compromised, could be disastrous for both cattle and wildlife in the region.

Finally, it is worth considering the role that poaching of wildlife has had in constraining a rinderpest outbreak and whether systematic culling could be used as a management tool to prevent or contain an outbreak. Evidence presented elsewhere (Campbell and Borner, chap. 6; Arcese, Hando, and Campbell, chap. 24; Campbell and Hofer, chap. 25), suggests that poaching has led to significant reductions in the size of several ungulate populations in the Serengeti region. If wildlife were the principal hosts of rinderpest, then reducing the size of their populations would reduce the possibility of a rinderpest outbreak. However, the available evidence suggests that cattle are the primary hosts of the virus and that successful control of the virus is entirely dependent upon high levels of vaccination coverage in cattle. The increased levels of poaching of wildlife have had no discernible effects on the incidence of rinderpest outbreaks. More significantly, Dobson and May (1986) have calculated that the width of a "cordon sanitaire" required to halt the spread of a rinderpest outbreak in wildlife would be on the order of 40 miles. This would require elimination of most of the wildlife in the Serengeti if an epidemic were spreading from west to east. This and all the above calculations suggest that the best way to control rinderpest in the Serengeti is to prevent an outbreak by maintaining comprehensive vaccination coverage of cattle continuously throughout the region.

ACKNOWLEDGMENTS

I would like to thank the Ngorongoro Conservation Area Authority, Serengeti Wildlife Research Institute, and Tanzania National Parks for permission to work in the Ngorongoro Conservation Area and Serengeti. Support for this project was provided by Wildlife Conservation International (New York Zoological Society). All of the work benefited considerably from discussions with Karim Herji, John Mosha, Titus Mlengeya, Emmanuel Chausi, Noah Tluway, Sebastian Chuwa, Nick Georgiadis, Paul Rossiter, Bryan Grenfell, Robert May, Chas McLellan, Pete Wallen, Paul Symons, Hugh Lamprey, Patricia Moehlman, and Jonah Western. I would also like to thank Bjorn Figenschou, Veronica Barrett, Ken Campbell, Bill Newmark, Tony Sinclair, Roger Burrows, and Joy Corbett for their hospitality during fieldwork in Tanzania.

REFERENCES

Anderson, R. M., Jackson, H. C., May, R. M., and Smith, A. M. 1981. Population dynamics of fox rabies in Europe. *Nature* 289:765–71.

Anderson, E. C., Jago, M., Mlengeya, T., Timms, C., Payne, A., and Hirji, K. 1990. A serological survey of rinderpest antibody in wildlife and sheep and goats in Northern Tanzania. *Epidemiol. Infect.* 105:203–14.

Anderson, R. M., and May, R. M. 1982. Directly transmitted infectious diseases: Control by vaccination. *Science* 215:1053–60.
———. 1985. Age related changes in the rate of disease transmission: Implication for the design of vaccination programmes. *J. Hygiene* 94:365–436.
———. 1986. The invasion, persistence, and spread of infectious diseases within animal and plant communities. *Phil. Trans. R. Soc. Lond. B* 314:533–70.
———. 1991. *Infectious Diseases of Humans.* Oxford University Press.
Anderson, R. M., and Trewhella, W. 1985. Population biology of the badger (*Meles meles*) and the epidemiology of bovine tuberculosis (*Mycobacterium bovis*). *Phil. Trans. R. Soc. Lond. B* 310:327–81.
Begon, M., Bowers, R. G., Kadianakis, N., and Hodgkinson, D. E. 1992. Disease and community structure: The importance of host self-regulation in a host-host-pathogen model. *Am. Nat.* 139:1131–50.
Branagan, D., and Hammond, J. A. 1965. Rinderpest in Tanganyika: A review. *Bull. Epizoot. Dis. Afr.* 13:225–46.
Clutton-Brock, J. 1987. *A natural history of domesticated mammals.* University of Texas Press, Austin.
Dobson, A. P. 1988. Behavioral and life history adaptations of parasites for living in desert environments. *J. Arid Environ.* 17:185–92.
Dobson, A. P., and May, R. M. 1986. Disease and conservation. In *Conservation biology: The science of scarcity and diversity,* ed. M. E. Soulé, 345–65. Sunderland, Mass.: Sinauer Associates.
Dobson, A. P., and Miller, D. 1989. Infectious disease and endangered species management. *Endangered Species Update* 6:1–5.
Estes, R. D., and Small, R. 1981. The large herbivore population of Ngorongoro Crater. *E. Afr. Wildl. J.* 19:175–86.
Grzimek, B., and Grzimek, M. 1960. *Serengeti shall not die.* London: Hamish Hamilton.
Holt, R. D., and Pickering, J. 1985. Infectious disease and species coexistence: A model of Lotka-Volterra form. *American Naturalist* 126:196–211.
Homewood, K. M., and Rodgers, W. A. 1991. *Maasailand Ecology.* Cambridge: Cambridge University Press.
Kermack, W. O., and McKendrick, A. G. 1927. A contribution to the mathematical theory of epidemics. *Proc. R. Soc. Lond. A* 115:700–721.
Kruuk, H. 1972. *The spotted hyena: A study of predation and social behavior.* Chicago: University of Chicago Press.
May, R. M. 1988. Conservation and disease. *Conserv. Biol.* 2: 28–30.
McCullogh, K. C., Sheshberadaran, H., Norrby, E., Obi, T. U., and Crowther, J. R. 1986. Monoclonal antibodies against morbilliviruses. *Rev. Sci. Tech. Off. Int. Epizoot.* 5:411–27.
Norrby, E., Sheshberadaran, H., McCullough, K. C., Carpenter, W. C., and Orvell, C. 1985. Is rinderpest virus the archevirus of the *Morbillivirus* genus. *Intervirology* 23:228–32.
Plowright, W. 1982. The effects of rinderpest and rinderpest control on wildlife in Africa. *Symp. Zool. Soc. Lond.* 50:1–28.
———. 1985. La peste bovine aujourd'hui dans le monde. Controle et possibilite d'eradication par la vaccination. *Ann. Med. Vet.* 129:9–32.

Plowright, W., and McCullough, B. 1967. Investigations on the incidence of rinderpest virus infection in game animals of N. Tanganyika and S. Kenya 1960/63. *J. Hygiene* 65:343–58.

Plowright, W., and Taylor, W. P. 1967. Long-term studies of immunity in East African cattle following inoculation with rinderpest culture vaccine. *Res. Vet. Sci.* 8:118–28.

Prins, H. H. T., and Weyerhauser, F. J. 1987. Epidemics in populations of wild ruminants; Anthrax and impala, rinderpest and buffalo in Lake Manyara National Park, Tanzania. *Oikos* 49:28–38.

Rossiter, P. B. et al. 1983. Re-emergence of rinderpest as a threat in East Africa since 1979. *Vet. Rec.* 113:459–61.

Rossiter, P. B. et al. 1987. Continuing presence of rinderpest virus as a threat in East Africa, 1983–1985. *Vet. Rec.* 120:59–62.

Schaller, G. 1972. *The Serengeti lion: A study of predator-prey relations.* Chicago: University of Chicago Press.

Schenzle, D. 1984. An age-structured model of pre- and post-vaccination measles transmission. *IMA J. Math. Appl. Med. Biol.* 1:169–92.

Scott, G. R. 1964. Rinderpest. *Adv. Vet. Sci.* 9:113–224.

Simon, N. 1962. *Between the sunlight and the thunder: The wildlife of Kenya.* London: Collins.

Talbot, L. M., and Talbot, M. H. 1963. *The wildebeest in western Masailand.* Wildlife Monographs no. 12. The Wildlife Society.

Taylor, W. P., and Watson, R. M. 1967. Studies on the epizootiology of rinderpest in blue wildebeest and other game species of Northern Tanzania and Southern Kenya, 1965–7. *J. Hygiene,* 65:537–45.

Yilma, T., Hsu, D., Jones, L., Owens, S., Grubman, M., Mebus, C., Yamanaka, M., and Dale, B. 1988. Protection of cattle against rinderpest with vaccinia virus recombinants expressing the HA or F gene. *Science* 242:1058–61.

TWENTY-FOUR

Historical and Present-Day Anti-Poaching Efforts in Serengeti

Peter Arcese, Justine Hando, and Ken Campbell

Many protected areas around the world are currently under immense pressure due to the illegal exploitation of the plant and animal species within them. Although not restricted to the poorer areas of the world, these pressures are frequently most severe where depressed economic conditions both increase the attractiveness of engaging in illegal harvest and decrease the ability of local authorities to provide sufficient enforcement against it. One emerging conservation ethic addresses the first of these issues by promoting the managed, legal use of wildlife resources in order to enhance local economies and encourage sustained harvest strategies (Western 1982; Bell 1986b, 1987; Lewis, Keweche, and Mwenya 1990; Mbano et al., chap. 28). In tandem with such plans, a few studies have also addressed the efficiency of enforcement techniques and the minimum levels of funding required to protect particular wildlife populations (Bell 1986a; Leader-Williams and Albon 1988; Leader-Williams, Albon, and Berry 1990; Milner-Gulland and Leader-Williams 1992).

In this volume, Mbano et al. (chap. 28) discuss a plan for regional development and the sustained use of wildlife resources in the Serengeti ecosystem, and Campbell and Hofer (chap. 25) explore spatial aspects of illegal meat hunting within Serengeti National Park. As a complement to those chapters, we provide some details of the history of the anti-poaching effort in Serengeti, offer a preliminary analysis of some factors affecting the efficiency of anti-poaching patrols, and explore some possible effects of illegal hunting on ungulate populations. We show that with a small investment in additional effort and organization, routine data collection by anti-poaching patrols could play a key role in the understanding of trends in ungulate populations within the Serengeti-Mara ecosystem and elsewhere.

METHODS

Historical Records

Since its inception as a national park, wardens and rangers in Serengeti have routinely compiled monthly reports on the number of individuals arrested while hunting illegally in the park. Data were also available for a variable fraction of all years on the number of wire snares collected and on the number of elephant, rhinoceros, wildebeest, and zebra carcasses found killed. We use these data to describe some long-term patterns in the anti-poaching effort in relation to ivory prices (Caldwell 1988; cited in Leader-Williams, Albon, and Berry 1990), time of year, the number of visitors to Serengeti, and the total number of rangers employed within the park and their average monthly salary in U.S. dollars (based on the average annual bank rate in Tanzania). Records of annual fuel consumption, available vehicles, and total number of patrols conducted by month or year were unavailable.

We obtained figures on the size of the human population living outside the park from the results of national censuses conducted in 1957, 1967, 1978, and 1988. These data provide a baseline index of the potential for changes in human hunting pressure inside the park due to changes in human population size alone. The data we use here were taken from the areas corresponding to the present boundaries of Tarime, Serengeti, Bunda, Bariadi, and Meatu Districts. These districts border the western edge of Serengeti and are the main regions of origin for hunters arrested in Serengeti (Turner 1988; Magombe and Campbell 1989; Campbell and Hofer, chap. 25; Mbano et al., chap. 28). The best linear fit to the data for 1957–1988 indicated an annual rate of population increase of 2.9% in the area along the western edge of the park, where most hunting takes place. As there was no suggestion of an accelerating rate of growth of the human population from a visual analysis of the data, a constant growth rate was assumed for subsequent analyses.

Surveys of Ranger Patrols

Beginning in June 1991, we distributed questionnaires to ranger posts in Serengeti in order to obtain standardized data on anti-poaching techniques and their success and on the number and species of animals captured by hunters operating illegally within the park. Questionnaires were filled out by the officer in charge upon completion of a patrol and compiled in Seronera. We analyzed data from 149 patrols conducted from June 1991 to February 1992, mainly in the central woodlands and western corridor areas of the park.

Information collected from the questionnaires included: the post of origin, number of rangers, officers, and vehicles on patrol, time of departure on and return from patrol, number of people observed, number of people arrested, type and number of weapons confiscated, and species and number of animals found killed. For the purpose of analysis, we divided patrols into three types: (1) foot patrols, (2) vehicle patrols, on which rangers spotted people or other signs of illegal activity solely from the vehicle while driving cross-country, and (3) mixed foot and vehicle patrols, on which rangers were transported by vehicle to an area that was subsequently searched on foot (e.g., when bush or watercourses prevented vehicle use).

Analyses

Statistical Analyses. We employed standard parametric techniques in most of our statistical comparisons (Sokal and Rohlf 1982). We tested for normality in the distributions of data using graphical analyses (e.g., probability plots), and we conducted tests using transformed values when data were poorly distributed (e.g., by arcsine—square root in the case of percentages, or by $\log_{10}$ for normalizing distributions of abundance and body mass). For analyses involving percentages, we conducted parallel nonparametric tests (e.g., Kruskal-Wallis ANOVA), but in no case did these yield markedly different levels of statistical significance. We thus present only the results of parametric tests of these data. All graphs are plotted using untransformed data, since these are more easily interpreted. All probability values reported are two-tailed.

Preference Indices and the Probability of Mortality. We used three common preference indices to explore the effect of illegal hunting on wildlife species in relation to their abundance, body size, and main habitats occupied: the "forage ratio," "rank preference index," and "Manly's alpha." These indices are typically used to estimate relative preference for specific food items by individual foragers. We use them to determine relative preference for and success in capturing various wildlife species by illegal hunters in Serengeti. Krebs (1989) describes each of these indices in detail, and he recommends the rank preference index and Manly's alpha in particular for situations in which there are large differences in the abundance of the species being selected. We found that preference scores obtained using these three indices were, nevertheless, very highly correlated (e.g., each of the three simple correlation coefficients between these indices exceeded 0.90; $P < .001$, $N = 13$ in each case). We therefore use only Manly's alpha for statistical analyses, since it is continuously distributed and easily normalized by arcsine—square root transformation.

We tested the sensitivity of Manly's alpha to the inclusion of abundant species that were rarely captured by excluding Thomson's gazelle and recalculating our results. We also repeated this process by excluding topi, a species of average abundance that was frequently killed. Neither of these alterations substantially influenced our results with respect to preference or to the relative risk of mortality (see below).

To determine whether individual species were killed significantly more or less often than expected given their relative abundances, we also calculated chi-square statistics for single-category goodness of fit tests. We hypothesized that if each species were equally likely to be killed, each should appear in the killed sample in proportions equal to those in the live sample (see also Marks 1976). To obtain the number of individuals of each species expected to have been killed, we multiplied the species' proportional abundance in the cumulative live estimate for all species by the total number of animals of the species recovered by rangers. We then calculated chi-square values for each species by comparing the expected number killed with that observed. Because of the number of tests conducted on these data, we reduced the critical value of alpha to 0.05/13, or 0.004.

We relied on figures from Campbell and Borner (chap. 6) to estimate the relative abundance of nonresident wildlife species in Serengeti (e.g., wildebeest, zebra, Thomson's and Grant's gazelles, ostrich, and eland; see table 6.6 in chap. 6). For residents, we used figures for the central and western portions of the park only (table 6.2 and 6.4 in chap. 6), since these were the areas where all patrols reporting killed animals were undertaken. However, it is unlikely that all species were equally available to hunters throughout the sample period. Many migrants, for example, spend much of the year on the short-grass plains, where they are likely to be immune from hunters using common capture methods (e.g., snares and pits). This difference in availability undoubtedly has an influence on our results that we cannot control with our present data. Another potential problem with our analyses of preference and mortality risk arises because wildlife surveys are prone to bias. This can be due to differences in the detectability of species that vary in size or habitat (Norton-Griffiths 1978; Krebs 1989; Campbell and Borner, chap. 6). Our estimates of preference and relative mortality should therefore be viewed as preliminary.

HISTORICAL PATTERNS OF ANTI-POACHING EFFORTS

Trophy Hunting

Few examples of the illegal exploitation of wildlife are more widely known than the hunting of African elephants and rhinos, whose populations have now collapsed over much of Africa (reviews in Douglas-Hamilton 1987; Leader-Williams 1990; Leader-Williams, Albon, and Berry 1990). Dublin and Douglas-Hamilton (1987) describe elephant populations in Serengeti as undergoing an initial increase in the 1970s, then suffering a decline during the 1980s. Both of these trends were at least partly attributed to illegal trophy hunting: the increase due to hunting outside the protected area causing immigration, and the decrease due to a rapid increase in the illegal exploitation of elephants within the park.

The near-disappearance of elephants occurred primarily from 1975 to 1986, and the local extinction of black rhinoceros in Serengeti from 1975 to 1980, as indicated by sharp peaks in the number of fresh carcasses found by rangers (fig. 24.1a). Records of rhino and elephant carcasses in Serengeti are unavailable for the period prior to 1975, but Turner (1988) noted that illegal hunting of these species was uncommon during this period. A. R. E. Sinclair (pers. comm.) suggested that elephant carcasses observed during a routine census of the northwest of Serengeti in May 1973 may have signaled the beginning of large-scale trophy hunting for ivory in the park.

Peaks in the number of trophy carcasses discovered in Serengeti corresponded to escalations in the world prices of both ivory and rhino horn (e.g., fig. 24.1b; see also Douglas-Hamilton 1987; Leader-Williams 1990) and to a sudden decline in tourist visits to the park (fig. 24.1b). By 1986, reductions in tourist revenue and operating budgets led to only a single vehicle being available to park staff for anti-poaching patrols. Acting together, these factors created a nearly impossible situation for enforcement and likely combined to exacerbate the decline of elephant and rhino populations in the park.

After a low count of 467 individuals in 1986, the 1989 census indicated that elephants are again increasing in Serengeti (table 6.3 in chap. 6). This may be due to the rebuilding of the enforcement capability in Serengeti since 1986, and more recently to the world ban on ivory trading. Rhinoceros populations remain negligible within Serengeti itself, however, and given the small size of adjacent populations in the Masai Mara National Reserve and Ngorongoro Conservation Area, they appear unlikely to recover in the near future.

Meat Hunting

Since Serengeti was gazetted as a park, the main duty of park rangers has been the curtailment of illegal meat hunting. Although hunting by local people is allowed on public lands outside the park (with hunting permits and firearms only), overexploitation of wildlife (likely due to illegal hunting), the conversion of land to agricultural use, and the relative abundance of ungulates inside the park have led to Serengeti being an attractive hunting area, despite the risk of arrest (Campbell and Hofer, chap. 25; Mbano et al., chap. 28).

Turner (1988) informally documented the early efforts of anti-poaching patrols in Serengeti and provided several examples of the techniques used by hunters to kill, prepare, and transport meat. The main

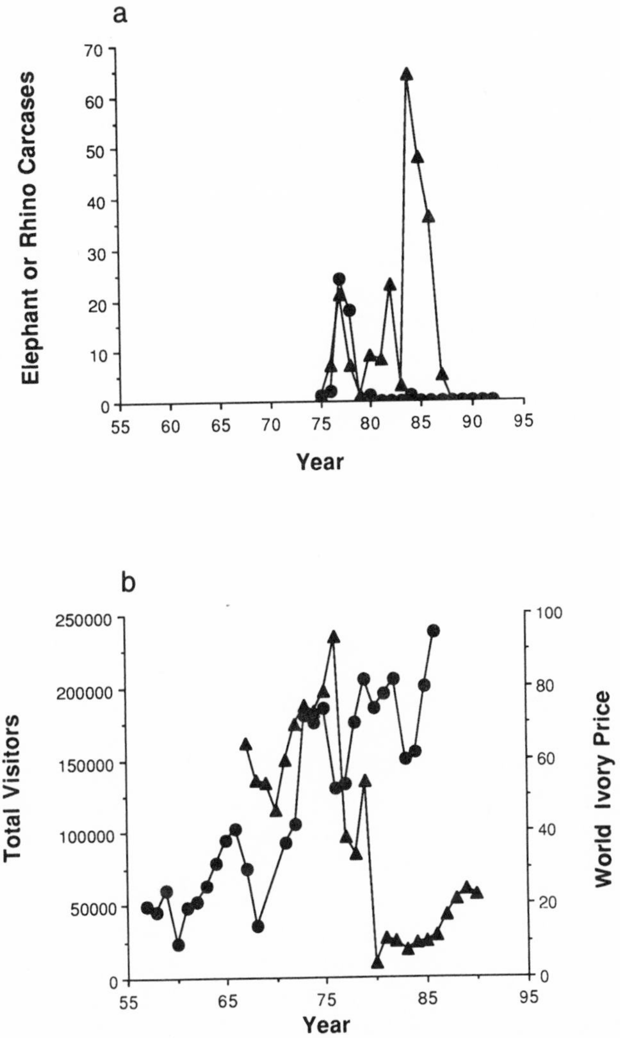

Figure 24.1 (*a*) The number of elephant (triangles) and rhinoceros (circles) killed illegally and found by rangers from 1975 to 1992. (*b*) The total number of visitors to Serengeti (triangles) from 1966 to 1991 and the world price of ivory ($U.S./kg; circles) from 1957 to 1985. (Data on ivory prices are from Caldwell 1988; cited and replotted in Leader-Williams, Albon, and Berry 1990.)

methods have not changed, and include setting wire snares in thickets, occasionally in lines of one hundred or more; digging pitfall traps, often near frequently used river crossings; and less often, the use of muzzle-loading rifles or other firearms. Spears and poison-tipped arrows are typically carried by hunters, but are mainly used to dispatch animals otherwise captured in snares or pitfalls.

Although several factors affect species preference among hunters (e.g., Marks 1976), the overwhelming dominance of the wildebeest population in Serengeti, their tendency to form large herds, and their willingness to enter thickets during migration has made them the focus of hunters, and of rangers deciding where to patrol (e.g., Turner 1988). Illegal hunting of wildebeest may have increased since 1975, as indicated by an approximate fourfold increase in the number of carcasses found killed (fig. 24.2). Although we were unable to statistically correct for variation in patrol effort over this period, we suggest that the upward trend (thought possibly not its magnitude) accurately reflects real changes in illegal wildebeest harvest over this period.

In comparison with wildebeest, the number of zebra carcasses found remained about constant from 1975 to 1992 (fig. 24.2). We note, however, that the wildebeest population increased by about 50% between 1975 and 1990, while zebra numbers remained stable (Sinclair, Dublin, and Borner 1985, Campbell 1989). Thus an alternative interpretation of the data in figure 24.2 is that harvest effort has not changed for wildebeest or zebra, but that the total number of wildebeest carcasses has risen in response to an increase in their abundance. However, this explanation predicts a much smaller increase in the number of wildebeest carcasses found than the one that was observed (e.g., ca. 0.5-fold vs. ca. 4-fold; fig. 24.2).

In contrast to the deleterious effects of illegal hunting on elephant and rhino populations in Serengeti, there is less evidence that hunting has had a significant negative effect on other ungulates in the park. For example, both wildebeest and zebra populations have remained approximately stable since 1977 (Campbell 1989; Campbell and Borner, chap. 6), even though these are two of the species most commonly taken illegally (Turner 1988; Magombe and Campbell 1989).

In contrast, Dublin et al. (1990) attributed 90% and 50% declines in local buffalo populations in the northwestern and western corridor areas of the park, respectively, to illegal hunting after ruling out the effects of disease. Sinclair (1977) described how the herding behavior of buffalo is exploited by hunters to eradicate local herds in cooperative drives into snare lines. Buffalo are also a preferred species among hunters in the northwest of the park (Sinclair 1977; pers. obs.), because of the quality and abundance of their meat and their mystique among local hunters. These factors may have acted together to increase the vulnerability of buffalo to hunting in Serengeti, and they suggest that factors other than commercial value may affect the stability of ungulate populations faced with illegal hunting.

Aside from elephants and rhinos, published evidence that illegal hunting has reduced ungulate populations in protected areas elsewhere

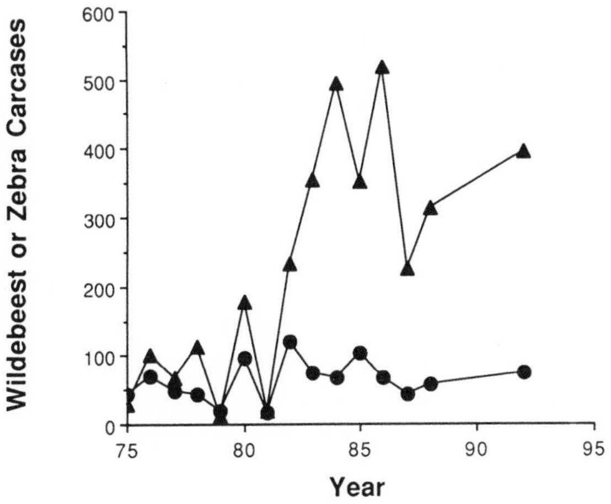

Figure 24.2 The number of zebra (circles) and wildebeest (triangles) killed illegally and found by rangers from 1975 to 1992.

in East Africa is sparse. This may be because early estimates of species abundance are not typically available. However, Edroma and Kenyi (1985) show that, especially in small areas, even small-bodied species such as reedbuck are potentially at risk of local extinction from illegal hunting. In North America, overhunting, which as in Africa is often done illegally, is the leading cause of endangerment and extinction among mammals (Hayes 1991).

Patterns of Arrests

Three main factors influenced the number of arrests made by rangers in Serengeti from 1957 to 1991: season, year, and the number of rangers employed. The number of arrests peaked during August through November, when the migratory wildebeest herds are typically in the northern portion of Serengeti or returning southward to the plains (fig. 24.3). At this time of the year, the mobility of rangers and their vehicles is greatest because it is the dry season, when vehicles can gain access to areas often impassable from December through June, and because grazers and fires remove much of the tall grass.

Arrests were made with about equal frequency in each of the months from January through July, but were lower than in the peak months of August through November (fig. 24.3). This probably occurred because during much of the former period the main wildebeest, zebra, and gazelle herds typically occupy the short-grass plains, where they are relatively immune from hunting. Turner (1988), whose anti-poaching efforts were primarily oriented to protecting the wildebeest herd, stated that once the wildebeest were on the plains, patrols were conducted much less often.

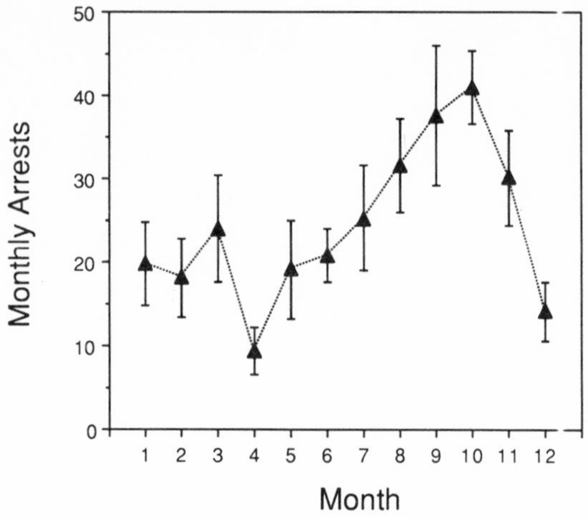

Figure 24.3 The mean number of arrests made in each month by rangers from 1975 to 1990. Vertical bars denote the standard errors of the estimates. Variation across months was statistically significant ($F_{11,168} = 122.5$, $P < .001$; ANOVA).

In April and December arrests were particularly few (fig. 24.3). In some areas of the park rainfall prevents effective patrols at these times. As a result, these are also the months when many rangers take their annual leave, resulting in a further reduction in patrol effort.

From 1957 to 1991, the annual number of arrests in Serengeti increased from about 100 to about 600 per year (fig. 24.4). This increase could have resulted from changes in (1) the number of rangers employed, (2) the efficiency of anti-poaching patrols, (3) the number of hunters entering the park, or (4) a combination of these factors. To explore these possibilities, we undertook several analyses with the following assumptions. First, we assumed that the total number of rangers employed was an index of anti-poaching effort. Although we later show that vehicles affect arrest rate, we had too little information on the number of vehicles available over the period to include this in our analysis. Second, we assumed that the number of hunters entering the park was a constant function of the size of the human population along the park's western edge, where most hunters originate (Magombe and Campbell 1989; Mbano et al., chap. 28). This assumption is probably flawed, however, since tendency to enter the park probably varies with such factors as local cultural traditions, economic conditions, seasonal changes in the availability of wildlife, and the intensity of the anti- poaching effort and thus the perceived risk of arrest. Nevertheless, as we have little information on these factors, population size is our best estimate of the number of potential hunters.

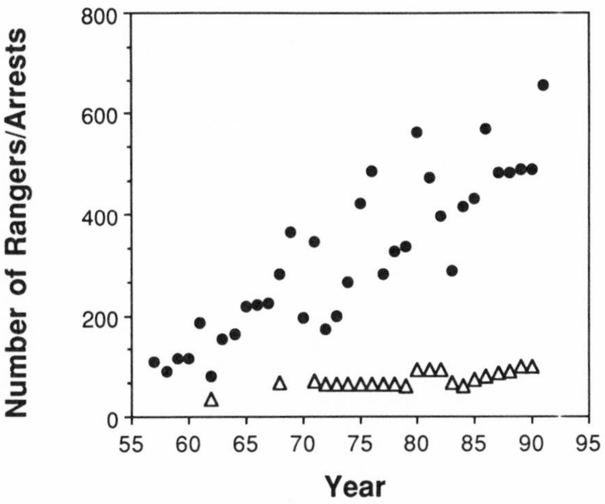

Figure 24.4 The annual number of arrests (solid circles) from 1957 to 1991 and the number of rangers employed (open triangles) from 1963 to 1991. For some of the years the data were unavailable.

We found that the observed increase in the annual number of arrests (ca. 5% per year) was greater than expected given an annual human population growth rate of 2.9% (see Methods) (fig. 24.5a). This suggests that increases in human population size alone cannot explain the increase in arrests (fig. 24.4). However, the number of rangers employed in Serengeti also approximately doubled from 1962 to 1991 (fig. 24.4). Increases in the number of arrests might also be expected due to this factor alone. We therefore standardized for the number of rangers employed in the park by dividing the total number of arrests by the number of rangers employed for each year to obtain the annual number of arrests per ranger. Plotting this figure by year revealed an approximate twofold increase in arrests made per ranger from 1962 to 1991 (ca. 3 to 6.5; fig. 24.5b). This shows that there was a substantial rise in the productivity of rangers over this period.

An increase in the annual number of arrests made per ranger could have resulted from at least two factors. Increases in the number of hunters entering the park could lead to more frequent encounters between rangers and hunters, and this might cause an increase in arrests even with no change in the number or efficiency of anti-poaching patrols. Alternatively, rangers could have become more efficient at searching for and capturing hunters due to improvements in equipment, transport, techniques, or training. In this case, we should observe an increase in the rate of arrests given a constant number of hunters entering the park.

We tested for an increase in the efficiency of rangers by comparing

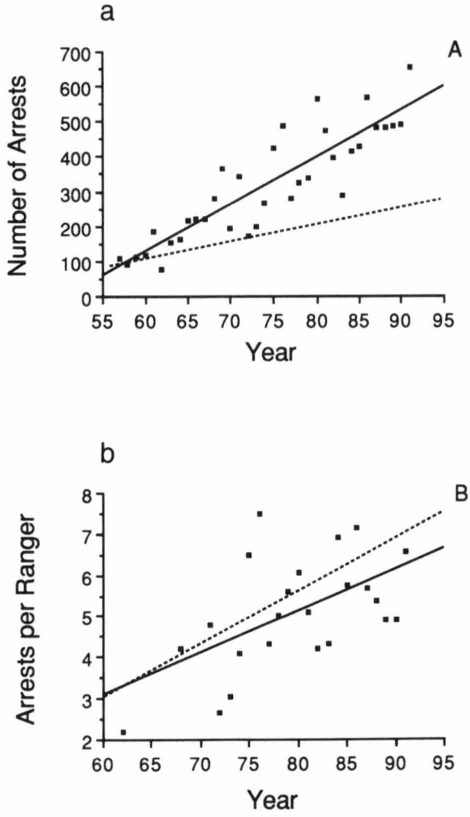

Figure 24.5 (*a*) The observed number of arrests from 1957 to 1991 (solid line: $Y = -685 + 13.54X$, $r = .89$), and the expected number given a population growth rate of 2.9% (dotted line; see text). The observed increase in the rate of arrests was significantly greater than expected by population growth (t test, $P < .01$). (*b*) The observed number of arrests per ranger from 1962 to 1991 (some missing years; solid line: $Y = -3.03 + 0.10X$, $r = .56$), and the expected number as calculated above. The observed increase in the rate of arrests per ranger was similar to that expected by population growth (t test, $P > .50$).

the annual number of arrests per ranger with the increase in arrests expected given a human population growth rate outside the park of 2.9% Again, we assumed that population growth reflects historical changes in the number of hunters in the park. Under this assumption, an increase in efficiency would be indicated if increases in the arrest rate exceed 2.9% per annum. However, fig. 24.5b shows that there was a close match between the increase in the number of arrests per ranger and that expected from human population growth outside the park. This suggests that encounter rate alone could account for the observed increase in the number of arrests made per ranger since 1962.

As discussed above, however, we have as yet no way to assess the number of hunters entering Serengeti, and this remains a serious obstacle to evaluating the effectiveness of anti-poaching patrols. Campbell and Hofer (chap. 25) present some possible alternatives, and in doing so they use an annual rate of population increase of 4% for people living in a narrow, 5 km strip along the western park border. Incorporating their estimate of population growth into our figures 24.5a and b has the effect of steepening the expected increase in the number and rate of arrests. This is consistent with the idea that historical increases in the annual number of arrests have resulted at least in part from higher encounter rates between rangers and hunters from 1962 to 1991.

In contrast to the factors discussed above, we found that monthly salary was unrelated to either the number of arrests or the number of snares collected annually per ranger, despite large fluctuations in salary over the years (fig. 24.6a–d). Reductions in the number of snares returned, as implied in figure 24.6b, could be linked to changes in hunting technique. They could also mean, however, that illegal hunters remove and take with them, or hide for future use, a greater proportion of snares than was done earlier. This might be expected given that a reward (1 Tsh, equaling ca. $0.15) was initiated in 1971 to encourage rangers to collect snares. Prior to this point, many snares were thought to have been left uncollected.

Although the reward appears to have had little effect on the overall pattern of snares returned from 1957 to 1983 (fig. 24.6b), it may also be that there were fewer snares available for the reasons mentioned above. At the time the reward was instituted, rangers earned approximately 7 Tsh per day in regular wages, suggesting that the rewards were relatively substantial. More recently, rewards for snares (200 Tsh), arrests (500 Tsh), and the recovery of firearms (10,000 Tsh) have reportedly had a marked positive effect on rangers, both in terms of their morale and their willingness to carry snares back from the field (M. Borner, Frankfurt Zoological Society, pers. comm.; see also Leader-Williams, Albon, and Berry 1990). In this case, rewards amounted to a substantial fraction of the daily salary of rangers in Serengeti, which averaged about 100 Tsh per day from 1986 to 1991. However, this incentive scheme, taken over by Tanzania National Parks in 1991, was curtailed in 1992 for lack of funds.

PRESENT-DAY PATTERNS OF ANTI-POACHING EFFORTS

Hunting Techniques

Our figures suggest that the use of snares is still the most popular technique for illegal hunters in Serengeti (e.g., Turner 1988). On the patrols covered by our questionnaires, rangers collected 2,157 wire snares (mean

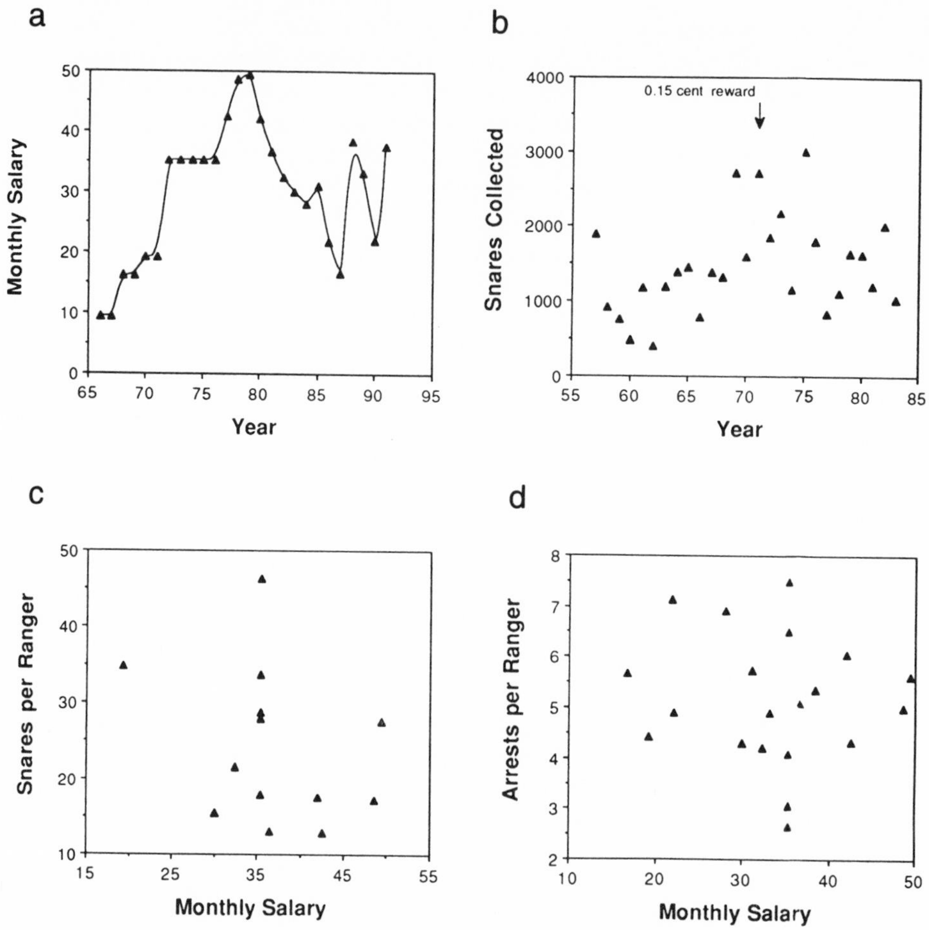

Figure 24.6 (*a*) The approximate monthly salary of rangers in $U.S. from 1966 to 1991. (*b*) The number of snares collected by rangers from 1957 to 1984. In 1971 a U.S. $0.15 reward was instituted for each snare returned (arrow). (*c*) The number of snares collected per ranger versus monthly salary (from *a*). (*d*) The number of arrests made per ranger versus monthly salary (as above).

per arrest = 11.3, SE = 2.0), 133 bows with arrows (mean bows per arrest = 0.60, SE = 0.06; mean arrows per bow = 6.7, SE = 0.5), 23 spears, and many knives, machetes, and axes. No firearms were recovered or reported. By contrast, Marks (1976) reported that hunters in the Bisa Valley of Zambia used muzzle-loading or other firearms almost exclusively.

Effects on Wildlife

Species Captured. Over all patrols, rangers reported 661 individuals of 20 species that were killed illegally inside the park. Five main species, however, accounted for 92% of all kills (wildebeest, 395; impala, 74; zebra, 52; topi, 46; and buffalo, 40), and wildebeest alone accounted for over half (59%). Aside from ungulates, rangers reported kills of three ostrich and one each of lion, cheetah, ratel, and monitor lizard. Other species included 10 giraffe, 9 eland, 6 warthog, 6 reedbuck, 6 waterbuck, 5 Grant's gazelle, 3 kongoni, 2 Thomson's gazelle, 1 hippopotamus, and 1 bushbuck. Given that no firearms were reported, and since snares were recovered in each instance where animals were killed, we assume that most of these animals were captured in snares.

Probability of Mortality. We obtained estimates of population abundance from wildlife surveys for thirteen of twenty species recorded killed by hunters (see Methods and Campbell and Borner, chap. 6). We compared the abundance of each of these species in the killed sample to their abundance as estimated in the live sample, assuming no bias in the latter (see Methods for a discussion of some problems inherent in these comparisons). We found that the probability of capture was not equal across the thirteen species considered (fig. 24.7, table 24.1). In particular, impala, topi, buffalo, giraffe, warthog, and waterbuck were killed significantly more often than expected given their abundances (fig. 24.7; see Methods). Six other species were killed about as often as expected, and only Thomson's gazelle were killed significantly less often than expected. Although the inclusion of Thomson's gazelle appears to have had a marked effect on our analysis, a reanalysis of the data after omitting them did not alter the statistical significance of the results (see Methods).

Preference by Abundance, Body Mass, and Habitat Use. We employed Manly's alpha as an index of preference by hunters for animals in the killed sample (see Methods; Krebs 1989). This index is distributed between zero and one, and it sums to equal one. Preference is indicated by values of alpha exceeding $1/N$, where N equals the number of species being considered. In this case, preference is indicated by values exceeding 1/13, or 0.077; avoidance is indicated by values lower than this. Relative preference for species in the killed sample is shown in figure 24.8.

Hunters in Serengeti primarily use wire snares placed in thickets to capture wildlife. The main habitat occupied by the thirteen species considered was thus, not surprisingly, found to be correlated with preference among hunters (fig. 24.9). Moreover, both abundance (fig. 24.10a) and body mass (fig. 24.10b) were also significantly related to preference, al-

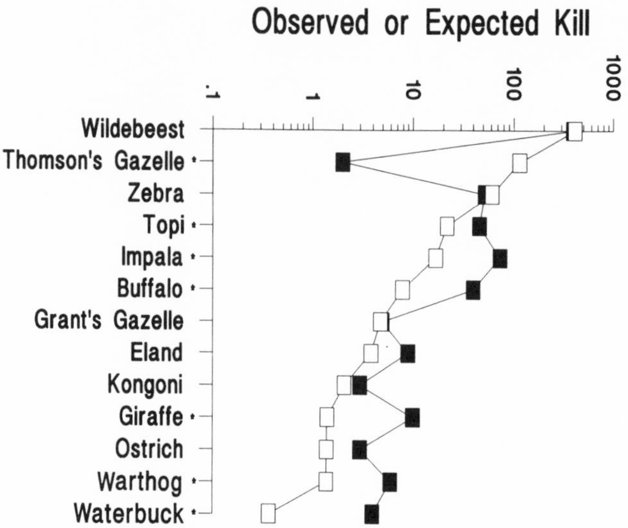

Figure 24.7 The observed number of individual animals captured by species (solid squares), and the expected number given their relative abundances (open squares; see Methods). Asterisks indicate a statistically significant difference between observed and expected number of animals killed at $P < .004$ (see Methods).

Table 24.1 Results of single-category goodness of fit tests (using chi-square) for comparisons of the observed and expected numbers of thirteen species of wildlife illegally killed in Serengeti, and three indices of relative preference for these species.

Species	Chi-square	Forage Ratio	Rank Preference	Manly's Alpha
Wildebeest	0.64	0.96	0.0	0.025
Thomson's gazelle	110.38	0.02	−11.0	0.001
Zebra	1.45	0.84	0.0	0.022
Topi	26.59	2.09	0.0	0.056
Impala	193.63	4.39	3.0	0.126
Buffalo	130.33	5.14	1.0	0.151
Grant's gazelle	0.01	1.10	−2.0	0.029
Eland	6.77	2.31	1.0	0.063
Kongoni	0.40	1.54	−2.5	0.041
Giraffe	52.28	7.71	4.0	0.245
Ostrich	1.87	2.31	−0.5	0.063
Warthog	15.34	4.62	4.0	0.134
Waterbuck	36.08	6.16	3.0	0.187

Note: $N = 649$; see Methods. Species are listed in order of decreasing abundance. For each index, preference is indicated by values exceeding, and avoidance indicated by values less than, a critical value: 1.0 for the forage ratio; 0.0 for the rank preference index; and 0.077 for Manly's alpha (see Methods; Krebs 1989).

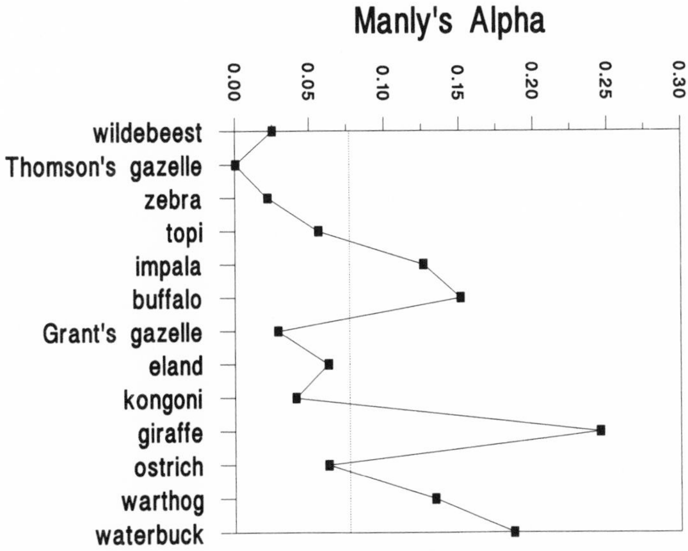

Figure 24.8 Manly's alpha index of preference for thirteen species in a sample of animals killed illegally (solid squares). Species are listed in decreasing order of abundance. Values of alpha exceeding 0.077 (dotted line) indicate preference; those below 0.077 indicate avoidance (see Methods).

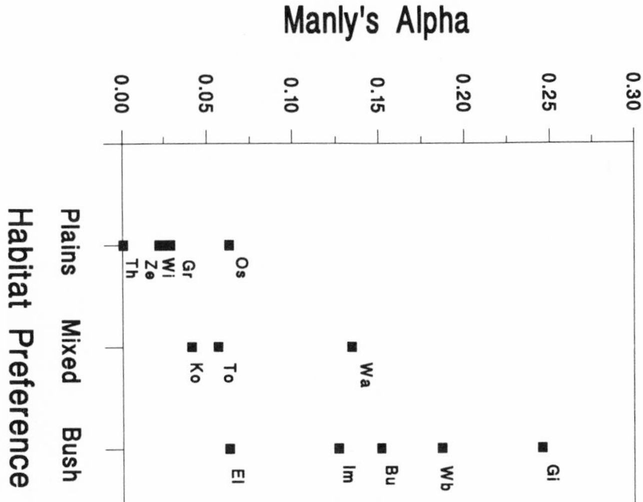

Figure 24.9 Manly's alpha index of preference for thirteen species (see Methods) versus the main habitats they occupy. Habitat was significantly related to relative hunter preference ($F_{2,10} = 9.06$, $P = .006$, ANOVA with Manly's alpha transformed by arcsine square root); species are grouped by the primary habitats occupied (e.g., Lamprey 1963; Sinclair 1977; Jarman and Sinclair 1979; Jarman and Jarman 1979). Species names are given in fig. 24.8 and table 24.1.

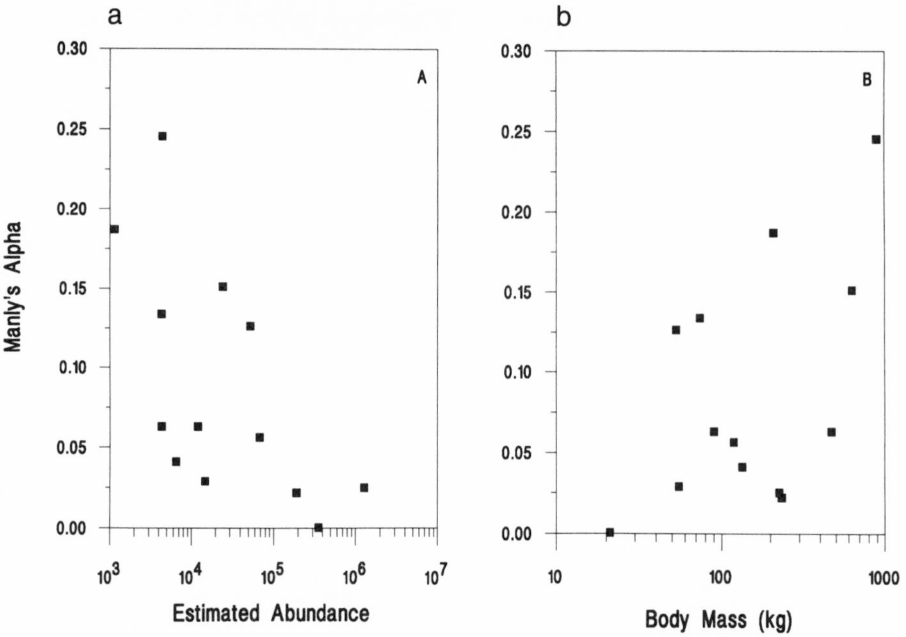

Figure 24.10 (*a*) Preference versus estimated abundance on a log scale ($Y = 0.74 - 0.106$ [SD = 0.035], $r^2 = .40$, $F_{1,11} = 9.03$, $P = .012$; abundance transformed by $\log_{10}$). (*b*) Preference versus body mass on a log scale ($Y = -0.09 + 0.169$ [SD = 0.075], $r^2 = .26$, $F_{1,11} = 5.11$, $P = .045$; body mass transformed by $\log_{10}$). Species are given in figure 24.8 and table 24.1.

though the relationship of body mass to preference was weaker than those of abundance and habitat to preference. The exclusion of Thomson's gazelle from these analyses did not alter the statistical significance of these results.

Although our estimates of the relative risk of poaching mortality and the preference of illegal hunters for wildlife species in Serengeti present a consistent picture, they are also preliminary. Probability of capture clearly depends on differences in the relative availability of species, as well as on differences in their susceptibility to various hunting techniques. At present we cannot disentangle the effects of such factors in our present data. This is because many species in Serengeti are migratory (e.g., wildebeest, zebra), others are typically resident (e.g., impala, topi, waterbuck, kongoni), and all show seasonal changes in habitat use and daily movement or social grouping patterns (e.g., Gosling 1974; Duncan 1975; Sinclair 1977; Jarman and Sinclair 1979; Jarman and Jarman 1979). Our comparisons may also suffer from sampling bias in estimates of species abundances and from uneven effort across areas in which hunters operate. To address these problems, future comparisons should use harvest and live animal counts taken from the same location, during the same period of time, and they should be conducted throughout the Serengeti area.

Preference and Population Trends. In general, we found no consistent relationship between current wildlife population trends and selectivity by illegal hunters. For example, topi may be experiencing an overall increase in abundance in the park (Campbell 1989; Campbell and Borner, chap. 6; Sinclair, chap. 9), but were killed by hunters more often than expected at random (fig. 24.7, table 24.1; but see fig. 24.8). Impala populations, however, appear to be approximately stable over the park as a whole (see refs. above), even though they also appear to be under heavy poaching pressure (fig. 24.7 and 24.8; table 24.1). By contrast, buffalo, giraffe, and waterbuck have undergone at least local recent declines (Campbell 1989; Campbell and Borner, chap. 6; Sinclair, chap. 9). These are expected given their high relative rates of capture (figs. 24.7 and 24.8; table 24.1).

It may be unreasonable, however, to expect correlations between relative kill rates and population trends under most circumstances. Several factors undoubtedly combine to affect the population size of each species in the park, and these are unlikely to have equal effects across all areas (e.g., Dublin et al. 1990; Sinclair, chap. 9; Hofer and East, chap. 16). In particular, because of uneven sampling, our results do not reflect the situation in the whole park.

We suggest that the number of each species killed illegally should be recorded as part of monitoring in Serengeti. This may be especially important for less common species, given that selectivity appears to be biased toward their more frequent capture (fig. 24.10). In support of this finding, Marks (1976) also reported that some of the least common species in his study area experienced the highest relative hunting pressure.

For rare species, losses of even a few individuals could have deleterious consequences for the survival of local populations (e.g., Caro and Durant, chap. 21). One anecdotal account of an adult roan antelope taken illegally within Serengeti (B. Mbano, pers. comm.) shows how important the illegal offtake could be for rare species: estimates of the roan antelope population within Serengeti range from 10 to 20 individuals (Sinclair, chap. 9). Interestingly, Turner (1988) reports that roan antelope were once regular, although uncommon, residents in Serengeti. Hofer and East (chap. 16) describe an analogous situation with potentially grave consequences for populations of predators.

Efficiency of Patrols

To explore what factors influenced the efficiency of anti-poaching patrols, we examined the total number of hunters observed illegally inside the park as well as the total number and proportion of these individuals that were eventually arrested. We consider each of these measures because they represent different aspects of success. For example, fig. 24.11 gives the distributions for the number of individuals observed and arrested on all patrols. Overall, a high proportion (50%–66%) of all patrols were successful in either arresting or observing illegal hunters (fig. 24.11). There was, however, evidence of a slight decline in the proportion of individuals sighted that were eventually arrested as the number sighted increased (fig. 24.11). This suggests that factors other than the total number of individuals sighted affected capture efficiency. Finally, we present the percentage of sighted individuals that were eventually arrested in two ways: (1) from those patrols on which at least one sighting was made, and (2) from all patrols. In the second case, patrols that failed to observe anyone were included as zeros. The latter figure is thus a composite of the efficiency of sighting individuals and of subsequently capturing them; we consider this a measure of overall capture success.

Seasonal Variation in Arrests. The number of monthly arrests was approximately equal across the 8 months for which we had data, except that it was slightly higher in July and August, and lower in October and December, than in other months (fig. 24.12a). There was slightly more variation by month in the total number of illegal hunters observed (fig. 24.12a). Compared with the pattern of arrests observed in the long-term data (see fig. 24.3), our current results show less seasonal variation. This may reflect regional effects on the frequency of arrests that occur within the park. Our current data were obtained primarily from posts in the central and western portions of the park, whereas the main migratory herds of wildebeest and zebra typically reside north of these areas from

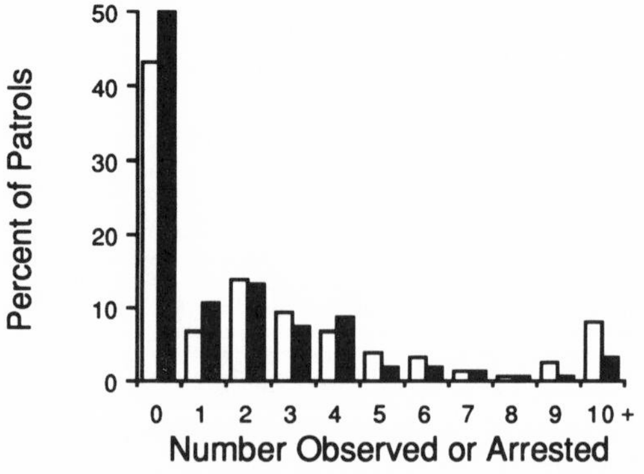

Figure 24.11 Percentage of patrols recording the observation (open bars) or arrest (solid bars) of from zero to ten or more illegal hunters inside Serengeti. The two distributions are marginally significantly different ($P = .06$, $D_{max} = 0.133$, $N = 149$; Kolmogorov-Smirnov two-sample test).

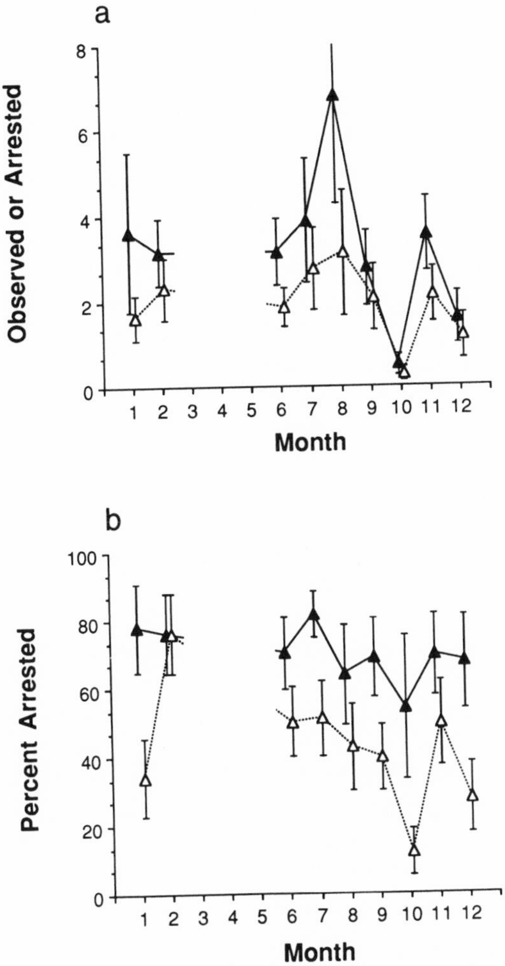

Figure 24.12 (*a*) Seasonal variation in the mean number of hunters observed (solid triangles; $F_{8,141} = 1.80$, $P = .08$; ANOVA) or arrested (open triangles; $F_{8,141} = 1.19$, $P = .31$) by rangers on patrol (sample sizes: Jan., 16; Feb., 7; June, 21; July, 16; Aug., 15; Sept., 26, Oct., 18; Nov., 14; Dec., 17). (*b*) The same data as in *a* replotted as the percentage of observed hunters that were later arrested (solid triangles; $F_{8,76} = 0.29$, $P = .97$; data transformed by arcsine square root; sample sizes: Jan., 7; Feb., 7; June, 15; July, 10; Aug., 10; Sept., 15; Oct., 4; Nov., 10; Dec., 7), or replotted as the percentage arrested, including patrols on which no sightings were made as zeros (open triangles; see text for explanation; $F_{8,141} = 2.07$, $P = .04$; conventions as in *a*). Vertical bars give the standard errors of the untransformed mean values.

August through October (e.g., Murray, chap. 11). In comparison, the long-term data reflect arrests made throughout the park.

There was little monthly variation in the percentage of all hunters observed that were subsequently captured (fig. 24.12b). This shows that once sighted, hunters were about equally likely to be arrested in all 8 months for which we had data. In contrast, the overall success of patrols in making arrests varied markedly (fig. 24.12b). Success peaked in February, when a high proportion of patrols observed illegal hunters and subsequently arrested most of them (fig. 24.12b). In October, however, eighteen patrols for which we had data averaged less than one observation of an illegal hunter per patrol (fig. 24.12a). Overall, these results suggest that although the largest number of arrests are typically made during the dry season from July to September (figs. 24.3, 24.12a), a high proportion of patrols might also succeed in the wet season if they were conducted as often.

Effect of Patrol Type and Number of Vehicles. Mixed foot and vehicle patrols were two to three times more effective at observing and arresting illegal hunters than vehicle-only patrols, and they were three to five times more successful than foot patrols (fig. 24.13a). Vehicle patrols lacking dedicated foot searches were only marginally more successful than rangers on foot (fig. 24.13a). These results should be interpreted carefully, however, since some rangers that initiated foot patrols only after observing evidence of hunting may have incorrectly recorded their patrols as dedicated, mixed foot-vehicle patrols. Although we do not believe this occurred often, it would have the effect of reducing the apparent success of "vehicle only" patrols and accentuating that of foot and vehicle patrols. Nevertheless, we suggest that vehicles substantially improve the effectiveness of anti-poaching patrols, particularly when combined with rangers working on foot.

Even though we found a strong positive effect of vehicles on success, we also found that two vehicles were no more successful than one (fig. 24.13b). And in terms of the number of individuals arrested, there was a suggestion that having more than one vehicle on patrol may have decreased effectiveness (fig. 24.13b). This result might be expected if hunters were more often alerted to the proximity of rangers by the sound of more than one vehicle, or by the increased probability of seeing rangers approaching the area.

The effects of patrol type and number of vehicles on the percentage of individuals arrested were similar to but much less striking than those in fig. 24.13. When only patrols making at least one sighting were considered, we found a slight but significant increase in effectiveness across foot, vehicle, and mixed patrols of approximately 20% overall (fig. 24.14a).

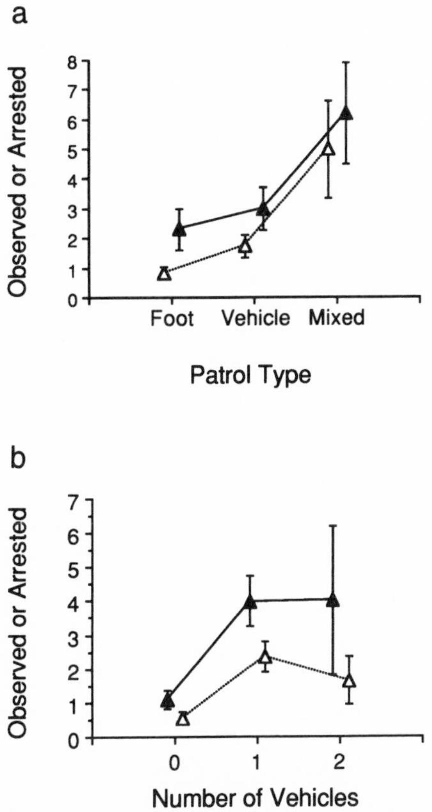

Figure 24.13 (*a*) The mean number of hunters observed (solid triangles; $F_{2,121} = 2.04$, $P = .14$; ANOVA) or arrested (open triangles; $F_{2,121} = 8.69$, $P < .001$) by rangers using one of three patrol techniques (sample sizes for patrol type: foot, 47; vehicle, 67; mixed, 10; see Methods). (*b*) The mean number of hunters observed (solid triangles; $F_{2,121} = 4.00$, $P = .02$) or arrested (open triangles; $F_{2,121} = 4.95$, $P = .009$) by rangers on foot ($N = 42$), or with one ($N = 68$) or two vehicles ($N = 14$). Vertical bars give the standard errors of the mean values.

The number of vehicles had a small and statistically insignificant effect on the percentage of individuals arrested (fig. 24.14b). By including all patrols, however, we found patterns very similar to those in figures 24.13a and b: mixed patrols and patrols with only one vehicle were more successful than others (fig. 24.14b). These comparisons show that the main effect of vehicles on the success of anti-poaching patrols is to increase the probability of observing hunters inside the park, rather than to affect the fraction of those individuals that are subsequently arrested.

Effect of Patrol Size and Departure Time. Rangers in Serengeti undertook anti-poaching patrols with as few as two and as many as twenty-

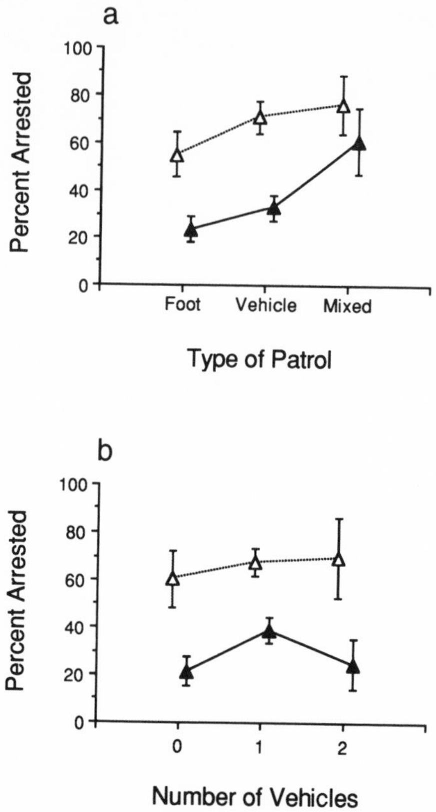

Figure 24.14 (*a*) The mean percentage of hunters observed that were arrested on all patrols (solid triangles; $F_{2,121} = 3.36$, $P = .038$; ANOVA; data transformed by arcsine square root; $N = 47$, 67, and 10 for foot, vehicle, and mixed, respectively), or only those on which at least one hunter was sighted (open triangles; $F_{2,56} = 1.28$, $P = .29$; $N = 15$, 39, and 5 for foot, vehicle, and mixed respectively), versus patrol type (see Methods). (*b*) The mean percentage of hunters observed that were arrested on all patrols (solid triangles; $F_{2,121} = 2.41$, $P = .09$; data transformed as above; $N = 47$, 67, and 10 for zero, one, and two vehicles respectively), or only those on which at least one hunter was sighted (open triangles; $F_{2,56} = 0.22$, $P = .81$; data transformed as above; $N = 15$, 39, and 5 for zero, one, and two vehicles respectively), versus the number of vehicles used on patrol. Vertical bars give the standard errors of the untransformed mean values.

one rangers and officers participating (fig. 24.15). We found, however, that group sizes of rangers and officers in the range of nine to fifteen were somewhat more successful than others when all patrols were considered (fig. 24.15). Similar relationships were found between the group size of rangers and both the total number of arrests made and the percentage of individuals arrested when at least one sighting was made. These results suggest that the most successful patrols were between nine and fifteen in group size.

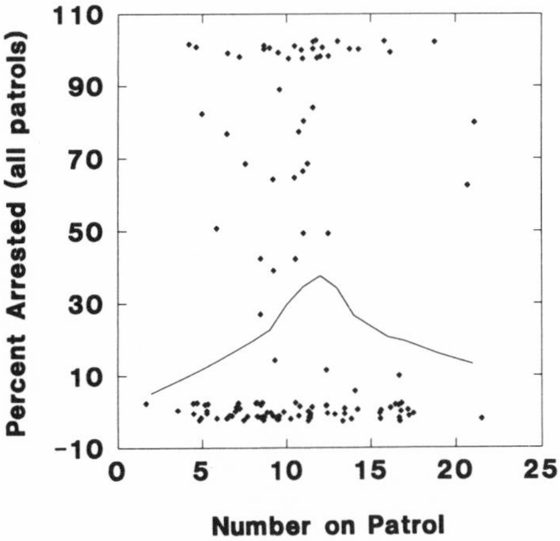

Figure 24.15 The number of rangers and officers participating in patrols versus their relative success (estimated as the percentage of individuals observed that were later arrested on patrols, including all patrols in which no one was seen as zeros; see text). The solid line represents a locally weighted sums of squares (LOWESS) best fit to the data (tension = 0.75, $N = 123$, data were jittered to reveal overlying points; Wilkinson 1989), and it indicates that group sizes between nine and fifteen were somewhat more successful than others. The tails of the relationship are not well estimated due to small sample size.

Rangers departed on patrol most often between 0600 and 0700 hours (fig. 24.16). The relationship between patrol success and the time of departure, however, showed an overall decreasing trend, with few patrols departing after 0900 hours having any likelihood of success (fig. 24.16). Although our data for early departures was sparse, the results suggest that the optimum departure time for anti-poaching patrols was about 0500 hours—slightly ahead of the most common departure time and about half an hour prior to daybreak. The result seems reasonable, given that illegal hunters probably anticipate the arrival of anti-poaching patrols during the early daylight hours, but not before (fig. 24.16). Patrols that departed unusually early may therefore have more often caught hunters unaware.

CONCLUSION

Killing wildlife for meat is the most serious form of poaching now occurring in Serengeti. This activity appears to have increased since the park's inception, as indicated by a sixfold increase in the number of arrests made from 1957 to 1991. However, the size of the ranger force has also doubled since 1963. Thus in order to understand what changes have occurred with respect to the efficiency of anti-poaching efforts in Serengeti, we must be able to estimate the number of people that enter the park to hunt. Research designed with this goal in mind is clearly required.

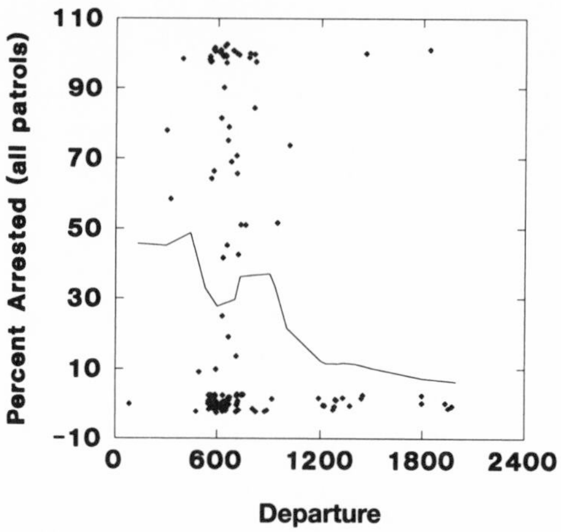

Figure 24.16 Departure time of anti-poaching patrols versus their success (estimated as for fig. 24.15). The solid line gives the LOWESS best fit to the data (conventions as for fig. 24.15), and it indicates that the chances of a departing patrol having a successful day declined rapidly with later departure times. The tails of the relationship are not well estimated due to small sample size. Many data points are overlaid by others ($N = 123$).

The most popular technique for killing wildlife is the setting of wire snares in thickets. Our results suggest that this practice causes the disproportionately frequent capture of species that use woodlands and thickets as their main habitat. Because these species also tend to be less abundant than those that primarily inhabit grasslands in Serengeti, their populations are more likely to suffer declines in the face of illegal harvest. The relative preferences for wildlife species among hunters were not well correlated with trends in the population sizes of individual species over the park as a whole. However, three species that were killed more often than expected given their relative abundances have experienced at least local declines in population size: these are buffalo, giraffe, and waterbuck. Populations of these and other species that appear to be heavily exploited (e.g., impala, warthog, and topi) should be carefully monitored for local or general declines in size. Efforts should also be undertaken to routinely monitor the number, sex, and age of all species that are illegally killed. With accurate estimates of the local abundance of species, the compilation and analysis of these data could play a key role in our understanding of trends in wildlife populations in Serengeti and elsewhere.

Our analyses of anti-poaching patrols suggest two main conclusions regarding their effectiveness. First, the frequency with which patrols are undertaken is the main factor affecting the number of arrests made. Both the probability of observing and of capturing hunters were about equal across the 8 months for which we had data. However, the total number of arrests varied significantly by month from 1975 to 1990. Periods of few arrests corresponded to months when patrols are often made difficult by rain and tall grass, to when rangers take their annual leave, and to when the migratory herds reside on the short-grass plains. This suggests that the success of the anti-poaching effort in Serengeti could be increased by (1) improvements in the road network, particularly in remote regions of the park; (2) staggering work schedules to avoid periods when there are too few rangers on hand to undertake patrols; and (3) conducting frequent patrols into areas where nonmigratory wildlife occur, even when the main wildebeest, zebra, and gazelle herds occupy the plains region of the park.

Second, we found that vehicles have a marked positive effect on the probability that anti-poaching patrols will encounter hunters and make arrests, particularly when they are combined with rangers working on foot. Vehicle patrols augmented by dedicated foot searches were three to five times more successful at observing and arresting hunters than foot patrols alone. However, routine financial commitments in Serengeti and elsewhere in Tanzania limit the number of anti-poaching vehicles that can be purchased and maintained.

In conclusion, we suggest the following as worthwhile goals for external funding organizations interested in aiding the anti-poaching effort in Serengeti: (1) providing and maintaining vehicles for anti-poaching patrols; (2) providing and maintaining equipment to expand the network of anti-poaching roads; (3) aiding in the development of a routine program for monitoring anti-poaching activities and wildlife abundances; and (4) providing expertise and training in the analysis, interpretation, and application of data gathered by such programs to improve and economize anti-poaching programs.

ACKNOWLEDGMENTS

We thank D. Babu for encouraging us to undertake these analyses, as well as the rangers and officers that faithfully completed survey forms after difficult and dangerous patrols. W. Summay, Warden in Charge of Serengeti, made several suggestions that helped refine our techniques and encouraged our effort. G. Sabuni, Serengeti Wildlife Research Institute, aided us in obtaining permits and housing for work in Serengeti. Frankfurt Zoological Society provided support for K. C., and P. A. was variously supported by the Natural Sciences and Engineering Research Council of Canada, Simon Fraser University, the University of Wisconsin-Madison, and by a grant from the Committee for Research and Exploration of the National Geographic Society. P. A. owes special thanks to R. Ydenberg for providing an office during the analysis and write-up of the initial draft of this chapter, and to T. Corfield of Kerr and Downey Safaris for providing assistance in the field. Two anonymous reviewers provided many helpful comments.

REFERENCES

Bell, R. H. V. 1986a. Monitoring of illegal activity and law enforcement in African conservation areas. In *Conservation and wildlife resources in Africa,* ed. R. H. V. Bell and E. McShane-Caluzi, 317–51. Washington, D.C.: Peace Corps.

———. 1986b. Traditional use of wildlife resources within protected areas. In *Conservation and wildlife resources in Africa,* ed. R. H. V. Bell and E. McShane-Caluzi, 297–315. Washington, D.C.: Peace Corps.

Campbell, K. L. I. 1989. Serengeti ecological monitoring programme. In *Serengeti Wildlife Research Center Bi-annual Report: 1988–89.*

Douglas-Hamilton, I. 1987. African elephants: Population trends and their causes. *Oryx* 21:11–24.

Dublin, H. T., and Douglas-Hamilton, I. 1987. Population trends of elephants in the Serengeti-Mara region. *Afr. J. Ecol.* 25:19–33.

Dublin, H. T., Sinclair, A. R. E., Boutin, S., Anderson, E., Jago, M., and Arcese, P. 1990. Does competition regulate ungulate populations? Further evidence from Serengeti, Tanzania. *Oecologia* 82:283–88.

Duncan, P. 1975. Topi and their food supply. Ph.D. dissertation, University of Nairobi.

Edroma, E. L., and Kenyi, J. M. 1985. Drastic decline of bohor reedbuck (*Redunca redunca* Pallas 1877) in Queen Elizabeth National Park, Uganda. *Afr. J. Ecol.* 23:53–55.

Gosling, M. 1974. The social behaviour of Coke's hartebeest *(Alcelaphus buselaphus cokei).* In *The behaviour of ungulates and its relation to management,* ed. V. Geist and F. Walther, n. s., no. 24, 488–511. Morges, Switzerland: IUCN.

Hayes, J. P. 1991. How do mammals become endangered? *J. Wildl. Mgmt.* 19:210–15.

Jarman, P. J., and Jarman, M. V. 1979. The dynamics of ungulate social organization. In *Serengeti: Dynamics of an ecosystem,* ed. A. R. E. Sinclair and M. Norton-Griffiths, 185–224. Chicago: University of Chicago Press.

Jarman, P. J., and Sinclair, A. R. E. 1979. Feeding strategy and the pattern of resource-partitioning in ungulates. In *Serengeti: Dynamics of an ecosystem,* ed. A. R. E. Sinclair and M. Norton-Griffiths, 130–63. Chicago: University of Chicago Press.

Krebs, C. J. 1989. *Ecological methodology.* New York: Harper and Row.

Lamprey, H. F. 1963. Ecological separation of large mammal species in the Tarangire Game Reserve, Tanganyika. *E. Afr. Wildl. J.* 1:63–92.

Leader-Williams, N. 1990. Allocation of resources for conserving African pachyderms. *Transactions of the 19th IUGB Congress,* Trondheim 1989, 633–39.

Leader-Williams, N., and Albon, S. D. 1988. Allocation of resources for conservation. *Nature* 336:533–35.

Leader-Williams, N., Albon, S. D., and Berry, P. S. 1990. Illegal exploitation of black rhinoceros and elephant populations: Patterns of decline, law enforcement, and patrol effort in Luangwa Valley, Zambia. *J. Appl. Ecol.* 27:1055–87.

Lewis, D., Keweche, G. B., and Mwenya, A. 1990. Wildlife conservation outside protected areas: Lessons from an experiment in Zambia. *Conserv. Biol.* 4:171–80.

Magombe, J. K., and Campbell, K. 1989. Poaching survey. In *Serengeti Wildlife Research Center Bi-annual Report: 1988–89.*

Marks, S. A. 1976. *Large mammals and a brave people.* Seattle: University of Washington Press.

Milner-Gulland, E. J., and Leader-Williams, N. 1992. A model of incentives for the illegal exploitation of black rhinos and elephants: Poaching pays in Luangwa Valley, Zambia. *J. Appl. Ecol.* 29:388–401.

Norton-Griffiths, M. 1978. *Counting animals.* 2d ed. Handbook no. 1. Techniques in African Wildlife Ecology. Nairobi: African Wildlife Foundation.

Sinclair, A. R. E. 1977. *The African buffalo: A study of resource limitation of populations.* Chicago: University of Chicago Press.

Sinclair, A. R. E., Dublin, H. T., and Borner, M. 1985. Population regulation of Serengeti wildebeest: A test of the food hypothesis. *Oecologia* 53:364–69.

Sokal, R. R., and Rohlf, F. J. 1982. *Biometry.* San Francisco: W. H. Freeman.

Turner, M. 1988. *My Serengeti years,* ed. B. Jackman. London: Elm Tree Books.

Western, D. 1982. Amboseli National Park: Enlisting landowners to conserve migratory wildlife. *Ambio* 11:302–8.

Wilkinson, L. 1989. *SYGRAPH.* Evanston, Ill.

TWENTY-FIVE

People and Wildlife: Spatial Dynamics and Zones of Interaction

Ken Campbell and Heribert Hofer

A mutual understanding and minimization of conflict between managers in conservation areas and local communities in adjacent areas is widely seen as critical to the continued existence of many protected regions (Myers 1972, 1979; Anderson and Grove 1987; Western and Pearl 1989; Robinson and Redford 1991; SRCS 1991). Integral to this process is an understanding of the mechanisms and levels of community benefits from wildlife and other natural resources, whether obtained legally or illegally (Myers 1972; Cumming 1981; Homewood and Rodgers 1984; Lindsay 1987; Hough 1991; Shaw 1991; Newby 1992).

By modeling the spatial dynamics of interactions between illegal meat hunters and wildlife, this study pursues two aims: first, to provide information of relevance to the management of the system, and second, to emphasize the fact that natural systems even as large as the Serengeti ecosystem do not exist in a vacuum. Trends in wildlife populations and the effects of hunting have previously been considered on a temporal scale (Sinclair 1973, 1979; Campbell 1989; Dublin et al. 1990), while the wider spatial dimensions of these trends, incorporating human populations outside the park, have largely been ignored. However, many management strategies contain explicit spatial components. These include the location of ranger posts and the structure of law enforcement work; conservation education and extension services; potential village hunting, licensing, and utilization schemes; and the siting of roads and infrastructure for both tourism and law enforcement work. Human demands on the ecosystem are likely to vary locally and are spread unevenly around and within protected areas. Cultural differences between the largely pastoralist Masai living to the east of the Serengeti and the agricultural and agro-pastoralist people living to the west result in major differences in their interactions with wildlife. As a result, illegal offtake of wildlife is almost exclusively the domain of hunters originating from areas to the west of the protected area, and the following analysis and discussion of wildlife meat hunting therefore concentrates on this area. Future management of wildlife resources will benefit from careful consideration of the spatial variation in the requirements of local communities and the pressures exerted by them, as well as the implications of ecological and biogeographic principles (Harris 1984; Lovejoy et al. 1986; Turton 1987; Shafer 1991; West and Brechin 1991).

Local communities adjacent to the protected area provide both the supply of illegal meat hunters and the market for their products. We develop a series of models to derive rough estimates of the extent of community benefits from illegal hunting of *all large herbivores* in the protected area. As movements of migratory herds vary substantially between years, the analysis of spatial patterns was restricted to *large resident herbivores* for which systematic aerial survey data were available. For these we appraise ecological factors affecting their distributions and investigate how spatial patterns of meat hunting may modify those distributions. This analysis permits us to identify areas with populations of large resident herbivores that currently experience high illegal hunting pressure or are likely to experience it in the future, and which may require additional law enforcement infrastructure.

METHODS

Study Area

The study area was defined as a 300 by 350 km region enclosing (1) the "protected area" (Serengeti National Park, Maswa Game Reserve, and the proposed Ikorongo and Grumeti Game Reserves); (2) seven administrative districts to the west of the protected area; and (3) Ngorongoro District (including the Ngorongoro Conservation Area) to the east. It thus included both the wildlife areas constituting the Tanzanian section of the Serengeti ecosystem and areas occupied by adjacent local communities (fig. 25.1). Boundaries of conservation areas were digitized from 1:50,000 scale topographic maps and checked against descriptions given by the official Government Gazette. In the case of Maswa GR, where the gazette has not been updated since boundary changes, borders were mapped during discussions with the reserve's manager. These were checked during a reconnaissance flight utilizing a GPS receiver to record coordinates of the de facto boundary between cultivated land and the reserve (R. Lamprey, pers. comm.). Parts of the Grumeti and Ikorongo Game Controlled Areas were the subject of proposals for upgrading to GR status and in these cases the proposed boundaries were used. Coordinates of 651 villages representing National Census enumeration areas were obtained from 1:50,000 and 1:250,000 scale topographic and thematic maps.

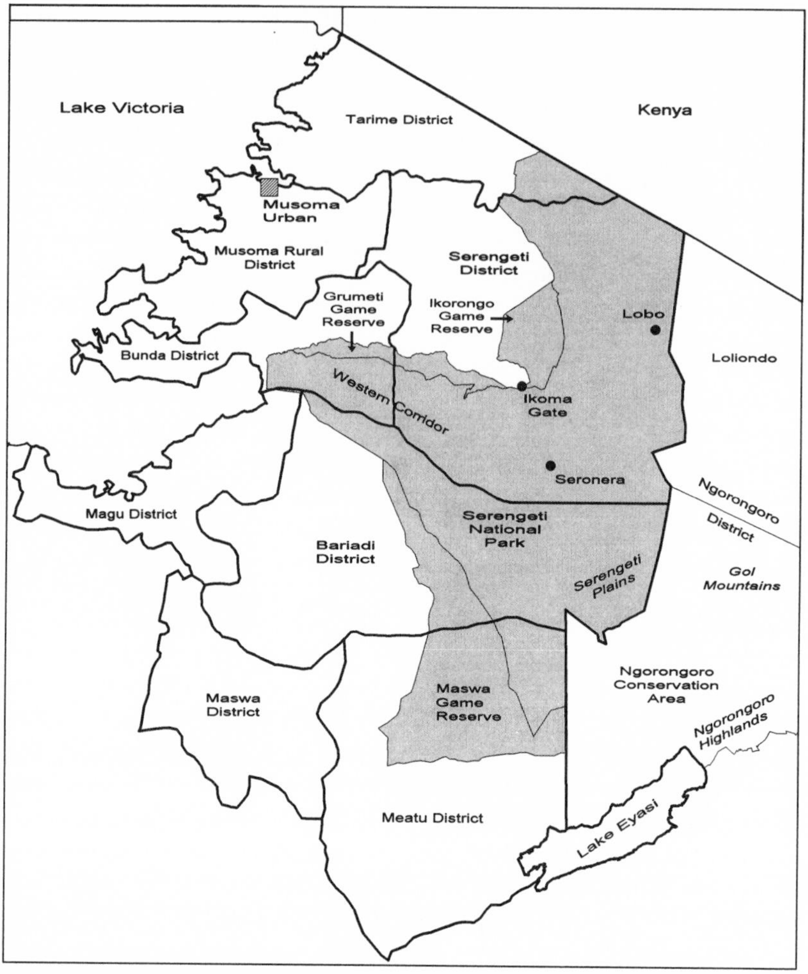

Figure 25.1 Base map of the study area showing conservation area boundaries, administrative districts, and locations mentioned in the text.

Environmental Gradients

Elevation was obtained from 1:50,000 topographic maps by recording altitude at intersections of a regular 1.25 km grid, resulting in 16 points per 5 km UTM grid square. The standard deviation of all points within one grid square provided an index of local relief.

Rainfall data have been collected in Serengeti National Park since the early 1960s using a network of daily and monthly rain gauges (Pennycuick and Norton-Griffiths 1976; Campbell 1989). These data were supplemented by Meteorological Department records from sites outside of the protected area. Average annual rainfall was calculated for the period

1970 to 1989 from a total of 125 meteorological stations, including 15 in Kenya, that had a minimum of 5 calendar years of complete data. Surfer (Golden Software 1990) was used to generate isohyets by interpolation of these point data as well as values for each 5 km UTM grid square (kriging, three nearest points and quadrant search pattern).

The Normalised Difference Vegetation Index (NDVI), derived from NOAA satellite data, is widely used for monitoring green vegetation over large areas (Justice et al. 1985) and provides an index of primary productivity. Monthly NDVI images for 1990 were calculated from 10-day composite images using IDA (Pfirmin 1989) and averaged to produce an image summarizing the distribution of vegetation greenness. Data for three decades prior to the 1991 wildlife census provided an image expressing the response to rainfall prior to the survey.

Land Cover and Wildlife

The distributions and population sizes of major herbivore species within the Serengeti ecosystem have been the subject of monitoring by aerial sample surveys since 1971 (Campbell and Borner, chap. 6). We used data from the most recent (1991) systematic aerial wildlife survey, which included Aerial Point Sampling (APS: Norton-Griffiths 1988) covering a survey area of 26,084 km^2 and 3,155 aerial photo samples. Missing data due to camera failure in 28 grid cells were replaced by 1988 APS data. Selected land cover information included data on total woody canopy, total grass cover, and animal tracks for each 5 km UTM grid square. The following eight large resident herbivores were considered: African buffalo, giraffe, Grant's gazelle, impala, kongoni, topi, warthog, and waterbuck. These species constitute the majority of the resident large herbivores in Serengeti (Campbell and Borner, chap. 6). For the analysis of distribution patterns we also included olive baboons, a common primate. Some other species, such as bushbuck and Bohor reedbuck, are present in the ecosystem, but aerial censuses provide poor estimates of the distribution and size of their populations, so they were excluded from the analysis. Species were classified as browsers or grazers according to McNaughton and Georgiadis (1986). Population estimates of wildebeest, the most common migratory herbivore, are from Campbell and Borner (chap. 6), while those of eland and zebra are from the 1991 aerial census.

Local Communities

Early National Census data (1948, 1957) were derived from sample surveys, whereas more recent censuses (1967, 1978, 1988) represent a total enumeration at a village level throughout the country (Central Statistics Bureau 1963, 1969; Bureau of Statistics 1978, 1988). Population parameters for 1948 to 1967 were available as summaries per district. Detailed data at a village enumeration area level were available for 1978 and 1988 from the seven districts within the study area (Bureau of Statistics 1978, 1988, 1991, and pers. comm.), except for Musoma Rural District, where 1978 data were compiled at a ward level. Some enumeration areas included more than one closely associated physical village.

Wildlife Meat Hunting

Information on home villages of wildlife meat hunters and the numbers of hunting trips per year was derived from replies to systematic questioning of 102 hunters arrested in the national park by anti-poaching patrols (Jan. to Dec. 1988 and Feb. to May 1992). Home villages were matched with known village locations. Numbers of wildlife killed per hunter and the relative proportions of each species were estimated from law enforcement patrol records (Jan. to Sept. 1988 and June 1991 to May 1992; see also Arcese, Hando, and Campbell, chap. 24). These questionnaires formed part of a continuing program to develop improved law enforcement monitoring information.

Data Processing and Analysis

All spatially referenced data were imported to a Geographic Information System (GIS) summarized on 5 km grid cells and analyzed using Idrisi (Eastman 1990). Distances were calculated from each village to the protected area boundary and from the center of each grid cell in the protected area to the nearest village. The size of the human population, its average annual rate of change, and size of the land area within each 5 km distance class from the protected area boundary were extracted from the GIS. Statistical analyses were performed using SYSTAT 5.0 (Wilkinson 1990). Because our analysis aimed to elucidate the influence of external factors on the distribution of large resident herbivores *inside the protected area,* statistical tests were confined to the grid cells from inside the protected area. GIS was also used as an integral component of supply and demand models (see Appendix 25.1 for details).

We explored whether wildlife and environmental data deviated from a random distribution in space (i.e., exhibited spatial autocorrelation) to (1) assess the scale of distribution patterns and (2) adjust the degrees of freedom and variances of test statistics (Cliff and Ord 1981). We calculated Moran's I as the coefficient of spatial autocorrelation (under the assumption of randomization) for grid cells inside the protected areas. As all wildlife distributions (except waterbuck) and environmental factors were significantly spatially autocorrelated, we adjusted the degrees of freedom and variances of all test statistics using correction factors derived from simulations by Bivand (1980).

RESULTS

Environmental Gradients and Wildlife

Here we discuss the spatial distribution of large resident herbivores. We summarize the background information necessary to (1) understand the large-scale patterns of the spatial distribution of large resident herbivores and (2) determine which populations or regions are key areas for large resident herbivores.

The Serengeti ecosystem lies within a rainfall gradient stretching from the southeast to the northwest (fig. 25.2). The lowest rainfall of 450 mm annually occurs in the east, in the rain shadow of the Crater (Ngorongoro) Highlands. Rainfall increases northward and westward and reaches 950 mm in the western corridor near Lake Victoria and 1,150 mm in the extreme north of the national park near the international border. This gradient was closely reflected by average NDVI for 1990 (correlation between rainfall and NDVI inside the protected area, Spearman's $r = .818$, $P < .0001$, $N = 672$). NDVI decreased sharply outside the boundary of the protected area south of the western corridor and west of Maswa GR. This is due to a decrease in woody canopy caused by extensive land clearance and fuelwood extraction and because low NDVI values are recorded during periods when cultivated land has little or no crop cover. A similar effect to the northwest of the protected area may be masked by the higher rainfall in that region. The distribution of the average vegetation index during the month prior to the 1991 wet season wildlife census was more even than that of the 1990 average but was dominated by the same southeast-northwest gradient (correlation of average for 1990 with average for the month prior to census, $r = .556$, $P < .0001$, $N = 672$).

Grassland forms the most extensive land cover throughout the study area (fig. 25.3), falling below 70% only in areas of extensive woodlands. Areas of low total woody canopy cover (0% to < 5%) are present in the southeast on the Serengeti plains and in parts of the north and west of the national park, generally coinciding with flat or gently undulating ground (correlation of relief with total woody canopy cover, $r = 0.35$, $P < .0001$, $N = 672$). Patches with more than 25% woody canopy cover occur in several distinct areas, the largest of which lie along the border between the national park and Maswa GR in the south, and in a band stretching from Ikoma to the eastern border of the national park. High relief is largely confined to parts of the west and southwest of the Serengeti and to the Crater and Loliondo Highlands in the east.

Except for buffalo, which show seasonal movements, and giraffe, most resident herbivores have small home ranges (table 25.1) compared with the size of each grid cell (25 km^2). We would therefore expect

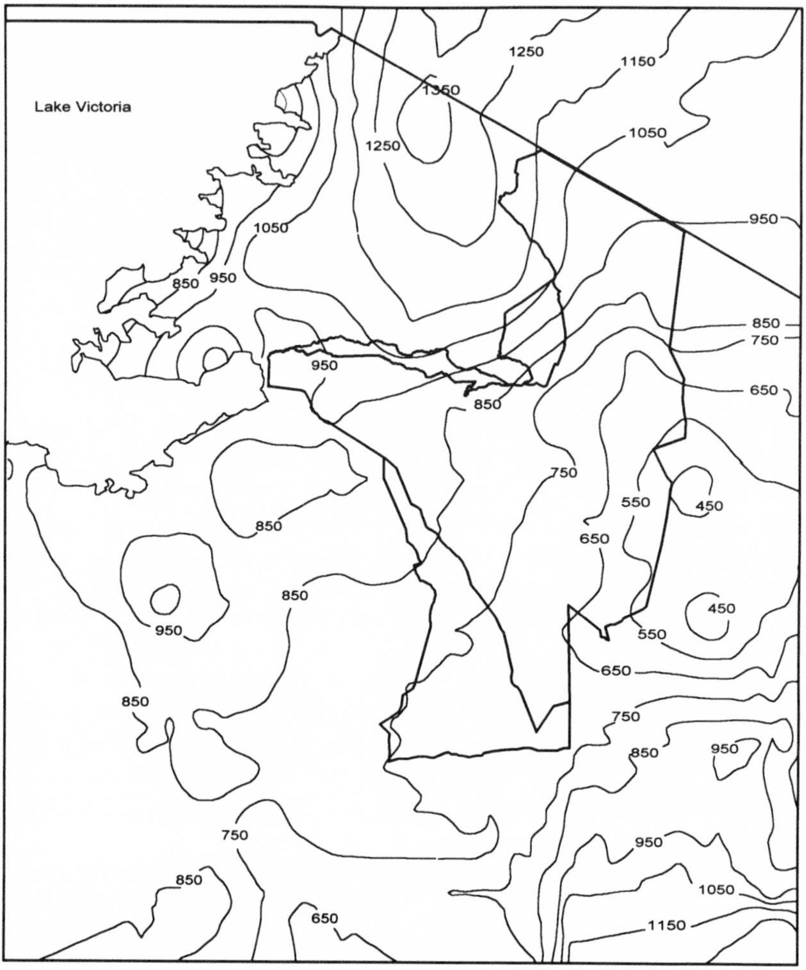

Figure 25.2 Spatial distribution of average rainfall 1970–1989. Isohyets in mm per annum.

patches of equal wildlife density to have a size equivalent to only a few grid cell units. Spatial correlograms for each species showed that with the exception of giraffe (two significant lags), species densities per cell were correlated only with the densities of immediately adjacent grid cells. Spatial interdependence therefore operated on a local scale.

Table 25.1 gives the population size of each large resident herbivore species. Figure 25.4 displays the spatial distribution of the combined densities of large resident herbivores in the ecosystem (1991). Large resident herbivores are present in the open woodland and wooded grassland habitats in the north and west outside the Serengeti plains, and their density

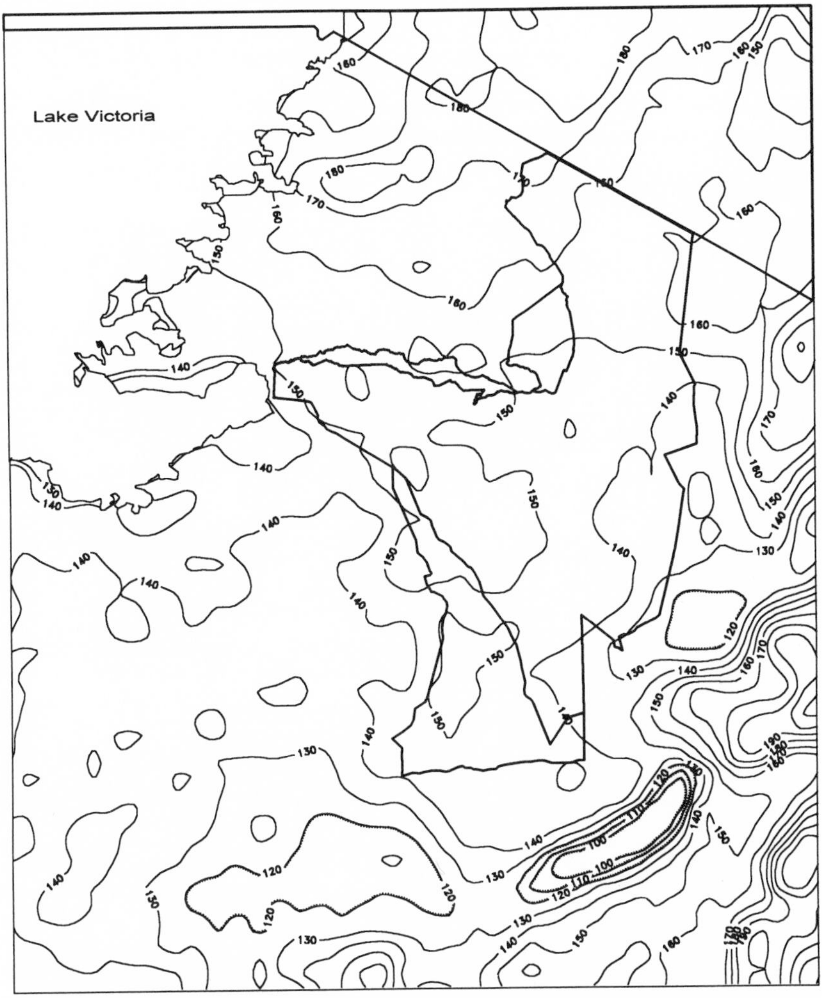

Figure 25.3 Contours showing the percentage of grass cover of the Serengeti ecosystem.

is principally dependent on rainfall (table 25.2). The Serengeti plains constitute the major wet season habitat of migratory ungulates (Maddock 1979). The only exception to the general pattern of resident herbivore distribution is Grant's gazelle, which occurs in significant numbers on the Serengeti plains throughout much of the year (Campbell 1988). Grant's gazelle is also the only species with a significant negative association with relief, implying a preference for flat, open grasslands. Several species respond significantly to rainfall and/or total woody canopy cover, suggesting specific habitat requirements. Thus, regions with moderate to high values of rainfall, relief, primary productivity, and total woody canopy

Table 25.1 Population parameters of major wildlife species within the protected area and the estimated annual offtake of wildlife by local communities in the 45 km wildlife demand zone west of the protected area.

Species[a]	Population[b]	Total grid cells[c]	Home range[d] (km^2)	Killed wildlife found by patrols[e] Number	(%)	Estimated annual offtake population	Offtake as % of estimate	Annual recruit-ment[d] (%)	Adult mortality[d] (%)
Kongoni (G)	11,716	81	0.3–10	3	0.4	872	7.4	21	10
Topi (G)	95,037	179	0.25–4	67	9.3	19,481	20.5	ca. 20	10
Buffalo (G)	64,111	49	>2,000	43	5.9	12,503	19.5	12–16	8
Warthog (G)	7,151	106	0.6–3.7	6	0.8	1,745	24.4	?	15
Waterbuck (G)	2,466	25	4.5	8	1.1	2,326	94.3	?	12
Grant's gaz. (M)	25,483	101	0.08–20	5	0.7	1,454	5.7	?	15
Impala (M)	79,098	208	2.9	78	10.8	22,679	28.7		10
Giraffe (B)	7,853	84	85–100	8	1.1	2,326	29.6	ca. 20	7
Baboon (F)	8,700	21	28	0	0	0	0	—	—
Bushbuck (MF)	(5,790)	—	<0.06	1	0.1	291	(5.0)	?	?
Reedbuck (G)	(28,320)	—	0.1–0.6	7	1.0	2,035	(7.2)	?	?
Wildebeest (G)*	1,300,000	migr.	migr.	409	56.5	118,922	9.2	15	12
Zebra (G)	146,867	migr.	migr.	64	8.8	18,609	12.7	?	10
Eland (M)	9,416	migr.	migr.	10	1.4	2,908	30.9	?	8
Other species	—	—	—	15	2.1	4,361	—	—	—
TOTAL	—	—		724	100.0	210,512	—	—	—

[a]Species feeding ecology: G, grazer; M, mixed grazer/browser; B, browser; F, fruits, seeds, flowers.
[b]1991 aerial survey estimates for the entire survey area excluding Ngorongoro Crater, uncorrected for survey bias. For estimates in parentheses see text; wildebeest from Campbell and Borner, chap. 6, rounded to the nearest 100,000; buffalo numbers recorded by the 1992 total count were 40,735 (Campbell and Borner, chap. 6); the higher figure reported here from the 1991 sample census probably results from sample error associated with a small number of very large herds.
[c]The number of grid cells in which a species was found inside the protected area.
[d]Home range sizes and recruitment rates for East African populations are from Kingdon 1982; Foster and Dagg 1972; Jewell 1972; Jarman and Jarman 1973; Sinclair 1977; Jarman 1979; Stammbach 1986; mortality in the Serengeti from Houston 1979.
[e]Numbers of carcasses recorded by a sample of patrols in Serengeti National Park.

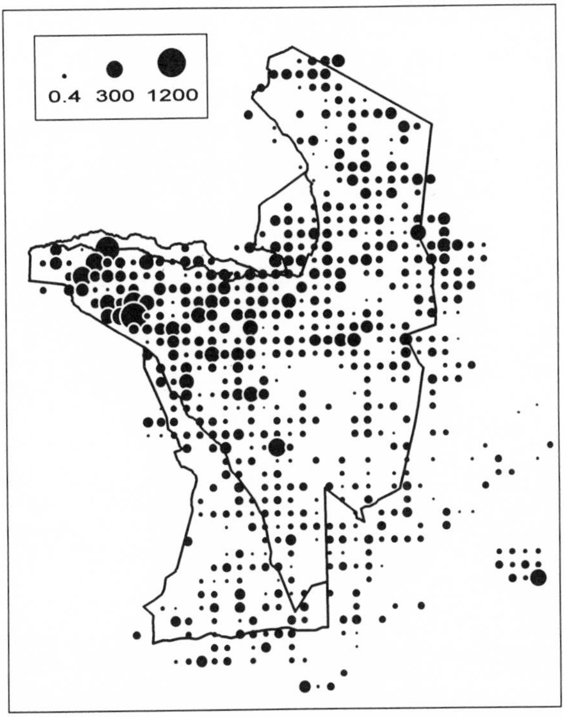

Figure 25.4 Spatial distribution of large resident herbivores, given as the density per km^2 for each 5 km grid square, from 1991 aerial survey data.

cover (all factors significantly and positively related to one another; see above) must be considered important areas for the community of large resident herbivores inside the protected area. Kongoni was the only species whose density increased with the distance of a grid cell from the nearest village (partial correlation coefficient of distance and kongoni density, $c_P = .148$, $P < .001$, holding woody canopy cover constant).

Low densities of, or an absence of, large resident herbivores were evident (1) in the northwest of the park close to the boundary, where there is no buffer zone between the park and cultivated land, and (2) in Maswa GR, which itself constitutes a buffer zone. This finding is corroborated by the highly significant increase in the percentage of area covered by animal tracks per grid cell as the distance of the grid cell from the nearest village outside the protected area increased ($c_P = .223$, $P < .0001$, holding relief, rainfall, grass, and woody cover constant). In contrast, there were no marked changes in the density of large resident herbivores outside the eastern boundary of the national park (i.e., in Ngorongoro District).

Table 25.2 Environmental and vegetation variables significantly correlated with, and the direction of their effect on, wildlife presence.

Wildlife presence	Food niche	Relief	Rainfall	NDVI for the month prior to the survey	Total woody canopy cover	Total grass cover
Animal tracks	—	Increases $p < .02$	Decreases $p < .0001$	NS	NS	Increases $p = .061$
Total resident herbivores	—	NS	Increases $p < .008$	NS	NS	NS
Topi	Grazer	NS	Increases $p = .041$	NS	NS	NS
Waterbuck	Grazer	NS	Increases $p = .04$	NS	NS	NS
Kongoni	Grazer	NS	NS	NS	Increases $p = .072$	NS
Grant's gazelle	Mixed	Decreases $p = .031$	NS	NS	NS	NS
Impala	Mixed	Increases $p = .046$	Increases $p < .007$	NS	Increases $p = .015$	NS
Giraffe	Browser	Increases $p = .02$	NS	NS	Increases $p = .04$	NS
Baboon	Fruit	NS	Increases $p < .002$	NS	NS	NS

Note: Values are *P* values of partial correlation coefficients adjusted for spatial autocorrelation between the index of wildlife presence and a particular variable while holding the four other independent variables constant. Because densities were predicted to increase (1) with rainfall and (2) for browsers and mixed grazer/browsers with total woody cover, *P* values are one-tailed for these categories. Species not listed (African buffalo and warthog, both grazers) showed no significant associations with any independent variable.

Local Communities

The human population in seven districts west of Serengeti has grown continuously since 1957 and reached a total of 1,733,958 in 1988. The considerable variation in annual rates of change between 1978 and 1988 per 5 km grid square (range -13.3% to +15.6%, SD 3.3, $N = 503$) indicated that population data summarized at a district level mask substantial changes occurring at the local level.

The maximum recorded distance of the home village of an arrested hunter from the boundary of the protected area was 45 km. This distance defined a "catchment area" surrounding the western edge of the protected area. Within this catchment area there was a total population of 1,055,910 in 1988 in 394 village enumeration areas, and average household size was 7.03. The average rate of population increase from 1978 to 1988 of 2.83% was similar to both the regional and national averages of 2.9% (table 25.3). Within the catchment area the rate varied with distance from the protected area boundary. A region of high increase occurred close to the protected areas (< 10 km) with a region of lower increase extending up to 25 km, followed by another area of above aver-

Table 25.3 Size and rate of increase of local communities west of the Serengeti in Tarime, Serengeti, Musoma Rural, Bunda, Bariadi, Maswa, and Meatu Districts, and Kalamela and Mkula Wards in Magu District, within 50 km from the boundary of the protected area.

Distance category (km)	Area (km²)	1978 population	% of 1978 total	1978 density (/km²)	1988 population	% of 1988 total	1988 density (/km²)	Mean annual % rate of increase
0 to < 5	3,429	62,302	6.92	18.17	92,767	7.80	27.05	4.06
5 to < 10	3,355	99,600	11.07	29.69	134,085	11.28	39.97	3.02
10 to < 15	3,289	111,737	12.42	33.98	136,954	11.52	41.65	2.06
15 to < 20	3,312	103,487	11.50	31.24	128,647	10.82	38.84	2.20
20 to < 25	3,338	76,319	8.48	22.86	96,913	8.15	29.04	2.42
25 to < 30	3,420	68,567	7.62	20.05	92,301	7.76	26.99	3.02
30 to < 35	3,444	92,300	10.26	26.80	129,836	10.92	37.70	3.47
35 to < 40	3,422	97,844	10.87	28.59	127,496	10.72	37.26	2.68
40 to < 45	3,449	83,648	9.30	24.25	116,911	9.83	33.90	3.40
45 to < 50	3,325	104,041	11.56	31.29	133,241	11.20	40.07	2.50
0 to 50	33,782	899,845	100	26.64	1,189,151	100	35.2	2.83

age increase. However, the rates of population change per village were unevenly spread within each distance class. Distinct regional patterns emerged, with areas of substantial decline and areas of rapid population increase most noticeably occurring in the southwest and northwest. The 1988 population distribution and 1978 to 1988 rates of change are displayed in figures 25.5a and b. Some settlements are on the boundary of the protected area, suggesting that encroachment by cultivation, fuelwood collection, livestock grazing, and other activities will become an increasing concern to the management of the protected area. An intriguing possibility is that rural migration and a reduction of the human population northwest of the protected area might be related to local depletion of wildlife (cf. fig. 25.4 with fig. 25.5b) and that high rates of increase in some areas close to the boundary may be related to the greater availability of wildlife in areas suitable for hunting.

Hunting and Wildlife

We developed a series of models to explore three related topics (fig. 25.6). We first consider how many people are involved in illegal wildlife meat hunting (referred to below simply as hunting), where they come from, what they hunt, and how much wildlife they kill per year, in order to provide a rough estimate of the size of the market and total demand for wildlife products by the communities west of the protected area. We then use the locations of villages and ranger posts and the physical characteristics of the protected area to estimate the spatial distribution of hunting, assuming that hunters seek to *maximize returns by minimizing the cost of hunting* (in terms of logistics, travel, and possible legal proceedings if arrested). This "profitability" model of hunting identifies areas where

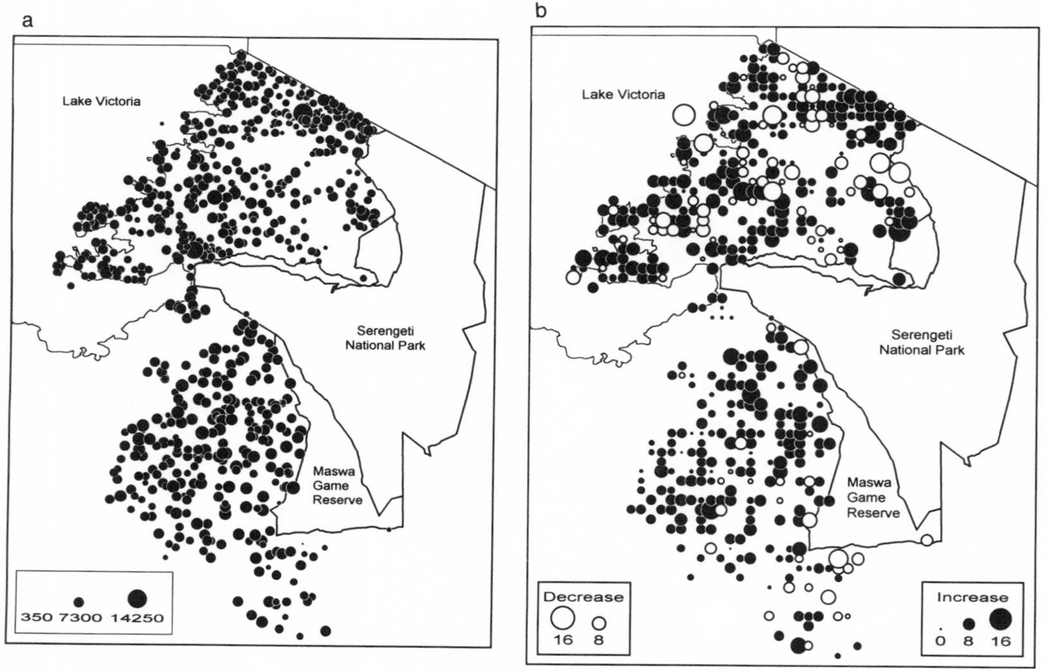

Figure 25.5 Spatial distribution of size and rate of change in the size of villages in seven districts west of the Serengeti. (*a*) Population size in 651 village enumeration areas in 1988. (*b*) Average annual percentage rate of population change, 1978–1988 summarized by 5 km grid square.

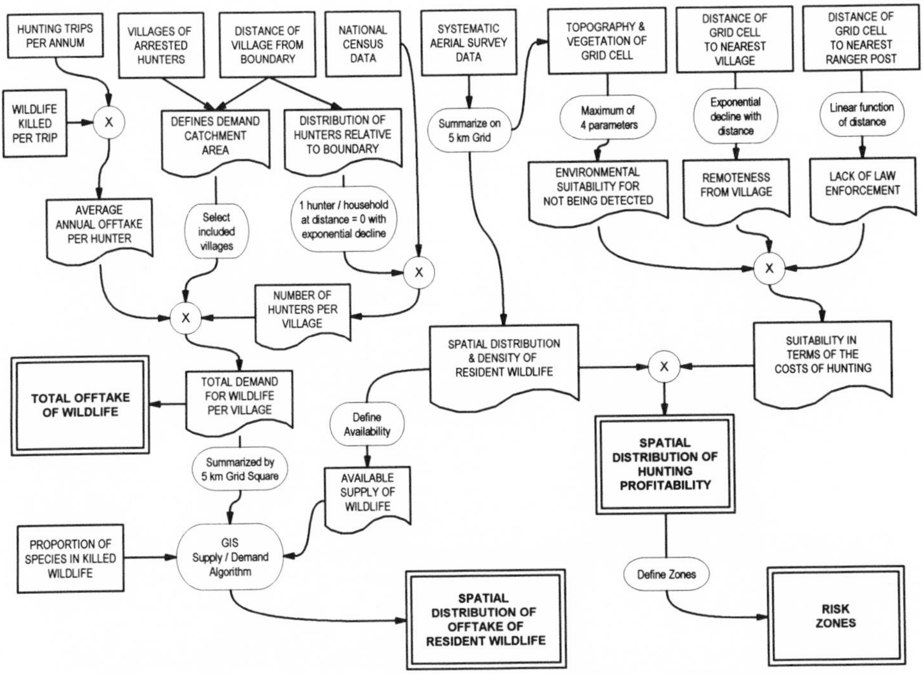

Figure 25.6 Flow diagram illustrating the "demand" and "profitability" models of wildlife offtake.

hunting should be most *rewarding,* and therefore identifies *centers of potential impact.* We then develop a "demand" model that ignores the cost of hunting but takes into account the estimated spatial distribution of the demand for wildlife in the communities outside the protected area as derived from the distribution of local communities.

Wildlife Meat Hunters. The majority of hunters in the sample arrested by anti-poaching patrols came from villages close to the protected area boundary, the proportion of hunters declining exponentially with the distance of their home village from the protected area (fig. 25.7). Following discussions with wardens and others involved in wildlife management both inside and outside the national park, we assumed that at zero distance from the protected area boundary an average of one person per household engages in hunting. According to figure 25.7 this value should decline exponentially to zero at the maximum distance from the protected area boundary of home villages from which hunters were arrested. For each village we estimated the total proportion of hunters per household based on the distance of the village from the protected area boundary and then multiplied this figure by the number of households

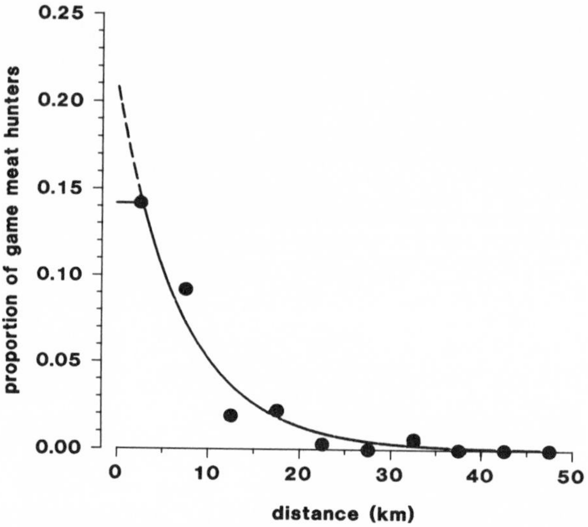

Figure 25.7 Proportion of local residents per household west of the Serengeti engaged in meat hunting in the protected area as a function of distance of the home village from the boundary of the protected area: $y = 0.208 \times e^{(-0.139 \times x)}$, $r^2 = .97$, slope $t_8 = 12.4$, $P < .001$. The solid line represents values used in calculating the proportion of hunters per village.

(villages closer than 2.5 km to the boundary were assumed to lie at 2.5 km, figure 25.7). On the basis of 1988 National Census data, this gave an estimated number of hunters per average household within the catchment area of 0.21 (median 0.084, SD 0.27, $N = 394$ villages) and a total of 31,655 people originating from areas west of the Serengeti engaged in hunting (table 25.4).

Patrol records showed that a total of 548 people, all male, were encountered with a total of 724 killed animals (see table 25.1), an average of 1.3 animals killed per person per hunting trip. In sample of arrested hunters questioned by anti-poaching patrols, the number of hunting trips per annum varied from 1 to 24 ($N = 84$). As the charge against arrested hunters may partly depend on their reply to this question, these figures probably represent an underestimate of the actual number of trips. Magombe and Campbell (1989) estimated 5 trips per annum from replies to questions on the time spent hunting and the average number of days per hunting trip, and we used this figure. This results in an average annual offtake per hunter of 6.5 animals. Total wildlife offtake by all villages within the catchment area can then be estimated as the number of hunters per village times the average annual harvest per hunter. We calculated that a total of 75,000 resident and 135,000 migratory wildlife were taken annually by hunters.

Table 25.4 Modeled number of meat hunters originating from areas west of the Serengeti ecosystem at increasing distance from the boundary of the protected area.

Distance class	Estimated number of hunters, 1988	Percentage of total population, 1988	Percentage of total meat hunters, 1988	Estimated number of hunters, 1978	Annual % rate of increase of hunters, 1978–1988
0 to < 5	12,987	14.00	41.0	8,443	3.99
5 to < 10	9,125	6.81	28.8	7,263	2.96
10 to < 15	5,166	3.77	16.3	4,070	2.01
15 to < 20	2,553	1.98	8.1	2,067	2.22
20 to < 25	908	0.94	2.9	753	2.39
25 to < 30	415	0.45	1.3	322	3.03
30 to < 35	278	0.21	0.9	215	3.32
35 to < 40	143	0.11	0.5	102	2.84
40 to < 45	80	0.07	0.3	59	3.47
TOTAL	31,655	2.66	100.0	23,294	3.11

Species Targeted by Hunters. The preferred method of hunting is snaring (Arcese, Hando, and Campbell, chap. 24). Large resident and migratory herbivores either wander into or are driven into snare lines. Other wildlife not specifically targeted by hunters are also caught, some of them in substantial numbers (Hofer, East, and Campbell 1993). We estimated species-specific offtake by hunters of herbivore species for which aerial survey estimates of density were available. We investigated hunting selectivity by regressing the $\log_{10}$ of the percentage of each species in the sample of wildlife killed by hunters on the $\log_{10}$ of its percentage of the total wildlife density derived from aerial censuses (fig. 25.8). As the slope of 0.82 ± 0.12 was not significantly different from 1 ($F_{1,9} = 2.16$, NS), illegal hunting within the entire protected area was in general considered to be nonselective, a result consistent with the largely passive nature of snaring. However, more concentrated hunting effort in specific localities is likely to result in the relative overexploitation of some species (Arcese, Hando, and Campbell, chap. 24).

Estimated annual offtake as a percentage of the population size estimated from aerial surveys varied among species (see table 25.1). The estimated relative offtake for waterbuck is clearly nonsensical. However, as aerial censuses of waterbuck are likely to underestimate actual population size (Campbell and Borner, chap. 6), it is currently impossible to provide a more reasonable figure.

The "Profitability" Model of Wildlife Offtake. We modeled the suitability for hunting of a given grid cell inside the protected area as a function of (1) its distance from the nearest village and (2) the likelihood of

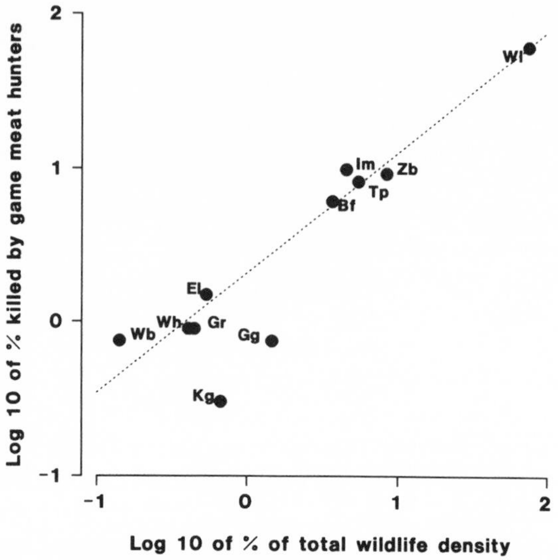

Figure 25.8 Percentage of each wildlife species killed by meat hunters as a function of the percentage of that species in the total wildlife density estimated by the 1991 aerial survey ($\log_{10} y = 0.22 + 0.82 \times \log_{10} x$, $r^2 = .83$, $F_{1,9} = 43.9$, $P < .001$). Bf, African buffalo; El, Eland; Gg, Grant's gazelle; Gr, Giraffe; Im, Impala; Kg, Kongoni; Tp, Topi; Wb, Waterbuck; Wh, Warthog; Wl, Wildebeest; Zb, Zebra.

its being patrolled by anti-poaching forces. The resulting index, $P_{H,}$ scaled to vary between 0 and 1, expressed the suitability of a location for hunting solely in terms of the costs of hunting, that is, costs due to legal action if arrested and the logistics of traveling to and from the hunting area. P_H was calculated as $P_L \times P_T \times P_{R,}$ where P_L is an index of the likelihood of a cell *not* being patrolled due to distance from the nearest ranger post, P_T is an index of the likelihood of *not* being detected in a cell due to topographic and vegetation features, and P_R is an index of the desirability of a cell for hunting based on the distance of the cell from the nearest village.

P_L was defined as a linear function of the distance of a grid cell from the nearest ranger post and scaled to the maximum distance. P_T was determined as the maximum of (1) relief, (2) total woody canopy cover, (3) quantity of riparian woodland or forest cover, and (4) rocky and stony ground, all factors scaled between 0 and 1. We hypothesized that an increase in the value of any of these factors would make the detection of hunters by patrols more difficult. P_R was set to decline exponentially with increasing distance, $V_{C,}$ of the grid cell from the nearest village, and scaled by setting P_R to 1 at distance 0 and to 0.1 at a distance of 30 km, or 2 days' walk, ($P_R = e^{[-0.07675 \times V_c]}$). No account was taken of the presence of roads or anti-poaching tracks, which may decrease the suitability of a cell for hunting by increasing the risk of arrest, nor of the presence of major rivers, which form a barrier to anti-poaching forces.

The suitability for hunting was highest in the central parts of Maswa GR, and high over most of the area of the proposed Ikorongo GR (fig. 25.9). Substantial areas with high values of P_H also occurred inside the national park: (1) south of the eastern end of the proposed Grumeti GR, (2) east of the southern end of the proposed Ikorongo GR, (3) in the northwest, and (4) in smaller pockets in the western corridor.

We then created an index of *the profitability of hunting* for each grid cell by multiplying its P_H with the total resident herbivore density in the cell. We called this the "profitability" model of hunting, as it describes those areas where hunting should be *most rewarding*. Accordingly, hunting would be most rewarding in the western corridor, and less rewarding in areas beyond an arc extending from southeast of the proposed Ikorongo GR to the midpoint of the border between the national park and Maswa GR (fig. 25.10).

The "Demand" Model of Wildlife Offtake. We estimated the distribution of wildlife offtake based on (1) the estimated demand for wildlife products per village (see section on wildlife meat hunters above) and (2) the "potential supply" of wildlife to hunters expressed as a percentage of the total wildlife density recorded per grid cell (see appendix 25.1 for details). This "demand" model took into account where outside the protected area most of the wildlife is wanted, but ignored the cost of hunting. It did, however, consider the remoteness of a cell in that it satisfied the demand by hunting first close to the home village and going farther only if the demand could not be met by wildlife in the grid cells close to the home village.

In the model simulations, total demand for resident herbivores decreased (fig. 25.11) as the distance of villages from the boundary of the protected area increased, as implied by the decline in the proportion of hunters coming from distant areas. Demand was predicted to peak (1) in Tarime District at the northwest corner of the national park, (2) in Bunda District in the area west of the proposed Grumeti GR, (3) in Bariadi District where the boundaries of the national park and Maswa GR merge, and (4) in Meatu District at the southwestern corner of Maswa GR. Peaks of total offtake of large resident herbivores (fig. 25.11) were predicted to occur in the western corridor, especially close to the boundary with the proposed Grumeti GR. High offtake was also predicted to occur near the northern section of Maswa GR. In the north, offtake would be concentrated in the central section of northern Serengeti.

The predicted pattern of offtake was calculated for each of eight resi-

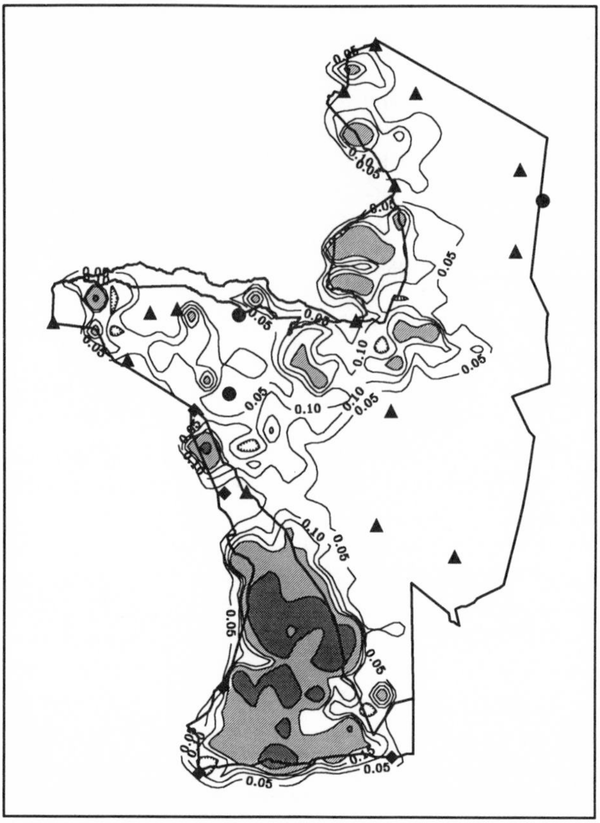

Figure 25.9 Spatial distribution of the index of the estimated suitability for hunting in a particular location P_H: the higher the index, the lower the cost of hunting (triangles, established National Park ranger posts; circles, National Park ranger posts planned and under construction; diamonds, existing Maswa GR ranger posts). Regions with a high suitability for hunting are shaded (>0.4 darkest, >0.2 second darkest). Regions with lower suitability for hunting are indicated by contours (0.05, 0.1, and 0.15).

dent herbivore species separately and combined to create an estimated distribution of the total offtake of residents. This resulted in some offtake being predicted for areas where it is thought unlikely to occur at present—for example, parts of the Serengeti plains—an effect due to the differing spatial availability of each species.

Zoning Hunting Impact. Several distinct zones of potential hunting impact emerged when we compared the estimates of the costs of hunting (index of suitability) with wildlife distribution and law enforcement activities. We defined five risk zones (fig. 25.12):

1. *Overexploited areas:* regions suitable for hunting ($P_H \geq 0.1$) but

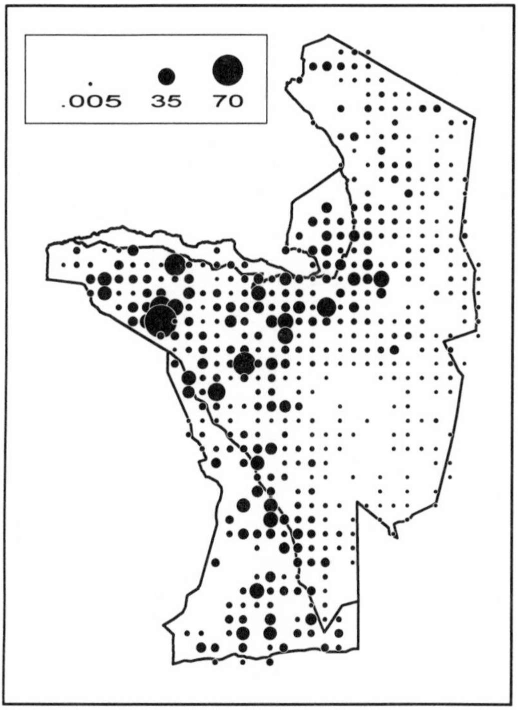

Figure 25.10 Estimated spatial distribution of the profitability of hunting expressed as total resident wildlife density per km² at risk: the higher the density, the more rewarding hunting is predicted to be.

containing few wildlife (density of all large resident herbivores < 1/km²). Examples are areas in the northwest of the park along the boundary, parts of the proposed Ikorongo GR, and large parts of Maswa GR.

2. *Endangered areas:* localities suitable for hunting ($P_H \geq 0.1$) containing medium to high densities of large resident herbivores (≥ 10/km²). Future hunting may concentrate in these areas. Examples are the western and central sections of the western corridor, the northwestern corner of Maswa GR and adjacent areas in the park, and the central and eastern sections of the proposed Ikorongo GR.

3. *Areas of escalation and conflict:* localities containing high wildlife densities (≥ 10/km²) but less suitable for hunting because of their greater chance of being patrolled (P_H 0.005–0.1). These are areas where hunters and patrols have a higher chance of meeting and where hunters may enter in larger and better armed groups to reduce the chances of arrest.

4. *Areas of future expansion:* areas suitable for hunting but not yet or only marginally exploited because demand can currently be satisfied by the supply in safer grid cells closer to the home village; possibly endangered in the future.

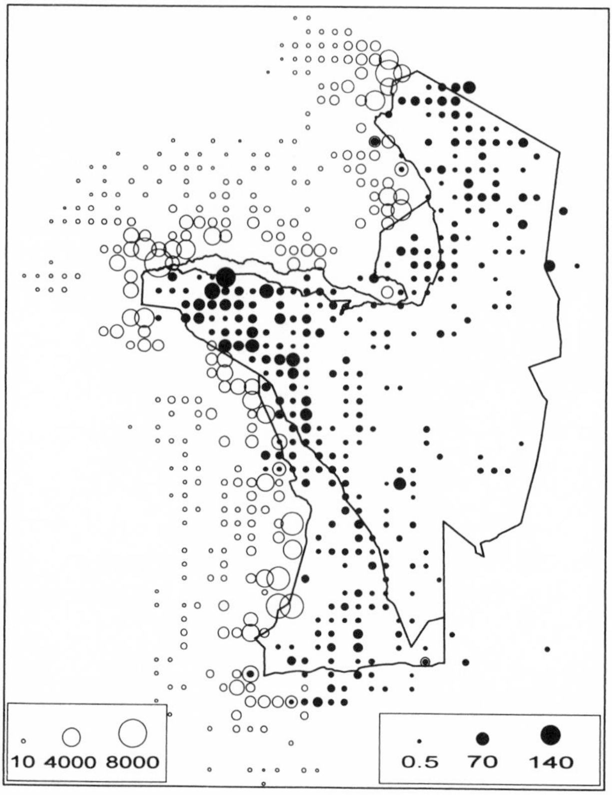

Figure 25.11 Predictions of the "demand" model. Open circles, total estimated demand for large resident herbivores (numbers per annum per 5 km grid cell) by local communities in the 45 km catchment area adjacent to the western boundary of the protected area. Solid circles, offtake of resident herbivores predicted by the demand model (expressed as density/km²).

5. *Untouched areas:* areas where wildlife presence is highly seasonal, suitability for hunting is low ($P_H < 0.005$), and the high chance of detection by patrols (or tourists) outweighs benefits from hunting. This zone largely encompasses the Serengeti plains.

Testing the Models. Our models are principally prospective as they seek to establish a framework that (1) identifies information critical to accurate assessments of hunting impact and (2) can be used to provide more accurate estimates with an improved database. In principle, most of our assumptions are testable, as are some of the implications of the models. Here we summarize our attempts at validating some of the assumptions and testing some of the implications.

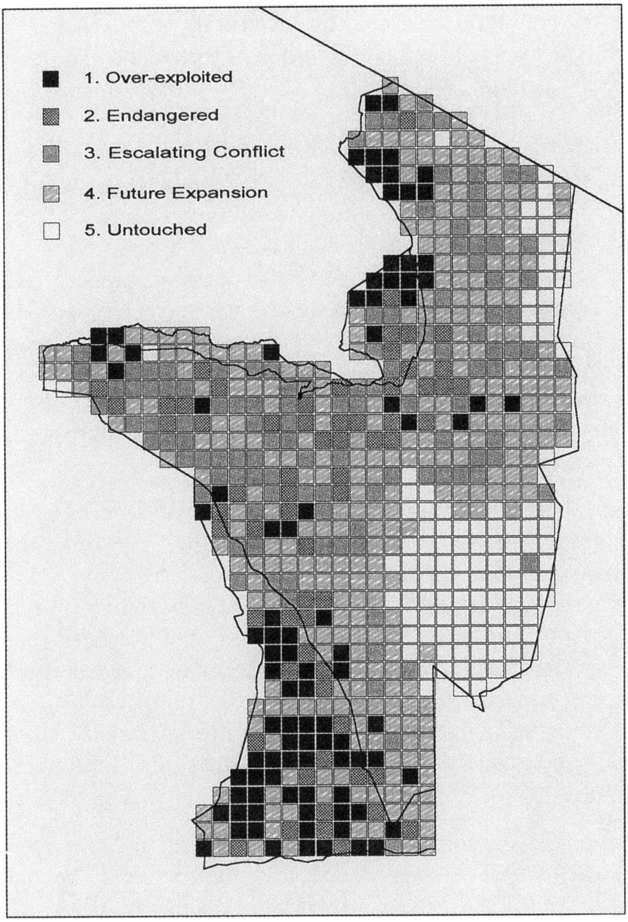

Figure 25.12 Zones of interaction of illegal meat hunting, law enforcement activities, and wildlife distribution. Zone 1, *overexploited,* comprises grid cells suitable for hunting (in terms of the costs of hunting) but containing very low resident wildlife densities. Zone 2, *endangered,* comprises localities suitable for hunting that contain medium to high densities of resident herbivores. Zone 3, *escalating conflict,* comprises cells containing medium to high wildlife densities but which are less suitable for illegal meat hunting due to increased chances of arrest. Zone 4, *future expansion,* comprises localities that are less suitable for game meat hunting due to the high chance of arrest and/or where wildlife densities and habitat constraints indicate lower probabilities of hunting. Zone 5, *"untouched,"* comprises largely areas where wildlife presence is highly seasonal, suitability for game meat hunting is low, and the chance of detection by patrols greatly outweighs the possible benefits from hunting.

We tested our model of wildlife offtake by comparing its estimate with estimates of wildlife offtake obtained independently by personnel of the Serengeti Regional Conservation Strategy project during workshops held with eleven villages in Serengeti District northwest of the national park (SRCS 1991; M. Maige, pers. comm.). For these villages our model estimated a total annual offtake of 11,564 animals of all wildlife species. The number of animals killed annually, estimated by the villages themselves, was 11,871. The close agreement between the two figures indicates that (1) the model provides a reasonable order of magnitude estimate of offtake, and (2) the assumption that on average one person per household engages in hunting at distance zero from the boundary of the protected area is reasonable.

If our cost model of hunting profitability is realistic and if hunting in the past has already had an appreciable effect on resident wildlife, then we would expect that

1. there should be a measurable decline in total resident wildlife density over time in the overexploited zone (risk zone 1) while other zones should show no such decline;
2. there should be a negative relationship between a grid cell's index of suitability for hunting and its total resident wildlife density in zones 1 to 4 (the Serengeti plains, zone 5, is not a useful area to test this);
3. for a particular species, the stronger the negative relationship between the index of suitability for hunting and its density, the less frequently the species should occur in the sample of wildlife recorded as killed by hunters.

Campbell and Borner (chap. 6, table 6.7) calculated average densities of each resident species for risk zones 1 (overexploited), 2 (endangered), and 3 (escalation of conflict) from wildlife censuses in 1988, 1989, and 1991. There was a strong, significant decline in resident wildlife densities from 1989 to 1991 in risk zone 1 (Wilcoxon matched sample test, $z = 2.37$, $n = 8$, $P = .014$, one-tailed) while in zones 2 and 3 there was a nonsignificant increase in resident wildlife densities (zone 2: $z = 1.86$, $P = .063$; zone 3: $z = 1.69$, $P = .091$, both two-tailed).

We tested the second prediction by correlating total resident wildlife density with the index P_H of suitability for hunting for grid cells belonging to zones 1 to 4. There was a significant decline of wildlife density with increased suitability for hunting per cell (Spearman $r = -.115$, $z = 2.4$, $P = .009$, one-tailed).

We tested the third prediction by correlating P_H with the species-specific density per grid cell and calculating residuals for each species from the linear regression in fig. 25.8. Negative residuals indicate "underselection" by hunters, while positive residuals indicate "overselection."

Species with negative correlation coefficients (for suitability versus density) had significantly lower residuals, that is, were underselected, compared with species with positive correlation coefficients (mean ± SEM -0.16 ± 0.15 versus $+0.26 \pm 0.07$, Mann-Whitney $U = 0$, $P < .013$, one-tailed, fig. 25.13). The confirmation of each prediction suggests that the models with their current estimates form a useful means of looking at human-wildlife interactions, and have the right order of magnitude in value and distribution.

DISCUSSION

The main findings of our study are:

1. Large resident herbivores are distributed throughout the protected area with the exception of the Serengeti plains in the southeast, the only area not experiencing hunting pressure.

2. Distribution patterns of resident herbivores are primarily determined by rainfall, woody canopy cover, and a topographic/edaphic gradient.

3. There is a large human population adjacent to the western bound-

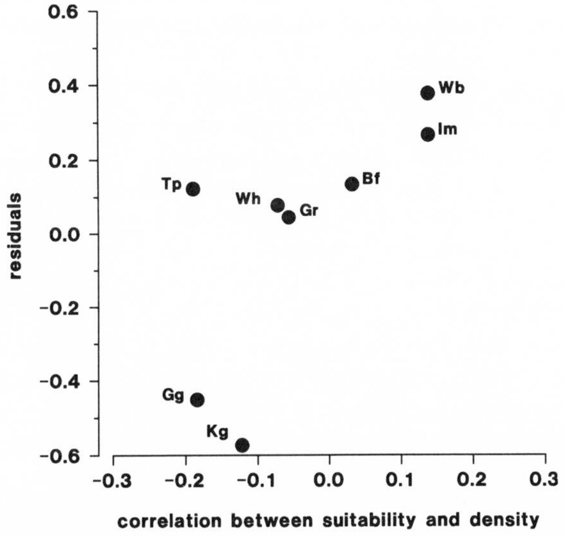

Figure 25.13 The relationship between the overall degree of selection of each prey species by hunters as indicated by the residuals from the relationship calculated in fig. 25.8 and the degree of correlation between the index of suitability for hunting (P_H) and the species-specific density per grid cell. Species whose densities are negatively correlated with P_H were underselected by hunters compared with those with positive correlations with P_H (see text). Species codes as in fig. 25.8.

ary of the protected area, with greater than average rates of increase within 10 km of the protected area boundary.

4. A large number of people currently engage in illegal meat hunting, and these people remove approximately 75,000 resident herbivores and 135,000 migratory herbivores annually from the protected area.

5. Gaps in the distribution of large resident herbivores, particularly in the west and northwest of the protected area, coincide with areas estimated to be highly suitable for hunting, suggesting that past hunting has already led to local depletion of all resident wildlife, and possibly to subsequent rural population migration and local reductions in settlements.

6. Existing and proposed Game Reserves acting as buffer zones between local communities and the national park will continue to suffer from unsustainable hunting pressure and will not prevent wide-scale hunting throughout the western and northern sections of the national park.

The models developed here are a first step toward a more comprehensive exploration of the effects of meat hunting on wildlife in the Serengeti ecosystem. Inevitably, there are shortcomings in the procedures and data used in our predictions. In the following sections, we discuss the basis of distribution patterns of large resident herbivores, assess the validity of the assumptions of the models, discuss the effects of hunting on wildlife, and point out some consequences of our predictions for management.

Spatial Dynamics of Resident Herbivores

The eight species of large resident herbivores we studied cover a wide spectrum of feeding habits, habitat preferences, body sizes, water requirements, and niche breadth among African herbivores. Resource partitioning (Vesey-Fitzgerald 1960; Lamprey 1963), water requirements (Western 1975), and possibly predation (Sinclair 1985) have been shown to modulate the distribution of herbivores in African ecosystems. Three factors principally determine coarse-scale patterns of herbivore distribution (Lamprey 1963; Ferrar and Walker 1974; McNaughton and Georgiadis 1986): (1) differentiation of feeding strategies, principally categorized by the proportion of woody species in the diet (the grazer-browser continuum); (2) habitat, typically defined as a gradient ranging from pure grassland to woodland or forest, and in our study represented by the percentage of land covered by grass versus woody canopy; and (3) a topographic/edaphic gradient, here represented by a measure of relief. In the Serengeti, feeding strategies vary from grazing (buffalo, kongoni, topi, warthog, waterbuck) to species that seasonally change the extent of their browsing (Grant's gazelle, impala) to pure browsing (giraffe). Digestibility of the diet (McNaughton 1987) and the energetic costs associated with specialized diets (Murray 1991) vary between herbivores, and in conjunction with the mineral heterogeneity of plant matter (McNaughton 1988, 1989, 1990), exert a considerable influence on the local distribution of herbivores, its patchiness, and its seasonal variation (McNaughton 1988, 1990; Murray 1991). In our study, these factors can be considered to be only partially represented by rainfall, grass and woody cover, vegetation index, and relief. Only by considering each factor explicitly will it be possible to disentangle the contributions of past exploitation by hunters and the ecology of each species to the current distribution patterns of some species, such as African buffalo, of which a significant proportion has been exterminated from the northwest of the national park (Dublin et al. 1990); kongoni, whose densities were positively related to woody canopy cover and the distance from the nearest village; and warthog, whose niche breadth (Ferrar and Walker 1974) may lessen the species' dependence on specific environmental gradients.

The substantial influence of plant mineral concentration on the spatial distribution of herbivores has only recently been fully appreciated (McNaughton 1988, 1989, 1990) but has important implications for conservation research and management decisions. By beef cattle standards, large areas of the Serengeti grassland are deficient in calcium, magnesium, phosphorus, and sodium (McNaughton 1989, 1990). A comprehensive mapping of mineral concentrations in plant matter would greatly assist in identifying areas in the Serengeti *critical for the successful reproduction of large resident herbivores* (McNaughton 1990), as lactating and pregnant females and young animals require higher concentrations of minerals than other age-sex classes (McDowell 1985).

There is a long-standing debate in the geographic and biological literature on how to identify the optimal spatial resolution to describe a particular process (Cliff and Ord 1981; Greig-Smith 1983; Haining 1990). Our records of the distribution of wildlife were based on the results of one aerial survey analyzed by 5 km UTM grid square. In an ideal situation one would require a representative picture of wildlife distribution derived from a series of surveys covering both wet and dry seasons. If seasonal movements of resident herbivores greatly exceeded the area of a grid cell, some of our analyses would become unreliable on a local scale, while patterns on a coarser scale should be unaffected. One indicator of the scale of local movement is the size of home ranges of social units. Information on home range sizes from the literature (see table 25.1) suggests that the area of 25 km^2 per grid cell is adequate for all resident herbivores except for African buffalo and possibly giraffe. Spatial autocorrelation of specific resident herbivore densities showed that they were significantly correlated only with those of immediate neighboring grid cells. Our choice of grid size was the minimum represented by the aerial survey data and was sufficiently small to preserve the patterns of variation in animal distribution as well as environmental and vegetation gradients and also provide an acceptable resolution for model predictions.

Models of Hunting

The models considering the demand for wildlife products by local communities outside the protected area hinged on three parameters: (1) the estimated number of hunters, (2) the number of wildlife killed per hunter per hunting trip, and (3) the number of hunting trips per annum. The estimated number of hunters is based on a relationship between the proportion of arrested hunters and the distance of their home villages from the protected area boundary, with the assumption of an average of one person per household being involved in hunting activities at 0 km from the boundary. The number of animals killed per trip probably represents a minimum estimate. It is unlikely that anti-poaching patrols arrest hunters only at the end of their trips (Magombe and Campbell [1989] found that in the western corridor hunters were arrested after a mean of 2.8 days, while the average stated length of a hunting trip extended to 5.2 days). The figure of 1.3 animals per hunter may therefore be low, and our models may underestimate the "true" extent of hunting. The number of hunting trips per annum is difficult to assess with confidence. In this case, replies to the question "how many hunting journeys in a year?" were not used because it was felt that they might be biased. The true figure is likely to vary (1) seasonally, (2) annually according to socioeconomic conditions, (3) regionally, and (4) with distance of home villages from the protected area boundary and from market centers. If the replies given to the question are taken at face value and the numbers of animals per hunter are corrected for the full length of a hunting trip, the annual harvest per hunter remains the same. Monthly arrests provide an indication of hunting effort but suffer from an uneven temporal and spatial distribution of law enforcement effort (Arcese, Hando, and Campbell, chap. 24). Do hunters from different areas prefer different wildlife species, or vary in the degree to which they depend on hunting? What are the economics of meat hunting, and could alternative sources of protein (beef, poultry, livestock, fish) replace wildlife? Are there local differences in dietary preferences that affect how feasible these alternatives might be?

Our confidence in the assumptions of the model based on estimated demand is enhanced by the similarity of the model's estimates and the independent estimates of offtake provided by a sample of villages (SRCS 1991). Other tests carried out on the models showed that: (1) the zoning of risks due to hunting of resident herbivores is realistic and that significant declines in wildlife density have indeed occurred within the zone at greatest risk; (2) the density of resident herbivores decreased with an increase in the index of suitability for hunting; and (3) expectations of over- and underselection by hunters were confirmed by law enforcement patrol and aerial survey data.

Our models assume that meat hunting is the exclusive province of local communities outside the boundary of the protected area. However, hunting by communities living inside the national park may become a serious problem if unchecked, as settlement sizes will expand due to the growth of existing staff villages attached to tourist lodges and national park headquarters, as well as planned staff villages and an expected influx of casual labour associated with the development of new tourist facilities.

Temporal and Spatial Trends in Hunting

Although we calculated offtake on an annual basis, hunting may be a seasonal activity for at least some of the local communities, as reflected by the seasonal changes in arrests by anti-poaching patrols (Arcese, Hando, and Campbell, chap. 24). With additional data on movements of hunters and patrols, it should become possible to track seasonal changes in hunting effort and examine seasonal changes in the effects of hunting. These data would also assist in elucidating how resident herbivores respond to changes in hunting pressure and human disturbance.

The population increase in communities near the boundary of the protected area between 1978 and 1988 outstripped the average rate of increase in the seven districts considered, suggesting that immigration into areas close to the boundary forms a substantial component of population development. This raises a number of important socioeconomic questions: Is this change merely a consequence of the abandonment of *ujamaa villagization* of the late 1960s and early 1970s, when people were sometimes forcefully moved? Alternatively, do people migrate toward the boundary of the protected area in order to improve their nutritional and economic status (1) by meat hunting, (2) because the soil is better for cultivation, or (3) due to increased availability of land, fuelwood, and other natural resources? How important is wildlife as a buffer in times of economic hardship? As hunting appears to be an exclusively male occupation, what alternative sources of occupation are available? What do men do when they do not hunt? If people move *away* from areas that lie at some distance from the boundary of the protected area, what measures are available to improve the nutritional and economic status of those communities? The movements of people between 1978 and 1988 from the northwestern border of Serengeti (an area with a significant decrease in resident wildlife numbers) to areas farther south but still adjacent to the protected area boundary (see fig. 25.5b) suggest that local availability of wildlife meat plays an important role in rural migration.

The "profitability" model was based on an index of a location's suitability for hunting that reflected the putative cost of hunting. This index is largely time-independent and can therefore be used to reconstruct patterns of hunting in the past as well as to predict the future. Gaps in the distribution of resident wildlife coincide with areas estimated to be highly suitable for meat hunting, suggesting that past hunting has already led to local depletion of resident wildlife. A striking example is the extermination of buffalo from the northwest in the late 1970s and early 1980s (Dublin et al. 1990; Sinclair, chap. 1; Campbell and Borner, chap. 6). A comparison of the predictions of the "profitability" (fig. 25.10) and "demand" models (fig. 25.11) suggests that in the future hunters will have to operate farther east and beyond former hunting grounds closer to home to satisfy demand—that is, hunt in locations that are both farther away from the home village and more likely to be checked by anti-poaching patrols.

In order to evaluate some of these predictions, data are required on the spatial distribution of patrolling effort and locations where hunters operate. Systematic surveys of the distribution of hunting effort (e.g., by location of current and old hunter camps) are required in addition to data on the locations of arrests and sightings of hunters by patrols. Anecdotal information suggests that hunters have already extended their hunting grounds, especially in the central sections of the north of the national park where the "discrepancy" between the predictions of the profitability and demand models is particularly conspicuous. Here, an increase in the number of encounters between hunters and anti-poaching patrols is likely and should lead to an escalation of conflict between hunters and patrols, as hunters will travel in larger, better armed groups in order to reduce the chance of being caught. It appears that this trend has already been observed in the north.

The Impact of Hunting on Wildlife

Table 25.1 compares information on recruitment and mortality with estimated total annual offtake. The heterogeneity of the sources, the unknown confidence limits for estimates of recruitment, mortality, and offtake, and the incompleteness of the information make any conclusion preliminary and tentative. These problems also show that systematic studies of basic population processes are urgently needed for a number of species. Assessment of the impact of hunting on the large resident herbivores may be complicated by seasonal changes in hunting intensity and seasonal movements of migratory herds. Hunting of wildebeest, the most commonly killed herbivore, is strongly seasonal and largely confined to periods when the herds leave the Serengeti plains during the dry season. Intensive hunting of the migratory herds has complex consequences for resident herbivores. The presence of wildebeest in an area may result in increased hunting effort. The higher number of snares and traps might increase the chances of mortality of residents, but the large number of migratory herbivores that get caught (and effectively "block" snares) might also decrease the chances of mortality of residents compared with areas without wildebeest. Snares, however, are often left behind and thus catch wildlife even after hunters have gone. The magnitudes of these effects probably depend on the movements of the herds, determined by the distribution and amount of dry season rainfall (Maddock 1979), and the ability of hunters to switch effort between areas, and thus they probably vary between years.

It is clear that the predicted offtake affects different herbivores in a variable manner (table 25.1). In some species, such as Grant's gazelle and kongoni, estimated offtake is small compared with recruitment or natural mortality, and hunting is expected to have little influence on population dynamics. This is presumably a consequence of the distribution patterns of both species, kongoni being restricted to sections inside the protected area far away from villages and Grant's gazelle preferring habitat predominantly occurring in the east, well away from the western boundary of the protected area. In other species (buffalo, giraffe, impala, topi, warthog, eland) offtake must be considered high and, if realized over a number of years, *unsustainable.*

Assessment of the effect of hunting on the population dynamics of the most commonly killed herbivore, the wildebeest, is complex. The sex ratio of animals killed by hunters was biased toward males (88%–98%, Georgiadis 1988; Hofer, East, and Campbell 1993), potentially reducing the impact of hunting. Furthermore, intensive hunting of wildebeest early in the dry season precedes the period when intraspecific competition for scarce high-quality dry season forage would lead to increased mortality (Sinclair 1979, 1985), and thus reduces the intensity of intraspecific competition. In addition, annual removal by hunters of 8% (Hofer, East, and Campbell 1993) of the spotted hyena population, the most important predator of wildebeest, has led to a decline in the hyena population (Hofer and East, chap. 16) and thus reduced the contribution of predators to wildebeest mortality. The main Serengeti predators (hyenas and lions) formerly removed roughly 2%–3% of the wildebeest population annually (Kruuk 1972; Schaller 1972), compared with the 12% of wildebeest removed by hyenas in the Ngorongoro Crater (Kruuk 1972). The difference between these figures is equivalent to an estimated 9% annual offtake of wildebeest by local communities (see table 25.1) and, with a stable wildebeest population, suggests that estimates of current hunting offtake of wildebeest are of the correct magnitude. Peterson and Casebeer (1972) recommended a "safe" annual harvest of 6% for the wildebeest population on the Athi-Kapiti plains south of Nairobi, an area comparatively depauperate in predators. This implies that the estimated annual offtake of wildebeest in the Serengeti by illegal hunters may exceed a "safe" level.

The general nonselectivity of snaring suggests the use of the proportion of a species in the sample killed by hunters as one method of obtaining order of magnitude estimates of its population size within the ecosystem. This might be particularly useful for species living in thick woodland and riparian habitats for which estimates from aerial or ground censuses are especially unreliable, such as bushbuck and reedbuck. Population estimates for these two species, as shown in table 25.1, were calculated using the observed percentage in the sample killed by hunters, the regression in figure 25.8, and the wildlife numbers counted in the 1991 aerial survey.

Spatial Zoning

Zoning is an important tool that can assist in the establishment of development and management priorities. We present a zoning scheme in which the protected area is stratified according to the estimated risk to resident wildlife from meat hunting activities (fig. 25.12). Clearly, the areas *most severely affected* (overexploited, zone 1) are: (1) about 44% of Maswa GR, (2) part of the western corridor and Grumeti GR, and (3) the northwest of the protected area. These areas are likely to continue to be severely affected by hunting and may also suffer from an increase in fuelwood collection and illegal livestock grazing. In the northwest, the Tabora ranger post is thought to have an "effective patrol area" extending over a width of only 3 grid squares (15 km). On this basis, the northern part of the protected area requires ranger posts at 15 km intervals along the boundary—a total of four new ranger posts. Areas where resident wildlife appear to be *most endangered* include those localities where suitability for hunting remains high and where wildlife numbers are also high (zone 2). Two such areas are (1) in the west, an area stretching from the northern end of Maswa GR toward the eastern end of Grumeti GR, and (2) large parts of Maswa GR. Both areas consist of undulating or hilly ground covered by woodland and experience medium rainfall. In these areas, wildlife could be protected by increased patrolling effort and by careful fire management designed to prevent destruction of forage (Stronach and McNaughton 1989). Escalation of conflict between hunters and law enforcement (zone 3) is likely to occur in several parts of the western corridor and in central parts of northern Serengeti.

The western corridor currently carries high densities of resident wildlife and is probably crucial to the continued existence of sizable populations of a number of resident herbivores. In particular, the plains in the southwestern parts of the western corridor support large herds of topi and buffalo. Is it possible that these herds have so far escaped large-scale hunting because they live on large open plains? These herds pose obvious targets for meat hunters and deserve particular attention by management.

CONCLUSIONS

Spatial models incorporating (1) wildlife distributions and densities, (2) the distribution and amount of demand for wildlife products by human populations outside the protected areas, and (3) estimates of the suitability of different areas for illegal meat hunting were shown to be of value in examining the problem of illegal offtake in the Serengeti. The total illegal offtake of wildlife from the Serengeti ecosystem was shown to be on the order of 200,000 animals annually. This offtake was driven by the demand for wildlife products by over 1 million people living in a 45 km wide "catchment" zone to the west of the protected areas.

In order to reduce such large-scale illegal hunting inside the protected area it will be necessary to provide viable alternative sources of meat capable of satisfying the current demand for wildlife meat and, in addition, to provide alternative sources of income for over 30,000 hunters. The time may have come for a thorough reappraisal of the problem of meat hunting in the protected areas and of regulations concerning licensed hunting in areas next to the national park. The scale of the problem is considered far too great to be solved by licensed hunting by local communities, and wider-ranging measures are required. In the interim, however, hunting efforts by local communities might be channeled in the direction necessary for the continued conservation of the Serengeti ecosystem by:

1. providing local communities with benefits derived from the conservation of the protected areas and with limited hunting licenses based on *sustainable quotas* on land *adjacent* to the protected areas, *provided that*
2. a comprehensive mechanism of monitoring is established both for setting sustainable quotas that take account of illegal offtake and for checking that quotas are not exceeded;
3. providing opportunities for conservation education that are focused on those communities that supply most of the hunters; and
4. *substantial* increases of law enforcement activities and infrastructure to curtail unlicensed hunting and effectively reduce hunting in the national park.

Since the current estimated offtake from within the GRs is clearly *unsustainable*, measures to replace wildlife meat in the diet are urgently required. We suspect, however, that only a substantial effort aimed at developing the nutritional and economic status of local communities adjacent to the protected area and at reducing high rates of population increase in the demand zone will secure a permanent improvement in relations between local communities and protected area management and a decline of unchecked large-scale meat hunting.

ACKNOWLEDGMENTS

We are grateful to the Tanzania Commission of Science and Technology for permission to conduct the study. We thank the Director General, Tanzania National Parks, the Director of the Wildlife Division, the Conservator of Ngorongoro Conservation Area, and the Director General of the Serengeti Wildlife Research Institute under whose auspices this work was carried out. We thank the Frankfurt Zoological Society (K. C.) and the Fritz-Thyssen-Stiftung and Max-Planck-Gesellschaft (H. H.) for financial support. We also thank the management and wardens of the Serengeti National Park, who have always proved receptive to new ideas and without whose dedication and hard work the Serengeti would not be the wildlife spectacle that it is today. In particular, the Principal Park Warden, W. Summay, and warden in charge of law enforcement, J. Hando, supported this work with enthusiasm. We thank S. A. Huish and M. East for logistic support and fruitful discussions, and two anonymous reviewers for useful comments. We are grateful to P. Arcese, who assisted in the development of the initial patrol record forms. Special thanks are due to Serengeti National Park rangers who conscientiously completed patrol record cards, interviewed arrested hunters, and filled out questionnaires.

APPENDIX 25.1 THE DEMAND MODEL OF WILDLIFE OFFTAKE

The estimated demand for wildlife in all villages in the 45 km catchment zone in seven districts west of the Serengeti was summarized for each 5 km UTM grid square, resulting in 291 demand centers. We modeled the spatial distribution of offtake for each species by running the supply and demand routine Hntrland in Idrisi. Supply images, providing the maximum potential offtake of each species per grid cell, were calculated as 50% of the surveyed density of each species for that cell. By setting the available supply at 50% we balanced two effects: (1) setting the supply higher would imply an almost immediate extinction of wildlife in grid cells close to the boundary of the protected area; (2) trial runs with the supply set at 25% suggested that at least in some species the estimated demand could not be met, and that meat hunting extended farther east into sections of the park where currently there is little evidence of intensive hunting. The Hntrland procedure satisfied demand by removing the supply radially outward from each demand center while balancing the supply to competing neighboring demand centers. The resulting image of remaining supply subtracted from the original supply provided an offtake image. Total offtake of resident wildlife was calculated as the sum of the offtake of each of eight resident wildlife species considered.

Postscript

Hofer et al. (1994), using a larger sample of 452 questionnaires, updated parameter estimates and confirmed that the spatial predictions of the profitability and demand models matched the spatial distribution of hunter activity as evidenced by location of patrol arrests. The modified estimates were 17,850 individuals operating as hunters or porters, killing 8.95 wildlife per person per year and a total of 159,800 wildlife per year, of which 87,000 were wildebeest (6.8% of the total wildebeest population). The magnitude of these estimates was shown to be correct by demonstrating that the estimate of offtake of hyaenas from evidence in poacher camps (375 animals per year) was equivalent to the estimate based on the fate of individually recognized study animals (423 animals per year). The larger sample size led Hofer et al. (1994) to more accurate estimates but did not change the main results and conclusions presented here.

Hofer, H., Campbell, K. L. I, East, M. L., and Huish, S. A. 1994. The impact of game meat hunting on target and non-target species in the Serengeti. Symposium "The exploitation of Mammals," ed. N. Dunstone and V. Taylor, London, November 1994 (in preparation).

REFERENCES

Anderson, D., and Grove, R. 1987. *Conservation in Africa: People, policies, and practice.* Cambridge: Cambridge University Press.

Bivand, R. 1980. A Monte Carlo study of correlation coefficient estimation with spatially autocorrelated observations. *Quaest. Geogr.* 6:5–10.

Bureau of Statistics. 1978. *Population census: Preliminary report.* Dar es Salaam, Tanzania: Ministry of Finance and Planning.

———. 1988. *Population census: Preliminary report.* Dar es Salaam, Tanzania: Ministry of Finance, Economic Affairs and Planning.

———. 1991. *Population census, regional profile: Shinyanga.* Dar es Salaam, Tanzania: President's Office, Planning Commission.

Campbell, K. L. I. 1988. *Programme report, March 1988.* Arusha, Tanzania: Serengeti Ecological Monitoring Programme.

———. 1989. *Programme report, September 1989.* Arusha, Tanzania: Serengeti Ecological Monitoring Programme.

Central Statistics Bureau. 1963. *African census report, 1957.* Dar es Salaam, Tanzania: Government Printer.

———. 1969. *1967 Population census.* Dar es Salaam, Tanzania: Ministry of Economic Affairs and Development Planning.

Cliff, A. D., and Ord, J. K. 1981. *Spatial processes: Models and applications.* London: Pion.

Cumming, D. H. M. 1981. The management of elephant and other large mammals in Zimbabwe. In *Problems in the management of locally abundant wild animals,* ed. P. A. Jewell and S. Holt, 91–118. New York: Academic Press.

Dublin, H. T., Sinclair, A. R. E., Boutin, S., Anderson, E., Jago, M., and Arcese, P. 1990. Does competition regulate ungulate populations? Further evidence from Serengeti, Tanzania. *Oecologia* 82:283–88.

Eastman, R. 1990. *Idrisi, a grid-based geographic analysis system, version 3.2, November 1990.* Clark University, Graduate School of Geography.

Ferrar, A. A., and Walker, B. H. 1974. An analysis of herbivore/habitat relationships in the Kyle National Park, Rhodesia. *J. S. Afr. Wildl. Mgmt. Assoc.* 4:137–47.

Foster, J. B., and Dagg, A. I. 1972. Notes on the biology of the giraffe. *E. Afr. Wildl. J.* 10:1–16.

Georgiadis, N. 1988. *Efficiency of snaring the Serengeti migratory wildebeest.* Unpublished manuscript deposited with SEMP, Serengeti.

Golden Software. 1990. *Surfer, version 4.* Golden, Colo.: Golden Software Inc.

Greig-Smith, P. 1983. *Quantitative plant ecology.* 3d ed. Oxford: Blackwell Scientific.

Haining, R. 1990. *Spatial data analysis in the social and environmental sciences.* Cambridge: Cambridge University Press.

Harris, L. D. 1984. *The fragmented forest.* Chicago: University of Chicago Press.

Hofer, H., East, M. L., and Campbell, K. L. I. 1993. Snares, commuting hyaenas, and migratory herbivores: Humans as predators in the Serengeti. *Symp. Zool. Soc. Lond.* 65:347–66.

Homewood, K., and Rodgers, W. A. 1984. Pastoralism and conservation. *Human Ecol.* 12:431–41.

Hough, J. 1991. Michiru Mountain Conservation Area: Integrating conservation with human needs. In *Resident peoples and national parks. Social dilemmas and strategies in international conservation,* ed. P. C. West and S. R. Brechin, 130–37. Tucson: University of Arizona Press.

Houston, D. C. 1979. The adaptations of scavengers. In *Serengeti, Dynamics of an ecosystem,* ed. A. R. E. Sinclair and M. Norton-Griffiths, 263–86. Chicago: University of Chicago Press.

Jarman, M. V. 1979. Impala social behaviour: Territory, hierarchy, mating, and the use of space. *Adv. Ethol.* 21:1–92.

Jarman, P. J., and Jarman, M. V. 1973. Social behaviour, population structure and reproductive potential of impala. *E. Afr. Wildl. J.* 11:329–38.

Jewell, P. 1972. Social organization and movement of topi, *Damaliscus korrigum,* during the rut at Ishasha Queen Elizabeth Park, Uganda. *Zool. Afr.* 7:233–55.

Justice, C. O., Townshend, J. R. G., Holben, B. N., and Tucker, C. J. 1985. Analysis of the phenology of global vegetation using meteorological satellite data. *Int. J. Remote Sensing* 6:1271–1318.

Kingdon, J. 1982. *East African mammals: An atlas of evolution in Africa,* vols. IIIB–D. London: Academic Press.

Kruuk, H. 1972. *The spotted hyena: A study of predation and social behavior.* Chicago: University of Chicago Press.

Lamprey, H. F. 1963. Ecological separation of the large mammal species in the Tarangire Game Reserve, Tanganyika. *E. Afr. Wildl. J.* 1:63–92.

Lindsay, W. K. 1987. Integrating parks and pastoralists: Some lessons from Amboseli. In *Conservation in Africa: People, policies, and practice,* ed. D. Anderson and R. Grove, 149–67. Cambridge: Cambridge University Press.

Lovejoy, T. E., Bierregaard, R. O., Rylands, A. B., Malcolm, J. R., Quintela, C. E., Harper, L. H., Brown, K. S., Powell, A. H., Powell, G. V. N., Schubart, H. O. R., and Hays, M. B. 1986. Edge and other effects of isolation on Amazon forest fragments. In *Conservation biology: The science of scarcity and diversity,* ed. M. E. Soulé, 257–85. Sunderland, Mass.: Sinauer Associates.

Maddock, L. 1979. The "migration" and grazing succession. In *Serengeti, Dynamics of an ecosystem,* ed. A. R. E. Sinclair and M. Norton-Griffiths, 104–29. Chicago: University of Chicago Press.

Magombe, J., and Campbell, K. L. I. 1989. Poaching survey. In *Serengeti Wildlife Research Centre, Biennial Report, 1988–89,* 69–70. Arusha, Tanzania: Serengeti Wildlife Research Centre

McDowell, L. R. 1985. *Nutrition of grazing ruminants in warm climates.* New York: Academic Press.

McNaughton, S. J. 1987. Adaptation of herbivores to seasonal changes in nutrient supply. In *The nutrition of herbivores,* ed. J. B. Hacker and J. H. Ternouth, 391–408. Sydney: Academic Press.

———. 1988. Mineral nutrition and spatial concentration of African ungulates. *Nature* 334:343–45.

———. 1989. Interactions of plants of the field layer with large herbivores. *Symp. Zool. Soc. Lond.* 61:15–29.

———. 1990. Mineral nutrition and seasonal movements of African migratory ungulates. *Nature* 345:613–15.

McNaughton, S. J., and Georgiadis, N. J. 1986. Ecology of African grazing and browsing mammals. *Annu. Rev. Ecol. Syst.* 17:39–65.

Murray, M. G. 1991. Maximizing energy retention in grazing ruminants. *J. Anim. Ecol.* 60:1029–45.

Myers, N. 1972. National parks in savanna Africa. *Science* 178: 1255–63.

———. 1979. *The sinking ark.* Oxford: Pergamon Press.

Newby, J. E. 1992. Parks for people—a case study from the Aïr Mountains of Niger. *Oryx* 26:19–28.

Norton-Griffiths, M. 1988. Aerial point sampling for land use surveys. *J. Biogeogr.* 15:149–56.

Pennycuick, L., and Norton-Griffiths, M. 1976. Fluctuations in the rainfall of the Serengeti ecosystem, Tanzania. *J. Biogeogr.* 3:125–40.

Peterson, J. C. B., and Casebeer, R. K. 1972. *Distribution, population status, and group composition of wildebeest,* Connochaetes taurinus, *and zebra,* Equus burchelli, *on the Athi-Kapiti plains, Kenya.* Project Working Document 2, Wildlife Management in Kenya. Nairobi: FAO.

Pfirmin, E. S. 1989. *IDA, Image Display and Analysis, Users Guide.* Arlington, Va.: Price, Williams and Associates.

Robinson, J. G., and Redford, K. H. 1991. *Neotropical wildlife use and conservation.* Chicago: University of Chicago Press.

Schaller, G. B. 1972. *The Serengeti lion: A study of predator-prey relations.* Chicago: University of Chicago Press.

Shafer, C. L. 1991. *Nature reserves, island theory, and conservation practice.* Washington, D.C.: Smithsonian Institution Press.

Shaw, J. H. 1991. The outlook for sustainable harvests of wildlife in Latin America. In *Neotropical wildlife use and conservation,* ed. J. G. Robinson and K. H. Redford, 24–34. Chicago: University of Chicago Press.

Sinclair, A. R. E. 1973. Population increases of buffalo and wildebeest in the Serengeti. *E. Afr. Wildl. J.* 11:93–107.

———. 1977. *The African buffalo: A study of resource limitation of populations.* Chicago: University of Chicago Press.

———. 1979. The eruption of the ruminants. In *Serengeti: Dynamics of an ecosystem,* ed. A. R. E. Sinclair and M. Norton-Griffiths, 82–103. Chicago: University of Chicago Press.

———. 1985. Does interspecific competition or predation shape the African ungulate community. *J. Anim. Ecol.* 54:899–918.

SRCS. 1991. *Serengeti Regional Conservation Strategy, a plan for conservation and development in the Serengeti region: Phase II final report and Phase III action plan.* United Republic of Tanzania, Ministry of Tourism, Natural Resources and Environment.

Stammbach, E. 1986. Desert, forest, and montane baboons: Multilevel societies. In *Primate societies,* ed. B. B. Smuts, D. L. Cheney, R. M. Seyfarth, R. W. Wrangham, and T. T. Struhsaker, 112–20. Chicago: University of Chicago Press.

Stronach, N. R. H., and McNaughton, S. J. 1989. Grassland fire dynamics in the Serengeti ecosystem, and a potential method of retrospectively estimating fire energy. *J. Appl. Ecol.* 26:1025–33.

Turton, D. 1987. The Mursi and National Park development in the Lower Omo Valley. In *Conservation in Africa: People, policies, and practice,* ed. D. Anderson and R. Grove, 169–86. Cambridge: Cambridge University Press.

Vesey-Fitzgerald, D. F. 1960. Grazing succession among East African game animals. *J. Mammal.* 41:161–72.

West, P. C., and Brechin, S. R. 1991. *Resident peoples and national parks: Social dilemmas and strategies in international conservation.* Tucson: University of Arizona Press.

Western, D. 1975. Water availability and its influence on the structure and dynamics of a savannah large mammal community. *E. Afr. Wildl. J.* 13:265–86.

Western, D., and Pearl, M. 1989. *Conservation for the twenty-first century.* New York: Oxford University Press.

Wilkinson. L. 1990. *SYSTAT: The System for Statistics.* Evanston, Ill.: SYSTAT, Inc.

TWENTY-SIX

Multiple Land Use in the Serengeti Region: The Ngorongoro Conservation Area

Scott Perkin

The Ngorongoro Conservation Area (NCA) lies immediately to the east of Serengeti National Park (SNP). Encompassing an area of nearly 8,300 km^2, the NCA was created in 1959 as a multiple land use area dedicated to the promotion of both natural resource conservation and human development. Today, the NCA combines wildlife protection and management with pastoralism, catchment forestry, paleontology, and tourism; in 1987–1988, nearly 25,000 Masai pastoralists resided within the NCA, together with some 286,000 head of livestock (Bureau of Statistics 1991; Perkin 1987).

During the three decades following its creation, the NCA has been overshadowed by its more famous neighbor and has received relatively little scientific attention or international support. Yet there are many reasons why the NCA is deserving of greater recognition from the conservation community. First, the NCA is an important protected area in its own right. It supports a rich diversity of habitats ranging from semi-arid grasslands to montane forest, a breeding population of the endangered black rhinoceros, several endemic plant species, and the dense wildlife populations of Ngorongoro Crater—which have now become Tanzania's foremost tourist attraction. In addition, the NCA contains a number of internationally significant paleontological sites, including Olduvai Gorge and Laetoli; the fossils and artifacts from these sites have produced a record of human evolution spanning nearly 4 million years. In recognition of these many different values, the NCA was inscribed upon the UNESCO World Heritage List in 1979.

Second, the NCA is an important buffer zone for Serengeti National Park, with which it shares a common boundary of over 100 km. In addition, the plains of the NCA form part of the Serengeti ecosystem, providing grazing and calving grounds for the annual migration of wildebeest and zebra. As such, the management of the NCA has direct implications not only for the future of SNP, but also for the migratory wildlife of the ecosystem as a whole.

Third, the NCA is the only multiple land use area of its kind in East Africa. For over 30 years, the NCA has been integrating conservation with human development. This experience provides insights that can be applied to the establishment and management of other multiple land use areas, not only adjacent to Serengeti National Park, but also in other parts of East Africa.

In light of these factors, this chapter takes a closer look at the performance of the Ngorongoro Conservation Area over the last three decades. In particular, the chapter aims to (1) evaluate the extent to which the NCA has fulfilled its principal conservation and development objectives; (2) assess the effectiveness of the NCA as a buffer zone for Serengeti National Park; and (3) analyze briefly the suitability of the multiple land use model for application elsewhere in East Africa. In carrying out these assessments, I have relied primarily upon the technical reports and findings of the Ngorongoro Conservation and Development Project (NCDP), which ran from 1987 to 1990.

EFFECTIVENESS OF THE NCA AS A CONSERVATION AND DEVELOPMENT INSTITUTION

Almost by definition, evaluation requires criteria against which performance can be measured. In the case of the NCA, the most appropriate criteria for such an evaluation are to be found in a policy statement issued by the Ngorongoro Conservation Area Authority (NCAA 1992). This policy document, intended to guide the long-term management of the NCA, establishes a comprehensive set of conservation and development objectives. In the sections below, the most important of these objectives are presented as criteria against which to evaluate the effectiveness of the multiple land use system between 1959 and 1990.

Conservation Objectives

For conservation, the main objectives are: (1) the protection of critical wildlife habitat; (2) the maintenance of species diversity in both flora and fauna populations; and (3) the maintenance of wildlife populations in sufficient numbers to ensure both the continuation of the natural spectacle and long-term genetic viability (NCAA 1992).

Critical Wildlife Habitat. Boshe (1988) has identified the following as critical wildlife habitat: the short-grass plains, Ngorongoro Crater, the migration corridors between Ngorongoro Crater and the lowlands, and the Northern Highlands Forest Reserve.

The Short-Grass Plains. The short-grass plains of the NCA play a vital role in the Serengeti ecosystem as the wet season grazing and calving grounds of the migratory herds of wildebeest and zebra. Because these plains lie outside Serengeti National Park, their status and long-term future have often been the cause of considerable concern for conservationists, who have feared that inappropriate development or overgrazing might cause irreversible ecological damage (e.g., Pearsall 1957). However, NCDP found that these areas had among the lowest human and livestock densities in the NCA. The low and erratic rainfall of the plains, coupled with the lack of permanent water supplies, had proven to be a strong disincentive to development. Although the Masai of several villages bordering the plains were using thorn fencing to protect grazing resources from wildlife, these efforts were localized and affected only a small proportion of the total plains area.

Concerns about increases in livestock grazing pressure (e.g., Makacha and Frame 1986) also appeared to be unfounded. The NCDP ground census in 1987 showed that the livestock population had remained well within previously recorded numbers (Perkin 1987). Cattle, for example, had fallen from some 161,930 in 1960 (Dirschl 1966) to 137,398 in 1987; sheep and goats had also shown a decrease, dropping from 243,632 animals in 1977 (Makacha and Frame, 1986) to 137,389. In addition, the risk that cattle might become infected with malignant catarrh fever—a fatal viral disease of cattle transmitted by wildebeest, which are asymptomatic carriers of the virus—was forcing most pastoralists to withdraw their cattle from the plains for a majority of the wet season (Machange 1988; Rossiter, Jessett, and Karstad 1983).

Concerns about soil and vegetation degradation (e.g., Anderson and Talbot 1965) also appeared to be unwarranted. Despite predictions that erosion hollows on the plains were coalescing to form a "virtual desert" (King 1981), regular monitoring over a 2-year period revealed no significant increases in the size of selected hollows or erosion benches (Perkin 1993). A time series analysis of Landsat imagery from 1972 to 1979 revealed little evidence of livestock-induced erosion (King 1981). Similarly, long-term monitoring of vegetation plots on the plains of SNP has revealed few changes to suggest that pasture quality is deteriorating, despite dramatic increases in the wildebeest population over the same period (Belsky 1985; Sinclair, chap. 5). These observations have led researchers such as Homewood and Rodgers (1991, 118) to conclude that the short-grass plains "are very resilient in terms of their tolerance to disturbance and their tendency to return to original species composition and vegetation structure. There seems to be little cause for concern over environmental degradation." In a similar but somewhat more colorful fashion, Aikman and Cobb (1989, 26) have concluded that "despite their battleground appearance at the end of the dry season, the short-grass plains appear to be both surprisingly stable and resilient."

In summary, NCDP concluded that the short-grass plains remained as accessible to the Serengeti wildlife migration at the end of the 1980s as they did at the time of the NCA's creation in 1959; there was little evidence to suggest that the plains were being degraded by wildlife or livestock grazing pressure.

Ngorongoro Crater. Ngorongoro Crater is technically a large caldera, covering some 250 km^2. Despite the formidable appearance of the caldera walls, which rise over 600 m, the Crater is accessible to both people and animals by a variety of routes. A dense and diverse array of wildlife inhabits the grasslands, wetlands, and woodlands of the caldera floor. Although the Crater occupies only a small proportion of the NCA, its spectacular scenery and easily viewable wildlife populations have made it the focus of much of the area's tourism, management, and administration.

In the late 1970s, permanent Masai settlements on the caldera floor were evicted under the *ujamaa* policies of the government of Tanzania, which sought to amalgamate dispersed households into clearly defined village centers. Since that time, management of the Crater has been oriented almost exclusively toward conservation and tourism, although some livestock grazing continued to take place on a day-permit basis at the time of NCDP's studies. With the exception of a small research cabin, a ranger station, camping facilities, and a water-pumping system (for the NCA staff and lodges on the Crater rim), no permanent development had been permitted within the Crater.

Nevertheless, the Crater had begun to experience a number of threats to its conservation values by the end of the 1980s. Increasing tourist pressure had led to off-road driving, vegetation damage, and the harassment of some wildlife species such as rhinoceros and lion. The Ngorongoro Ecological Monitoring Programme had suggested that the lack of a regular burning program was leading to progressively poorer pasture quality (NEMP 1989a). Lastly, there were concerns that the construction of a lodge on the north side of the Crater rim could result in a reduction of water to the caldera floor, since the lodge planned to draw its supplies directly from one of the Crater's principal water sources.

The Ngorongoro Conservation and Development Project concluded that the Crater continued to provide prime wildlife habitat. The Project stressed, however, that careful monitoring and management would be required if the Crater's values were to be safeguarded over the long term.

Ngorongoro Crater Migration Corridors. Ngorongoro Crater is not a self-contained ecosystem. Many species, including wildebeest, zebra, eland, and Thomson's gazelle, use the western flank of Ngorongoro Crater as a corridor to and from the plains (Estes and Small 1981). Such corridors are now recognized worldwide as being important for conservation in general by providing emigration and immigration flow. At the time of the assessment carried out by NCDP, this corridor remained largely accessible to wildlife; however, there were indications that increasingly dense pastoral settlement in the area, coupled with water development projects and the expansion of illegal cultivation, might eventually obstruct this route. NCDP highlighted the need for land use planning and the careful siting of development projects to keep the corridor open.

Northern Highlands Forest Reserve. The Northern Highlands Forest Reserve (NHFR) spans some 890 km^2 of the Crater Highlands. The forest is an important botanical and wildlife refuge, supporting a wide diversity of flora and fauna. It is also an important catchment area, providing water to the surrounding, densely settled villages of Karatu, Mbulumbulu, and others.

In 1989, the Ngorongoro Ecological Monitoring Programme (NEMP) examined changes in the extent of canopy cover in the NHFR. Using aerial photographs from 1959 and 1972, NEMP investigated changes within a 190 km^2 study area of the NHFR. Approximately a quarter of this area had lost more than 75% of its canopy cover; less than 2% of the forest had remained intact over the time period in question (NEMP 1989a). Similarly, Misana (1989) used multitemporal Landsat imagery to investigate changes in the NCA's vegetation between 1979 and 1987; she reported that considerable forest destruction had occurred in the eastern section of the NHFR, where some 1,400 ha had been lost over the time period examined.

In sum, despite the importance of the NHFR, the evidence available up to 1990 suggested that the forest was deteriorating. The major cause of forest destruction appeared to have been the illegal removal of fuelwood and building poles by the agricultural communities adjacent to the NHFR boundaries; although the conservation authorities had been successful at preventing agricultural encroachment per se, demand for wood products was so high that effective control of illegal harvesting had been difficult to achieve. Burning and grazing by the NCA's pastoralists also appeared to be playing a role in forest degradation. Although grazing within the NHFR was permitted only upon receipt of a permit, enforcement had been erratic and livestock numbers had proven difficult to control. Fires lit by the Masai to improve pasture conditions within forest glades often appeared to spread into adjacent forest areas, causing considerable damage (Struhsaker et al. 1989). Large wild mammals, such as elephant and buffalo, might also have been contributing to forest degradation

Species Diversity. An accurate assessment of the extent to which the NCA has been successful at maintaining its full complement of flora and fauna was made difficult by an acute shortage of information. At the time of the NCDP review, there were few data on the area's small mammals, insects, amphibians, or reptiles, and the area's plant species were still being inventoried. Within these limitations, however, NCDP found little evidence to suggest that plant or animal species had become extinct within the NCA since the time of its establishment.

A possible exception is the lesser kudu. Small numbers of this species were sighted in the Endulen/Kakesio area during the aerial survey conducted by EcoSystems in 1980; the total population at that time was estimated to be 37 (EcoSystems 1980). However, subsequent aerial surveys by NCDP did not report kudu in this area or elsewhere in the NCA. It may be that lesser kudu are not normally resident within the area, but rather, are sporadic and temporary visitors (as are oryx). On the other hand, the figures may represent the extinction of a small resident population. Further information is required on the status of this species.

Wildlife Population Sizes. As with species diversity, an appraisal of the extent to which the NCA has succeeded at maintaining wildlife population sizes was hampered by a lack of long-term data, even for the larger and more "popular" mammals. Although the migratory animals such as wildebeest and zebra had been regularly monitored as part of the Serengeti's research programs, there was little information on trends in the NCA's resident animal populations.

One exception to this picture was the long-term monitoring of the wildlife populations of Ngorongoro Crater itself, which had been censused since the mid-1960s. Table 26.1 shows that the Crater continued to support an exceptionally high density of animals over the period 1980–1989, although there may have been declines in several species, including wildebeest, kongoni, and eland. The factors underlying these changes were unclear, but may have included a decrease in pasture quality because of the lack of burning (NEMP 1989a) or illegal hunting (Campbell 1986). In contrast, buffalo appeared to be increasing; it may be that the buffalo population was expanding (or moving down from the adjacent forests) to fill the niche that was formerly occupied by Masai cattle.

The NCA's black rhinoceros population had also been monitored on an intermittent basis. In the 1960s, Goddard (1967) identified 68 individuals in the vicinity of Olduvai Gorge; by the time of NCDP's inception, this population had been exterminated through poaching. Within the

Table 26.1 10-year means of the wildlife populations of Ngorongoro Crater.

Period	Wildebeest	Zebra	Buffalo	Grant's gazelle	Thomson's gazelle	Kongoni	Eland	Waterbuck	Elephant
1970–1979[a]	14,497	4,042	505	1,577	3,291	163	306	43	22
1980–1989[b]	9,620	3,553	1,852	1,309	3,437	113	72	31	22

[a]Data for 1970–1978 from Estes and Small 1981; data for 1979 from Campbell 1986.
[b]Data from Campbell 1986; NEMP 1989b. Elephant data are for 1986, 1988, and 1989 only.

Crater itself, numbers had fallen from some 108 in the 1960s (Goddard 1967) to fewer than 30 by the end of the 1980s (Homewood and Rodgers 1991); in 1989, NEMP identified only 16 individuals (NEMP 1989b). Nevertheless, the Crater population remained one of the largest and most viable black rhinoceros populations in northern Tanzania; the maintenance of these animals in the face of heavy poaching pressure was an important conservation achievement.

Additional information on wildlife population trends was obtained by comparing the results of an aerial census of the NCA carried out by the consulting group EcoSystems in 1980 with the results of an aerial count carried out by NCDP in 1987 (table 26.2). This comparison suggested that the NCA continued to support significant populations of resident wildlife, but that a number of species, particularly buffalo and elephant, had declined.

The interpretation of these results was made difficult because both surveys were "once-off snapshots" without long-term, supporting data. The interpretation was also complicated by the poor confidence limits of many of the 1987 estimates and a number of methodological problems with the 1987 count (Campbell and Borner 1987). Nevertheless, the reduction in buffalo numbers was marked and was likely to have been significant; later surveys (e.g., Campbell and Borner, chap. 6) provided further data to support this trend. The complete absence of elephants outside Ngorongoro Crater in 1987 was also striking (although Campbell and Borner [chap. 6] later reported the presence of some 259 animals). In both cases, the declines were almost certainly attributable to poaching. The EcoSystems survey, for example, noted an abnormally high number of elephant carcasses in the Endulen area in 1980. Similarly, the effects of illegal hunting on the buffalo populations of the Serengeti National Park have been well documented (e.g., Dublin et al. 1990).

In summary, NCDP concluded that poaching had reduced the numbers of rhinoceros, buffalo, and elephant, but that longer-term monitoring of the areas outside Ngorongoro Crater would be required to determine population trends in other resident species. Within the Crater, the apparent declines in herbivores such as wildebeest suggested that a more active program of pasture management (particularly burning) might be

Table 26.2 Wildlife populations of the NCA (excluding Ngorongoro Crater), as estimated by wet season aerial surveys carried out in 1980 and 1987.

Species	1980[a]	95% c.l.[b]	1987[c]	95% c.l.
Buffalo	8,751	13	388	175
Eland[d]	4,385	12	5,372	16
Elephant[e]	687	26	0	—
Giraffe	2,834	14	1,666	9
Grant's gazelle	9,957	11	6,715	7
Impala	2,228	13	3,301	67
Kongoni	901	10	275	87
Ostrich	1,282	13	2,217	43

[a]Data from EcoSystems 1980. The estimates have been derived by combining the numbers given for the NCA with those given for the Endulen Game Controlled Area, which the EcoSystems survey mistakenly considered as a separate administrative unit, then subtracting the estimated populations for Ngorongoro Crater.

[b]Confidence limits are for Arusha Region as a whole, not specifically for the NCA.

[c]Data from Perkin and Campbell 1988; K. L. I. Campbell, pers. comm.

[d]Eland are migratory animals, but are included here for comparative purposes.

[e]Although no elephants were seen in 1987, subsequent surveys provided an esimate of 259 animals outside the Crater (Campbell and Borner, chap. 6).

required. Despite these problems, it was evident that the NCA continued to support a wildlife spectacle of international importance.

Development Objectives

Development priorities for the NCA are: (1) the provision of water supplies to meet the needs of both people and livestock; (2) the provision of livestock development assistance, including veterinary services, breeding programs, and marketing facilities; (3) the provision of economic opportunities for residents, particularly through employment; (4) the promotion of social services to enhance health, education, and food security; and (5) the active involvement of the area's residents in management decisions (NCAA 1992).

The extent to which the NCA had fulfilled each of these objectives up to 1990 is briefly examined below.

Water Supplies. An assessment of the NCA's water supply network was carried out by Aikman and Cobb in 1988. They reported that a majority of the systems they inspected were non-functional; that many domestic supplies were shared with livestock and were, as a result, at risk of contamination; and that the principal factor contributing to the breakdown of supplies had been a general lack of maintenance. They concluded that the system was inadequate to meet the needs of the human and livestock populations of the NCA (Aikman and Cobb 1989).

Veterinary Services. Veterinary services and livestock development programs were appraised by Field, Moll, and ole Sonkoi (1988). A full-time veterinarian had recently been appointed, following an absence of many years; essential livestock drugs were scarce; few cattle dips were in working order; breeding programs had largely collapsed; and only one livestock market was functioning in the district. Tick-borne diseases were common, and had led to reduced productivity and high rates of mortality in cattle. In short, despite the central importance of livestock to the pastoral economy, this sector had been neglected for many years.

Economic Opportunities. Although the Ngorongoro Conservation Area Authority employed over 300 people in 1989, only 13 positions were occupied by Masai residents of the NCA; similarly, the various hotels and lodges relied almost exclusively upon staff from other areas of Tanzania. There were no revenue-sharing schemes in place through which the Masai might have benefited from the NCA's sizeable tourist trade, nor had communities been assisted in establishing alternative income-generating schemes.

Social Services.

Education and Health. The provision of education and health services within the NCA is the responsibility of the Ngorongoro District authorities and the relevant parent ministries, and does not fall within the mandate of the Ngorongoro Conservation Area Authority. Nevertheless, as stressed by Maro (1990), there are many ways in which the NCAA could collaborate with the district to improve these services, particularly through the provision of material and financial assistance (for example, helping to maintain kerosene supplies for vaccine refrigerators). Up to 1990, such collaboration was limited, and the provision of both education and health services was constrained by a wide range of factors.

Food Security. McCabe et al. (1989) raised a number of concerns in relation to nutritional status and food security. In particular, they noted that, partly as a result of the poor quality of animal health services, the number of livestock had failed to keep pace with the growth in the human population. Although livestock numbers had fluctuated around a relatively stable mean, the human population had increased nearly threefold, rising from some 8,700 people in 1966 to almost 25,000 people in 1988 (Dirschl 1966; Bureau of Statistics 1991). These factors had led to a severe decline in the number of livestock per capita, and many families owned too few livestock on which to survive. Anthropometric measurements carried out by McCabe's team indicated that levels of malnourishment were high and increasing by international standards; some 19% of the children under 5 years of age were malnourished, as were 12% of the men and 15% of the women.

In other parts of East Africa, pastoralists who have experienced similar declines in livestock to human ratios have turned to wage labor or cultivation in order to meet their nutritional requirements. Both of these options, however, were unavailable to the Masai of the NCA at the time of the NCDP review; prospects for employment, as discussed previously, were poor, and cultivation had been prohibited since 1975 because of conservation concerns (although illegal cultivation had become widespread). These factors made the NCA Masai uniquely dependent upon their livestock for both food and income, and their nutritional situation was correspondingly precarious. To improve this situation, NCDP recommended that a multifaceted strategy be adopted, aimed at improving livestock health services, increasing the availability of grain and other foodstuffs, diversifying the economic base, and stabilizing the human population.

In late 1992, the proscription on cultivation was lifted by a decree from the Prime Minister. This was intended as a temporary measure to assist the Masai while the NCAA developed alternative strategies for improving food security. Although cultivation was still being practiced in 1994, it remained the NCAA's intention to impose increasingly strict limitations on this land use and eventually to phase out cultivation altogether.

Community Involvement in Management. At the time of the Ngorongoro Conservation and Development Project, there were no formal structures to ensure the representation or involvement of the Masai in the NCA's management. Although a Masai representative from the District Council was a member of the NCA's Board of Directors, there was no legal requirement for such involvement (Forster and Malecela 1989). Similarly, with the exception of a small extension team initiated by NCDP, there was no forum for discussion between the NCAA and residents, nor were there any mechanisms in place through which the Masai could contribute to day-to-day decisions about the running of the area.

In conclusion, an examination of the period 1959–1990 suggests that the NCA was moderately successful at fulfilling its principal conservation objectives, although forest degradation and poaching of several species were perennial management problems. The promotion of human socioeconomic objectives, however, had lagged, and there was evidence of a decline in both the quality and availability of development services; food security had become increasingly tenuous, and had emerged as the dominant concern in virtually all communities. The factors responsible for these trends, and their implications for the NCA's effectiveness as a buffer zone, are explored in the following sections.

EFFECTIVENESS OF THE NCA AS A BUFFER ZONE FOR THE SERENGETI

Despite its frequent usage, the term "buffer zone" remains ambiguous in most conservation literature. In general terms, however, most conservationists would agree with Garratt's (1984, 69) stipulation that buffer zones should "permit limited levels of use while not imposing the strict limitations of the protected area or the more liberal land uses of the general region." As such, buffer zones are essentially transition zones, designed to avoid the "hard edges" that can be formed when a national park or protected area abuts an area of very different and incompatible land use (e.g., intensive cultivation or dense settlement).

In maintaining "soft edges" as opposed to "hard" ones, at least five management objectives become implicit: (1) the maintenance of habitats and resources that are used by migratory wildlife from the "core" protected area; (2) the maintenance of wildlife migration routes to and from these resources; (3) the reduction of conflict between incompatible land uses; (4) the preservation of the conservation assets of the buffer zone itself; and (5) the sustainable fulfillment of the socioeconomic requirements of the resident human population, both as a means of reducing pressure on the core protected area and as a goal in its own right.

From the perspective of these objectives, it can be concluded from the evaluation carried out above that the NCA has acted as a successful buffer zone for Serengeti National Park. Up to 1990, habitats and resources critical to the long-term survival of SNP's migratory wildlife populations remained intact. Wildlife migration routes remained unimpeded. Inappropriate development projects and land use activities incompatible with wildlife conservation had been largely avoided. At the same time, the NCA had also been moderately successful at protecting those features that make the area an important protected unit its own right; the principal failures—forest degradation and the poaching of species such as buffalo and elephant—were common problems throughout the East African region, and were not unique to the NCA or its multiple-use system of management.

It is only the fifth objective, then, that remained largely unfulfilled. However, in discussing the development status of the NCA's resident communities, it is necessary to place the NCA within the context of other remote pastoral areas in Tanzania and Kenya (McCabe, Perkin, and Schofield 1992). With the important exception of food security, the socioeconomic status of the NCA Masai in the 1980s probably did not differ significantly from that of pastoralists in many other parts of East Africa. Maro (1990), for example, has argued that education and health services within the NCA at the time of his study were roughly on a par with those offered in the adjacent pastoral areas of Sale and Loliondo. Similarly, McCabe, Perkin, and Schofield (1992) have drawn attention to the declines in the ratio of livestock to people that have occurred among the Samburu, while Nestel (1986) has reported high rates of malnourishment from Masai group ranches in Kenya. What is clear, however, is that the NCAA could have done much more than it has. As one former Masai Member of Parliament has said: "The people of the NCA should be like a calf with two mothers, cared for by both the Authority and the District; instead, they have become like orphans" (M. S. ole Parkipuny, pers. comm.).

Perkin (1993) has analyzed the reasons for this relative lack of attention to socioeconomic requirements and has identified a number of contributory causes. The most important factor appears to have been a lack of commitment to the NCA's multiple-use philosophy on the part of the Ministry of Natural Resources, which often perceived the NCA as being little different from a conventional national park. As a result, the majority of the NCA's staff were drawn from traditional wildlife management backgrounds and had little experience with the social sciences or development activities; in addition, most staff members came from agricultural areas and thus may have held less empathy for pastoral land use systems.

As a result of these policies, the majority of the NCA's resources and funding were devoted to traditional wildlife conservation activities such as anti-poaching. Because of the lack of Masai representation and involvement, there was no internal system of checks and balances, and hence there were few opportunities for residents of the area to redress this management bias. The situation was exacerbated by a number of additional factors, including the lack of a clear statement of management policy, the absence of a management plan with detailed conservation and development objectives, and the lack of a formal land use zoning scheme.

SUITABILITY OF THE MULTIPLE LAND USE APPROACH FOR THE EAST AFRICA REGION

The NCA has demonstrated that multiple-use areas can be relatively successful within an East African context. In many parts of the region, it is becoming increasingly difficult to establish conventional national parks in which no human utilization is permitted because of moral, economic, and political pressures (West and Brechin 1991). In these areas, multiple land use systems may be better options for the conservation of natural areas, and the NCA could serve as a potential management model.

However, the NCA's experiences suggest that the multiple land use model may not be easily transferred to other areas of East Africa. In particular, it appears difficult for multiple-use areas to maintain a balance between conservation and development activities, given the prevailing socioeconomic and political conditions of the region: interdisciplinary institutions are scarce, mechanisms for ensuring public participation are largely in their formative stages, and effective systems of checks and balances remain elusive. These factors seem to push multiple-use systems toward one management extreme or the other on the conservation-development spectrum. Within the NCA, management activities over the first three decades were oriented largely toward conservation at the expense of development efforts; it is possible that the reverse could have occurred. For example, in 1968–1969, when the NCA was under the jurisdiction of the Ministry of Agriculture, the NCA was nearly dissolved and large sections of the area were nearly converted to intensive cultivation and livestock ranching (Arhem 1985).

The NCA also highlights a number of additional problems that are likely to be faced by other multiple-use areas in the region, including (1) the difficulty of maintaining effective control of natural resource utilization systems (as demonstrated by the NCA's difficulties in enforcing the forest grazing permit system); (2) the difficulty of assessing the environmental impact of people and livestock (and hence, of designing sustainable resource utilization systems) in the absence of long-term ecological monitoring; and (3) the difficulty of coping with the high rates of human population growth that characterize much of East Africa (Perkin 1993).

CONCLUSION

The experience gained from the NCA's three decades of multiple-use management provides insights for the establishment of other multiple land use areas. The lessons learned—although often clear with hindsight—were not so apparent in 1959. New multiple-use areas should address the following points (Perkin and Mshanga 1992; Perkin 1993):

1. A clear statement of management policy, supported by unambiguous legislation
2. A management plan with clear conservation and development objectives
3. A land use zoning scheme
4. The technical and legal capacity to plan and control development
5. A commitment to meeting the needs of local communities
6. The recruitment of staff from both the natural and the social sciences
7. Management systems that involve local people and operate on a power- and revenue-sharing basis
8. Integration of the multiple land use area with the administrative district of which it is part

9. Comprehensive ecological monitoring
10. Population policies, aimed first at clearly defining those groups with a right to use the natural resources of the area

Although an outstanding network of national parks and reserves has been established throughout East Africa, many ecologically important areas remain unprotected. These include wildlife dispersal areas adjacent to national parks (such as the Loliondo Game Controlled Area in the Serengeti region) and areas of high biological diversity, such as wetlands and forests, that have been neglected because of the overemphasis placed on conserving large mammals and their savanna habitats. As land use pressures within the East African region continue to mount, it will become increasingly difficult to establish additional national parks to protect these areas; despite their inherent management problems, multiple-use systems could become the most viable conservation option in many instances.

To date, more attention has been given to conventional national parks and strictly protected reserves. However, the NCA's experiences suggest that multiple land use areas can maintain natural ecosystems while allowing both human settlement and the utilization of essential resources. This approach should now be expanded to incorporate other areas in East Africa that might otherwise be lost to conservation.

ACKNOWLEDGMENTS

The Ngorongoro Conservation and Development Project was a collaborative endeavor between the Ngorongoro Conservation Area Authority, the Tanzanian Ministry of Lands, Natural Resources, and Tourism, and IUCN. Funding for the project was provided by the Norwegian Agency for International Development and the Food Aid Counterpart Fund of the EEC. The opinions expressed in this chapter, however, are those of the author alone.

I am indebted to many people and organizations for the assistance provided to me during the course of the project. In particular, I would like to thank the former Director of Wildlife, C. Mlay, and the former Chair of the NCA Board of Directors, E. M. Malecela, for their constant support. I would also like to thank the following members of the Ngorongoro Conservation Area Authority: J. Kayera, former Conservator of Ngorongoro; P. Mshanga; and L. ole Mariki. E. Chausi, A. Kijazi, J. ole Kuwai, S. Makacha, M. Meng'oriki, J. Mosha, J. ole Koromo, L. ole Sayalel, E. ole Sella, M. Shoo, and N. Tluway also contributed a great deal to the project.

The Masai community offered a tremendous amount of information and encouragement. I am grateful to: P. ole Kasiaro, Councillor for Ngorongoro Ward; M. S. ole Parkipuny, formerly the Member of Parliament for Ngorongoro District; and F. S. ole Naingisa, Chairman of the Ngorongoro District Council.

My thanks are also extended to B. Mbano, M. Borner, K. Campbell, S. Huish, S. Lelo, J. ole Kuwai, R. Malpas, J. Mgombi, and J. Root for making the aerial surveys possible. Lastly, I am grateful to R. Malpas and the staff of the IUCN Regional Office in Nairobi for their administrative and logistical support.

REFERENCES

Aikman, D. I., and Cobb, S. M. 1989. *Water development in the Ngorongoro Conservation Area.* Ngorongoro Conservation and Development Project Technical Report no. 8. Nairobi: IUCN Regional Office for Eastern Africa.

Anderson, G. D., and Talbot, L. M. 1965. Soil factors affecting the distribution of the grassland types and their utilization by wild animals on the Serengeti plains, Tanganyika. *J. Ecol.* 53:33–56.

Arhem, K. 1985. *Pastoral man in the Garden of Eden: The Maasai of the Ngorongoro Conservation Area, Tanzania.* University of Uppsala, Department of Cultural Anthropology.

Belsky, A. J. 1985. Long term vegetation monitoring in the Serengeti National Park, Tanzania. *J. Appl. Ecol.* 22:449–60.

Boshe, J. I. 1988. *Important ecological aspects of the wildlife populations of the Ngorongoro Conservation Area.* Ngorongoro Conservation and Development Project Technical Report no. 6. Nairobi: IUCN Regional Office for Eastern Africa.

Bureau of Statistics. 1991. *Tanzania Sensa 1988. Population Census Regional Profile, Arusha.* Dar-es-Salaam: Bureau of Statistics, President's Office, Planning Commission.

Campbell, K. L. I. 1986. Census of wildlife in Ngorongoro Crater, 26–27 August, 1986. Seronera: Serengeti Ecological Monitoring Programme. Mimeograph.

Campbell, K. L. I., and Borner, M. 1987. Ngorongoro Conservation Area. Livestock and Wildlife Census, 12–16 April 1987. Preliminary Results of SRF Survey. Seronera: Serengeti Ecological Monitoring Programme. Mimeograph.

Dirschl, H. J. 1966. *Management and development plan for the Ngorongoro Conservation Area.* Dar-es-Salaam, Tanzania: Ministry of Agriculture, Wildlife, and Forests.

Dublin, H. T., Sinclair, A. R. E., Boutin, S., Anderson, E., Jago, M., and Arcese, P. 1990. Does competition regulate ungulate populations? Further evidence from Serengeti, Tanzania. *Oecologia* 82:283–88.

EcoSystems. 1980. The status and utilisation of wildlife in Arusha Region, Tanzania. Nairobi: EcoSystems Ltd.

Estes, R. D., and Small, R. 1981. The large herbivore population of Ngorongoro Crater. *Afr. J. Ecol.* 19:175–85.

Field, C. R., Moll, G., and ole Sonkoi, C. 1988. *Livestock development in the Ngorongoro Conservation Area.* Ngorongoro Conservation and Development Project Technical Report no. 1. Nairobi: IUCN Regional Office for Eastern Africa.

Forster, M., and Malecela, E. M. 1989. *Review of the legislation governing the Ngorongoro Conservation Area.* Ngorongoro Conservation and Development Project Technical Report no. 7. Nairobi: IUCN Regional Office for Eastern Africa.

Garratt, K. 1984. The relationship between adjacent lands and protected areas: Issues of concern for the protected area manager. In *National parks, conservation, and development: The role of protected areas in sustaining society,* ed. J. A. McNeely and K. R. Miller, 65–71. Washington, D.C.: Smithsonian Institution Press.

Goddard, J. 1967. Home range, behaviour, and recruitment rates of two Black Rhinoceros (*Diceros bicornis* [L.]) populations. *E. Afr. Wildl. J.* 5:133–50.

Homewood, K. M., and Rodgers, W. A. 1991. *Maasailand ecology: Pastoralist development and wildlife conservation in Ngorongoro, Tanzania.* Cambridge: Cambridge University Press.

King, R. B. 1981. Landform and erosion in the Ngorongoro Conservation Area. Background paper for the new development and management plan for Ngorongoro. Dar-es-Salaam: Bureau of Resource Assessment and Land Use Planning, University of Dar-es-Salaam. Mimeograph.

Machange, J. 1988. *Livestock/wildlife interactions in the Ngorongoro Conservation Area.* Ngorongoro Conservation and Development Project Technical Report no. 4. Nairobi: IUCN Regional Office for Eastern Africa.

Makacha, S., and Frame, G. W. 1986. Population trends and ecology of Maasai pastoralists and livestock in Ngorongoro Conservation Area, Tanzania. Serengeti Wildlife Research Institute Contribution no. 338. Mimeograph.

Maro, W. E. 1990. *Education and health services within the Ngorongoro Conservation Area.* Ngorongoro Conservation and Development Project Technical Report no. 13. Nairobi: IUCN Regional Office for Eastern Africa.

McCabe, J. T., Perkin, S., and Schofield, E. C. 1992. Can conservation and development be coupled among pastoral people? An examination of the Maasai of the Ngorongoro Conservation Area, Tanzania. *Hum. Org.* 51:353–66.

McCabe, J. T., Schofield, E. C., Pedersen, G. N., Lekule, A., and Tumaini, A. 1989. *Food security and nutrition among the Maasai of the Ngorongoro Conservation Area.* Ngorongoro Conservation and Development Project Technical Report no. 10. Nairobi: IUCN Regional Office for Eastern Africa.

Misana, S. B. 1989. *An assessment of vegetation change in the Ngorongoro Conservation Area.* Ngorongoro Conservation and Development Project Technical Report no. 9. Nairobi: IUCN Regional Office for Eastern Africa.

NCAA (Ngorongoro Conservation Area Authority). 1992. Management and development plan 1992/93–1996/97. Policy Statement for the Ngorongoro Conservation Area Authority. Ngorongoro Crater: Ngorongoro Conservation Area Authority. Mimeograph.

NEMP (Ngorongoro Ecological Monitoring Programme). 1989a. *Annual report.* 1989. Ngorongoro Crater: Ngorongoro Ecological Monitoring Programme.

———. 1989b. *Semi-annual report,* April 1989. Ngorongoro Crater: Ngorongoro Ecological Monitoring Programme. 58 pp.

Nestel, P. 1986. A society in transition: Developmental and seasonal influences on the nutrition of Maasai women and children. *Food Nutr. Bull.* 8 (1):2–18.

Pearsall, W. H. 1957. Report on an ecological survey of the Serengeti National Park, Tanganyika. *Oryx* 4:71–136.

Perkin, S. 1987. 1987 wet season ground census: Preliminary report. Nairobi: IUCN Regional Office for Eastern Africa. Mimeograph.

———. 1993. Integrating conservation and development: An evaluation of multiple land-use in the Ngorongoro Conservation Area, Tanzania. Ph.D. thesis, School of Development Studies, University of East Anglia.

Perkin, S., and Campbell, K. L. I. 1988. Ngorongoro Conservation Area: Aerial survey of livestock and wildlife, 23–25 October 1987. Preliminary results of SRF Survey. Nairobi and Seronera: IUCN Regional Office for Eastern Africa and Serengeti Ecological Monitoring Programme. Mimeograph.

Perkin, S., and Mshanga, P. 1992. Ngorongoro: Seeking a balance between conservation and development. Paper presented at the IV World Congress on National Parks and Protected Areas, Caracas, 10–21 February, 1992.

Rossiter, P. B., Jessett, D. M., and Karstad, L. 1983. Role of wildebeest fetal membranes and fluids in the transmission of malignant catarrhal fever virus. *Vet. Rec.* 113:150–52.

Struhsaker, T. T., Odegaard, A., Ruffo, C., and Steele, R. 1989. *Forest conservation and management in the Ngorongoro Conservation Area.* Ngorongoro Conservation and Development Project Technical Report no. 5. Nairobi: IUCN Regional Office for Eastern Africa.

West, P. C., and Brechin, S. R., eds. 1991. *Resident peoples and national parks: Social dilemmas and strategies in international conservation.* Tucson: University of Arizona Press.

TWENTY-SEVEN

Economic Incentives to Develop the Rangelands of the Serengeti: Implications for Wildlife Conservation

M. Norton-Griffiths

The rangelands that surround the northern and eastern boundaries of Serengeti National Park are home to the Masai, who still largely follow a pastoral way of life. Although the majority of their income is derived from livestock management, an increasing share comes from wildlife-based tourism and from agriculture. The income from agricultural production is generated by the conversion of rangeland to farmland and ranchland, at the ultimate expense of wildlife conservation. While pastoralism is generally compatible with wildlife, agriculture and ranching are generally not.

This chapter examines the conflicts between national conservation objectives, the maintenance of the traditional Masai way of life, and the equitable distribution of revenues from wildlife-based tourism. Specifically, data from the Mara area in Kenya are used to compare the revenues and profits that the Masai are currently earning from using their land for traditional livestock management, augmented by tourism and agriculture, with the revenues and profits that they could potentially earn from using this same land to its full agricultural potential.

There are two critical issues to be assessed. First, are the differentials between current and potential revenues large enough to create economic incentives for the Masai to develop their land for agriculture—at the expense of wildlife? Second, can revenues from wildlife-based tourism or other sources reduce these incentives to the extent that Masai landowners will elect to leave their land largely undeveloped—to the benefit of wildlife?

THE SERENGETI ECOSYSTEM

Regional Perspective

The Serengeti ecosystem holds three distinct types of land: (1) the formal conservation estate of Serengeti National Park (SNP), Masai Mara National Reserve (MMNR), and the Maswa Game Reserve, in which the interests of wildlife are paramount; (2) the multiple land use area of the Ngorongoro Conservation Area (NCA), in which the interests of both wildlife and pastoral Masai are of equal importance; and (3) extensive agricultural and rangeland areas, under a variety of more or less formal land tenure systems, in which the interests of the landowners and/or land users are paramount. Wildlife, both resident and migratory, is found on all of this land.

The migratory wildlife is of specific concern here and is the focus of national, regional, and international conservation interest. The almost 2 million migratory animals (1.3 million wildebeest, 0.4 million gazelle, 0.2 million zebra) use areas far larger than that of the formal conservation estate in their annual migrations. During the rainy season (Maddock 1979) the migrants make extensive use of the short grasslands of the Ngorongoro Conservation Area, while during their northern and southern movements they pass through the agricultural areas to the northwest of Serengeti National Park, around Ikoma, and the eastern rangeland areas around Loliondo. However, of most importance are the rangeland areas surrounding the northern boundary of the Masai Mara National Reserve in Kenya, which, along with the MMNR itself, protect essential dry season grazing resources.

The Mara Area

The Mara area in Kenya (fig. 27.1), as defined by Douglas-Hamilton (1988), epitomizes the conflicts between pastoralism, tourism, and agriculture. The 1,368 km^2 Masai Mara National Reserve is a formal conservation estate owned by the government of Kenya and managed by the Narok District Council (Talbot and Olindo 1992). Land use within the reserve is restricted to wildlife tourism. Its major conservation value is the protection of resident wildlife communities and the provision of critical dry season grazing resources for migratory populations. It is the premier wildlife attraction in Kenya, generating 8% of national tourist revenues, 10% of all tourist bednights, and some \$20 million in foreign exchange (table 27.1).

The MMNR is surrounded by 4,566 km^2 of group ranches that are under private ownership, either by a group of families (group ownership) or by one individual family (individual ownership). The group ranches contain year-round communities of resident wildlife, but migrants also spill out onto them during the dry season. These dry season grazing resources on the group ranches are important to the migrant wildebeest, for losing them could reduce numbers by 30% (see below). The ranches generate \$10 million of foreign exchange through tourist activities (table 27.1), as well as \$2.4 million through traditional livestock management and a further \$3.8 million through large-scale agriculture.

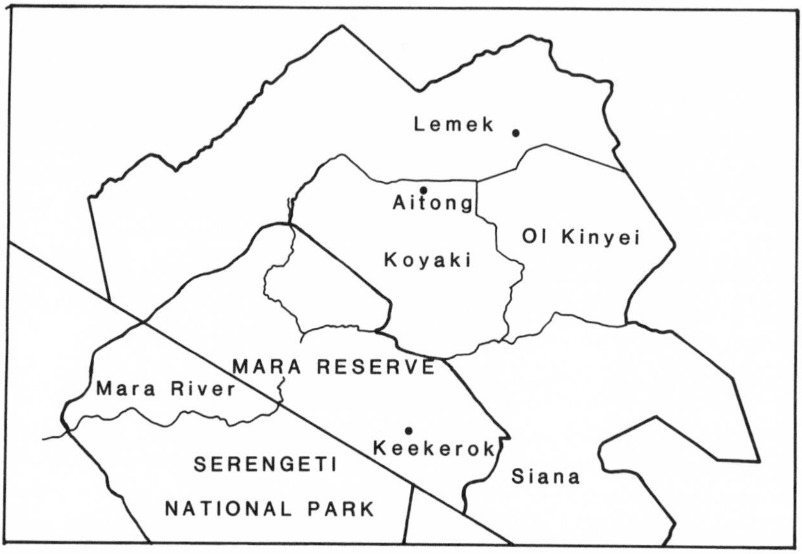

Figure 27.1 The Mara region, showing the group ranches and the Mara Reserve.

Conflicts of Interest

This close juxtaposition and overlap of people, livestock, and crops with migratory and resident wildlife leads inevitably to strong conflicts of interest. On the one side, national conservation authorities point to the very tangible benefits from the conservation estate and its wildlife, including the substantial earnings from tourism and hunting, as justification for their activities. On the other, private landowners and land users are becoming increasingly aware of the hidden costs of supporting wildlife on their land, even though they share to some extent in the benefits arising from it.

With wildlife revenues far exceeding those from livestock or agriculture, it might appear at first sight to be mutually beneficial to all parties in the Mara area to maintain the status quo. However, significant areas of the group ranches around the MMNR have high agricultural potential, and private landowners are realizing that current profits compare unfavorably with the potential profits from developing their land.

THE VALUE OF AGRICULTURAL PRODUCTION

It is this conversion of privately owned rangelands to agriculture and to ranching that is of most concern to wildlife managers and conservationists. To understand why this is happening it is necessary to compare the revenues and profits to landowners from the present mix of land uses on the group ranches with their potential earnings if the land were devel-

Table 27.1 Land use and total revenues in the Mara area.

	Masai Mara National Reserve	Surrounding group ranches
Area (km²)	1,368	4,566
Land tenure	State-owned conservation estate	Group or individual ownership
Total revenues[a]		
Wildlife-based tourism[b]	20.0	10.0
Nomadic pastoralism and livestock trading[b]	Nil	2.4
Large-scale agriculture[c]	Nil	3.8
TOTAL	20.0	16.2

[a]In millions of U.S. dollars: 1989 base year.
[b]1987 data from Douglas-Hamilton (1988), multiplied by a GDP inflator of 1.185 for the period 1987–1989 (IBRD 1992).
[c]1987 data from Douglas-Hamilton (1988), multiplied by 1.163 wheat price inflator for the period 1987–1989 (IBRD 1992).

oped. These potential earnings will of course be influenced strongly by the basic agricultural capabilities of the land, for land with good rainfall and soils is inherently more productive than is drier land with poorer soils.

Short and Gitu (1990) classified six land potential zones in Kenya on the basis of elevation, rainfall, and temperature (table 27.2), each of which affects crop and livestock production. They also gave the total hectares of each zone, not including the areas set aside as national park or forest reserve and which are therefore not available for agricultural use or development (table 27.3). Norton-Griffiths and Southey (1993) then estimated the revenues and profits from agricultural and livestock production within each of these land potential zones from (1) estimates of crop hectares (broken down by crop types) and of livestock densities (broken down by livestock type and management) and (2) the revenues and profits from individual crop and livestock production activities.

Measures of the hectares of different crops and the densities of livestock within each zone, originated from land use surveys carried out in Kenya between 1981 and 1986 (EcoSystems 1987). Crop cover data were obtained from aerial point sampling (Norton-Griffiths 1988) and are strictly comparable within and between the zones (table 27.3). Livestock data for the higher-potential zones 1–4 came from the same surveys but were augmented from Bekure et al. (1991) for zone 5 and from EcoSystems (1985b) for zone 6. Livestock were classified in each zone into grade cattle, indigenous cattle, and smallstock, and management was classified as zero grazing (stall fed), open grazing, and ranching.

Revenues and profits from individual crop and livestock production activities within each zone were taken from farm level budgets based on

Table 27.2 Environmental characteristics of the land potential zones.

Zone	Elevation (meters)	Rainfall (mm)	Temperature (°C)
Zone 1: Per Humid	2,500	> 2,000	2–15
Zone 2: Humid	1,700	1,600	15–21
Zone 3: Sub-humid	1,400	1,400	21–24
Zone 4: Transitional	1,100	700	21–24
Zone 5: Semiarid	700	600	24–31
Zone 6: Arid	< 700	400	> 31

Table 27.3 Land uses in each land potential zone.

Zone	Area (km^2)[a]	Cultivation (ha/km^2)	% cash crops[b]	Livestock (no./km^2)
Zone 1: Per Humid	680	17.1	25%	33.8
Zone 2: High Potential	19,210	49.9	45%	146.5
Zone 3: Medium potential	62,290	36.6	38%	122.6
Zone 4: Arable	83,730	20.7	23%	128.7
Zone 5: Ranching	132,940	2.9	46%	90.7
Zone 6: Pastoral	216,600	0.0		22.9

[a]Area of each zone less the land set aside as national parks, national reserves, or formal forest estate.
[b]Crops were classified into individual mono- and intercrop areas.

the policy analysis matrix (PAM) approach of Monke and Pearson (1989). The Policy Analysis for Rural Development program in Kenya (PARD 1991; Sellen 1991) interviewed a cross-section of smallholder and commercial farmers in seven districts, concentrating always on the dominant agro-ecological conditions and the dominant crop and livestock systems, to develop these budgets. Each PAM budget might represent the averaged results from thirty or more farmer interviews, and each can be assigned to one individual land potential zone.

These PAM crop budgets are all expressed in 1989 Kenya shillings and itemize total revenues and total costs on a per hectare basis. Total revenues include the value of a crop (and its residues) at the prevailing district prices or, if relevant, at the prevailing national prices (e.g., for estate-grown crops). Total costs differentiate between fixed and intermediate inputs, and include all direct and indirect costs as well as working capital and marketing. Costs are also based on the conditions prevailing in a district at the time of the survey. Family labor, for example, was valued at its opportunity cost, which was taken to be the local unskilled wage.

Although PAM budgets were available for the more profitable of the livestock production systems (intensive zero grazing, semi-zero grazing, and extensive grazing) in zones 1–4, data from Bekure and Chabari (1991) were used for ranching and pastoral production in the lower-potential zones 5 and 6. The International Livestock Commission for Africa (ILCA), working in Kajiado Masailand (zone 5) between 1981 and 1983 (Bekure et al. 1991), measured the output per hectare from the consumption and sales of cattle and smallstock to be KSh 59.95 per hectare per year (Bekure and Chabari 1991, their table 9.1), or $6.18/ha/year using the GDP deflator for 1981–1989 and the appropriate exchange rate. This is in close agreement with three other estimates for zone 5: first, the theoretical long-term sustainable output for the pastoral Masai livestock system of KSh 59.00/ha/year calculated from the ILCA livestock system model (table 10.1 in ILCA 1991); second, the figure of $5.19/ha/year calculated by Douglas-Hamilton (1988) for cattle alone in Masailand; and third, average revenues of $5.12/ha/year and $4.89/ha/year from the 1989 annual accounts of two large commercial cattle ranches. There were no comparable data for zone 6, so the estimated revenue per hectare for zone 5 was scaled simply on the basis of the observed stocking densities.

While ILCA (Bekure et al. 1991) provided data to estimate total revenues per hectare from livestock, they provided no leads to estimating a profit equivalent to the PAM budget profits. The profitability of the zone 4 extensive grazing (31%) was therefore applied to the estimated revenues for zones 5 and 6. Note that these livestock budgets reflect only the value of sales and consumption of livestock and livestock products. They do not reflect the capital value of the herd, nor the value of herd growth.

The final step was to multiply the crop hectares and livestock densities within each zone by the relevant PAM budgets to get revenues and profits for all crop and livestock production activities. This gives a revenue and profit aggregated and averaged across the whole area of each zone, rather than values for each cultivated hectare. Profits are calculated in the PAM budgets as total revenues less total costs and represent the private profit to the landowner. They are not strictly net since they leave out the cost of land, and they therefore show returns to land rather than returns to capital.

Revenues and profits (table 27.4) are meager in zone 1, where the high elevations and low temperatures restrict land use mainly to forest use with little cultivation. Zone 2 is the land of highest potential and holds the majority of estate and smallholder tea and coffee, as well as the highest densities of rural populations and livestock. Almost half of all crops are cash crops, and revenues and private profits average a substantial $412.00 and $151.00 per hectare per year. Intensive smallholder cultivation and cash cropping is also found throughout zone 3 (most crop production in Kenya takes place within these two zones), and revenues and profits remain high there. Zone 4 is more marginal for cropping, with fewer cash crops and more modest revenues and profits, whereas zone 5 is generally unsuitable for agriculture, though suitable for ranching. Zone 6 represents the vast arid rangelands suitable only for livestock.

Table 27.4 Total revenues and private profits for agricultural and livestock production in the land potential zones.

	Total revenues			Private profits		
	Agriculture	Livestock	Total	Agriculture	Livestock	Total
Zone 1	60.0	58.4	118.4	17.6	20.7	38.3
Zone 2	320.8	90.9	411.7	115.7	35.0	150.7
Zone 3	183.6	48.4	232.0	70.0	20.7	90.7
Zone 4	104.0	45.4	149.4	34.6	19.6	54.2
Zone 5	15.0	6.2	21.2	3.7	1.6	5.3
Zone 6	0.0	1.6	1.6	0.0	0.6	0.6

Note: U.S. dollars ha/yr: 1989 base year. 1989 K.Shs converted at U.S. $1 = KShs 20.6 (IBRD 1992).

It is now possible to estimate the potential revenues and profits that could be earned from developing the group ranches and converting the land to agricultural and livestock production by multiplying the areas of the land potential zones making up the group ranches (table 27.5) by the revenues and profits for each zone (table 27.4). Strictly speaking, this procedure estimates the total revenues and profits from agricultural and livestock production on land of exactly similar agricultural potential to the group ranches, but elsewhere in Kenya and outside of Masailand.

ECONOMIC INCENTIVES TO DEVELOP RANGELANDS

Current Profits to Landowners

The profits earned by the landowners of the group ranches bear little relationship to the total revenues generated on their land (table 27.6). The landowners control the livestock trade, from which they therefore receive all profits, but both agriculture and tourism are in the hands of outside operators. Of the total revenues of $16 million generated each year on the group ranches, only 8% ($3.00 per hectare) go to the landowners as profits. Of these profits, 72% are from livestock, 16% from agriculture, and 14% from tourism.

The great majority of the tourist revenues from the Mara area—$20 million from the Masai Mara National Reserve and $10 million from the group ranches—are created by and are channelled through the operators of packaged tours, tented safaris, balloons, lodges, and airlines. Some revenues do go directly to landowners through employment and through bednight and visitor fees. The Kenya Wildlife Service (KWS) is also trying to implement a more equitable distribution of revenues, and luxury safari operators are negotiating significant sole use concessions directly with group ranch owners. Nonetheless, in 1989 the landowners received only 1.6% of the tourist revenues generated on their land. (A similar figure holds for Amboseli National Park, where the landowners of the wet season dispersal zone received in 1990 only 1% of the $15 million of wildlife revenues generated there).

Table 27.5 Area (km²) of land potential zones within the group ranches surrounding the Masai Mara National Reserve.

	All group ranches around MMNR	Koyaki and OlKinyei group ranches	Other group ranches
Zone 1	0	0	0
Zone 2	0	0	0
Zone 3	772	0	772
Zone 4	2,192	63	2,129
Zone 5	1,602	1,602	0
Zone 6	0	0	0
Total	4,556	1,665	2,901

Table 27.6 Current and potential revenues and profits to landowners on the group ranches surrounding the Masai Mara National Reserve.

	Current land use		Developed land use	
	Revenues	Profits	Revenues	Profits
Tourism	10.00	0.16	Nil	Nil
Livestock	2.37	0.91	14.68	6.15
Agriculture	3.80	0.20	39.37	13.58
Total	16.17	1.27	54.05	19.73
U.S. $/Ha	35.41	2.78	118.39	43.21

Note: In millions of U.S. dollars: 1989 base year. Ratio of current to potential profits: 1:15.

Agriculture presents a similar set of circumstances to the landowner. The current agricultural revenues (Douglas-Hamilton 1988) are from large-scale wheat farming rather than smallholder production. The wheat farmers are not Masai but are commercial operators who lease land directly from landowners. Lease payments go to the Masai landowners and all other revenues and profits go to the operators. In 1989 these lease payments amounted to 5.3% of the $3.8 million in revenues.

Potential Profits to Landowners

The potential profits to the landowners from developing their land fully for agricultural and livestock production are very much greater than are any of the returns under current land uses (table 27.6). Revenues from tourism would naturally disappear, since fully developed land is incompatible with resident and migratory wildlife. Landowners would instead receive the full profits of $20 million out of revenues of $54 million from agricultural and livestock production, equivalent to $43.00 per hectare per year—15 times the profit per hectare under present land usage.

The group ranches do not all have the same agricultural potential, so they will not all benefit from development to the same extent. Two group ranches in particular, Koyaki and OlKinyei, lie almost entirely in zone 5 (table 27.5) and have relatively poor agricultural potential. These two ranches border the MMNR (fig. 27.1) and are important overspill areas for the migratory populations. They are also the present focus of wildlife-based tourist game viewing around the Mara Reserve. The potential revenues from developing Koyaki and OlKinyei could be as much as $26.00/ha/year (table 27.7) with profits of $7.00/ha/year, broadly comparable to the existing profits of $3.00/ha/year. In contrast, the potential revenues on the other group ranches that have much higher agricultural potential are some $171.00/ha/year, with expected profits of $64.00/ha/year–23 times greater than at present.

Economic Incentives

The differentials between current and potential profits amount to $4.37/ha/year on Koyaki and OlKinyei ranches and $61.13/ha/year on the other group ranches (table 27.7). These inequalities represent the opportunity costs to the landowners of not developing their land, and they are large enough to create economic incentives for the owners of group ranches to develop their land for agriculture, especially land of good agricultural potential.

There are indications enough to show that Masai rangelands have been under such pressure for some years and that the process of conversion to agricultural use is well under way. Within Narok District as a whole, over 15,000 hectares of the high-potential land along the western boundaries with Kericho, Kisii, and South Nyanza Districts had been converted to agricultural use by 1984 (EcoSystems 1985a), and was being farmed by both smallholders and commercial operators.

Elsewhere, large-scale wheat and maize farming is spreading south and east into the Loita and Rift Valley parts of the district. For example, 20 years ago the Kedong ranch in the Rift Valley was the focus of wildlife ranching experiments; it was inconceivable then that it could become the wheat farm it is today. Narok District now produces a major share of Kenya's wheat, and this conversion of land has been in response both to growing markets and to advances in farm technology and plant breeding. Zero tillage and new germplasms make farming in these formerly marginal lands a paying proposition. What may have been marginal land 20 years ago is now potentially productive.

Within the Mara area, Douglas-Hamilton (1988) reported both recent sales of land to Kisii smallholders and an increase in wheat hectares from 18,000 in 1973 to more than 27,000 in 1987. This wheat land is all being leased from the Masai. Standards of land husbandry are often poor, especially on land with short leases. Wheat farmers who either own land in Narok or hold long-term land leases practice better land husbandry and get higher yields.

Table 27.7 Potential revenues and profits on group ranches with higher or lower agricultural potential.

	Koyaki and OlKinyei group ranches (lower potential)		All other group ranches (higher potential)	
	Revenues	Profits	Revenues	Profits
Agriculture	3.06	0.81	36.32	12.77
Livestock	1.28	0.38	13.40	5.77
Totals	4.38	1.19	49.72	18.54
Per hectare	26.05	7.15	171.38	63.91
Current profits per hectare (table 27.6)		2.78		2.78
Ratio current: potential profits		1:3		1:23
Inequality: potential − current profits		4.37		61.13

Note: In millions of U.S. dollars: 1989 base year.

These same incentives encourage the conversion of rangeland of lower agricultural potential into commercial ranches. While this is not yet a major factor in Narok, it is in other pastoral districts of Kenya, such as Laikipia, which has been largely alienated from pastoralists and converted to ranchland. The Galana Ranch, east of Tsavo National Park, is a more recent case. Two thousand square kilometers were alienated from the pastoral Orma and converted into a livestock ranch. In each case, subsistence pastoral (i.e., dairy) production systems were changed into beef production systems.

Rangelands in Tanzania are coming under the same kinds of pressures as those in Kenya, and for the same reasons. In Loliondo, east of the Serengeti, a significant area was recently alienated for the production of barley. In Arusha region, vast areas of Masailand have been alienated for wheat farms—the Hanang scheme—and for bean farms. It is inevitable that the rangeland areas in the Tanzanian parts of the Serengeti ecosystem will sooner, rather than later, be subjected to these kinds of economic incentives and pressures.

Sadly, governments rarely look kindly on nomadic pastoralists, who are invariably seen as backward, uncontrollable, and a drain on the national economy. These attitudes, along with the usually weak land tenure systems, mean that rangelands are all too easily alienated from pastoralists to the benefit of others, resulting in the pastoralists becoming ever more marginalized (Homewood, Rodgers, and Arheim 1987; Homewood and Rodgers 1991; McCabe, Perkin, and Schofield 1992).

IMPLICATIONS OF DEVELOPING THE RANGELANDS

Conservation Interests

Conservation interests will not be well served by further development in the Mara area. Apart from a reduction in biodiversity, habitat types, and plant and animal populations, there is the possibility of a serious impact on the Serengeti migratory wildebeest, for the group ranches surrounding the MMNR protect their critical dry season grazing resources. The population model described by Hilborn et al. (chap. 29) predicts a 12% reduction in wildebeest numbers from their steady state following the loss of dry season grazing on the 2,900 km^2 of group ranches with high agricultural potential (table 27.5), and a further reduction in numbers of 21% if the grazing on the additional 1,665 km^2 of the Koyaki and OlKinyei group ranches is lost as well. The loss of all the group ranches in the Mara area could therefore trigger a population decline of some 30%.

National wildlife managers must assess carefully the effects of losing greater or lesser tracts of these rangelands and answer some very hard questions. How many wildebeest should there be for the Serengeti ecosystem to continue functioning as it has in past decades? How much biodiversity and habitat diversity should be preserved, and why? Policymakers must be much more proactive in the face of these challenges. If they sit idly by, these areas will be lost by default.

The Rights and Interests of the Masai

Policy initiatives in response to this challenge must not ignore the rights and interests of the Masai. There is much talk of maintaining traditional lifestyles and value systems among ethnic groups, and while no one can argue against this in principle, people cannot be condemned to a permanent poverty trap just to safeguard tradition. While elder Masai undoubtedly value their traditional lifestyles, younger and better educated Masai want change.

It would also be a mistake to base policy initiatives on the romantic notion that traditional people invariably make wise guardians of their natural resources and wildlife. Most cultures exploit their resource base to the limits of their technology. The conversion of rangeland to agricultural use suggests that Masai pastoralists are no different in this respect, and that they do not refrain from developing their land at the ultimate expense of the wildlife on it (mainly by fencing) once they have the capital and technology to do so. Population growth among the Masai is another important factor that will lead to the subdivision of the group ranches into smaller and smaller units of production. Whether these rangelands are developed into agricultural lands or ranches, or whether wildlife-generated revenues can be made large enough to remove the incentives to do so, the Masai must participate fully and equitably in the development process and in the ensuing benefits. Yet already in Narok District, the agents of change are not the Masai themselves, but outsiders purchasing and leasing land.

Kenya faces major problems from rising human populations, and relatively little land remains free for development. And while it is unlikely that the government will ever degazette national parks and reserves such as the Masai Mara National Reserve—international response, especially from donors, would be too intense—it is equally unlikely that land of high agricultural potential that is not under formal conservation estate can remain undeveloped for much longer. The continuing dilemma facing the government is how to reconcile the benefits from conservation and from international tourism with the increasing need to provide food and land for its population.

POLICY OPTIONS TOWARD THE GROUP RANCHES

If national wildlife managers wish to preserve wildlife on the group ranches, then they must reduce the economic incentives for the private landowners to develop their land. This can be done only by matching their opportunity costs, namely, the profits forgone. In other words, wildlife managers must compensate the Masai for not developing their land.

Revenues from wildlife-based tourism could be used for the direct compensation of landowners. To be effective, however, these revenues must go directly to the landowners of the group ranches, not as at present to the county council (where they are plundered) or even toward "community projects." Thus, revenues must be clearly seen as "profits" from maintaining the existing land use system.

The full compensation due to the landowners in the Mara area for not developing their land is $18.5 million each year at 1989 prices (table 27.6): this matches their forgone profits. This $18.5 million represents a 62% increase on current (1989) tourist revenues from the area and is equivalent to an additional $56.00 per visitor day (on average some 150,000 people visit the Mara area each year [Douglas-Hamilton 1988] and stay for 2.2 nights). The tourist trade might find it hard to support this increase, even if it were gradually phased in over a number of years.

However, the distribution of compensation would depend on the agricultural potential of the land. The owners of the low-potential Koyaki and OlKinyei group ranches would share approximately $0.7 million, with the remaining $17.8 million being shared among the other group ranches. Compensating only Koyaki and OlKinyei ranches ($0.728 million per year) would represent an increase of only 2.4% in tourist revenues, equivalent to an additional fee of $2.20 per visitor day—a much more feasible target.

Efforts in this direction are already under way, although the transfers are nowhere near large enough yet. The KWS has started to divert revenues directly to landowners, on the order of $0.50 per visitor day, which generates approximately $165,000 each year among the group ranches. Some group ranches are also entering into direct deals with tour companies. For example, the owners of the 877 km^2 Koyaki ranch have signed a sole use contract with private safari operators for $40,000 per year, a bit short of the $300,000 due to them for full compensation. Similarly, the high-potential Lemek group ranch (497 km^2) is unilaterally keeping all wildlife revenues generated on its land from daily entry fees and from bednight fees, but is also being sued by the Narok County Council. These revenues amount to some $500,000 a year, but they are significantly less than the $3 million due for full compensation.

An alternative approach might be to lease back from the Masai the most important parts of the group ranches for conservation, basing the rent per hectare on the agricultural potential of the land. In return for annual lease payments, the Masai would undertake to limit settlements, livestock densities, and livestock development within the leased areas. Such policies need not of course be implemented overnight, but could be phased in over a period of years.

Compensating only Koyaki and OlKinyei ranches, through direct payments or leases, and letting the others select their own development path would appear to hold out the promise of an equitable or even optimal solution. National conservation interests would be broadly maintained, with probably acceptable effects on migratory populations, biodiversity, and habitats. The private landowners would reap the same benefits as if they had developed their land and be able to follow traditional lifestyles if they so wished. The government would see tourist revenues maintained and land of higher potential developed.

While the principle of paying landowners not to develop land, or to produce less on already developed land, may be new to Kenya, it is well established elsewhere. In the United States and Europe, for example, governments make substantial payments to farmers not to produce crops, or to leave land fallow. In the EC countries the "set aside" program pays farmers some $250.00/ha/year for setting land aside from production. Indeed, farmers can qualify for other farm subsidies only by joining the "set aside" program. Sources other than tourist revenues could therefore be sought to develop a "set aside" program in Kenya.

Unfortunately, the weakness of all the options discussed above is the sensitivity of tourist revenues to political change and turmoil, and to global economic trends. A second concern is the continuing escalation of land values and the returns from agriculture. Land values in Kenya are rising in real terms, as is the value of agricultural production (Government of Kenya 1992). These rates of increase will themselves intensify as a function of increasing population. Wildlife-generated revenues are unlikely to respond in the same way. Although they may track global inflation, mass market tourism is relatively elastic—especially in the face of competition from other countries.

COMMUNITY-BASED CONSERVATION PROGRAMS

Today, wildlife managers and conservationists are much more closely attuned to the rights and interests of landowners and land users surrounding formal conservation reserves than they were in the past. In response, they are implementing and supporting community-based Integrated Conservation and Development Projects (Kiss 1992; Wells and Brandon 1992) throughout the Tanzanian and Kenyan parts of the Serengeti ecosystem in an effort to help convince landowners to conserve wildlife on their land.

In Tanzania, local communities all around Serengeti National Park are involved through these ICDPs in consumptive and nonconsumptive wildlife management, primarily within buffer zones, and share in the ensuing benefits. In Kenya, as discussed above, the Kenya Wildlife Service is striving toward a more equitable distribution of benefits from tourist earnings to make "ecotourism" (Lindberg 1991) more profitable for private landowners. Furthermore, tour and hunting operators in both Kenya and Tanzania are negotiating sole use contracts with individual landowners and community groups.

Godoy (1992) describes how dependency on natural resource–based production systems is linked to low levels of economic development. As economic well-being improves, the opportunity cost of using natural resources rather than engaging in other economic activities gradually increases until it is no longer worthwhile. Thus a rich farmer living alongside a national park has much better things to do than hunt an animal or cut down a tree. In contrast, it is worth the while of a poor farmer to do this. This may be why meat hunting (or poaching) is so much a feature of Tanzania rather than of Kenya. Indeed, efforts to formalize the use of park resources (for example, setting local hunting quotas for wildebeest in the Serengeti) merely exacerbate dependency on them. In the Serengeti, every wildebeest harvested (or poached) from the park is one less head of livestock traded. Far better to trade 80,000 livestock, with all the ensuing benefits and economic effects, than harvest 80,000 wildebeest. By its very nature a dead wildebeest can only ever be a dead wildebeest: in contrast, a traded cow creates cash, goods, services, and economic activity.

There is indeed a perverse risk that ICDPs will engender the very changes they are trying to prevent. By increasing dependency on park resources, these programs foster and maintain uneconomic and primitive activities that create a poverty trap and hinder development.

CONCLUSION

Within the Serengeti ecosystem, there are large differentials between the revenues and profits that Masai landowners are currently earning from using their land for traditional livestock management, augmented by tourism and agriculture, and the revenues and profits that they can potentially earn from using this same land to its full agricultural potential. These differentials range from 1:3 on land with relatively poor agricultural potential to 1:23 on land with good agricultural potential.

These differentials create strong economic incentives for the landowners of group ranches to develop their land for agriculture and ranching at the expense of wildlife. This calls into question the viability of conservation policies for rangeland areas based on a mix of traditional livestock management and wildlife-based tourism.

It is clear that the current revenues to landowners from wildlife-based tourism are simply not adequate to stop development. It is equally clear that the community-based ICDPs in the Serengeti ecosystem do not generate the scale of revenues needed to prevent the development of land. The odd primary school, cattle dip, bednight fee, and social center simply will not suffice. If national wildlife managers and conservationists wish to preserve wildlife on the group ranches, then they must reduce the economic incentives to private landowners to develop their land. This can be done only by matching their opportunity costs, namely, the profits forgone. In other words, wildlife managers must compensate the Masai for not developing their land. The sums involved are not trivial: within the Mara area alone they amount to some $18.5 million each year.

Unless the compensation to landowners for not developing their land is raised substantially, significant areas of habitat will disappear, and the migratory wildebeest populations might decline by some 30%. The principle of paying landowners not to develop or use their land is well established worldwide—the EC currently pays $250.00/ha/year to farmers for "set aside." Funds other than tourist revenues could therefore be sought to develop a "set aside" program in Kenya.

ACKNOWLEDGMENTS

This work was carried out while visiting the Harvard Institute for International Development, Cambridge, Mass, U.S.A. in 1992–1993. I am most grateful to the Director and staff of HIID for the kindness, facilities, and advice afforded to me.

REFERENCES

Bekure, S., and Chabari, F. 1991. An economic analysis of Masai livestock production. In *Masai herding: An analysis of the livestock production system of Masai pastoralists in eastern Kajiado District, Kenya,* ed. S. Bekure, P. N. de Leeuw, B. E. Grandin, and P. J. H. Neate, 115–25. Nairobi: International Livestock Centre for Africa.

Bekure, S., de Leeuw, P. N., Grandin, B. E., and Neate, P. J. H. 1991. *Masai herding: An analysis of the livestock production system of Masai pastoralists in eastern Kajiado District, Kenya.* Nairobi: International Livestock Centre for Africa.

Douglas-Hamilton, I. 1988. *Identification study for the conservation and sustainable use of the natural resources in the Kenyan portion of the Mara-Serengeti Ecosystem.* Nairobi: European Development Fund of the European Economic Community.

EcoSystems. 1985a. *Integrated land use survey of Western Kenya.* Nairobi: Government of Kenya, Lake Basin Development Authority, Ministry of Energy and Regional Development.

———. 1985b. *Turkana District resources survey: Final report,* vol. 1. Nairobi: Government of Kenya, Ministry of Energy and Regional Development.

———. 1987. *Integrated land use database for Kenya.* Nairobi: Government of Kenya, Long Range Planning Unit, Ministry of Planning and National Development.

Godoy, R. 1992. The effect of income on the extraction of non-timber tropical forest products among the Sumu Indians of Nicaragua: A preliminary study. Cambridge: Harvard Institute for International Development. Mimeograph.

Government of Kenya. 1992. *Economic Survey 1992.* Nairobi: Government of Kenya, Central Bureau of Statistics, Ministry of Planning and National Development.

Homewood, K., and Rodgers, W. A. 1991. *Masailand ecology: Pastoral development and wildlife conservation in Ngorongoro Tanzania.* Cambridge: Cambridge University Press.

Homewood, K., Rodgers, W. A., and Arheim, K. 1987. Ecology of pastoralism in the Ngorongoro Conservation Area, Tanzania. *J. Agric. Sci.* (Camb.) 108:47–72.

IBRD. 1992. *African development indicators.* Washington, D.C.: World Bank.

Kiss, A., ed. 1992. *Living with wildlife: Wildlife resource management with local participation in Africa.* Technical Paper 130. Washington, D.C.: World Bank.

Lindberg, K. 1991. *Policies for maximising nature tourism ecological and economic benefits.* Washington, D.C.: World Resources Institute.

Maddock, L. 1979. The migration and grazing succession. In *Serengeti: Dynamics of an ecosystem,* ed. A. R. E. Sinclair and M. Norton-Griffiths, 104–29. Chicago: University of Chicago Press.

McCabe, J. T., Perkin, S., and Schofield, C. 1992. Can conservation and development be coupled among pastoral people? An examination of the Masai of the Ngorongoro Conservation Unit, Tanzania. *Hum. Org.* 51(4):353–66.

Monke, E. A., and Pearson, S. R. 1989. *The policy analysis matrix for agricultural development.* Ithaca and London: Cornell University Press.

Norton-Griffiths, M. 1988. Aerial point sampling for land use surveys. *J. Biogeogr.* 15:149–56.

Norton-Griffiths, M., and Southey, C. 1993. *The opportunity costs of biodiversity conservation: a case study of Kenya.* CSERGE Working Paper GEC 93–21. Norwich: University of East Anglia.

PARD (Policy Analysis for Rural Development). 1991. *Farm budgets in selected districts of Kenya: Policy analysis for rural development.* Working paper series 14. Nairobi: Egerton University.

Sellen, D. 1991. *Representative farms and farm incomes for seven districts in Kenya.* Research Training in Agricultural Policy Analysis Project. Nairobi: USAID.

Short, C., and Gitu, K. W. 1990. *Land use and agricultural potential in Kenya.* Technical Paper 90–02. Nairobi: Government of Kenya, Long Range Planning Unit, Ministry of Planning and National Development.

Talbot, L., and Olindo, P. 1992. The Masai Mara and Amboseli reserves. In *Living with wildlife: Wildlife resource management with local participation in Africa,* ed. A. Kiss. 67–74. Technical Paper 130. Washington, D.C.: World Bank.

Wells, M., and Brandon, K. 1992. *People and parks: Linking protected area management with local communities.* Washington, D.C.: World Bank.

TWENTY-EIGHT

The Serengeti Regional Conservation Strategy

B. N. N. Mbano, R. C. Malpas, M. K. S. Maige,
P. A. K. Symonds, and D. M. Thompson

The Serengeti-Mara ecosystem of northern Tanzania and southwestern Kenya is one of the most important wildlife regions in the world. An outstanding network of parks and reserves has been established to protect this natural and cultural heritage; the most important protected areas are Serengeti National Park itself, the Maswa Game Reserve, the Ngorongoro Conservation Area, and the Masai Mara National Reserve in Kenya (see fig. 25.1).

There is, however, growing concern about the long-term viability of the ecosystem. Land use pressures in the region are escalating rapidly, bringing human populations into increasing conflict with the protected areas. In the west, settlements and farms are appearing on the perimeter of the national park. This habitation is affecting the areas immediately surrounding the park that act as "buffer zones." There have been increases in poaching, unplanned fires, and illegal tree cutting. Similar problems are being experienced in Maswa Game Reserve. In Ngorongoro, the issue of the compatibility of pastoralism and conservation continues to raise pressing questions. In Kenya, large agricultural schemes are leading to a reduction in the land available for grazing (Norton-Griffiths, chap. 27). From the standpoint of local human populations, wildlife is seen as a source of livestock diseases, as competition for grazing, and as a threat to crops. Finally, the protected areas themselves are faced with internal management problems, such as poaching and wildfires, that stem from a lack of financial and material resources.

The Serengeti Regional Conservation Strategy (SRCS) was launched at a workshop at Seronera in Serengeti National Park in December 1985, with the goal of identifying and implementing long-term solutions to the resource use conflicts threatening conservation of the ecosystem. The basic premise of the workshop was that conservation and human development in the Serengeti region can no longer proceed in isolation from one another. In particular, it was recognized that the long-term future of the region's protected areas can be assured only if the protected area authorities find solutions to such problems as poaching and encroachment. Such an approach requires an understanding of the needs of neighboring communities and the development and implementation of mechanisms to ensure that those needs are met without detriment to the region's resource base.

The overall goal of the SRCS is to design a new approach toward the management and utilization of the Serengeti region's natural resources, in which:

first, human development needs and natural resource conservation requirements in the region are reconciled with one another through the cooperation of all resource users and managers;
second, the protected areas, and the wildlife resource in particular, play a central role in the economic development of the region;
third, local communities are committed to the conservation of the Serengeti region's wildlife resource through being directly involved in its management and utilization and through receiving direct benefits;
fourth, local communities achieve sustainable use of other natural resources in the region through ownership of land and village-generated land use plans, thereby reducing pressures on the resources of the protected areas.

To achieve these objectives, the SRCS needed to promote action across a diversity of fronts. An integrated approach in which all the concerned parties combined their efforts to identify and implement appropriate solutions was essential if the conservation and development problems of the region were to be tackled in a rational way. The immediate priorities were to create an awareness of present resource use problems and their underlying causes, to investigate existing land uses and trends, to design activities to resolve resource use conflicts, and to instigate policy and administrative changes to facilitate these activities.

In the future, the SRCS will require a framework for the planning of resource use and the provision of technical and material support to the various government and private natural resource managers and users in the Serengeti region. This framework will enable them to implement solutions to rapidly changing land use pressures and problems. We do not see the SRCS as producing a single plan that can then be used by the region's existing authorities to harmonize human development and conservation in the region. Rather, we have placed emphasis on the SRCS providing a planning framework through which the links between protected area authorities and the national, regional, district, and village governments can be developed. While we have stressed that the implementation of changes in resource use policy and practice must be the responsibility of the relevant authorities, we envision the SRCS as a necessary instrument to provide this coordination framework, as well as to pioneer new approaches to sustainable development of the region's natural resources. In addition, we anticipate that the SCRS, on the basis of the experience it gains, will provide technical advice to the responsible authorities on conservation and development activities in the medium to long term.

There are three phases of the SRCS: phase I, problem recognition; phase II, awareness building, specific problem identification and quantification, the design of solutions and initial implementation, and the development of coordination and technical advisory capacity; and phase III, full implementation. The Seronera workshop in December 1985 fulfilled phase I of this process. Phase II was initiated in January 1989 under the auspices of the Ministry of Tourism, Natural Resources, and Environment, co-financed by the Norwegian Agency for International Development (NORAD), the Frankfurt Zoological Society (FZS), and the European Community Food Aid Counterpart Fund (FACF), and with the technical and managerial support of IUCN, the World Conservation Union.

This chapter describes phase II of the SRCS for the period 1989–1991, and what is intended for phase III in terms of philosophy and technical framework.

Phase II focused on:

1. Developing an understanding of the region's resource base, present resource utilization practices, and the constraints to sustainable use
2. Identifying priority areas for intervention to ensure the effective management of the region's protected areas
3. Determining the principal reasons underlying current unsustainable land use practices as a basis for the design and implementation of programs to promote sustainable resource use
4. Developing activities aimed at improving the interaction between the protected areas and the local communities
5. Implementing the SRCS program. This focus included planning and monitoring the promotion of SRCS in the local community, and establishment of communication with local people.

THE DATA BASE FOR CONSERVATION AND DEVELOPMENT

A data base for conservation and development is needed to identify priority needs, to plan appropriate interventions, and to monitor their effects. Data on regional resource use and supporting infrastructure were collected from a variety of sources, including aerial and ground surveys and district questionnaires. Data were obtained on wildlife numbers and distribution, human demography, agriculture, livestock, water, forestry, roads, and community welfare.

So far data collection has been quantitative for the wildlife surveys only, and other data have been more qualitative. In the future more quantitative methods should be used by district resource managers. These data will be designed for storage in computers that can be used by both staff of the SRCS and resource managers working in protected areas. The principal activities proposed for phase III to meet these needs are the convening of a district workshop to design methods for collecting data, a consultancy to design information and retrieval methods, and the subsequent installation of the data base itself.

The wildlife monitoring data will be used to detect changes in natural populations and the effects of pilot wildlife utilization schemes, to set and adjust harvest quotas, and to monitor trends in poaching activities. With the exception of monthly ground surveys in Bunda and Serengeti Districts, the data collection methods used in phase II were not directed at the wildlife utilization areas outside the park and therefore did not produce sufficient information for the region. We intend, in phase III, to implement regular aerial surveys focused on the wildlife utilization areas, and to expand the ground surveys.

The district resource surveys revealed the generally unsustainable use of resources in the region and the dilapidated state of support services and extension infrastructure such as roads and health services. Baseline information on resources and supporting infrastructure has now been entered into the data base, and this will be used to monitor the effects of SRCS activities in the future. Techniques to be used to update this data base in phase III will include remote sensing and ground censuses.

The status of the region's road networks is an important concern. The data collected have been essential in identifying the major areas for road improvement in the protected areas during phase III. Communities neighboring the protected areas were not, however, covered in phase II; such areas are a priority for the future.

PRIORITIES FOR THE MANAGEMENT OF PROTECTED AREAS

The management of the protected areas remains a top priority. Our immediate responsibilities are to ensure the integrity of the boundaries against human encroachment and to secure the natural ecosystems and wildlife populations on which the SRCS is founded. We have, therefore, focused on examining the protected area boundaries and on methods for law enforcement. We are also studying how protected areas can generate more revenue to help underwrite the cost of management and to enable benefits to be provided to local communities. Part of this effort has concentrated on ways of improving park interpretation facilities and tourist roads.

Surveys on the ground showed that protected area boundaries are indistinct and need marking, and that more precise description is needed for legal documents. The SRCS will help the area authorities to find funds for equipment for boundary grading, as well as technical advice on the proper boundary alignment.

The roads used for anti-poaching patrols within all the protected areas are in a state of disrepair, and there is a need for renovations and construction of new patrol roads, particularly in the northern Serengeti, the Maswa Game Reserve, and the game controlled areas. In conjunction with the boundary marking planned for phase III, the SRCS will help with road work, once again through fund-raising for equipment and technical advice.

Surveys showed that ranger infrastructure (patrol posts, field gear, transport, radio communication, etc.) is inadequate in all the region's protected areas. SRCS will work with Tanzania National Parks to find the funds for field equipment, transport, and vehicle maintenance facilities, and for the rehabilitation and construction of ranger housing. Similar activities are planned in the Maswa Game Reserve and the proposed new game reserves, focusing on the construction of fully equipped, secure guard posts serviced by airstrips in areas where poaching and encroachment are presently high and law enforcement inadequate.

In collaboration with the Serengeti Tourism, Education, and Extension Program (STEEP), priorities for improving park interpretation facilities, particularly for local tourism, have also been identified. In conjunction with STEEP, we will help the park staff to provide education in natural history and wildlife management among the local villages, and to bring local people into the national park so that they can appreciate the wildlife spectacle.

SUSTAINABLE USE OF RESOURCES

The development of sustainable resource use in the local communities outside the protected areas is the only effective way of relieving human population pressures on the protected areas in the long term. However, if local communities are to accept new practices, it is essential that changes be made voluntarily by the residents, not through coercion or enforced controls. The SRCS has so far introduced methods to improve crop production, to provide positive incentives for farmers to use sustainable methods of cultivation, and to involve the farmers themselves in discussing problems and identifying solutions.

Land tenure is an essential prerequisite to promoting effective land husbandry, since villagers are unlikely to invest time, money, and labor in the careful utilization of resources that they do not own. At the commencement of phase II, no village in the region had title to its land. An immediate priority was therefore to provide assistance to villages and district authorities in demarcating villages and securing the title deeds. This work has now been completed in the Loliondo division of Ngorongoro District, and will soon commence in other areas in the region. Once villages have title to their land, they will be assisted in assessing their own resources and in developing a land use management plan. The SRCS will start in Loliondo, where land tenure is secured, and follow on in other areas as soon as land titles are issued, or where input by the SRCS is specifically requested by the relevant authorities.

The SRCS has carried out a few pilot projects designed to improve the use of resources such as timber. For example, in some villages tree nurseries have been established, which will provide the poles for building houses. Training has begun in tree nursery management, and village authorities have been very supportive. The expansion of these pilot schemes will focus on Serengeti and Bunda Districts, and will concentrate on training and material support through the Community Conservation Centres (see below). These projects will include seed and tree nurseries, the promotion of new crop varieties, training in soil conservation techniques, livestock health and improvement activities, and water development.

INTERACTIONS OF PEOPLE AND PROTECTED AREAS

Poaching of wildlife by local communities in the region is presently extensive, and is the main source of conflict and polarization between park authorities and local communities. Methods that encourage communities to understand the value of the wildlife resource and which provide direct and sustainable benefits to the communities have to be developed to address both the development needs of the people and the conservation needs of the protected areas.

In phase II we have made some progress toward this end. First, we have established communication channels between wildlife managers and the local communities. Second, we have developed new programs aimed at increasing the benefits to local communities from wildlife utilization. Third, we have suggested how existing buffer zones might be secured and extended, and what types of land uses in buffer zones would be acceptable for conservation.

Establishing Communication Channels

There is currently a lack of communication among the authorities in different protected areas. To improve communication between local people and the park, we convened two village workshops in Serengeti and Bunda Districts. The workshops were attended by village and district representatives and protected area managers.

These workshops provided a forum for the discussion of issues of common concern to park managers and local communities. Besides their primary objective of establishing a dialogue, the workshops provided information on village development needs, and on the people's perspective on wildlife utilization options. As a result, more workshops will be organized in phase III, covering all seven districts adjoining the national park. Although the workshops are a good way of communicating, they are infrequent, and so other methods are needed. To this end we will set up Community Conservation Centres to provide support for the Serengeti National Park extension warden.

The Community Conservation Centres will be established by the local communities themselves with support from members of the SRCS. Their purpose is not only conservation education but also to provide a focal point for collaborative activities between park managers and villagers, as well as for demonstrations, training, and other extension activities. Initially, three Centres will be established in the top priority areas of Serengeti and Bunda Districts, with later expansion of the program depending on the availability of funding and on the interest generated in the villages themselves.

Benefits from Wildlife for Local Communities

We have introduced schemes by which local people would receive tangible benefits from the protected areas so that they would then support conservation. We looked at four methods of providing these benefits. First, we started community wildlife schemes in which the local people carried out hunting and sold the products themselves. Second, we examined how the benefits from government and commercial utilization schemes might be diversified. Third, we looked at how revenues from the protected areas could be shared with local people, and fourth, we explored the implementation of community development programs by the protected area authorities.

Two methods were used to test the feasibility of community wildlife utilization schemes: first, information was gathered from village workshops on the current amount and type of catch, and second, experiments were designed to identify suitable traditional hunting methods. Both approaches have provided information on the traditional hunting methods used, the number of animals taken, and the success rate. We will determine in phase III which of these methods is the most appropriate to use on the pilot utilization schemes. We will also determine how village quotas for wildlife utilization should be established and regulated, how benefits to the villagers can be equitably distributed, and how they can be maximized, for example, through the manufacture of wildlife artifacts and through effective marketing structures. Monitoring systems will measure the effects of the pilot schemes on the wildlife populations. The pilot schemes will initially be in Serengeti and Bunda Districts, and will later be expanded to other areas depending on their success, requests from village authorities, availability of funding, and government approval.

To investigate mechanisms by which local people could benefit from government and commercial schemes, members of SRCS collected information on existing schemes operating in the region. Sport hunting fees are the largest source of income from wildlife utilization in the region, although this income is far from optimal because of the failure of hunting companies to use their hunting concessions effectively. We have suggested that there should be policy changes to increase the income from these concessions. We have worked with private sport hunting companies to find ways of testing specific mechanisms for distributing benefits to the communities. The Cullman Reward Scheme is a project that is funded by the voluntary donations of hunting clients and gives rewards in exchange for anti-poaching work and the collection of snares and weapons. It has been particularly successful and will now be expanded and duplicated.

We also examined other possibilities for revenue sharing with the local communities. However, further work requires an economic study to determine the percentage of revenue that could be diverted. We also need to develop mechanisms both for transferring funds to villages or district councils and for managing those funds.

During the village workshops, we looked at how the protected areas could use their infrastructure to assist in village development activities. For example, road construction equipment owned by the protected area authorities could be used to build and maintain village access roads. Currently these roads are in a poor state and receive almost no maintenance. The use of protected area infrastructure in this way makes good economic sense, as well as having positive benefits for park/people relations. These possibilities will be investigated further in collaboration with the SNP extension service. In addition, the SRCS will support training and education of national park wardens in community extension skills through STEEP.

Establishment of Buffer Zones

One of our top priorities was to examine the status of the areas adjacent to the national park and to determine ways in which a system of buffer zones could be established. Land adjacent to the national park is being subjected to a variety of land uses, and different areas have differing legal status. It is not practical to establish a single uniform category of buffer zone under these circumstances, and we have recommended a more flexible system of zones in which different legal statuses, management systems, and land uses operate. Three main categories of buffer zones are proposed:

1. Mandatory buffer zones: areas that are already gazetted, or are in the process of being gazetted, to protected area status. Two classes of mandatory zone are distinguished: class 1, where protected area status already exists—this class includes Maswa Game Reserve and the Ngorongoro Conservation Area—and class 2, where an upgrading of the legal status of the area is in progress—this class includes the Grumeti, Ikorongo, and Kijereshi Proposed Game Reserves.
2. Voluntary buffer zones: areas without protected area status where land uses compatible with conservation will be established voluntarily by local communities as part of village land use plans. This category includes part of Loliondo Game Controlled Area and those parts of the Ikorongo and Grumeti areas that will not be upgraded to game reserve status.
3. Hard edges: areas where land adjacent to the park is heavily developed for agriculture and settlement and where the introduction of an effective buffer zone is (in the short term) impractical. This category includes land adjoining the park in Serengeti and Tarime Districts to the northwest and in Bunda District at the end of the western corridor.

We have developed a number of land use prescriptions within each zone category, as well as a program of implementation to secure the buffer zones and to enable them to fulfill their function. These prescriptions include strengthening law enforcement, increasing revenues, improving physical infrastructure, and developing management plans for mandatory buffer zones; promoting village land use plans and pilot village wildlife utilization schemes for voluntary buffer zones; and promoting sustainable development activities and the removal of land uses that are incompatible with conservation from hard edge zones.

BUILDING A REGIONAL STRATEGY

The various initiatives that we have proposed above will evolve over time, and so our long-range strategy must also evolve. Therefore, the SRCS will develop in a number of ways to support the activities described above. First, we will establish planning and evaluation mechanisms. Second, we must generate broader support in the local community for our strategy, and finally, we must establish better communication and coordination mechanisms.

Planning mechanisms are needed to provide the technical and philosophical framework of the SRCS and to ensure that the discrete activities implemented under the umbrella of the SRCS are coordinated. We will need to examine and adjust objectives and activities as the strategy itself evolves.

So far, the principal mechanism by which SRCS activities have been planned and evaluated is its Steering Committee, which has responsibility for overseeing all activities of SRCS. With the expansion of activities in phase III, and the growing international as well as national significance of the program, additional planning mechanisms will be introduced. These will include a technical workshop and a series of specialist technical papers for examination of issues on which to base the interventions of the SRCS. We intend the second SRCS technical workshop, like its predecessor held at Seronera in 1985, to involve international as well as national experts, local community representatives, and both Tanzanian and Kenyan natural resource managers.

We realize that current wildlife exploitation by local communities may prove to be unsustainable. Better systems have to be developed to provide greater security for conservation, and to do this we must enlist central, regional, and local governments, the protected area managers, and the local communities themselves. So far this process has only just started, but a number of activities are proposed. These include the implementation of a national awareness campaign, the production of high-quality promotional and audiovisual materials, the regular publication of a quarterly newsletter, the design of an SRCS custom logo, and when opportunity arises, promotion of the strategy through national and international forums.

We need to establish communication and coordination mechanisms through which problems can be discussed by the resource users and managers, solutions designed, and activities coordinated. Effective communication and coordination will become increasingly necessary as the SRCS unfolds. The principal coordination mechanism has been provided by the SRCS Steering Committee, which includes central and regional government resource managers and planners, under the chairmanship of the Director of Wildlife. The Steering Committee has provided technical guidance and political support throughout the implementation of phase II, and has provided an effective mechanism for promoting cooperation and communication between strategy partners through which sectoral concerns can be aired. The Committee has met six times in phase II, convening in different localities around the region.

The Steering Committee will continue to provide the principal coordination mechanism for the SRCS in phase III. In addition, because of the increased complexity of the program and the growing number of strategy partners, we will need an additional coordination mechanism, the Steering Committee technical subcommittees. These committees will include representatives of other cooperating organizations and specialists and will provide a mechanism for promoting wider involvement in SRCS activities.

PROGRAM ORGANIZATION

The SRCS has accomplished the majority of the objectives established for phase II. Major regional resource issues have been identified and quantified, priorities have been set for the management of protected areas, and a start has been made on identifying and testing approaches to promoting sustainable development in the local communities. A range of activities has begun to promote positive park/people interactions. In addition, the SRCS itself has been established on a firm footing through awareness of the problems and the potential solutions, and through the construction of a planning and coordination framework. The stage has been set to implement the various proposals of phase III that have been identified in phase II.

Although the majority of these proposals will be implemented by government agencies and other cooperating organizations, the SRCS itself will need to have sufficient technical, administrative, and financial resources to ensure an adequate level of support. A program coordination unit (PCU) will therefore be created in phase III under the overall direction of a program director seconded by the Ministry of Tourism, Natural Resources, and Environment. The technical program will be supervised by a chief technical advisor provided by IUCN. A number of other technical officers and advisors and support staff will complete the PCU staff structure. Offices and staff housing for the PCU will be established outside the national park at Nyabuta in Serengeti District, one of the top priority areas for the project.

Because the SRCS covers a very large area with poor communication facilities, discrete, semiautonomous program components will be established in different geographic localities, with the responsibility for implementing the phase III proposals in that area. Coordination of these program components and administrative and logistical support will be provided by the PCU. The first such component will be launched in Serengeti and Bunda Districts by the PCU itself, and will provide a basis for testing many of the concepts and pilot activities mentioned above prior to their being extended to new program components.

ACKNOWLEDGMENTS

We thank all those who helped develop the SRCS, too many to mention. In particular, we thank C. A. Mlay, Director of Wildlife; D. Babu, Director of Tanzania National Parks; B. Maragesi and W. Summay, Chief Park Wardens of Serengeti National Park; H. M. Nyka, Acting Director of Serengeti Wildlife Research Centre; and M. Borner of the Frankfurt Zoological Society. This strategy has been funded by the Norwegian Agency for International Development (NORAD), the Frankfurt Zoological Society, and the European Community Food Aid Counterpart Fund.

TWENTY-NINE

A Model to Evaluate Alternative Management Policies for the Serengeti-Mara Ecosystem

R. Hilborn et al.

The Serengeti has been the site of dozens of research projects since the 1960s, and there have often been as many as fifteen researchers active in the park at any time. This research is well known in both popular and scientific circles around the world, and the contribution of these research programs to our scientific understanding has been great. The list of major books, scientific papers, and films resulting from research in the Serengeti is unrivaled. Yet in the 1990s researchers are under increasing pressure to justify the relevance of their work to the needs of the people living in and around the Serengeti, and to the nation of Tanzania.

The Serengeti ecosystem is affected by many actions of a governmental and nongovernmental nature. Anti-poaching patrols, burning policy, hotel development, animal vaccination in areas surrounding the park, and changing land uses in surrounding areas have become management issues. Further, changes in rainfall pattern, international tourism, and growth of the human population in areas surrounding the park will all have major implications for the plants and animals living in the Serengeti ecosystem. The decision makers in Tanzanian government agencies are actively considering their management options, and it is the responsibility of researchers who work in the Serengeti to ensure that the knowledge they have gained is available to these government officials to help them predict the likely consequences of their management actions.

To assist in this transfer of scientific knowledge from researchers to managers, a workshop was held at the Serengeti Wildlife Research Centre (SWRC) on 7–11 December 1991.* The purposes of the workshop were

*The other participants in the workshop, in random order, are N. Georgiadis, J. Lazarus, J. M. Fryxell, M. D. Broten, B. N. N. Mbano, M. G. Murray, A. R. E. Sinclair, S. M. Durant, B. Mwasaga, M. K. S. Maige, P. Arcese, S. Albon, H. Hofer, M. Kapela, A. Dobson, M. East, H. Nkya, H. T. Dublin, C. Packer, K. L. I. Campbell, S. C. Gascoyne, S. R. Creel, P. Hetz, N. M. Creel, and T. M. Caro.

threefold: (1) to synthesize current knowledge in order to evaluate possible management actions; (2) to bridge the gap between researchers and managers; and (3) to identify critical uncertainties and major data needs.

The workshop used a method developed at the University of British Columbia in the 1970s (Walters 1974; Holling 1978) in which the construction of a simulation model is used as the mechanism to achieve the three objectives listed above. The key elements in this workshop process are to have managers and decision makers specify what questions they would like to ask, and to have scientists use their knowledge to build a policy evaluation tool in the form of a model. This approach has been widely used on perhaps 150 environmental management problems (Holling 1978; Walters 1986).

In the previous Serengeti book (Sinclair and Norton-Griffiths 1979) Hilborn and Sinclair (1979) constructed a model of the Serengeti ecosystem. This model focused exclusively on the interaction of rainfall, grass, ungulates, and predators, and was designed as a research, rather than a management, tool. The Hilborn and Sinclair model did predict quite accurately how the wildebeest would respond to the rainfall regime that has occurred in the 14 years since the model was constructed.

DESIGNING THE MODEL

Types of Models

The most important step in building a model is deciding what the model will exclude. No model can possibly be as complex as any natural system (or even a single cell), so modelers must begin by bounding the problem—what will be included and what will be left out. What a model will include depends upon the reason for building the model; some models include only one or two variables, others include thousands. Below I list frequent uses for models and comment on the appropriate model size for each use.

Prediction: Models used to make predictions are usually very small, and many phenomena are compressed into simple relationships (Linhart and Zucchini 1986).

Research coordination: Models used to help coordinate researchers and managers are generally complex enough to include the factors that each researcher studies.

Policy evaluation: While similar to predictive models, policy evaluation models are more frequently used to explore scenarios by simulation and must be complex enough to include the array of management actions under consideration.

Institutional memory: This type of model is a statement of how those who construct it believe the system works. As such, the model can serve as memory for later managers or scientists who need to know the understanding of their predecessors.

Management training: Models are often used to help train decision makers in the dynamic response of the systems they manage. Aircraft flight simulators are very sophisticated versions of management training models, but models have seen considerable use in the training of natural resource managers (Hilborn and Walters 1987). These models are similar to policy evaluation models but usually are highly refined to meet user needs.

Optimization: Models can be used in formal optimization algorithms to calculate the "best" actions to achieve some specified objective.

Hypothesis testing: Models can serve as a formal statement of a scientific hypothesis that can be used for quantitative comparison with other models and data. Such models are generally small and simple.

The model constructed during the Serengeti workshop was built primarily for research coordination and policy evaluation. Thus, when deciding what to include and what to exclude, we generally opted to include the species and/or research areas of all the participants, but due to the short duration of the workshop we often had to make the model simpler than most researchers would have preferred.

Building a model for research coordination involves many compromises. Each individual researcher frequently has detailed knowledge of one component of the system. As the purpose is not to look in detail at any researcher's knowledge and data, but rather to foster communication between researchers and managers, the workshop technique uses several devices to make sure the model is simple enough to be understood by all participants and that it addresses the key interdisciplinary connections that are often missed in individual research programs.

Issues the Model Should Address

We begin by considering what issues the model should address; those chosen by the participants were:

Population change in humans
Visitor capacity
Adjacent land use
Hunting and poaching
Climate change
Disease outbreak
Vegetation change
Species loss
Economics and cash flow
International tourism
Population dynamics of herbivores and carnivores

Indicators of Performance. Once we have chosen the general areas to include, we then list the key indicators of system performance. These indicators are the outputs from the computer model that we will use to evaluate how well any particular management policy performs. The indicators chosen were:

Animal population sizes
Tourist numbers
Tourist satisfaction
Revenue to national parks
Employment
Vegetation condition
Illegal harvest
Household income
Livestock per family
Encroachment on national parks land
Local population health care level

Management Actions to Consider. The next step is to decide what management actions we hope to be able to evaluate. These are the analogues of the flight controls on an aircraft flight simulator; those chosen were:

Poaching enforcement
Hotel construction
Livestock vaccination
Burning and/or suppression of burning
Road construction
Adjacent land uses
Reintroductions
Improvement of water supply and infrastructure

Spatial and Temporal Resolution. The lists of actions and indicators above can be thought of as the design criteria for our model. They tell us what the model should be able to accept as inputs and what it should produce as outputs. Next we must decide what spatial and temporal scales the model will use. There are a number of options for spatial scale, including (1) a grid pattern, (2) areas of arbitrary size and definition, (3) an explicit model in which each organism has a location in space, and (4) no spatial resolution or implicit spatial resolution.

After some discussion and consideration of computer limitations, we decided on option 2, a model with ten spatial areas as shown in figure 29.1. The areas were chosen to reflect the annual migratory pattern of wildebeest, zebra, and Thomson's gazelle, and to include the areas of significant human impact. Many participants felt that smaller spatial units would be appropriate, but considering that we had only 4 days to build

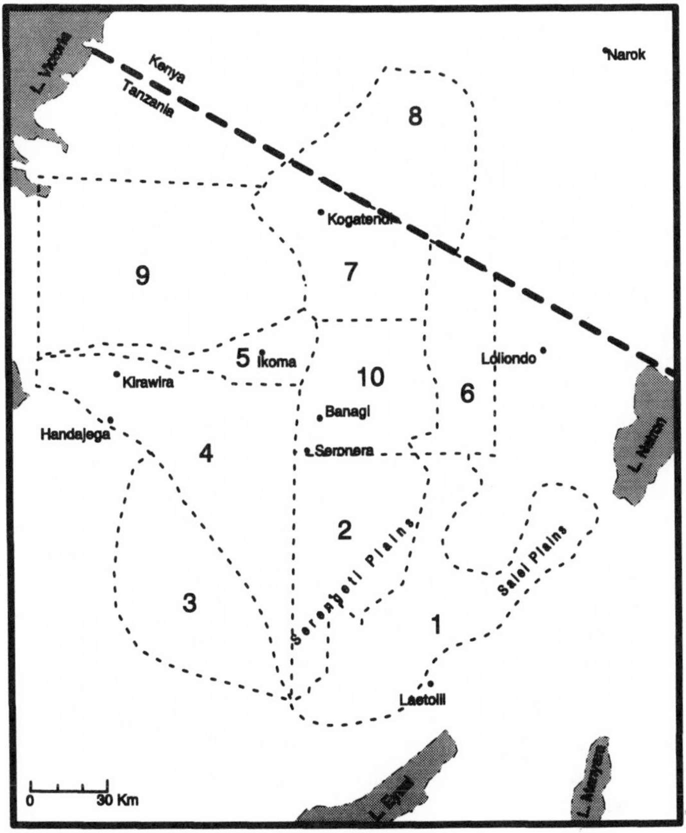

Figure 29.1 Map of the Serengeti ecosystem showing the ten spatial areas of the model: 1, Plains East; 2, Plains; 3, Maswa; 4, Western Corridor; 5, Ikoma; 6, Loliondo; 7, Northern Park; 8, Mara; 9, Northwest; 10, Seronera.

the model and that its purpose was research coordination, we used the areas shown.

The appropriate temporal resolution was also a compromise between conflicting objectives. We elected to consider two seasons, an 8-month wet season and a 4-month dry season. Some parts of the model, such as human population dynamics, did not need to operate on this intra-annual step, but others, particularly vegetation, ungulates, and predators, did need such a division.

Model Components

Subgroups. A key element in the workshop process is to break the participants into groups according to areas of disciplinary specialty, but to use the lists of actions and indicators to ensure that these subgroups spend their time building a submodel that is appropriate to the other groups, rather than building a detailed model of their own specialty that is not appropriate for interfacing with the rest.

We divided into five subgroups; group titles and participants are listed in table 29.1.

Information Flow. The key technique for ensuring that each subgroup provides the information needed by other groups is the information flow table, often called the "looking outward" table (Walters 1974). In this table each group specifies what inputs it needs from other groups in order to simulate the indicators listed earlier and evaluate the actions. The ultimate construction of the submodels is thus dictated by the demands of other groups. This technique was developed to help counter the trend of subgroups to build models that were of internal interest (looking inward) rather than being of interest to the other groups (looking outward).

As this table is constructed, the entire workshop gains an understanding of what the variables are and in what units the model will be built. We attempt, when filling in this table in what is normally a long and tiring exercise, to specify the exact units and spatial scale of each variable. Table 29.2 is the result of this process.

Generally, the model as defined at this stage remains a bit more complex and detailed than the final product. As the groups work on their own submodels, they either realize that some of the factors listed earlier are too detailed to be considered, that the process can be implicitly incorporated in another model, or that time or data are unavailable.

In the following sections we discuss the model as actually constructed during the workshop.

MODEL STRUCTURE

The model began with the year 1960, when research began in the Serengeti and the wildebeest eruption took place. We normally ran a 60-year scenario, with the year 2020 being the last year of simulation. For each year, the submodels were called in the following order: (1) vegetation, (2) ungulates, (3) predators, (4) inside park, (5) outside park. Finally, for the

Table 29.1 Group assignments.

Vegetation	Ungulates	Predators	Inside/park	Outside park
N. Georgiadis	J. Lazarus	J. M. Fryxell	M. D. Broten	B. N. N. Mbano
M. G. Murray	A. R. E. Sinclair	S. M. Durant	B. Mwasaga	M. K. S. Maige
P. Arcese	S. Albon	H. Hofer	M. Kapela	A. Dobson
		M. East	H. Nkya	H. T. Dublin
		C. Packer	K. L. I. Campbell	S. C. Gascoyne
		S. R. Creel	P. Hetz	
		N. M. Creel		
		T. M. Caro		

Table 29.2 Information flow table.

	To group				
From group	Vegetation	Ungulates	Predators	Inside park	Outside park
Vegetation	Grass Brush Small trees Large trees	Same	Same	Same	Same
Ungulates	# wildebeest # zebras # tommies # brn animals # giraffes # elephants	Same	Same	Same	
Predators		# eaten by species	# lions # hyenas # cheetahs # wild dogs # leopards # jackals	Same	
Inside park	Off-road trips % burned		# road kills	# visitor nights Hotel capacity Roads	Employment
Outside park	Land use Livestock	Kill by species Land available Disease Deaths	# killed	Poaching effort Livestock	Land use Hunters Human pop Livestock Diseases Amenities

end of each year, a graphics routine was called to plot the scenario on the computer screen. The program was written in QuickBasic, and run on MS-DOS–compatible computers. It takes approximately 3 minutes on a 20 MHz 386 computer to run the model from the year 1960 to 2020.

Vegetation Submodel

The vegetation group built a model that included grass, bushes, and several size classes of trees, similar to the model published in Dublin, Sinclair, and McGlade (1990). However, due to time constraints, the full vegetation model was not incorporated into the simulations performed at the workshop. Here we describe only the parts of the vegetation model concerned with two classes of grass: dry season grass and wet season grass. The key variables in the vegetation model are:

$Rain_{dry:y}$ = dry season rain in year y in mm
$Rain_{wet:y}$ = wet season rain in year y in mm
$Grazed_{y,i}$ = percentage of area grazed in dry season, year y, area i
$Burned_{y,i}$ = percentage of area burned in year y, area i
$Tallgrass_{y,i}$ = amount of dry season grass in year y, area i (kg/ha)
$Green_{y,i}$ = amount of dry season new growth in year y, area i (kg/ha)

$Woods_{y,i}$ = percentage of area i that is woodland in year y
$Cultiv_{y,i}$ = percentage of area i that is cultivated in year y
$Wild_y$ = total wildebeest population in year y (in thousands)
$pburn_{y,i}$ = the proportion of area i not in cultivation that is burned in year y

The equations for the vegetation model are:

$$Grazed_{y,i} = Wild_y/(1{,}000 + Wild_y)$$

This simply says that the percentage of the area grazed increases as the wildebeest population increases, with 50% of the area grazed when there is a wildebeest population of 1 million. Note that this percentage is determined by the total wildebeest population size because we assume that the distribution of wildebeest is the same in every dry season.

$$Burned_{y,i} = pburn\ (1 - Grazed_{y,i})$$

The variable *pburn* is determined in the parks management submodel, and here we simply assume that areas that have been intensively grazed by wildebeest will not burn.

$$Tallgrass_{y,i} = (7.7 \times Rain_{wet,y} - 202) \times (1 - Burned_{y,i} - Woods_{y,i} - Cultiv_{y,i})$$

$$Green_{y,i} = -800 + Rain_{dry,y} \times 8,$$

or if $Green_{y,i}$ is predicted to be less than 50, then $Green_{y,i} = 50$.

In both of these equations the grass production is assumed to be largely determined by rainfall.

The key interactions of the vegetation with other components of the system are that as the wildebeest population changes, the proportion of the area burned will change, and as land is cultivated or converted to woodland, the amount of grass available will be reduced.

Ungulate Submodel

The following definitions are used in the ungulate submodel:

$N_{y,i,j}$ = numbers of ungulate species j, year y, area i
$Metwt_j$ = metabolic weight of species j
$Resid_{s,i,j}$ = proportion of migratory species j resident in area i in season s
$HalfSat_j$ = the logarithm of dry season grass per animal that produces 50% survival (half saturation)

There are three migratory species in the model—wildebeest, zebra, and Thomson's gazelle—and three nonmigratory species—elephant, buffalo, and "brown animal," which collectively refers to topi, impala, and kongoni. The key relationships in the population dynamics of the ungulates can be summarized as

$$N_{y+1} = N_y\text{SurvDry} + \mathit{Calves} - \mathit{Hunterkill} - \mathit{Disease\ deaths} - \mathit{Predator\ kills}.$$

The number of calves born is simply proportional to the population size:

$$\mathit{Calves} = N_y \times \mathit{Calving\ rate}.$$

The key dynamic factor is the dry season survival, which is assumed to be related to the amount of dry season food per individual by the following relationship:

$$\mathit{SurvDry} = \log(\mathit{GrassPerAnimal})/[\log(\mathit{GrassPerAnimal}) + \mathit{HalfSat}]$$

where the grass per animal is

$$\mathit{GrassPerAnimal} = (\mathit{Grass} \times \mathit{area}_i \times 100)/(N_{ij} \times \mathit{Metwt}_j \times 1{,}000 \times 120).$$

The numerator is *grass* (kg/ha) times 100 ha per km^2 times the number of km^2 in area *i*. For wildebeest and Thomson's gazelle *grass* is dry season grass. For zebra *grass* is wet season grass/4.5, which reflects the use of long grass during the dry season by zebra but the lower value of the long wet season grass during the dry season. For the nonmigratory species *grass* is the wet season grass divided by 3. The denominator is the number of individuals present, times their metabolic weight, times 1,000 to convert from numbers in thousands to numbers times 120 days in the dry season. The parameters for the starting numbers, metabolic weights, survival, and *HalfSat* values of the ungulates are given in table 29.3.

At the beginning of each season, the migratory species are allocated to the ten areas based on the residence proportions *Resid* shown in table 29.4 and the following equation:

$$N_{y,s,i,j} = N_{y,j}\,\mathit{Resid}_{s,i,j}.$$

Predator Submodel

The predator submodel has two major components: calculation of the kill of prey items, and the population dynamics of the predators. The kill of prey is calculated from the multiprey type II functional response, whose form is

$$\mathit{Kill}_i = \frac{\mathit{Density}_i \times \mathit{pAttack}_i}{1 + \Sigma_j\, \mathit{Handle}_j\, \mathit{Density}_j\, \mathit{pAttack}_j}$$

where

$\mathit{Density}_i$ = density of the prey in numbers per ha
Handle_i = handling time for a single predator to consume one prey *i*
$\mathit{pAttack}_i$ = probability of successful attack on species *i*
Kill_i = number of prey items of species *i* killed per unit time

Table 29.3 Parameters for the ungulate submodel.

Species	1960 numbers (in thousands)	Metabolic weight	Calf survival	Half saturation
Wildebeest	250	53	0.2	0.12
Zebra	200	80	0.15	1.5
Thomson's gazelle	50	9.5	0.2	0.5
Elephant	2.5	503	0.05	30
Buffalo	225	105	0.15	110
Brown animals	100	32	0.25	20

Table 29.4 Residence proportions of migratory ungulates in different areas in the wet and dry seasons.

	Wet season			Dry season		
Area	Wildebeest	Zebra	Thomson's gazelle	Wildebeest	Zebra	Thomson's gazelle
1 Plains East	0.5	0.4	0.6	0	0	0
2 Plains	0.4	0.3	0.4	0	0	0.1
3 Maswa	0.1	0.1		0	0	0
4 Western corridor		0.05		0.3	0.2	0.4
5 Ikoma		0		0.1	0.1	0
6 Loliondo		0.1		0	0	0.2
7 Northern park		0		0.2	0.2	0
8 Mara		0		0.2	0.3	0
9 Northwest		0		0.	0.	0
10 Seronera		0.05		0.2	0.2	0.3

Note that the subscripts are given only for prey species; when calculating, these calculations are made for each predator species in each area in each season. The handling times for each predator and prey species are given in table 29.5. The probability of successful attack is given in table 29.6.

The population dynamics of the predators are summarized as follows:

$$N_{y+1} = N_y\,(1 + \mathit{Recruitment} - \mathit{Mortality}) - \mathit{Number\ poached}.$$

The number poached applies only to lions and hyenas and for a single area is simply

$$\mathit{Number\ poached} = \mathit{Predators\ present} \times \mathit{Poacher\ trips} \times \mathit{Poacher\ efficiency}.$$

The mortality of predators is assumed to be linearly related to predator density as follows:

$$\mathit{Mortality} = \mathit{Base\ mortality} \times \mathit{Slope} \times \mathit{Predator\ density}.$$

The base mortalities and slopes are given in table 29.7. For cheetahs the mortality is assumed to increase with hyena density as well as cheetah density.

Table 29.5 Relative handling times for different predator and prey species.

Prey species	Lion	Hyena	Cheetah	Leopard	Wild dog
Wildebeest	3.14	8.6	4.3		8.1
Zebra	4	10			
Thomson's gazelle	0.46	1.2	0.27	1	0.5
Elephant	200				
Buffalo	12				
Brown animals	1.43	3.7	2.4	3.3	4.5

Note: These numbers are multiplied by 4.0 for use in the model.

Table 29.6 Relative probability of successful attack used as an indicator of prey preference.

Prey species	Lion	Hyena	Cheetah	Leopard	Wild dog
Wildebeest	1	0.2	0.1		0.05
Zebra	1	0.2			
Thomson's gazelle	0.1	1	0.1		1
Elephant					
Buffalo	0.7				
Brown animals	1	0.2	1	1	0.1

Note: These numbers are multiplied by 0.002 for use in the model.

The recruitment rate is assumed to decrease exponentially as a function of density using the following equation:

$$\textit{Recruitment rate} = \textit{Max recruitment} \times \exp(-\textit{constant} \times \textit{density}).$$

For lions, hyenas, and leopards, *Max recruitment* is proportional to the abundance of their food, averaged over the wet and dry season. Food is the numerator of the prey kill equation times the weight of the individual prey items.

$$\textit{Max recruitment} = \textit{Food available} \times \textit{Recruitment maximum parameter}$$

For cheetah and wild dogs, the maximum recruitment is assumed to be prey independent. Table 29.7 gives the population dynamics parameters for each species. Table 29.8 gives the initial numbers of predators by area.

All species except hyenas are assumed to be resident. Hyenas are assumed to "commute" to feeding areas, so that in each season the "feeding area" for the hyenas in each area is specified. In practice, this means that hyenas on the edge of the plains commute to the woodlands to feed in the dry season, and woodland hyenas commute to the plains to feed in the wet season.

Inside Park Submodel

The model of inside-park activities has two major components. Most important is the assessment of tourism quality and tourism growth, which is closely linked to the abundance of major species and tourist density.

Table 29.7 Population dynamics parameters for predator model.

Species	Lion	Hyena	Cheetah	Leopard	Wild dog
Base mortality	0.02	0.06	0.05	0.02	
Slope of mortality	0.3	0.0	0.076	0.0	
Mortality per poacher trip	1/100,000	1/100,000			
Recruitment maximum parameter	0.052	0.08	0.6	0.0048	0.15
Recruitment slope parameter	−5.31	−0.866	−6	−16	−4.55

Table 29.8 Predator numbers in 1991.

Area	Lion	Hyena	Cheetah	Leopard	Wild dog
1 Plains East	100	2,000	100	0	0
2 Plains	100	1,500	200	0	0
3 Maswa	300	300	35	130	15
4 Western corridor	700	1,800	100	180	20
5 Ikoma	100	700	0	70	10
6 Loliondo	100	200	40	100	12
7 Northern park	300	300	40	110	14
8 Mara	400	1,500	50	130	15
9 Northwest	0	0	0	0	0
10 Seronera	700	1,200	35	120	15

Note: 1960 numbers are assumed to be 60% of 1991 numbers for lion and hyena.

The second component is a large number of calculations of employment and revenues.

The key relationships are as follows:

$$TourismQuality = AnimalQuality \times (1 - Crowding)$$

Tourism quality goes up as more animals are seen, and down as more tourists are present.

$$Crowding = Tourists/(150{,}000 + Tourists)$$

Crowding is an increasing function of the numbers of tourists.

$$AnimalQuality = UngulateQuality + PredatorQuality$$

The more ungulates and predators seen, the better for tourism quality.

$$UngulateQuality = (Wildebeest + Zebra + BrownAnimals)/10$$

Wildebeest, zebra, and brown animals are so much more abundant than other species that they dominate what is seen, but since their numbers are in thousands and we divide by 10, we are saying, in effect, that ten thousand wildebeest are equal to one cheetah.

$$PredatorQuality = 0.5 \times Lions + Cheetahs + Leopards$$

Lions are considered half as valuable as cheetahs and leopards. Hyenas were not considered a tourist attraction.

$$TourismGrowthRate = 1 + Sensitivity \times (TourismQuality - 1{,}800)$$

This is a linear relationship between *TourismQuality* and the growth rate in tourism. When *TourismQuality* is greater than 1,800, tourism will increase; when it is less than 1,800, tourism will decrease. The sensitivity parameter determines how quickly it will increase or decrease, and a value of 4/40,000 was used for base runs. The constants 1,800 and 4/40,000 were selected by trial and error to make the simulations roughly mimic the real system behavior.

The following calculations were made for employment and parks revenue:

$$TotalTourists_{y+1} = TotalTourists_y \times TourismGrowthRate$$
$$TotalTouristNights = TotalTourists \times TouristStayDuration$$

Tourist stay duration is 2 nights.

$$BedsInPark = BedsInPark \times BedGrowthRate$$

The growth rate of beds in the park is specified as a control variable for different scenarios.

$$TotalRevenues = (TotalTouristNights \times DailyFee) + (TotalTouristNightsInHotel \times HotelFeetoParks) + (TotalCampNights \times CampFee)$$

The sources of revenue are a daily park fee (\$15.00), a fee from the hotel per night spent in hotel (\$5.00), and a fee per camper night (\$15.00).

Some fraction (50% or 75%) of the total revenue is allocated to the park's operating revenue; the rest is passed on to the Tanzanian government.

The park's residual funds are computed as follows:

$$ResidualFunds = ResidualFunds + OperatingRevenue - OperatingBudget$$

The operating budget has two components:

$$OperatingBudget = AntiPoachingBudget + CapitalImprovements + ParksManagement$$

Employment is assumed to be dependent on the number of tourist nights as follows:

$$ResidentEmployees = 0.6 \times TotalTouristNights/365$$
$$Dependents = ResidentEmployees \times 5$$

Resident employees include both hotel staff and park staff.

Outside Park Submodel

The three major components of the submodel for human activities outside the park were poacher effort and kill, human population growth, and changes in land use outside the park.

The poaching model assumes that residents of each area will go on poaching trips in their own area and an adjacent area. The number of potential trips made by residents of an area depends upon the number of people in the area and a local factor we call *TripsPerPerson.* Thus, as the human population grows, the number of potential poacher trips increases.

We assumed that anti-poaching patrols decrease the number of poacher trips, or at least their effectiveness. The number of poacher trips is calculated as:

$$PoacherTrips = Population \times TripsPerPerson \times Reduction$$
$$Reduction = 1 - PatrolDays/(120 + PatrolDays)$$

Thus, when the number of patrol days is 120, *PoacherTrips* will be 0.5 times the potential poacher trips (*Reduction* = 0.5). High levels of enforcement will decrease *PoacherTrips* to close to zero. These calculations are made for each area in each year.

The poacher kill of species *i* is:

$$PoacherKill_i = Number_i \times PoacherTrips \times VulnerabilityToPoachers_{i.}$$

The human population is assumed to grow by a constant rate, which can be changed. Table 29.9 lists the areas and human-related parameters of each area.

Land in wilderness is assumed to decrease in proportion to the human population and to be converted to cultivated land. The number of domestic animals is assumed to be three per resident.

PREDICTIONS AND SCENARIOS

We have used the model to explore six scenarios that reflect a variety of natural and human-induced changes in the Serengeti. It cannot be over-

Table 29.9 Habitats, and human parameters of the ten areas.

Area	Spatial area (sq. km)	Human population (1,000s)	Hunting trips per 1,000 people	% area cultivated	% area wilderness	% area urban	Livestock
1 Plains East	4,000	10	50	1	98	1	130
2 Plains	3,000	0.1	10	0	100	0	0
3 Maswa	2,600	50	50	0	100	0	0
4 Western corridor	3,600	1	10	0	100	0	0
5 Ikoma	1,500	100	50	0	100	0	0
6 Loliondo	2,000	20	10	2	97	2	90
7 Northern park	2,200	0.1	10	0	100	0	0
8 Mara	2,600	1	0	12	86	2	330
9 Northwest	4,000	1,000	10	30	86	4	337
10 Seronera	2,400	0.1	10	0	100	0	0

emphasized that these scenarios are extremely tentative and reflect a level of resolution that was chosen for the convenience of a 4-day workshop rather than the most appropriate level for prediction. The purpose of the workshop was to facilitate communication between researchers and National Parks staff. Thus, these scenarios reflect what the model says will happen, rather than being a best scientific estimate. Further, very little effort was made to adjust model parameters so that model outputs were close to current or historical values; thus most of the quantitative values of population sizes, tourist numbers, and so forth may be significantly different from what is actually observed in the Serengeti. Given all these caveats, it is nevertheless interesting to explore the predicted consequences of different decisions and natural events.

Scenario 1: Current Actions

Figure 29.2 shows the "base scenario" without any changes in current actions. The figure is divided into four panels, which show the values for key indicators of different components of the model.

The model run begins in 1960, and we can see that the increase in the animal populations mimics reasonably well the changes from 1960 to 1990. Wildebeest increase dramatically, zebras relatively little. The predators show slight increases, with hyenas showing the most. The closure of the border in 1977 is reflected in the drop in tourists and a drop in tourism quality due to lack of infrastructure.

Poaching effort increases constantly throughout the scenario, and poacher kill increases rapidly in the 1960s and 1970s, peaks in the 1980s, and then declines as the wildebeest and zebra are declining.

Scenario 2: Increased Anti-Poaching Patrols

Figure 29.3 shows a scenario in which anti-poaching patrols are increased fivefold in 1990. This increase causes a dramatic reduction in poacher kill, which arrests the decline of the ungulates in the late 1990s. Essentially this scenario shows that the decline in the ungulates seen in scenario 1 could be reversed by poaching control. The poacher kill in scenario 1 peaked at about 100,000 (wildebeest and zebra), while in scenario 2 the poacher kill never exceeded 50,000. Note also that the decline in hyenas is less severe in scenario 2 than it was in scenario 1.

Scenario 3: Greatly Reduced Anti-Poaching Patrols

Scenario 3 (fig. 29.4) explores the consequences of almost total elimination of anti-poaching patrols in 1990. The model predicts that this would lead to a rapid decline in the ungulates so that both wildebeest and zebra would be nearly gone by the year 2000.

The hyenas and lions respond to the decline of the ungulates by severe

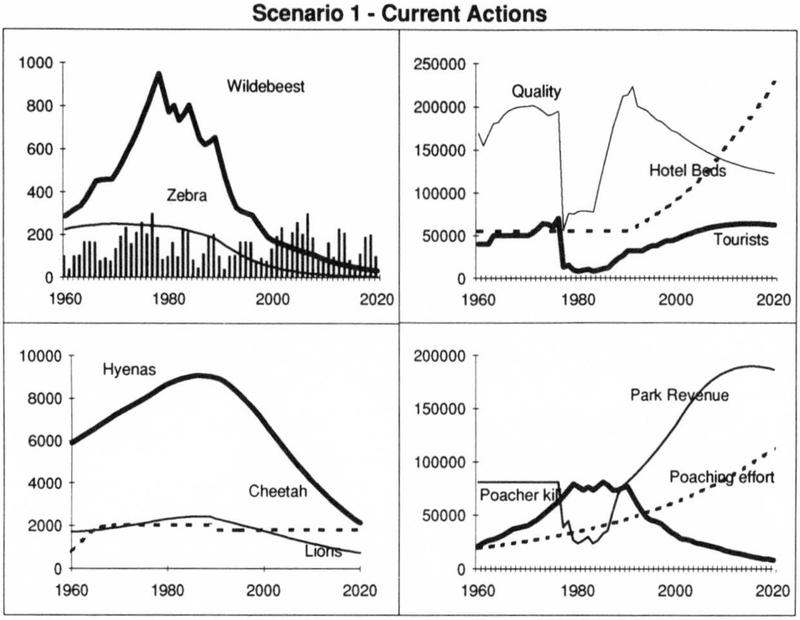

Figure 29.2 Output from scenario 1, the current situation of animal populations, park management, and poaching. Shown in upper left panel are wildebeest population (thick solid line) and zebra population (thin solid line), both in thousands, and dry season rainfall in mm (vertical bars). Shown in the upper right panel are number of tourist nights (thick solid line), number of hotel beds (dashed line), and tourism quality (thin solid line). Shown in the lower left panel are numbers of hyenas (thick solid line), numbers of lions (thin solid line), and numbers of cheetahs (dashed line). Shown in the lower right panel are poacher kill (thick solid line), park revenue (thin solid line), and poaching effort (dashed line).

declines of their own. Cheetahs are unaffected because they feed primarily on Thomson's gazelles, which are not severely poached.

Scenario 4: Poor Rainfall

In scenario 4 (fig. 29.5) we assume that rainfall after 1990 will be 66% of that seen in scenario 1. There is reasonably little difference between this scenario and scenario 1, except that the decline in ungulates is more rapid.

Scenario 5: Human Population Decline

In scenario 5 (fig. 29.6) we explore the consequences of a severe decline in the human population, a possible consequence of the current AIDS epidemic or some similar catastrophe. From the year 1990 onward we let the human population decline by 5% each year. This is rapidly reflected in a reduction of poaching effort and poacher kill, and a rebuilding of the wildebeest population.

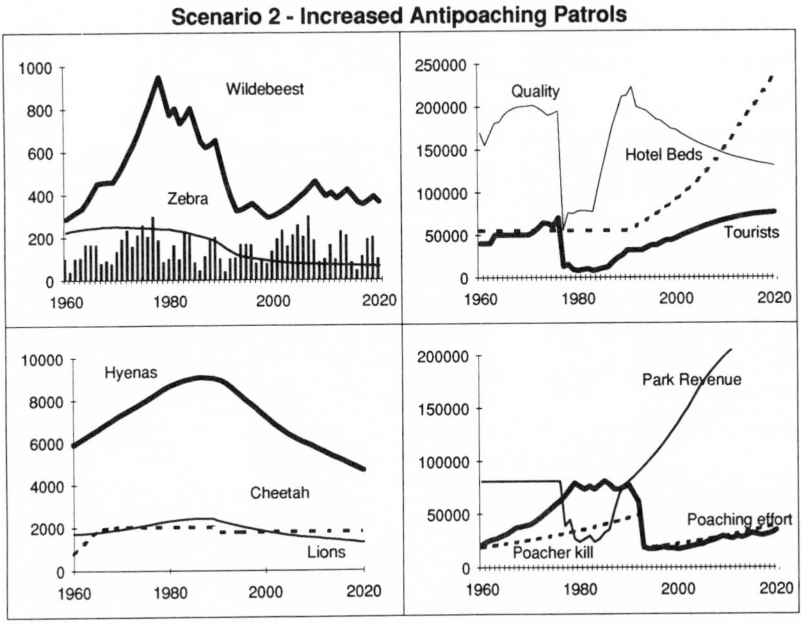

Figure 29.3 Output from scenario 2, a fivefold increase in anti-poaching patrols. See figure 29.2 for description of indicators.

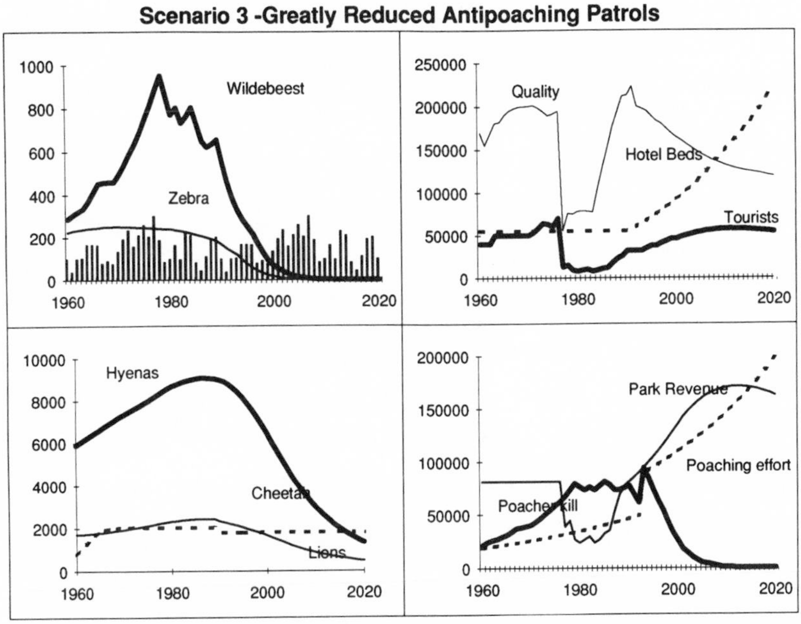

Figure 29.4 Output from scenario 3, a near elimination of anti-poaching patrols in 1990. See figure 29.2 for description of indicators.

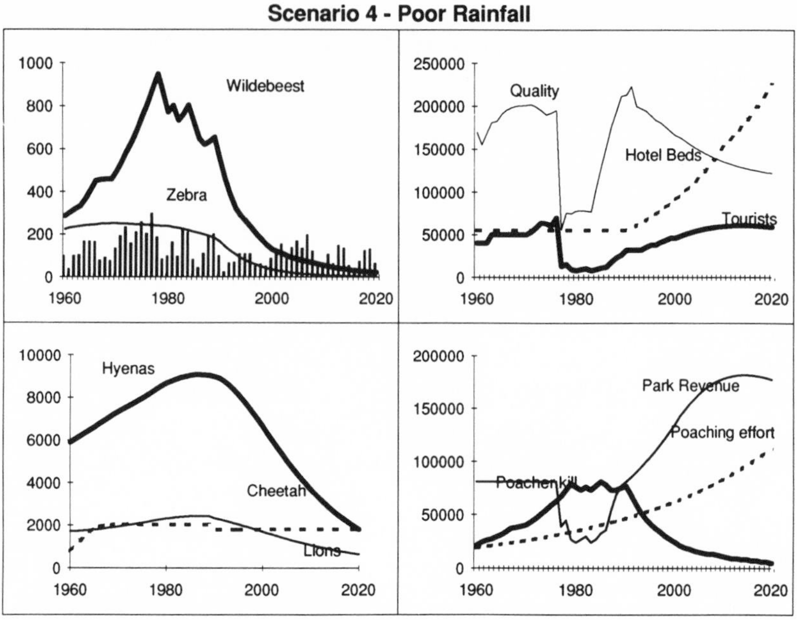

Figure 29.5 Output from scenario 4, a 44% decrease in rainfall from 1990. See figure 29.2 for description of indicators.

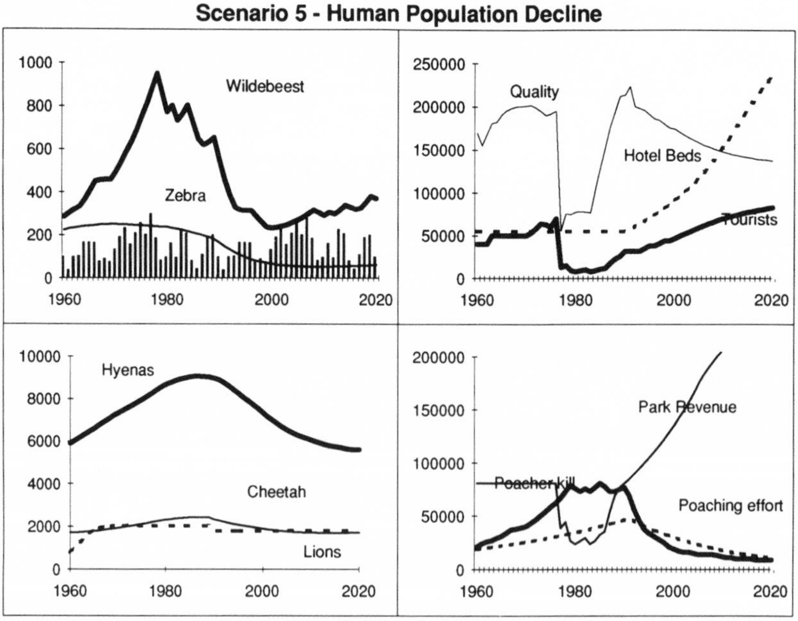

Figure 29.6 Output from scenario 5, an annual decrease in human population of 5% from 1990. See figure 29.2 for description of indicators.

Scenario 6: Rinderpest Epidemic

Scenario 6 (fig. 29.7) examines the effects of an outbreak of rinderpest. We have simulated a rinderpest epidemic by adding an additional 10% mortality each year, the consequence being a very rapid decline in wildebeest numbers by 1990.

CONCLUSION

The primary purposes of this model were to facilitate communication and to identify research priorities. The scenarios shown above provide a tentative exploration that can be used to guide more detailed data collection and modeling. There were few dramatic surprises in the scenarios explored, except perhaps the slight impact of reduced rainfall. It is clear that if the levels of mortality due to poaching are as high as we have assumed in the model, then the fate of the wildebeest and buffalo depends primarily on the control of poaching.

Research Priorities

We have identified three major areas in need of further research, the results of which would make the predictions of any model more reliable. The fate of the ungulates appears to depend primarily on the level of the

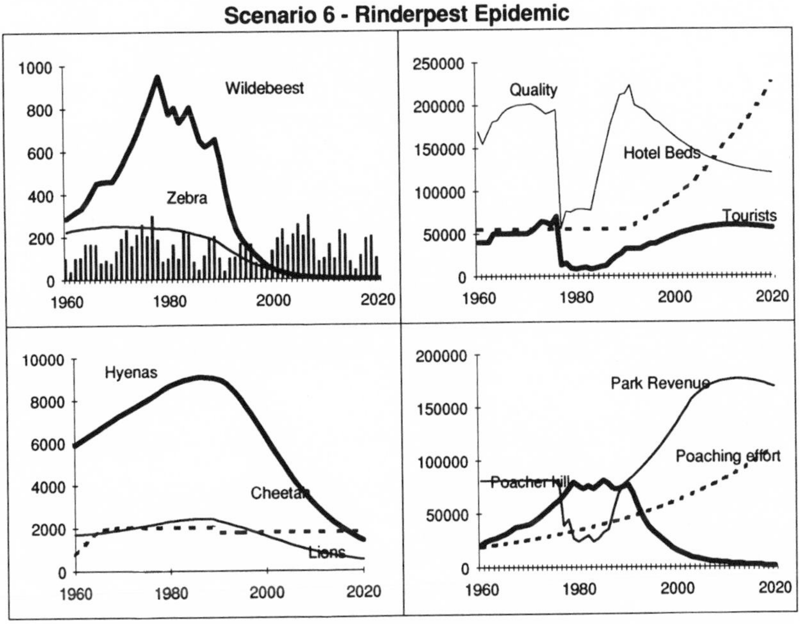

Figure 29.7 Output from scenario 6, an increase in wildebeest mortality due to rinderpest from 1990. See figure 29.2 for description of indicators.

poacher kill. The estimates we have used are tentative, and much more research is needed. Further, the effectiveness of anti-poaching patrols needs to be understood. It is clear that the wildebeest and zebra populations can withstand some level of poaching; the question is how much is too much.

Since tourism is the major source of income for the parks, we need to understand what determines the quality of tourism. This could be explored by various tourist surveys and by comparisons among different parks to see what factors are related to the number of tourists.

It is widely believed that the predators—lions, leopards, and cheetahs—are major components of tourism quality. The population-level responses of these predators are not well understood. The predator modeling group had considerable difficulty in formulating the components of recruitment and mortality necessary for predicting the effects of changes in the park.

Cautions

A final word of caution: This model does not represent the state of the art; it is merely one realization (4 days worth) of how things might work. Each individual group could build a better model. A research group could build a better combined model. This model provides a first look at how the system might work and should be viewed as a starting point for further exploration. The process of the workshop was much more important than the product.

ACKNOWLEDGMENTS

This workshop was made possible by the cooperation and funding of the Serengeti Wildlife Research Centre, Tanzania National Parks, the Frankfurt Zoological Society, and a Natural Sciences and Engineering Research Council of Canada grant to A. R. E. Sinclair. The assistance of Ken Campbell in supplying data, computers, and other support is greatly appreciated. As in any cooperative modeling process, everyone who participated in the workshop made a valuable contribution.

REFERENCES

Dublin, H. T., Sinclair, A. R. E., and McGlade, J. 1990. Elephants and fire as causes of multiple stable states in the Serengeti-Mara woodlands. *J. Anim. Ecol.* 59:1147–64.

Hilborn, R., and Sinclair, A. R. E. 1979. A simulation of the wildebeest population, other ungulates, and their predators. In *Serengeti: Dynamics of an ecosystem,* ed. A. R. E. Sinclair and M. Norton-Griffiths, 287–309. Chicago: University of Chicago Press.

Hilborn, R., and Walters, C. J. 1987. Microcomputer simulation for training and teaching. *Environ. Softw.* 1:156–63.

Holling, C. S., ed. 1978. *Adaptive environmental assessment and management.* New York: Wiley.

Linhart, H., and Zucchini, W. 1986. *Model selection.* New York: Wiley.

Sinclair, A. R. E., and Norton-Griffiths, M., eds. 1979. *Serengeti: Dynamics of an ecosystem.* Chicago: University of Chicago Press.

Walters, C. J. 1974. An interdisciplinary approach to development of watershed models. *Tech. Forecasting Soc. Change* 6:299–323.

———. 1986. *Adaptive management of renewable resources.* New York: Macmillan.

APPENDIX A

SCIENTIFIC AND COMMON NAMES OF THE LARGER MAMMAL SPECIES OF THE SERENGETI-MARA REGION

Order Primates

Erythrocebus patas	Ikoma patas monkey
Cercopithecus aethiops	Vervet monkey
Cercopithecus mitis	Blue monkey
Colobus abyssinicus	Black and white colobus
Papio cynocephalus	Olive baboon

Order Pholidota

Manis temmincki	Ground pangolin

Order Lagomorpha

Lepus capensis	Cape hare
Lepus crawshayi	Crawshay's hare

Order Rodentia

Hystrix crisata	African porcupine
Hystrix africaeaustralis	Cape porcupine
Pedetes capensis	Spring hare

Order Carnivora

Panthera leo	Lion
Panthera pardus	Leopard
Acinonyx jubatus	Cheetah
Felis serval	Serval
Felis caracal	Caracal
Felis sylvestris	African wildcat
Canis aureus	Golden jackal
Canis mesomelas	Black-backed jackal
Canis adustus	Side-striped jackal
Lycaon pictus	African wild dog
Otocyon megalotis	Bat-eared fox
Ictonyx striatus	Zorilla
Poecilogale albinucha	African striped weasel
Melivora capensis	African honey badger, ratel
Viverra civetta	African civet
Nandinia binotata	Palm civet
Genetta genetta	Common genet
Herpestes ichneumon	Great grey mongoose
Herpestes sanguineus	Slender or black-tipped mongoose
Helogale parvula	Dwarf mongoose
Atilax paludinosus	Marsh mongoose
Mungos mungo	Banded mongoose
Ichneumia albicauda	White-tailed mongoose
Proteles cristatus	Aardwolf
Crocuta crocuta	Spotted hyena
Hyaena hyaena	Striped hyena

Order Tubulidentata

Orycterpus afer	Aardvark

Order Proboscidea

Loxodonta africana	African elephant

Order Hyracoidea

Dendrohyrax arboreus	Tree hyrax
Heterohyrax brucei	Bush hyrax
Procavia johnstoni	Rock hyrax

Order Perissodactyla

Equus burchelli	Burchell's zebra
Diceros bicornis	Black rhinoceros

Order Artiodactyla

Potamochoerus porcus	Bushpig
Phacochoerus aethiopicus	Warthog
Hippopotamus amphibius	Hippopotamus
Giraffa camelopardalis	Giraffe
Sylvicarpa grimmia	Common duiker
Raphicerus campestris	Steinbok
Ourebia ourebi	Oribi
Oreotragus oreotragus	Klipspringer

Order Artiodactyla (continued)

Rynchotragus (Madoqua) kirkii	Kirk's dikdik
Redunca redunca	Bohor reedbuck
Redunca fulvorufula	Mountain reedbuck
Kobus ellipsiprymnus	Defassa waterbuck
Aepyceros melampus	Impala
Gazella thomsoni	Thomson's gazelle
Gazella granti	Grant's gazelle
Hippotragus equinus	Roan antelope
Damaliscus korrigum	Topi
Alcelaphus buselaphus	Kongoni or Coke's hartebeest
Connochaetes taurinus	Wildebeest
Tragelaphus scriptus	Bushbuck
Tragelaphus strepsiceros	Greater kudu
Tragelaphus imberbis	Lesser kudu
Taurotragus oryx	Eland
Oryx gazella	Oryx
Syncerus caffer	African buffalo

APPENDIX B

SCIENTIFIC AND COMMON NAMES OF THE BIRDS OF THE SERENGETI-MARA REGION

This list of birds in the Serengeti-Mara ecosystem is based on unpublished records of A. R. E. Sinclair (1965–1993) and those published in Sinclair 1978, Schmidl 1982, Short, Horne, and Muringo-Gichuki 1990, and B. W. Finch (undated). Records for the Ngorongoro Crater Highlands and Loita Hills are not included. Some doubtful records have been omitted. Latin nomenclature follows Short et al. with two exceptions: (1) both the yellow-shouldered widowbird (*Euplectes macrocerus*) and the yellow-mantled widowbird (*E. macrourus*) occur at least parapatrically, and (2) both the grey-headed sparrow (*Passer griseus*) and Swahili sparrow (*P. suahelicus*) are common, distinct, sympatric geographically and ecologically and do not interbreed. These birds behave as true biological species and so in both cases the names assigned by Mackworth-Praed and Grant (1955) are used. Common names generally follow Short et al. except where more familiar usage follows Mackworth-Praed and Grant (1952, 1955).

Latin Name	Common Name
Struthionidae	**Ostriches**
Struthio camelus	Ostrich
Podicipedidae	**Grebes**
Tachybaptus ruficollis	Little grebe, dabchick
Phalacrocoracidae	**Cormorants**
Phalacrocorax carbo	Great cormorant
Phalacrocorax africanus	Long-tailed cormorant
Anhinga rufa	African darter
Pelecanidae	**Pelicans**
Pelecanus onocrotalus	Great white pelican
Pelecanus rufescens	Pink-backed pelican
Ardeidae	**Herons, Egrets**
Ixobrychus minutus	Little bittern
Ixobrycgys sturmii	Dwarf bittern
Gorsachius leuconotus	White-backed night heron
Ardeola ralloides	Squacco heron
Ardeola idae	Madagascar squacco heron
Bubulcus ibis	Cattle egret
Butorides striatus	Green-backed heron
Egretta ardesiaca	Black heron
Egretta gargetta	Little egret
Egretta intermedia	Yellow-billed egret
Casmerodius albus	Great white egret
Ardea purpurea	Purple heron
Ardea cinerea	Grey heron
Ardea melanocephala	Black-headed heron
Ardea goliath	Goliath heron
Scopidae	**Hammerkop**
Scopus umbretta	Hammerkop
Ciconiidae	**Storks**
Mycteria ibis	Yellow-billed stork
Anastomus lamelligerus	Open-billed stork
Ciconia nigra	Black stork
Ciconia abdomii	Abdim's stork

Latin Name	Common Name
Ciconia episcopus	Wooly-necked stork
Ciconia ciconia	White stork
Ephippiorhynchus senegalensis	Saddle-billed stork
Leptoptilos crumeniferus	Marabou stork
Threskiornithidae	**Ibises, spoonbills**
Plegadis falcinellus	Glossy ibis
Bostrychia hagedash	Hadada
Threskiornis aethiopicus	Sacred ibis
Platalea alba	African spoonbill
Phoenicopteridae	**Flamingos**
Phoenicopterus ruber	Great flamingo
Phoeniconaias minor	Lesser flamingo
Pandionidae	**Osprey**
Pandion haliaetus	Osprey
Accipitridae	**Vultures, Hawks, Eagles**
Aviceda cuculoides	African cuckoo-hawk
Machaerhamphus alcinus	Bat-hawk
Elanus caeruleus	Black-shouldered kite
Milvus migrans	Black kite
Haliaetus vocifer	Fish eagle
Aegypius tracheliotus	Lappet-faced vulture
Aegypius occipitalis	White-headed vulture
Necrosyrtes monachus	Hooded vulture
Gyps rueppellii	Ruppell's griffon vulture
Gyps africanus	White-backed vulture
Neophron percnoptreus	Egyptian vulture
Gypaetus barbatus	Lammergeier
Circaetus gallicus	Short-toed snake-eagle
Circaetus cinereus	Brown snake-eagle
Circaetus cinerascans	Banded snake-eagle
Terathopius ecaudatus	Bateleur
Polyboroides typus	Harrier-hawk
Melierax poliopterus	Pale chanting goshawk

Latin Name	Common Name
Melierax metabates	Dark chanting goshawk
Melierax gabar	Gabar goshawk
Kaupifalco monogrammicus	Lizard-buzzard
Butastur rufipennis	Grasshopper-buzzard
Circus macrourus	Pallid harrier
Circus pygargus	Montague's harrier
Circus aerieginosus	Eurasian marsh-harrier
Circus ranivorus	African marsh-harrier
Accipiter tachiro	African goshawk
Accipiter badius	Shikra
Accipiter ovampensis	Ovampo sparrowhawk
Accipiter minullus	Little sparrowhawk
Accipiter melanoleucus	Great sparrowhawk
Buteo buteo	Common buzzard
Buteo augur	Augar buzzard
Aquila pomarina	Lesser spotted eagle
Aquila rapax	Tawny and Steppe eagle
Aquila verreauxi	Verreaux's eagle
Hieraaetus wahlbergi	Wahlberg's eagle
Hieraaetus spilogaster	African hawk-eagle
Hieraaetus pennatus	Booted eagle
Hieraaetus bellicosus	Martial eagle
Spizaetus occipitalis	Long-crested hawk-eagle
Spizaetus ayresii	Ayres's hawk-eagle
Sagittariidae	**Secretary Bird**
Sagittarius serpentarius	Secretary bird
Falconidae	**Falcons**
Polihierax semitorquatus	Pygmy falcon
Falco naumanni	Lesser kestrel
Falco tinnunculus	Kestrel
Falco rupicoloides	Grester kestrel
Falco alopex	Fox-kestrel
Falco ardosiaceus	Grey kestrel
Falco amurensis	Eastern red-footed falcon
Falco concolor	Sooty falcon
Falco subbuteo	Eurasian hobby
Falco cuvieri	African hobby

Latin Name	Common Name
Falco biarmicus	Lanner falcon
Falco peregrinus	Peregrine falcon
Anatidae	**Ducks, Geese**
Dendrocygna bicolor	Fulvous whistling-duck
Dendrocygna viduata	White-faced whistling-duck
Alopochen aegyptiacus	Egyptian goose
Plectropterus gambensis	Spur-wing goose
Sarkidiornis melanotos	Knob-billed goose
Nettapus auvitus	Pygmy goose
Anas crecca	Green-winged teal
Anas capensis	Cape wigeon
Anas platyrhynchos	Mallard
Anas undulata	Yellow-billed duck
Anas acuta	Pintail
Anas erythrorhyncha	Red-billed duck
Anas hottentota	Hottentot teal
Anas querquedula	Garganey
Anas clypeata	Shoveler
Netta erythrophthalma	Southern pochard
Phasianidae	**Francolins, Spurfowl**
Guttera pucherani	Crested guineafowl
Numida meleagris	Helmeted guineafowl
Coturnix coturnix	Quail
Coturnix delagorguei	Harlequin quail
Francolinus coqui	Coqui francolin
Francolinus shelleyi	Shelley's francolin
Francolinus sephaena	Crested francolin
Francolinus afer	Red-necked spurfowl
Francolinus rufopictus	Grey-breasted spurfowl
Francolinus leucoscepus	Yellow-necked spurfowl
Turnicidae	**Button-Quail**
Turnix sylvatica	Button-quail
Gruidae	**Cranes**
Balearica regulorum	Crowned-crane
Rallidae	**Rails, Crakes**
Rallus caerulescens	African water-rail
Sarothrura elegans	Pygmy crake
Crex egregia	African crake
Amaurornis flavirostris	Black crake
Gallinula chloropus	Moorhen
Fulica cristata	Red-knobbed coot
Heliornithidae	**Finfoot**
Podica senegalensis	African finfoot
Otididae	**Bustards**
Neotis denhami	Jackson's (Denham's) bustard
Ardeotis kori	Kori bustard
Eupodotis senegalensis	Senegal (white-bellied) bustard
Eupodotis melanogaster	Black-bellied bustard
Jacanidae	**Jacana**
Actophilornis africana	Jacana
Rostratulidae	**Painted-snipe**
Rostratula benghalensis	Painted-snipe
Recurvirostridae	**Stilt, avocet**
Himantopus himantopus	Black-winged stilt
Recurvicrostra avosetta	Avocet
Burhinidae	**Thicknees, Stone-Curlews**
Burhinus vermiculatus	Water thicknee
Burhinus capensis	Spotted stone-curlew
Glareolidae	**Coursers**
Cursorius temminckii	Temminck's courser
Cursorius africanus	Two-banded courser
Cursorius cinctus	Heuglin's courser
Cursorius chalcopterus	Violet-tipped courser
Glareola pratincola	Common pratincole
Charadriidae	**Plovers**
Vanellus senegallus	Wattled plover
Vanellus armatus	Blacksmith plover
Vanellus spinosus	Spur-winged plover
Vanellus buguleris	Senegal plover
Vanellus melanopterus	Black-winged plover

Latin Name	Common Name
Vanellus coronatus	Crowned plover
Charadrius hiaticula	Ringed plover
Charadrius pecuarius	Kittlitz's sandplover
Charadrius tricollaris	Three-banded plover
Charadrius pallidus	Chestnut-banded sandplover
Charadrius asiaticus	Caspian plover
Scolopacidae	**Sandpipers, Snipes**
Numerius arquata	Curlew
Limosa limosa	Black-tailed godwit
Tringa erythropus	Spotted redshank
Tringa totanus	Redshank
Tringa nebularia	Greenshank
Tringa stagnatilis	Marsh sandpiper
Tringa ochropus	Green sandpiper
Tringa glareola	Wood sandpiper
Actitis hypoleucos	Common sandpiper
Phalaropus lobatus	Red-necked (northern) phalarope
Gallinago gallinago	Common snipe
Gallinago nigripennis	African snipe
Calidris alba	Sanderling
Calidris minuta	Little stint
Calidris ferruginea	Curlew sandpiper
Calidris alpina	Dunlin
Philomachus pugnax	Ruff
Laridae	**Gulls**
Larus cirrocephalus	Grey-headed gull
Larus ridibundus	Black-headed gull
Sternidae	**Terns**
Sterna nilotica	Gull-billed tern
Chlidonias leucopterus	White-winged black tern
Pteroclidae	**Sandgrouse**
Pterocles exustus	Chestnut-bellied sandgrouse
Pterocles decoratus	Black-faced sandgrouse
Pterocles gutturalis	Yellow-throated sandgrouse
Columbidae	**Pigeons, doves**
Treron calva	Green pigeon

Latin Name	Common Name
Turtur tympanistria	Tambourine dove
Turtur afer	Blue spotted wood dove
Turtur chalcospilos	Emerald spotted wood dove
Oena capensis	Namaqua dove
Columba arquatrix	Olive pigeon
Columba guinea	Speckled pigeon
Streptopelia lugens	Pink-breasted (dusky turtle) dove
Streptopelia decipiens	Mourning dove
Streptopelia semitorquata	Red-eyed dove
Streptopelia capicola	Ring-necked dove
Streptopelia senegalensis	Laughing dove
Psittacidae	**Parrots**
Agapornis fischeri	Fischer's lovebird
Agapornis personata	Yellow-collared lovebird
Poicephalus meyeri	Brown parrot
Musophagidae	**Turacos**
Crinifer zonurus	Eastern grey plaintain-eater
Corythaixoides personata	Bare-faced go-away bird
Musophaga rossae	Ross's turaco
Turaco schalowi (persa)	Schalow's turaco
Cuculidae	**Cuckoos, Coucals**
Clamator glandarius	Great spotted cuckoo
Clamator jacobinus	Black-and-white-cuckoo
Clamator levaillantii	Levaillant's cuckoo
Cuculus solitarius	Red-chested cuckoo
Cuculus clamosus	Black cuckoo
Cuculus canorus	Eurasian cuckoo
Cuculus gularis	African cuckoo
Chrysococcyx cupreus	Emerald cuckoo
Chrysococcyx klaas	Klaas's cuckoo
Chrysococcyx caprius	Didric cuckoo
Centropus grillii	Black coucal
Centropus superciliosus	White-browed coucal

Latin Name	Common Name
Tytonidae	**Barn-owls**
Tyto alba	Barn owl
Tyto capensis	Cape grass owl
Strigidae	**True Owls**
Otus senegalensis	African scops owl
Otus leucotis	White-faced scops owl
Bubo africanus	Spotted eagle-owl
Bubo lacteus	Verreaux's eagle-owl
Scotopelia peli	Pel's fishing-owl
Strix woodfordii	African wood-owl
Glaucidium perlatum	Pearl-spotted owlet
Asio capensis	African marsh-owl
Caprimulgidae	**Nightjars**
Caprimulgus inornatus	Plain nightjar
Caprimulgus tristigma	Freckled nightjar
Caprimulgus fraenatus	Dusky nightjar
Caprimulgus europaeus	Eurasian nightjar
Caprimulgus clarus	Slender-tailed nightjar
Caprimulgus fossii	Gabon nightjar
Macrodipteryx vexillaria	Pennant-winged nightjar
Apodidae	**Swifts**
Cypsiurus parvus	Palm swift
Apus barbatus	Black swift
Apus niansae	Nyanza swift
Apus apus	Eurasian swift
Apus affinis	Little swift
Apus caffer	White-rumped swift
Apus horus	Horus swift
Tachymarptis melba	Alpine swift
Coliidae	**Mousebirds**
Urocolius macrourus	Blue-naped mousebird
Colius striatus	Speckled mousebird
Trogonidae	**Trogons**
Apaloderma narina	Narina's trogon
Alcedinidae	**Kingfishers**
Halcyon leucocephala	Grey-headed kingfisher
Halcyon senegalensis	Woodland kingfisher
Halcyon chelicuti	Striped kingfisher

Latin Name	Common Name
Ceyx pictus	Pygmy kingfisher
Corythornis cristata	Malachite kingfisher
Megaceryle maxima	Giant kingfisher
Ceryle rudis	Pied kingfisher
Meropidae	**Bee-eaters**
Merops pusillus	Little bee-eater
Merops oreobates	Cinnamon-chested bee-eater
Merops albicollis	White-throated bee-eater
Merops superciliosus	Madagascar bee-eater
Merops apiaster	Eurasian bee-eater
Coraciidae	**Rollers**
Coracias naevia	Rufous-crowned roller
Coracias garrulus	Eurasian roller
Coracias caudata	Lilac-breasted roller
Eurystomus glaucurus	Broad-billed roller
Phoeniculidae	**Wood-Hoopoes**
Phoeniculus purpureus	Green wood-hoopoe
Phoeniculus damarensis	Violet wood-hoopoe
Phoeniculus cyanomelas	Scimitarbill
Phoeniculus minor	Abyssinian scimitarbill
Upupidae	**Hoopoe**
Upupa epops	Hoopoe
Bucerotidae	**Hornbills**
Bucorvus cafer	Ground hornbill
Tochus erythrorhynchus	Red-billed hornbill
Tochus deckeni	Von der Decken's hornbill
Tochus alboterminatus	Crowned hornbill
Tochus nasutus	Grey hornbill
Ceratogymna subcylindricus	Black-and-white casqued hornbill
Capitonidae	**Barbets**
Gymnobucco bonapartei	Grey-throated barbet
Pogoniulus psisillus	Red-fronted tinkerbird
Pogoniulus bilineatus	Golden-rumped tinkerbird

Latin Name	Common Name
Tricholaema diademata	Red-fronted barbet
Tricholaema lacrymosa	Spotted-flanked barbet
Lybius leucocephalus	White-headed barbet
Lybius bidentatus	Double-toothed barbet
Trachyphonus erythrocephalus	Red-and-yellow barbet
Trachyphonus darnaudii	D'Arnaud's barbet
Indicatoridae	**Honeyguides**
Prodotiscus regulus	Wahlberg's honeyguide
Indicator variegatus	Scaly-throated honeyguide
Indicator indicator	Black-throated honeyguide
Indicator minor	Lesser honeyguide
Picidae	**Woodpeckers**
Jynx ruficollis	Red-breasted wryneck
Campethera nubica	Nubian woodpecker
Campethera abingoni	Golden-tailed woodpecker
Campethera cailliautii	Little spotted woodpecker
Dendropicos fuscescens	Cardinal woodpecker
Dendropicos namaquis	Bearded woodpecker
Dendropicos goertae	Grey woodpecker
Alaudidae	**Larks**
Mirafra cantillans	Singing bushlark
Mirafra albicauda	Northern white-tailed bushlark
Mirafra africana	Rufous-naped lark
Mirafra rufocinnamomea	Flappet lark
Mirafra africanoides	Fawn-colored lark
Eremoptreix leucopareia	Fischer's sparrowlark
Calandrella cinerea	Red-capped lark
Calandrella somalica	Athi short-toed lark
Galerida fremantlii	Short-tailed lark
Hirundinidae	**Swallows**
Riparia paludicola	Brown-throated sand-martin

Latin Name	Common Name
Riparia cincta	Banded martin
Hirundo griseopyga	Grey-rumped swallow
Hirundo fuligula	Rock-martin
Hirundo rustica	Eurasian swallow
Hirundo smithii	Wire-tailed swallow
Hirundo senegalensis	Mosque swallow
Hirundo daurica	Red-rumped swallow
Hirundo abyssinica	Striped swallow
Delichon urbica	House-martin
Psalidoprocne holomelaena	Black rough-wing swallow
Psalidoprocne albiceps	White-headed rough-wing swallow
Motacillidae	**Wagtails, pipits**
Motacilla flava	Yellow wagtail
Motacilla clara	Mountain wagtail
Motacilla alba	Pied wagtail
Macronyx croceus	Yellow-throated longclaw
Macronyx ameliae	Rosy-breasted longclaw
Anthus novaeseelandiae	Richard's pipit
Anthus leucophrys	Plain-backed pipit
Anthus similis	Long-billed pipit
Anthus caffer	Bush (Little tawny) pipit
Anthus cervinus	Red-throated pipit
Anthus trivialis	Tree pipit
Campephagidae	**Cuckooshrikes**
Campephaga flava	Black cuckooshrike
Coracina caesia	Grey cuckooshrike
Pycnonotidae	**Bulbuls**
Pycnonotus barbatus	Yellow-vented bulbul
Andropadus latirostris	Yellow-whiskered greenbul
Andropadus gracilirostris	Slender-billed greenbul
Phyllastrephus cerviniventris	Grey-olive greenbul
Phyllastrephus fischeri	Fischer's greenbul
Phyllastrephus cabanisi	Olive greenbul
Prionopidae	**Helmetshrikes**
Eurocephalus rueppelli	White-crowned shrike

Latin Name	Common Name
Prionops plumata	Straight-crested (white) helmetshrike
Prionops poliolopha	Grey-crested helmetshrike
Malaconotidae	**Bush-shrikes, boubous**
Nilaus afer	Northern brubru
Dryoscopus cubla	Black-backed puffback shrike
Tchagra senegala	Black-headed bush-shrike
Tchagra australis	Brown-headed bush-shrike
Laniarius erythrogaster	Black-headed gonolek
Laniarius ferrugineus	Tropical boubou
Laniarius funebris	Slate-coloured boubou
Malaconotus sulfureopectus	Sulphur-breasted bush-shrike
Malaconotus blanchoti	Grey-headed bush-shrike
Laniidae	**Shrikes**
Corvinella melanoleuca	Magpie-shrike
Lanius collurio	Red-backed shrike
Lanius isabellinus	Red-tailed shrike
Lanius minor	Lesser grey shrike
Lanius excubitoroides	Grey-backed fiscal
Lanius dorsalis	Taita fiscal
Lanius collaris	Fiscal
Lanius senator	Wood-chat shrike
Turdidae	**Thrushes**
Erithacus luscinia	Sprosser
Erithacus megarhynchos	Nightingale
Irania gutturalis	White-throated robin
Cossypha heuglini	White-browed robin-chat
Cossypha natalinsis	Red-capped robin-chat
Cichladusa guttata	Spotted morning-warbler
Erythropygia leucophrys	Red-backed scrub-robin
Phoenicurus phoenicurus	Redstart
Saxicola rubetra	Whinchat

Latin Name	Common Name
Oenanthe isabellina	Isabelline wheatear
Oenanthe pileata	Capped wheatear
Oenanthe oenanthe	Eurasian wheatear
Oenanthe pleschanka	Pied wheatear
Cercomela familiaris	Familiar chat
Myrmecocichla aethiops	Anteater chat
Myrmecocichla nigra	Sooty chat
Myrmecocichla cinnamomeiventris	Cliff-chat
Monticola saxatilis	Rock-thrush
Turdus libonyanus	Kurrichane thrush
Timaliidae	**Babblers**
Turdoides rubiginosus	Rufous chatterer
Turdoides melanops	Black-lored babbler
Turdoides jardinei	Arrow-marked babbler
Sylviidae	**Warblers**
Bradypterus baboecala	Little rush warbler
Melocichla mentalis	Moustached warbler
Schoenicola platyura	Fan-tailed warbler
Acrocephalus schoenobaenus	Sedge warbler
Acrocephalus scirpaceus	Eurasian reed warbler
Acrocephalus palustris	Marsh warbler
Acrocephalus arundinaceus	Great reed warbler
Acrocephalus gracidirostrius	Lesser swamp warbler
Cisticola erythrops	Red-faced cisticola
Cisticola woosnami	Trilling cisticola
Hippolais pallida	Olivaceous warbler
Cisticola aberrans	Rock-loving cisticola
Cisticola chiniana	Rattling cisticola
Cisticola galactotes	Winding cisticola
Cisticola robusta	Stout cisticola
Cisticola natalensis	Croaking cisticola
Cisticola fulvicapilla	Tabora cisticola
Cisticola brachyptera	Siffling cisticola
Cisticola juncidis	Zitting cisticola
Cisticola aridula	Desert cisticola

Latin Name	Common Name
Cisticola brunnescens	Pectoral-patch cisticola
Cisticola ayresii	Wing-snapping cisticola
Prinia subflava	Tawny-flanked prinia
Apalis flavida	Black-breasted apalis
Apalis cinerea	Grey apalis
Phyllolais pulchella	Buff-bellied warbler
Camaroptera brachyura	Grey-backed camaroptera
Calamonastes simplex	Grey bush-warbler
Eremomela icteropygialis	Yellow-bellied eremomela
Eremomela scotops	Green-capped eremomela
Sylvietta whytii	Red-faced crombec
Hypergerus lepidus	Grey-capped warbler
Hyliota australis	Southern yellow-bellied hyliota
Phylloscopus trochilus	Willow warbler
Phylloscopus sibilatrix	Wood warbler
Parisoma boehmi	Banded tit-flycatcher (parisoma)
Sylvia atricapilla	Blackcap
Sylvia borin	Garden warbler
Sylvia communis	Whitethroat
Muscicapidae	**Flycatchers**
Bradornis pallidus	Pale flycatcher
Bradornis microrhynchus	Grey flycatcher
Bradornis semipartitus	Silverbird
Melaenornis fischeri	White-eyed slaty flycatcher
Melaenornis pammelaina	Southern black flycatcher
Muscicapa striata	Spotted flycatcher
Muscicapa aquatica	Swamp flycatcher
Muscicapa adusta	Dusky flycatcher
Muscicapa caerulescens	Ashy flycatcher
Platysteiridae	**Puffback flycatchers**
Batis molitor	Chin-spot puffback flycatcher
Platysteira cyanea	Wattle-eye

Latin Name	Common Name
Monarchidae	**Paradise Flycatchers**
Elminia longicauda	Blue flycatcher
Terpsiphone rufiventer	Black-headed paradise flycatcher
Terpsiphone viridis	Paradise flycatcher
Remizidae	**Penduline-Tits**
Anthoscopus caroli	African penduline-tit
Paridae	**Titmice**
Parus albiventris	White-bellied tit
Parus fringillinus	Red-throated tit
Nectariniidae	**Sunbirds**
Anthreptes collaris	Collared sunbird
Nectarinia olivacea	Olive sunbird
Nectarinia amethystina	Amethyst sunbird
Nectarinia senegalensis	Scarlet-chested sunbird
Nectarinia venusta	Variable sunbird
Nectarinia pulchella	Beautiful sunbird
Nectarinia kilimensis	Bronze sunbird
Nectarinia reichenowi	Golden-winged sunbird
Nectarinia mariquensis	Mariqua sunbird
Nectarinia bifasciata	Little purple-banded sunbird
Zosteropidae	**White-eyes**
Zosterops senegalensis	Yellow white-eye
Emberizidae	**Buntings**
Emberiza tahapisi	Cinnamon-breasted rock bunting
Emberiza flaviventris	Golden-breasted bunting
Fingillidae	**Finches**
Serinus sulphuratus	Brimstone canary
Serinus burtoni	Thick-billed seedeater
Serinus citrinelloides	African citril
Serinus atrogularis	Yellow-rumped seedeater
Serinus dorsostriatus	White-bellied canary
Serinus mozambicus	Yellow-fronted canary

Latin Name	Common Name
Estrildidae	**Waxbills**
Pytilia melba	Green-winged pytilia
Amadina fasciata	Cut-throat
Lagonosticta senegala	Red-billed firefinch
Lagonosticta rubricata	African firefinch
Uraeginthus bengalus	Red-cheeked cordon-bleu
Uraeginthus cyanocephalus	Blue-capped cordon-bleu
Uraeginthus ianthinogaster	Purple grenadier
Estrilda melanotis	Yellow-bellied waxbill
Estrilda erythronotos	Black-cheeked waxbill
Estrilda rhodopygea	Crimson-rumped waxbill
Estrilda astrild	Waxbill
Estrilda subflava	Zebra waxbill
Ortygospiza atrocollis	African quailfinch
Lonchura cantans	Silverbill
Lonchura griseicapilla	Grey-headed silverbill
Lonchura cucullata	Bronze mannikin
Lonchura bicolor	Black-and-white mannikin
Ploceidae	**Weavers**
Bubalornis niger	Red-billed buffalo-weaver
Dinemellia dinemelli	White-headed buffalo-weaver
Plocepasser mahali	White-browed sparrow-weaver
Histurgops ruficauda	Rufous-tailed weaver
Pseudonigrita arnaudi	Grey-headed social weaver
Sporopipes frontalis	Speckle-fronted weaver
Amblyospiza albifrons	Grosbeak weaver
Ploceus baglafecht	Reichenow's weaver
Ploceus ocularis	Spectacled weaver
Ploceus nigricollis	Black-necked weaver
Ploceus xanthops	Holub's golden weaver
Ploceus velatus	Vitelline masked weaver
Ploceus intermedius	Masked weaver

Latin Name	Common Name
Ploceus jacksoni	Golden-backed weaver
Ploceus spekei	Speke's weaver
Ploceus cucullatus	Black-headed weaver
Ploceus rubiginosus	Chestnut weaver
Malimbus rubniceps	Red-headed weaver
Quelea cardinalis	Cardinal quelea
Quelea erythrops	Red-headed quelea
Quelea quelea	Red-billed quelea
Euplectes gierowii	Black bishop
Euplectes hordeaceus	Black-winged red bishop
Euplectes orix	Southern red bishop
Euplectes capensis	Yellow bishop
Euplectes albonotatus	White-winged widowbird
Euplectes macrocercus	Yellow-shouldered widowbird
Euplectes macrourus	Yellow-mantled widowbird
Euplectes ardens	Red-collared widowbird
Euplectes axillaris	Fan-tailed widowbird
Euplectes jacksoni	Jackson's widowbird
Anomalospiza imberbis	Parasitic weaver
Vidua chalybeata	Indigobird
Vidua funerea	Variable (dusky) indigobird
Vidua hypocherina	Steel-blue whydah
Vidua macroura	Pin-tailed whydah
Vidua fischeri	Straw-tailed whydah
Vidua paradisaea	Paradise whydah
Passeridae	**Sparrows**
Passer motitensis	Rufous sparrow
Passer eminibey	Chestnut sparrow
Passer suahelicus	Swahili sparrow
Passer griseus	Grey-headed sparrow
Petronia pyrgita	Yellow-spotted petronia
Sturnidae	**Starlings**
Onychognathus morio	Red-winged starling
Lamprotornis chalybeus	Blue-eared glossy starling

Latin Name	Common Name
Lamprotornis purpuropterus	Ruppell's long-tailed glossy starling
Spreo superbus	Superb starling
Spreo hildebrandti	Hildebrandt's starling
Cinnyricinclus leucogaster	Violet-backed starling
Creatophora cinerea	Wattled starling
Buphagus africanus	Yellow-billed oxpecker
Buphagus erythrorhychus	Red-billed oxpecker

Latin Name	Common Name
Oriolidae	**Orioles**
Oriolus oriolus	Eurasian golden oriole
Oriolus auratus	African golden oriole
Oriolus larvatus	Black-headed oriole
Dicruridae	**Drongos**
Dicrurus adsimilis	Drongo
Corvidae	**Crows**
Corvus capensis	Cape rook
Corvus albus	Pied crow
Corvus albicollis	White-necked raven

REFERENCES

Finch, B. W. (n.d.). Birds of the Masai Mara checklist. Oak Brook, Ill.: Friends of Conservation.

Mackworth-Praed, C. W., and Grant, C. H. B. 1952, 1955. *Birds of Eastern and North Eastern Africa.* Vols. 1&2.

Schmidl, D. 1982. *The birds of the Serengeti National Park, Tanzania.* Brit. Ornithol. Union, Checklist No. 5.

Short, L. L., Horne, J. F. M., and Muringo-Gichuki, C. 1990. *Annoted check-list of the birds of East Africa.* Proc. Western Foundation of Vertebrate Zoology, Los Angeles, Calif. 3:61–246.

Sinclair, A. R. E. 1978. Factors affecting the food supply and breeding season of resident birds and movement of palaearctic migrants in a tropical African savannah. *Ibis* 120:480–97.

CONTRIBUTORS

Peter Arcese
Department of Wildlife Ecology
University of Wisconsin
Madison, Wisconsin 53706
United States of America

F. F. Banyikwa
Biological Research Laboratories
Syracuse University
130 College Place
Syracuse, New York 13244–1220
United States of America

Markus Borner
Frankfurt Zoological Society
Serengeti National Park
P. O. Box 3134, Arusha
Tanzania

Michael D. Broten
Department of Resource Surveys and Remote Sensing
Ministry of Planning and National Development
P. O. Box 47146, Nairobi
Kenya

Roger Burrows
Department of Continuing and Adult Education
University of Exeter
Cotley, Streatham Rise
Exeter EX4 4PE
United Kingdom

J. D. Bygott
Box 1501, Karatu
Tanzania

Ken Campbell
Tanzania Wildlife Conservation Monitoring
Serengeti Wildlife Research Centre
P. O. Box 3134, Arusha
Tanzania

Tim Caro
Center for Population Biology
Department of Wildlife and Conservation Biology
University of California
Davis, California 95616
United States of America

Emmanuel B. Chausi
Ngorongoro Conservation Area Authority
Box 1, Ngorongoro
Tanzania

Nancy M. Creel
Wildlife Conservation Research Unit
Department of Zoology
University of Oxford
South Parks Road
Oxford OX1 3PS
United Kingdom

Scott R. Creel
Department of Biological Sciences
Purdue University
West Lafayette, Indiana 47907
United States of America

Andy Dobson
Ecology and Evolutionary Biology
Princeton University
Princeton, New Jersey 08544–1003
United States of America

Holly T. Dublin
Worldwide Fund for Nature
Regional Office for Eastern Africa
P. O. Box 62440, Nairobi
Kenya

Sarah Durant
Serengeti Wildlife Research Institute
P. O. Box 661, Arusha
Tanzania

Marion East
Max-Planck-Institut für Verhaltensphysiologie
Abteilung Wickler
D–82319 Seewiesen Post Starnberg
Germany

Lee F. Elliott
Department of Biological Sciences
Purdue University
West Lafayette, Indiana 47907
United States of America

Clare D. FitzGibbon
Large Animal Research Group
Department of Zoology
University of Cambridge
Cambridge CB2 3EJ
United Kingdom

Laurence G. Frank
Psychology Department
University of California
Berkeley, California 94720
United States of America

John M. Fryxell
Department of Zoology
University of Guelph
Guelph, Ontario, N1G 2W1
Canada

Nicholas Georgiadis
The Wildlife Conservation Society
Bronx Zoo
Bronx, New York 10460
United States of America

J. P. Hanby
Box 1501, Karatu
Tanzania

Justine Hando
Tanzania National Parks
P.O. Box 3134
Arusha
Tanzania

R. Hilborn
School of Fisheries WH–10
University of Washington
Seattle, Washington 98195
United States of America

Heribert Hofer
2 Max-Planck-Institut für Verhaltensphysiologie
Abteilung Wickler
D–82319 Seewiesen Post Starnberg
Germany

Kay E. Holekamp
Department of Psychology
Department of Zoology
Michigan State University
East Lansing, Michigan 48824
United States of America

M. Karen Laurenson
Department of Zoology
University of Cambridge
Downing Street
Cambridge CB2 3EJ
United Kingdom

John Lazarus
Department of Psychology
University of Newcastle upon Tyne
Newcastle upon Tyne NE1 7RU
United Kingdom

M. K. S. Maige
Serengeti Regional Conservation Strategy
P. O. Box 1541, Arusha
Tanzania

R. C. Malpas
IUCN, World Conservation Union
P. O. Box 68200, Nairobi
Kenya

B. N. N. Mbano
Wildlife Division
P.O. Box 1994,
Dar es Salaam,
Tanzania

S. J. McNaughton
Biological Research Laboratories
Syracuse University
130 College Place
Syracuse, New York 13244–1220
United States of America

Simon A. R. Mduma
Centre For Biodiversity Research
Department of Zoology
University of British Columbia
Vancouver V6T 1Z4
Canada

Patricia D. Moehlman
The Wildlife Conservation Society
Bronx, New York 10460
United States of America

Martyn G. Murray
Research Group in Mammalian Ecology and Reproduction
Physiological Laboratory
University of Cambridge
Downing Street
Cambridge CB2 3EG
United Kingdom

M. Norton-Griffiths
Centre for Social and Economic Research on the Global Environment
University College London, Gower Street
London WC1E 6BT
United Kingdom

C. Packer
Department of Ecology, Evolution and Behavior
University of Minnesota
1987 Upper Buford Circle
St. Paul, Minnesota 55108
United States of America

Scott Perkin
School of Development Studies
University of East Anglia
Norwich NR4 7TJ
United Kingdom

Victor A. Runyoro
Ngorongoro Conservation Area Authority
Box 1, Ngorongoro
Tanzania

Mohammed Said
Department of Resource Surveys and Remote Sensing
Ministry of Planning and National Development
P. O. Box 47146, Nairobi
Kenya

D. Scheel
Department of Ecology, Evolution, and Behavior
University of Minnesota
1987 Upper Buford Circle
St. Paul, Minnesota 55108
United States of America

A. R. E. Sinclair
Centre for Biodiversity Research
Department of Zoology
University of British Columbia
Vancouver V6T 1Z4
Canada

Laura Smale
Department of Psychology
Department of Zoology
Michigan State University
East Lansing, Michigan 48824
United States of America

P. A. K. Symonds
Tally Wood, Limpsfield Chart
Oxted, Surrey RH8 0TF
United Kingdom

D. M. Thompson
18 Brodie Road
Guilford, Surrey GU1 3RZ
United Kingdom

Peter M. Waser
Department of Biological Sciences
Purdue University
West Lafayette, Indiana 47907
United States of America

INDEX